Illustrated Generic Names of Fungi

Etymology, Descriptions, Classifications, and References

Featuring More Than 1,000 Original Watercolors

Miguel Ulloa and Elvira Aguirre-Acosta

with the technical assistance of
Samuel Aguilar Ogarrio

Departamento de Botánica
Instituto de Biología

Universidad Nacional Autónoma de México

The American Phytopathological Society
St. Paul, Minnesota, U.S.A.

Front cover and all internal images: Courtesy Miguel Ulloa—© APS.

This book was developed and produced by Dr. Ulloa with technical assistance in formatting and editing from Samuel Aguilar. The book was submitted in final form to APS PRESS. No editing or proofreading was done by PRESS staff.

To the extent permitted under applicable law, neither The American Phytopathological Society (APS) nor its suppliers and licensors assume responsibility for any damage, injury, loss, or other harm that results from an error or omission in the content of this publication in any of its forms and/or of any of the products derived from it. This publication and its associated products do not address every possible situation, and unforeseen circumstances may make the recommendations they contain inapplicable in some situations. Users of these materials cannot assume that they contain all necessary warning and precautionary measures and that other or additional information or measures may not be required. Users must rely on their own experience, knowledge, and judgment in evaluating or applying any information. Given rapid advances in the sciences and changes in product labeling and government regulations, APS recommends that users conduct independent verifications of diagnoses and treatments.

APS reserves the right to change, modify, add, or remove portions of these terms and conditions at its sole discretion at any time and without prior notice. Please check the Terms and Conditions of Use stated on APSnet periodically for any modifications.

Reference in this publication to a trademark, proprietary product, or company name by personnel of the U.S. Department of Agriculture or anyone else is intended for explicit description only and does not imply approval or recommendation to the exclusion of others that may be suitable.

Library of Congress Control Number: 2018967612
International Standard Book Numbers:
Print: 978-0-89054-618-5
Online: 978-0-89054-619-2

© 2020 by The American Phytopathological Society
All rights reserved.

No portion of this book may be reproduced in any form, including photocopy, microfilm, information storage and retrieval system, computer database, or software, or by any means, including electronic or mechanical, without written permission from the publisher.

Printed in the United States of America on acid-free paper.

The American Phytopathological Society
3340 Pilot Knob Road
St. Paul, Minnesota 55121, U.S.A.

This book is dedicated to

Dr. Teófilo Herrera, IBUNAM (1924–)
and Dr. Richard T. Hanlin, University of Georgia (1931–),
both emeritus professors and mentors, who positively influenced
the academic life of Miguel Ulloa

—The authors

Presentation of the book *Etimología e Iconografía
de Géneros de Hongos,* IBUNAM, 1994

With love to my children, Miguel, Alejandro,
Jonathan, Daniela, and Marlon

—M. U.

To my children, José Luis and Elvira

—E. A. A.

To my wife, Su,
and my children, Samuel and Diana

—S. A.

Preface

To members of the general public, including most students, scientific names are difficult-to-pronounce words that have no particular meanings. To anyone working with living organisms, however, scientific names provide the means by which one can communicate precisely what one is working with, regardless of language or cultural barriers. This accurate communication among biologists is essential for the understanding and advancement of scientific research in all fields.

A scientific name consists of two words: the genus and the specific epithet. Because this binomial is Latin, it is always written in italic or bold type or underlined to distinguish it from other text. The generic name is always capitalized, whereas the species name is not, at least in more recent literature. In older publications, the species name was often capitalized if it was derived from a proper noun, following the custom in English and its Germanic origins.

The present book deals only with generic names. Ideally, the name reflects some biological characteristic of the organism being described; however, the only requirement is that the name be Latinized. In addition to fungi being named for morphological characteristics, some are named after geographical locations or habitats or to honor mycologists or other individuals, and some are even given whimsical names. Traditionally, names were derived from Greek or Latin roots, but in recent years, names have been derived from words from other languages.

This book is basically an expanded version of the *Etimología e Iconografía de Géneros de Hongos,* written by Miguel Ulloa and Teófilo Herrera and published in 1994 by the Instituto de Biología, Universidad Nacional Autónoma de México (IBUNAM). In that book, genera were arranged in the taxonomic order used at that time, and 795 black-and-white images illustrated the text (406 drawings and 389 photographs). To the 807 generic names in that original work, an additional 784 have been added in this new book, bringing the total number of genera to 1,592.

These generic names are arranged alphabetically in the mode of a dictionary, and the current taxonomic classifications have been added. As in the original work, for each generic name, the etymological derivation is provided, along with the bibliographic citation in which the name was originally proposed. This citation is followed by comments on characteristics of the fungus. Older works often did not provide the derivations of new names, so it was necessary to decipher this information from the publication, which in some instances was not possible. In recent years, it has become customary to provide this etymological information in papers describing new taxa, making the task much easier.

With the goal of making the work more appealing and useful, watercolor illustrations of species belonging to 1,052 of the genera have been created by Miguel Ulloa specifically for this volume. All the illustrations were improved using the program Adobe Photoshop; the color, balance, background, brightness, and other aspects were enhanced to obtain maximum quality. In addition to adding to the book's scientific value, the beauty of these illustrations gives the book an aesthetic interest. It is the authors' hope that students of mycology and related fields will find this volume a useful supplement to the usual mycology texts. Knowing the meanings of generic names aids in remembering the organisms themselves and enhances the appreciation of these fascinating organisms.

Most of the genera included in this book belong to the kingdom Fungi. In addition, genera are included from two other kingdoms previously regarded as fungi: the Chromista and the Protozoa. These nonfungal genera are treated in mycology courses and studied by mycologists. The 1,592 generic names of these three kingdoms are arranged alphabetically to facilitate the search of any genus. The entries provide basic biological phenomena among these organisms, which exhibit an impressive richness of sizes, shapes, and colors, as well as different somatic and reproductive structures, diverse physiological and biochemical activities adapted to distinct habitats, and varied life cycles as saprobes, symbionts, and parasites in the living world.

Like the *Illustrated Dictionary of Mycology, Second Edition* (written by Miguel Ulloa and Richard T. Hanlin and published in 2012 by The American Phytopathological Society), which included more than 5,000 technical terms and 2,708 color illustrations (2,036 photographs and 672 drawings), the present work was formatted using Adobe InDesign, a program that provides better spacing of the words and an overall improvement in the appearance of the text. All the generic and species names of fungi treated, as well as the abbreviations of the journals and books cited in the references, were checked in *Index*

Fungorum and *Mycobank* to make them as current as possible. Many generic names of fungi have one or more synonyms, as cited in the entries. The classification of the treated generic and species names depends on the period of time that this work was completed. Taxonomic variations, cited as *–see* (for referring to the valid names) and *syn.* (to indicate a synonym), are part of the constant process of study and evaluation of these groups of organisms to achieve a phylogenetic classification of the fungi; the outcome of this course of events will result in more changes of names and their classifications.

According to the *Dictionary of the Fungi, Tenth Edition* (written by P. M. Kirk et al. and published in 2008), the kingdom Fungi includes 8,283 genera; Chromista, 126 genera; and Protozoa, 122 genera. Thus, the total number of recognized genera treated in this work (1,592) is less than 18% of the number of genera reported in the mycological literature. Given the small number of genera described in the two latter kingdoms, genera are lacking for some letters of the alphabet.

We are grateful to our students, colleagues, and others for providing helpful comments, suggestions, and criticisms, which have contributed to improving this volume. Dr. Richard Hanlin initiated the translation of the book *Etimología* in 1994 for making the English version, but other tasks (especially the preparation of the dictionaries of fungi by Ulloa and Hanlin) resulted in a delay of this new book written in English. Our thanks go to biologist Abraham Medina, Curator of Photography, Comisión Nacional Para el Conocimiento y Uso de la Biodiversidad, Mexico City, who helped search the literature for preparing some of the basidiomycete entries.

Finally, we wish to present in this preface a summary of the characteristics of the title and subtitles of this volume: *Illustrated Generic Names of Fungi: Etymology, Descriptions, Classifications, and References, Featuring More Than 1,000 Original Watercolors.* These elements explain the etymological analyses of generic names in all entries; the original bibliographic reference in which the generic name was first published; the original description of the structure of the fungus and the biological importance of each generic name; and the updated classifications from genus to the three kingdoms studied by mycologists (Fungi, Chromista, and Protozoa) for the total number of entries.

We also wish to provide special recognition of the valuable editorial review and corrections of the whole text made by Dr. Amy Rossman, Senior Editor for APS PRESS. She performed with professionalism and enthusiasm for several months to produce an improved book of illustrated generic names of fungi. We also thank the technical assistance and support of Samuel Aguilar, who contributed mainly in helping to prepare the images for insertion into the text and in the final formatting of the book.

We acknowledge the editorial advice of Antonio Bolívar, Managing Editor of Redacta, S.A., Mexico City, and the interest and help of Dr. Víctor Manuel Sánchez Cordero, Director of IBUNAM, in the undertaking of this project. This institution provided financial support to APS to reduce the cost of producing the book.

Miguel Ulloa
Elvira Aguirre-Acosta
Mexico City

Introduction

The primary purpose of *Illustrated Generic Names of Fungi* is to help readers obtain answers in pursuing biological information for any fungal genus, including the authority or authorities who first discovered and described the genus; the current classification of the genus, from family to kingdom; the etymology or meaning of the name of the genus; a basic description of the genus; and an image of a species in the genus.

One thousand five hundred and ninety-two generic names of fungi are considered in this book. In some cases, the generic name was described many years ago; for example, A. von Haller published the name *Trichia* in *Historia stirpium indigenarum helvetiae* in 1768, more than 250 years ago. Other generic names have been described only recently; *Juglanconis* was described by Voglmayr and Jaklitsch in 2017 (*Persoonia* 38:142). Each entry includes the original bibliographic citation of the generic name, followed by the current classification from genus to kingdom. Most genera are in the kingdom Fungi or one of the other nonfungal kingdoms, Chromista or Protozoa, with organisms that have been studied by mycologists. The genera included are from all over the world, describing fungal diversity from five continents. Brief diagnostic morphological characteristics of each genus are presented. These descriptions also mention the ecology and mode of nutrition of each genus as parasites on organisms ranging from protozoa to vertebrate hosts, including human mycoses caused by fungi, and symbionts with insects and many groups of arthropod hosts, as well as those fungi associated with plants (decomposers, mycorrhizae, and pathogens).

An important aspect of the content of this book is etymology, which is the study of the origins and meanings of the words. For the generic names of fungi, the etymology rests on knowledge of the origins, structures, and transformations of words. The etymological meaning of any word is an essential part of linguistics—indispensable to speaking and writing with the accuracy, clarity, precision, and elegance that distinguishes academic and scientific language. The technical and scientific vocabulary are derived from ancient words, mainly Greek and Latin, from which innumerable words or neologisms have been derived. Etymology is a natural and constant orthographic norm. It expresses the essential basis of the definition, providing a powerful aid to memory to more easily recall the meaning of a word.

Another key aspect of this book is the nature of the illustrations. More than 1,000 original watercolors of fungal species are included, all created with passion and fervor by Miguel Ulloa. Not only do these illustrations give the book a sense of artistic value, but they also serve to distinguish it from the usual mycology texts.

The authors hope that the features combined in this work will be well received and useful to students of this fascinating group of organisms.

Preface to the Spanish Edition of
Etimología e Iconografía de Géneros de Hongos
by Miguel Ulloa y Teófilo Herrera (1994)

In this book are presented the etymological analyses of 807 generic names of fungi, arranged taxonomically. These names and their classification correspond to those included in the book *El reino de los hongos: micología básica y aplicada*, by Teófilo Herrera and Miguel Ulloa (1990), and they belong to 80 orders and 21 classes distributed among the three taxonomic divisions, Myxomycota, Eumycota and Lichens, into which the kingdom Fungi is divided.

The primary purpose of this work is to bring together, in a single work, the annotated etymology of each generic name considered, including, besides the author(s), the original bibliographic citation of the description of the genus, the Greek or Latin roots (in the majority of cases, both, or occasionally of another language) from which they have been derived, and a brief commentary on the etymology or true significance of the name, since this can be related to some biological characteristic of the organism, whether morphological, physiological or ecological, with its mode of living, habitat, or some other peculiarity that distinguishes it. Some generic names are dedicated to a person or can be derived from the place where it was found. In addition, some data on the description or other aspects of the biology of each genus are included. To facilitate understanding the Greek roots of the generic names analyzed, the Greek characters have been transliterated into Spanish.

The bibliographic citations follow B-P-H (*Botanica Periodicum Huntianum*), edited by Lawrence et al., 1968, except for journals not included in this catalog, which are cited as they were found. In a few cases it was not possible to find the title of the work in which the genus was described, so only the name of the journal where it was published is indicated.

Much care has been taken to make the book a reliable and well documented source that can be consulted by those desiring to know the significance of the generic names of the many fungi that are usually encountered in mycological research, teaching and publication. When the respective authors have indicated the origin, significance and application of the names in their works, the task is relatively simple, but when these important clues are not included, which unfortunately is the most common situation, it is necessary to analyze the etymology of the generic names by referring to many other bibliographic sources (such as dictionaries of Latin and Greek, dictionaries of mycology and in general all the specialized mycological works that it was possible to consult), to try to arrive at a coherent explanation of their true significance and application. For this reason, and because in many cases it was not possible to find the works with the original description of the genus, for some generic names only an approximate explanation could be given and it is possible to have inadvertently committed some errors or omissions. Therefore, the users of this book are asked to bring to the attention of the authors any clarification or irregularity so that necessary corrections can be made in case future editions are made.

It is important to point out the advantage of analyzing and commenting on the etymology of the generic names of fungi since, in the majority of cases, these names (like those of the other living beings described by science) include useful information, including the spelling, which contributes to achieving a more correct, intelligent and professional use of these names, so that one can avoid, where possible, memorization without understanding their true meaning. There were relatively few generic names of fungi for which the authors gave at least the etymology, and in the great majority of cases reviewed neither the etymology nor any commentary was found. For them, in the present book, it was endeavored to make the etymological analysis, incorporating a succinct explanation of the etymology of all of the generic names analyzed, including those in which the meaning appeared obvious. However, in a few cases it was not possible to determine their etymology, probably because of their remote historic origin or due to the name being assigned capriciously, but even in these cases an attempt was made to decipher and provide some commentary about their morphology, biology and importance that characterizes or relates them to the fungus in question.

Of the works consulted that bring together the greatest number of generic names of fungi, for those that in many cases give at least the Latin etymology, is the *Sylloge Fungorum* (1882-1913) by Pier Andrea Saccardo, although the etymology is not annotated or the commentary is generally very sparse, in addition to which the work lacks illustrations and is written entirely in Latin. It should be noted that previously it was a little more common for authors to follow the norm of explaining the origin of scientific names used, including genera and species, a norm that unfortunately very few follow at present.

The present work is intended to make up for these types of omissions and contribute to filling a void which, in the opinion of the authors, has existed in mycology from our time in Mexico as well as in other Spanish speaking countries, and probably throughout the world, since this *Etimología e iconografía de géneros de hongos* contains a considerable quantity of generic names, with the corresponding illustrations, representing almost all of the orders and classes of the kingdom Fungi. In this respect it is fitting to point out the recent edition of the work of Miguel Oltra, *Origen etimológico de los nombres científicos de los hongos*, which the Mycological Society of Madrid published in November, 1991, as the first number of the *Monographs* of this Society. It is hoped that Oltra's work will complement the one published here, and vice versa.

With the object of making the book more useful and attractive, 795 illustrations have been included. The discrepancy of this number with the 807 genera contained in the book is due to the fact that it was not possible to find the information necessary to illustrate some genera, and figures were not included for a few that are considered synonyms of others, or whose morphology was very similar to that of other genera illustrated; moreover, for some, more than one figure was included. In the majority of the 795 figures the drawings or photos are of a known species that is representative of the genus in question (i.e., in a few cases the figures refer to unidentified species of the respective genus, which are cited as sp.) which shows, when possible, the characteristic related to the applications of the name of that genus. All of the drawings included in this book (406) are originals made by Miguel Ulloa in black and white, water color, and some with ball-point pen. Of the 389 photographs, 229 are by the same author (MU), although the coauthor of the work, Teófilo Herrera, and various collaborators and colleagues also contributed valuable photographs. The initials in the captions of the figures indicate their authors; the names corresponding to these initials, as well as the institutions where these authors work, are presented in alphabetical order in the credits of photographs at the end of the book. The captions for the figures give the scale at which they are reproduced.

We hope that the incorporation of numerous graphic elements in the book provides valuable information for its use. The inquietude of undertaking and including these graphic elements arises from the firm conviction of Miguel Ulloa that images enrich in a notable manner, or even surpass, written information, as has been reflected in several of his mycological works, such as the *Atlas of Introductory Mycology* (with coauthor Richard T. Hanlin, 1988), *El reino de los hongos: micología básica y aplicada* (with coauthor Teófilo Herrera, 1990) and the *Diccionario ilustrado de micología* (as sole author, 1991). Several of the figures included in this book have been utilized to illustrate some concepts contained in these three works, but it should be made clear that the great majority of the illustrations in the present book were especially made for it. Thus, e.g., the 703 figures contained in *El reino de los hongos* represent only 269 genera of fungi which were illlustrated in some manner, so that more than 540 genera included in that work lack an illustration, and in this book of etymology and iconography the major part of the 807 genera that it contains were illustrated with at least one figure.

It is appropriate and necessary to recognize the valuable and special collaboration of the following colleagues in the Institute of Biology, UNAM, in various stages of the work. Biologist Calixto Benavides participated actively in arranging the figures in the book, in obtaining photographs, in the preparation of various figure captions and calculating scales for them, among other tasks. MSc. Elvira Aguirre-Acosta and Dr. Patricia Lappe collaborated with great care and shyness in the search of mycological references and the typing of the book text. Elvira Aguirre-Acosta developed the index included in the book and she helped in the selection of herbarium specimens for taking photographs. Also recognized is the participation of Biologist Margarita Villegas and MSc. Joaquín Cifuentes of the Faculty of Sciences of UNAM, who helped in the initial stages of the work, in searching the bibliographic information that was later used in the etymological analyses, mainly of genera belonging to various Agaricales, Aphyllophorales and other basidiomycetes.

The authors also express their gratitude to the following persons and institutions: Dr. Evangelina Pérez-Silva of the Institute of Biology, UNAM, and Biologists Leticia Montoya Bello and Víctor Bandala Muñoz, as well as to Dr. Gastón Guzmán, of the Institute of Ecology, Xalapa, Veracruz, who furnished herbarium material for taking pictures, as did Biologists Luis López Eustaquio and Daniel Portugal of the Autonomous University of the State of Morelos, Biologist Rosario Vázquez of the National School of Biological Sciences of IPN, and Biologist Margarita Villegas of the Faculty of Sciences of UNAM. Dr. Robert W. Lichtwardt of the University of Kansas, USA, sent useful bibliographic information on the etymology of some generic names of Trichomycetes. MSc. Angélica Calderón-Villagómez, of the Institute of Biology, UNAM, collaborated in locating some bibliographic citations. Dr. Rupert Barneby, of the New York Botanical Garden, and Fernando Chiang, of the Institute of Biology, UNAM, carefully reviewed the work.

Finally, we sincerely thank Dr. Antonio Lot Helgueras, Director of the Institute of Biology, UNAM, for his support and interest in the undertaking of this book, as well as to the personnel of the publisher for their accurate and professional editorial work.

Miguel Ulloa
Teófilo Herrera
Mexico City

Table of Contents

Preface	v
Introduction	vii
Preface to *Etimología e Iconografía de Géneros de Hongos* **(1994)**	ix
Kingdom Fungi	1
A	3
B	39
C	59
D	105
E	123
F	141
G	149
H	167
I	189
J	193
K	197
L	205
M	223
N	251
O	259
P	269
Q	311
R	313
S	325
T	359
U	379
V	387
W	393
X	397
Y	403
Z	407
Kingdom Chromista	411
Kingdom Protozoa	429
Abbreviations and symbols	449
Bibliography	451

Kingdom Fungi

Pilobolus, Ascotricha, Psilocybe, Monochaetia, Diplocarpon, Crucibulum, Arachnopeziza

Auricularia, Cladonia, Uncinocarpus, Petriella, Entoloma, Zasmidium

A

Armillaria ostoyae

A

Abgliophragma R. Y. Roy & Gujarati
Wiesneriomycetaceae, Inc. sed., Inc. sed., Inc. sed., Pezizomycotina, Ascomycota, Fungi
Roy R. Y., S. Gujarati. 1966. *Abgliophragma setosum*, gen. nov. et sp. nov. from soil. *Trans. Br. mycol. Soc.* 49:363-365.
From L. prefix *ab-*, from, away + genus *Gliophragma*, to indicate similarity to this genus. Hyphae black, tuberculate, bearing sporodochia surrounded by black, septate, setae; conidiophores subhyaline, unbranched, arranged in a palisade-like manner with sterigmatoid structures at tips; conidia hyaline, several-celled, cells connected by short isthmi, produced singly. Isolated from soil, Banaras Hindu University Campus, Varanasi, India.

Abgliophragma setosum R. Y. Roy & Gujarati: sporodochia with setae, conidiophores, and conidia, x 200.

Abrachium Baseia & T. S. Cabral
Phallaceae, Phallales, Phallomycetidae, Agaricomycetes, Agaricomycotina, Basidiomycota, Fungi
Cabral, T. S., et al. 2012. *Abrachium*, a new genus in the Clathraceae, and *Itajahya* reassessed. *Mycotaxon* 119:419-429.
From L. *a-*, absence < Gr. *brachiôn*, upper arm > L. *brachium*, arm, i.e., without arms. Fructification when immature an "egg", subglobose, epigeous, white, with basal mycelial cords. Stipe cylindrical, spongy; receptacle spongy, sunflower-shaped, without arms, with central disc covered by a gelatinous gleba. Spores cylindrical to bacilloid, hyaline, smooth. The genus is separated from *Aseroë* on the basis of molecular and morphological characteristics. In the Brazilian Atlantic rainforest and northeastern Brazil.

Abrachium floriforme (Baseia & Calonge) Baseia & T. S. Cabral: basidiocarp, on soil, x 0.5.

Absidia Tiegh.
Cunninghamellaceae, Mucorales, Inc . sed., Inc . sed., Mucoromycotina, Zygomycota, Fungi
van Tieghem, P. 1878. Troisième mémoire sur les mucorinées. *Annls Sci. Nat., Bot.*, sér. 6,4(4):312-398.
From L. *absidis* < Gr. *apsís*, joint or keystone of an arch + L. suf. *-ia*, which denotes quality or state of, referring to the arched, semicircular shape of the sporangium as well as to the columella. Common saprobe in soil, grows on plant detritus. Some species are pathogenic to animals.

Acaulospora

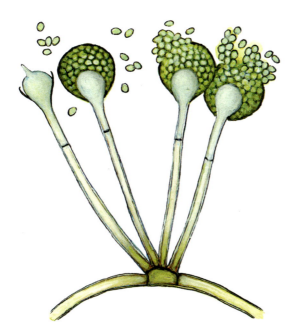

Absidia spinosa Lendn.: sporangiophores, sporangia, and columella, x 400.

Acarellina Bat. & H. Maia
Inc. sed., Inc. sed., Inc. sed., Inc. sed., Pezizomycotina, Ascomycota, Fungi
Batista, A. C., H. da Silva Maia. 1960. *Acarellina psidii* - tipo de um novo gênero de fungos Leptostromataceae. *Publicações Inst. Micol. Recife 246*:1-9.
From genus *Acarella* + L. suf. *-ina*, denoting likeness, for the similarity of the fruiting bodies. Superficial mycelium absent. Pycnostromata superficial, dimidiate-scutate, pale brown, smooth, with a circular central ostiole and pelliculose borders. Upper wall of prosenchyma, composed of radiate hyphae. Basal wall thin, subhyaline. Conidiophores in a basal layer, cylindrical, simple, hyaline. Conidia acrogenous, hyaline, ellipsoid, 1-celled, smooth. Collected on leaves of *Psidium albidum* in Vitoria, Pernambuco, Brazil.

Acarocybiopsis J. Mena, et al.
Inc. sed., Inc. sed., Inc. sed., Inc. sed., Pezizomycotina, Ascomycota, Fungi
Mena-Portales, J., et al. 1999. *Acarocybiopsis*, a new genus of synnematous hyphomycetes from Cuba. *Mycol. Res.* 103:1032-1034.
From genus *Acarocybe* + Gr. suf. *-ópsis*, aspect, i.e., similar to this genus. Conidiomata synnematous, indeterminate, brown. Stipe mostly of parallel, smooth, septate hyphae, of two kinds: an ascending or erect hypha, brown, and paler descending hyphae. Capitulum terminal, composed of a single conidium. Conidiogenous cells integrated, terminal, percurrent, with a lateral, uncinate protuberance that bends downwards and produces descending hypha, with percurrent enteroblastic apical proliferations. Conidia holoblastic, solitary, dry, acrogenous, ellipsoidal to obovoid, brown, smooth, secession schizolytic. Collected on fallen branches in the Sierra de Cubitas, Camagüey, Cuba.

Acarospora A. Massal.
Acarosporaceae, Acarosporales, Acarosporomycetidae, Lecanoromycetes, Pezizomycotina, Ascomycota, Fungi
Massalongo, A. B. 1852. *Ric. auton. lich. crost.* (Verona):27.
From Gr. *ákari*, harvest-mite, mite + *sporá*, spore, for the shape of the spores, oblong-ellipsoid, like the shape of a mite. Thallus crustaceous, squamulose-areolate, frequently lobed toward margin. Apothecia immersed or superficial, with a differentiated thalloid excipulum. Grows preferentially on acid rocks, in deserts as well as in alpine and subalpine forests. Photobiont *Protococcus*.

Acarospora sinopica (Wahlenb.) Körb.: crustaceous thallus on rock, x 10.

Acaulopage Drechsler
Zoopagaceae, Zoopagales, Inc. sed., Inc. sed., Zoopagomycotina, Zygomycota, Fungi
Drechsler, C. 1935. Some non-catenulate conidial phycomycetes preying on terricolous amoebae. *Mycologia* 27(2):185-186.
From Gr. *a*, without + *kauléo*, to grow a stalk + *páge*, loop, snare. Merosporangia sessile or nearly so, directly on hyphae, i.e., without distinctive sporangiophores, trapping amoebae that adhere by means of a sticky substance, and then invading them by haustoria. Hyphae and sporangiophores formed externally.

Acaulospora Gerd. & Trappe
Acaulosporaceae, Diversisporales, Inc. sed., Glomeromycetes, Glomeromycotina, Glomeromycota, Fungi
Gerdemann, J. W., J. M. Trappe. 1974. The Endogonaceae in the Pacific Northwest. *Mycol. Mem.* 5:31-36.
From Gr. *a*, without + *kauléo*, to grow a stem + *sporá*, spore, i.e., a spore without a stem or column, sessile. Azygospores sessile. Endomycorrhizogenous fungi associated with a variety of plants (grasses, legumes, palms and others).

Acervus

Acaulopage tetraceros Drechsler: hyphae and haustoria within a terricolous amoeba, and external hyphae with obconic spores, x 430.

Acaulospora colombiana (Spain & N.C. Schenck) Kaonongbua, et al.: development of the spore from the vesicle or saccule; the vesicle becomes empty, due to its contents fills out the mature spore, x 200.

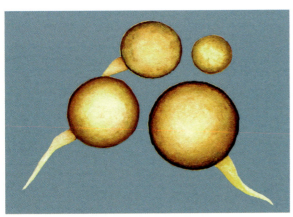

Acaulospora laevis Gerd. & Trappe: azygospores, x 260. Described in Oregon, U.S.A.

Acervus Kanouse

Pyronemataceae, Pezizales, Pezizomycetidae, Pezizomycetes, Pezizomycotina, Ascomycota, Fungi
Kanouse, B. B. 1938. Notes on new or unusual Michigan Discomycetes. V. *Pap. Mich. Acad. Sci.* 23:149-154.
From L. *acervus*, heap, for the shape of the clustered ascomata. Apothecia sessile, densely cespitose in clusters varying from few to many, much contorted from mutual pressure, arising from a black, rubbery, sclerotiform base, externally slightly verrucose, soft leathery when fresh, somewhat friable when dry. Hymenium orange-colored. Asci cylindric, inoperculate, 8-spored. Ascospores ellipsoid, simple, smooth. Paraphyses clavate, curved, orange-colored. On soil.

Acervus epispartius (Berk. & Broome) Pfister: ascomata on soil, x 1.5.

Acetabula (Fr.) Fuckel—see Helvella L.

Helvellaceae, Pezizales, Pezizomycetidae, Pezizomycetes, Pezizomycotina, Ascomycota, Fungi
Fuckel, K. W. G. L. 1870. Symbolae mycologicae. Beiträge zur Kenntnis der rheinischen Pilze. *Jb. nassau. Ver. Naturk.* 23-24:330.
From L. *acetabulum*, vinegar bottle, cup or glass for vinegar.

Achaetobotrys Bat. & Cif.

Antennulariellaceae, Capnodiales, Dothideomycetidae, Dothideomycetes, Pezizomycotina, Ascomycota, Fungi
Batista, A. C., R. Ciferri. 1963. Capnodiales. *Saccardoa* 2:1-296.
From Gr. *a*, which expresses deprivation or negation of something + *chaíte*, crest > L. *chaeta*, bristle, or *chaete*, long hair, mane + L. *botrys*, derived from Gr. *bótrys*, raceme of grapes. Fructifications glabrous, globose, sessile, grouped in ascostromata as in a raceme. Ascospores brown, 1-septate. Saprobic on the surface of leaves, where it develops in the sugary secretions of parasitized insects.

Achaetomiella Arx—see Chaetomium Kunze

Chaetomiaceae, Sordariales, Sordariomycetidae, Sordariomycetes, Pezizomycotina, Ascomycota, Fungi
Arx, J. A. von. 1970. *Gen. Fungi Sporul. Cult.* (Lehre), p. 247.
From Gr. *a*, without + genus *Chaetomium* Kunze (this from Gr. *chaíte*, crest, long hair, mane + L. dim. suf. *-ium*) + L. dim. suf. *-ella*.

Acrocalymma

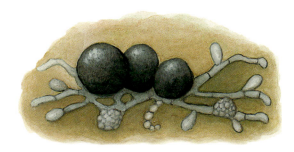

Achaetobotrys affinis (L. R. Fraser) Bat. & Cif.: moniliform hyphae and obovoid, dark pseudothecia, x 500.

Achaetomium J. N. Rai, et al.
Chaetomiaceae, Sordariales, Sordariomycetidae, Sordariomycetes, Pezizomycotina, Ascomycota, Fungi
Rai, J. N., et al. 1964. *Achaetomium*, a new genus of Ascomycetes. *Can. J. Bot.* 42(6):693-697.
From Gr. *a*, absence of something + genus *Chaetomium* Kunze, i.e, like *Chaetomium*, but lacking the prominent hairs of the latter genus. Ascomata superficial, scattered or gregarious. Perithecia subglobose to obpyriform, ostiolate, papillate or with a short ostiolar neck, yellowish to gray when young, becoming dark brown at maturity, appearing glabrous, but covered by a light-colored or yellowish hyphal tomentum. Wall of perithecium thick. Paraphyses lacking. Asci unitunicate, cylindrical or subclavate, undifferentiated at apex, 8-spored, evanescent at maturity. Ascospores 1-celled, globose to ellipsoidal, limoniform or rhomboid, dark brown, smooth, with a single lateral or terminal germ pore, extruded in a cirrhus. On soils.

Aciculoconidium D. S. King & S. C. Jong
Inc. sed., Saccharomycetales, Saccharomycetidae, Saccharomycetes, Saccharomycotina, Ascomycota, Fungi
King, D. S., S. C. Jong. 1976. *Aciculoconidium*: a new hyphomycetous genus to accommodate *Trichosporon aculeatum*. *Mycotaxon* 3(3):401-408.
From L. *acicula*, point, needle < Gr. *akídos* + NL. *conidium* < Gr. *kónis*, dust, for the pointed terminal cell of the conidia. Mycelium hyaline, composed of branched, septate hyphae. Blastoconidia catenulate, candida-like, subspherical to ellipsoid; terminal conidia needle-shaped, rounded at proximal end and pointed at distal end. Isolated from *Drosophila pinicola* in California.

Acremonium Link
Inc. sed., Hypocreales, Hypocreomycetidae, Sordariomycetes, Pezizomycotina, Ascomycota, Fungi
Link, J. H. F. 1809. Observationes in ordines plantarum naturales. *Mag. Ges. Naturf. Freunde, Berlin* 3(1-2):15.
From Gr. *akrémon*, thicket + L. dim. suf. -*ium*, referring to the branching of the conidiophores. Conidiophores compound in basal part, produces phialoconidia in viscous heads or in dry chains. Saprobic fungus in soil, although it is also found in the rhizosphere of plants and as a secondary invader of decomposing vegetable remains. One species causes maduromycosis in humans.

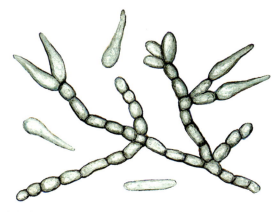

Aciculoconidium aculeatum (Phaff, M. W. Mill. & Shifrine) D. S. King & S. C. Jong: septate hyphae, blastoconidia, and terminal needle-shaped conidia, x 1,500.

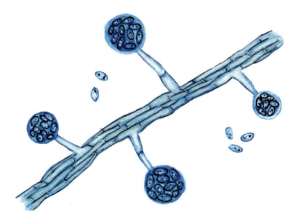

Acremonium alternatum Link: hyphae package with conidiophores and gloeoid conidial heads, x 500.

Acrocalymma Alcorn & J. A. G. Irwin
Acrocalymmaceae, Pleosporales, Pleosporomycetidae, Dothideomycetes, Pezizomycotina, Ascomycota, Fungi
Alcorn, J. L., J. A. G. Irwin. 1987. *Acrocalymma medicaginis* gen. et sp. nov. causing root and crown rot of *Medicago sativa* in Australia. *Trans. Br. mycol. Soc.* 88(2):163-167.
From Gr. *akrón* > L. *acros*, at the tip, extremity + Gr. *kalymma* > L. *calymma*, hood, for the polar helmet-shaped mucilaginous appendages on the conidia. Conidiomata pycnidial, separate, globose, papillate or rostrate, unilocular, ostiolate. Conidiogenous cells hyaline, discrete, cylindrical to lageniform, smooth, determinate, phialidic, arising from inner cells of wall. Conidia cylindrical to fusoid, straight, hyaline, aseptate, at length 1-3 septate, pale brown, smooth, with polar helmet-shaped mucilaginous appendages originating from a sheath around young conidium. On stems of *Medicago sativa* in Queensland, Australia.

Acrotheca

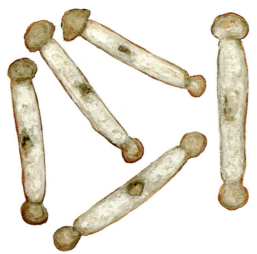

Acrocalymma medicaginis Alcorn & J. A. G. Irwin: conidia, x 2,100.

Acrotheca Fuckel—see **Ramularia** Unger
Mycosphaerellaceae, Capnodiales, Dothideomycetidae, Dothideomycetes, Pezizomycotina, Ascomycota, Fungi
Fuckel, K. W. G. L. 1860. *Jb. nassau. Ver. Naturk.* 15:43.
From Gr. *ákros*, at the end, apex + *théke*, box, receptacle.

Actidium Fr.
Mytilinidiaceae, Mytilinidiales, Inc. sed., Dothideomycetes, Pezizomycotina, Ascomycota, Fungi
Fries, E. M. 1815. *Observ. mycol.* (Havniae) 1:190.
From Gr. *aktís*, genit. *aktínos*, spokes of a wheel + L. dim. suf. *-idium*, for the star-shaped ascostromata (hysterothecia). Ascostromata hysterothecioid, black, carbonaceous, partially immersed, superficial or situated on a subiculum. Ascospores 1-septate, generally brown. Saprobic on wood or tree branches.

Actidium hysterioides Fr.: stellate hysterothecia on decorticated wood of *Pinus*, with a longitudinal slit on each arm, x 30.

Actinomucor Schostak.
Mucoraceae, Mucorales, Inc. sed., Inc. sed., Mucoromycotina, Zygomycota, Fungi
Schostakowitsch, W. 1898. *Actinomucor repens* n. g. n. sp. *Ber. dt. bot. Ges.* 16:155-158.
From Gr. *aktís*, genit. *aktínos*, radius + L. *mucor*, mold < *muceo*, to become moldy, to spoil. Sporangiophore with a subterminal verticil, radial branches producing sporangia. Saprobic in soil. Used commercially in the production of tofu or Chinese cheese from soybean.

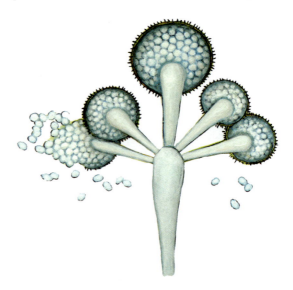

Actinomucor elegans (Eidam) C. R. Benj. & Hesselt.: verticillate sporangiophore with echinulate sporangia, x 850.

Adomia S. Schatz
Xylariaceae, Xylariales, Xylariomycetidae, Sordariomycetes, Pezizomycotina, Ascomycota, Fungi
Schatz, S. 1985. *Adomia avicenniae*: a new ascomycetous genus from Red Sea and Australian mangroves. *Trans. Br. mycol. Soc.* 84(3):555-559.
From Hebrew *adom*, red + L. des. *-ia*, in reference to the Red Sea location of the holotype specimen. Ascomata solitary, ampulliform to subglobose, immersed, ostiolate, periphysate, clypeate, hyphae of ostiole fused at tip, forming an orange-colored, brittle cover. Peridium prosenchymatous, brown below, appearing black on surface. Paraphyses originating from base and apex of centrum. Asci 8-spored, cylindrical, unitunicate, with a persistent refractive cap in apex, developing from base of centrum. Ascospores ellipsoid, unicellular, brown with a single appendage. On pneumatophores of *Avicennia marina*, Sinai-Red Sea, Egypt.

Aessosporon Van der Walt —see **Sporobolomyces** Kluyver & C. B. Niel
Sporidiobolaceae, Sporidiobolales, Inc. sed., Microbotryomycetes, Pucciniomycotina, Basidiomycota, Fungi
van der Walt, J. P. 1970. The perfect and imperfect states of *Sporobolomyces salmonicolor*. *Antonie van Leeuwenhoek* 36:49-55.
From Gr. *aísso*, to throw + *spóros*, *spóron*, spore.

Agaricus L.
Agaricaceae, Agaricales, Agaricomycetidae, Agaricomycetes, Agaricomycotina, Basidiomycota, Fungi

Linnaeus, C. von, 1753. *Sp. pl.* 2:1171.

From Gr. *agarikón*, mushroom, excrescence, a name applied since the time of Dioscorides to a particular type of fungus of probable origin in Agari or Agaroi, a Scythian town of Agaria in Sarmatia, on the northern coast of the Sea of Azov, where there existed an expert in medicine who probably used the fungus of the same name, later called "*agaricum*" in Latin. Basidiocarps medium or large, fleshy, with free gills, and a cottony or membranaceous veil. Spores chocolate brown. Some species are edible and of important commercial value, while others are somewhat toxic. *Agaricus bisporus* (J. E. Lange) Imbach is the mushroom produced commercially; *A. campestris* L. is the common meadow mushroom.

Agaricus benesii (Pilát) Pilát: basidiocarp, x 1.

Agarwalomyces R. K. Verma & Kamal

Inc. sed., Inc. sed., Inc. sed., Inc. sed., Pezizomycotina, Ascomycota, Fungi

Verma, R. K., Kamal. 1987. *Agarwalomyces indicus* gen. et sp. nov., a fructicolous synnematous hyphomycete from Uttar Pradesh. *Trans. Br. mycol. Soc.* 89(4):596-599. Named in honor of the Indian mycologist *G. P. Agarwal* + connective -o- + L. *myces*, fungus, for his service as Head, Department of Biological Sciences, Rani Durgavati University, Jabalpur, India. Colonies black, mycelium immersed; hyphae branched, septate, hyaline to pale brown. Stromata pseudoparenchymatous, partially embedded in host tissue, black. Conidiomata arising from stromata, of scattered erect black synnemata with rounded heads, individual hyphae fused, parallel, dark brown to black, straight. Conidiophores septate, branched, rough, pale brown. Conidiogenous cells integrated on free ends of conidiophores, cylindrical, polyblastic, terminal, intercalary, with irregular sympodial proliferations, unthickened scars. Conidia holoblastic, dry, catenate or botryose, acropleurogenous, globose, aseptate, pale brown to dark brown, verruculose. On living fruits of undetermined Lythraceae, Chaubattia, Ranikhet, Almora, Uttar Pradesh, India.

Agrocybe Fayod

Strophariaceae, Agaricales, Agaricomycetidae, Agaricomycetes, Agaricomycotina, Basidiomycota, Fungi

Fayod, V. 1889. Padrome d'une histoire naturelle desagaricinées. *Annls Sci. Nat.*, Bot., sér. 7, 9:358.

From Gr. *agrós*, field, crop + *kýbe*, pileus, head, because it generally grows outside of the forest, in open areas. Pileus generally flat to planoconvex, glabrous, occasionally areolate; gills adherent, dull brown; stipe frequently with a membranaceous ring. Grows in open, humid areas, such as gardens, fields, or near forests. Several of the species have been reported as edible; some are cultivated in southeastern Europe.

Agrocybe praecox (Pers.) Fayod: basidiocarps growing in garden, x 1.

Agrogaster D. A. Reid

Bolbitiaceae, Agaricales, Agaricomycetidae, Agaricomycetes, Agaricomycotina, Basidiomycota, Fungi

Reid, D. A. 1986. New or interesting records of Australasian Basidiomycetes: VI. *Trans. Br. mycol. Soc.* 86(3):429-440.

From Gr. *agrós*, a field + *gastér*, belly, stomach. Sporophore agaricoid or secotioid, related to *Agrocybe*. Pileus conical to almost globose or convex, smooth, somewhat hygrophanus. Pileus border easily or distinctly separating from stipe. Lamellae regular or very irregular and anastomosing to form chambers. Cuticle of pileus cellular, of sphaerocysts. Cheilo- and pleurocystidia abundant. Spores smooth, thin-walled, subhyaline to brown, elliptical, with a more or less central apiculus. Collected under *Podocarpus dacrydioides* and *P. spicatus*, in Adhuriri and Kaituna Reserves, New Zealand.

Aigialus

Aigialus Kohlm. & S. Schatz
Aigialaceae, Pleosporales, Pleosporomycetidae, Dothideomycetes, Pezizomycotina, Ascomycota, Fungi
Kohlmeyer, J., S. Schatz. 1985. *Aigialus* gen. nov. (Ascomycetes) with two new marine species from mangroves. *Trans. Br. mycol. Soc.* 86(4):699-707.
From Gr. *aigialos*, coast, seashore, in reference to the habitat of the fungus. Ascomata subglobose in frontal view, laterally compressed, completely or mostly immersed in a black stroma, with a longitudinal furrow at top, ostiolate, apapillate, carbonaceous to coriaceous, black, gregarious. Peridium two-layered. Ostiole depressed or slightly projecting, in center of apical furrow. Pseudoparaphyses trabeculate, embedded in a gelatinous matrix. Asci 8-spored, cylindrical, pedunculate, thick-walled, fissitunicate, not bluing in IKI. Ascospores biseriate, ellipsoidal to broadly fusiform in frontal view, flat on one side, convex on the other, muriform, dark brown except for hyaline to pale brown apical cells, smooth, apical and subapical cells covered by a gelatinous sheath or cap. On submerged roots and branches of *Rhizophora mangle* in the Caribbean Sea.

Ajellomyces McDonough & A. L. Lew.—see **Histoplasma** Darling
Ajellomycetaceae, Onygenales, Eurotiomycetidae, Eurotiomycetes, Pezizomycotina, Ascomycota, Fungi
McDonough, E. S., A. L. Lewis. 1968. The ascigerous stage of *Blastomyces dermatitidis*. *Mycologia* 60(1):76-83.
Dedicated to the Brazilian medical mycologist *Libero Ajello* + L. suf. *-myces* < Gr. *mýkes*, fungus.

Akenomyces G. Arnaud ex D. Hornby
Inc. sed., Inc. sed., Inc. sed., Agaricomycetes, Agaricomycotina, Basidiomycota, Fungi
Hornby, D. 1984. *Akenomyces costatus* sp. nov. and the validation of *Akenomyces* Arnaud. *Trans. Br. mycol. Soc.* 82(4):653-664.
From L. *akene*, achena, a small dry fruit, hence, achene fungus + L. suf. *-myces* < Gr. *mýkes*, fungus. Sclerotia commonly pyriform, also subglobose, ellipsoidal, oval, obovoid, fusiform-obovoid, ovoid, obpyriform, or obclavate on short pedicels with inserted hyaline hairs that extend upwards over sclerotia, often imparting a veined appearance. Sclerotial surface a brown continuous membrane composed of closely united, parallel filaments, enclosing a loose, hyaline, sclerotic tissue of oil-rich filaments and irregular spaces. Subiculum with clamp-connexions.

Albahypha Oehl, et al.—see **Claroideoglomus** C. Walker & A. Schüßler
Glomeraceae, Glomerales, Inc. sed., Glomeromycetes, Inc. sed., Glomeromycota, Fungi
Oehl, F., et al. 2011. Revision of Glomeromycetes with entrophosporoid and glomoid spore formation with three new genera. *Mycotaxon* 117:297-316.
From L. *alba*, white + *hypha*, hypha, referring to the white, slightly funnel-shaped subtending hypha.

Akenomyces costatus D. Hornby: ellipsoidal sclerotium with a veined appearance, x 100. On roots of rotten *Triticum*, in Great Britain.

Albatrellus Gray
Albatrellaceae, Russulales, Inc. sed., Agaricomycetes, Agaricomycotina, Basidiomycota, Fungi
Gray S. F. 1821. *Nat. Arr. Brit. Pl. (London)* 1:645.
From L. *albus*, white + *ater*, atrum, dark, in reference to the colors of the fruiting bodies. Basidioma solitary or in clusters, with cap and stem, stem central or eccentric to lateral, with bases or cap margin fused. Context mostly tough-fleshy, white or becoming brightly colored. Hymenophore regularly poroid. Hyphal system monomitic, generative hyphae septate, with or without amyloid or non-amyloid walls. Terrestrial, widely distributed.

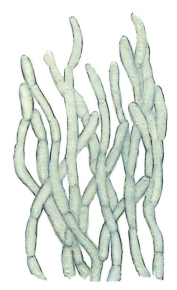

Albatrellus tibetanus H.D. Zheng & P.G. Liu: Pileipellis hyphae. Collected in *Picea* forest, in Tibet, China x 1,000.

Aldonata Sivan. & A. R. P. Sinha

Parmulariaceae, Asterinales, Dothideomycetidae, Dothideomycetes, Pezizomycotina, Ascomycota, Fungi

Sivanesan, A., A. R. P. Sinha. 1989. *Aldonata*, a new ascomycete genus in the Parmulariaceae. *Mycol. Res.* 92(2):246-249.

From genus *Aldona* + L. des. *-ta*, for the similarity to this genus. Leaf parasites forming diffuse, somewhat circular leaf spots. Stromata epiphyllous, subcuticular rarely penetrating epidermis or tissues below, composed of several layers of hyaline, elongated, pseudoparenchymatous cells. Ascomata irregularly aggregated to somewhat stellate in appearance, linear or radiate from a common center, at first subcuticular, later erumpent, black, opening by longitudinal dehiscence. Peridium thick, composed of pseudoparenchymatous cells. Hamathecium of pseudoparaphyses which are filiform, hyaline, septate and branched. Asci broadly clavate, bitunicate with fissitunicate dehiscence, mostly 8-spored, short-stalked. Ascospores clavate, hyaline, smooth, straight, dictyoseptate with 5-10 transverse septa, constricted at septa, often surrounded by a thin mucilaginous sheath, apical cell rounded, basal cell tapering, straight or curved, cylindrical, narrow. Collected on living leaves of *Pterocarpus dracaus* in Port Blair, India.

Alectoria Ach.

Parmeliaceae, Lecanorales, Lecanoromycetidae, Lecanoromycetes, Pezizomycotina, Ascomycota, Fungi

Acharius, E. 1809. *In*: Luyken, *Tent. Hist. Lich.*:95.

From Gr. *aléktor*, rooster, related to a popular belief connecting this genus with the rooster + L. suf. *-ia*, which denotes belonging to, e.g., in L. and Es. the name alectoria is given to the crystalline stone that forms in the liver of some old roosters, and which is supposed to have medicinal properties. Thallus fruticose, cylindrical or flattened cylindric, erect, extended or hanging, with dichotomous or subdichotomous branching. Photobiont *Protococcus*. Grows on conifers, old wood, rocks and soil.

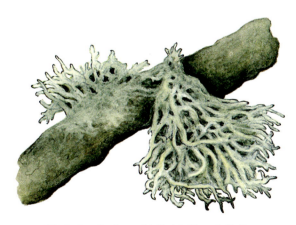

Alectoria ochroleuca (Hoffm.) A. Massal.: thalli on a conifer twig, x 1.

Aleuria Fuckel

Pyronemataceae, Pezizales, Pezizomycetidae, Pezizomycetes, Pezizomycotina, Ascomycota, Fungi

Fuckel, K. W. G. L. 1870. Symbolae mycologicae. Beiträge zur Kenntnis der rheinischen Pilze. *Jb. nassau. Ver. Naturk.*23-24:325.

From Gr. *áleuron*, wheat flour + L. suf. *-aria*, which indicates similarity, possession or connection, for the pruinose to furfuraceous ornamentation on the external surface of the apothecia. Apothecia cupuliform, sessile or pedicellate, covered by short, thin, hyaline hairs. Hymenium yellow, orange or red. Spores reticulate or with spiny bands and crests. Grows on soil, in open gardens and forests in temperate regions.

Aleuria aurantia (Pers.) Fuckel: sessile apothecium on soil, x 10.

Aliquandostipite Inderb.

Aliquandostipitaceae, Jahnulales, Inc. sed., Dothideomycetes, Pezizomycotina, Ascomycota, Fungi

Inderbitzin, P., et al. 2001. Aliquandostipitaceae, a new family for two new tropical ascomycetes with unusually wide hyphae and dimorphic ascomata. *Am. J. Bot.* 88 (1):52-61.

From L. *aliquando*, sometimes, at times + *stipite*, stipitate, because some ascomata are sessile on the mycelium, whereas others are borne at the apex of a stout hyphal stalk. Mycelium superficial on substrate, light to dark brown, hyphae up to 40 μm wide, immersed hyphae narrower. Sessile ascomata single, immersed to erumpent or superficial on substrate, globose to broadly ellipsoidal, papillate, pale brown when young, dark brown with age. Pseudoparaphyses present. Asci clavate, bitunicate, 8-spored. Ascospores oval, 1-septate, constricted at septum, pale brown, with a sheath. Stalked ascomata borne on broad hyphae arising singly or in groups, globose to oval, tapering to stalk. On decaying branches in tropical rainforest in Thailand.

Allantonectria Earle

Nectriaceae, Hypocreales, Hypocreomycetidae, Sordariomycetes, Pezizomycotina, Ascomycota, Fungi

Earle, F. S. 1901. Fungi, pp. 1-30 *In*: E. L. Greene, *Plant. Bak.* 2:1-42.

Allomyces

From Gr. *allantós*, a sausage + genus *Nectria*, for the shape of the ascospores. Perithecia densely cespitose, 12-20 or more perithecia united on a stroma; stromatic clusters erumpent, thickly scattered or subconfluent. Perithecia bright-red, becoming dark dull-red when dry, globose, smooth or slightly roughened, collapsing. Asci 8-spored, clavate, minute, aparaphysate. Ascospores 1-celled, allantoid-cylindric, curved, hyaline, distichous or inordinate, minute. On dead, withered leaves of *Yucca* in Hermosa, Colorado, U.S.A.

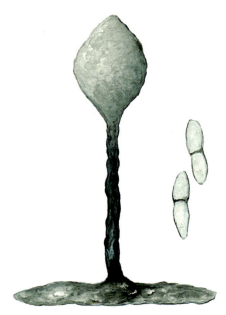

Aliquandostipite khaoyaiensis Inderb.: pedicellate ascoma, x , and free ascospores, x 500.

Allomyces E. J. Butler

Blastocladiaceae, Blastocladiales, Inc. sed., Blastocladiomycetes, Inc. sed., Chytridiomycota, Fungi
Butler, E. J. 1911. On *Allomyces*, a new aquatic fungus. *Ann. Bot.* (London) 25:1023-1034.
From Gr. *állos*, another, something foreign to the organism + *mýkes* > L. *myces*, fungus, a name probably derived from the peculiarity of having alternation of generations in its life cycle, with saprobic sporothalli and gametothalli attached to remains of animal and plant origin.

Alternaria Nees

Pleosporaceae, Pleosporales, Pleosporomycetidae, Dothideomycetes, Pezizomycotina, Ascomycota, Fungi
Nees von Esenbeck, C. G. D. 1817. *Syst. Pilze* (Würzburg):72.
From L. *alternare*, to alternate, take turns + suf. -*ia*, which denotes quality of or state of a being, referring to the shape of the conidia, which are alternately thick and slender. Conidia in simple or branched chains with acropetal succession, muriform porospores, dark, ovate or obclavate, with distal end tapering to a beak. Common in soil, on organic remains, includes plant pathogenic species that are primary or secondary invaders of plants. Spores most frequent in atmosphere and the cause of respiratory allergies in humans.

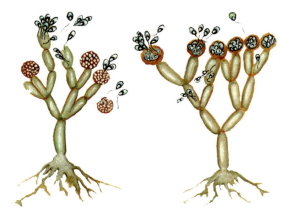

Allomyces macrogynus (R. Emers.) R. Emers. & C. M. Wilson: sporothallus and gametothallus, with rhizoids, x 400.

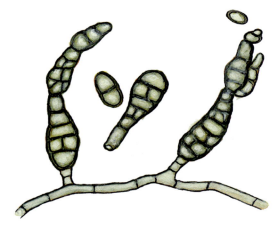

Alternaria alternata (Fr.) Keissl.: conidiophore and muriform porospores, x 500.

Alveariospora M. Silva, et al.

Inc. sed., Inc. sed., Inc. sed., Inc. sed., Pezizomycotina, Ascomycota, Fungi
Da Silva M., et al. 2012. *Alveariospora*, a new anamorphic genus from trichomes of *Dimorphandra mollis* in Brazil. *Mycotaxon* 119:109-116.
From L. *alvearium*, beehive, hive + *spora*, spores, for the shape of the conidia. Colonies on natural substrate effuse, hairy, brown, olivaceous or black. Mycelium superficial, immersed. Conidiophores distinct, single, unbranched, septate, brown or olivaceous, smooth. Conidiogenous cells integrated, terminal, at first with a single terminal conidiogenous locus, then indeterminate, polyblastic, with successive sympodial proliferation, rupturing outer wall around each scar. Conidiogenous loci evident, lenticular, protuberant, thickened, black, conidial secession schizolytic. Conidia solitary, ellipsoidal, oval to broadly navicular, dictyoseptate,

Amauroascus

distoseptate, verruculose or smooth, brown or dark brown, conspicuously cicatrized at base, with a cellular, cylindrical or subulate, brown apical appendage. Collected in the Brazilian Cerrado.

Amanita pantherina (DC.) Krombh.: basidiocarp on soil, x 1.

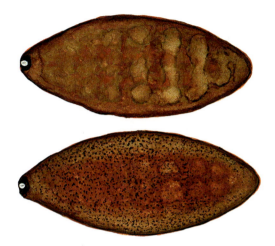

Alveariospora distoseptata Meir. Silva, et al.: ellipsoidal, distoseptate, and verruculose conidia, x 1,000. Isolated from trichomes of *Dimorphandra mollis*, in Minas Gerais, Brazil.

Amanita Pers.

Amanitaceae, Agaricales, Agaricomycetidae, Agaricomycetes, Agaricomycotina, Basidiomycota, Fungi
Persoon, C. H. 1797. *Tent. disp. meth. fung. (Lipsiae)*:65.
From Gr. *amanós*, referring to the name of a mountain in Southeast Asia Minor, between Syria and Cilicia. It is supposed that Galen obtained specimens of fungi from Mount Amano (in Gr. *Amanós* and in L. *Amanus*), to which he applied the Greek name. Possibly derived from the Gr. *amánores*, pustules, for the presence of these on the pileus of various species. Fructifications with universal veil and a partial veil, which disappear, leaving remnants on pileus and/or stipe, at first as warts, pustules or patches on pileus, and as a volva on base of stipe. Partial veil remains as a ring on upper part of stipe. Gills generally free. Spores white to cream-colored. Species lacking a ring previously regarded as *Amanitopsis* but now placed in *Amanita*. Terricolous species, edible as well as toxic; majority mycorrhizogenous, in forests of pines and other conifers and broad-leaved forests or in tropical jungles. Prior to Linnaeus, these fungi were designated under the Latin term *"boletus"*, which now is utilized for another group of fungi.

Amanitopsis Roze—see Amanita Pers.

Amanitaceae, Agaricales, Agaricomycetidae, Agaricomycetes, Agaricomycotina, Basidiomycota, Fungi
Roze, E. 1876. Essai d' une nouvelle classification des Agaricacées. *Bull. Soc. bot. Fr.* 23:45-54.
From genus *Amanita* Pers. + L. suf. *-opsis*, from Gr. *-ópsis*, aspect, appearance, due to the similarity to this genus.

Amanita perpasta Corner & Bas: basidiocarp on soil, x 1. Described in Singapore.

Amauroascus J. Schröt. (syn. Kuehniella G. F. Orr)

Onygenaceae, Onygenales, Eurotiomycetidae, Eurotiomycetes, Pezizomycotina, Ascomycota, Fungi
Schröter, J. 1893. Die Pilze Schlesiens. *In*: J. F. Cohn (ed.), *Krypt.-Fl. Schlesien* (Breslau) 3.2(1-2):210.
From Gr. *amaurós*, dark + *askós* > L. *ascus*, sac, ascus, an apparent reference to the asci with dark spores. Ascomata gymnothecia, spherical to irregular, composed of undifferentiated, loosely interwoven, thin-walled hyphae that surround asci. Gymnothecia formed in dense clusters on mycelium, globose to subglobose or ellipsoid, white, yellow, or brown. Peridial hyphae undifferentiated, but occasionally brownish with slightly thickened walls; appendages lacking or with spiral appendages at periphery. Asci formed in clusters, globose to ovoid, 8-spored, with

Amauroderma

evanescent walls. Ascospores globose, subglobose or elliptical, 1-celled, hyaline, yellow or reddish-brown to dark brown, smooth or appearing roughened under light microscopy, punctate or ridged under scanning electron microscopy. Asexual morphs of arthroaleuriospores and aleuriospores. Racquet hyphae present. On wood, soil, and keratinous debris.

leaves. Lacks a conidial morph. Causes black mildew or sooty mold, parasitic on tropical trees and shrubs in the tropical region of Amazonia, Brazil.

Amauroderma schomburgkii (Mont. & Berk.) Torrend: mature basidiocarps on wood, x 1.

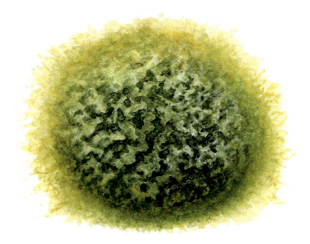

Amauroascus aureus (Eidam) Arx (syn. **Kuehniella aurea** (Eidam) Udagawa & Uchiy.): globose gymnothecium with spiral appendages on its surface, on soil, x 10.

Amauroderma Murrill (syn. Lazulinospora Burds. & M. J. Larsen)

Ganodermataceae, Polyporales, Inc. sed., Agaricomycetes, Agaricomycotina, Basidiomycota, Fungi

Murrill, W. A. 1905. The Polyporaceae of North America. *Bull. Torrey bot. Club* 32(7):366.

From Gr. *amaurós*, dark + *dérma*, skin, cutis, having a dark colored surface on the pileus. Subiculum arachnoid to byssoid, margin poorly delimited, concolorous with fertile areas or paler. Hyphal system monomitic, hyphae hyaline to pale yellow, septate, lacking clamps. Basidiocarps annual, pileate-stipitate, mucedinoid, arachnoid or byssoid, thinly adherent or loosely attached to substrate. Pileus circular to reniform, black to ochraceous-brown, often concentrically zonate, with an eccentric stipe. Basidia 4-spored. Basidiospores subglobose to ovoid, warted to sparsely echinulate, turning blue in aqueous KOH. On bark, roots and wood in warm temperate and tropical forests.

Amazonia Theiss.

Meliolaceae, Meliolales, Inc. sed., Sordariomycetes, Pezizomycotina, Ascomycota, Fungi

Theissen, F. 1913. Über einige Microthyriaceae. *Annls mycol.* 11(6):499.

From L. *Amazonia*, from the *Amazons*, warrior women of Scythia, and this name derived from *Amazon*, *Amazones* or *Amazonis* (Gr. *a*, without + *mazós*, wet-nurse, teat, i.e., without a teat). Ascomata flat, radiate, epiphytic on

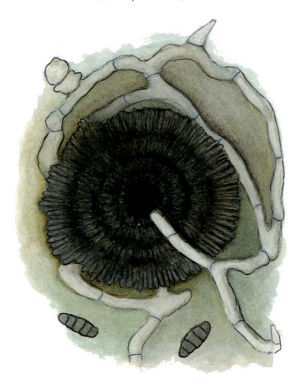

Amazonia palaquii Hosag. & P. J. Robin: radiate, flattened ascoma on the leaves of *Palaquium* sp. (Sapotaceae), and two free ascospores, x 500. Described in Kerala, India.

Ambispora C. Walker, et al.

Ambisporaceae, Archaeosporales, Inc. sed., Glomeromycetes, Glomeromycotina, Glomeromycota, Fungi

Walker, C., et al. 2007. Molecular phylogeny and new taxa in the Archaeosporales (Glomeromycota): *Ambispora fennica* gen. sp. nov., Ambisporaceae fam. nov., and emendation of *Archaeospora* and Archaeosporaceae. *Mycol. Res.* 111:137-153.

From L. *ambi*, both + *spora*, spore < Gr. *sporá*, seed, spore, for having two types of spores. Spores glomoid or acaulosporoid, hyaline to white at first, becoming pale ochraceous with age, formed on a large saccule that collapses and becomes detached at maturity. Some spores sessile, but others with persistent pedicel, similar to subtending hypha of a *Glomus* spore. Wall structure complex, consisting of three groups. Glomoid spores hyaline. variable in shape, globose to subglobose or ellipsoid, occasionally obovoid or irregular. Wall structure of two components in a single group. Spores open-pored i.e. lacking any occlusion, if closed, then by formation of a septum from laminated inner wall component. No reaction in Melzer's reagent.

Amoebophilus P. A. Dang.
Cochlonemataceae, Zoopagales, Inc. sed., Inc. sed., Zoopagomycotina, Zygomycota, Fungi
Dangeard, P. A. 1910. Études sur le développement et la structure des organismes inférieurs: (I) Les amibes. *Botaniste* 11:4-57.
From genus *Amoeba* < Gr. *amoibé*, change, transformation + *phílos*, to have an affinity for. Parasitizes terricolous amoebae, invades cells when merosporangia adhere to membrane, germinates, and haustorium extends into interior. New merosporangia formed on exterior of invaded cells.

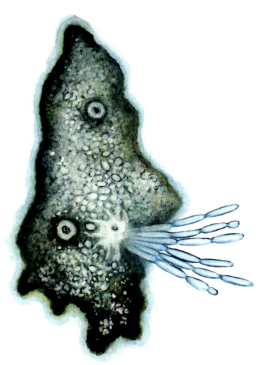

Amoebophilus simplex G. L. Barron: an amoeba with adhered conidia, and haustoria inside the host cell, x 900.

Amorosia Mantle & D. Hawksw.
Amorosiaceae, Pleosporales, Pleosporomycetidae, Dothideomycetes, Pezizomycotina, Ascomycota, Fungi

Mantle, P. G., et al. 2006. *Amorosia littoralis* gen. sp. nov., a new genus and species name for the scorpinone and caffeine-producing hyphomycete from the littoral zone in The Bahamas. *Mycol. Res.* 110(12):1371-1378.
Named in honor of the Caribbean scientist *Emmanuel Ciprian Amoroso* (1901-1982), for many years Professor of Physiology at the Royal Veterinary College, London. Colonies on PDA effuse, dark brown, reverse dark brown. Mycelium well-developed, superficial, composed of pale brown, thin-walled hyphae, frequently septate, smooth-walled. Setae and appressoria absent. Chlamydospores occasional, formed in short chains, arising from mycelium; individual chlamydospores subhyaline, broadly ellipsoid to subglobose, walls thick. Conidiophores micronematous to semi-macronematous, arising singly, pale brown, similar to mycelium. Conidiogenous cells integrated, terminal or intercalary, subhyaline to pale brown, smooth-walled. Conidia arising singly, dry, lateral, elongate-clavate, pale brown to brown, 3-4-septate when mature, evenly pigmented, constricted at septa, smooth-walled, lacking any gelatinous sheath or appendage, germinating apically. Isolated from Crooked Island in Southern Bahamas.

Amorphotheca Parbery
Amorphothecaceae, Inc. sed., Eurotiomycetidae, Eurotiomycetes, Pezizomycotina, Ascomycota, Fungi
Parbery, D. G. 1969. *Amorphotheca resinae* gen. nov., sp. nov.: the perfect state of *Cladosporium resinae*. *Aust. J. Bot.* 17:331-357.
From Gr. *ámorphos*, shapeless, amorphous + *théke*, box, depositary. Cleistothecia with peridium composed of an acellular material, amorphous on reaching maturity and dark due to a melanoid deposit. Grows on vegetable resins and in substrates derived from petroleum, such as creosote and kerosene.

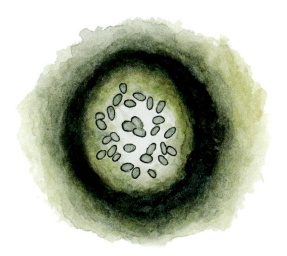

Amorphotheca resinae Parbery: transverse section of a mature ascoma, with the amorphous peridium, and containing ascospores, x 600.

Amphicypellus

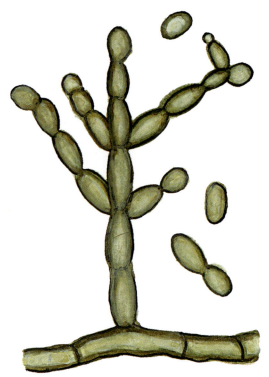

Amorphotheca resinae Parbery: conidial state, x 1,000. Isolated from soil, in Australia.

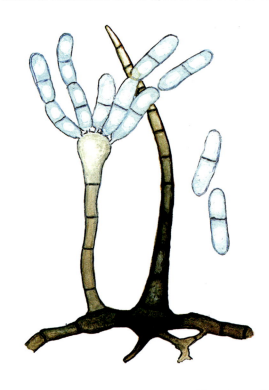

Ampullicephala setiformis (R. F. Castañeda) R. F. Castañeda, et al.: conidiophore, conidia, and seta, x 1,500.

Amphicypellus Ingold—see **Chytriomyces** Karling
Chytriomycetaceae, Chytridiales, Chytridiomycetidae, Chytridiomycetes, Inc. sed., Chytridiomycota, Fungi
Ingold, C. T. 1944. Studies on British chytrids II. A new chytrid on *Ceratium hirundinella* and *Peridinium. Trans. Br. mycol. Soc.* 27(1-2):93-97.
From Gr. *amphicypellus*, double cup, referring to the sporangium and apophysis.

Ampullicephala R. F. Castañeda, et al.
Inc. sed., Inc. sed., Inc. sed., Inc. sed., Pezizomycotina, Ascomycota, Fungi
Castañeda Ruiz, R. F., et al. 2009. Two setose anamorphic fungi: *Ampullicephala* gen. nov. and *Venustosynnema grandiae* sp. nov. *Mycotaxon* 109:275-288.
From L. *ampulla*, bottle, flask + Gr. *kephalé*, head > L. *cephala*, referring to the shape of the conidial head and conidiogenous cells. Colonies on natural substrate hairy to velvety, spreading, brown or white-brown. Mycelium superficial, immersed. Setae cylindrical, acerose or acuminate, erect, septate, simple or branched, smooth or verrucose, brown to black, arising on same hyphae near conidiophores. Conidiophores macronematous, mononematous, septate, brown to olivaceous, smooth or verrucose. Conidiogenous cells polyblastic, synchronous, ampulliform, terminal, determinate, with small, inconspicuous denticles. Conidia blastic-synchronous, cylindrical or oblong, hyaline, 1-septate or pluriseptate, smooth or verrucose, produced in short acropetal chains. Sexual morph unknown. Collected on decaying leaves of *Bauhinia cumanensis* in Soroa, Pinar del Rio, Cuba.

Amylomyces Calmette—see **Rhizopus** Ehrenb.
Inc. sed., Mucorales, Inc. sed., Inc. sed., Mucoromycotina, Zygomycota, Fungi
Calmette, L.C.A. 1892.Contribution à l'étude des ferments de l'amidon; la levûre chinoise. *Annls Inst. Pasteur, Paris* 6:604-620.
From Gr. *ámylon, ámylos*, starch + *mýkes* > L. *myces*, fungus, i.e., the starch fungus.

Amyloporiella A. David & Tortic
Polyporaceae, Polyporales, Inc. sed., Agaricomycetes, Agaricomycotina, Basidiomycota, Fungi
David, A., M. Tortic. 1984. *Amyloporiella* gen.nov. (Polyporaceae). *Trans. Br. mycol. Soc.* 83:659-667.
From genus *Amyloporia* Bourdot & Galzin + L. dim. suf. *-ella*. Fructifications resupinate, annual or perennial, trama more or less amyloid (rarely in *A. crassa*), bitter. Subiculum mostly chalky. Pore surface white, yellowish-brownish or golden yellow. Hyphal system dimitic to subtrimitic, generative hyphae with clamp-connections, skeletal hyphae amyloid, thick-walled to almost solid, mostly soluble in 10% KOH. Cystidioles numerous. Spores smooth, hyaline, allantoid to ovoid, inamyloid, acyanophilous. Associated with brown rot on conifers, particularly *Larix*.

Amylosporus Ryvarden (syn. **Rigidoporopsis** I. Johans. & Ryvarden)
Bondarzewiaceae, Russulales, Incertae sedis, Agaricomycetes, Agaricomycotina, Basidiomycota, Fungi
Ryvarden, L. 1973. New genera of the Polyporaceae. *Norw. J. Bot.* 20:1.
From Gr. *ámylon*, *ámylos*, starch + *spóros* < *sporá*, spore, referring to amyloid spores. Fruitbody annual, sessile, semicircular, single but some of them confluent and somewhat overlapping, forming a large agglomerate; stipe rudimentary. Context homogeneous but somewhat slightly zonate, white, watery and fleshy when fresh, very pale ochraceous and fragile when dry. Pore surface white when fresh, cream to ochraceous when dried, darker than the context, tubes up to 1 cm thick concolorous with the pore surface; pores round to angular 2-4 per mm with lacerate dissepiments. Hyphal system dimitic: generative hyphae hyaline, very thin-walled, with rare branches, simple septate but also with clamps that are single, double and sometimes verticillate; Skeletal hyphae thick-walled, aseptate, straight or more or less flexuous, contorted, with a narrow lumen to subsolid and typical branches. Cystidia none. Basidia clavate, hyaline, thin-walled, with a simple septum at the base, with 4 sterigmata. Basidiospores numerous, hyaline, broadly ellipsoid, firm walled, so finely echinulate that the ornamentations are not easily seen, amyloid.

Anamika K. A. Thomas, et al.—see **Hebeloma** (Fr.) P. Kumm.
Hymenogastraceae, Agaricales, Agaricomycetidae, Agaricomycetes, Agaricomycotina, Basidiomycota, Fungi
Thomas, K. A., et al. 2002. *Anamika*, a new mycorrhizal genus of Cortinariaceae from India and its phylogenetic position based on ITS and LSU sequences. *Mycol. Res.* 106(2):245-251.
From Skt., *anamika*, nameless, lass.

Anaptychia Körb.
Physciaceae, Teloschistales, Lecanoromycetidae, Lecanoromycetes, Pezizomycotina, Ascomycota, Fungi
Körber, G. W. 1848. *Grundriss Krypt.-Kunde*:197.
From Gr. pref. *ana-*, upwards, back, and again, similar to + *ptýx*, genit. *ptychós*, pleat, crease, sheet, layer. Thallus foliose, fruticose or subfruticose, occasionally with elongated, narrow, creased lobules, often forming a channel through lower part of each lobule. Apothecia with disc more or less concave and excipulum subentire, crenate or dentate-ciliate. Photobiont *Protococcus*. On rocks, soil and trees.

Anaselenosporella Heredia, et al.
Inc. sed., Inc. sed., Inc. sed., Inc. sed., Pezizomycotina, Ascomycota, Fungi
Castañeda Ruiz, R. F., et al. 2010. *Anaselenosporella sylvatica* gen. & sp. nov. and *Pseudoacrodictys aquatica* sp. nov., two new anamorphic fungi from Mexico. *Mycotaxon* 112:65-74.
From Gr. *ana-*, upwards, back, similar to + genus *Selenosporella*, for the similarity to this genus. Colonies on natural substratum effuse, hairy, brown or black. Mycelium superficial, immersed. Conidiophores macronematous, mononematous, erect or prostrate, septate, smooth or verruculose, brown. Conidiogenous cells polyblastic, lageniform, cylindrical to subulate, indeterminate with holoblastic sympodial proliferations, discrete. Conidial secession schizolytic. Conidiogenous loci flattened, lenticular or convex, lateral and apical, slightly melanized. Conidia solitary, acicular, filiform, fusiform to semi-circular, unicellular, hyaline, smooth or verruculose, dry or hygroscopic. Sexual morph unknown. Collected on decaying leaves of an unidentified plant in Huatusco, Veracruz, Mexico.

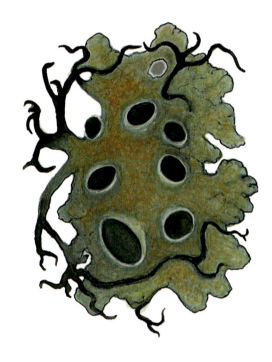

Anaptychia ciliaris (L.) Körb. ex A. Massal.: thallus with apothecia, and ciliate border, x 3.

Anaseptoidium R. F. Castañeda, et al.
Inc. sed., Inc. sed., Inc. sed., Inc. sed., Pezizomycotina, Ascomycota, Fungi
Castañeda-Ruiz, R. F., et al. 2012. Two new fungi from Mexico: *Anaseptoidium* gen. nov. and *Cylindrosympodium sosae* sp. nov. *Mycotaxon* 119:141-148.
From Gr. pref. *ana-*, upwards, back, and again, similar to + genus *Septoidium*, referring to the latter anamorphic genus. Colonies on natural substrate effuse, creeping, funiculose, brown. Mycelium superficial, immersed. Stomatopodia absent. Conidiophores macronematous,

Anastomyces

mononematous, erect, sometimes reduced to conidiogenous cells. Conidiogenous cells monoblastic, integrated, determinate. Conidial secession schizolytic. Conidia solitary, acrogenous, cylindrical to oblong, fimbriate, septate, smooth or verruculose, pale brown to brown. Sexual morph unknown. Collected on synnemata of *Phaeoisaria clavulata* in Agüita Fría, Veracruz, Mexico.

Anastomyces W. P. Wu, et al.
Inc. sed., Inc. sed., Inc. sed., Inc. sed., Agaricomycotina, Basidiomycota, Fungi
Wu, W., et al. 1997. Notes on three fungicolous fungi: *Anastomyces microsporus* gen. et sp. nov., *Idriella rhododendri* sp. nov. and *Infundibura adhaerens*. *Mycol. Res.* 101(1):1318-1322.
From Gr. *anastómosis*, a coming together + *mykés*, fungus > L. *myces*, for the fusion of two young conidia. Conidiomata fungicolous, sporodochial, hyaline to yellow-brown, conical, producing wet spore masses; basal stroma composed of hyaline to pale brown, septate, thin-walled, smooth, branched hyphae. Conidiophores hyaline, septate, branched irregularly, thin-walled, smooth, formed from stromatic hyphae. Conidiogenous cells integrated, determinate, cylindrical to subcylindrical, hyaline, thin-walled, smooth, with two loci on each conidiogenous cell; loci apical, persistent, protuberant, unthickened. Conidial ontogeny holoblastic, wall hologenous, delimited by one septum from conidiogenous cells, secession schizolytic. Conidia solitary, hyaline, aseptate, smooth, thin-walled, guttulate, ellipsoid to subcylindrical but soon becoming H-shaped due to anastomosis of two young conidia from same conidiogenous cell. Collected on ascomata of *Apiospora* sp. on a grass in Chengde, Hebei Prov., China.

Ancorasporella J. Mena, et al.
Inc. sed., Inc. sed., Inc. sed., Inc. sed., Pezizomycotina, Ascomycota, Fungi
Mena Portales, J., et al. 1998. *Ancorasporella*, a new genus of hyphomycetes from Mexico. *Mycol. Res.* 102(6):736-738.
From genus *Ancoraspora* + L. dim. suf. *-ella*, for the similarity to this genus. Colonies effuse, black, hairy. Mycelium mostly immersed. Stromata often formed. Conidiophores macronematous, mononematous, erect, straight or flexuous, septate, unbranched, smooth, dark brown to black, paler at apex. Conidiogenous cells monotretic or polytretic, integrated, terminal, cylindrical, sympodial, cicatrized. Conidia solitary, dry, acropleurogenous, triangular to fusiform and curved, rounded at ends, broadly fusiform in vertical view, symmetrical, attached at 90° to long axis of conidiogenous cell by central cell, septate, verrucose, brown, with a dark scar at point of attachment. Collected on dead fern frond in Rancho Guadalupe, Xalapa, Veracruz, Mexico.

Ancorasporella mexicana J. Mena, et al.: conidiophores and pluriseptate conidia, x 740.

Anellaria P. Karst.—see **Panaeolus** (Fr.) Quél.
Inc. sed., Agaricales, Agaricomycetidae, Agaricomycetes, Agaricomycotina, Basidiomycota, Fungi
Karsten, P. A. 1879. Rysslands, Finlands och den Skandinaviska halföns Hattsvampar. *Bidr. Känn. Finl. Nat. Folk* 32:517.
From L. *anellarius*, with rings + des. *-a-*, which are the remnants of the partial veil around the stipe once the pileus expands.

Angusia G. F. Laundon—see **Maravalia** Arthur
Chaconiaceae, Pucciniales, Inc. sed., Pucciniomycetes, Pucciniomycotina, Basidiomycota, Fungi
Laundon, G. F. 1964. Angusia (Uredinales). *Trans. Br. mycol. Soc.* 47(3):327-329.
Named for the collector of the fungus (Zambia), A. Angus + L. suf. *-ia*.

Aniptodera Shearer & M. A. Mill.
Halosphaeriaceae, Microascales, Hypocreomycetidae, Sordariomycetes, Pezizomycotina, Ascomycota, Fungi
Shearer, C. A., Miller, M. A. 1977. Fungi of the Chesapeake Bay and its tributaries V. *Aniptodera chesopeakensis* gen. et sp. nov. *Mycologia*. 69(5):893
From Gr. *ánisos*, unequal, dissimilar + *deré*, neck. Perithecia superficial or partially immersed, globose to subglobose, hyaline, membranous; ostiolate, neck elongated, cylindrical, periphysate. Asci in hymenium at base of perithecium, clavate, unitunicate, wall thickened below the apex, apex with pore, asci deliquescing at maturity. Paraphyses absent, catenophyses present. Ascospores 2-celled, hyaline, thick walled. Isolated from balsa wood blocks in Chalk Point, Md., U.S.A.

Aniptodera aquadulcis (S. Y. Hsieh, et al.) J. Campb., J. L. Anderson & Shearer (syn. **Halosarpheia aquadulcis** S. Y. Hsieh, et al.): perithecium with long ostiolar neck, on decomposing wood in fresh water, Taiwan, x 120; ascus with two-celled ascospores, x 450; free ascospore with apical, hook-shaped appendages, x 1,600.

Anisostagma K. R. L. Petersen & Jørg. Koch
Halosphaeriaceae, Microascales, Hypocreomycetidae, Sordariomycetes, Pezizomycotina, Ascomycota, Fungi
Petersen, K. R. L., J. Koch. 1996. *Anisostagma rotundatum* gen. et sp. nov., a lignicolous marine ascomycete from Svanemøllen Harbour in Denmark. *Mycol. Res.* 100(2):209-212.
From Gr. *ánisos*, unequal + *stágma*, drop, in reference to the ascospore, which contains one large globule surrounded by numerous droplets. Ascomata single or gregarious, globose to broadly ellipsoidal, immersed to partly immersed, ostiolate with a neck, periphysate, cream-colored to pale brown. Peridium two-layered. Catenophyses present. Asci 8-spored, clavate, pedunculate, thin-walled, unitunicate, without an apical apparatus, early deliquescing, developing from a small cushion at base of ascomatal venter. Ascospores globose to ellipsoidal, one-celled, with one large globule surrounded by numerous droplets, hyaline, without appendages. Inhabiting oak mooring posts.

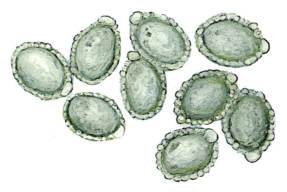

Anisostagma rotundatum K. R. L. Petersen & Jörg. Koch: ascospores, x 1,000.

Ankistrocladium Perrott—see **Casaresia** Gonz. Frag.
Dermateaceae, Helotiales, Leotiomycetidae, Leotiomycetes, Pezizomycotina, Ascomycota, Fungi
Perrott, E. 1960. *Ankistrocladium fuscum* gen. nov., sp. nov., an aquatic hyphomycete. *Trans. Br. mycol. Soc.* 43(3):556-558.
From Gr. *ankístron*, fishhook + L. *cladium* < Gr. *kládion*, small branch, for the hooked tip of the spores.

Annellophora S. Hughes
Inc. sed., Inc. sed., Inc. sed., Inc. sed., Pezizomycotina, Ascomycota, Fungi
Hughes, S. J. 1952. *Annellophora* nom. nov. (*Chaetotrichum* Syd. non Rabenh.). *Trans. Br. mycol. Soc.* 34(4):544-550.
From L. *annellus*, a ring + Gr. *phóros*, bearing, referring to the successive conidial scars on the conidiophore. Mycelium superficial, foliicolous, solitary or in association with Cyanophyceae, Asterineae or Meliolineae, composed of hyaline to brown septate, branched hyphae. Conidiophores simple, straight or curved, more or less cylindrical, septate, brown below, paler above, elongating by numerous successive cylindrical proliferations through successive conidial scars, finally giving upper part of conidiophore an annellate appearance; arising directly from a repent hypha as a lateral branch. Conidia obclavate to fusiform, 3-7-septate, pale brown to brown, paler at apex, with or without a single basal and lateral cellular appendage arising singly as blown-out ends of apex of conidiophore and its successive proliferations.

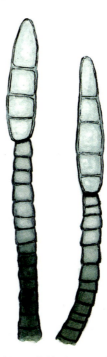

Annellophora africana S. Hughes: annellate conidiophores with terminal conidia, x 500. Described in Ghana.

Anomalemma Sivan.—see **Exosporiella** P. Karst.
Melanommataceae, Pleosporales, Pleosporomycetidae, Dothideomycetes, Pezizomycotina, Ascomycota, Fungi

Antarctomyces

Sivanesan, A. 1983. Studies on Ascomycetes. *Trans. Br. mycol. Soc.* 81(2):313-332.
From Gr. *anómos*, irregular + *lemma*, bark, rind, perhaps in reference to the intertwined hyphae of the stroma.

Antarctomyces Stchigel & Guarro
Thelebolaceae, Thelebolales, Leotiomycetidae, Leotiomycetes, Pezizomycotina, Ascomycota, Fungi
Stchigel, A. M., et al. 2001. *Antarctomyces psychrotrophicus* gen. et sp. nov., a new ascomycete from Antarctica. *Mycol. Res.* 105(3):377-382.
From L. *Antarcticus*, Antarctic + *myces* < Gr. *mýkes*, fungus, for the type locality. Mycelium mainly submerged, of septate, branched and unbranched, anastomosing, hyaline hyphae. Ascomata composed of naked asci, without excipulum. Asci ellipsoidal to subglobose, unitunicate, non-catenate, 8-spored. Paraphyses absent. Ascospores ellipsoidal to fusiform, hyaline, spinulose, without germ pores, 1-celled. Conidiophores thick, hyaline, with lateral cylindrical protuberances. Conidiogenous cells enteroblastic, integrated, intercalary, determinate. Conidia subglobose to irregularly cylindrical, hyaline, smooth, thick-walled, aggregated in slimy masses, 1-celled. Chlamydospores irregular, single or forming long chains, one or two-celled. Isolated from soil from King George Island, South Shetland Islands, Antarctica.

Antarctomyces psychrotrophicus Stchigel & Guarro: ellipsoidal and spinulose ascospore, x 6,000.

Antennatula Fr. ex F. Strauss (syn. **Hormisciella** Bat.)
Euantennariaceae, Incertae sedis, Incertae sedis, Dothideomycetes, Pezizomycotina, Ascomycota, Fungi
Strauss, F. 1850. *Flora*, Regensburg 33(Beil.):98,99.
From L. *antenna*, antenna, a movable segment organ of sensation on the head of insects, myriapods, and crustaceans + dim. suf. *-ula*. Colonies round or effuse, black. Mycelium superficial, septate, moniliform hyphae which taper only very slightly and gradually, composed of subglobose or oblong cells deeply constricted at septa, catenulate, prostrate on substrate. Conidiophores lacking. Conidia formed directly from hyphal cells, initially clavate, 1-celled and hyaline, later 3-5-septate, cylindric-fusoid, brown. On living leaves and needles of trees.

Antennatula dingleyae S. Hughes: torulose hyphae, cylindrical conidia, light brown to dark brown, with basal cells and the apex more clear, mucronate or rounded, with some conidia germinating, x 400. On branches of *Discaria tournatoa*.

Aphanotria Döbbeler
Bionectriaceae, Hypocreales, Hypocreomycetidae, Sordariomycetes, Pezizomycotina, Ascomycota, Fungi
Döbbeler, P. 2007. Ascomycetes on *Polytrichadelphus aristatus* (Musci). *Mycol. Res.* 111(12):1406-1421.
From Gr. *aphanés*, invisible + the last syllable of *Nec(tria)*, because of the hidden nectriaceous ascomata. Ascomata immersed within leaf nerves, perithecial, ellipsoid, apically attenuate, often ending in a rostrum outside leaf, hyaline, glabrous. Rostrum cylindrical or slightly tapering towards apex, often bent, hyaline, apically more or less truncate. Rostrum filled with numerous periphyses. Apical paraphyses inconspicuous, filamentous, cylindrical, without ramifications and anastomoses, single cells of varying size. Asci unitunicate, cylindrical, straight or slightly bent, when immature very thick-walled, apically rounded, without apical structures, foot gradually tapering, containing (3-7) 8 uniseriate spores, aborted spores usually visible, mature or immature asci easily detached. Iodine reaction negative. Ascospores fusiform, almost symmetrical, hyaline, with five transverse septa, not constricted at septa, epispore distinctly warted, cyanophilous, variable in size. Hyphae inconspicuous, hyaline, predominantly within walls of nerve cells and decomposing them, abaxial epidermal layer free of hyphae. Asexual morph not observed. Collected on *Polytrichadelphus aristatus* in Venezuela and Colombia.

Apinisia La Touche
Onygenaceae, Onygenales, Eurotiomycetidae, Eurotiomycetes, Pezizomycotina, Ascomycota, Fungi
La Touche, C. J. 1968. *Apinisia graminicola* gen. et sp. nov. *Trans. Br. mycol. Soc.* 51(2):283-285.
Named after British Dr. and Mrs. A. E. Apinis + L. des. *-ia*, for their work on gymnoascaceous fungi. Cleistothecia white, globose, when aggregated, globose to ovate, excluding appendages. Peridium composed of hyaline irregular hyphae with thin walls, but forming unequal chlamydospores in chains that become free at maturity. Ascospores yellow, globose with lightly or

finely echinulate walls. Isolated from rotting grass in a garden in Leeds, England.

Apiosordaria Arx & W. Gams (syn. **Echinopodospora** B. M. Robison)
Lasiosphaeriaceae, Sordariales, Sordariomycetidae, Sordariomycetes, Pezizomycotina, Ascomycota, Fungi
Arx, J. A. von, W. Gams. 1967. *Nova Hedwigia* 13:201.
From Gr. *ápion*, pear + genus *Sordaria* Ces. & De Not., for the shape of the mature ascospores. Ascomata ostiolate or nonostiolate perithecia, superficial or with base immersed in substrate, scattered or aggregated. Perithecia ovate to obpyriform, dark brown to brownish-black, glabrous or with flexuous, brown hairs on upper part. Wall of perithecium pseudoparenchymatous or membranaceous, transparent, thin. Centrum containing septate, filamentous paraphyses. Asci unitunicate, clavate, tapering to a short stipe, with an indistinct apical ring not bluing in iodine, 4-8-spored. Ascospores elliptical to obclavate when delimited, becoming 2-celled through formation of a septum along base of spore initial, cutting off a small basal cell. Larger upper cell of ascospore becoming dark brown at maturity, lower cell remaining hyaline. Mature ascospores angular to ellipsoid, ovate or clavate, with a minute apical hyaline appendage, in some species with small gelatinous appendages. Surface of upper cell ornamented with pits or striae that are often obscured in mature spores. Asexual morphs are cladorrhinum-like. In soil and plant debris.

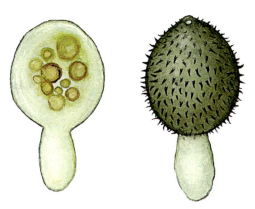

Apiosordaria jamaicensis (B. M. Robison) J. C. Krug, et al.: immature and mature ascospores, x 1,000. Described in Jamaica.

Apiosphaeria Höhn. (syn. **Oswaldina** Rangel)
Phyllachoraceae, Phyllachorales, Inc. sed., Sordariomycetes, Pezizomycotina, Ascomycota, Fungi
Höhnel, F. von. 1909. *Sber. Akad. Wiss., Math.-natur. Kl., Abt. I*,118:1218.
From Gr. *ápion*, pear + genus *Sphaeria* Haller, for the shape of the mature ascospores. Ascomata ostiolate perithecia, formed in circular lesions on host, immersed in leaf tissue, frequently formed along leaf veins. Perithecia globose to somewhat depressed, with erumpent ostiolar neck; ostiole lined with slender periphyses. Ostiole surrounded by a blackened clypeus formed in host epidermis; clypeus also formed in epidermis beneath perithecium. Perithecial wall pale brown, composed of several layers of flattened cells. Centrum containing filamentous paraphyses. Asci unitunicate, oval to subcylindrical, with a blunt apex and a distinct apical ring that stains blue in iodine, 8-spored. Ascospores obovoid, 2-celled, septate near lower end, constricted at septum, upper cell with a large oil globule; spores initially hyaline, with upper cell becoming yellowish-brown, lower cell remaining hyaline. Parasitic on living leaves of *Tabebuia* spp. and *Tecoma* spp. in the American tropics.

Apiosporina Höhn. (syn. **Dibotryon** Theiss. & Syd.)
Venturiaceae, Venturiales, Pleosporomycetidae, Dothideomycetes, Pezizomycotina, Ascomycota, Fungi
Höhnel, F. von, 1910. *Fragmente zur Mykologie*, No. 510. *Sber. Akad. Wiss. Wien. Math.-naturw. Kl., Abt. 1, 119*:439.
From Gr. *ápion*, pear + *sporá*, spore + L. suf. *-ina*, which indicates similarity. Pseudothecia dark, densely grouped on an extensive superficial mycelium that develops on hypertrophied tissues of host plant. Asci bitunicate. Ascospores pyriform or apiosporous, hyaline, with a septum near lower end. Found on forest and cultivated soils, rotted wood, dung, and as a pathogen of various plants.

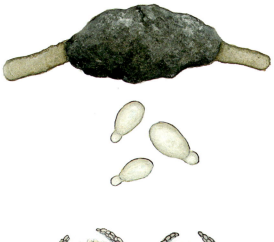

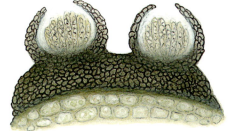

Apiosporina morbosa (Schwein.) Arx: pseudothecia on prune twig, x 1; longitudinal section of two pseudothecia with ascospores, x 50; free ascospores, x 750.

Apostemidium

Apostemidium (P. Karst.) P. Karst.—see **Vibrissea** Fr.
Vibrisseaceae, Helotiales, Leotiomycetidae, Leotiomycetes, Pezizomycotina, Ascomycota, Fungi
Karsten, P. A. 1871. Mycologia fennica I Discomycetes. *Bidr. Känn. Finl. Nat. Folk* 19:1-264.
From L. *apostema*, abscess + dim. suf. *-idium*, possibly for the nature of the ascoma.

Appendiculella Höhn.
Meliolaceae, Meliolales, Inc. sed., Sordariomycetes, Pezizomycotina, Ascomycota, Fungi
Höhnel, F. von. 1919. Über die Gattung *Meliola* Fr. *Sber. Akad. Wiss. Wien, Math.-naturw. Kl., Abt. 1, 128*:555-559.
From L. *appendicula*, dim. of *appendix*, appendage, something that projects from a larger part + dim. suf. *-ella*, in reference to. Ascomata spheroidal with larviform appendages, derived from black, epiphytic mycelium without setae. Obligate parasite on leaves of tropical trees and shrubs.

Appendiculella lozanellae Rodr. Just. & M. Piepenbr.: ascoma with larviform appendages, on *Lozanella enantiophylla*, x 300. Described in Panama.

Aqualignicola Ranghoo, et al.
Annulatascaceae, Inc. sed., Sordariomycetidae, Sordariomycetes, Pezizomycotina, Ascomycota, Fungi
Ranghoo, V. M., et al. 2001. *Brunneosporella aquatica* gen. et sp. nov., *Aqualignicola hyalina* gen. et sp. nov., *Jobellisia viridifusca* sp. nov. and *Porosphaerellopsis bipolaris* sp. nov. (ascomycetes) from submerged wood in freshwater habitats. *Mycol. Res.* 105(5):625-633.
From L. *aqua*, water, aquatic + *lignum*, wood + suf. *-icola*, inhabitant, reference to the aquatic habitat and the woody substrate where the fungus was found. Ascomata immersed, globose to subglobose, membranous, hyaline to brown, solitary to gregarious. Neck short, dark-brown, covered with setae. Setae lanceolate, solid, surrounding neck. Peridium composed of several layers of pale brown, pseudoparenchymatous cells. Paraphyses filiform, hyaline, simple to rarely branched, tapering towards apex. Asci 8-spored, cylindrical, pedicellate, unitunicate, apically truncate, with a large, J-, refractive, cylindrical apical ring. Ascospores uniseriate to biseriate, ellipsoidal-fusiform, hyaline, unicellular. Collected on submerged wood in the Lam Tsuen River, Tai Po, Hong Kong, China.

Aqualignicola vaginata D. M. Hu, L. Cai & K. D. Hyde: ascoma with setae surrounding the neck, x 200; ascus, x 330, and two free ascospores, x 1,000.

Aquamarina Kohlm., et al.
Inc. sed., Inc. sed., Inc. sed., Sordariomycetes, Pezizomycotina, Ascomycota, Fungi
Kohlmeyer, J., et al. 1995. Fungi on *Juncus roemerianus*. New marine and terrestrial ascomycetes. *Mycol. Res.* 100(4):393-404.
From L. *aquamarina*, sea water, sea-green, verging towards blue, referring to the color of the ascomata. Ascomata subglobose, immersed, ostiolate, with a long neck, periphysate, coriaceous, sea-green verging towards blue, single. Peridium two-layered. Hamathecium composed of simple paraphyses with a gel coating; centrum of ascomatal initials filled with a thin-walled pseudoparenchyma. Asci 8-spored, cylindrical, short-stalked, thin-walled, unitunicate, with an apical, non-amyloid ring, developing successively on ascogenous tissue at base of locule. Ascospores uniseriate, fusiform, 3-septate, hyaline.

Aquaphila Goh, et al.
Tubeufiaceae, Tubeufiales, Pleosporomycetidae, Dothideomycetes, Pezizomycotina, Ascomycota, Fungi
Goh, T. K., et al. 1998. *Aquaphila albicans* gen. et sp. nov., a hyphomycete from submerged wood in the tropics. *Mycol. Res.* 102(5):587-592.
From L. *aqua*, water + Gr. *phílos*, loving, in reference to the aquatic habitat where this fungus is found. Colonies on natural substrate effuse, non-dematiaceous. Mycelium partly immersed in woody substratum and partly superficial. Stromata, setae, and hyphopodia absent. Conidiophores semi-macronematous, mononematous, borne as lateral branches from superficial hyphae, hyaline, delicate, septate, simple or branched, flexuous and geniculate. Conidiogenous cells integrated, terminal

or intercalary, denticulate, monoblastic or polyblastic, proliferation sympodial, indeterminate. Conidial secession schizolytic. Conidia borne singly, acrogenous and becoming lateral as a result of conidiophore proliferation, hyaline, fusoid to falcate or sigmoid, broad, multi-euseptate, apedicellate. Growing on submerged angiosperm wood in the South Pacific.

Arachnion Schwein.
Agaricaceae, Agaricales, Agaricomycetidae, Agaricomycetes, Agaricomycotina, Basidiomycota, Fungi
Schweinitz, L.D. 1822. Synopsis fungorum Carolinae superioris. *Schr. naturf. Ges. Leipzig* 1:59.
From Gr. *aráchne*, spider, spider web + dim. suf. *-ion*, in reference to the mature fructification with intermixed filaments in the gleba. Fructifications small, globose to subglobose, epigeous, sessile. Peridium thin, fragile; gleba a mass of small peridioles with spores; spores subglobose to short ellipsoid, smooth, thick-walled, with a stump of a pedicel, rusty-brown in iodine. Rarely collected in Michigan, New York, and North Carolina, U.S.A.

Arachnion album Schwein.: external and internal views of basidiocarps, x 2.

Arachnopeziza Fuckel
Arachnopezizaceae, Helotiales, Leotiomycetidae, Leotiomycetes, Pezizomycotina, Ascomycota, Fungi
Fuckel, L. 1870. Symbolae Mycologicae. *Jb. nassau. Ver. Naturk.* 23-24:1-459.
From Gr. *aráchne*, spider, spider web + genus *Peziza* L., for the web-like subiculum. Apothecia gregarious, seated on a thin, spiderweb-like, white or yellowish mycelial subiculum, at first closed, rounded, later opening, becoming patellate, externally clothed with fine bristly hairs. Asci clavate, inoperculate, 8-spored. Ascospores ellipsoid to fusoid, clavate or filiform, becoming several-septate, often with an apiculus at each end, hyaline. Paraphyses filiform, usually enlarged above.

Arbusculina Marvanová & Descals
Inc. sed., Inc. sed., Inc. sed., Inc. sed., Pezizomycotina, Ascomycota, Fungi
Marvanová, L., E. Descals. 1987. New taxa and new combinations of Aquatic Hyphomycetes. *Trans. Br. mycol. Soc.* 89(4):499-507.
From L. *arbuscula*, a small tree + L. suf. *-ina*, likeness. Colony dark, aerial mycelium abundant, woolly; hyphae hyaline, thin-walled as well as brown, thick-walled. Minute, dark sclerotia. Conidiophores single, apical, seldom lateral, semimacronematous, flexuous, simple or sparsely branched, monilioid or cylindrical, often broadening distally, septate, subfuscous, paler towards apex. Conidiogenous cells single, apical, integrated, determinate or percurrent. Conidia single, apical, branched, hyaline to subfuscous, septate, cells torulose, obclavate or ampulliform when apical; stalk short, sometimes clavate, 1-2 celled; axis straight or slightly curved at branch insertions, base truncate to obtuse; branches irregularly arranged, (0-) several, apical or lateral, primary, secondary and rarely tertiary, alternate, straight to slightly curved. Secession schizolytic.

Arachnopeziza aurelia (Pers.) Fuckel: apothecium with a basal subiculum, on soil, x 3. Described in France.

Arcuadendron Siegler & J. W. Carmich.
Inc. sed., Inc. sed., Inc. sed., Inc. sed., Pezizomycotina, Ascomycota, Fungi
Siegler, L., J. W. Carmichael. 1976. Taxonomy of *Malbranchea* and some other hyphomycetes with arthroconidia. *Mycotaxon* 4(2):349-488.
From L. *arcuo*, curvature + Gr. *déndron*, tree, in reference to the branching of the conidiophores. Vegetative hyphae hyaline or yellowish-green, septate, differentiated conidiophores lacking. Conidiogenous cells integrated, serial, indeterminate, progressive, growing from apical conidium. Conidia formed in acropetal succession serial, alternate, separated by short hyphal segments that dissolve to release conidia. Conidia broadly ellipsoidal or triangular with truncate ends, hyaline or yellowish, smooth or verrucose. Originally isolated from soil in India.

Argopericonia

Argopericonia B. Sutton & Pascoe
Inc. sed., Inc. sed., Inc. sed., Inc. sed., Pezizomycotina, Ascomycota, Fungi
Sutton B. C., I. G. Pascoe. 1987. *Argopericonia* and *Tryssglobulus*, new Hyphomycete genera from *Banksia* leaves. *Trans. Br. mycol. Soc.* 88(1):41-46.
From Gr. *argós*, bright, shining + genus *Periconia*, referring to the shiny spores. Colonies saprotrophic. Mycelium superficial, hyaline, branched, euseptate. Stromata absent. Conidiophores arising from vegetative mycelium, macronematous, determinate, erect, dark brown, smooth, euseptate, producing a hyaline, apical conidiogenous head consisting of either a single apical cuneiform cell or two cells, lower one cuneiform and upper one smaller, short cylindrical with an obtuse apex. Conidiogenous cells integrated, hyaline, smooth, each producing a subapical ring of non-protuberant, unthickened conidiogenous loci. Conidia holoblastic, dry, shining, solitary or short catenate, hyaline, aseptate, smooth, ellipsodial, thin-walled, guttulate, flattened at base, occasionally conidiogenously forming additional apical to subapical conidia. Collected on living leaves of *Banksia marginata*, in Yaugher, Victoria, Australia.

Arkoola J. Walker & Stovold
Venturiaceae, Venturiales, Pleosporomycetidae, Dothideomycetes, Pezizomycotina, Ascomycota, Fungi
Walker J., G. E. Stovold. 1986. *Arkoola nigra* gen. et sp. nov. (Venturiaceae) causing black leaf blight of soybean in Australia. *Trans. Br. mycol. Soc.* 87(1):23-44.
From Australian aborigine word *arkoola*, hair, referring to the abundant black superficial mycelium and setose pseudothecia. Superficial mycelium abundant, dark brown to black, composed of large septate branched, dark brown hyphae. Pseudothecia mainly superficial, sometimes erumpent, large, black, setose, subglobose to broadly ovoid to obpyriform, uniloculate, ostiolate. Wall of pseudothecia composed of several layers. Asci cylindrical, short stipitate, 8-spored, bitunicate with fissitunicate dehiscence, ocular chamber present. Ascospores large, 1-septate, ellipsoidal to fusiform, pale greenish, surrounded by a thin gelatinous sheath. Pseudoparaphyses abundant, hyaline, filiform, branched, cellular. Parasitic on living plants and saprobic on fallen leaves and other dead organs of plants. Collected on dead leaves of soybean in Rydalmere, New South Wales, Australia.

Armillaria (Fr.) Staude (syn. **Armillariella** (P. Karst.) P. Karst.)
Physalacriaceae, Agaricales, Agaricomycetidae, Agaricomycetes, Agaricomycotina, Basidiomycota, Fungi
Staude, F. 1857. *Schwämme Mitteldeutschl.* 28:xxviii,130.
From L. *armilla*, ring, bracelet + L. suf. *-aria*, which indicates possession or connection, for the presence of a ring on the upper part of the stipe. Fructifications fleshy; pileus cinnamon-colored or bright brownish-orange, gills firmly united to stipe, not easily separated; gills adnexed-emarginate, sinuate or semi-free, sometimes decurrent; internal veil and ring well developed. Spore print white; spores typically binucleate, amyloid or inamyloid. Occasionally with black rhizomorphs. Grows on soil or wood.

Armillaria ostoyae (Romagn.) Herink: basidiocarps, x 0.5.

Armillariella (P. Karst.) P. Karst.—see **Armillaria** (Fr.) Staude
Physalacriaceae, Agaricales, Agaricomycetidae, Agaricomycetes, Agaricomycotina, Basidiomycota, Fungi
Karsten, G. K. W. H. 1881. Hymenomycetes fennici. *Acta Soc. Fauna Flora fenn.* 2(1):4.
From *Armillaria* (Fr.) Kummer + L. dim. suf. *-ella*, i.e., a small *Armillaria*.

Arrasia Bernicchia, et al.
Inc. sed., Inc. sed., Inc. sed., Inc. sed., Agaricomycotina, Basidiomycota, Fungi
Bernicchia, A., et al. 2011. *Arrasia rostrata* (Basidiomycota), a new corticioid genus and species from Italy. *Mycotaxon* 118:257-264.
Dedicated to the Italian mycologist *Luigi Arras*, for his mycological excursions across Sardinia. Basidiomata effuse, adnate, thin, white, smooth, finely farinaceous, with a distinct margin. Hyphal system monomitic, hyphae clamped. Dendrohyphidia filamentous, branched, clamped. Basidia suburniform at first, then flexuous, clavate to obclavate, basally clamped, with 4 sterigmata. Basidiospores broadly subfusiform to biapiculate, dis-

tal end elongating into a thick-walled rostrum, walls hyaline, slightly thickened, smooth, cyanophilous, inamyloid, nondextrinoid. Collected on bark of trunk and old branches of living *Juniperus phoenicea*, Nuoro Province, Sardinia, Italy.

Arthonia Ach.
Arthoniaceae, Arthoniales, Arthoniomycetidae, Arthoniomycetes, Pezizomycotina, Ascomycota, Fungi
Acharius, E. 1806. *Neues J. Bot. 1*(3):3.
From Gr. *árdo*, water, irrigate, to spray, sprinkle, to dust + L. suf. *-ia*, which indicates characteristic of. Thallus rudimentary, appears to have been sprinkled or dusted when apothecia form, crustous, occasionally areolate; lacks cortical tissues, formed, partially or totally, within substrate, appearing grayish. Apothecia tiny or small, round or irregular, linear or stellate, with a black, brown, grayish or reddish disc.

Arthonia radiata (Pers.) Ach.: crustose thallus with numerous apothecia, growing on *Carpinus* and *Fagus* bark, x 1. Described in Germany.

Arthothelium A. Massal.
Arthoniaceae, Arthoniales, Arthoniomycetidae, Arthoniomycetes, Pezizomycotina, Ascomycota, Fungi
Massalongo, A. B. 1852. *Ric. auton. lich. crost* (Verona):54.
From Gr. *árdo*, water, irrigate, to spray, to sprinkle, to dust + *théles* or *thelé*, nipple, the skin that covers the nipple, or membrane, for the thin, crustaceous-membranaceous thallus that appears to have been sprinkled when the apothecia form. Thallus rudimentariy and lacking cortical tissues. Apothecia small or large, round or irregular, more or less immersed, with a black or dark brown disc. On tree bark, which it partially penetrates.

Arthrinium Kunze
Apiosporaceae, Xylariales, Xylariomycetidae, Sordariomycetes, Pezizomycotina, Ascomycota, Fungi
Kunze, G. 1817. *Mykologische Hefte* (Leipzig)*1*:9-10.
From Gr. *árthron*, articulation + L. suf. *-inus*, similar to, belonging to + dim. suf. *-ium*, due to the presence of distinctive refringent and thick septa in the conidiophores, a characteristic that, together with the dark, unicellular, lenticular conidia with a hyaline germ slit, constitute the distinctive features of this fungus that is saprobic in soil and on decomposing grass. The presence of the refractive septa gives the appearance of knuckles to the hyphae, from which the name of the genus is derived. Sexual morph apiospora-like.

Arthothelium spectabile A. Massal.: crustaceous-membranaceous thallus, sprinkled with small apothecia, on tree bark, x 2.5.

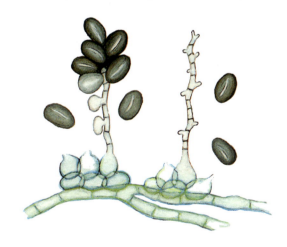

Arthrinium phaeospermum (Corda) M. B. Ellis: basauxic conidiophores, with meristematic blastospores, x 1,000.

Arthrobotrys Corda
Orbiliaceae, Orbiliales, Orbiliomycetidae, Orbiliomycetes, Pezizomycotina, Ascomycota, Fungi
Corda, A. C. J. 1839. *Pracht-Fl. Eur. Schimmelbild*:43.
From Gr. *árthron*, articulation, joint + *bótrys*, cluster, referring to the arrangement of the conidia, which form a cluster on each of the intercalary and terminal nodal cells along the conidiophore. Soil fungus characterized by developing traps to capture and destroy nematodes, besides being able to grow as a saprobe.

Arthrocladium

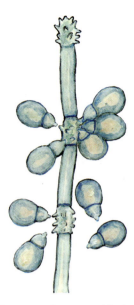

Arthrobotrys oligosporus Fresen.: conidiophore with verticillate groups of conidia, x 800.

Arthrocladium Papendorf
Inc. sed., Inc. sed., Inc. sed., Inc. sed., Pezizomycotina, Ascomycota, Fungi
Papendorf, M. C. 1969. New South African soil fungi. *Trans. Br. mycol. Soc.* 52(3):483-489.
From Gr. *árthron*, joint + *kládos*, dim. *kladion* > L. *cladion*, for the branched, jointed conidiophores and conidia. Hyphae septate, pale smoky brown. Conidiophores obsolete or distinct, pale brown-olivaceous. Conidia solitary on hyphae or single and in small groups terminally on conidiophores, filamentous with proximal articulate spore-body and septate, gradually tapering, tail-like distal extension, smooth, pale brown-olivaceous. Isolated from soil of *Acacia karroo* community, Potchefstroom, Transvaal, South Africa.

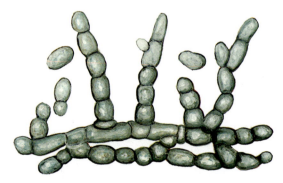

Arthrocladium tropicale M. M. F. Nascim., et al.: conidiophores with arthroconidia, x 1,000.

Arthroderma Curr.
Arthrodermataceae, Onygenales, Eurotiomycetidae, Eurotiomycetes, Pezizomycotina, Ascomycota, Fungi
Berkeley, M. J. 1860. *Outl. Brit. Fung.* (London). Lovell Reeve, London, p. 357.
From Gr. *árthron*, articulation, joint + *dérma*, skin, film, covering, because the peridium or covering of the cleistothecium has appendages composed of geniculate, phalangoid cells. Soil saprobe. *Trichophyton* Malmsten is not the sexual morph but a separate genus. Causes various superficial tineas or mycoses of the skin of humans and higher animals.

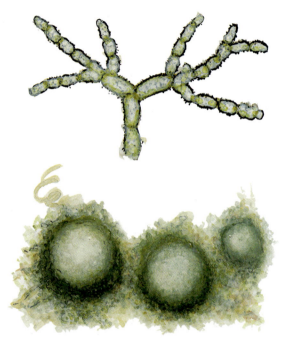

Arthroderma quadrifidum C. O. Dawson & Gentles: ascomata, x 50. Above there is a magnification of a phalangoid peridial appendage, x 1,500. Described in Great Britain.

Arthrographis G. Cochet ex Sigler & J. W. Carmich.
Eremomycetaceae, Inc. sed., Inc. sed., Dothideomycetes, Pezizomycotina, Ascomycota, Fungi
Siegler, L., J. W. Carmichael. 1976. Taxonomy of *Malbranchea* and some other hyphomycetes with arthroconidia. *Mycotaxon* 4(2):349-488.
From Gr. *árthron*, joint + *graphé* > *graphís*, pencil, lines of a drawing, possibly referring to the growth pattern in culture. Vegetative hyphae branched, hyaline. Conidiophores simple, hyaline, branched at apex, appearing arborescent. Fission arthroconidia formed by disjunction and segmentation of hyaline fertile branches borne at apex of the conidiophore, or by fragmentation of undifferentiated hyphae. Arthroconidia hyaline or yellow, smooth, cylindrical, appearing discoid in end view. Keratinolytic on hair and other substrates, including from clinical specimens.

Arthromyces T. J. Baroni & Lodge
Tricholomataceae, Agaricales, Agaricomycetidae, Agaricomycetes, Agaricomycotina, Basidiomycota, Fungi

Baroni, T. J., et al. 2007. *Arthromyces* and *Blastosporella*, two new genera of conidia-producing lyophylloid agarics (Agaricales, Basidiomycota) from the neotropics. *Mycol. Res.* 111(5):572-580.

From Gr. *árthron*, jointed + *mýkes*, fungus > L. *myces*, for the arthroconidia formed on the pileus. Species with a mycenoid or collybioid habit with extremely crowded, narrow, attached lamellae, surface of pileus and stipe producing chains of dark fuscous brown or dark olivaceous brown, ornamented arthroconidia, basidia with siderophilous/cyanophilous bodies and basidiospore walls cyanophilic. Collected in montane forest of Parque Armando Bermudez, Santiago Prov., Dominican Republic.

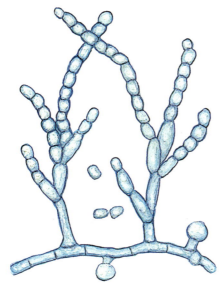

Arthrographis kalrae (R. P. Tewari & Macph.) Sigler & J. W. Carmich.: conidiophores with arthroconidia, x 150. Described in New York, U.S.A.

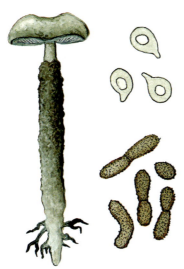

Arthromyces claviformis T. J. Baroni & Lodge: basidiocarp, x 1.5; basidiospores, x 1,000, and arthroconidia coming out from the pileus, x 1,000. On clay near *Eucalyptus* or under *Pinus occidentalis*.

Arthrowallemia R. F. Castañeda, et al.
Inc. sed., Inc. sed., Inc. sed., Inc. sed., Pezizomycotina, Ascomycota, Fungi

Castañeda Ruiz, R. F., et al. 1998. *Arthrowallemia*, a new genus of hyphomycetes from tropical litter. *Mycol. Res.* 102(1):16-18.

From Gr. *árthron*, jointed + genus *Wallemia* Johan-Olsen, the name of a similar fungus. Colonies effuse, hairy, brown to black. Mycelium mostly immersed. Conidiophores macronematous, mononematous, erect, cylindrical, septate, smooth or verrucose, subhyaline to brown, sometimes with percurrent proliferations, unbranched. Conidiogenous cells holothallic, terminal, determinate, integrated. Conidia thallic-arthric, in unbranched chains, forming by disarticulation of the conidiogenous hyphae at the septa, cylindrical to spathulate, septate, smooth-walled or verrucose, almost hyaline to black; conidial secession schizolytic. Sexual morph unknown. Collected on decaying leaves of an unidentified plant in Santiago de las Vegas, Ciudad de la Habana, Cuba.

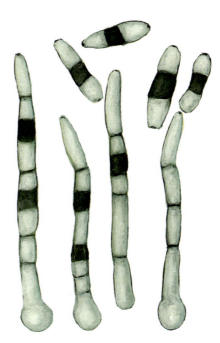

Arthrowallenia formosa R.F. Castañeda, et al.: conidiophores and conidiogenous cells, with conidia thallic-arthric, schizolytic, 0-3 septate, with smooth wall, x 500.

Arxiella Papendorf
Inc. sed., Inc. sed., Inc. sed., Inc. sed., Pezizomycotina, Ascomycota, Fungi

Papendorf, M. C. 1969. New South African soil fungi. *Trans. Br. mycol. Soc.* 52(1):483-489.

Named in honor of the Dutch mycologist J. A. von Arx + L. dim. suf. *-ella*. Hyphae septate, branched, thick-walled, hyaline to faintly olivaceous. Conidiophores one- to many-celled, continuous with hypha or basally septate, simple or branched, hyaline or faintly olivaceous, all cells conidiiferous; conidia apical or lateral on cells of conidiophore, sessile or on short or elongate sterigma-like projections, single or in short chains or in small irregular groups of interconnected members, reniform with ends obliquely cornute, hyaline to faintly colored, smooth, medianly 1-septate. Isolated from leaf-litter and top soil of *Acacia karroo* community.

Aschersonia Mont.—see **Moelleriella** Bres.
Clavicipitaceae, Hypocreales, Hypocreomycetidae, Sordariomycetes, Pezizomycotina, Ascomycota, Fungi
Montagne, J. F. A. 1848. Centurie i-ix de plantes cellulaires nouvelles tant indigènes qui' exotiques. *Annls Sci. Nat.*, Bot., sér. *3,10*:121.
Dedicated to the German Dr. *F. M. Ascherson* + L. suf. -*ia*, which denotes belonging to.

Ascobolus Pers.
Ascobolaceae, Pezizales, Pezizomycetidae, Pezizomycetes, Pezizomycotina, Ascomycota, Fungi
Persoon, C. H. 1792. *In*: J. F. Gmelin, Caroli Linné *Syst. Nat.*, Edn. 13, 2(2):1461.
From Gr. *askós*, sac or bag + *bállo* (L. *bolus*), to emit, to throw, hurl, shoot, because it discharges ascospores violently when the asci mature in the apothecia. Apothecia small (0.5-5 mm, occasionally up to 2.5 cm or more in diameter), dark and thick-walled. Ascospores dark brown with purple tones, occasionally black. Principally on the dung of herbivorous or carnivorous animals, but also on soil, wood or decomposing leaves.

Ascobolus sacchariferus Brumm.: apothecium with exserted mature asci, x 120. Described in the Netherlands.

Ascobotryozyma J. Kerrigan, et al. see—**Botryozyma** Shann & M.T. Sm.
Inc. sed., Saccharomycetales, Saccharomycetidae, Saccharomycetes, Saccharomycotina, Ascomycota, Fungi
Kerrigan, J., et al. 2001. *Ascobotryozyma americana* gen. nov. et sp. nov. and its anamorph *Botryozyma americana*, an unusual yeast from the surface of nematodes. *Antonie van Leeuwenhoek* 79(1):7-16.

From L. *ascus* < Gr. *askós*, wine skin, *ascus* + genus *Botryozyma*, i.e., a *Botryozyma* with asci. Asci originating from thallus of *Botryozyma* Shann & M.T. Sm . Asci globose or broadly ellipsoid, evanescent. Ascospores hyaline, lunate, smooth. Thalli attached to the nematode by basal cells. Basal cells obovoid, with bases branched. Other cells subcylindrical to subovoid, budding at the apex. Asexual morph *Botryozyma americana* J. Kerrigan, M.T. Sm. & J.D. Rogers. Collected in Spokane Co., Washington on *Panagrellus dubius* from *Populus tremuloides*.

Ascocoma H. J. Swart
Phacidiaceae, Helotiales, Leotiomycetidae, Leotiomycetes, Pezizomycotina, Ascomycota, Fungi
H. J. Swart. 1986. Australian leaf-inhabiting fungi XXIV. *Coma circularis* and its teleomorph. *Trans. Br. mycol. Soc.* 87(4):603-612.
From Gr. *askós*, wine skin > L. *ascus* + genus *Coma* Nag Raj & W.B. Kendr., i.e., a species of *Coma* with asci. Stromata subcuticular, confluent when close together and forming multilocular complexes, composed of vertical rows of cubical cells, surface layer black, internal tissue pale brown. Loculi variable in size, frequently confluent, opening irregularly with the surface layer breaking away, exposing the hymenium. Asci clavate to cylindrical with a rounded apex and a thick wall but apparently unitunicate, eight spored. Ascospores fusiform with rounded ends, one celled or unevenly two celled, hyaline in the ascus. Paraphyses numerous, unbranched, septate, with slightly swollen rounded tip, embedded in a gelatinous matrix. On living leaves of *Eucalyptus pauciflora*, Mt. Howitt, Australia.

Ascodesmis Tiegh.
Ascodesmidaceae, Pezizales, Pezizomycetidae, Pezizomycetes, Pezizomycotina, Ascomycota, Fungi
van Tieghem, P. E. L. 1876. Nouvelles observations sur le dèveloppement du fruit. *Bull. Soc. bot. Fr.* 23:275.
From Gr. *askós*, sac + *desmís*, genit. *desmídos*, fascicle, small sheaf for the arrangement of the asci in the ascomata. Fructification lacking a receptacle, small, composed of operculate, obovoid, oblong-obovoid, saculiform or broadly clavate asci. Ascospores multiseriate or irregular; generally eight-spored, rarely fewer in each ascus; brown or purple-brown episporic pigment that forms an irregular reticulum or even crests, spines or warts. On dung, some species only on the excrement of specific animals.

Ascodichaena Butin
Ascodichaenaceae, Rhytismatales, Leotiomycetidae, Leotiomycetes, Pezizomycotina, Ascomycota, Fungi
Butin, H. 1977. Taxonomy and morphology of *Ascodichaena rugosa* gen. et sp. nov. *Trans. Br. mycol. Soc.* 69(2):249-254.

From Gr. *askós* wine skin > L. *ascos*, ascus + genus *Dichaena* Fr., i.e., a *Dichaena* with asci. Ascomata blackish-brown, membranaceous, breaking out of the periderm, forming dense groups, appearing round to elongate hysterioid, opening by a simple, divided or stellate fissure, which expands under humid conditions; wall of the fruit body composed of more or less dark brown elongated or irregularly round cells. Asci unitunicate, shortly stipitate, broad clavate-cylindrical, splitting at the thin-walled apex, which lacks a pore and does not blue in Melzer's reagent. Paraphyses numerous, septate, apically clavate. Ascospores 4-6-8 per ascus, unicellular, ellipsoid to ovoid, in fresh material with granulate, hyaline protoplasm. Spermogonia developing next to the ascomata. Conidiophores arising from the innermost layer of cells, simple or divided, bearing spherical to ovate, hyaline microconidia (spermatia), under humid conditions forming drops of spores on the top of the pycnidia, not germinating in water. Pycnidia (macroconidial state) grow on extended stromata mixed with the ascomata, varying in shape and size, normally elongate hysterioid and bilabiate to oval, trilabiate or quadrilabiate. Macroconidia ovate to cylindrical hyaline. On hardwood branches in Europe.

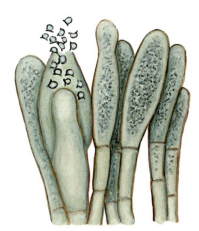

Ascoidea rubescens Bref.: immature asci, and one mature ascus liberating helmet-shaped ascospores, x 500.
Described from sap of damaged tree, in Germany.

Ascodesmis sphaerospora W. Obrist: immature and mature asci with ascospores, x 500; two free echinulate ascospores, x 750.
Described in Ohio, U.S.A.

Ascoidea Bref.
Ascoideaceae, Saccharomycetales, Saccharomycetidae, Saccharomycetes, Saccharomycotina, Ascomycota, Fungi
Brefeld, O., G. Lindau. 1891. *In*: O. Brefeld, Die Hemiasci und die Ascomyceten. *Unters. Gesammtgeb. Mykol.* (Leipzig) 9:94-108.
From Gr. *askós*, wine-skin, ascus + L. suf. *-oidea*, similarity. The name refers to the multispored asci unlike usual 8-spored asci. Asci proliferate internally. Asexual morph blastospores. Inhabits the mycangia of ambrosia beetles, which disseminate the spores, on which they feed.

Ascomauritiana Ranghoo & K. D. Hyde
Inc. sed., Inc. sed., Inc. sed., Inc. sed., Pezizomycotina, Ascomycota, Fungi
Ranghoo V. M., K. D. Hyde. 1999. *Ascomauritiana lignicola* gen. et sp. nov., an ascomycete from submerged wood in Mauritius. *Mycol. Res.* 103(8):938-942.
From L. *ascos*, ascus < Gr. *askós*, wine skin, ascus + *Mauritius*, for the country of origin. Ascomata solitary or aggregated, partly to completely immersed in the host substrate, perpendicular or horizontal to the host surface, dark brown to black, obpyriform, with a periphysate neck. Peridium comprising two strata, an inner stratum and an outer stratum. Paraphyses comprising short chains of globose cells, often breaking up into individual cells, early deliquescing. Asci 4-spored, cylindric-clavate, unitunicate, tapering at the base to a short stalk, rounded at the apex, lacking an apical apparatus, early deliquescing. Ascospores biseriate, broadly fusiform, somewhat curved, 5-septate, not constricted at the septa, at first hyaline, becoming brown to black at maturity, smooth-walled, with germ pores at each end, no mucilaginous sheath. Collected on submerged wood in Deep River, Beau Champ Sugar Estate, Beau Champ, Flacq District, Mauritius.

Ascomauritiana lignicola Ranghoo & K. D. Hyde: four-spored ascus, x 200, and free mature ascospores, x 300.

Ascosphaera

Ascosphaera L. S. Olive & Spiltoir
Ascosphaeraceae, Onygenales, Eurotiomycetidae, Eurotiomycetes, Pezizomycotina, Ascomycota, Fungi
Spiltoir, C. F., L. S. Olive. 1955. A reclassification of the genus *Pericystis* Betts. Mycologia 47(2):238-244.
From Gr. *askós*, wine-skin, sac, ascus + *sphaíra* > L. *sphaera*, sphere, because the asci are united in spherical groups (ascus balls) inside a sac-shaped, transparent, spherical structure (the sporocyst), which has been regarded as a cleistothecium. In the pollen of bee hives and as a parasite of the larvae of domesticated honey bees.

Ascosphaera apis (Maasen ex Claussen) L. S. Olive & Spiltoir: sporocyst with sporiferous balls, containing ascospores, x 550. Described in Germany.

Ascotaiwania Sivan. & H. S. Chang
Inc. sed., Inc. sed., Sordariomycetidae, Sordariomycetes, Pezizomycotina, Ascomycota, Fungi
Sivanesan, A., H. S. Chang. 1992. *Ascotaiwania*, a new amphisphaeriaceous ascomycete genus on wood from Taiwan. Mycol. Res. 96(6):481-484.
From L. *ascos*, ascus < Gr. *askós*, wine skin, ascus + *Taiwan* + L. des. *-ia*, for the country of origin. Perithecia solitary to aggregated, partly to completely immersed in the host substrate, oblique or horizontal, dark brown to black, globose, with a lateral short to long, erect, periphysate beak. Peridium composed of pseudoparenchymatous cells. Paraphyses filiform, hyaline, simple to rarely branched, deliquescing early. Asci cylindrical, 8-spored, pedicellate, functionally unitunicate, with a distinct nonamyloid apical annulus. Ascospores uniseriate to overlapping biseriate in the ascus, fusoid, transversely 7-septate, not constricted, straight to somewhat curved, smooth, with larger brown central cells and smaller hyaline to subhyaline end cells. Collected on wood in Wulae, Taipei, Taiwan.

Ascothailandia Sri-indr., et al.—see **Canalisporium** Nawawi & Kuthub.
Incertae sedis, Incertae sedis, Hypocreomycetidae, Sordariomycetes, Pezizomycotina, Ascomycota, Fungi
Sri-indrasutdhi, V., et al. 2010. Wood-inhabiting freshwater fungi from Thailand: *Ascothailandia grenadoidia* gen. et sp. nov., *Canalisporium grenadoidia* sp. nov. with a key to *Canalisporium* species (Sordariomycetes, Ascomycota). Mycoscience 51:411-420.
From Gr. *askós*, sac, ascus + *Thailand*, country of origin + L. des. *-ia*, i.e., an ascomycote from Thailand.

Ascotremella Seaver
Helotiaceae, Helotiales, Leotiomycetidae, Leotiomycetes, Pezizomycotina, Ascomycota, Fungi
Seaver, F. J. 1930. Photographs and descriptions of cup-fungi-X. *Ascotremella*. Mycologia 22(2):53.
From Gr. *askós*, wine skin > L. *ascos*, ascus + genus *Tremella* (from L. *tremulus*, trembling, from *tremo*, to tremble + dim. suf. *-ella*) for the tremelloid ascoma. Apothecia densely crowded or cespitose, tremelloid, sessile or substipitate. Asci cylindric, often much swollen so spores appear relatively small, 8-spored. Ascospores ellipsoid or more or less irregular, usually containing two small oil-drops, hyaline. Paraphyses slender, simple or branched. On wood.

Ascotremella faginea (Peck) Seaver: tremelloid ascomata on wood, among mosses, x 1.

Ascotricha Berk. (syn. **Dicyma** Boulanger)
Xylariaceae, Xylariales, Xylariomycetidae, Sordariomycetes, Pezizomycotina, Ascomycota, Fungi
Berkeley, M. J. 1838. Notes of British fungi. Ann. nat. Hist., Mag. Zool. Bot. Geol. 1, 1:257-264.
From Gr. *askós*, wine skin, sac, ascus + *thríx*, genit. *trichós*, hair. Ascomata perithecial with hairs, hairs with swollen, sympodial branching producing conidia. Conidiophores branched sympodially, at times dichotomously or trichotomously, similar to a cymose inflorescence, giving rise to polyblastic, denticulate conidiogenous cells. Conidia warted, olivaceous, unicellular. Grows as cellulolytic saprobe on seeds, vegetable detritus, soil, paper, cotton, and other cellulosic organic substrates.

Ascotrichella Valldos. & Guarro
Xylariaceae, Xylariales, Xylariomycetidae, Sordariomycetes, Pezizomycotina, Ascomycota, Fungi

Valldosera M., J. Guarro. 1988. Some coprophilous Ascomycetes from Chile. *Trans. Br. mycol. Soc.* 90(4):601-605.
From genus *Ascotricha* Berk. + L. dim. suf. *-ella*, i.e., similar to *Ascotricha*. Ascomata widely ovoid to subglobose, superficial, ostiolate, hairy, dark brown to black; peridium membranaceous. Paraphyses filiform, hyaline, septate. Asci cylindrical, unitunicate, non-amyloid, with an apical ring. Ascospores discoid, brown, simple, with a longitudinal germ slit, without a gelatinous sheath. Conidia ovoid to elliptic, developing on long ascomatal hairs, brown, verruculose, lateral, terminal or intercalary. Asexual morph humicola-like. Collected on unidentified dung in Frutono, Chile.

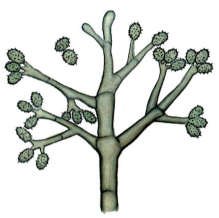

Ascotricha chartarum Beck. (syn. **Dicyma ampullifera** Boulanger): conidiophore branching dichotomously or trichotomously, terminating with unicellular and echinulate conidia, x 1,000.

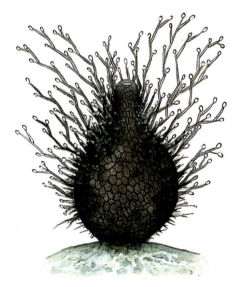

Ascotricha xylina L. M. Ames: perithecium covered with dichotomous branches that produce conidia, x 250. Described in Philippines.

Asellaria R. A. Poiss.
Asellariaceae, Asellariales, Inc. sed., Inc. sed., Kickxellomycotina, Zygomycota, Fungi

Poisson, R. 1932. *Asellaria caulleryii* n. g., n. sp., type nouveau d'entophyte, parasite intestinal des *Aselles* (crustacées isopodes). Description des stades connus et d'une partie de son cycle évolutif. *Bull. biol. Fr. Belg.* 66:232-254.
From the genus *Aselles*, the isopod crustacean that serves as its host + L. suf. *-aria*, which indicates connection or possession. Thallus branched, attached to *Aselles*; produces arthrospores in proctodeum of terrestrial and aquatic isopod crustaceans.

Asellaria aselli D. Scheer ex S. T. Moss & Lichtw.: endocommensal thallus adhered to the crustacean proctodeum (*Asellus aquaticus*), x 280. Described in United Kingdom.

Aseroë Labill.
Phallaceae, Phallales, Phallomycetidae, Agaricomycetes, Agaricomycotina, Basidiomycota, Fungi
Labillardière, J.J.H. 1800. *Relation du Voyage à la Recherche de La Pérouse*. H.J. Jansen. *Bull. Murith. Soc. Valais. Sci. Nat.* 1:145.
From Gr. *asê* (*asén*), disgust + *roê* (*roi*) juice; because it has a bad odor. Peridium of three layers, outer thin and furfuraceous, middle thick and gelatinous, inner mucilaginous. Receptacle hollow, cylindrical stem bearing apically a horizontal discoid expansion from margin from which arises a variable number of laterally arranged, simple or bifurcate arms. Spore mass mucilaginous, olivaceous, foetid, imposed upon upper part of disc, proximal portions of inner surfaces of arms. Spores elliptical, tinted, smooth. Growing solitary on ground and wood in Australia, Tasmania, New Zealand, China, Japan, Malaysia, South America, England and southern U.S.A.

Ashbya Guillierm.—see **Eremothecium** Borzi
Eremotheciaceae, Saccharomycetales, Saccharomycetidae, Saccharomycetes, Saccharomycotina, Ascomycota, Fungi

Aspergillus

Guilliermond, A. 1928. Recherches sur quelques ascomycetes inferieurs isoles de la stigmatomycose des grains de cotonnier. *Revue Gen. Botanique* 50:562.
Dedicated to the American mycologist S. F. Ashby + L. ending -*a*.

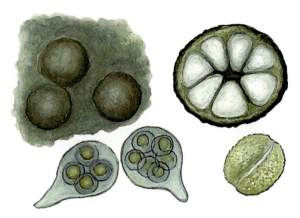

Aspergillus flavus Link (syn. **Petromyces flavus** B. W. Horn, et al.): three sclerotia on agar, x 12; vertical section of one sclerotium, showing seven ascocarps inside, x 30; asci with uniguttulate ascospores, x 1,000; tuberculate ascospore with the equatorial line, x 3,000. Derived from cross of asexual morphs isolated from seed of peanut and from soil, Georgia, U.S.A.

Aseroë rubra Labill.: basidiocarp on soil, x 1.

Aspergillus P. Micheli ex Link (syns. **Chaetosartorya** Subram., **Emericella** Berk., **Eurotium** Link, **Gymnoeurotium** Malloch & Cain, **Harpezomyces** Malloch & Cain, **Neosartorya** Malloch & Cain, **Petromyces** Malloch & Cain, **Sartorya** Vuill., **Sporophormis** Malloch & Cain)

Aspergillaceae, Eurotiales, Eurotiomycetidae, Eurotiomycetes, Pezizomycotina, Ascomycota, Fungi
Link, J. H. F. 1809. Observationes in ordines plantarum naturales. *Mag. Gesell. naturf. Freunde, Berlin* 3(1-2):16. From L. *aspergillo* < *asper*, rough + *gillo*, vessel, bladder container used in sprinking sacred water + dim. suf. -*illus*, referring to the vesicle of the conidial head. Colonies forming large compact masses of hyphae that develop into ascocarp-bearing stromata. Stromata superficial, ovoid to cylindrical, sclerenchymatous, containing one to several ascocarps. Ascocarp initials coiled, formed among vegetative hyphae, eventually cleistothecial often covered by Hülle cells, subglobose to globose or irregular, tomentose, thin-walled, hyaline. Asci produced in small clusters, irregularly arranged, subglobose to globose, evanescent, 8-spored. Ascospores oblate, hyaline, reddish-purple or bluish, smooth, sometimes with two equatorial crests or with a thin hyaline line around equator. Conidial heads covered with uniseriate phialides and conidia. Conidia small, globose. Ubiquitous, with a large number of saprobic species, common in soil as saprobe, parasites of plants, animals and humans, some osmophilic, others of industrial importance. Several species produce toxigenic compounds.

Aspergillus flavus Link: biserial conidial head, x 1,500.

Aspergillus galapagensis (Frisvad, et al.) Samson, et al.: cleistothecia on agar culture, x 300; asci with ascospores, x 400; bivalve ascospores, with two equatorial crests, x 1,200; conidiophore with phialides and conidia, x 500. Isolated from soil, Galapagos, Ecuador.

Asterina

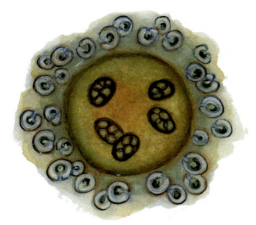

Aspergillus nidulans (Eidam) G. Winter (syn. **Emericella nidulans** (Eidam) Vuill.): cleistothecium covered by Hülle cells, with asci and reddish-purple ascospores, x 200.

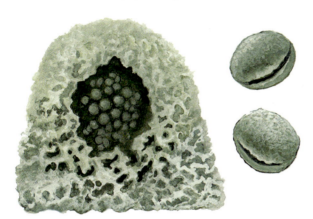

Aspergillus spinulosus Warcup (syn. **Sporophormis spinulosa** Malloch & Cain): hyphal cleistothecium, gymnoascus-like, with a central opening with ascospores inside, x 200; two liberated, equinulate ascospores, with two symmetrical valves, x 3,000.

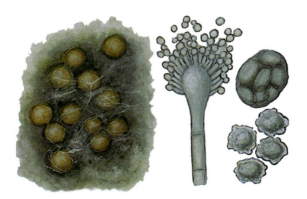

Aspergillus taklimakanense Abliz & Y. Horie: cleistothecia, x 100; conidiophore with phialides and conidia, x 500; ascus with ascospores, x 500; free ascospores, x 1,500. Isolated from desert soil, Xinjian, China.

Asteridiella McAlpine (syn. **Parasteridiella** H. Maia)
Meliolaceae, Meliolales, Inc. sed., Sordariomycetes, Pezizomycotina, Ascomycota, Fungi

McAlpine, D. 1897. *Proc. Linn. Soc. N.S.W.* 22(1):38. From genus *Asteridium* Speg. + L. dim. suf. *-ella*, and *Asteridium* from Gr. *astér*, star + L. dim. suf. *-idium*, for the starlike shape of the colonies. Black mildew or sooty mold, mycelium brownish-black, smooth, flat, with lateral capitate and mucronate hyphopodia. Ascomata globose, membranous-carbonaceous. Paraphyses present. Asci 2-spored, oblong, unitunicate. Ascospores brownish-black, broadly oblong, phragmosporous. Conidia lacking. Obligate parasite on living leaves of diverse tropical plants.

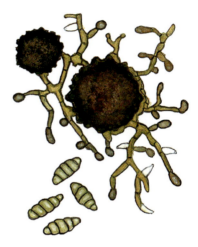

Asteridiella chowrirae Hosag., Thimm. & Jayash.: perithecia with mucronate and capitate hyphopodia, x 150; ascospores, x 300. On leaves of *Euphorbia pulcherrima*, from Karnataka, India.

Asterina Lév.
Asterinaceae, Asterinales, Dothideomycetidae, Dothideomycetes, Pezizomycotina, Ascomycota, Fungi
Léveillé, J. H. 1845. *Annls Sci. Nat., Bot.*, sér. 3,3:59.
From L. *aster*, star + suf. *-ina*, which indicates possession or appearance. Pseudothecia orbicular, open by a wide pore or by irregularly stellate or radial fissures; cells collapse or disintegrate, forming a mucilage; pseudothecia unilocular, developing from superficial mycelium. Asci cylindrical or obclavate; ascospores 1-septate, turn brown when mature.

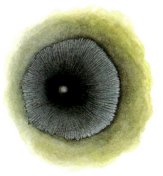

Asterina veronicae (Lib.) Cooke: orbicular, radiate pseudothecia, with a central pore, growing on the leaves of the host (*Veronica officinalis*), x 250.

Asterophora

Asterophora Ditmar
Lyophyllaceae, Agaricales, Agaricomycetidae, Agaricomycetes, Agaricomycotina, Basidiomycota, Fungi
Ditmar, L. P. F. 1809. *Duo genera fungorum. Neues J. Bot.* (Schrader) 3(3):55-57.
From Gr. *astér*, *astéros*, star + *phóros*, carrier. Pileus with lamellae often somewhat reduced, rather thick, obtuse, narrow, distant, producing few star-shaped basidiospores. Chlamydospores generally in abundance, brown, stellate, smooth, cyanophilous. Parasitic on Russulaceae in Europe, North America, Cuba, East Asia and New Guinea.

Asterophora lycoperdoides (Bull.) Ditmar: basidiocarps growing on the surface of a pileus of Russulaceae, x 1.

Asterostomopora Bat. & H. Maia
Inc. sed., Inc. sed., Inc. sed., Inc. sed., Pezizomycotina, Ascomycota, Fungi
Batista, A. C., et al. 1960. Novos fungos Asterinothyriaceae e Plenotrichaceae. *Publicações Inst. Micol. Recife* 221:1-22.
From Gr. *astér*, genit. *astéros*, star + *stoma*, mouth + L. *porus*, pore. Mycelium superficial, composed of sinuous, brown hyphae, irregularly branched, septate, with lateral hyphopodia, without setae. Pycnostromata dispersed, orbicular, dimidiate, smooth, with a central pore. Upper wall formed of rectangular cells radially arranged; basal wall inconspicuous. Conidiophores formed on underside of upper wall. Conidia olive-brown, ellipsoid, 1-celled, smooth. Collected on an unknown plant in St. Catherine, Jamaica.

Astraeus Morgan
Diplocystidiaceae, Boletales, Agaricomycetidae, Agaricomycetes, Agaricomycotina, Basidiomycota, Fungi
Morgan, A. P. 1889. North American Fungi. The Gasteromycetes. *J. Cincinnati Soc. Nat. Hist.* 12:8-22.
From Gr. *astér*, *ástron*, star + suf. *-eus*, belonging to; referring to the stellate shape of the fructification at maturity. Fructifications subglobose in button stage with coriaceous, cartilaginous exoperidium that fragments into segments similar to radii of a star. Endoperidium thin, sessile, membranaceous with dehiscence by means of an apical pore. Gleba of long, branched filaments (capillitium) and large, globose, finely verrucose, brown spores. Grows in semiarid pastures and scrub lands in transition zone to pines and oaks or in sandy soils in arid regions.

Astraeus hygrometricus (Pers.) Morgan: stellate basidiocarp on soil, x 1.

Athelia Pers. (syn. **Sclerotium** Tode)
Atheliaceae, Atheliales, Agaricomycetidae, Agaricomycetes, Agaricomycotina, Basidiomycota, Fungi
Persoon, C. H. 1922. *Mycol. Eur.* (Erlanga) 1:83.
Etymology not determined. In soil, capable of causing blight, stem and root rot of a variety of plants especially as soilborne pathogen *A. rolfsii* (Curzi) C.C. Tu & Kimbr.

Athelia rolfsii (Curzi) C. C. Tu & Kimbr. (syn. **Sclerotium rolfsii** Sacc.): dark sclerotia, capable of attacking a wide variety of plants, that develop on a white mycelium, x 20.

Aureobasidium

Atkinsonella Diehl
Clavicipitaceae, Hypocreales, Hypocreomycetidae, Sordariomycetes, Pezizomycotina, Ascomycota, Fungi
Diehl, W. W. 1950. *Balansia* and the Balansiae in America. *Agriculture Monogr.*, US Dept Agric. 4:1-82.
Named in honor of the American mycologist *George F. Atkinson* + L. dim. suf. *-ella*. Stroma developing around culm and leaf base, superficial, first forming a thin, flat hypothallus, grayish. Conidial morphs formed on hypothallus, followed by perithecia. Stromata pulvinate, erumpent through hypothallus, short-stalked to subsessile, separate, few in number, black. Perithecia obclavate, immersed in stroma in a single row, crowded, with erumpent ostiolar necks. Ostiole lined with slender periphyses. Perithecial wall composed of several rows of cells with slightly thickened, brown walls. Centrum with septate, lateral paraphyses. Asci long, subcylindrical, tapered toward base, unitunicate, with a thickened apical cap, not blueing in iodine, 8-spored. Ascospores filiform, parallel in ascus, multiseptate, hyaline. Asexual morphs both typhodial or sphacelia-like and ephelidial. Parasitic on living grasses in the genera *Danthonia* and *Stipa*.

Atractilina Dearn. & Barthol.
Inc. sed., Inc. sed., Inc. sed., Inc. sed., Pezizomycotina, Ascomycota, Fungi
Dearness, J. 1924. New and noteworthy fungi-III. *Mycologia* 16(4):143-176.
From Gr. *átraktos*, spindle, shaft + L. suf. *-ina*, denoting likeness. Mycelium forming erect synnemata on host lesions; synnemata brush-like in upper quarter, bearing conidiogenous cells. Conidia single or weakly catenulate, hyaline to subhyaline, 1-4 but mostly 3-septate, tapering to a narrow base. On *Callicarpa americana* in Florida, U.S.A.

Aulographina Arx & E. Müll.
Asterinaceae, Asterinales, Dothideomycetidae, Dothideomycetes, Pezizomycotina, Ascomycota, Fungi
Arx, J. A. von, E. Müller. 1960. Über die neue Ascomycetengattung *Aulographina*. *Sydowia* 14:330-333.
From genus *Aulographum* Lib. + L. suf. *-ina*, denoting likeness, i.e., a genus similar to *Aulographum*. Mycelium superficial on host, composed of brown, torulose, much branched hyphae. Ascomata sessile thyriothecia, shield-shaped, elongate, often branched, opening by a longitudinal slit. Pseudoparaphyses filiform. Asci numerous, bitunicate, 4-8-spored. Ascospores elongate, with a median septum, hyaline. On conifer needles and leaves of hardwoods.

Aulographum Lib.
Aulographaceae, Asterinales, Dothideomycetidae, Dothideomycetes, Pezizomycotina, Ascomycota, Fungi
Libert, M. A. 1834. *Pl. crypt. Arduenna, fasc.* (Liège) 3:no. 272.
From Gr. *aulón*, genit. *aulónos*, canal, aqueduct, tube, or from *aulós*, fistula, furrow, tube, bobbin + *graphé*, writing, engraving. Pseudothecia on a subcuticular mycelial membrane apothecioid, elongate or oblong opening by a longitudinal slit when mature, exposing asci. Asci short, ovoid or cylindrical. Ascospores 1-septate, hyaline or pale brown. On leaves and branches.

Aulographum hederae Lib.: apothecioid pseudothecia, on dry leaf of *Hedera helix*, from Belgium, x 200.

Atractilina alinae Melnik & U. Braun: erect synnemata with conidia, x 100. On decaying leaves from Vietnam.

Aureobasidium Viala & G. Boyer (syn. **Pullularia** Berkhout)

Auricularia

Saccotheciaceae, Dothideales, Dothideomycetidae, Dothideomycetes, Pezizomycotina, Ascomycota, Fungi
Viala, P., G. Boyer. 1891. *Rev. gén. Bot.* 3:371.
From L. *aureus*, golden < *aurum*, gold + *basidium* < Gr. *basídion*, dim. of *básis*, base. Hyphae yellowish, conidia borne on short denticles. Conidia ovoid, unicellular, hyaline or pigmented, produced in mucilage. Ubiquitous and omnivorous, on painted surfaces, wood, flower nectar, soil, decomposing fruits, paper pulp and many other substrates.

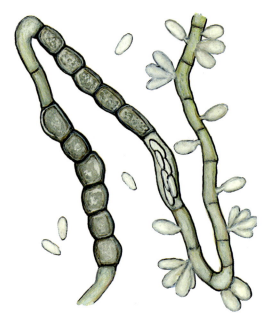

Aureobasidium pullulans (de Bary & Löwenthal) G. Arnaud: hyphae with conidia originating from denticles, and chlamydospores, x 800.

Auricularia Bull.

Auriculariaceae, Auriculariales, Inc. sed., Agaricomycetes, Agaricomycotina, Basidiomycota, Fungi
Bulliard, P. 1780. *Herb. Fr.* (Paris) 3:tab 290.
From L. *auris*, ear, from this *auricula*, ear + suf. *-aria*, which indicates connection or possession. Basidiomata gelatinous-leathery or subcartilaginous, with a lateral attachment to substrate, resembling ears, completely smooth or with alveoli, veins or hairs, on upper or lower surface, sessile, occasionally attached by small pedicel. On wood in temperate and tropical forests. Several species are consumed as food and can be cultivated under artificial conditions.

Auriculoscypha D. A. Reid &.Manim.

Septobasidiaceae, Septobasidiales, Inc. sed., Pucciniomycetes, Pucciniomycotina, Basidiomycota, Fungi
Reid, D. A., P. Manimohan. 1985. *Auriculoscypha*, a new genus of Auriculariales (Basidiomycetes) from India. *Trans. Br. mycol. Soc.* 85(3):532-535.
From L. *auricula*, the ear + Gr. *skýphos* > L. *scypha*, a cup, a cupulate auriculariaceous fungus. Sporophores cupulate, with central, dorsally attached stipe, woody, non-gelatinous texture; flesh golden-brown, cottony-fibrillose, non-gelatinous; hyphae of context almost hyaline toward margin, elsewhere pale brown to brown with thickened refractive walls, septate, lacking clamp-connexions. Basidia auricularioid, clavate or circinate. Spores hyaline, cylindrical or allantoid, septate or muriform. Collected on branches of cashew (*Anacardium occidentale*), Calicut University campus, North Kerala, India.

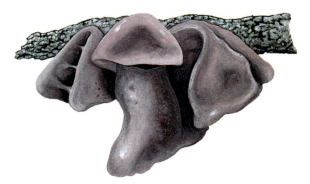

Auricularia auricula-judae (Bull.) Quél.: basidiocarps on dead wood, x 2.

Auriculoscypha anacardiicola D. A. Reid & Manim.: basidiomata on a branch of *Anacardium occidentale*, x 2.

Aurificaria D. A. Reid

Hymenochaetaceae, Hymenochaetales, Inc. sed., Agaricomycetes, Agaricomycotina, Basidiomycota, Fungi
Reid, D .A. 1963. New or interesting records of Australasian Basidiomycetes: V. *Kew Bull.* 17(2):278.
From L. *aureus*, *auri*, golden + *fic*, from *facio*, to make, in reference to the brilliant golden-brownish color of the context + suf. *-aria*, connection or possession. Sporophores centrally stipitate and often much lobed or sessile, dimidiate, imbricate and entire. Upper surface of pileus dark brown, often covered with rusty pruina. Context brownish-gold, texture hard, almost woody but with a relatively light weight. Hyphal structure monomitic, without clamp-connexions at septa. Setae and setoid hyphae lacking in internal surface of tubes. Spores small, globose to widely elliptic, with distinct dull-brown walls. Lignicolous or terrestrial.

Austroboletus

Aurificaria indica (Massee) D. A. Reid): lignicolous basidioma, x 0.5. Described in Gujurat, India.

Auriscalpium Gray
Auriscalpiaceae, Russulales, Inc. sed., Agaricomycetes, Agaricomycotina, Basidiomycota, Fungi
Gray, S. F. 1821. *Nat. Arr. Brit. Pl.* (London) 1:650.
From L. *auriscalpium*, ear cleaner, from *auris*, ear + *scalpo*, to dig, to scrape, to rub, for the shape of the basidioma. Basidiomata generally small, dark, and of a tough consistency. Pileus lobate-reniforme with an eccentric stipe. Growing on conifer remains, usually on fallen pine cones, also among mosses or on the remains of grasses. One of the most common species, *A. vulgare* S. F. Gray (=*Hydnum auriscalpium* L. ex Fr.), has extremely hairy dark brown, reddish, or violaceous fructifications with pale brown or grayish teeth covered by the hymenium.

Auriscalpium vulgare Gray: mature basidiocarps on soil, among mosses, x 1. Described from forests in Europe.

Aurosphaeria Sun J. Lee, et al.
Inc. sed., Inc. sed., Inc. sed., Inc. sed., Pezizomycotina, Ascomycota, Fungi
Lee, J. L., et al. 2009. *Aurosphaeria*, a novel coelomycetous genus. *Mycotaxon* 107:463-472.
From L. *aurum*, gold + *sphaera*, sphere < Gr. *sphaīra* + L. des. *-ia*, for the color and shape of the pycnidia. Mycelial colonies on agar whitish, with a yellow reverse. Pycnidia in clusters, erumpent, golden, globose, lacking an ostiole, thin-walled with tufts of pigmented hyphae on exterior. Conidiophores hyaline, branched at tips, with enteroblastic, phialidic, determinate, smooth conidiogenous cells. Conidia globose, hyaline, smooth, thin-walled, borne in chains. Produces biologically active compounds. Isolated as an endophyte from *Drosera montana* in the Bolivian Amazon.

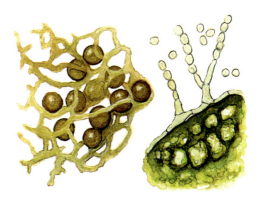

Aurosphaeria flaviradians Sun J. Lee, et al: clusters of pycnidia, x 400, and conidiophores with conidia, x 1,000.

Australopilus Halling & N. A. Fechner—see Royoungia Castellano, et al.
Boletaceae, Boletales, Agaricomycetidae, Agaricomycetes, Agaricomycotina, Basidiomycota, Fungi
Halling, R. E., et al. 2012. Affinities of the *Boletus chromapes* group to *Royoungia* and the description of two new genera, *Harrya* and *Australopilus*. *Australian Systematic Botany* 25(6):418-431.
From L. *austral*, southern, from Australia, + *pilus*, cap.

Austroboletus (Corner) Wolfe
Boletaceae, Boletales, Agaricomycetidae, Agaricomycetes, Agaricomycotina, Basidiomycota, Fungi
Wolfe, C.B. Jr. 1979. *Austroboletus* and *Tylopilus* subg. *Porphyrellus*, with emphasis on North American taxa. *Biblthca Mycol.* 69:1-132
From L. *austro*, austral, of the South + genus *Boletus* Tourn. Pileus viscid to dry with setoid cystidia sometimes present at pores. Stipe glabrous to subglabrous, faintly punctate to almost scabrous, reticulate or longitudinally ridged, elevated or with a glutinous sheet, finally with a subapical glutinous, narrow annulus; context unchanging or variously discolored, blackening at times by autoxidation; taste mild or more rarely slightly bitter. Hyphae without clamp connections, inamyloid. Spores with variable ornamentation visible in light microscope, rarely appearing smooth, distinctly verruculose in the SEM, mostly elongate-fusoid, inamyloid to pseudoamyloid. Ectomycorrhizal; growing in the north and south temperate regions as well as the neo- and paleotropics.

Autoicomyces

Autoicomyces Thaxt.

Ceratomycetaceae, Laboulbeniales, Laboulbeniomycetidae, Laboulbeniomycetes, Pezizomycotina, Ascomycota, Fungi

Thaxter, R. 1908. Contribution towards a monograph of the Laboulbeniaceae. Part II. Mem. Am. Acad. Arts Sci., ser. 2. 13:217-469.

From Gr. *autós*, same, by itself, own + *oíkos*, house or dwelling + *mýkes* > L. *myces*, fungus, for all the known forms inhabit species of a single genus, *Berosus*, an aquatic coleopteran. Receptacle consisting of three superposed cells, the lowest often involved by the blackened foot, the upper surmounted by a pair of cells giving rise to the single perithecium and the antheridial appendage respectively. Antheridial appendage consisting of a series of superposed cells producing ramigenous branches irregularly along its inner margin. Perithecium usually appendiculate, determinate, the wall-cells in rows.

Autoicomyces contortus (Thaxt.) I. I. Tav. (syn. **Ceratomyces contortus** Thaxt.): thallus adhered on aquatic coleoptera, x 600.

B

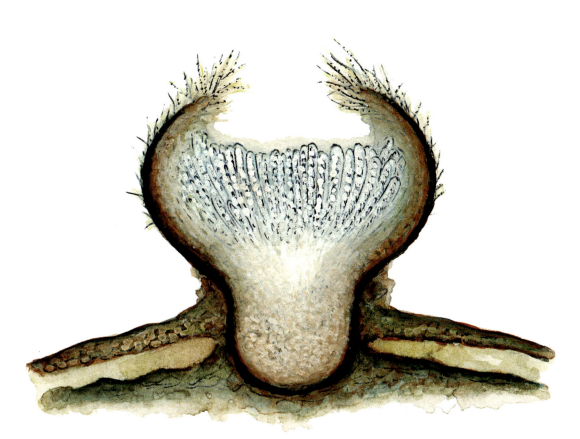

Bulbomollisia radiata

B

Babjeviella Kurtzman & M. Suzuki
Debaryomycetaceae, Saccharomycetales, Saccharomycetidae, Saccharomycetes, Saccharomycotina, Ascomycota, Fungi
Kurtzman, C. P., M. Suzuki. 2010. Phylogenetic analysis of ascomycete yeasts that form coenzyme Q-9 and the proposal of the new genera *Babjeviella*, *Meyerozyma*, *Millerozyma*, *Priceomyces*, and *Scheffersomyces*. *Mycoscience* 51:2-14.
Named in honor of the Russian zymologist *Inna P. Babjeva* + L. dim. suf. *-ella*, for her outstanding contributions to yeast systematics and ecology. Asci unconjugated, forming one to four hat-shaped ascospores released by ascus evanescence. Bulb-like structures sometimes form on brims of ascospores. Not determined if species homothallic or heterothallic. Asexual cell division by multilateral budding on a narrow base. Cells short ellipsoid to elongate. Neither pseudohyphae nor true hyphae formed. Sugars not fermented, but many of the sugars and polyols in standard yeast growth tests assimilated. Major ubiquinone CoQ-9 and mol% G + C content for type strain 49.9. Diazonium blue B test negative. *Babjeviella inositovora* (Golubev & Blagod.) Kurtzman & M. Suzuki is one of the few ascomycetous yeasts that can utilize inositol as a sole source of carbon.

Baeomyces Pers.
Baeomycetaceae, Baeomycetales, Ostropomycetidae, Lecanoromycetes, Pezizomycotina, Ascomycota, Fungi
Persoon, C. H. 1794. Einige Bemerkungen über die Flechten. *Ann. Bot.* (Usteri) 7:19.
From Gr. *baiós*, slender, slim, weak small, dry + *mýkes* > L. *myces*, fungus. Primary lichenized thallus crustose, inconspicuous, fragile. Short, erect, solid podetia with pink or brown apothecia. On soil, sand, and moist rocks.

Baeospora Singer
Marasmiaceae, Agaricales, Agaricomycetidae, Agaricomycetes, Agaricomycotina, Basidiomycota, Fungi
Singer, R. 1938. Notes sur quelques Basidiomycetes. *Revue Mycol.*, Paris 3:187-199.
From Gr. *baiós*, little + *sporá*, spore. Carpophores collybioid, rarely subpleurotoid; pileus with initially incurved margin, hygrophanous or subhygrophanous, brown to violet; hymenophore lamellate, lamellae narrow, subdecurrent to subfree, crowded, pallid to somewhat lilac. Spores to 6 µm, subglobose to cylindric, hyaline, amyloid, smooth.

Baeomyces rufus (Huds.) Rebent.: podetia with terminal, vinaceous apothecia, on soil, x 12.

Bagnisiella Speg.
Dothideaceae, Dothideales, Dothideomycetidae, Dothideomycetes, Pezizomycotina, Ascomycota, Fungi
Spegazzini, C. L. 1880. Fungi Argentini. *Anales Soc. cient. argent.* 10(4):146.
Dedicated to the Italian mycologist *C. Bagnis* + L. dim. suf. *-ella*. Ascostromata composed of erumpent, pulvinate pseudothecia, with asci forming a compact palisade exposed when upper part of pseudothecium breaks and becomes black, giving the mature ascostroma an apothecioid appearance. Spores hyaline lacking septa. On dead stems of plants.

Baipadisphaeria Pinruan
Nectriaceae, Hypocreales, Hypocreomycetidae, Sordariomycetes, Pezizomycotina, Ascomycota, Fungi
Pinruan, U., et al. 2010. *Baipadisphaeria* gen. nov., a freshwater ascomycete (Hypocreales, Sordariomycetes) from decaying palm leaves in Thailand. *Mycosphere* 1:53-63.

From Thai *baipad*, fan leaves of a palm + Gr. *sphaíra* (L. *sphaera*), sphere, referring to round ascomata. Ascomata solitary to scattered, subglobose, immersed, dark brown, coriaceous, ostiolate, lacking ascomatal setae. Paraphyses septate, hypha-like, unbranched. Asci apedicellate, clavate to ovoid, unitunicate, 8-spored, apically narrow and rounded, lacking an apical structure. Ascospores fusiform to subcylindrical, straight or curved, hyaline to pale brown, smooth-walled. On submerged trunk of the palm *Licuala longicalycata* in Thailand.

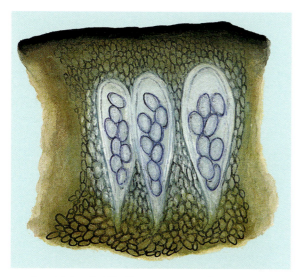

Bagnisiella mirabilis (Starbäck) Theiss.: vertical section of an ascostroma, showing three asci with ascospores, growing on plant detritus, x 500.

Balansia Speg.
Clavicipitaceae, Hypocreales, Hypocreomycetidae, Sordariomycetes, Pezizomycotina, Ascomycota, Fungi
Spegazzini, C. L. 1885. *Anales Soc. cient. argent.* 19(1):45.
Named in honor of the French botanist *Benedict Balansa* + L. des. *-ia*, a French botanist who collected in Argentina. Ascomata stromatic, with stroma originating from systemic mycelium in host tissues, first forming a thin, flat hypothallus on leaf surface. Conidia produced on hypothallus. Perithecial stroma forming later, over hypothallus, effuse to elongate-pulvinate or capitate and sessile or stipitate, dark brown to blackish on surface, interior hyaline, fleshy. Perithecia immersed in a single layer in periphery of stroma, crowded. Asci unitunicate, cylindrical with short stalk, with thickened apical cap, formed in a basal cluster, 8-spored. Ascospores filiform, hyaline, phragmosporous, often breaking into part-spores. Asexual state ephelis-like. Parasitic on grasses, causing diseases in temperate and tropical areas.

Banksiamyces G. W. Beaton
Helotiaceae, Helotiales, Leotiomycetidae, Leotiomycetes, Pezizomycotina, Ascomycota, Fungi

Beaton, G., G. Weste. 1982. *Banksiamyces* gen. nov., a discomycete on dead *Banksia* cones. *Trans. Br. mycol. Soc.* 79:271-277.
From the Australian plant genus *Banksia* + L. *myces*, fungus < Gr. *mykês*, fungus. Apothecia superficial, dark brown to black, tough; disk dark grey, depressed receptacle cupulate, stalked, when dry covered with a whitish meal formed from a hyaline, gelatinous exudate, margin inrolled, contorted or laterally compressed; ectal excipulum and stalk cortex of brown, thick- or thin-walled, globose or ellipsoidal cells; medullary excipulum and stalk medulla of brown, granular, thick- or thin-walled hyphae and a hyaline, gelatinous tissue of hyphae with staining lumina, walls faintly visible or not, gelatinous tissue always present in either medullary excipulum, stalk medulla or both. Asci cylindrical-clavate, 8-spored, with a positive iodine reaction in ascus plug, apices thick-walled, spores uniseriate, biseriate or rarely irregular. Ascospores ellipsoidal, tapering or slightly curved, hyaline, non-septate, most with two polar oil drops, covered in ascus with a highly refractive gelatinous or mucilaginous coating. Paraphyses cylindrical, septate, thick- or thin-walled, simple, branched or forked at tips; immersed, with tips of the asci, in a hyaline or brown matrix. On seed follicles of dead cones of *Banksia* in Australia.

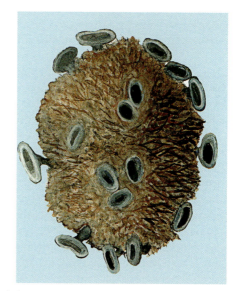

Banksiamyces toomansis (Berk. & Broome) G. W. Beaton: apothecia growing on the surface of a fruit of *Banksia quercifolia*, x 1.

Barbatosphaeria Réblová (syn. Tectonidula Réblová)
Inc. sed., Inc. sed., Inc. sed., Sordariomycetes, Pezizomycotina, Ascomycota, Fungi
Réblová, M. 2007. *Barbatosphaeria* gen. et comb. nov., a new genus for *Calosphaeria barbirostris*. *Mycologia* 99(5):727.
From L. *barbatus*, bearded + *sphaera*, sphere, refers to the conspicuous pubescence covering neck and perithecia.

Barnettella

Perithecia nonstromatic, immersed between bark and cortex in circular groups or solitary, venter globose to subglobose, necks elongating, converging, piercing cortex in a common point. Ostiolum periphysate. Perithecial wall leathery, two- or three-layered. Paraphyses septate, tapering towards tip, longer than asci. Asci unitunicate, cylindrical-clavate, with shallow, refractive apical ring and visible invagination, when young long-stipitate with bulbose base, at maturity short-stipitate, tapering towards base, floating freely. Ascospores allantoid, U- to horseshoe-shaped or 3-4 circular aseptate or septate, hyaline. Asexual states ramichloridium-like and sporothrix-like.

Barbatosphaeria barbirostris (Dufour) Réblová: perithecia with pubescent necks, x 60; an ascus with bicellular ascospores, x 1,300.

Barnettella D. Rao & P. Rag. Rao (syn. **Vinculum** R. Y. Roy, et al.)
Inc. sed., Inc. sed., Inc. sed., Inc. sed., Pezizomycotina, Ascomycota, Fungi
Rao D., P. R. Rao. 1964. *Barnettella*, a new member of Melanconiaceae. *Mycopath. Mycol. Appl.* 22:56.
Name dedicated to professor *H. L.Barnett*, of the West Virginia University, U.S.A. + dim. suf. *-ella*. Sporodochia black, globose to irregular, aggregating to form a large compact mass; conidia in long chains, interconnected with several hyaline cells arranged in a single layer, globose to barrel shaped, dark, smooth-walled, multicellular, cells arranged irregularly or in tiers, end cells hyaline. On decaying grass on the Banaras Hindu University campus, Varanasi, India.

Barnettozyma Kurtzman, et al.
Wickerhamomycetaceae, Saccharomycetales, Saccharomycetidae, Saccharomycetes, Saccharomycotina, Ascomycota, Fungi
Kurtzman, C. P., et al. 2008. Phylogenetic relationships among species of *Pichia*, *Issatchenkia*, and *Williopsis* determined from multigene sequence analysis, and the proposal of *Barnettozyma* gen. nov., *Lindnera* gen. nov. and *Wickerhamomyces* gen. nov. *FEMS Yeast Res.* 8:939-954.
Named in honor of the English mycologist *James A. Barnett* + connective *-o-* + Gr. *zýme*, yeast. Asci globose to ellipsoid, unconjugated or arising from conjugation between a cell and its bud or between independent cells. Some species heterothallic. Asci deliquescent or persistent with one to four ascospores that are hat shaped or spherical with an equatorial ledge. Cell division by multilateral budding on a narrow base; budded cells spherical, ovoid or elongate. Pseudohyphae in some species; true hyphae in *B. wickerhamii* (Van der Walt) Kurtzman, Robnett & Bas.-Powers. Glucose fermented by most species; some ferment other sugars. A variety of sugars, polyols, and other carbon sources assimilated by most species, but not methanol and hexadecane. Some species use nitrate. Where determined, predominant ubiquinone is CoQ-7. The diazonium blue B reaction is negative. The genus is phylogenetically circumscribed from analysis of LSU, SSU rRNA, and EF-1a gene sequences.

Barnettella indica (R. Y. Roy, et al.) P. Rag. Rao: black, subglobose sporodochium, x 70, on stems and pods of *Bothriochloa portusa*, *Cynodon dactylon* and *Dichantium annulatum*, in Uttar Pradesh, India, with long chains of barrel-shaped, multicellular, darkly pigmented conidia, interconnected with hyaline cells, x 700.

Basidiobolus Eidam
Basidiobolaceae, Basidiobolales, Inc. sed., Inc. sed., Inc. sed., Zygomycota, Fungi
Eidam, E. 1886. *Basidiobolus*, eine neue Gattung der Entomophthoraceen. *Beitr. Biol. Pfl.* 4:181-251.
From Gr. *basídion*, dim. of *básis*, base + *bállo* > L. *bolus*, to hurl, because phototropic sporangiophores at base of sporangiolum, forcefully casts it, includes one or a few spores. On dung, especially of amphibians and reptiles, and on decomposing vegetation; rarely a pathogen of animals, including humans.

Basidiobotrys Höhn.—see **Xylocladium** P. Syd. ex Lindau

Xylariaceae, Xylariales, Xylariomycetidae, Sordariomycetes, Pezizomycotina, Ascomycota, Fungi
Höhnel, F. X. R. von. 1909. Fragmente zur Mykologie (VI Mitteilung, Nr. 182 bis 288). *Sber. Akad. Wiss. Wien, Math.-naturw. Kl., Abt. I*, 118:420.
From Gr. *basídion*, dim. of *básis*, base + *bótrys*, cluster.

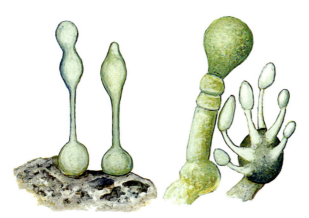

Basidiobolus ranarum Eidam: phototrophic sporangiophores, x 340, terminal pyriform sporangiolum x 500; and cluster of pedicellate unispored sporangiola (microspores), x 650, growing on dung of *Rana esculenta* and *R. oxyrrhina*. Described in Germany.

Basidioradulum Nobles—see **Xylodon** (Pers.) Gray
Schizoporaceae, Hymenochaetales, Inc. sed., Agaricomycetes, Agaricomycotina, Basidiomycota, Fungi
Nobles, M. K. 1967. Conspecifity of *Basidioradulum* (*Radulum*) *radula* and *Corticium hydnans*. *Mycologia* 59:192-211.
Described by Fries for *Hydnum radula* (1815) and *Radulum orbiculare* (1825), introducing in the root of the new combination the word basidio because it is a basidiomycete.

Battarrea Pers.
Agaricaceae, Agaricales, Agaricomycetidae, Agaricomycetes, Agaricomycotina, Basidiomycota, Fungi
Persoon, C. H. 1801. *Syn. meth. fung.*(Göttingen) 1:129.
Dedicated to the Italian mycologist *Giovanni Antonio Battarra* (1714–1789) + L. des. *-ea*, because Battarra ends in a vowel. Fructification a semi-woody squamulose-lacerate stipe with globose pileus with circumscissile dehiscence, i.e., in shape of a circular equatorial fissure from which it separates, in the manner of a helmet. Upper part of sporiferous sac exposes powdery gleba composed of a capillitium with elaters and globose or subglobose spores. In arid and semiarid regions.

Battarreoides T. Herrera
Agaricaceae, Agaricales, Agaricomycetidae, Agaricomycetes, Agaricomycotina, Basidiomycota, Fungi
Herrera, T. 1953. Un hongo nuevo procedente del estado de San Luis Potosí. *Battarreoides potosinus* gen. nov. sp. nov. *Anales Inst. Biol. Univ. Nac. México* 24(1):41-46.
From genus *Battarrea* Pers. + L. suf. *-oides* < Gr. *-oeídes*, similar to. Fructification similar to that of *Battarrea*, but dehiscence of pileus by means of pores, so that peridium permanent, not deciduous, as in *Battarrea*, with circumscissile dehiscence. In sandy soils and scrubland in arid and semiarid regions.

Battarrea phalloides (Dicks.) Pers.: mature basidioma, on soil, x 1.

Battarreoides diguetii (Pat. & Har.) R. Heim & T. Herrera: mature basidioma, on sandy soil, x 1.

Baudoinia J. A. Scott & Unter.
Inc. sed., Capnodiales, Dothideomycetidae, Dothideomycetes, Pezizomycotina, Ascomycota, Fungi
Scott, J. A., et al. 2007. *Baudoinia*, a new genus to accommodate *Torula compniacensis*. *Mycologia* 99:592-601.
Named in honor of the French pharmacist *Antonin Baudoin*, who first called attention to the fungus. Colonies darkly pigmented, sooty. Mycelium black, effused,

Bdellospora

velvety to crust-like; vegetative hyphae dark brown, thick-walled, often moniliform. Conidiophores lacking. Conidiogenous cells integrated within vegetative hyphae. Conidia dry, nonseptate or uniseptate at median, thick-walled, globose to barrel-shaped, brown to black, typically with coarse surface ornamentation, dehiscencing by schizolysis. Ramoconidia absent. Common in areas where ethanol vapor exists, such as distilleries, coating the building walls with fungal growth, causing a phenomenon known as "warehouse staining".

Bdellospora Drechsler
Cochlonemataceae, Zoopagales, Inc. sed., Inc. sed., Zoopagomycotina, Zygomycota, Fungi
Drechsler, C. 1935. Some conidial phycomycetes destructive to terricolous amoebae. *Mycologia* 27:25.
From Gr. *bdélla*, leech + *sporá*, spore. The name refers to characteristics of merosporangia, i.e., similar to leeches which adhere to animals on which they feed, adhering to cells of terricolous amoebae. Each merosporangium perforates membrane of amoeba developing an haustorium in interior, transforms into swollen body from which emerge sporangiophores and merosporangia. Sexual reproduction is by means of coiled zygophores.

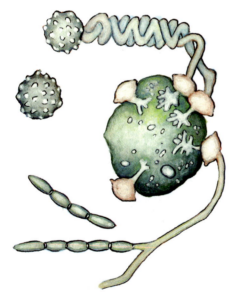

Bdellospora helicoides Drechsler: merosporangia infecting a terricolous amoeba (*Amoeba terricola*). Above, there are coiled zygophores with zygospores; bottom, there are chains of merosporangia, x 1,200.

Beauveria Vuill.
Cordycipitaceae, Hypocreales, Hypocreomycetidae, Sordariomycetes, Pezizomycotina, Ascomycota, Fungi
Vuillemin, P. 1912. *Beauveria*, nouveau genre de verticillacées. *Bull. Soc. bot. Fr.* 59:34-40.
Named in honor of the French mycologist *J. Beauver* (1874-1938) + L. suf. *-ia*, which denotes belonging to. Fungus with hyaline conidia borne in clusters on conidiogenous cells on a rachis that elongates sympodially. Parasites of insects, with a cordyceps-like sexual morph; also in soil. The muscardine disease of silk worm is an important example of the diseases caused by this fungus.

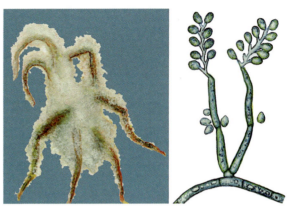

Beauveria bassiana (Bals.-Criv.) Vuill.: spider killed by the fungus, x 2; conidiophores with conidia, x 800.

Belemnospora P. M. Kirk
Inc. sed., Inc. sed., Inc. sed., Inc. sed., Pezizomycotina, Ascomycota, Fungi
Kirk, P. M. 1981. New or interesting microfungi II. Dematiaceous Hyphomycetes from Ester Common, Surrey. *Trans. Br. mycol. Soc.* 77:279-297.
From Gr. *belemnon*, dart + *sporá*, seed, spore > L. *spore*, spore. Colonies effuse, minutely hairy, pale brown to brown, often inconspicuous. Mycelium mostly superficial, composed of a network of anastomosing, pale brown to brown, smooth, septate hyphae. Conidiophores semi-macronematous to macronematous, mononematous, solitary, simple, pale brown to brown, smooth, septate. Conidiogenous cells holoblastic, monoblastic, integrated, terminal, cylindrical, percurrent. Conidia acrogenous, solitary, seceding schizolytically, dry, smooth, cylindrical, with base truncate and pointed apex, with or without septa. On rotten wood of *Quercus robur* in Surrey, U.K.

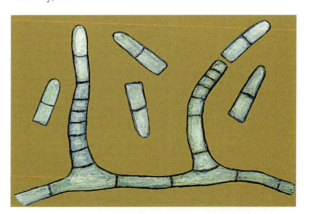

Belemnospora epiphylla P. M. Kirk: conidiophores and conidia, x 1,700.

Beltraniella

Belonioscypha Rehm—see **Cyathicula** De Not.
Helotiaceae, Helotiales, Leotiomycetidae, Leotiomycetes, Pezizomycotina, Ascomycota, Fungi
Rehm, H. 1892. Ascomyceten: Hysteriaceen und Discomyceten. *In*: *Rabenh. Krypt.-Fl.*, Edn 2 (Leipzig), Vol. 1., pt. 3:1-1,272.
From genus *Belonium* + L. *scyphus* < Gr. *skýphos*, a cup.

Belonopsis (Sacc.) Rehm (syn. **Trichobelonium** (Sacc.) Rehm.
Dermateaceae, Helotiales, Leotiomycetidae, Leotiomycetes, Pezizomycotina, Ascomycota, Fungi
Rehm, H. 1891. Rabenhorst's Kryptogamen-Flora, Pilze -Ascomyceten. *1*(3):401-608.
From Gr. *bélon*, a needle, any sharp point + suf. *-ópsis*, similar. Apothecia sessile, gregarious, fleshy to subgelatinous, light- or dark-colored, seated on a white or colored mycelial subiculum. Asci cylindric to clavate, usually 8-spored. Ascospores ellipsoid to fusoid, becoming 3-septate.

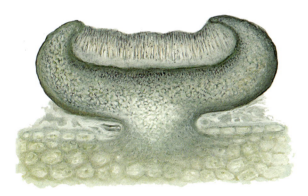

Belonopsis guestphalicum (Rehm) Nannf.: vertical section of an apothecium in host tissues (*Typha angustifolia*, English Lake, England), seated on a white subiculum, and showing the hymenium, excipulum, pedicel, interwoven asci with paraphyses, and ellipsoidal, 3-4 septate ascospores, x 300.

Beltrania Penz.
Beltraniaceae, Sordariales, Sordariomycetidae, Sordariomycetes, Pezizomycotina, Ascomycota, Fungi
Penzig, O. 1882. *Beltrania* un nuovo genere di Ifomiceti. *Nuovo Giornale Botanico Italiano* 16:72-75.
Named to honor the Italian mycologist, *Vito Beltrani*, who was dedicated to the study of Uredinales. Mycelium with setae dark, simple, pointed. Conidiophores simple or less often forked, brown, with conidia (sympodulospores) biconic, 1-celled, brown wih a paler middle band, borne singly on denticles or ovoid separating cells. Saprophytic.

Beltraniella Subram. (syn. **Ellisiopsis** Bat.)
Amphisphaeriaceae, Amphisphaeriales, Xylariomycetidae, Sordariomycetes, Pezizomycotina, Ascomycota, Fungi
Subramanian, C.V. 1952. Fungi Imperfecti from Madras-III. *Proc. Indian Acad. Sci.*, Sect. B, *36*:227.
From genus *Beltrania* Penz + dim. suf. *-ella*. Mycelium brown, superficial; hyphae septate, sparingly branched, lacking hyphopodia. Setae erect, simple, continuous, blackish. Conidiophores minute, conoid-oblong, denticulate, brownish, formed among setae. Conidia lageniform, 1-celled, subhyaline, with a single transverse hyaline band and a single, narrow basal projection where attached to conidiophore. On living and decaying leaves, widespread.

Beltrania mundkuri Piroz. & S.D. Patil: simple conidiophores with biconical conidia, and setae, x 1,000, in decaying leaves of *Syzygium cumini*, in India.

Beltraniella endiandrae Crous & Summerell: simple conidiophore with conidia, and setae, x 1,500, in decaying leaves of *Endiandra*, in New South Wales, Australia.

Bensingtonia

Bensingtonia Ingold
Agaricostilbaceae, Agaricostilbales, Inc. sed., Agaricostilbomycetes, Pucciniomycotina, Basidiomycota, Fungi
Ingold, C. T. 1986. *Bensingtonia ciliata* gen. et sp. nov., a ballistosporic fungus. *Trans. Brit. mycol. Soc.* 86:325-328.
Named in honor of the author's village of *Bensington* (now Benson), England + L. suf. *-ia*. Fungus ballistosporic, hyaline, ovoid, germinating by ballistospores or rarely forming hyphae. Colonies on agar limited, soon dominated by simple or branched aerial hyphae, for the most part derived from blastospores. Ballistospores unicellular, subovoid, formed terminally on apicula of aerial hyphae. Isolated from *Auricularia auricula-judae* var. *lacteae* Quél., in Lullingstone, Kent, U.K.

Bertia De Not.
Bertiaceae, Coronophorales, Hypocreomycetidae, Sordariomycetes, Pezizomycotina, Ascomycota, Fungi
De Notaris, G. 1844. Cenni sulla tribù dei Pirenomiceti sferiacei e descrizione di alcuni generi spettanti alla medesima. *G. bot. ital.* 1(1):334.
Dedicated to the Italian botanist *J. Berti* + *a*, because the name ends in a vowel. Ascomata solitary or aggregated, frequently on a compact stroma. Peridium dark, carbonaceous, without ostiole, opening at maturity by disintegration of apex where gelatinous cells that are transformed into a mucilaginous mass, i.e., swollen body or Quellkörper. Asci pedicellate, claviform; ascospores uniseptate, fusiform, hyaline. Parasitic on the wood of trees, rarely lichenicolous.

Bertia moriformis (Tode) De Not.: gregarious perithecia, on the surface of *Fagus* bark, x 35.

Bettsia Skou (syn. **Pericystis** Betts)
Ascosphaeraceae, Onygenales, Eurotiomycetidae, Eurotiomycetes, Pezizomycotina, Ascomycota, Fungi
Skou, J. P. 1972. Ascosphaerales. *Friesia* 10:5.
The name is given in honor of *Annie D. Betts* + L. des. *-ia*, a British apiculturist, who first described the fungus. The sac (sporocyst) contains spherical groups of asci, which has been considered a cleistothecium. Causes pericystic mycosis or chalk disease of the larvae of domesticated honey bees.

Bettsia alvei (Betts) Skou: sporocysts with dark ascospores, x 500.

Bicornispora Checa, et al.
Coryneliaceae, Coryneliales, Eurotiomycetidae, Eurotiomycetes, Pezizomycotina, Ascomycota, Fungi
Checa, J., et al. 1996. *Bicornispora exophiala*, a new genus and species of the Coryneliales and its black yeast anamorph. *Mycol. Res.* 100:500-504.
From L. *bi*, two + *cornu*, horn + *spora*, spore < Gr. *sporá*, seed, spore, for the horn-like extensions on the spore. Ascomata scattered, globose, short-stalked, dark brown, iridescent, glabrous, non-ostiolate. Peridium pseudoparenchymatous, brown. Paraphyses lacking. Asci unitunicate, 8-spored, ovoid to ellipsoidal, non-amyloid, without obvious apical apparatus, soon evanescent. Ascospores one-celled, dark brown, smooth, reniform, with an acute, horn-like extension at each end, smooth-walled. On fallen branches of *Cytisus purgans* in Mazanares el Real, Madrid, Spain.

Bicornispora seditiosa Checa, et al.: ascoma on plant detritus, x 50; horned ascospores, x 1,400.

Bidenticula Deighton—see **Fusarium** Link
Nectriaceae, Hypocreales, Hypocreomycetidae, Sordariomycetes, Pezizomycotina, Ascomycota, Fungi

Deighton, F. C. 1972. Four leaf-spotting Hyphomycetes from Africa. *Trans. Br. mycol. Soc.* 59:419-427.
From L. *bi*, two + *denticula*, toothed, referring to paired denticles on conidiophores.

Bifusella Höhn.
Rhytismataceae, Rhytismatales, Leotiomycetidae, Leotiomycetes, Pezizomycotina, Ascomycota, Fungi
Höhnel, F. 1917. Mycologische Fragmente. *Annls mycol.* 15:293-383.
From L. *bi*, two + *fusus*, spindle + dim. suf. *-ella*, for the shape of the ascospores. Apothecia subcuticular, elliptical, black. Hymenium flat; subhymenium thin. Paraphyses lacking or disappearing early. Asci clavate, unitunicate. Ascospores bifusiform or more or less rod-like. Pycnidia large, covered by a thin pseudoparenchymatous layer attached to cuticle. Spermatia large, bacillar. On conifer needles, causing lesions.

Bifusepta Darker
Rhytismataceae, Rhytismatales, Leotiomycetidae, Leotiomycetes, Pezizomycotina, Ascomycota, Fungi
Darker, G. D. 1963. A new phragmosporous genus of the Hypodermataceae. *Mycologia* 55:812-818.
From L. *bis*, two, twice + *fusus*, spindle + *septum*, partition, for the phragmosporous ascospores. Apothecia elliptical, golden-brown to grayish-black, immersed, subcuticular, erumpent at maturity, opening by a longitudinal slit. Covering layer of dark pseudoparenchyma alternating with wedges of parenchyma. Subhymenium flat, thin, hyaline, seated on a darkened basal layer. Paraphyses simple, filiform. Asci clavate, 4-spored. Ascospores long, bifusiform, phragmosporous, surrounded by a gelatinous sheath. Pycnidia small, black, circular; spores bacillar. On conifer needles.

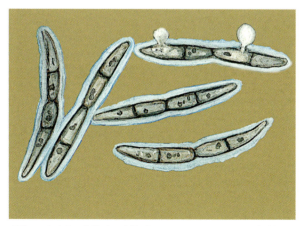

Bifusepta tehonii Darker: bifusiform ascospores, surrounded by a gelatinous sheath, x 600.

Bipolaris Shoemaker (syn. Cochliobolus Drechsler)
Pleosporaceae, Pleosporales, Pleosporomycetidae, Dothideomycetes, Pezizomycotina, Ascomycota, Fungi

Shoemaker, R. A. 1959. Nomenclature of *Drechslera* and *Bipolaris*, grass parasites segregated from *Helminthosporium*. *Can. J. Bot.* 37:879-887.
From L. *bis*, twice + *polaris*, polar, belonging to or relative to the pole, because of the germination of the conidia, which produce a germ tube at each pole. Pseudothecia dark. Asci with ascospores arranged in a tight spiral forcefully discharged from ascus. Ascospores filiform. Conidia fusoid, straight or curved, septate, originating from pores in wall of geniculate, brown conidiophores. Sexual state cochliobolus-like. Generally parasitic on grasses.

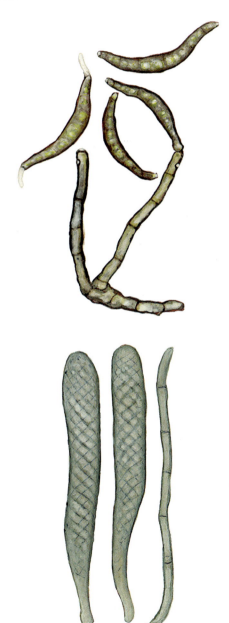

Bipolaris maydis (Y. Nisik. & C. Miyake) Shoemaker (syn. **Cochliobolus heterostrophus** (Drechsler) Drechsler): conidiophore and conidia, x 300; asci with ascospores arranged in tight spirals, x 900; and free septate ascospore, x 750.

Biscogniauxia

Biscogniauxia Kuntze (syn. **Nummularia** Tul. & C. Tul.)

Xylariaceae, Xylariales, Xylariomycetidae, Sordariomycetes, Pezizomycotina, Ascomycota, Fungi

Kuntze, O. 1891. *Revis. gen. pl.* (Leipzig) 2:398.

From L. *bis*, twice, double + genus *Cogniauxia*, which was named in honor of the Belgian botanist *Alfred C. Cogniaux* + L. des. *-ia*. Stromata erumpent through bark at maturity, discrete, discoid or cupulate, applanate or shallow-concave, without perithecial outlines, black, dark brown, grey or greyish-lilac. Stromata carbonaceous, bipartite, with an outer layer that covers ostioles when young and sloughs off at maturity to expose them. Upper part of stroma composed only of fungal tissue, with basal portion consisting of either fungal tissue or both host and fungal tissues. Perithecia formed in a single layer in the periphery of stroma, with slightly to conspicuously papillate ostioles, ovate-cylindrical, narrowing abruptly above to form a distinct neck. Centrum containing asci intermixed with paraphyses. Paraphyses filiform, septate. Asci unitunicate, cylindrical, short-stipitate, with a flattened apical ring that stains blue in iodine, 8-spored. Ascospores smooth, pitted, pleated, or striate, of two types: 1-celled, ellipsoid to subglobose and nearly symmetrical to asymmetrical, dark brown, with a germ slit, or 2-celled, upper cell large, brown, with a germ slit, and lower cell smaller, hyaline. Asexual states nodulisporium-like and periconiella-like. On trunks and branches of angiospermous trees.

Biscogniauxia nummularia (Bull.) Kuntze: perithecial stromata on beech bark, x 1.

Bisporella Sacc.

Helotiaceae, Helotiales, Leotiomycetidae, Leotiomycetes, Pezizomycotina, Ascomycota, Fungi

Saccardo, P. A. 1884. Conspectus generum Discomycetum husque cognitorum. *Botan. Zbl.* 18:213-220.

From genus *Bispora* Corda + L. dim. suf. *-ella*, i.e., a small *Bispora*. Apothecia gregarious, sessile or nearly so, shallow cup-shaped, tapering to a small base, tough, smooth. Disc bright to pale yellow, yellow-orange, cream-colored or white, receptacle paler to darker, minutely downy in some species. Ectal excipulum gelatinous. Paraphyses filiform, branched, septate, containing yellow oil droplets. Asci unitunicate, inoperculate, cylindric-clavate, 4-8-spored, J+. Ascospores elliptical to elliptic-fusiform, biseriate or obliquely uniseriate in ascus, 1-celled, often becoming 1-3-septate, hyaline, with an oil droplet at each end. Asexual states in bloxamia-like, cystodendron-like or eustilbum-like. On bark or dead wood or old fruiting bodies of fungi.

Bisporella citrina (Batsch) Korf & S. E. Carp: apothecia on dead bark, x 10.

Bitrimonospora Sivan., et al.—see **Monosporascus** Pollack & Uecker

Diatrypaceae, Xylariales, Xylariomycetidae, Sordariomycetes, Pezizomycotina, Ascomycota, Fungi

Sivanesan, A., et al. 1974. *Bitrimonospora indica* gen. et sp. nov., a new Loculoascomycete from India. *Trans. Br. mycol. Soc.* 63:595-596.

From Gr. *bi*, two + *tri*, three + *mono*, single + *sporá*, seed, spore, referring to the presence of 1 but also, 2 or 3 ascospores in each ascus.

Blakeslea Thaxt.

Choanephoraceae, Mucorales, Inc. sed., Inc. sed., Mucoromycotina, Zygomycota, Fungi

Thaxter, R. 1914. New or peculiar zygomycetes. III. *Blakeslea* n. g., *Dissophora* n. g., *Haplosporangium* n. g. *Bot. Gaz.* 58:353-366.

Named in honor of *A. F. Blakeslee*, a mycologist from the United States. Two types of sporangia on separate sporangiophores, one large, many-spored sporangia and also small, few-spored sporangia. Sporangiospores fusiform, striate, with several long, hyaline polar appendages. In soil, and plant substrates.

Blastocladia Reinsch

Blastocladiaceae, Blastocladiales, Inc. sed., Blastocladiomycetes, Inc. sed., Chytridiomycota, Fungi

Reinsch, P. F. 1877. Beobachtungen über einige neue Saprolegnieae. *Jahrb. wiss. Bot.* 11:283-311.

From Gr. *blastós*, germ, bud, shoot + *kládos*, branch + *-ia*, L. suf. that denotes quality or state, due to thallus of cylindrical basal cell that expands at apex to form short, clavate lobes or branches bearing reproductive organs. Saprobic on decomposing branches and fruits.

Blakeslea trispora Thaxt.: sporangiophore with three-spored sporangiola, x 270.

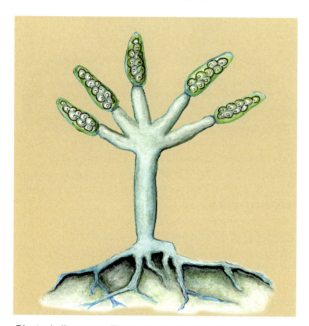

Blastocladia ramosa Thaxt.: sporangiophore with sporangia, x 60.

Blastocladiella V. D. Matthews—see **Clavochytridium** Couch & H. T. Cox

Blastocladiaceae, Blastocladiales, Inc. sed., Blastocladiomycetes, Inc. sed., Chytridiomycota, Fungi

Matthews, V. D. 1937. A new genus of the Blastocladiaceae. *J. Elisha Mitchell scient. Soc.* 53:191-195.

From genus *Blastocladia* Reinsch + L. dim suf. *-ella*. Resembles *Blastocladia* but differs in its simpler thallus.

Blastomyces Gilchrest & W. R. Stokes

Ajellomycetaceae, Onygenales, Eurotiomycetidae, Eurotiomycetes, Pezizomycotina, Ascomycota, Fungi

Gilchrest C., W. R. Stokes 1898. A case of pseudo-lupus vulgaris caused by a blastomyces. *J. Exp. Med.* 3:76.

From Gr. *blastós*, bud, shoot + *mýkes* > L. *myces*, fungus, referring to blastospores. Ascomata cleistothecial, ajellomyces-like. Causing blastomyces, a systemic disease that starts as a pulmonary infection in tissues of humans and other animals.

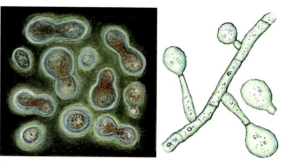

Blastomyces dermatitidis Gilchrist & W. R. Stokes: parasitic phase with budding cells in human lung tissue, x 800. At the right, saprobic phase with aleuriospores, x 1,500.

Blumeria Golovin ex Speer

Erysiphaceae, Erysiphales, Leotiomycetidae, Leotiomycetes, Pezizomycotina, Ascomycota, Fungi

Speer, E. O. 1975. Untersuchungen zur Morphologie und Systematik der Erysiphaceen I. Die Gattung *Blumeria* Golovin und ihre Typusart *Erysiphe graminis* DC. *Sydowia* 27:1-6.

Named in honor of the Swiss mycologist *Samuel Blumer* + L. des. *-ia*. Mycelium hyaline, superficial, forming white colonies on host tissues, producing abundant conidia. Haustoria with finger-like lobes. Older mycelium becoming pale brown and forming numerous erect, curved, hyaline, thick-walled setae as secondary mycelium. Conidia produced in a chain with basipetal maturation from a mother cell, oldest conidium at apex. Ascomata enclosed chasmothecium, with a basal ring of mycelioid appendages. Ascomatal wall thick. Asci several, unitunicate, clavate to subcylindrical or ellipsoid, with a short stalk, usually 8-spored, rarely 4-spored. Ascospores ellipsoid, hyaline. Asexual state oidium-like. Worldwide obligate parasite on living leaves and stems of grasses, causing powdery mildew disease. The only species is *Blumeria graminis* (DC.) Speer.

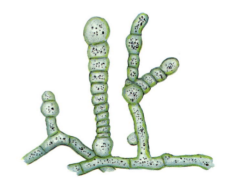

Blumeria graminis (DC.) Speer (syn. **Acrosporium monilioides** Nees): meristematic arthrospores, growing in host tissue, x 150.

Blumeriella

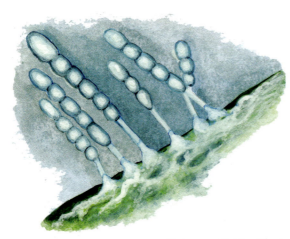

Blumeria graminis (DC.) Speer (syn. **Oidium monilioides** (Nees) Link): conidiophores with egg-shaped conidia, x 230.

Blumeriella Arx
Dermateaceae, Helotiales, Leotiomycetidae, Leotiomycetes, Pezizomycotina, Ascomycota, Fungi
Arx, J. A. von. 1961. Über *Cylindrosporium padi*. *Phytopath. Z.* 42:161-166.
From genus *Blumeria* Golovin ex Speer + L. dim. suf. *-ella*. Apothecia immersed in host tissues, overlying tissues erumpent, but apothecia remaining depressed; apothecia flat to shallow-convex; ectal excipulum of brown-walled textura globulosa or textura angularis. Asci unitunicate, subclavate. Ascospores 1-celled, hyaline, ellipsoid. Asexual states in phloeosporella-like or cylindrosporium-like. Parasitic on living leaves of *Prunus* species.

Bolbitius Fr.
Bolbitiaceae, Agaricales, Agaricomycetidae, Agaricomycetes, Agaricomycotina, Basidiomycota, Fungi
Fries, E. 1838. *Epicrisis Systematis Mycologici*, p.253.
From. Gr. *bólbiton*, cow dung + L. *-ius*, because the species grow in manure or in fields fertilized with this excrement. Basidiocarp with lamellate hymenophore. Margin of pileus sulcate, plicate or pectinate; spores smooth with broad truncate germ pore; spore print brown or bright ferruginous; lamellae sometimes weakly deliquescent. On dung, humus or earth, also on rotten trunks of trees, and swampy or sandy soils and sawdust.

Boletellus Murrill
Boletaceae, Boletales, Agaricomycetidae, Agaricomycetes, Agaricomycotina, Basidiomycota, Fungi
Murrill, W. A. 1909. The Boletaceae of North America-1. *Mycologia* 1(1):4-18.
From genus *Boletus* L. + L. dim. suf. *-ellus*. Pileus scaly or naked, dry or viscid; hymenophore with yellow colors, sometimes with red pores. Spore print deep olivaceous to olive-brown; spores smooth or with short spines, also reticulate in some species, always elongate, except in some species with reticulate ornamentation, hyphae with or without clamp connections. On soil rarely at base of trees or on very decayed wood. In North America and pantropical, also in Japan and in Europe. Some species are edible, facultatively or obligately ectomycorrhizal and therefore of importance in forestry.

Bolbitius titubans (Bull.) Fr.: basidiomata growing in herbivorous dung, x 1.

Boletochaete Singer
Boletaceae, Boletales, Agaricomycetidae, Agaricomycetes, Agaricomycotina, Basidiomycota, Fungi
Singer, R. 1944. New genera of fungi. I. *Mycologia* 36:358-360.
From genus *Boletus* L. + L. *chaete*, a bristle < Gr. *chaité*, long flowing hair, seta, i.e., a seta-bearing bolete. Pileus with subepithelial or trichodermical-palisade epicutis; pores small similar to *Boletus*, at first white or whitish, adnate to deeply depressed around stipe; hymenophoral trama of *Boletus* subtype; spores pale cinnamon, pink or pale-ochraceous, i.e. pink as in *Tylopilus* P. Karst.; spores small, inamyloid, ellipsoid, pseudocystidia numerous; stipe often attenuated downwards or bulbous, pruinose or scurfy, central, solid; context slightly bitter. Paleotropical.

Boletus L.
Boletaceae, Boletales, Agaricomycetidae, Agaricomycetes, Agaricomycotina, Basidiomycota, Fungi
Linnaeus, C. 1753. *Sp. pl.* 2:1176.
From Gr. *bolítes*, an old name that referred, according to Galen, to certain roots, as well as to edible mushrooms; it has the same root as *bōlos*, lump, gleba, ball or lump of soil, for the shape and color of the pileus, similar to a

mound of earth. Basidiomata fleshy, small to large, variable in color, often changing color when context injured or exposed. Hymenium develops on interior surface of tubes, where basidiospores produced; spores escape through pores at base of tubes comprising hymenophore. Terricolous, with ectomycorrhizal associations. Several of the species are edible, but others are toxic.

Boletus pinophilus Pilát & Dermek: basidioma, on soil, x 0.5. Described in Czechoslovakia.

Bombardia (Fr.) P. Karst.

Lasiosphaeriaceae, Sordariales, Sordariomycetidae, Sordariomycetes, Pezizomycotina, Ascomycota, Fungi

Karsten, P. A. 1873. Mycologica Fennica. Pars secunda. Pyrenomycetes. *Bidr. Känn. Finl. Nat. Folk* 23:1-252.

From ML. *bombarda*, stone-throwing engine + L. des. *-ia*, for the apparent resemblance of the ascomata to stones thrown by this structure. Perithecia ostiolate, narrowly ellipsoidal to clavate, lacking a neck, aggregated, brown to black, tapering to an elongated base. Ascomatal wall 4-layered. Paraphyses abundant, filiform, branched. Asci unitunicate, clavate, with a truncate apex and a simple, non-amyloid apical ring. Ascospores cylindrical, sigmoid when formed, caudate at both ends, becoming septate, with upper cell swelling and becoming dark brown, ellipsoidal to ovoid, with a truncate base and an upper germ pore; lower cell remaining hyaline, cylindrical or bent, finally collapsing. On dead wood.

Bombardia bombarda (Batsch) J. Schröt.: ostiolate perithecia on dead wood, x 15.

Boothiella Lodhi & Mirza (syn. **Thielaviella** Arx & T. Mahmood)

Chaetomiaceae, Sordariales, Sordariomycetidae, Sordariomycetes, Pezizomycotina, Ascomycota, Fungi

Lodhi, S. A., F. Mirza 1962. A new genus of the Eurotiales. *Mycologia* 54(2):217.

The name is dedicated to the British mycologist Dr. *C. Booth*. Ascomata immersed or becoming superficial, spherical or nearly so, non-ostiolate, with a wall composed of a few layers of large, hyaline cells; asci cylindrical, unitunicate, 4-spored, borne at base, deposited in a hymenial layer, deliquescent; ascospores ellipsoidal or subspherical, continuous, dark brown, with a basal germ pore. Isolated from agricultural soil in Lahore, Pakistan.

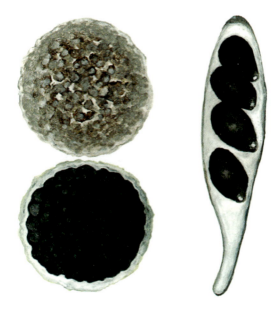

Boothiella tetraspora Lodhi & J.H. Mirza: at the left, cleistothecial ascomata (above immature, below mature dark brown, with ascospores), x 130; at right, an ascus unitunicate with 4-spores, x 1,500.

Botryandromyces I. I. Tav. & T. Majewski

Laboulbeniaceae, Laboulbeniales, Laboulbeniomycetidae, Laboulbeniomycetes, Pezizomycotina, Ascomycota, Fungi

Tavares, I. I., T. Majewski. 1976. *Siemaszkoa* and *Botryandromyces*, two segregates of *Misgomyces* (Laboulbeniales). *Mycotaxon* 3:193-208.

From Gr. *bótrys*, cluster of grapes + *andrós*, male + *mýkes* > L. *myces*, fungus, for the clustered antheridia. Perithecium with two adjacent vertical rows of wall cells consisting of three cells of unequal length and the other two rows of cells consisting of four cells of unequal length. Secondary stalk cell and basal cells of perithecium with well-defined walls. Lower tier of wall cells forming venter of perithecium. Antheridia simple phialides, arising in a cluster from small cells just above spore septum. On Heteroceridae.

Botryoderma

Botryandromyces heteroceri (Thaxt.) I. I. Tav. & T. Majewski: parasitic thallus adhered to the insect host (*Heterocerus obsoletus*), x 400.

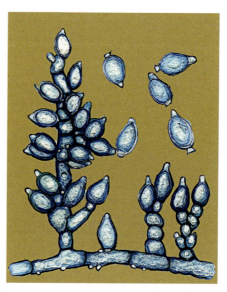

Botryoderma rostratum Papendorf & H. P. Upadhyay: conidiophores and conidia, x 1,200. Described in Brazil.

Botryoderma Papendorf & H. P. Upadhyay
Inc. sed., Inc. sed., Inc. sed., Inc. sed., Pezizomycotina, Ascomycota, Fungi
Papendorf, M. C., H. P. Upadhyay. 1969. *Botryoderma lateritium* and *B. rostratum* gen. et spp. nov. from soil in South Africa and Brazil. *Trans. Br. mycol. Soc.* 52:257-265.
From Gr. *bótrys*, cluster of grapes + *derma*, skin, referring to the clusters of conidiophores on the surface of the substrate. Hyphae branched, septate, thin-walled, hyaline. Conidiophores variable, sometimes lacking, when present uni- or multicellular, subglobose to shortly filamentous and partly irregular or torulose, thin-walled, hyaline. Sporogenous cells produced laterally on fertile hyphae or laterally and terminally on conidiophores, usually inflated, subglobose, obpyriform, clavoid or subcylindrical, thin-walled, hyaline. Conidia aleuriospores, either sessile or on short sterigmata, produced typically in small groups in basipetal succession on sporogenous cells, one-celled, elliptical, oblong-elliptical, ovoid, globose or subglobose, with a basal or sublateral and often slightly stipitate annular secession scar, smooth, hyaline or faintly colored, thick-walled, without germ slit or pore. Isolated from soil in an *Acacia karoo* community.

Botryophialophora Linder—see **Myrioconium** Syd. & P. Syd.
Sclerotiniaceae, Helotiales, Leotiomycetidae, Leotiomycetes, Pezizomycotina, Ascomycota, Fungi
Linder, D. L. 1944. I. Classification of the marine fungi. Pp. 401-433, *In*: Barghoorn, E. S., D. L. Linder, Marine fungi: their taxonomy and biology. *Farlowia* 1:395-467.
From Gr. *bótrys*, cluster of grapes + genus *Phialophora* Medlar.

Botryosphaeria Ces. & De Not.
Botryosphaeriaceae, Botryosphaeriales, Inc. sed., Dothideomycetes, Pezizomycotina, Ascomycota, Fungi
Cesati, V., G. De Notaris. 1863. Sferiacei italici. *Comment. Soc. crittog. Ital.* 1:211.
From Gr. *bótrys*, cluster of grapes + *sphaíra* > L. *sphaera*, sphere, in reference to pseudothecia grouped together in host tissue. Pseudothecia medium to large, erumpent, sunken in a pulvinate stroma. Ascospores hyaline, rarely brown, ovoid or ellipsoid, lacking septa. Generally, on dead woody stems and wood, causes regressive death of plants and rot of citrus fruits. Asexual morphs fusicoccum-like.

Botryotinia Whetzel— see **Botrytis** P. Micheli ex Pers.
Sclerotiniaceae, Helotiales, Leotiomycetidae, Leotiomycetes, Pezizomycotina, Ascomycota, Fungi
Whetzel, H. H. 1945. A synopsis of the genera and species of the Sclerotiniaceae, a family of stromatic inoperculate Discomycetes. *Mycologia* 37:648-714.
From Gr. *bótrys*, cluster + L. des. *-inia*, nature of something, for the clusters of conidia.

Botryozyma Shann & M. T. Smith (syn. **Ascobotryozyma** J. Kerrigan, et al.)
Trigonopsidaceae, Saccharomycetales, Saccharomycetidae, Saccharomycetes, Saccharomycotina, Ascomycota, Fungi
Smith, M. T., et al. 1992. *Botryozyma nematodophila* gen. nov., spec. nov. (Candidaceae). *Antonie van Leeuwenhoek* 61:277-284.
From Gr. *bótrys*, cluster of grapes + *zýme*, yeast, for the origin of the fungus. Yeast cells cylindrical, reproducing by multilateral, holoblastic budding. Pseudohyphae formed, producing branch-like structures on their

terminal cells. Cell wall ascomycetous as shown by ultrastructure, lacking D-xylose and L-fucose. Asci originating from thallus, globose or broadly ellipsoid, evanescent. Ascospores hyaline, lunate, smooth. Thalli attached to the nematode by basal cells. Isolated from the nematode *Panagrellus zymophilus* on sour-rot grapes from Verona, Italy.

Botrytis P. Micheli ex Pers. (syn. **Botryotinia** Whetzel)
Sclerotiniaceae, Helotiales, Leotiomycetidae, Leotiomycetes, Pezizomycotina, Ascomycota, Fungi
Persoon, C. H. 1794. *Neues Mag. Bot.* 1:120.
From Gr. *bótrys*, cluster of, grapes + L. suf. *-ítis*, like, conidia arranged in clusters (botryoblastospores) on ampullae or globose vesicles on conidiophore branches. Stroma consisting of black, flattened or irregularly hemispherical sclerotium, usually formed just beneath cuticle of epidermis, finally erumpent, remaining firmly attached to host. Spermatia globose, borne on branching spermatiophores, enveloped in a mucilaginous matrix. Conidia on short sterigmata in dense clusters, smooth, subglobose or pyriform. Cosmopolitan, pathogenic on plants, common in the phylloplane, causing a blight or rotting of the leaves, flowers, and fruits, the so-called gray mold of great economic importance. Sexual state botryotinia-like, sclerotioid apothecia.

Bovista Pers.
Agaricaceae, Agaricales, Agaricomycetidae, Agaricomycetes, Agaricomycotina, Basidiomycota, Fungi
Persoon, C. H. 1794. *Neues Mag. Bot.* 1:86.
From G. *Bovista*, puffball, this from a German dialect *bofist*, wind, because on being pressed the sporiferous sac ejects a mass of spores; probably also derived from L. *bos, bovis*, ox + des. *-ista*, habit, that which practices something, related to, probably because some species grow in grasslands where oxen graze. Fructifications globose or subglobose, with central base of attachment, and a double peridium of which only the smooth, bright, flexible internal layer remains at maturity, at times with remains of superficial layer. Opens by an apical pore through which powdery spores exit. Gleba composed of capillitium that is free from internal wall of peridium, composed of separate units that are branched irregularly or dichotomously and end as pointed filaments. In grasslands and open woods.

Brachybasidium Gäum.
Brachybasidiaceae, Exobasidiales, Exobasidiomycetidae, Exobasidiomycetes, Ustilaginomycotina, Basidiomycota, Fungi
Gäumann, E. A. 1922. Über die Gattung *Kordyana* Rac. *Ann. mycol.* 20:269.
From Gr. *brachýs*, short + L. *basidium*, derived from Gr. *basídion*, dim. of *básis*, base, basidium, for shape of basidia. Fructification of an hymenial layer of bispored basidia that emerge in groups through stomata on underside of leaves. Basidia developing in sori covered by protective host tissue and maturing in irregular succession, only a few basidia sporiferous at the same time, thus production of spores continuous for a long period of time. Basidia not produced directly on dikariotic vegetative cells; first, probasidia with thick walls formed, and from each one of these, emerges a cylindrical metabasidium, establishing a constriction between persistent probasidium and metabasidium. On members of palm family (Arecaceae).

Botrytis cinerea Pers.: conidiophore and conidia, x 1,000.

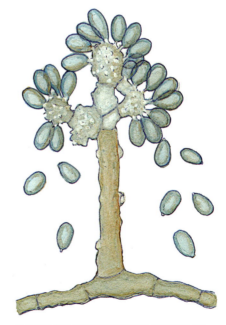

Botrytis ficariarum Hennebert: conidiophore with clusters of conidia, x 2,000. Described in Belgium.

Brasiliomyces

Bovista nigrescens Pers.: mature basidiomata on soil, x 1.

Brachybasidium pinangae (Racib.) Gäum.: basidia with basidiospores, coming through a stoma in the leaf of pinang palm, x 300.

Brasiliomyces Viégas
Erysiphaceae, Erysiphales, Leotiomycetidae, Leotiomycetes, Pezizomycotina, Ascomycota, Fungi
Viégas, A. P. 1944. Alguns fungos de Brasil II. Ascomicetos. *Bragantia* 4:1-392.
From Por. *Brasil*, the country of origin, Brazil + L. *myces* < Gr. *mýkes*, fungus. Mycelium superficial, hyaline, forming white colonies on host tissues, bearing erect, tapered, hyaline setae in one species. Ascomata globose to subglobose, non-ostiolate, seated on surface of mycelium; appendages lacking, but basal attachment hyphae often present. Ascomatal wall thin, yellowish or pale brown with age. Asci unitunicate, saccate, in fascicles of 2-5 per ascoma, each ascus containing 3-8 ascospores. Ascospores hyaline, one-celled, oblong to oval, with rounded ends. On living leaves of angiosperms in the tropics and subtropics.

Brasiliomyces chiangmaiensis To-Anun & S. Takam.: globose perithecium, with asci and ascospores, x 1,000. Growing on leaves of *Dalbergia cultrata*, from Thailand.

Brefeldiella Speg.
Brefeldiellaceae, Inc. sed., Inc. sed., Dothideomycetes, Pezizomycotina, Ascomycota, Fungi
Spegazzini, C. 1889. Fungi puiggariani pugillus. *Boln. Acad. nac. Cienc. Córdoba* 11:558.
Dedicated to the German mycologist *O. Brefeld* + L. dim. suf. *-ella*. Pseudothecia apothecioid developing as circular swellings beneath an orbicular, membranous thallus composed of cells arranged radially. Internal mycelium inconspicuous or nonexistent. Pseudothecia with short, clavate or cylindrical asci and hyaline ascospores. On leaves and branches of trees in tropical regions.

Bridgeoporus T. J. Volk, et al.
Inc. sed., Inc. sed., Inc. sed., Agaricomycetes, Agaricomycotina, Basidiomycota, Fungi
Burdsall, H. H., Jr., et al. 1996. *Bridgeoporus*, a new genus to accommodate *Oxyporus nobilissimus* (Basidiomycotina, Polyporaceae). *Mycotaxon* 60:387-395.
Named in honor of the American mycologist *William Bridge Cooke* + connective *-o* + L. *porus*, pore, who originally described the fungus. Basidiomata perennial, sessile, ungulate, imbricate, or usually substiptate. Texture fibrous, rubbery, tough. Pileus surface a dense mat of white mycelial fibers, becoming brown with age. Hyphal system monomitic, lacking clamps. Context white, tough, fibrous, cinnamon-buff to ochraceous. Pores concolorous with context, round. Basidia pyriform, 4-spored, septate at base. Basidiospores broadly ovoid, hyaline, smooth, thin-walled, J-.

Bridgeoporus nobilissimus (W. B. Cooke) T. J. Volk, et al.: ungulate basidioma, x 0.05.

Brunneosporella Ranghoo & K. D. Hyde
Annulatascaceae, Inc. sed., Sordariomycetidae, Sordariomycetes, Pezizomycotina, Ascomycota, Fungi
Ranghoo, V. M., et al. 2001. *Brunneosporella aquatica* gen. et sp. nov., *Aqualignicola hyalina* gen. et sp. nov., *Jobellisia viridifusca* sp. nov. and *Porosphaerellopsis bipolaris* sp. nov. (ascomycetes) from submerged wood in freshwater habitats. *Mycol. Res.* 105:625-633.

From L. *brunneus*, brown + *spora*, spore < Gr. *sporá*, seed, spore + L. dim. suf. *-ella*, for the brown ascospores. Ascomata immersed, globose to subglobose, membranous, dark brown to black, solitary or gregarious. Neck dark brown, periphysate. Peridium composed of several layers of pale brown, elongate, angular pseudoparenchymatous cells. Paraphyses simple to rarely branched, hyaline, septate, tapering towards apex. Asci 8-spored, cylindrical, pedicellate, thin-walled, unitunicate, apically truncate, with a J-, refractive, discoid apical ring. Ascospores uniseriate to biseriate, ellipsoidal-fusiform, brown, 1-septate, smooth-walled. On submerged wood in Plover Cove Reservoir, Hong Kong, China.

Buellia De Not.
Caliciaceae, Caliciales, Lecanoromycetidae, Lecanoromycetes, Pezizomycotina, Ascomycota, Fungi
De Notaris, G. 1846. Fragmenti lichenografici. *G. bot. ital.* 2(1):195.
Dedicated to *Buell* + L. suf. *-ia*, which denotes belonging to. Lichenized thallus crustous, granulose, verrucose or areolate; apothecia hard, immersed or sessile, with disc flat or convex, commonly black, with a fleeting excipulum of same color. On trees, old wood, mosses, soil, and mainly rocks.

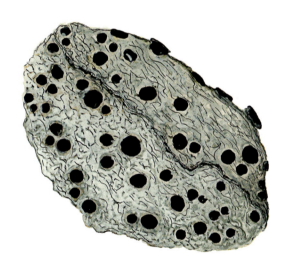

Buellia elegans Poelt: crustous thallus with apothecia, x 4. Described in Italy.

Buerenia M. S. Reddy & C. L. Kramer
Protomycetaceae, Taphrinales, Taphrinomycetidae, Taphrinomycetes, Taphrinomycotina, Ascomycota, Fungi
Reddy, M. S., C. L. Kramer. 1975. A taxonomic revision of the Protomycetales. *Mycotaxon* 3:1-50.
Named in honor of the Swiss mycologist G. von Büren + L. des. *-ia*, for his contributions to knowledge of the protomycetaceous fungi. Ascogenous cells formed throughout tissues in swellings on stems and leaves of host; spherical to elliptical, pale yellowish-brown, smooth-walled, germinating without a rest period. Ascospores formed within ascogenous cells; vesicles not formed. Parasitic on Apiaceae.

Buerenia myrrhidendri Döbbeler: galls (synasci) on fruits of *Myrrhidendron donnell-smithii*, x 5. Described in Costa Rica.

Bulbocatenospora R. F. Castañeda & Iturr.
Inc. sed., Inc. sed., Inc. sed., Inc. sed., Pezizomycotina, Ascomycota, Fungi
Castañeda, R. F., et al. 2000. *Bulbocatenospora*, a new hyphomycete genus from Venezuela. *Mycol. Res.* 104:107-109.
From L. *bulbus*, bulb + *catena*, chain + *spora*, spore < Gr. *sporá*, spore, seed, for bulbous spores formed in chains. Colonies effuse, black. Conidiophores inconspicuous, mostly reduced to conidiogenous cells, monotretic, determinate, intercalary, with a well-defined pore. Conidial secession schizolytic. Conidia muriform, ovoid, triangular to irregular, complanate, with dark brown to brown bulbous cells formed in unbranched, acropetal chains. On fallen leaves in Parque Nacional Henry Pitier, Aragua, Venezuela.

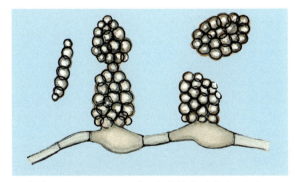

Bulbocatenospora complanata R. F. Castañeda, et al.: conidiophores with clusters of conidia, seen in dorsal and lateral views, x 1,000.

Bulbomollisia Graddon
Dermateaceae, Helotiales, Leotiomycetidae, Leotiomycetes, Pezizomycotina, Ascomycota, Fungi

Bulgaria

Graddon, W. D. 1984. Some new Discomycete species: 6. *Trans. Br. mycol. Soc.* 83:377-382.
From L. *bulbus*, bulb + genus *Mollisia* (Fr.) P. Karst, for bulbous base of apothecium. When fresh apothecia a slate-colored disk with a brilliant white margin, immersed base of small-celled hyaline textura prismatica, changing to slender vertically aligned hyphae in apothecium; outermost cells pale brown continue as outer cells of apothecium. Excipular cells small, outermost brown, inner hyaline; towards margin giving rise to slender, thin-walled, septate, hyaline hairs next to hymenium. Asci 8-spored, slightly clavate, with pore bluing in iodine. Ascospores 1-2-celled, hyaline, narrowly clavate, biguttulate.

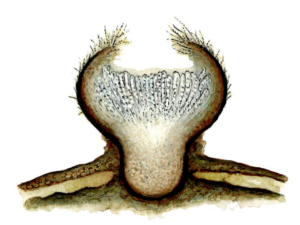

Bulbomollisia radiata Graddon: longitudinal section of an apothecium, containing asci and ascospores, on branches of *Rubus fruticosus*, x 100. Described in Great Britain.

Bulgaria Fr. (syn. **Phaeobulgaria** Seaver)
Phacidiaceae, Phacidiales, Leotiomycetidae, Leotiomycetes, Pezizomycotina, Ascomycota, Fungi
Fries, E. M. 1822. *Syst. mycol., sistens fungorum ordines, genera, et species* (Lund) 2:166.
From L. *bulga*, leather bag, sac + suf. -*aria*, which denotes similarity, resemblance, connection with something, pertaining to, or place of something, due to shape of saccate-turbinate apothecia, structure more or less complex, composed of differentiated tissues. Apothecia medium large, singly or cespitose, gelatinous within, externally brownish. Hymenium black or bluish-black. Asci clavate, inoperculate, with an apical pore, 8-spored. Ascospores ellipsoid, unicellular, unequal-sided, brown. Paraphyses filiform. Common in forests, principally on bark of dead wood, but occasionally parasitizing hardwood trees such as oaks.

Bulleromyces Boekhaut & A. Fonseca
Bulleraceae, Tremellales, Inc. sed., Tremellomycetes, Agaricomycotina, Basidiomycota, Fungi
Boekhout T., et al. 1991 *Bulleromyces* genus novum (Tremellales), a teleomorph for *Bullera alba*, and the occurrence of mating in *Bullera variabilis*. *Antonie van Leeuwenhoek* 59(2):91.
Dedicated to the Canadian mycologist A. H. R. Buller (1874-1944) + L. *myces*, fungus. One of the "mirror yeasts", so-called because colonies in Petri dish cultures produce a mirror image of colonies on cover; due to active discharge of ballistospores, which adhere to objects near colonies. Inhabiting the phyllosphere of plants, particularly on leaves parasitized by rusts.

Bulgaria inquinans (Pers.) Fr.: gregarious apothecia, x 1.

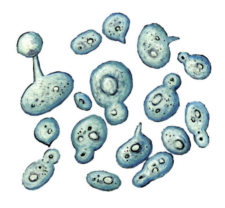

Bulleromyces albus Boekhout & Á. Fonseca (syn. **Bullera alba** (W. F. Hanna) Derx): vegetative budding cells, and ballistospores on sterigmata ("mirror yeast"), x 1,350. Isolated in agar culture, in Florida, U.S.A.

Byssochlamys Westling—see **Paecilomyces** Bainier
Trichocomaceae, Eurotiales, Eurotiomycetidae, Eurotiomycetes, Pezizomycotina, Ascomycota, Fungi
Westling, R. 1909. *Byssochlamys nivea*, on föreningslänk mellen familjerna Gymnoascaceae och Endomycetaceae. *Svensk. bot. Tidskr.* 3:125-137.
From Gr. *býssos*, cloth of very fine flax of antiquity + *chlamýs*, clammy, clothing.

Byssomerulius Parmasto
Phanerochaetaceae, Polyporales, Inc. sed., Agaricomycetes, Agaricomycotina, Basidiomycota, Fungi
Parmasto E. 1967. Corticiaceae U.R.S.S. IV. Descriptiones taxorum novorum. Combinationes novae. *Izv. Akad. Nauk Estonsk. SSR, Ser. Biol.*16:383.

From Gr. *býssos*, fine and soft fabric + similar to genus *Merulius*, but softer like a linen cloth, in reference to fruiting surface, with pitlike depressions or shallow tubes. Basidiocarp resupinate or resupinate-reflexed, suberose, membranous, not gelatinous, hymenophore merulioid with pores circular or angular. Hymenium white to yellow, sometimes slightly rose or red. Basidia clavate. Spores hyaline, ellipsoid or cylindric. On wood of angiosperm trees and conifers.

Byssosphaeria Cooke (syn. **Macbridella** Seaver)
Melanommataceae, Pleosporales, Pleosporomycetidae, Dothideomycetes, Pezizomycotina, Ascomycota, Fungi
Cooke M.C., Plowright C.B. 1879. British Sphaeriacei. *Grevillea* 7:84.

From Gr. *býssos*, a fine yellowish flax, fine thread + *sphaira* > L. *sphaera*, sphere, refers to the fine subiculum around the perithecia. Perithecia in dense cespitose clusters seated on stroma, bright colored, reddish or yellowish, becoming darker with age, globose to sub-cylindrical, collapsing or entire. Asci cylindrical-clavate, 8-spored. Ascospores elliptical or fusoid, 1-septate, at first hyaline, becoming smoky-brown to brownish-black. On tree bark.

Byssomerulius corium (Pers.) Parmasto: resupinate basidiocarp, on wood, x 1,500.

Byssosphaeria schiedermayeriana (Fuckel) M.E. Barr : globose perithecia embedded in a villose subiculum, x 20; apical part of a bitunicate ascus with fusiform ascospores x 300.

C

Chaetomium crispatum

C

Caliciopsis Peck

Coryneliaceae, Coryneliales, Eurotiomycetidae, Eurotiomycetes, Pezizomycotina, Ascomycota, Fungi
Peck, C. H. 1883. *Ann. Rep. N.Y. St. Mus. nat. Hist.* 33:32.
From *Calicium* Pers., a genus of lichens (from L. *calix*, *calicis*, cup + *-ium*, dim. suf.) + L. suf. *-opsis*, from Gr. *-ópsis*, aspect, appearance, in reference to the shape of its ascocarps, which are ascostromal and transitional between the Euascomycetes and the Loculoascomycetes. Ascomata lobed, with a particular dehiscence, in which an infundibuliform perforation forms that permits the expulsion of asci from terminal ascigerous locule. *Caliciopsis pinea* Peck produces its erumpent ascostromata on the bark of white pine (*Pinus strobus*).

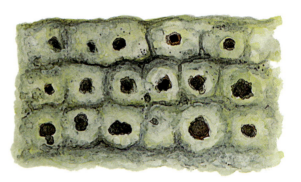

Calicium notarisii (Tul.) M. Prieto & Wedin (syn. **Cyphelium notarisii** (Tul.) Blomb. & Forssell): crustous thallus of this lichen, with superficial apothecia, x 10.

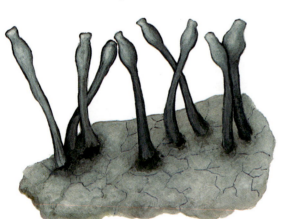

Caliciopsis pinea Peck: lobed, erumpent ascostromata on the bark of white pine (*Pinus strobus*), x 20. Described in New York, U.S.A.

Calicium salicinum Pers.: crustous thallus with long stipitate apothecia, x 30.

Calicium Pers.

Caliciaceae, Teloschistales, Lecanoromycetidae, Lecanoromycetes, Pezizomycotina, Ascomycota, Fungi
Persoon, C. H. 1794. *Ann. Bot. (Usteri)* 7:20.
From L. *calix*, *calicis*, cup + *-ium*, dim. suf., for the shape of its ascocarps. Thallus crustous, generally rudimentary, lacks areoles, with rhizoids. Apothecia on long, slender, dark-colored stipes. Spores bicellular, dark brown. Photobiont *Cystococcus*. Growing on old wood and decorticated tree trunks in shady places.

Callistosporium Singer

Tricholomataceae, Agaricales, Agaricomycetidae, Agaricomycetes, Agaricomycotina, Basidiomycota, Fungi
Singer, R. 1944. New genera of fungi. I. *Mycologia* 36(4):363.
From Gr. *kállistos*, superlative of *kalós*, beautiful + *spóros*, *sporá*, seed, spore + L. dim. suf. *-ium*, meaning with small, beautiful spores, which are brightly colored. Pileus hygrophanous or non-hygrophanous, with a cuticle consisting of repent, elongate hyphae,

sometimes with peg-like hyphal branches arising from them; hymenophore lamellate; lamellae subdecurrent to narrowly adnexed or emarginate; stipe central, thin, fleshy fragile to subcartilaginous; context consisting of hyphae without clamps. spore print white when fresh; spores ellipsoid, smooth, inamyloid, bright colored in dried specimens, scarcely or weakly cyanophilic; basidia normal but some pigmented as spores. On the base of palm trees, on various kinds of wood, even on mosses and on the soil. In the Asiatic subtropic and tropics and in America from the tropics to the boreal region (Canada), south to Argentina, also two species in Europe.

Calloria Fr.

Dermateaceae, Helotiales, Leotiomycetidae, Leotiomycetes, Pezizomycotina, Ascomycota, Fungi
Fries, E. M. 1849. *Summa veg. Scand., Sectio Post.* (*Stockholm*):359.
From Gr. *kallós*, a beauty > L. *callos* + des. *-ia*, quality of. Apothecia minute, sessile or subsessile, occasionally with a short, stem-like base, superficial or suberumpent, dark-grayish, greenish or more frequently bright-colored, red, yellow or purplish, externally smooth or nearly so. Asci clavate, typically 8-spored. Ascospores ellipsoid or fusoid, normally 1-septate, rarely 3-septate, hyaline. Paraphyses filiform, simple or branched, ends either free or agglutinated, forming an epithecium. On dead plant materials.

Calloria neglecta (Lib.) B. Hein: multiple reddish apothecia, on plant detritus, x 1.

Calocera (Fr.) Fr.

Dacrymycetaceae, Dacrymycetales, Inc. sed., Dacrymycetes, Agaricomycotina, Basidiomycota, Fungi
Fries, E. M. 1828. *Elench. fung.* (*Greifswald*) 1:233.
From Gr. *kalós*, pretty, beautiful > L. pref. *calo-*, + Gr. *kéras*, horn. Basidiomata corniculate or clavate, simple, occasionally branched, generally brightly colored, yellow or orange, consistency gelatinous-ceraceous, leathery and firm. In humid forests on wood, in particular on conifers.

Calocera cornea (Batsch) Fr.: yellowish basidiomata, on dead wood, x 1.

Calogaya Arup, et al.

Teloschistaceae, Teloschistales, Lecanoromycetidae, Lecanoromycetes, Pezizomycotina, Ascomycota, Fungi
Arup, U., et al. 2013. A new taxonomy of the family Teloschistaceae. *Nordic Journal of Botany*. *31*(*1*):16-83.
From Gr. *kalós*, beautiful > L. pref. *calo-* + *Gaya* < Dr. *Ester Gaya* (English mycologist), to appreciate her contribution of the study of this group of fungi.

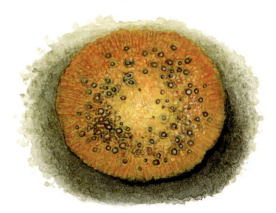

Calogaya pusilla (A. Massal.) Arup, et al. (syn. **Caloplaca saxicola** (Hoffm.) Nordin): crustous, brightly colored thallus, with apothecia, on dead wood, x 1.

Calonectria De Not.

Nectriaceae, Hypocreales, Hypocreomycetidae, Sordariomycetes, Pezizomycotina, Ascomycota, Fungi
De Notaris, G. 1867. Nuove reclute per la pirenomicetologia italica. *Comm. Soc. crittog. Ital.* 2 (fasc. 3):477.
From Gr. *kalós*, beautiful > L. pref. *calo-* + genus. *Nectria* (Fr.) Fr., for the attractiveness of the ascomata. Ascomata ostiolate perithecia, solitary to aggregated, superficial, brightly colored, turning purple in KOH; ascomatal wall composed of two regions. Centrum

Caloplaca

containing apical paraphyses. Asci unitunicate, broadly clavate to obovate, undifferentiated at apex. Ascospores elliptic to fusiform, often somewhat curved, 1- to several-septate, hyaline. Asexual morph cylindrocladium-like. On vascular plants, often causing serious diseases.

Caloplaca Th. Fr.
Teloschistaceae, Teloschistales, Lecanoromycetidae, Lecanoromycetes, Pezizomycotina, Ascomycota, Fungi
Fries, Th. M. 1860. *Lich. Arctoi.* 3:218.
From Gr. *kalós*, beautiful > L. pref. *calo-* + Gr. *pláx*, genit. *plakós*, round plate, tablet, broad surface, table, sheet. Thallus crustous, generally with intense, bright color, e.g., orange or red, with concolorous apothecia; marginal lobulation on thallus, occasionally finely fruticulose. Grows on rocks and trunks and branches of trees.

Calosphaeria Tul. & C. Tul.
Calosphaeriaceae, Calosphaeriales, Sordariomycetidae, Sordariomycetes, Pezizomycotina, Ascomycota, Fungi
Tulasne, L. R., C. Tulasne. 1863. *Select. fung. carpol.* (Paris)2:108.
From Gr. *kalós*, beautiful > L. pref. *calo-* + genus *Sphaeria* Haller, a beautiful spherical fungus. Ascomata ostiolate perithecia, solitary, scattered or arranged in valsoid groups with convergent ostiolar necks, formed free on substrate or beneath loose bark, occasionally immersed in substrate, with ostiolar necks not extending far beyond surface of substrate. Perithecia black, with subglobose to obpyriform venter and long ostiolar neck that is laterally or obliquely attached. Perithecial wall composed of several layers of flattened cells. Centrum consisting of a hymenial layer of asci and paraphyses lining inner wall of perithecium. Paraphyses broad, elongate, tapered, septate. Asci unitunicate, clavate with a narrow tapered stipe, or oblong, with a blunt rounded apex, 8-spored; ascus apex undifferentiated, not bluing in iodine. Ascospores 1-celled or occasionally 2-celled, hyaline, allantoid or sometimes straight and oblong or ellipsoid, sometimes budding inside ascus. Asexual morphs ramichloridium-like and sporothrix-like. On hardwood branches, often associated with other fungi.

Calostoma Desv.
Calostomataceae, Boletales, Agaricomycetidae, Agaricomycetes, Agaricomycotina, Basidiomycota, Fungi
Desvaux, N. A. 1809. Observationes sur quelques genres à établir dans la famille des champignons. *J. Bot.* (Desvaux) 2:94.
From Gr. *kalós*, beautiful > L. pref. *calo-* + Gr. *stóma*, mouth. Fructifications epigeous at maturity, differentiated into a fragile, alveolate-reticulate, whitish semi-subterraneous foot and a globose, reddish-orange foot, constituted of four layers: an external gelatinous layer beneath which are pigmented, transparent, and the internal or membranous layers. Gleba composed of a capillitium with annular swellings, which eventually disintegrates. Spores large, ornamented with long spines or reticulum. Grows in subtropical and oak forests.

Calosphaeria pulchella (Pers.) J. Schröt.: ostiolate perithecia, on hardwood branch, x 20.

Calostoma cinnabarinum Desv.: mature basidioma, with reddish peristome, on soil, x 1.

Calvatia Fr.
Agaricaceae, Agaricales, Agaricomycetidae, Agaricomycetes, Agaricomycotina, Basidiomycota, Fungi
Fries, E. M. 1849. *Summa veg. Scand., Sectio Post.* (Stockholm):442.
From L. *calva*, cranium, or from *calvus*, bald, smooth + des. *-atia*, from L. suf. *-atus*, which indicates similarity or possession. Fructifications at maturity smooth like a cranium or bald head, due to partial or total falling off of exoperidium, when this persists, smooth, areolate or composed of flat scales. Gleba lilac to olivaceous-yellow at maturity; base sterile fibrous or alveolate; capillitium composed of long, septate or continuous, fragile filaments, commonly fragmented, united to internal wall of endoperidium. Spores globose or somewhat elliptical, smooth or echinulate. Grows in grasslands, sandy meadows, low forests and oak groves.

Calvatia cyathiformis (Bosc) Morgan: mature basidioma, with broken peridium, on soil, x 0.4.

Calycella (Fr.) Boud.
Helotiaceae, Helotiales, Leotiomycetidae, Leotiomycetes, Pezizomycotina, Ascomycota, Fungi
Boudier, J. L. E. 1885. Nouvelle classification des discomycètes charnus. *Bull. Soc. mycol. Fr.* 1:112.
From Gr. *kályx*, genit. *kálykos* > L. *calyx, calicis*, chalice, stemmed glass + L. dim. suf. *-ella*, for the shape of the fructifications. Apothecia turbinate, caliciform, sessile or with a short stipe, yellow or orange, with walls that become gelatinous, not easy to distinguish hyphae, which have translucid or crystalline walls at maturity. Ascospores hyaline, usually 1-septate. Common on wood in humid and temperate forests.

Calycina Nees ex Gray
Hyaloscyphaceae, Helotiales, Leotiomycetidae, Leotiomycetes, Pezizomycotina, Ascomycota, Fungi
Gray, S. F. 1821. *Nat. Arr. Brit. Pl.* (London) 1:669.
From Gr. *kályx*, cup > L. *calyx* + suf. *-ina*, likeness. Apothecia medium-large, stipitate or sessile, length of stem variable, light-colored or dull, externally smooth, or with poorly developed hair-like structures. Asci cylindric or clavate, usually 8-spored. Ascospores ellipsoid to fusoid, hyaline, for a long time simple, later becoming one to several septate. On wood and branches.

Camaropella Lar. N. Vassiljeva
Boliniaceae, Boliniales, Sordariomycetidae, Sordariomycetes, Pezizomycotina, Ascomycota, Fungi
Vassiljeva, L. N. 1997. *Camarops pugillus* (Schw.: Fr.) Shear in the Russian Far East. *Mikol. Fitopatol.* 31(1):5-7.
From Gr. *kamára*, a vaulted chamber + *pellos*, dusky, in reference to the shape of the ascostroma. Perithecia erumpent from wood, eutypelloid clusters of confluent perithecial beaks. Asci cylindrical, paraphysate. Ascospores uniseriate, unicellular, ellipsoid, brownish.

Camaropella lutea (Alb. & Schwein.) Lar. N. Vassiljeva: sagittal section of an ascostroma, which shows embedded perithecia in a dusky vaulted chamber, on wood, x 5.

Camarophyllus (Fr.) P. Kumm.—see **Hygrophorus** Fr.
Hygrophoraceae, Agaricales, Agaricomycetidae, Agaricomycetes, Agaricomycotina, Basidiomycota, Fungi
Kummer, P. 1871. *Führ. Pilzk.* (Zerbst) 8:26.
From Gr. *kamára*, arc, arc dome + *phýllon*, leaf, in reference to the form of the gills.

Camillea Fr.
Xylariaceae, Xylariales, Xylariomycetidae, Sordariomycetes, Pezizomycotina, Ascomycota, Fungi
Fries, E. M. 1849. *Summa veg. Scand., Sectio Post.* (Stockholm):322.
Dedicated to the illustrious French scientist *Camille Montagne*. Stromata cylindrical, dark, carbonaceous with embedded perithecia, with only ostioles projecting on surface. Saprobic on wood.

Camillea leprieurii (Mont.) Mont.: carbonaceous perithecial stromata, on wood, x 3.

Canalisporium Nawawi & Kuthub. (syn. **Ascothailandia** Sri-indr., Boonyuen, Sivichai & E. B. G. Jones)
Inc. sed., Inc. sed., Hypocreomycetidae, Sordariomycetes, Pezizomycotina, Ascomycota, Fungi

Canariomyces

Nawawi A., A.J. Kuthubutheen 1989. *Canalisporium*, a new genus of lignicolous hyphomycetes from Malaysia. *Mycotaxon* 34:477.
From L. *canalis*, canal, channel, conduit + *spora*, spore + dim. suf. -ium, referring to the cell lumina of transversal or longitudinal septa connected by narrow canal
Ascomata globose or subglobose, dark brown to black perithecia each with a long neck, ostiolate, scattered, immersed, semi-immersed or superficial. Neck brown or black, variable in length, straight or curved, periphysate. Paraphyses hyphal, tapering. Asci unitunicate, 8-spored, long cylindrical, truncate at apex, with a J-apical ring, persistent. Ascospores uniseriate or overlapping, fusiform, straight or curved, 3-septate, with large dark central cells and small lighter end cells, smooth. Asexual morph canalisporium-like. Collected on submerged wood and bark in Asia.

Canariomyces Arx

Microascaceae, Microascales, Hypocreomycetidae, Sordariomycetes, Pezizomycotina, Ascomycota, Fungi
Arx, J. A. von. 1984. *Canariomyces notabilis*, a peculiar ascomycete from the Canary Islands. *Persoonia* 12(2):185-187.
From L. *canario*, island > Sp. *Canario*, Canary Islands + L. *myces*, fungus < Gr. *mykés*, i.e., a fungus from the Canary Islands. Ascomata perithecia superficial, globose, smooth, blackish, nonostiolate. Ascomatal wall thin, dark brown. Asci arranged irregularly on branched ascogenous hyphae, botryose, sessile, spherical or broadly ovate, 8-spored. Ascospores ellipsoidal or broadly fusiform, 1-celled, hyaline and dextrinoid when young, becoming brown, with 2-3 darker striations at maturity. Isolated from roots of *Phoenix canariense* in the Canary Islands.

Candelaria A. Massal.

Candelariaceae, Candelariales, Inc. sed., Lecanoromycetes, Pezizomycotina, Ascomycota, Fungi
Massalongo, A. B. 1852. Synopsis lichenum Blasteniosporum. *Flora, Regensburg* 35:567.
From L. *candela*, candle < *candeo*, to be red-hot, burning + suf. -aria, which indicates connection or possession, for bright yellow color. Thallus foliose, bright yellow, with small, aggregated lobules, confused with a crustous lichen, with cortex and rhizines on lower surface. Apothecia generally small, with a yellow disc. Grows on tree bark and on rocks.

Candelariella Müll. Arg.

Candelariaceae, Candelariales, Inc. sed., Lecanoromycetes, Pezizomycotina, Ascomycota, Fungi
Müller, J. 1894. Conspectus systematicus lichenum Novae Zelandiae. *Bull. Herb. Boissier* 2(app. 1):11.
From genus *Candelaria* A. Massal. + L. dim. suf. -ella; the latter genus. from L. *candela*, candle < *candeo*, to be candescent, red-hot + suf. -aria, which indicates connection or possession, for the intense yellow color of the majority of the species of both genera. Thallus yellow, crustous, minutely squamulose, arranged in rosettes or dispersed, with small apothecia. Grows on rocks, soil and old wood. Photobiont *Protococcus*.

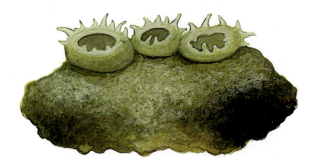

Candelaria fibrosa (Fr.) Müll. Arg.: foliose thallus, with yellowish apothecia, on tree bark, x 8.

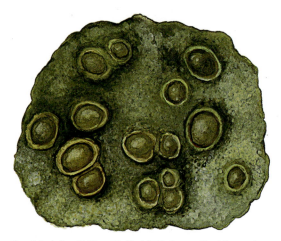

Candelariella vitellina (Hoffm.) Müll. Arg.: yellowish, crustous thallus, with apothecia, on wood, x 1.

Candelina Poelt

Candelariaceae, Candelariales, Inc. sed., Lecanoromycetes, Pezizomycotina, Ascomycota, Fungi
Poelt, J. 1974. Systematik der Flechten. *Phyton* Horn 16(1-4):194.
From L. *candela*, candle < *candeo*, to be red-hot, burning + suf. -ina, which indicates possession or similarity, for the intense yellow color of species in the genus.

Candida Berkhout

Inc. sed., Saccharomycetales, Saccharomycetidae, Saccharomycetes, Saccharomycotina, Ascomycota, Fungi
Berkhout, C. M. 1923. De schimmelgeschlachten *Monilia*, *Oidium*, *Oospora* und *Torula*. Thesis, Utrecht, 72 pp.
From L. *candida*, white, for the color of its colonies. Ascosporogenous yeast with worldwide distribution and many species. Includes pathogens such as candidiasis in humans and other animals and saprobes, some of industrial importance.

Capnodium

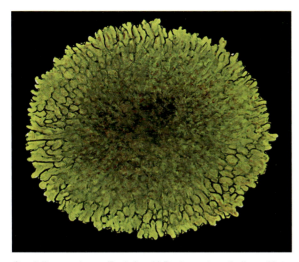

Candelina mexicana (B. de Lesd.) Poelt: crustous thallus, with an intense yellow color, on rock, x 2.

Candida albicans (C. P. Robin) Berkhout: giant, white colonies on agar, x 1.

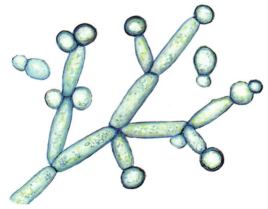

Candida albicans (C. P. Robin) Berkhout: pseudomycelium with blastospores, and thick-walled chlamydospores, x 1,500.

Cannonia Joanne E. Taylor & K. D. Hyde
Xylariaceae, Xylariales, Xylariomycetidae, Sordariomycetes, Pezizomycotina, Ascomycota, Fungi
Taylor, J. E., K. D. Hyde. 1999. *Cannonia* gen. nov., from palms in the Southern Hemisphere. *Mycol. Res.* 103(11):1398-1402.
Named in honor of the British mycologist *Paul Cannon* + L. des. -ia. Stromata poorly developed. Clypeus composed of host cells filled with brown, thick-walled hyphae. Ascomata immersed, black beak-like to papillate ostioles, singular or tightly clustered; in vertical section, globose to pyriform, asci peripheral. Ostiole central, aperiphysate. Peridium comprising two strata of brown-walled, compressed cells. Asci (1-)8 multispored, clavate to broadly cylindrical, unitunicate, short pedicellate, with apical apparatus lacking, sometimes deliquescent. Ascospores overlapping uni-, bi- or multiseriate, broadly fusiform or irregularly-shaped, laterally compressed, unicellular, mid yellow-brown, biguttulate, with a straight, indistinct, full length germ slit. Paraphyses hypha-like, sometimes lacking at maturity.

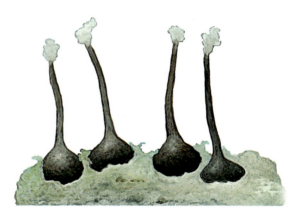

Cannonia australis (Speg.) Joanne E. Taylor & K. D. Hyde: dark perithecia on tissues of *Syagrus* palm, x 30.

Cantharellus Adans. ex Fr.
Cantharellaceae, Cantharellales, inc. sed., Agaricomycetes, Agaricomycotina, Basidiomycota, Fungi
Fries, E. M. 1821. *Syst. mycol.* (Lundae) 1:316.
From L. *cantharus* < Gr. *kántharos*, wine glass, beverage glass, pitcher + L. dim. suf. *-ellus*, for the shape of the fructification. Basidiocarps fleshy with thick, branched gills forming veins or pleats, radially arranged, decurrent toward lower part of stipe; spores white or pale ochraceous. *Cantharellus cibarius* Fr., commonly called chanterelle, is a highly valued edible species. Grows on soil of conifer forests.

Capnodium Mont.
Capnodiaceae, Capnodiales, Dothideomycetidae, Dothideomycetes, Pezizomycotina, Ascomycota, Fungi
Montagne, J. P. F. C. 1849. *Cryptogamia guyanensis. Annls. Sci. Nat., Bot., sér.* 3,11:233.
From Gr. *kapnós*, smoke, vapor > *kapnódes*, smoky + L. dim. suf. *-ium*. Ascostromata dark forming a black subiculum. Pseudothecia glabrous or with hyphal pro-

Carpenteles

jections; asci with muriform ascospores. Epiphytic on leaves, a saprobe on the sugary exudates of the insect parasites of various plants, referred to as sooty molds.

Cantharellus cibarius Fr.: yellow, mature basidioma, on soil in pine forest in Sweden, x 1.

Capnodium citri Berk. & Desm.: black ascostromata, on the surface of the leaf of citrus plant, x 1.

Carpenteles Langeron—see **Penicillium** Link
Trichocomaceae, Eurotiales, Eurotiomycetidae, Eurotiomycetes, Pezizomycotina, Ascomycota, Fungi
Langeron, M. 1922. Utilité de deux nouvelles coupures générique dans les perisporiaces: *Diplostephanus* n. g. et *Carpenteles* n. g. *C. r. hebd. Séanc. Mém. Soc. Biol.* 87:343-345.
In honor of the American mycologist *C. W. Carpenter* + Gr. suf. *-es*, agent or doer.

Caryospora De Not.
Caryosporaceae, Pleosporales, Pleosporomycetidae, Dothideomycetes, Pezizomycotina, Ascomycota, Fungi
De Notaris, G. 1855. *Micr. Ital. Novi* 9:7.

From Gr. *káryon*, nut > L. *caryon* + *spora*, spore < Gr. *sporá*, seed, spore, referring to the original substrate of stone fruit endocarps. Ascomata pseudothecia, discrete, scattered, erumpent to superficial with base rounded or flattened. Pseudothecia large, up to 1 mm diameter, mammiform, with pore in papillate apex; pseudothecial wall carbonaceous, composed of several layers of compressed, darkly pigmented cells. Centrum containing numerous pseudoparaphyses. Asci few in number, bitunicate, oblong to broadly ellipsoid, 1-8-spored. Ascospores 2-celled, reddish-brown to dark brown, broadly ellipsoid, tapering to pointed or rounded ends, smooth; primary septum median, usually constricted at septum, sometimes with small secondary septa at ends; young ascospore surrounded by a thin gelatinous sheath that forms rough areas on spore surface when dry. Asexual morph asterostomella-like. Saprobic on woody substrates, stone fruit endocarps, and woody herbaceous and graminicolous stems.

Caryospora obclavata Raja & Shearer: erumpent pseudothecium on dead wood, x 90; an ascus with eight bicellular ascospores, x 500.

Casaresia Gonz. Frag. (syn. **Ankistrocladium** Perrott)
Dermateaceae, Helotiales, Leotiomycetidae, Leotiomycetes, Pezizomycotina, Ascomycota, Fungi
González Fragoso, R. 1920. Nuevo género y especies de hifal sobre hojas de *Sphagnum*. *Boln Real Soc. Españ. Hist. Nat., Biologica* 20:112.
The name is given in honor of Dr. *Antonio Casares-Gil*. Spanish Botanist + L. des. *-ia*. Mycelium septate, fuscus. Conidiophores simple, short, dark, one-celled, bearing a single apical conidium. Conidia dark brown, multicellular, each of a long, thick-walled, septate, curved, fusoid axis, hooked at tip, bearing at base two pairs of lateral branches of same shape as axis. One lower branch frequently branches again to produce a hexa-radiate spore; occasionally with seven or nine branches. On submerged decorticated twigs and plant material.

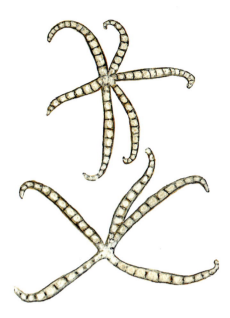

Casaresia sphagnorum Gonz. Frag. (syn. **Ankistrocladium fuscum** Perrott): penta- and hexaradiate conidia, x 1,000. Described in Great Britain.

Cataractispora K. D. Hyde, et al.

Annulatascaceae, Inc. sed., Sordariomycetidae, Sordariomycetes, Pezizomycotina, Ascomycota, Fungi

Hyde, K. D., et al. 1999. *Cataractispora* gen. nov. with three new freshwater lignicolous species. *Mycol. Res.* 103(8):1019-1031.

From L. *cataracta*, waterfall, river + *spora*, spore < Gr. *sporá*, seed, spore, in reference to the freshwater habitat of this species. Ascomata globose, subglobose or ellipsoidal, immersed, semi-immersed or superficial, coriaceous, ostiolate, beaked, solitary, brown or black. Paraphyses hypha-like, septate, tapering distally. Asci 8-spored, cylindrical, unitunicate, with a relatively massive J- refractive, bipartite apical ring. Ascus wall bipartite. Ascospores uniseriate or overlapping uniseriate, fusiform or ellipsoidal, hyaline, aseptate to multiseptate, with episporial verruculose wall ornamentation, with or without polar chambers; appendage derived from episporial projections at tip, accumulates within polar chambers or ascospore polar cells, released to form long, thread-like bipolar appendages in water. Collected on submerged wood in fresh water.

Catenaria Sorokin

Catenariaceae, Blastocladiales, Inc. sed., Blastocladiomycetes, Inc. sed., Chytridiomycota, Fungi

Sorokin, N. W. 1889. *Revue mycol., Toulouse* 11(43):139.

L. *catena*, chain + suf. *-aria*, which indicates connection or possession > *catenaria*, having chains. Zoosporangia arranged in a linear series, connected by a narrow isthmus. Develop as parasites or saprobes of nematodes, eggs of small animals, rotifers, fungi and algae.

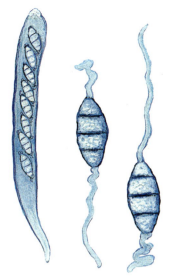

Cataractispora receptaculorum W. H. Ho, et al.: an ascus with eight ascospores, x 420; two septate ascospores with polar appendages, x 900.

Catenaria allomycis Couch: resting spores and rhizoids within a hypha of **Allomyces** E.J. Butler, x 950. Described in Texas, U.S.A.

Catinella Boud.

Inc. sed., Inc. sed., Inc. sed., Dothideomycetes, Pezizomycotina, Ascomycota, Fungi

Boudier, É. 1907. *Hist. Class. Discom. Eur. (Paris)*, 1-223.

From L. *catinus*, a bowl + dim. suf. *-ella*, for the shape of the ascoma. Apothecia sessile, solitary or several together, attached to substrate by numerous dark brown fibers, patellate or nearly so, dark greenish, subgelatinous, smooth or with many short, crowded hairs. Asci cylindric or subcylindric, 8-spored. Ascospores simple, greenish. Paraphyses filiform, granular, septate. On moist rotten wood and old leaves.

Cenangiomyces Dyko & B. Sutton

Inc. sed., Inc. sed., Inc. sed., Agaricomycetes, Agaricomycotina, Basidiomycota, Fungi

Dyko, B. J., B. C. Sutton. 1979. Two new and unusual deuteromycetes. *Trans. Br. mycol. Soc.* 72(3):411-417.

From genus *Cenangium* Fr. + L. *myces*, fungus. Conidiomata cupulate, erumpent, separate, light colored, base and walls composed of textura oblita. Ostiole absent. Conidiophores hyaline, branched, septate. Conidiogenous cells percurrently proliferating, with clamp connections,

Cenangium

integrated. Conidia holoblastic, filiform to subulate, septate, apical and basal cells becoming evacuated, remaining as appendages. Collected on needles of *Pinus* in Sussex, England.

Catinella olivacea (Batsch) Boud.: sessile, patellate apothecium, on wood, x 15.

Cenangium Fr.
Helotiaceae, Helotiales, Leotiomycetidae, Leotiomycetes, Pezizomycotina, Ascomycota, Fungi
Fries, E. M. 1818. *K. svenska Vetensk-Akad. Handl.*, ser. 3, 39:360.
From Gr. *kenós*, empty, hollow > L. *cenos* + Gr. *angeíon*, vessel, receptacle > L. *angium*. Ascomata apothecia, erumpent through bark, single or in small clusters, dark brown to black. Apothecia sessile to substipitate, when dry elongate or sometimes triangular and strongly inrolled at margins, hard and brittle, becoming fleshy, circular, shallowly cupulate when moist. Ectal excipulum sharply differentiated from medullary excipulum. Medullary excipulum composed of hyaline, interwoven hyphae as textura intricata embedded in gel. Hypothecium consisting of a thin layer of thin-walled, lightly pigmented, interwoven hyphae. Hymenium smooth, yellowish-brown. Paraphyses filamentous, septate, not forming an epithecium. Asci unitunicate, inoperculate, clavate to cylindric-clavate, short-stipitate, undifferentiated at apex but with wall slightly thickened, not bluing in iodine, 8-spored. Ascospores hyaline, 1-celled, fusoid to ellipsoid, smooth. Pathogenic on dying branches of *Pinus* spp.

Cenococcum Moug. & Fr.
Gloniaceae, Mytilinidiales, Pleosporomycetidae, Dothideomycetes, Pezizomycotina, Ascomycota, Fungi
Mougeout, J. B., E. M. Fries. 1829. *Syst. mycol.* (*Lundae*), Vol. 3(1):65.
From Gr. *kenós*, empty, hollow + *kókkos* < *kokkíon*, little ball, seed, grain. Sclerotia globose, at first fleshy, then suberose or corneous, with a hollow center. On soil.

Cenangium ferruginosum Fr.: ascomata on young branch of wild pine (*Pinus sylvestris*), x 2.

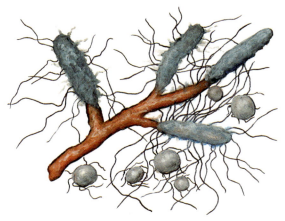

Cenococcum graniforme (Sowerby) Ferd. & Winge: sclerotia of this mycorrhizal fungus, x 4.

Cephaloascus Hanawa
Cephaloascaceae, Saccharomycetales, Saccharomycetidae, Saccharomycetes, Saccharomycotina, Ascomycota, Fungi
Hanawa, S. 1920. Studien über die auf gesunder und kranken Haut angesiedelten Pilzkeime. *Jap. J. Derm. Urol.* 20(germ. sec.):14.
From Gr. *kephalé*, head + *askós* > L. *ascus*, wine sac, ascus. Asci formed by budding on apex of diploid ascophores. Each ascus with four, rarely eight, hat-shaped ascospores. The only species of the genus, *C. fragans* Hanawa, isolated from a human ear and decomposing wood.

Cephalosporium Corda—see Sarocladium W. Gams & D. Hawksw.
Inc. sed., Hypocreales, Hypocreomycetidae, Sordariomycetes, Pezizomycotina, Ascomycota, Fungi
Corda, A. C. J. 1839. *Icon. fung.* (Prague) 3:11-12.
From Gr. *kephalé*, head + *sporá*, spore + L. dim. suf. *-ium*.

Ceratobasidium

Cephaloascus fragans Hanawa: ascophores with terminal groups of four-spored asci, x 1,250. Isolated from a human patient, in Japan.

Cephalotrichum stemonitis (Pers.) Nees (syn. **Doratomyces stemonitis** (Pers.) F. J. Morton & G. Sm.): lance-shaped synnemata, which at the apex form heads of conidia, on soil, x 50.

Cephalotheca Fuckel

Cephalothecaceae, Inc. sed., Sordariomycetidae, Sordariomycetes, Pezizomycotina, Ascomycota, Fungi
Fuckel, K. W. G. L. 1869. *Symbolae mycologicae. Beiträge zur Kenntnis der rheinischen Pilze. Vol. 1. Jb. nassau. Ver. Naturk.* 25-26:297.
From Gr. *kephalé*, head + *théke*, box, deposit. Cleistothecia black, when young covered with hyphae or a thick, yellowish down, at maturity, cleistothecia with apically glabrous heads. Saprobic in soil.

Cephalotheca sulfurea Fuckel: dark cleistothecia, on rotting plants of oak on damp soil, in Austria, x 18.

Cephalotrichum Link

Microascaceae, Microascales, Hypocreomycetidae, Sordariomycetes, Pezizomycotina, Ascomycota, Fungi
Link, H. F. 1809. Observationes in ordines plantarum naturales. Dissertatio I. Magazin der Gesellschaft Naturforschenden Freunde Berlin. 3(1):3-42.
From Gr. *kephalé*, a head + *thríx*, genit. *trichós*, because of the head of the spores. Produces its spores in a dry head at the apex of a complex, erect conidiophore or synnema, which is often up to a mm in height. The synnemata produce chains of powdery conidia with a "bottle brush" or "feather" appearance. The ovoid conidia are produced from annellidic conidiogenous cells covering the sporogenous area.

Ceramothyrium Bat. & H. Maia

Chaetothyriaceae, Chaetothyriales, Chaetothyriomycetidae, Eurotiomycetes, Pezizomycotina, Ascomycota, Fungi
Batista, A. C., H. da Silva Maia. 1956. *Ceramothyrium* a new genus of the family Phaeosaccardinulaceae. *Atti dell' instituto Botanico e Laboratorio Crittogamico dell' Università di Pavia* 14:1-32.
From Gr. *kéranos*, a vessel, earthen pot + *thyreós*, oblong shield + L. suf. -*ium*, perhaps in reference to the color and shape of the ascoma. Mycelium superficial, easily detached, olivaceous to blackish-brown, in small, effuse, thin pelliculose to membraneous colonies composed of septate, reticulate to irregularly branched hyphae. Setae and hyphopodia lacking. Ascomata subglobose to globose or somewhat flattened laterally, yellowish-brown to brown, ostiolate, unilocular, membranous to subcarnose, developing beneath mycelial pellicle. Asci clavate to saccate, 8-spored. Ascospores hyaline, clavate-fusoid, phragmosporous. Collected on living leaves of *Paivea langsdorfii* in São Lourenço, Pernambuco, Brazil.

Ceratobasidium D. P. Rogers

Ceratobasidiaceae, Cantharellales, Inc. sed., Agaricomycetes, Agaricomycotina, Basidiomycota, Fungi
Rogers, D. P. 1935. Notes on the lower Basidiomycetes. *Univ. Iowa Stud. nat. Hist.* 17(5):4.
From Gr. *kéras*, genit. *kératos*, horn + L. *basidium* (derived from Gr. *basídion*, dim. of *básis*, base) basidium, for the horn-shaped sterigmata on the basidium. Fructifications resupinate, corticoid, waxy, waxy-gelatinous, with aseptate, subspherical probasidia. Metabasidia aseptate, wide, ovoid to spheropedunculate, forming inflated-elongate, horned sterigmata that produce basidiospores that germinate by repetition. Includes species mycorrhizal with terrestrial orchids as well as species parasitic on plants. *Ceratobasidium fibrillosum* (Burt) D. P. Rogers

Ceratocystis

& H. S. Jacks., is a species distributed from Colombia to Mexico; the fructification is waxy-gelatinous and fibrillose, which on drying remains as a grayish layer.

Ceratobasidium calosporum D. P. Rogers: longitudinal section of basidioma, showing basidia and basidiospores, x 1,000.

Ceratocystis Ellis & Halst. (syn. **Thielaviopsis** Went)
Ceratocystidaceae, Microascales, Hypocreomycetidae, Sordariomycetes, Pezizomycotina, Ascomycota, Fungi
Ellis, J. B., B. D. Halsted. 1890. *New Jersey Agric. Coll. Exp. Sta. Bull.* 78:14.
From Gr. *kéras*, genit. *kératos*, horn + *kýstis*, bladder, vesicle, sac. Ascomata perithecioid, more or less globose at base with a long neck. Phialides subcylindrical with deep collarettes. Conidia unicellular, subhyaline, in basipetal chains also thallic-arthric phase with subhyaline or brown conidia, single or in chains. Parasitic and saprobic on diverse plants and wood. Asexual morph chalara-like.

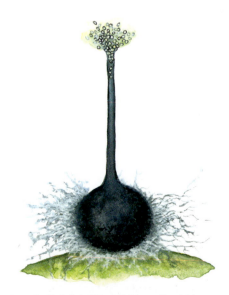

Ceratocystis fimbriata Ellis & Halst.: perithecioid ascocarp, on tubers of *Batatas edulis*, in New Jersey, U.S.A., releasing ascospores through the ostiole, x 240.

Ceratomyces Thaxt.
Ceratomycetaceae, Laboulbeniales, Laboulbeniomycetidae, Laboulbeniomycetes, Pezizomycotina, Ascomycota, Fungi
Thaxter, R. 1892. Contributions from the cryptogamic laboratory of Harvard University XVII. Further additions to the North American species of Laboulbeniaceae. *Proc. Amer. Acad. Arts & Sci.* 27:34.
From Gr. *kerátion*, dim. of *kéras*, genit. *kératos*, horn + *mýkes* (L. *myces*), fungus. Thallus composed of a more or less massive receptacle, multicellular with 9-60 columns of cellular walls, turbinate, with a cupuliform distal depression, bordered by numerous sterile appendages and spermatiophorous branches. Parasitic on aquatic coleopterans.

Cercophora Fuckel
Lasiosphaeriaceae, Sordariales, Sordariomycetidae, Sordariomycetes, Pezizomycotina, Ascomycota, Fungi
Fuckel, K. W. G. L. 1870. Symbolae mycologicae. Beiträge zur Kenntniss der Rheinischen Pilze. *Jb. nassau. Ver. Naturk.* 23-24:1-459.
From Gr. *kérkos*, tail + *phóros*, bearer, for the caudate ascospores. Ascomata ostiolate perithecia, obpyriform to conical or subglobose, dark-brown to black, usually covered with septate hairs or setae; wall 3-4-layered. Paraphyses filamentous to ventricose. Asci unitunicate, clavate, with non-amyloid apical ring, 8-spored. Ascospores at first cylindrical, straight, caudate at both ends, becoming sigmoid, septate, upper cell swelling becoming brown, ellipsoidal to ovoid, truncate at base, with upper germ pore, smooth; lower cell remaining hyaline, collapsing at maturity. Asexual morph lacking. On dung, wood, or in soil.

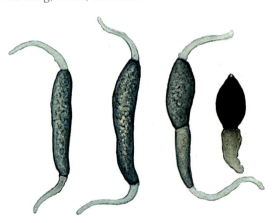

Cercophora palmicola Hanlin & Tortolero: young and mature ascospores, on rotten stems of Palmae, in Venezuela, x 1,200.

Cercospora Fresen. ex Fuckel
Fresenius, J. B. G. W. 1863. *Beitr. Mykol.* 3:91-93.
From Gr. *kérkos* > L. *cerco*, tail, worm + *sporá*, spore, referring to the phragmospores. Conidia elongate,

cylindrical or filiform, multiseptate, hyaline, usually truncate at point of attachment with pores of geniculate conidiophores. Sexual morph mycosphaerella-like. Parasitic on leaves, stems and fruits of plants, rarely saprobic in soil.

Ceriosporopsis Linder
Halosphaeriaceae, Microascales, Hypocreomycetidae, Sordariomycetes, Pezizomycotina, Ascomycota, Fungi
Linder, D. L. 1944. I. Classification of the marine fungi, pp. 401-433, *In*: Barghoorn, E. S., D. L. Linder, Marine fungi: their taxonomy and biology. *Farlowia* 1(3):395-467.
From genus *Ceriospora* Niessl + Gr. suf. *-ópsis*, aspect, appearance, for the similarity to this genus. Perithecia black or dark brown, membranous, collapsing on drying, ovoid to elongate-ellipsoid, ostiolate; ostiole eccentric, rounded-conoid or papilliform. Paraphyses absent. Asci not seen but probably 8-spored since the spores occur frequently in bundles of eight. Ascospores hyaline or dilutely colored, two-celled, with a stout hyaline, tapering elongate appendage at each end, appendages dehiscent or deliquescent, then ascospores truncate at each end. Mostly buried in substratum becoming exposed by erosion of the decayed immersed test blocks.

Cerrena Gray
Polyporaceae, Polyporales, Inc. sed., Agaricomycetes, Agaricomycotina, Basidiomycota, Fungi
Gray, S. F. 1821. *Nat. Arr. Brit. Pl.* (London) 1:649.
From It. *cerrena*, the popular name of the fungus in Italy; referring to the shape of small hills. Fructifications shell-shaped, without a stipe, generally gregarious, forming imbricated bodies. Hymenophore disposed in fused tubes developing in labyrinthiform structures finally with teeth. Spores elliptic, smooth. On dead wood of trees but not of conifers.

Cetraria Ach.
Parmeliaceae, Lecanorales, Lecanoromycetidae, Lecanoromycetes, Pezizomycotina, Ascomycota, Fungi
Acharius, E. 1803. *Methodus*, Sectio post. (Stockholmiæ):292.
From L. *cetra*, a type of leather shield + suf. *-aria*, which indicates connection or possession. Thallus foliose, subfoliose or fruticose. Apothecia marginal or terminal with flat, concave or convex disc, pale brown, chestnut brown or black. The photobiont is *Protococcus*. Grows on trees, rocks, soil and old wood.

Chaetoceratostoma Turconi & Maffei—see **Chaetoceris** Clem. & Shear
Inc. sed., Hypocreales, Hypocreomycetidae, Sordariomycetes, Pezizomycotina, Ascomycota, Fungi
Turconi, M., L. Maffei. 1918. Note micologiche e fitopatologiche. *Atti Ist. bot. R. Univ. Pavia, 2 Sér.* 15:144.
From Gr. *chaíte*, crest, mane + *kératos*, from *kéras*, horn + *stóma*, mouth.

Cetraria islandica (L.) Ach.: foliose thallus with apothecia, x 1.

Chaetoceris Clem. & Shear (syn. **Chaetoceratostoma** Turconi & Maffei)
Inc. sed., Hypocreales, Hypocreomycetidae, Sordariomycetes, Pezizomycotina, Ascomycota, Fungi
Clements, F. E., C. L. Shear 1931. *The Genera of Fungi.* Hafner Publ. Co. New York, p. 262.
From Gr. *chaíte*, crest, mane + *kératos*, from *kéras*, horn. Perithecia with a long horn-like neck, with hairs or setae whole length of neck, ending in fimbriate ostiole. Asci evanescent. Ascospores dark, apiculate, released in a mucilaginous cirrhus. On plant substrates.

Chaetocladium Fresen.
Mucoraceae, Mucorales, Inc. sed., Inc. sed., Mucoromycotina, Zygomycota, Fungi
Fresenius, G. 1863. *Beitr. Mykol.* 3:97.
From Gr. *chaíte*, crest, pony tail, mane + *kládos*, branch + L. dim. suf. *-ium*, i.e., branches with manes. Sporangiola borne on verticillate fertile branches whose apices terminate in stiff, sterile crests or spines. Saprobic in soil and dung, often a parasite of other Mucorales.

Chaetomidium (Zopf) Sacc.
Chaetomiaceae, Sordariales, Sordariomycetidae, Sordariomycetes, Pezizomycotina, Ascomycota, Fungi
Saccardo, P. A. 1882. *Syll. fung.* (Abellini) 1:39.
From the genus *Chaetomium* Kunze (this from Gr. *chaíte*, crest, head of hair, mane, coma + L. dim. suf. *-ium*) + L. dim. suf. *-idium*. Similar to *Chaetomium*. Perithecia dark with hairs, lacking an ostiole. Cellulolytic saprobe on vegetable detritus, soil, wood, dung, and other cellulosic substrates.

Chaetomium

Chaetocladium brefeldii Tiegh. & G. Le Monn.: sporangiophore, with branches that show the verticillate arrangement of sporangiola, x 900.

Chaetomidium heterotrichum R. J. Mey.: perithecium, whose peridium is surrounded with different-colored hairs, from bark of *Quercus falcata*, Georgia, U.S.A., x 90.

Chaetomium Kunze (syn. **Achaetomiella** Arx)
Chaetomiaceae, Sordariales, Sordariomycetidae, Sordariomycetes, Pezizomycotina, Ascomycota, Fungi
Kunze, G., J. K. Schmidt. 1817. *Mykologische Hefte* (Leipzig) *1*:15.
From Gr. *chaíte*, crest, head of hair, mane, coma + L. dim. suf. *-ium*. Perithecia covered with or without hairs. Cellulolytic saprobe on diverse organic substrates, such as wood, cloth, straw, and dung.

Chaetonaevia Arx
Dermateaceae, Helotiales, Leotiomycetidae, Leotiomycetes, Pezizomycotina, Ascomycota, Fungi
Arx, J. A. von. 1951. Eine neue Discomycetengattung aus Skandinavien. *Antonie van Leeuwenhoek 17*:85-89.
From NL. *chaeta*, bristle < Gr. *chaíte*, long hair, mane + genus *Naevia* Fr., i.e., a *Naevia*-like fungus with bristles on the ascoma. Ascomata small apothecia developing as a closed, globose structure, becoming erumpent. Apothecial margin bearing a ring of long, rigid, straight, thick-walled, hyaline setae. Paraphyses numerous. Asci broadly clavate to cylindrical, rounded at apex, with a thick wall, 4-spored. Ascospores parallel in ascus, fusoid, tapering toward both ends, inaequilateral to slightly curved, initially hyaline, 1-celled, later olivaceous to brown, with a median septum. Collected on dead leaves of *Arctostaphylus alpina* in Sweden.

Chaetomium crispatum (Fuckel) Fuckel: perithecium, with straight and contorted hairs, and a cirrhus of ascospores outside the ostiole, on a substrate rich in cellulose, x 110.

Chaetosartorya Subram.—see **Aspergillus** P. Micheli ex Haller
Trichocomaceae, Eurotiales, Eurotiomycetidae, Eurotiomycetes, Pezizomycotina, Ascomycota, Fungi
Subramanian, C.V. 1972. The perfect states of *Aspergillus*. *Curr. Sci. 41(21)*:761.
From Gr. *chaíte*, crest, pony tail, mane + genus *Sartorya* Vuill. < named in honor of A. Sartory + the des. *-a*.

Chaetosphaerella E. Müll. & C. Booth
Müller, E., C. Booth. 1972. Generic position of *Sphaeria phaeostroma*. *Trans. Br. mycol. Soc. 58(1)*:73-77.
From Gr. *chaíté*, long hair, mane + *sphaíra* > L. *sphaera*, sphere + L. dim. suf. *-ella*. Perithecia densely aggregated, superficial, sitting on and in a subiculum composed of dark, thick hyphae and fastened in the substrate by a foot-like base, globose, provided with dark setae and ostiolate. Asci unitunicate containing a simple plate-like apical apparatus, 8-spored. Ascospores cylindric, curved, broadly rounded at the ends, 4-celled, the longer median cells are brown, the end cells hyaline, paraphyses filiform, often present only in young ascomata.

Chaetothyriomyces Pereira-Carv., et al.
Chaetothyriaceae, Chaetothyriales, Chaetothyriomycetidae, Eurotiomycetes, Pezizomycotina, Ascomycota, Fungi
Pereira-Carvalho, D., et al. 2009. *Chaetothyriomyces*: a new genus in family Chaetothyriaceae. *Mycotaxon 107*:483-488.

From Gr. *chaíte*, long hair, mane + *thyreós*, oblong shield + L. *myces*, fungus, a fungus similar to members of the family Chaetothyriaceae. Mycelial colonies effuse, superficial, that form a net-like pellicle. Ascomata form beneath the pellicle, globose or subglobose, dark brown, unilocular, smooth. Asci broadly clavate, bitunicate with 16 ascospores. Ascospores elliptical, 2-celled, hyaline. On living leaves of *Qualea grandiflora* in Mato Grosso do Sul, Brazil.

Chaetosphaerella phaeostroma (Durieu & Mont.) E. Müll. & C. Booth: vertical section of a perithecium, with setae and asci, seated on a subiculum and on black stromatic stroma, developing as a saprobe in dead wood, x 100; liberated, curved ascospores, with two dark-colored central cells, and hyaline end cells at both extremes, and free, ellipsoidal conidia, with their cells arranged in a similar way to the ascospores, x 500.

Chalara (Corda) Rabenh. (syn. **Hughesiella** Bat. & A. F. Vital)

Inc. sed., Microascales, Hypocreomycetidae, Sordariomycetes, Pezizomycotina, Ascomycota, Fungi
Rabenhorst, L. 1844. *Deutschl. Krypt.-Fl.* (Leipzig) 1:38.
From Gr. *chalára*, chain. Conidia in basipetal chains originating within phialide or conidiogenous cell. Phialides ventricose with an elongated collarette. Conidia unicellular, cylindrical, with truncate ends, hyaline or lightly pigmented. Chlamydospores dark. Causing blue stain on wood, also saprobic in soil.

Chalaropsis Peyronel

Inc. sed., Microascales, Hypocreomycetidae, Sordariomycetes, Pezizomycotina, Ascomycota, Fungi
Peyronel, B. 1916. Una nuova mallattia del lupino prodotta da *Chalaropsis thielavioides* Peyr. nov. gen. et nov. sp. *Staz. Sper. Agric. Ital.* 49:583-596.
From genus *Chalara* < Gr. *chalára*, chain + L. suf. *-opsis* < Gr. *-ópsis*, appearance, aspect, i.e., similar to the genus *Chalara*. Distinguished from *Chalara* by presence of septate aleuriospores borne individually or in short chains. Saprobic in the soil.

Cheiromoniliophora Tzean & J. L. Chen

Inc. sed., Pleosporales, Pleosporomycetidae, Dothideomycetes, Pezizomycotina, Ascomycota, Fungi
Tzean, S. S., J. L. Chen. 1990. *Cheiromoniliophora elegans* gen. et sp. nov. (Hyphomycetes). *Mycol. Res.* 94(3):424-427.
From Gr. *cheir*, genit. *cheirós*, hand + L. *monile*, necklace + Gr. *phóros*, bearing, referring to the cheiroid conidia borne in chains. Conidiophores semi-macronematous, macronematous, simple or irregularly branched, septate, smooth or verrucose. Conidiogenous cells discrete, catenulate, proximally clavate, pyriform to obovoid, distally spherical, subspherical, blastic, smooth or occasionally verrucose, hyaline to subhyaline. Conidia acrogenous or pleurogenous, cheiroid, in one plane, composed of 2-3 arms, arising from a more or less triangular basal cell. Arms multiseptate, slightly constricted at septa, consisting of 1-5 cells, brown or dark brown, smooth, thick-walled. Isolated from fallen, decaying leaves from Kukwang, Nantou, Taiwan.

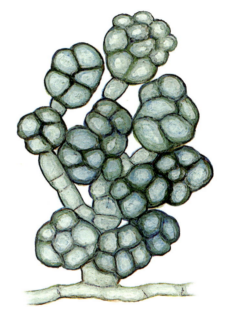

Cheiromoniliophora gracilis R. F. Castañeda, et al.: conidiophore with cheiroid conidia, on fallen leaves of *Nectandra coriacea*, Cuba, x 1,400.

Chiodecton Ach.

Roccellaceae, Arthoniales, Arthoniomycetidae, Arthoniomycetes, Pezizomycotina, Ascomycota, Fungi
Acharius, E. 1814. *Syn. meth. lich.* (Lund):108.
From Gr. *chión*, snow, *chióneos*, color of snow + *dektós*, *dektón*, pleasant, acceptable. Thallus whitish, grayish-white, reddish-white, or yellowish-white, rugose, verrucose or smooth to granulose; often cracked, with coral red border, thin or moderately thick. Apothecia small or medium sized, immersed or superficial, grouped in a stroma, with a well developed proper excipulum, brown or black. On tree bark.

Chlamydoabsidia

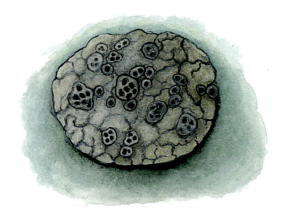

Chiodecton natalense Nyl.: crustous thallus, with apothecia, on tree bark, x 1.

Chlamydoabsidia Hesselt. & J. J. Ellis
Cunninghamellaceae, Mucorales, Inc. sed., Inc. sed., Mucoromycotina, Zygomycota, Fungi
Hesseltine, C. W., J. J. Ellis. 1966. Species of *Absidia* with ovoid sporangiospores. I. *Mycologia* 58(5):761-785.
From Gr. *chlamýs, chlamýdos*, cape, clothes, covering + the genus *Absidia* Tiegh. (< L. *absidis*, joint or keystone of an arch + L. suf. *-ia*, which denotes quality or state of). Similar to *Absidia* but forming large, dark chlamydospores at apices of aerial hyphae. Isolated from roots of pea.

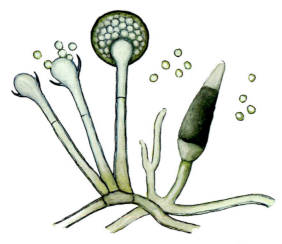

Chlamydoabsidia padenii Hesselt. & J. J. Ellis: apophysate sporangiophore, sporangiospores, and dark chlamydospore, x 350. Described in Washington, U.S.A.

Chlamydopus Speg.
Agaricaceae, Agaricales, Agaricomycetidae, Agaricomycetes, Agaricomycotina, Basidiomycota, Fungi
Spegazzini, C. 1898. Fungi Argentini novi vel critici. *Anal. Mus. nac. Hist. nat. B. Aires* 6:189.
From Gr. *chlamýs*, genit. *chlamýdos*, cloak, cape, mantle + *poús*, foot, referring to the thick stipe of the fruitbody. Basidiocarps consisting of a long stem with a two-layer peridium upon its dilated apex. Exoperidium fragile, breaking away; endoperidium tough, membranous, persistent, dehiscing by an apical pore that enlarges with age. Stem solid, enlarged, somewhat inflated apically, with a fibrillose cupulate "volva". Gleba of spores and capillitium of simple or sparingly branched threads intermixed with clusters of persistent fasciculate basidia. Basidia bearing apically 1-4 spores on short sterigmata. Solitary in sandy soils of Australia, North and South America.

Chlamydozyma Wick.—see **Metschnikowia** T. Kamienski
Metschnikowiaceae, Saccharomycetales, Saccharomycetidae, Saccharomycetes, Saccharomycotina, Ascomycota, Fungi
Wickerham, L. J. 1964. A preliminary report on a perfect family of exclusively protosexual yeasts. *Mycologia* 56:257.
From Gr. *chlamýs*, genit. *chlamýdos*, clothing, covering, exterior membrane, sheath + *zýme*, ferment, yeast.

Chlorencoelia J. R. Dixon
Hemiphacidiaceae, Helotiales, Leotiomycetidae, Leotiomycetes, Pezizomycotina, Ascomycota, Fungi
Dixon, J. R. 1975. *Chlorosplenium* and its segregates. II. The genera *Chlorociboria* and *Chlorencoelia*. *Mycotaxon* 1(3):193-237.
From Gr. *chlorós*, green + genus *Encoelia* (Fr.) P. Karst. for the similarity to this genus. Apothecia superficial, solitary to gregarious, blue-green, disc shallow cupulate, becoming expanded when mature. Asci 8-spored, cylindric-clavate with long tapering stalks, inoperculate, strongly J+, apex rounded to subconic. Ascospores ellipsoid to cylindric-oblong or allantoid, 1-2-celled, hyaline. Paraphyses filiform, septate. On decayed wood.

Chlorencoelia torta (Schwein.) J. R. Dixon: mature apothecia, on dead wood, x 4.

Chlorociboria Seaver ex C. S. Ramamurthi, et al.
Chlorociboriaceae, Helotiales, Leotiomycetidae, Leotiomycetes, Pezizomycotina, Ascomycota, Fungi
Ramamurthi, C. S., et al. 1957. A revision of the North American species of *Chlorociboria* (Sclerotiniaceae). *Mycologia* 49(6):854-863.

Chlorosplenium

From Gr. *chlóros*, green + genus *Ciboria* Fuckel, i.e., a green *Ciboria*. Ascomata apothecia superficial, solitary to caespitose, fleshy to coriaceous, centrally or eccentrically stipitate; disc shallow cupulate to infundibuliform, orange-yellow to bluish-green; receptacle and stipe of same color as disc, glabrous or finely tomentose, sometimes with dark greenish pustules. Ectal excipulum of gelatinized hyphae. Hymenium of asci interspersed with paraphyses. Paraphyses filiform, with blunt apices, septate, branched near base. Asci unitunicate, cylindric to narrowly clavate, with long tapering stipe, apex rounded to subconic and amyloid, 8-spored. Ascospores fusiform to fusiform-ellipsoid or suballantoid, 1-celled or occasionally 1-septate, hyaline or greenish. Asexual morph dothiorina-like. On decayed wood to which it imparts a bright to dark green stain.

From Gr. *chlorós*, green, greenish-yellow + *splen*, spleen + NL. *-ium*. Apothecia sessile or short-stipitate, with margin upturned, yellowish. Hymenium concave, becoming olivaceous-green. Asci clavate, 8-spored. Ascospores simple, hyaline. Paraphyses filiform, slightly clavate. On wood.

Chlorophyllum agaricoides (Czern.) Vellinga (syn. **Endoptychum agaricoides** Czern.): agaricoid, semi-sessile fruit bodies; the gleba (in the sectioned fruit body) composed of superimposed gills, on arid soil, x 0.5.

Chlorociboria aeruginascens (Nyl.) Kanouse ex C. S. Ramamurthi, et al.: stipitate apothecia, on soil, among mosses, x 10.

Chlorophyllum Massee (syn. **Endoptychum** Czern.)
Agaricaceae, Agaricales, Agaricomycetidae, Agaricomycetes, Agaricomycotina, Basidiomycota, Fungi
Massee, G. E. 1898. Fungi exotici, I. *Bull. Misc. Inf., Kew* 138:113-136.

From Gr. *chlorós*, green or yellow-green + *phýllon*, leaf, lamina, for having green or greenish-gray gills, due to the color of the spores. Similar to *Lepiota* (Pers.) Gray, but differs in having white gills when young, which turn dark greenish-gray with age, due to accumulation of green spores. Fructifications agaricoid or with gleboid interior of folded, anastomosed and superimposed gills, globose or subglobose, sessile or semi-sessile, some species with a gleba and capillitium. Spores green, dark, smooth. Epigeous in meadows or open, humid places or subhypogeous in arid zones. Toxic.

Chlorophyllum molybdites (G. Mey.) Massee ex P. Syd.: basidiocarp with the ring in the stipe, and greenish gills, on soil, x

Chlorosplenium Fr.
Dermateaceae, Helotiales, Leotiomycetidae, Leotiomycetes, Pezizomycotina, Ascomycota, Fungi
Fries, E. M. 1849. *Summa veg. Scand., Sectio Post.* (Stockholm):356.

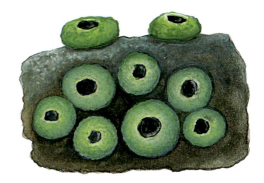

Chlorosplenium chlora (Schwein.) M. A. Curtis: greenish-yellow apothecia, on wood, x 8.

Choanephora

Choanephora Curr.
Choanephoraceae, Mucorales, Inc. sed., Inc. sed., Mucoromycotina, Zygomycota, Fungi
Currey, F. 1873. Note on *Cunninghamia infundibulifera*. *J. Linn. Soc., Bot.* 13:578.
From Gr. *choáne*, crucible, funnel, cavity of the brain, tube + *phóros*, bearer < *phéro*, to carry. Sporangia many-spored, with persistent wall that undergoes circumscissile dehiscence upon maturing, leaving infundibuliform-like receptacles. Sporangiola as heads of one-celled sporangia. Spores with polar piliform appendages. Parasitic on plants causing wet rot, especially of cucurbits or saprobic in soil.

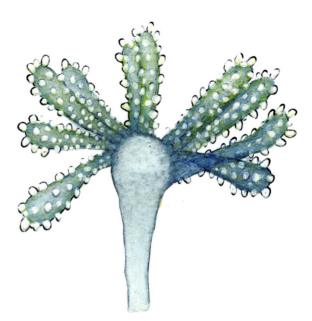

Chromelosporium fulvum (Link) McGinty, et al.: conidiophore, and conidiogenous cells covered with botryoblastospores, x 1,100.

Choanephora cucurbitarum (Berk. & Ravenel) Thaxt.: sporangiophore with heads of sporangiola, x 400.

Chromelosporium Corda
Pezizaceae, Pezizales, Pezizomycetidae, Pezizomycetes, Pezizomycotina, Ascomycota, Fungi
Corda, A. C. J. 1833. In: Sturm, *Deutschl. Fl.*, 3 Abt. (Pilze Deutschl.) 3(13):81.
From Gr. *chrôma*, color + *spóros*, spore + L. dim. suf. *-ium*. Colonies initially white, rapidly turning yellow, then cinnamon with age. Conidia botryoblastospores on ampullae. Apothecia brown, peziza-like. Common saprobe in soil, especially in sterilized flowerpots in greenhouses.

Chromocrea Seaver—see **Trichoderma** Pers.
Hypocreaceae, Hypocreales, Hypocreomycetidae, Sordariomycetes, Pezizomycotina, Ascomycota, Fungi
Seaver, F. J. 1910. The Hypocreales of North America-III. *Mycologia* 2(2):48-92 + plates XXIXXII.
From Gr. *chrôma*, color + *kreas* > L. *creas*, flesh, referring to the colored stroma.

Chromocreopsis Seaver—see **Thuemenella** Penz. & Sacc.
Xylariaceae, Xylariales, Xylariomycetidae, Sordariomycetes, Pezizomycotina, Ascomycota, Fungi
Seaver, F. J. 1910. The Hypocreales of North America-III. *Mycologia* 2(2):48-92.
From genus *Chromocrea* Seaver + Gr. suf. *-ópsis*, appearance.

Chrysomphalina Clémençon
Hygrophoraceae, Agaricales, Agaricomycetidae, Agaricomycetes, Agaricomycotina, Basidiomycota, Fungi
Clémençon, H. 1982. Kompendium der Blätterpilze Europäische omphalinoide Tricholomataceae. *Zeitschrift für Mykologie.* 48(2):195-237.
From Gr. *chrysós*, gold + *omphalós*, navel + L. suf. *-ina*, which denotes similarity, for the umbilicate or omphaloid pileus. Pileus generally brightly colored, yellow, green, purple or opaque, gray or almost black, more or less hygrophanous, smooth or scaly near center. Gills decurrent. Stipe fleshy to subcartilaginous, hollow at maturity. On sandy or rocky soil, dead wood or among mosses, occasionally lichenized.

Chrysosporium Corda
Onygenaceae, Onygenales, Eurotiomycetidae, Eurotiomycetes, Pezizomycotina, Ascomycota, Fungi
Corda, A. C. J. 1833. Deutschlands Flora, Abt. III. Die Pilze Deutschlands. 3-13:65-96.
From Gr. *chrysós*, gold + *sporá*, spore + dim. suf. *-ium*, referring the golden color of the spores. Colonies white at first, then rosy buff from the centre outwards. 1-celled, large and tuberculate conidia are formed on

undifferentiated hyphae, and may be sessile, or borne laterally on short stalks or producing in an intercalary position. Associated with bat dwellings.

Chrysomphalina chrysophylla (Fr.) Clémencon (syn. **Omphalina chrysophylla** (Fr.) Murrill): tubaeform basidiomata, on soil, x 1.

Chrysosporium chiropterorum Beguin & Larcher: terminal and lateral golden conidia, 1-celled, borne sessile or on short pedicles, and tuberculate conidia with digitiform projections, x 750. Isolated from bat guano (*Rhinolophus ferrumequinum*) in France.

Chytridium A. Braun

Chytridiaceae, Chytridiales, Chytridiomycetidae, Chytridiomycetes, Inc. sed., Chytridiomycota, Fungi

Braun, A. 1851. *Betracht. Erschein. verjüng. Natur* (Leipzig), p. 198.

From Gr. *chytrídion*, dim. of *chytrís*, earthern jar, pot, flower vase + L. dim. suf. *-ium*. Thallus epi- and endobiotic. Parasitic on marine algae or saprobic in freshwater.

Chytridium olla A. Braun: epi- and endobiotic thallus, in oogonia of a chlorophyceous alga, *Oedogonium*, in Germany, with a zoosporangium releasing uniflagellate zoospores, x 700.

Chytriomyces Karling (syn. **Amphicypellus** Ingold)

Chytriomycetaceae, Chytridiales, Chytridiomycetidae, Chytridiomycetes, Inc. sed., Chytridiomycota, Fungi

Karling, J. S. 1945. Brazilian chytrids. VI. *Rhopalophlyctis* and *Chytriomyces*, two new chitinophyllic operculate genera. *Am. J. Bot.* 32(7):362-369.

From Gr. *chytrís*, pot, earthern jar, flower vase + L. *myces*, fungus; as in other genera of chytrids, its name derives from the shape of the thallus. Thallus monocentric, extramatrical consisting of a sporangium, apophysis and rhizoidal system. Sporangium spherical, dehiscing by a subapical lid. Apophysis globose. Rhizoidal system consisting of three or four main axes arising laterally from apophysis; each main axis tapering and distally divided into numerous fine branches. Epibiotic portion forms sporangium; endobiotic portion produces rhizoidal system. Saprobic on plant and animal detritus, parasitic on diatoms.

Chytriomyces hyalinus Karling: epi- and endobiotic thallus, with a prosporangium containing young uniflagellate zoospores, x 1,000. Collected in Brazil, and Connecticut, New York & Virginia, U.S.A.

Ciboria

Ciboria Fuckel
Sclerotiniaceae, Helotiales, Leotiomycetidae, Leotiomycetes, Pezizomycotina, Ascomycota, Fungi
Fuckel, K. W. G. L. 1870. Symbolae mycologicae. Beiträge zur Kenntniss der Rheinischen Pilze. *JJb. nassau. Ver. Naturk.* 23-24:1-459.
From Gr. *kibórium*, pl. *kibória*, seed vessel of the Egyptian lotus, drinking cup > L. *ciborium*, drinking cup, for the shape of the ascomata. Ascomata stipitate apothecia arising from a mummiform sclerotium. Apothecia cupulate to shallow saucer-shaped, up to 14 mm diam, usually brown, but rarely red, yellow or whitish; stipe slender, up to 30 mm long. Paraphyses filiform, with slightly enlarged tips. Asci unitunicate, cylindric or subclavate, inoperculate, pore bluing in iodine, 4-8-spored; ascospores 1-celled, hyaline, ellipsoid to subellipsoid, sometimes inaequilateral, smooth or minutely roughened. On flowers and fruits of flowering trees; apothecia on fallen, overwintered tissues.

Ciboria coryli (Schellenb.) N. F. Buchw.: cupulate apothecia, developed from a mummiform sclerotia on plant tissues, x 2.

Ciboriella Seaver—see **Hymenoscyphus** Gray
Helotiaceae, Helotiales, Leotiomycetidae, Leotiomycetes, Pezizomycotina, Ascomycota, Fungi
Seaver. F. J. 1951. *North American Cup-fungi* (Inoperculates) New York, 428 pp.
From genus *Ciboria* Fuckel + L. dim. suf. *-ella*.

Ciborinia Whetzel
Sclerotiniaceae, Helotiales, Leotiomycetidae, Leotiomycetes, Pezizomycotina, Ascomycota, Fungi
Whetzel, H. H. 1945. A synopsis of the genera and species of the Sclerotiniaceae, a family of stromatic, inoperculate discomycetes. *Mycologia* 37(6):648-714.
From Gr. *kibórium*, pl. *kibória*, seed vessel of the Egyptian lotus, drinking cup > L. *ciborium*, drinking cup + euphonic connective *n* + des. *-ia*, for the shape of the ascomata. Ascomata stipitate apothecia, arising from a sclerotium; sclerotium discoid, black, circular, subcircular to occasionally elongated, thin and flat. Apothecia cupulate to saucer-shaped or discoid, up to 10 mm diam, yellow-brown to brown, reddish or whitish. Paraphyses filamentous, usually swollen at apex. Asci unitunicate, cylindrical to clavate or subclavate, inoperculate, with amyloid apical ring, 8-spored. Ascospores 1-celled, hyaline, ellipsoid to fusoid. Parasitic in living leaves of vascular plants, with apothecia forming on fallen sclerotia.

Ciboriopsis Dennis—see **Moellerodiscus** Henn.
Sclerotiniaceae, Helotiales, Leotiomycetidae, Leotiomycetes, Pezizomycotina, Ascomycota, Fungi
Dennis, R. W. G. 1962. New or interesting British Helotiales. *Kew Bull.* 16(2):317-327.
From genus *Ciboria* Fuckel + Gr. *-ópsis*, denoting appearance, likeness, i.e., a fungus similar in appearance to *Ciboria*.

Circinella Tiegh. & G. Le Monn.
Syncephalastraceae, Mucorales, Inc. sed., Inc. sed., Mucoromycotina, Zygomycota, Fungi
van Tieghem, P., G. Le Monnier. 1873. Recherches sur les mucorinées. *Annls. Sci. Nat., Bot., sér.* 5, 17:261-399.
From L. *circino* < *circinus*, circle, to make round, to put in the shape of a circle + dim. suf. *-ella*. Sporangiophores with circinate branches, often umbellate, bearing sporangia at tips. Saprobic in soil, dung and seeds, especially Brazil nuts.

Circinella umbellata Tiegh. & G. Le Monn.: sporangiophore, with circinate branches terminating on sporangia, x 350.

Citeromyces Santa María
Saccharomycetaceae, Saccharomycetales, Saccharomycetidae, Saccharomycetes, Saccharomycotina, Ascomycota, Fungi

Cladosporium

Santa María, J. 1957. Un nuevo género de levaduras: *Citeromyces*. *Bol. Inst. Nac. Invest. Agron.* 17:269-276.
From L. *citer, citerior*, denoting proximity in time + *myces* < Gr. *mýkes*, fungus, referring to the hypothesis that this genus of yeasts is of recent origin. Multilateral budding, without pseudohyphae or true mycelium. Asci globose with a thick wall and a single warted ascospore. Cultures brown or dark brown, with abundant sporulation. Capable of assimilating nitrate and nitrite and of vigorously fermenting some sugars. In sweetened condensed milk and fruits preserved in syrup.

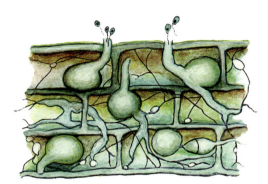

Cladochytrium tenue Nowak.: polycentric thallus, with the rhizoidal system that forms sporangia and resting spores, in decomposing plant tissues of *Acorus calamus* and *Iris pseudacorus*, in Poland, x 380.

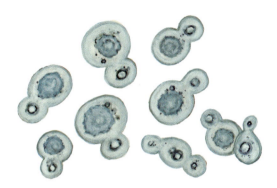

Citeromyces siamensis Nagats., et al.: vegetative budding cells, and asci containing a single warted ascospore, x 1,700. Isolated from dry salted sepia, in Thailand.

Cladonia P. Browne
Cladoniaceae, Lecanorales, Lecanoromycetidae, Lecanoromycetes, Pezizomycotina, Ascomycota, Fungi
Browne, P. 1756. *Prim. fl. Holsat.* (Kiliae):90.
From Gr. *kládos*, branch, bud, shoot + L. suf. *-ia*, which denotes quality or state of a thing. Primary thallus small, squamulose, persistent or fugacious. Secondary thallus of more or less branched, hollow podetia. Apothecia varied in shape and size, borne on terminal part of branches or cups of podetia. Mainly on soil, often with mosses, and base of trees, old wood and rocks.

Civisubramaniania Vittal & Dorai
Inc. sed., Inc. sed., Inc. sed., Inc. sed., Pezizomycotina, Ascomycota, Fungi
Vittal B. P. R., M. Dorai. 1986. *Civisubramaniania eucalypti* gen. et sp. nov. *Trans. Br. mycol. Soc.* 87(3):482-485.
Dedicated to the Indian mycologist *C. V. Subramanian* + L. des. *-ia*, for his significant contributions to knowledge of hyphomycetes. Colonies pulvinate, consisting of numerous macronematous, branched, verrucose, closely undulating conidiophores compact at base, intermixed with helically twisted brown sterile hyphae. Conidiogenous cells discrete, clustered at tips of conidiophores; fertile conidiogenous cells flask-shaped with long necks bearing conidia from different loci at tip. Conidia holoblastic, solitary, dry, hyaline, ellipsoidal to subspherical, echinulate. Collected on dead leaves of *Eucalyptus tereticornis*, in Vandalur, Madras, India.

Cladochytrium Nowak.
Cladochytriaceae, Chytridiales, Chytridiomycetidae, Chytridiomycetes, Inc. sed., Chytridiomycota, Fungi
Nowakowski, L. 1877. Beitrag zur Kenntniss der Chytridiaceen. *In*: J. F. Cohn, *Beitr. Biol. Pfl.* 2:92.
From Gr. *kládos*, branch + *chytrís*, kettle, receptacle + L. dim. suf. *-ium*. Thallus polycentric, rhizoidal system with many branches producing sporangia and resting spores. On decomposing plant tissues, as well as gelatinous covering of aquatic algae.

Cladonia cristatella Tuck.: podetia (secondary thallus) with reddish apothecia, on soil, among mosses, x 12.

Cladosporium Link (syn. **Hormodendrum** Bonord.)
Cladosporiaceae, Capnodiales, Dothideomycetidae, Dothideomycetes, Pezizomycotina, Ascomycota, Fungi
Link, H. F. 1816. Observationes in ordines plantarum naturales. *Mag. Gessell. naturf. Freunde*, Berlin 7:37.
From Gr. *kládos*, branch, shoot + *spóros*, spore + L. dim. suf. *-ium*, i.e., a little branch with spores. Conidiophores

Claroideoglomus

erect, pigmented, dendritically branched, producing acropetal chains of blastoconidia with dark, attachment scars on ends where connected to other conidia, ramoconidia or conidiophores. Common saprobe on organic remains, pathogenic on plants and animals, and in humans may cause respiratory allergies.

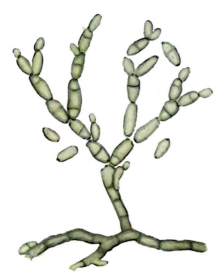

Cladosporium cladosporioides (Fresen.) G. A. de Vries: conidiophore with acropetal chains of blastoconidia, x 900.

Claroideoglomus C. Walker & A. Schüßler (syn. **Albahypha** Oehl, et al.)
Claroideoglomeraceae, Glomerales, Inc. sed., Glomeromycetes, Glomeromycotina, Glomeromycota, Fungi
Schüßler, A., C. Walker. 2010. The Glomeromycota. A species list with new families and new genera (Gloucester), 56 pp.
From L. clarus, clear + Gr. suf. -oidéos, like + genus Glomus Tul. & C. Tul., for the color of the subtending hyphae. Spores generally formed singly from white, funnel-shaped, subtending hyphae, hyaline to white, rarely subhyaline, wider at spore base than their width 10-20 μm distance from spore, conspicuously bill-shaped. Spores generally single, with 1-4 wall layers, pore closure at spore base often with a septum arising from structural layer, from an adherent innermost (semi-) flexible layer, or from both layers. In soil or rarely in roots.

Clasterosphaeria Sivan.
Magnaporthaceae, Magnaporthales, Sordariomycetidae, Sordariomycetes, Pezizomycotina, Ascomycota, Fungi
Sivanesan, A. 1984. Teleomorphs of Clasterosporium anomalum and C. cyperi. Trans. Br. mycol. Soc. 83(4):710-718.
From genus Clasterosporium Schwein. + Sphaeria Haller. Mycelium amphigenous, effuse, superficial, setose, hyphopodiate, composed of branched, septate, reticulate hyphae. Perithecia superficial, globose, dark brown to black, collapsing at maturity, ostiolate. Peridium two-layered. Asci 8-spored, thick-walled, unitunicate with a distinct, non-amyloid, refractive apical ring-like structure. Ascospores fusiform, 3-septate, central septum often just below middle, hyaline to subhyaline. Paraphyses hyaline, strap-shaped, septate, deliquescent. Collected on leaves of Cyperus haspan in Selangor, Malaysia.

Claroideoglomus hanlinii Błaszk., Chwat & Góralska: light yellow spores, x 330, proceeding from the roots of Phoenix dactylifera, growing in maritime dunes near Varadero, Hicacos Peninsula, Cuba.

Clathrus P. Micheli ex L. (syns. **Colonnaria** Raf., **Lindera** C. Cunn.)
Phallaceae, Phallales, Phallomycetidae, Agaricomycetes, Agaricomycotina, Basidiomycota, Fungi
Linnaeus, C. 1753. Sp. pl. 2:1179.
From L. clathrus < Gr. kleíthron, iron grating, in reference to the clathrate fructification, i.e., with the shape of a lattice work or trellis. Peridium subglobose, globose or obovate, with a thin external layer and a thick, gelatinous middle layer; receptaculum of arms united at apex that generally form a hollow, spherical reticulum. Gleba or masses of spores on internal surface of arms, mucilaginous, olivaceous, fetid. Spores elliptical, smooth. Grows in soils rich in humus or on rotten wood, usually in tropical regions.

Claussenomyces Kirschst. (syn. **Dendrostilbella** Höhn.)
Tympanidaceae, Helotiales, Leotiomycetidae, Leotiomycetes, Pezizomycotina, Ascomycota, Fungi
Kirschstein, W. 1923. Ein neuer märkischer Discomycet. Verh. Bot. Ver. Prov. Brandenb. 65:122.
The name is given in honor of the German professor Peter Claussen + Gr. mýkes > L. myces, fungus. Apothecia turbinate, pulvinate, slightly longitudinally prolonged, sessile or sub-stalked, growing in groups of numerous ascocarps. Hymenium greenish-white to blackish-green, in some cases reddish-brown. Outer surface smooth,

concolorous with the hymenium, margin regular. Flesh jelly and elastic, deep greenish or brownish. In some species apothecia are found together with their conidial stages (*Dendrostilbella* Höhn.). Asci clavate, inoperculate, 8-spored. Ascospores fusiform or subcylindrical, sometimes slightly curved, with 3-16(-21) more or less marked transversal septa, and sometimes one or more longitudinal ones, smooth, hyaline, irregularly arranged in the ascus. Paraphyses cylindrical, enlarged and forked in the upper part. On decaying and decorticated deciduous or coniferous wood, on cones or resinous exudates.

Clathrus ruber P. Micheli ex Pers.: clathrate fructification, with the reddish external layer, and the gelatinous internal layer where the gleba is formed, on soil, x 0.5.

Clavaria Vaill. ex L.

Clavariaceae, Agaricales, Agaricomycetidae, Agaricomycetes, Agaricomycotina, Basidiomycota, Fungi
Linnaeus, C. 1753. *Sp. pl.* 2:1182.
From L. *clava*, club, nightstick + L. suf. *-aria*, which indicates connection or possession, in reference to the club shape. Basidiomata simple or branched, of variable color, with monomitic hyphae. Spores white, smooth. Lignicolous or terricolous. Many species segregated in the genera *Clavariadelphus* Donk, *Clavicorona* Doty, *Clavulina* J. Schröt., and *Ramaria* (Fr.) Bon.

Clavaria zollingeri Lév.: branched basidioma, on soil, x 0.5.

Clavariadelphus Donk

Clavariadelphaceae, Gomphales, Phallomycetidae, Agaricomycetes, Agaricomycotina, Basidiomycota, Fungi
Donk, M. A. 1933. Revision der Niederländischen Homobasidiomycetae- Aphyllophoraceae II. *Med. Bot. Mus. Herb. R. Univ. Utrecht* 9:72.
From genus *Clavaria* Vaill. ex L. + Gr. *adelphós*, brother, for its similarity to the genus *Clavaria*. Basidiomata clavate, generally with dull colors, brown to yellow or purple. Terricolous. Edible, probably mycorrhizal. Most common are *C. pistillaris* (Fr.) Donk, frequent in oak and subtropical forests but rare in conifer forests, and *C. truncatus* (Quél.) Donk in forests of fir and oak.

Clavariadelphus ligula (Schaeff.) Donk: clavate basidiomata, on soil, x 0.5.

Claviceps Tul. (syn. Sphacelia Lév.)

Clavicipitaceae, Hypocreales, Hypocreomycetidae, Sordariomycetes, Pezizomycotina, Ascomycota, Fungi
Tulasne, L. R., 1853. Nouvelles recherches sur l'appareil reproducteur des champignons. *Annls Sci. Nat.*, *Bot.*, sér. 3, 20:43.
From L. *clavus*, *clavi*, nail, or from *clava*, club, walking stick, or nightstick, and from NL. *ceps*, derived from L. *caput*, *capitis*, head, for the club shape of the ascocarps. Stromata well developed in which perithecia immersed. Asci unitunicate, long, cylindrical, with a thickened apex perforated by a pore through which ascospores expelled, 8-spored. Ascospores hyaline, filiform. Mycelial mat on flowers produces sporodochia with unicellular conidia arranged on conidiophores. Conidia giving rise to a pseudoparenchymatous mass that is transformed into a sclerotium that replaces cereal grain head. Sclerotium falls to ground to produce stromata. Parasitizes grains including rye in which it produces the disease known as ergot caused by *C. purpurea* (Fr.) Tul. with an asexual morph as synonym *Sphacelia segetum* Lév.

Clavicorona

Claviceps purpurea (Fr.) Tul.: pedicellate stromata, with embedded perithecia, growing from a sclerotium formed in a rye grain, x 50.

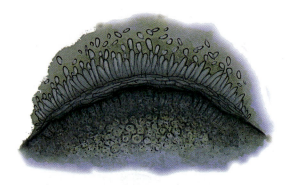

Claviceps purpurea (Fr.) Tul. (syn. **Sphacelia segetum** Lév.): sporodochium with numerous unicellular, hyaline conidia, x 250. The sporodochia are plant pathogens that cause gangrene or inflammation in the heads of the cereals that they parasitize.

Clavicorona Doty

Auriscalpiaceae, Russulales, Inc. sed., Agaricomycetes, Agaricomycotina, Basidiomycota, Fungi
Doty, M. S. 1947. Proposals and notes on some genera of clavarioid fungi and their types. *Lloydia* 10:38-44.
From L. *clava*, club, nightstick, from this the pref. *clavi-*, + *corona*, crown, for the coronate apex of the small branches or clubs. Basidiomata with few or many branches, of light colors. Grows on decayed wood in oak and subtropical forests. The peculiar *Clavicorona pyxidata* (Fr.) Doty produces branched fructifications that end in miniature crowns.

Clavispora Rodr. Mir.

Metschnikowiaceae, Saccharomycetales, Saccharomycetidae, Saccharomycetes, Saccharomycotina, Ascomycota, Fungi
Rodrigues de Miranda, L. 1979. *Clavipsora*, a new yeast genus of the Saccharomycetales. *Antonie van Leeuwenhoek* 45:479-483.
From L. *clava*, club + *spora*, < Gr. *sporá*, spore, for the clavate ascospores. Cells reproduce by multipolar budding; pseudomycelium often formed. Conjugation of mating types precedes ascus formation. Ascospores clavate, smooth, 1-4 per ascus, sometimes containing a small oil droplet, easily liberated from ascus. Metabolism oxidative and fermentative. External vitamin source not required. Isolated from citrus peel juice in Israel.

Clavicorona subechinulata Gardezi: basidioma with many branches, on soil in coniferous and hardwood forests, in Jammy-Kashmir, India, x 0.5.

Clavispora lusitaniae Rodr. Mir.: an ascus with four clavate and verrucose ascospores, x 7,000.

Clavochytridium Couch & H. T. Cox (syn. **Blastocladiella** V. D. Matthews)

Blastocladiaceae, Blastocladiales, Inc. sed., Blastocladiomycetes, Inc. sed., Chytridiomycota, Fungi
Couch, J., H.T. Cox. 1939. *J. Elisha Mitchell scient. Soc.* 55(2):389.
From L. *clava*, club + genus *Chytridium* A. Braun, for the club-like appearance of the sporangiophores. Thallus monocentric, eucarpic, with an intramatrical rhizomycelium and an extramatrical zoosporangium. Zoosporangia sessile or stalked, with one or more exit papillae. Zoospores posteriorly uniciliate, with two to several refractive bodies, emerging fully formed and swimming after a few seconds. Rhizoidal system well developed, sometimes septate or constricted, delimited from the zoosporangium by a cross wall at maturity.

Clitopilus

Clavochytridium simplex Doweld: sporangiophores with sporangia, x 50. Described from dead fly in pool of fresh water, Virginia, U.S.A.

Clavulina J. Schröt.
Clavulinaceae, Cantharellales, Inc. sed., Agaricomycetes, Agaricomycotina, Basidiomycota, Fungi
Schröter, J. 1889. Die Pilze Schlesiens. *In*: J. F. Cohn, *Krypt.-Fl. Schlesien (Breslau)*3.1(25-32):442.
From L. *clavula*, club, or nightstick of small size + des. -*ina*, which indicates similarity or relation. Basidiomata similar to *Clavaria*, simple or branched. Terricolous, rarely lignicolous in temperate and tropical forests. Several species edible, most common *C. cinerea* (Fr.) J. Schröt. in conifer and oak forests, fructifications with large branches, ash-gray or violaceous, with a short, grayish-white base.

Clavulina amethystina (Bull.) Donk: multibranched basidioma, on soil, among mosses, x 1.

Climacocystis Kotl. & Pouzar
Polyporaceae, Polyporales, Inc. sed., Agaricomycetes, Agaricomycotina, Basidiomycota, Fungi

Kotlaba F., Z. Pouzar. 1958. Polypori novi vel minus cogniti Cechoslovakiae III., *Ceská Mykol.* 12(2):103.
From Gr. *klímax*, genit. *klimátos*, a ladder, slope + *kýstis*, a bladder, pouch. Basidiocarps annual, whitish, dimidate, semicircular, without a stipe, pulvinate, hispid or floccose at first, then smooth, with whitish, separable tubes. Monomitic hyphal system with clamp connections, acute, thick-walled and ventricose cystidia, spores broadly ellipsoid hyaline, smooth and non-amyloid.Causing white rot on living or dead deciduous trees. It is widely distributed in the northern hemisphere.

Climacocystis borealis (Fr.) Kotl. & Pouzar (syn. **Spongipellis borealis** (Fr.) Pat.): mature basidiome with a spongy pileus, on soil, Sweden, x 0.3.

Clitocybe (Fr.) Staude
Tricholomataceae, Agaricales, Agaricomycetidae, Agaricomycetes, Agaricomycotina, Basidiomycota, Fungi
Kummer, P. 1871. *Führ. Pilzk.* (Zerbst):26.
From Gr. *klitós*, side, slope, incline, inclination, hillside + *kýbe*, head. Basidiomata solitary, cespitose or forming fairy rings, white, grayish, brown, yellowish, greenish, orangish, or reddish, fleshy, regular or irregular, incurved on margin, with central depression of pileus; stipe central, fibrous on outer part; gills decurrent, rarely adnate, with an acute, curved border. Spores white. Both toxic and edible species. The edible greenish *C. odora* (Bull.) Fr. has an agreeable odor similar to anise; several with white fructifications, such as *C. cerussata* Fr. and *C. rivulosa* (Pers.) Fr., are poisonous. Common in soil, rare on wood, in forests and grasslands.

Clitopilus (Fr. ex Rabenh.) P. Kumm.
Entolomataceae, Agaricales, Agaricomycetidae, Agaricomycetes, Agaricomycotina, Basidiomycota, Fungi
Kummer, P. 1871. *Führ. Pilzk.* (Zerbst):23.
From Gr. *klitós*, inclination, slope + *pîlos*, hat, pileus, for the silouette of the pileus. Basidiomata with pileus generally depressed or infundibuliform, at times irregular

Clonostachys

or asymmetrical, similar to that of *Clitocybe*, medium to large, fleshy, gills pink due to spores. Mealy odor, a mild or bitter taste. Grows on wood or insects, in humus or in sandy soils; some species are edible.

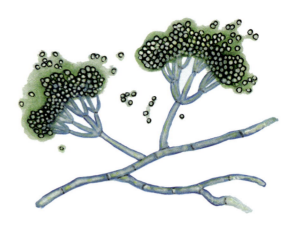

Clonostachys rosea (Link) Schroers, et al. (syn. **Gliocladium roseum** Bainier): penicillia with phialides and conidia, that accumulate in gloeoid masses on the penicillia, x 230.

Clitocybe odora (Bull.) P. Kumm.: mature basidiomata, showing laminal hymenophore, on soil, x 0.5.

Coccidiascus Chatton

Eremotheciaceae, Saccharomycetales, Saccharomycetidae, Saccharomycetes, Saccharomycotina, Ascomycota, Fungi

Chatton, E. R. 1913. *Coccidiascus legeri* n. g., n. sp., levure ascosporée parasite des cellules intestinales of *Drosophila funebris* Fabr. *Compt. Rend. Séances Mém. Soc. Biol.* 75:117-120.

From L. *coccidium* < Gr. *kokkíon*, little ball + *ascus*, wineskin, ascus. Yeast without mycelium, develops as budding cells. Asci with fusiform or helicoidal ascospores. Intestinal endoparasite of flies, *Drosophila funebris*.

Clitopilus prunulus (Scop.) P. Kumm.: basidioma with decurrent lamellae, on soil, x 1.

Coccidiascus legeri Chatton: endoparasitic asci with ascospores in the intestinal cells of the *Drosophila funebris* fly, in France, x 1,000; below, the free ascospores, x 2,000.

Clonostachys Corda (syn. **Bionectria** Speg.)

Bionectriaceae, Hypocreales, Hypocreomycetidae, Sordariomycetes, Pezizomycotina, Ascomycota, Fungi
Corda, A. C. J. 1840. *Icon. fung.* (Prague) 4:30.
From Gr. *klón*, dim. *klónion*, a branch, twig + *stáchys*, an ear of grain, spike, for the ascomata developed on a branch. Ascomata nectrioid, light colored, fleshy. Asci unitunicate, 8-spored. Ascospores generally 1-septate, hyaline. Conidiophores branching to form penicillia; phialides produce glutinous phialoconidia that accumulate in gloeoid masses. Sexual morph bionectria-like. Fungicolous, used in biological control, also common in soil.

Coccidioides G. W. Stiles

Onygenaceae, Onygenales, Eurotiomycetidae, Eurotiomycetes, Pezizomycotina, Ascomycota, Fungi
Stiles, G. W. 1896. *In*: Rixford & Gilchrist, *Rep. Johns Hopkins Hosp.* 1:243.
From L. *coccidium* < Gr. *kokkíon*, little ball + L. suf. *-oides* < Gr. *-oeídes*, similar to, i.e., similar to the coccidian. Spheres with endospores in its parasitic phase initially

Coccomyces

interpreted as sporiferous structures of coccidia, sporozoans (protozoans) that live as endocellular parasites of animals. Enteroarthric conidia in its saprobic mycelial phase. Causal agent of coccidioidomycosis of humans and other animals, an infection that begins in respiratory tract and lungs, but can become systemic.

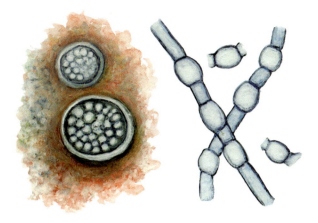

Coccidioides immitis G. W. Stiles: parasitic phase showing the spheres with endospores, x 350; on the right, the saprobic phase with enteroarthric conidia, x 1,100.

Coccodiella Hara

Phyllachoraceae, Phyllachorales, Inc. sed., Sordariomycetes, Pezizomycotina, Ascomycota, Fungi

Hara, K. 1910. *Bot. Mag.*, Tokyo 25:224.

From Gr. *kókkos*, grain, berry > L. *coccus*, abl. *cocco* + connective *-di-* + dim. suf. *-ella*, because the ascomata resemble tiny berries. Stromata hypophyllous, initially subepidermal, becoming erumpent, appearing superficial, solitary or aggregated in clusters, pulvinate to hemispherical, attached to host tissues by a basal hypostroma. Tissues of stroma pseudoparenchymatous, fleshy to subcoriaceous. Surface of stroma bearing one to usually many perithecial lobes, each containing a single perithecium. Exterior of stroma blackish-brown, internal tissues dark yellow to yellow-orange, exposed when perithecial lobes removed. Perithecia immersed in stroma, subglobose to obpyriform, with an ostiolar papilla or short neck lined with periphyses. Perithecial wall composed of several layers of flattened cells. Centrum containing numerous filamentous, septate paraphyses. Asci unitunicate, clavate to cylindrical, short-stipitate, with an undifferentiated apex, J-, 8-spored. Ascospores 1-celled, hyaline, sometimes brownish with age, ellipsoidal, oblong or fusoid, smooth. Asexual morph hemidothis-like. Parasitic on living leaves, especially of Lauraceae and Melastomataceae in the tropics.

Coccomyces De Not.

Rhytismataceae, Rhytismatales, Leotiomycetidae, Leotiomycetes, Pezizomycotina, Ascomycota, Fungi

De Notaris, G. J. 1847. Prime linee di una nuova disposizione dei Pirenomiceti Isterini. *G. bot. ital.* 2:38.

From Gr. *kókkos*, kernel, grain > L. *coccus* + Gr. *mýkes* > L. *myces*, fungus, an apparent reference to the shape of the ascoma. Ascomata apothecia, immersed in an intramatrical stroma, in some species a bleached area bordered by a black line. Stromata orbicular to angular, consisting of layers of carbonized hyphae with thickened walls, often pseudoparenchymatous, opening irregularly or along preformed lines of dehiscence, covering layer splitting into 3-5 teeth. Apothecium composed of an hymenium of asci and paraphyses resting on a subhymenium and several layers of pseudoparenchymatous cells. Disc concave to convex, usually yellowish, darkening on drying. Paraphyses simple, apices inflated, sometimes branched or circinate, tips free or surrounded by a gelatinous epithecium. Asci thin-walled, cylindrical and subsessile or clavate and long-stalked, with a pointed apex not blue in iodine, 4-8-spored. Ascospores hyaline, cylindrical, fusiform or filiform, usually with a thin gelatinous sheath, 1-celled or septate, smooth. Asexual morph leptothyrium-like. Saprobic on twigs, bark, or wood, or parasitic on leaves and stems, usually of dicotyledonous plants.

Coccodiella arundinariae Hara: hypogeous stromata on the leaf of *Dendrocalamus latiflorus*, x 1; sagittal section of perithecia, x 40, ascus with eight ascospores, x 380.

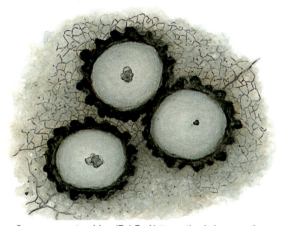

Coccomyces tumidus (Fr.) De Not.: apothecia immersed in an intramatrical stroma, x 13.

Cochliobolus

Cochliobolus Drechsler—see **Bipolaris** Shoemaker
Pleosporaceae, Pleosporales, Pleosporomycetidae, Dothideomycetes, Pezizomycotina, Ascomycota, Fungi
Drechsler, C. 1934. Phytopathological and taxonomic aspects of *Ophiobolus*, *Pyrenophora*, *Helminthosporium* and a new genus, *Cochliobolus*. Phytopathology 24:973.
From Gr. *kochlías*, snail, spiral + *bállo* > L. *bolus*, to throw, discharge.

Cochlonema Drechsler
Cochlonemataceae, Zoopagales, Inc. sed., Inc. sed., Zoopagomycotina, Zygomycota, Fungi
Drechsler, C. 1935. Some conidial phycomycetes destructive to terricolous amoebae. Mycologia 27(1):18-19.
From Gr. *kóchlos*, *kóchlon*, snail + *néma*, thread, filament, i.e., a filament in the shape of a snail. Vegetative hyphae of this predatory fungus of terricolous amoebae endozoic, thick and compactly coiled. When the host cell dies, sporangiophores and zygophores emerge through membrane of amoeba giving rise to multispored merosporangia and zygospores.

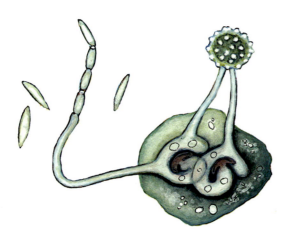

Cochlonema euryblastum Drechsler: vegetative hyphae inside a terricolous amoeba, giving rise to sporangiophores and zygophores, x 570. Described in Virginia, U.S.A.

Codinaeopsis Morgan-Jones
Chaetosphaeriaceae, Chaetosphaeriales, Sordariomycetidae, Sordariomycetes, Pezizomycotina, Ascomycota, Fungi
Morgan-Jones, G. 1976. Notes on hyphomycetes. X. *Codinaeopsis* gen. nov. Mycotaxon 4(1):166-170.
From genus *Codinaea* Maire + Gr. suf. *-ópsis*, aspect. Colonies broadly effuse, appressed, whitish to brown, setose. Mycelium partly superficial, partly immersed in substrate. Hyphae septate, branched, smooth, subhyaline to pale brown. Conidiophores macronematous, mononematous, erect, straight or slightly curved, septate, smooth, dark brown with paler apices, bearing short encircling collar hyphae. Conidiogenous cells borne terminally and laterally on collar hyphae or terminally on main axis, cylindrical, monophialidic or polyphialidic, with collarettes. Conidia 1-celled, cylindrical, curved, hyaline, with a basal scar, bearing a single setula at each end, aggregated in a mucilaginous mass. Collected on dead stems of *Rubus* in Alabama, U.S.A.

Coelomomyces Keilin
Coelomomycetaceae, Blastocladiales, Inc. sed., Blastocladiomycetes, Inc. sed., Chytridiomycota, Fungi
Keilin, D. 1921. On a new type of fungus: *Coelomomyces stegomyiae* n. g., n. sp., parasitic in the body-cavity of the larva of *Stegomyia scutellaris* Walker (Diptera, Nematocera, Culicidae). Parasitology 13:225-234.
From Gr. *koilóma*, coeloma, cavity < *koîlos*, hollow + suf. *-oma*, which is used for the generalization of a concept + *mýkes* > L. *myces*, fungus. Obligate parasite that lives inside cavity of body or coeloma of insects, e.g., the larvae of mosquitos.

Coelomomyces lativittatus Couch & H. R. Dodge ex Couch: sporangia with uniflagellate zoospores, inside the insect coeloma, x 1,000. Described in Georgia, U.S.A.

Coemansia Tiegh. & Le Monn.
Kickxellaceae, Kickxellales, Inc. sed., Inc. sed., Kickxellomycotina, Zygomycota, Fungi
van Tieghem, P., G. Le Monnier. 1873. Recherches sur les mucorinées. Annls. Sci. Nat., Bot., sér. 5, 17:392.
Dedicated to the Belgian botanist E. Coemans + L. suf. *-ia*, which denotes belonging to. Sporangiophores erect with lateral sporocladia forming pseudophialides on lower surface. Each pseudophialide forms a single unispored sporangiolum (merosporangium) on lower surface. On soil, dung and occasionally dead insects.

Cokeromyces Shanor
Mycotyphaceae, Mucorales, Inc. sed., Inc. sed., Mucoromycotina, Zygomycota, Fungi
Shanor, L., et al. 1950. A new genus of the Choanephoraceae. Mycologia 42(2):271-278.
Dedicated to the American botanist W. C. Coker + Gr. *mýkes* > L. *myces*, fungus. Sporangiola columellate on contorted pedicels originatiing from a fertile vesicle. Saprobic on rabbit dung.

Colletotrichum

Coemansia spiralis Eidam: sporangiophore with lateral sporocladia, x 500; sporocladium with merosporangia, x 2,000.

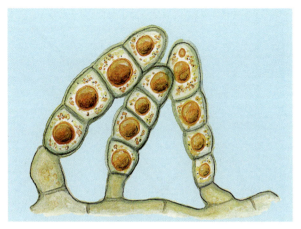

Coleosporium plumeriae Pat.: multicellular teliospores, on the underside of the leaves of *Plumeria alba*, in Guadeloupe, x 1,000.

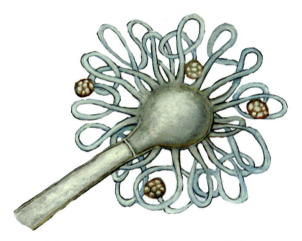

Cokeromyces recurvatus Poitras: sporangiophore with a terminal vesicle, which originates sporangiola on contorted pedicels, x 600. Described in Illinois, U.S.A.

Coleosporium Lév.

Coleosporiaceae, Pucciniales, Inc. sed., Pucciniomycetes, Pucciniomycotina, Basidiomycota, Fungi
Léveillé, J. H. 1847. *Ann. Sci. Nat. Bot., sér. 3*,8:373.
From Gr. *koleós*, covering, sheath + *sporá*, spore + L. dim. suf. *-ium*. Teliospores uni- or pluricellular, with crusty layers and a thin, pale wall with thick, gelatinous layer on the apical portion. Heteroecious subepidermal rusts with aecia on leaves of *Pinus* (some autoecious) and telia on angiosperms, such as Apocynaceae, Asteraceae, Begoniaceae, and Caprifoliaceae.

Collema Weber ex F. H. Wigg.

Collemataceae, Peltigerales, Lecanoromycetidae, Lecanoromycetes, Pezizomycotina, Ascomycota, Fungi
Wiggers, F. H. 1780. *Prim. fl. Holsat* (Kiliae):89.
From Gr. *kóllema*, that which is stuck or glued < *kólle* or *kólla*, glue, gum < *kolláo*, to glue, paste + L. suf. *-ma*. Thallus gelatinous due to photobiont algae with viscous sheaths of *Nostoc*. Dendricolous, saxicolous and terricolous, frequently associated with mosses.

Colletotrichum Corda (syns. Glomerella Spauld. & H. Schrenk, Peresia H. Maia)

Glomerellaceae, Glomerellales, Hypocreomycetidae, Sordariomycetes, Pezizomycotina, Ascomycota, Fungi
Corda, A. C. I. 1831. *In: Sturm, Deutschl. Fl., 3 Abt.* (Pilze Deutschl.) 3(12):41.
From Gr. *kolletós*, stuck together, agglutinated < *kólle*, *kólla*, gelatin, glue + *thríx*, genit. *trichós*, hair, bristle + L. *-um*. Mycelium intramatrical, hyphae olivaceous, septate, irregularly branched. Fructifications acervular, subcuticular or subepidermal, circular, flat, separate or confluent, with long, dark brown bristles mixed with compact conidiophores. Conidiophores of two types: cylindrical, parallel, short, 1-celled, simple, hyaline; also long, setose, brownish-black, with a subglobose apex, obtuse or lanceolate, septate, straight or slightly curved, simple. Conidia hyaline, 1-celled, acrogenous, ellipsoid, ovoid, or cylindrical, catenulate or not. Perithecia in groups within a stroma, often poorly developed, glomerella-like. Causing anthracnose and decay on numerous wild and cultivated plants.

Colletotrichum gloeosporioides (Penz.) Penz. & Sacc. (syn. Glomerella cingulata (Stoneman) Spauld. & H. Schrenk): perithecia within a stroma, x 225.

Collybia

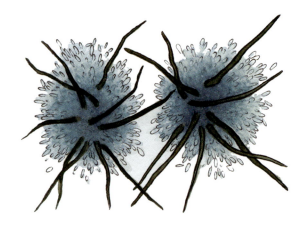

Colletotrichum kahawae J. M. Waller & Bridge: acervuli with dark brown bristles, and conidia, on berries of *Coffea arabica*, Kenya, x 200.

Collybia (Fr.) Staude
Tricholomataceae, Agaricales, Agaricomycetidae, Agaricomycetes, Agaricomycotina, Basidiomycota, Fungi
Staude, F. 1857. *Schwämme Mitteldeutschl.* 28:xxviii,119.
From Gr. *kóllybos*, a small coin, due to the shape and size of the pileus. Basidiomata generally small or medium, solitary or cespitose, discoidal, convex to campanulate at first, but later extended, at times umbonate, with an involute margin in young specimens. Stipe central, cartilaginous, fibrous, hollow. Gills adnate, adnexed or free. Spores white, rarely yellowish, greenish or brownish-red. Grows on soil or wood in forests and grasslands region. Several edible species, but not highly prized. Certain species, such as *C. dryophila* (Bull. ex Fr.) Quél., form "fairy rings" in grasslands or in clearings in forests.

Colonnaria Raf.—see **Clathrus** P. Micheli ex L.
Phallaceae, Phallales, Phallomycetidae, Agaricomycetes, Agaricomycotina, Basidiomycota, Fungi
Rafinesque, C. S. 1808. Prospectus of Mr. Rafinesque Schmaltz's Two Intended Works on North American Botany. *Med. Repos.*, ser. 2, 5:355.
From L. *columna > colonna*, column + suf. *-aria*, to contain or have something.

Colpoma Wallr.
Rhytismataceae, Rhytismatales, Leotiomycetidae, Leotiomycetes, Pezizomycotina, Ascomycota, Fungi
Wallroth, C. F. W. 1833. *Fl. crypt. Germ.* (Norimbergae) 2:422.
From Gr. *kólpos*, bosom, bay, gulf > L. *colpos*; the relevance of the name is unclear. Ascomata narrow, elongate apothecia, immersed in host tissues, often parallel to long axis of substrate, becoming erumpent at maturity, folding back overlying host tissues. Apothecia black, oblong to oblong-elliptical or oval, often curved or branched, opening by a longitudinal slit to expose hymenium. Exterior margin consisting of a blackened clypeus composed of host cells permeated by hyphae, intergrading into layers of hyaline cells on interior. Hymenium yellow, composed of asci and paraphyses. Paraphyses filiform, unbranched, aseptate, coiled or contorted at apex. Asci clavate to cylindrical, unitunicate, inoperculate, undifferentiated at apex, 8-spored. Ascospores filiform, rounded at upper end and tapered below, hyaline, 1-celled, but sometimes forming septa late, surrounded by a thin gelatinous sheath, fasciculate in ascus. Asexual morphs conostroma-like. On woody stems and branches.

Colpoma quercinum (Pers.) Wallr.: hysterothecia, growing on woody substrates, x 10.

Coltricia Gray (syn. **Polystictus** Fr.)
Hymenochaetaceae, Hymenochaetales, Inc. sed., Agaricomycetes, Agaricomycotina, Basidiomycota, Fungi
Gray, S.F. 1821. *A natural arrangement of British plants.* 1:1-82.
Italian name of mushroom. Basidiocarps annual, stipitate, soft and tough when fresh, hard and brittle when dry; pileus yellowish to deep rusty brown, tomentose to silky with appressed hairs. Pore surface cinnamon to rusty brown, pores angular, medium to large, Stipe usually central and concolorous with the pileus. All parts of basidiocarps black with KOH. Hyphal system monomitic, generative hyphae with simple septa, hyaline to pale rusty brown, narrow to wide with thickened walls, setae absent generally. Spores cylindrical to ellipsoid, at maturity golden yellow to rusty brown. On the ground or well decayed wood.

Completoria Lohde
Completoriaceae, Entomophthorales, Inc. sed., Inc. sed., Entomophthoromycotina, Zygomycota, Fungi

Coniochaeta

Lohde, G. 1874. Über einige neue parasitische Pilze. *Tagebl. Versamm. dt. naturf. Ärzte* (Breslau) 47:203-206. From L. *completoris*, that which completes, completed, to complete, to finish. Intracellular parasite of fern prothalli destroys cellular contents filling each infected cell with a compact group of hyphal bodies from which emerge sporangiophores with forcibly discharged sporangiola.

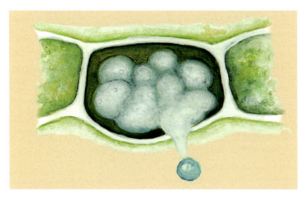

Completoria complens Lohde: intracellular hyphal bodies, parasitic of fern prothallus of *Aspidium falcatum*, Poland; one of these hyphal body gives out a sporangiophore with sporangiolum, x 550.

Condylospora Nawawi

Inc. sed., Inc. sed., Inc. sed., Inc. sed., Pezizomycotina, Ascomycota, Fungi

Nawawi, A. 1976. *Condylospora* gen.nov., a hyphomycete from a foam sample. *Trans. Br. mycol. Soc.* 66(2): 363-365. From Gr. *kondylos*, knuckle > L. *condylos* + *spora*, spore, referring to the knuckle-like bend in the conidia. Mycelium hyaline, branched, septate. Conidiophores as lateral branches of hyphae, either short or long, simple, septate, geniculate, with denticulate scars. Conidiogenous cells holoblastic, terminal, sympodial. Conidia hyaline, filiform, multiseptate, bent in middle. Isolated from foam, in Malaysia.

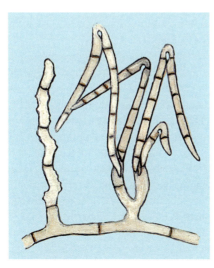

Condylospora spumigena Nawawi: aquatic conidiophore with multicellular, v-shaped, conidia, x 900. Described in Peninsular Malaysia.

Conidiobolus Bref.

Ancylistaceae, Entomophthorales, Inc. sed., Inc. sed., Entomophthoromycotina, Zygomycota, Fungi

Brefeld, O. 1884. *Conidiobolus utriculosis* und minor. *In*: J. Schröter, *Mykol. Untersuch.* 4:35-78.

From Gr. *kónis*, dust + dim. suf. *-ídion* > L. *conidium* + Gr. *bállo* > L. *bolus*, to hurl, i.e., it hurls the conidium. Sporangiola unispored, discharged violently from phototrophic sporangiophores. Saprobic in soil, dung and vegetable detritus, also parasitize insects and mammals, including humans.

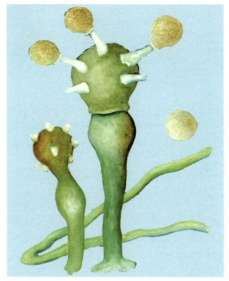

Conidiobolus coronatus (Costantin) A. Batko: sporangiophores with unispored sporangiola, x 900.

Coniochaeta (Sacc.) Cooke (syns. **Ephemeroascus** Emden, **Poroconiochaeta** Udagawa & Furuya)

Coniochaetaceae, Coniochaetales, Sordariomycetidae, Sordariomycetes, Pezizomycotina, Ascomycota, Fungi

Cooke, M. C. 1887. Synopsis *Pyrenomycetum. Grevillea* 16(77):16-19.

From Gr. *kónis*, dust, powder + *chaíte*, mane, long hair > L. *chaeta*, bristle, for the setae on the ascoma. Ascomata ostiolate or non-ostiolate perithecia, single or aggregated, confluent in stromata or non-stromatic, superficial or partially immersed, subglobose to obpyriform, brown to dark brown, glabrous or bearing brown setae or stiff hairs. Ostiole papillate, lined with periphyses. Perithecial wall thin, membranous, 2-layered of thick-walled, rounded cells. Centrum containing paraphyses. Asci unitunicate, clavate to cylindrical, stalked, short pedicellate, with a small, indistinct, non-amyloid apical ring, formed in a basal fascicle, evanescent at maturity, 8- spored, or sometimes 4-spored. Paraphyses filiform to subvesiculose, septate. Ascospores 1-celled, olivaceous brown to dark brown, ellipsoidal, discoid-lenticular or reniform, smooth or with rounded pits, with a longitudinal germ slit, often

Coniophora

inconspicuous, often surrounded by a thin, gelatinous sheath. Phialoconidia present. On wood and herbaceous stems or isolated from soil.

Coniochaeta tetraspora Cain: perithecium with setae, x 150; ascus with four ascospores, two of them with germ slit, x 800.

Coniophora DC.
Coniophoraceae, Boletales, Agaricomycetidae, Agaricomycetes, Agaricomycotina, Basidiomycota, Fungi
De Candolle, A. P., J. de Lamarck. 1815. *Fl. franç., Edn 3, 6*:34.
From Gr. *kónis*, dust, ash + suf. *-phóras* (L. *-phora*), from *phéro*, to wear, for the generally dusty to granulose hymenophore. Fructifications fleshy, waxy, subcoriaceous or membranous, resupinate, effuse. Hymenium smooth, subundulate, tuberculate or granulose. Spores elliptical, navicular or subfusiform, smooth. Lignicolous or terricolous. *Coniophora puteana* (Schum.) Karst. is a destructive species of wood whose fructification is pale yellowish or olivaceous-brown, with a white margin and a smooth hymenium, subundulate or gyrose, often subtuberculate, dusty; it causes decay of tree stumps, fallen trunks and worked wood.

Coniophora puteana (Schumach.) P. Karst.: resupinate basidiomata, on tree bark, x 1.

Conioscyphopsis Goh & K. D. Hyde
Inc. sed., Inc. sed., Inc. sed., Inc. sed., Pezizomycotina, Ascomycota, Fungi
Goh, T. K., K. D. Hyde. 1998. A new hyphomycete genus, *Conioscyphopsis*, from wood submerged in a freshwater stream and a review of *Conioscypha*. *Mycol. Res. 102*(2):308-312.
From genus *Conioscypha* Höhn. + Gr. *-ópsis*, appearance, for the similarity between the two genera. Colonies effuse, black. Mycelium mostly immersed, composed of branched, septate, smooth, subhyaline to pale brown hyphae. Stromata none. Setae and hyphopodia absent. Conidiophores micronematous. Conidiogenous cells dematiaceous, discrete, ampulliform to subglobose, thick-walled, smooth, arising laterally or terminally from hyphae, sessile or sometimes short-stalked, monoblastic; when aged, regenerating percurrently. Conidia exogenous, acrogenous, solitary, dry, dematiaceous, more or less ovate, aseptate, asetulate, thick-walled, smooth, with a thickened hilum. Conidial ontogeny holoblastic by apical wall building; conidial secession schizolytic; proliferation enteroblastic without periclinal thickening. Isolated from submerged, decayed angiosperm wood in North Queensland, Australia.

Conocybe Fayod
Bolbitiaceae, Agaricales, Agaricomycetidae, Agaricomycetes, Agaricomycotina, Basidiomycota, Fungi
Fayod, M.V. 1889. Prodrome d'une histoire naturelle des Agaricinés. *Annls Sci. Nat., Bot., sér. 7, 9*:357.
From Gr. *kónos* > L. *conus*, cone + Gr. *kýbe* > L. *cybe*, head, for the cone-shaped pileus. Pileus hygrophanous, glistening when dry. Lamellae usually at first strongly ascendant; veil none or slight on margin, none on stipe. Spore print rust color; spores smooth or faintly verruculose, verrucose in some tropical species. On mosses and grasses, or on decayed wood and dung.

Cookeina Kuntze
Sarcoscyphaceae, Pezizales, Pezizomycetidae, Pezizomycetes, Pezizomycotina, Ascomycota, Fungi
Kuntze, O. 1891. *Revis. gen. pl.* (Leipzig) 2:849.
Dedicated to the English mycologist *Mordecai Cubitt Cooke* (1825-1914) + L. suf. *-ina*, which indicates pertaining to or possession. Apothecia urnulate, with long or short pedicels, sometimes sessile, generally of brilliant red, orange or yellow colors. On fallen trunks or branches in wet jungle or tropical forests.

Coprinellus P. Karst. (syn. Ozonium Link)
Psathyrellaceae, Agaricales, Agaricomycetidae, Agaricomycetes, Agaricomycotina, Basidiomycota, Fungi
Link, J. H. F. 1809. Observationes in ordines plantarum naturales. *Mag. Ges. Naturf. Freunde* 3:21.
From Gr. *kópros*, dung, excrement + suf. dim. *-ellus*, a

small *Coprinus*. Pileipellis composed of an epithelium layer of globose cells, i.e., a cystoderm; pileus setules present or absent; veil present or absent; lamellae adnate to sub-free. On herbivorous dung, soil or wood.

Cookeina tricholoma (Mont.) Kuntze: cupuliform apothecia, on wood, x 1.

Coprinopsis P. Karst.
Psathyrellaceae, Agaricales, Agaricomycetidae, Agaricomycetes, Agaricomycotina, Basidiomycota, Fungi
Karsten, P. A. 1881. Hymenomycetes Fennici enumerati. *Acta Societatis pro Fauna et Flora Fennica.* 2(1):1-40.
From Gr. *kópros*, dung, excrement + suf. *-ópsis*, aspect, for the similarity to the genus *Coprinus*. Pileus conical to campanulate, becoming planar and revolute, margin splitting in age, plicate, fragile, shiny, almost cristalline-like, with white patches of cottony veil remnants. Stipe subbulbous, tomentose, shaggy, or woolly, sometimes glabrous with age, white. Basidiospores ellipsoid to amygdaliform, apiculus visible, with a central germ pore, smooth, dark earth to chocolate brown or nearly black. Pileipellis a cutis of radially arranged, elongated to somewhat inflated hyphae; veil always present; setules on pileus always absent. Veil composed of floccose to granular remnants that readily slough off from the pileus; annulus absent; pleurocystidia present; sterigmata not plugged. Common on lawn, sand, herbivorous dung, or wood.

Coprinus Pers.
Agaricaceae, Agaricales, Agaricomycetidae, Agaricomycetes, Agaricomycotina, Basidiomycota, Fungi
Persoon, C. H. 1797. *Tent. disp. meth. fung.* (Lipsiae):62.
From Gr. *kópros*, dung, excrement + L. suf. *-inus*, which indicates relation, possession or similarity, due to the coprophilous habitat. Basidiomata with autodeliquesing gills at maturity, disintegrating into a black ink due to mass of black spores. Veil composed of tenacious, floccose remnants that do not easily wash away from the pileus surface; annulus present; pleurocystidia absent; sterigmata plugged with gold-colored material. Grows in open places, on dung, soil, dead wood, or other fungi. Some species edible, capable of being cultivated.

Coprinopsis lagopus (Fr.) Redhead, et al. (syn. **Coprinus lagopus** (Fr.) Fr.): basidiomata, on herbivorous dung, x 1.5.

Coprotiella Jeng & J. C. Krug
Thelebolaceae, Thelebolales, Leotiomycetidae, Leotiomycetes, Pezizomycotina, Ascomycota, Fungi
Jeng, R. S., J. C. Krug. 1976. *Coprotiella*, a new cleistocarpous genus of the Pyronemataceae with ascospores possessing de Bary bubbles. *Mycotaxon* 4(2):545-550.
From genus *Coprotus* Korf & Kimbr. + L. dim. suf. *-ella*, for the similarity between the two genera. Ascocarps scattered or gregarious, non-stromatic, non-ostiolate, subglobose to globose, hyaline at first, becoming creamy or yellowish, glabrous. Peridium membranous. Asci unitunicate, nonamyloid, broadly clavate, evanescent. Ascospores 1-celled, smooth, hyaline, thin-walled, with conspicuous de Bary bubble when mature. Gelatinous sheath lacking. Isolated from horse dung from Tucuman, Argentina.

Coprotinia Whetzel
Rutstroemiaceae, Helotiales, Leotiomycetidae, Leotiomycetes, Pezizomycotina, Ascomycota, Fungi
Whetzel, H. H. 1944. A new genus of the Sclerotiniaceae. *Farlowia* 1:483-487.
From Gr. *kópros*, dung > L. *copros*, for the habitat of the fungus. Stromata not observed in nature, of indefinite form, consisting of one to several layers of black hyphae. Spermatia and conidial morph not seen. Apothecia gregarious, numerous, long, slender-stipitate, some shade of brown, relatively small. Stem hair-like, minutely roughened by ends of hyphal tips. Asci small, clavate. Ascospores 2-seriate, crowded near end of ascus, hyaline. On dung.

Cora Fr.
Inc. sed., Inc. sed., Agaricomycetidae, Agaricomycetes, Agaricomycotina, Basidiomycota, Fungi

Cordyceps

Fries, E. M. 1825. *Syst. orb. veg.* (Lundae) *1*:300.
From Gr. *kóre*, pupil of the eye, young, virgin, beautiful woman, maiden, doll, for the colorful reniform, circular or semicircular thallus. Thallus of loosely interwoven hyphae that extend over substrate and photobiont (*Scytonema*) to form a delicate pileus. Hymenium in ventral part of pileus of basidia and cystidia. On mossy soils, old wood and trees.

Cora cyphellifera Dal-Forno, et al.: thalli, on branches and twigs of trees, Ecuador, x 0.8.

Cordyceps Fr. (syn. **Phytocordyceps** C. H. Su & H. H. Wang)
Cordycipitaceae, Hypocreales, Hypocreomycetidae, Sordariomycetes, Pezizomycotina, Ascomycota, Fungi
Fries, E. M. 1824. *Observ. Mycol.* (Havniae) 2:316.
From Gr. *kordýle*, stick, club, bar + NL. *ceps*, derived from L. *caput, capitis*, head. Mycelium transformed into a pseudosclerotium composed of tissues of parasitized fungus or host. Stromata arising from masses of mycelium, clavate, yellow to orange, stalk simple, cylindric, apical fertile portion clavate. Perithecia superficial or with base scarcely immersed, pyriform, crowded in a stellate-radial arrangement. Asci unitunicate, cylindric, with an enlarged, non-amyloid, hemispheric apical cap penetrated by a fine pore, 8-spored. Paraphyses lacking. Ascospores consisting of two elongate, septate, fusoid units linked by a filiform connective, not separating into part-spores. On insects, arachnids or other fungi, rarely on seeds.

Coriolopsis Murrill
Polyporaceae, Polyporales, Inc. sed., Agaricomycetes, Agaricomycotina, Basidiomycota, Fungi
Murrill, W. A. 1905. The Polyporaceae of North America: XI. A synopsis of the brown pileate species. *Bull. Torrey bot. Club* 32(7):358.
From genus *Coriolus* Quél. + L. suf. *-opsis* < Gr. *-ópsis*, aspect, similar to, like *Coriolus*. Hymenophore thin, flexible or rigid, annual, epixylous, sessile, dimidiate, often largely resupinate; surface pale brown, zonate, hairy, margin and context thin, coriaceous to woody, pale ferruginous to almost white; hymenium concolorous, tubes small, regular, thin-walled, entire; spores smooth, hyaline. Widely distributed.

Coriolus Quél.—see **Trametes** Fr.
Polyporaceae, Polyporales, Inc. sed., Agaricomycetes, Agaricomycotina, Basidiomycota, Fungi
Quélet, L. 1886. *Enchir. fung.* (Paris):175.
From L. *corium*, leather, skin, hide (from Gr. *chórion*) + dim. suf. *-olus*, for the leathery consistency of the fructification.

Corollospora Werderm. (syn. **Peritrichospora** Linder)
Halosphaeriaceae, Microascales, Hypocreomycetidae, Sordariomycetes, Pezizomycotina, Ascomycota, Fungi
Werderman 1922. *Notizbl. Bot. Gart. Berlin-Dahlem* 8: 248.
From L. *corona*, dim. *corolla*, a crown + Gr. *sporá*, spore, > L. *spora*, for the ring of cilia around the center of the ascospore. Perithecia solitary or scattered, mostly part immersed in substratum, globose, carbonaceous, black, ostiolate; ostiole extremely small, papillate, central. Paraphyses absent. Asci fusoid or broadly fusoid, 8-spored. Ascospores fusoid, biappendiculate, hyaline becoming brownish around central septum, either 5-septate or 1-septate, with 1-3 pseudosepta on each side of septum, polytrichous around septum; terminal appendages hyaline, subcylindrical, with obtuse apices. On sand grains, shells or marine animals, or immersed test blocks.

Corollospora portsaidica Abdel-Wahab & Nagah.: vertical section of ascoma, globose, papillate, shiny black, carbonaceous, adhered to dead floating wood, x 240, in Port Said, Mediterranean Sea, Egypt; a free fusiform ascospore, 1-septate, with an appendage in each pole and an equatorial double appendages, x 830.

Corylomyces

Coronophora Fuckel
Coronophoraceae, Coronophorales, Hypocreomycetidae, Sordariomycetes, Pezizomycotina, Ascomycota, Fungi
Fuckel, K. W. G. L. 1864. *Fungi rhenani exsic., fasc.* 10:961.
From L. *corona*, crown, garland, diadem + Gr. suf. *-phóros*, bearer, from *phéro*, to carry, produce, sustain, referring to the manner in which the mature ascocarp opens. Ascocarps globose or cupulate, in small groups, lacking a true ostiole, forms a mass of cells at apex that later gelatinize, swell and disintegrate to form an opening, described as a Quellkörper (swollen body), coronate. Parasitic on branches, trunks or other pyrenomycetes.

Coronophora gregaria Fuckel: ascocarps, on bark of *Sorbus aucuparia*, x 25; the magnification of one ascocarp, x 50.

Corticium Pers.
Corticiaceae, Corticiales, Inc. sed., Agaricomycetes, Agaricomycotina, Basidiomycota, Fungi
Persoon, C. H. 1794. Neuer Versuch einer systematischen Einteilung der Schwämme. *Neues Mag. Bot.* 1:110.
From L. *cortex, corticis*, bark, outer layer + L. suf. *-ium*, which indicates connection or similarity. Basidiocarps resupinate, small, thin layer, effuse. On wood or bark of vascular plants. Some species are plant pathogens, e.g., *C. salmonicolor* Berk. & Broome, whose specific name refers to the color of the fructification, and causes pink disease of tropical plants of economic importance, such as cacao, cinchona, citrus, rubber, and tea.

Cortinarius (Pers.) Gray
Cortinariaceae, Agaricales, Agaricomycetidae, Agaricomycetes, Agaricomycotina, Basidiomycota, Fungi
Gray, S. F. 1821. *Nat. Arr. Brit. Pl.* (London) 1:627.
From L. *cortina*, curtain, veil + suf. *-aria*, which indicates connection or possession. Basidiomata with context fleshy, fibrous, with partial veil or "spiderweb" that covers gills, tends to disappear as pileus expands, leaving only an annular zone of fibers on stipe. Spore print reddish-brown. Obligately ectomycorrhizal with trees. Some species edible, others with orellanine, a compound toxic to humans and other mammals.

Corticium meridioroseum Boidin & Lanq.: resupinate and effuse basidiomata, on wood of *Pyrus amygdalus*, France, x 1.

Cortinarius terpsichores Melot: basidioma with the curtain, on soil in deciduous woodland, x 1.

Cortinellus Roze—see **Tricholoma** (Fr.) Staude
Tricholomataceae, Agaricales, Agaricomycetidae, Agaricomycetes, Agaricomycotina, Basidiomycota, Fungi
Roze, E. 1876. *Bull. Soc. Bot. Fr.* 23:50.
From genus *Cortinarius* (Pers.) Gray, this from L. *cortina*, curtain, veil + dim. suf. *-ellus*.

Corylomyces Stchigel, et al.
Inc. sed., Sordariales, Sordariomycetidae, Sordariomycetes, Pezizomycotina, Ascomycota, Fungi
Stchigel, A. M., et al. 2006. *Corylomyces*: a new genus of Sordariales from plant debris in France. *Mycol. Res.* 110(11):1361-1368.
From L. *Corylus*, genus of the hazel tree + L. *myces*, fungus, for the host on which the fungus was found. Ascomata superficial, ostiolate, tomentose, with a long neck composed of hairs. Peridium of textura angularis. Hairs hypha-like. Paraphyses and periphyses absent. Asci cylindrical, 8-spored, evanescent, short stipitate. Ascospores one- or two-celled, dark brown, lunate, bilaterally compressed, with an apical germ pore. Collected in Saint Pé de Bigorre, France, from a hazelnut (*Corylus avellana*) decomposing on soil.

Corynascus

Corynascus Arx
Chaetomiaceae, Sordariales, Sordariomycetidae, Sordariomycetes, Pezizomycotina, Ascomycota, Fungi
Arx, J. A. von. 1973. Ostiolate and nonostiolate pyrenomycetes. *Proc. K. Ned. Akad. Wet., Ser. C, Biol. Med. Sci.* 76(3):289-296.
From Gr. *koryné*, club, club-shaped + *askós*, sac > L. wine skin, ascus, for the club-shaped asci. Ascomata perithecia, solitary, superficial to immersed, globose, nonostiolate, brown, smooth. Ascomatal wall composed of pseudoparenchyma cells. Asci clavate to ellipsoidal, unitunicate, 8-spored, with deliquescent walls. Ascospores ellipsoidal to fusoid, 1-celled, brown, with a germ pore at each end. Conidial morphs chrysosporium-like.

Corynelia clavata (L. f.) Sacc.: lobate ascostromata; this species is parasitic on tissues of *Podocarpus coriaceous*, x 3.

Corynascus similis Stchigel, et al.: globose ascoma, on soil, x 400; ascus with young ascospores, and two free mature ascospores, x 1,000. Isolated from soil of Rajasthan, India.

Coryneliospora fructicola (Pat.) Fitzp.: clavate, erumpent ascostromata, on fruit tissues of *Rapanea melanophleos*, x 10.

Corynelia Fr.
Coryneliaceae, Coryneliales, Eurotiomycetidae, Eurotiomycetes, Pezizomycotina, Ascomycota, Fungi
Fries, E. M. 1818. *Observ. mycol.* (Havniae) 2:343.
From Gr. *korýne*, club, nightstick + L. suf. *-ia*, which denotes pertaining to, for the shape of its ascostromata. Ascostromata erect, often lobed, with a line of dehiscence as infundibuliform perforation, through which asci exposed. Asci clavate, pedicellate. Ascospores globose. Parasitic on *Podocarpus*, a tropical conifer.

Coryneliospora Fitzp.
Coryneliaceae, Coryneliales, Eurotiomycetidae, Eurotiomycetes, Pezizomycotina, Ascomycota, Fungi
Fitzpatrick, H. M. 1942. Revisionary studies in the Coryneliaceae. *Mycologia* 34(4):485.
From Gr. *korýne*, club, nightstick + *sporá*, spore, for the clavate shape of the ascostromata. Ascostromata erumpent through bark, undergoing dehiscence to expose ascigerous stratum. Asci clavate, pedicellate. Ascospores globose, echinulate. On bark of *Podocarpus*, tropical conifers.

Corynespora Güssow (syn. **Corynesporasca** Sivan.)
Corynesporascaceae, Pleosporales, Pleosporomycetidae, Dothideomycetes, Pezizomycotina, Ascomycota, Fungi
Güssow, H.T. 1906. *Z. PflKrankh.* 16:13.
From Gr. *koryné*, a club, club-shaped shoot + *sporá*, spore, for the form of conidia. Conidiophores mostly hypophyllous, erect, simple, sparingly septate, dark brown, with successive cylindrical proliferations, lighter in colour towards the apex. Conidia usually solitary, club-shaped, obclavate to cylindrical, straight or often slightly curved, tapering towards the apex, pale olivaceous brown, smooth, 4-20 pseudoseptate, hilum conspicuous. Causes a severe leaf spot disease on more than 70 host plant species from tropical and subtropical countries.

Corynesporasca Sivan.—see **Corynespora** Güssow
Corynesporascaceae, Pleosporales, Pleosporomycetidae, Dothideomycetes, Pezizomycotina, Ascomycota, Fungi
Sivanesan, A. 1996. *Corynesporasca caryotae* gen. et sp. nov. with a *Corynespora* anamorph, and the family Corynesporascaceae. *Mycol. Res.* 100(7):783-788.
From genus *Corynespora* Güssow + L. *ascus*, sac.

Crepidotus

Corynesporopsis P. M. Kirk
Inc. sed., Inc. sed., Inc. sed., Inc. sed., Pezizomycotina, Ascomycota, Fungi
Kirk, P. M. 1981. New or interesting microfungi II. Dematiaceous hyphomycetes from Esher Common, Surrey, England. *Trans. Br. mycol. Soc.* 77(2):279-297.
From genus *Corynespora* Güssow + Gr. suf. *-ópsis*, aspect, appearance, i.e., similar to *Corynespora*. Colonies effuse, blackish-brown to black. Mycelium partly superficial, partly immersed, composed of branched, septate, pale brown, smooth hyphae. Conidiophores macronematous, mononematous, arising singly or fasciculate. Conidia shortly catenate, arising through a pore at apex of conidiogenous cell, terminal conidium, dry, smooth, ellipsoid to cylindrical, euseptate. On decayed hardwoods.

Corynetes Hazsl.—see **Geoglossum** Pers.
Geoglossaceae, Geoglossales, Inc. sed., Geoglossomycetes, Pezizomycotina, Ascomycota, Fungi
Hazsl.1881. *Természettud. Köréböl Magyar Tud. Akad. Értek.* 11(19):7.
From Gr. *kórynetes*, club-bearer > L. *corynetes*, for the shape of the ascomata.

Cotylidia P. Karst.
Inc. sed., Inc. sed., Inc. sed., Agaricomycetes, Agaricomycotina, Basidiomycota, Fungi
Karsten P. A. 1881. Enumeratio Thelephorearum Fr. et Clavariearum Fr. Fennicarum, systemate novo dispositarum. *Revue mycol.*, Toulouse 3(9):21-23.
From Gr. *kotýle* > L. *cotyle*, cup-shaped + dim. suf. *-idia*, in reference to the shape of the fructification. Basidiomata infundibuliform or spatulate, stipitate, light-colored, hymenophore smooth or rugose, flesh pale. Basidiospores ellipsoid, hyaline, smooth, inamyloid. Lignicolous, muscicolous, or terricolous. Frequent in the tropics, some species reported from temperate regions, including Europe.

Craterellus Pers.
Cantharellaceae, Cantharellales, Inc. sed., Agaricomycetes, Agaricomycotina, Basidiomycota, Fungi
Persoon, C. H. 1825. *Mycol. eur.* (Erlanga) 2:1-214.
From Gr. *kratér* > L. *crater*, cup, the mouth of a volcano + L. dim. suf. *-ellus*, i.e., little cup or little crater. Pileus fleshy or membranous, infundibuliform, or flabelliform, tubiform or trumpet-shaped, with opening recurved, fertile undersurface smooth to irregularly-veined, flesh thin, brittle. Spores broadly elliptic or subglobose. On the ground or rotten twigs or prostrate branches. Some species are edible sometimes with a fragrant odor similar to the odor of peach and apricot.

Crauatamyces Viégas
Inc. sed., Inc. sed., Inc. sed., Dothideomycetes, Pezizomycotina, Ascomycota, Fungi
Viégas, A. P. 1944. Alguns fungos de Brasil II. Ascomicetos. *Bragantia* 4(1-6):1-392.
From Tupi *caraguatá*, *Bromelia* spp. with thick, fibrous leaves and densely clustered fruits > *crauata*, a cluster of fruits + L. *myces*, fungus < Gr. *mykês*, because the dense clusters of ascomata resemble the clusters of mature caraguatá fruits. Mycelium forming a black, erumpent, pseudoparenchymatous stroma giving rise to uniloculate pseudothecia. Pseudothecia black, obpyriform, smooth, densely aggregated on a common stroma, each with a cylindrical base. Locules globose, opening by an apical pore, containing filiform pseudoparaphyses. Asci clavate, with a thickened apex, 8-spored. Ascospores filiform, brown, phragmosporous, with rounded ends. Collected on stems of *Eupatorium* sp. in São Paulo State, Brazil.

Creonectria Seaver—see **Nectria** (Fr.) Fr.
Nectriaceae, Hypocreales, Hypocreomycetidae, Sordariomycetes, Pezizomycotina, Ascomycota, Fungi
Seaver, F. J. 1909. The Hypocreales of North America-II. *Mycologia* 1(5):177-207.
From Gr. *kréas* > L. *creos*, flesh + genus *Nectria* (Fr.) Fr., i.e., a fleshy *Nectria*.

Crepidotus (Fr.) Staude
Inocybaceae, Agaricales, Agaricomycetidae, Agaricomycetes, Agaricomycotina, Basidiomycota, Fungi
Staude, F. 1857. *Schwämme Mitteldeutschl* 25:71.
From Gr. *crepís*, genit. *crepídos*, a kind of shoe for humans, with the shape of sandal + *ous*, *otós*, ear, flap, in reference to the pileus without a stipe; ear-shaped. Pileus without a stipe, with indefinite or well differentiated cuticle, yellow, pink, red, rusty brown or white; lamellae variously attached to base, often decurrent if stipe present, not connected by anastomoses; spores often perforated or with cylindric spines or warts, frequently spores smooth, globose with oblique hylar appendage, oblong or elliptic; cheilocystidia always present. On wood, on herbaceous stems, mosses, ferns and palms; occasionally parasitic on living trees. Cosmopolitan.

Crepidotus fimbriatus Hesler & A. H. Smith: basidiomes cupulate, becoming kidney to shell-shaped, margin incurved, fringed with a cottony tomentum. Fruiting on the undersurface of downed conifer branches, in North America, x 1.

Crinipellis

Crinipellis Pat.
Marasmiaceae, Agaricales, Agaricomycetidae, Agaricomycetes, Agaricomycotina, Basidiomycota, Fungi
Patouillard, N.T. 1889. Fragments mycologiques. Notes sur quelques champignons de la Martinique. *J. Bot., Paris* 3:335-343.
From L. *crinis*, crinate, crest of hair + *pellis*, skin, meaning covered with hair. Pileus and usually stipe covered with thick-walled, hair-shaped elements, pseudoamyloid to almost amyloid, stipe central or eccentric. Hymenophore lamellate. Spore print white or nearly so; spores hyaline, inamyloid, of various shapes, smooth, thin-walled but often becoming somewhat thick-walled, cyanophilic. Hyphae with clamp connections. On dead and living plants, especially Poaceae such as bamboo.

Crinitospora B. Sutton & Alcorn
Inc. sed., Inc. sed., Inc. sed., Inc. sed., Pezizomycotina, Ascomycota, Fungi
Sutton, B. C., J. L. Alcorn. 1985. Undescribed species of *Crinitospora* gen. nov., *Massariothea*, *Mycoleptodiscus* and *Neottiosporina* from Australia. *Trans. Br. mycol. Soc.* 84(3):437-445.
From L. *crinitus*, with hair, bearded + *spora*, spore, i.e., a spore with hairs. Mycelium immersed, branched, septate, hyaline to pale brown. Conidiomata eustromatic, immersed, epidermal, separate, brown, acervular, dehiscence by breakdown of overlying host tissue. Conidiophores hyaline, septate towards base, unbranched. Conidiogenous cells integrated or discrete, determinate or indeterminate, smooth, cylindrical to lageniform. Conidia holoblastic, produced in succession or solitary, medianly 1-euseptate, rarely 0-2-euseptate, smooth, ellipsoid, obtuse at apex, broadly truncate at base, with 4-10 apically inserted, unbranched, divergent cellular appendages. On branches of *Mangifera indica*, Bowen, Queensland, Australia.

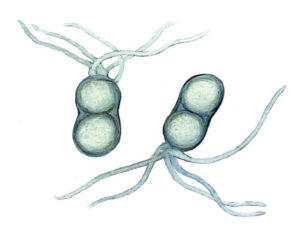

Crinitospora pulchra B. Sutton & Alcorn: bicellular conidia, with several appendages on the base, from branch of *Mangifera indica*, Queensland, Australia, x 850.

Cristelloporia I. Johans. & Ryvarden (syn. **Murrilloporus** Ryvarden)
Hydnodontaceae, Trechisporales, Inc. sed., Agaricomycetes, Agaricomycotina, Basidiomycota, Fungi
Johansen I., L. Ryvarden. 1979. Studies in the Aphyllophorales of Africa VII. Some new genera and species in the Polyporaceae. *Trans. Br. mycol. Soc.* 72(2):189-199.
From Gr. *krystallos*, clear ice, glass > L. *cristello* + genus *Poria* P. Browne, referring to the crystals formed by the fungus. Hyphal system dimitic, generative hyphae hyaline, thin-walled, with clamps at septa; skeletal hyphae yellow to golden, thick-walled, solid, narrow, dextrinoid, often with secondary simple septa, needle-like crystals present among hyphae. Fructifications annual, resupinate, sessile to pileate, cottony to coriaceous, reddish-brown to chestnut at base, with thin cuticle. Pore surface white when fresh, cream to yellow, pores round to angular or irregular, tubes concolorous with pore surface. Basidia small, clavate to urniform. Cystidia hyphoid, hyaline, thin-walled. Spores broadly ellipsoid, hyaline, thin-walled, asperulate, non-amyloid, non-dextrinoid. Collected in Ashanti region of Ghana.

Crocicreas Fr. (syn. **Belonioscypha** Rehm)
Helotiaceae, Helotiales, Leotiomycetidae, Leotiomycetes, Pezizomycotina, Ascomycota, Fungi
Fries, E. M. Summa veg. Scand., Sectio Post. (Stockholm):418.
From Gr. *krókis*, the woof of weft, loose thread, lint + *kréas*, flesh, referring to the flesh of the apothecia. Small, stipitate to subsessile apothecia, whitish to brownish hymenium, scattered or gregarious, at first rounded, expanding and becoming campanulate or turbinate, light colored, externally smooth. Asci clavate or cylindrical, inoperculate, 4-8-spored, with non septate to multiseptate ascospores ellipsoid, ovoid, obpyriform, fusiform or filiform in shape. Paraphyses filiform. Growing on plant materials.

Cronartium Fr.
Cronartiaceae, Pucciniales, Inc. sed., Pucciniomycetes, Pucciniomycotina, Basidiomycota, Fungi
Fries, E. M. 1815. *Observ. mycol.* (Havniae) 1:220.
Probably from the Gr. *chrónnymi* and *chronnýo*, to stain, to contaminate + *ártios*, capable, suitable, for the spots it produces on the plants that it parasitizes. Aecia on branches, trunks or cones of pines; telia on Fagaceae or Scrophulariaceae. Causes rusts on plants of economic importance including blister rust of white pine, rust of currant, fusiform rust of pines, and rust of oak leaves. It produces galls on the stems and cones of adult pine and dwarfing in young pines.

Crucibulum Tul. & C. Tul.
Agaricaceae, Agaricales, Agaricomycetidae, Agaricomycetes, Agaricomycotina, Basidiomycota, Fungi

Tulasne, L. R., C. Tulasne. 1844. Recherches sur l'organization et le mode de fructification des champignons de la tribu des nidulariées. *Annls. Sci. Nat., Bot., sér. 3,* 1:89.

From ML. *crucibulum,* crucible, for the shape of the basidiocarp, similar to a cauldron or receptacle, like those employed to smelt metals. Fructifications solitary or gregarious, with pale, lenticular peridioles attached to internal wall of peridium by means of a funiculum. Spores hyaline, elliptical or oval-elliptical, smooth. On wood.

Cronartium quercuum (Berk.) Miyabe ex Shirai: aecia on *Pinus aristata* trunk, x 0.5. On *Quercus nigra* and *Q. tinctoria,* in South Carolina & Pennsylvania, U.S.A.

Crucibulum laeve (Huds.) Kambly: mature basidiomata with white peridiole, on wood, x 4.

Crumenulopsis J. W. Groves (syn. **Digitosporium** Gremmen)

Helotiaceae, Helotiales, Leotiomycetidae, Leotiomycetes, Pezizomycotina, Ascomycota, Fungi

Groves, J. W. 1969. *Crumenulopsis,* a new genus name to replace *Crumenula* Rehm. *Can. J. Bot.* 47:47-51.

From genus *Crumenula* De Not. + Gr. suf. *-ópsis,* aspect, appearance, i.e., a genus similar in appearance to *Crumenula.* Ascomata apothecia, erumpent, single or in small clusters, dark brown or purplish to black, with furfuraceous exterior. Apothecia sessile to substipitate, when dry elongate or sometimes triangular and strongly inrolled at margins, hard and brittle, becoming circular and shallowly cupulate when moist, fleshy. Hymenium smooth, yellowish-brown. Paraphyses filamentous, septate, clavate or scarcely enlarged at tips, not forming an epithecium. Asci unitunicate, inoperculate, clavate to cylindric-clavate, short-stipitate, undifferentiated at apex, not bluing in iodine, 4-8-spored. Ascospores hyaline, 1-celled, fusoid, oval or lacrimiform, smooth, with or without a thin gelatinous sheath. Asexual morph digitosporium-like. On dying branches of *Pinus* spp.

Cryphonectria (Sacc.) Sacc. & D. Sacc. (syns. **Endothia** Fr., **Endothiella** Sacc.)

Cryphonectriaceae, Diaporthales, Sordariomycetidae, Sordariomycetes, Pezizomycotina, Ascomycota, Fungi

Saccardo, P. A. 1905. *Syll. fung.* (Abellini) 17:783.

From Gr. *krýphos,* a hiding place > L. *crypho* + genus *Nectria* (Fr.) Fr., referring to the perithecia immersed in a stroma. Stromata valsoid, brightly colored, reddish to brownish orange, with erumpent ectostromal disk. Ascomata ostiolate perithecia, subglobose, clustered in stroma, oblique, with long ostiolar necks converging and extending to surface. Ascomatal wall dark brown. Asci unitunicate, ellipsoid to clavate, with nonamyloid, refractive apical ring, base of ascus often deliquescent, asci lying free at maturity, 8-spored. Ascospores hyaline, 2-celled, with median septum, ellipsoid or ovoid. Asexual morph in pseudstromatic pycnidial, separate or aggregated, reddish to brownish-orange, unilocular or complex and multilocular, endothiella-like. Causing cankers on deciduous trees. *Cryphonectria parasitica* (Murrill) M. E. Barr causes the destructive chestnut blight disease.

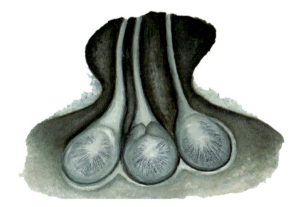

Cryphonectria parasitica (Murrill) M. E. Barr (syn. **Endothia parasitica** (Murrill) P. J. Anderson & H. W. Anderson): perithecial stromata within the tissues of the host plant, causing cankers in chestnuts and other trees; each perithecium forms a long neck towards the erumpent ectostroma, x 60. On living or recently killed branches of *Castanea dentata,* New York, U.S.A.

Cryptascoma

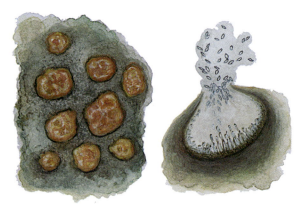

Cryphonectria parasitica (Murrill) M.E. Barr (syn. **Endothiella parasitica** Roane): Pseudostromatic pycnidial conidiomata, causing oak canker, x 1, and sagittal section of a pycnidium with conidia, on *Castanea dentata*, New York, U.S.A., x 125.

Cryptascoma Ananthap.
Valsaceae, Diaporthales, Sordariomycetidae, Sordariomycetes, Pezizomycotina, Ascomycota, Fungi
Ananthapadmanaban, D. 1988. *Cryptascoma*, a new genus of the Valsaceae. *Trans. Br. mycol. Soc.* 90(3):479-481.
From Gr. *kryptós*, hidden > L. *cryptos* + ascoma, a fruiting body with asci. Stroma absent. Perithecia in groups, immersed, vertically oriented, partially erumpent, ostiolate. Peridium two-layered. Ostiole periphysate. Asci unitunicate, thin-walled, ellipsoid, becoming free in perithecial cavity, nonamyloid, 8-spored. Ascospores fusiform, l-septate, with an appendage at each end of cell. Appendage simple, filiform. Paraphyses absent. Collected on dead branches in Berijam, Kodaikanal, Madurai Dt, Tamil Nadu State, India.

From Gr. *kryptós*, hidden + *kókkos* < *kokkíon*, small ball, due to the cellular shape of this yeast, which is surrounded by a mucilaginous capsule. Basidia originate in groups on terminal part or at times on lateral part of thin, nonseptate hyphae with clamp connections and swollen at apex. Basidia on terminal part produce small, sessile basidiospores with thin walls in basipetal chains. No persistent probasidium. Also forms blastospores laterally on hyphae. Causes pulmonary, meningeal or systemic cryptococcosis in humans and other higher animals, also saprobic and pathogenic on other animals. On bird dung and soil enriched with this dung. Some species live in the gummy exudate of Cactaceae and the fruit juice of plants.

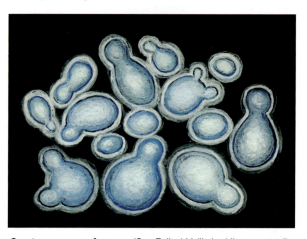

Cryptococcus neoformans (San Felice) Vuill.: budding yeast cells, surrounded by a mucilaginous capsule, from fermenting fruit juice, x 4,000.

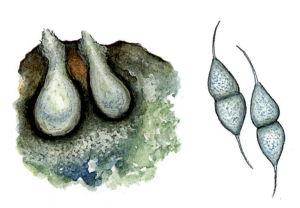

Cryptascoma bisetulum Ananthap.: perithecia immersed in the dead tree branch, x 25; and fusiform, one-septate, ascospores, with an appendage at each end of the cell, x 1,000.

Cryptococcus Vuill. (syn. **Filobasidiella** Kwon-Chung)
Tremellaceae, Tremellales, Inc. sed., Tremellomycetes, Agaricomycotina, Basidiomycota, Fungi
Vuillemin, P. 1901. Les blastomycètes pathogènes. *Rev. Gén. Sci. Pures Appl.* 12:732-751.

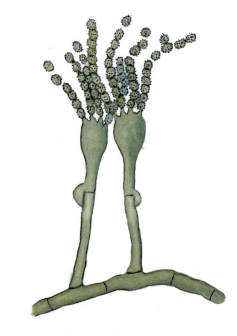

Cryptococcus neoformans (San Felice) Vuill.) (syn. Filobasidiella neoformans Kwon-Chung): basidia with small, sessile basidiospores in basipetal chains, x 500. Isolated in Maryland, U.S.A.

Cryptocoryneopsis B. Sutton

Inc. sed., Inc. sed., Inc. sed., Inc. sed., Pezizomycotina, Ascomycota, Fungi

Sutton, B. C. 1980. *Cryptocoryneopsis umbraculiformis* gen. et sp. nov. from Australia. *Trans. Br. mycol. Soc.* 74(2):393-398.

From genus *Cryptocoryneum* Fuckel + Gr. suf. -ópsis, aspect, for the similarity to that genus. Colonies sparse, widely effuse. Mycelium mostly superficial, composed of branched, septate, pale brown, anastomosing, smooth hyphae. Conidiophores macronematous, mononematous, unbranched, straight, smooth, septate, dark brown, thick-walled. Conidiogenous cells integrated, terminal holoblastic, single on each conidiophore. Conidia acrogenous, dry, solitary, umbrella-shaped, with a basal central point of secession and body of conidium comprised of 4-8 pale brown cells at periphery of which each produces a cheiroid group of cells; the latter consist of 1-4 upper dark brown more or less globose cells and 2-4 downwardly directed, 2-4 septate pale brown pendulous branches. Collected on dead leaves of *Banksia oblongifolia* in Queensland, Australia.

Cryptomyces Grev.

Cryptomycetaceae, Rhytismatales, Leotiomycetidae, Leotiomycetes, Pezizomycotina, Ascomycota, Fungi

Greville, R. K. 1825. *Scott. crypt. fl.* (Edinburgh) 4:206.

From Gr. *kryptós*, hidden, covered + *mýkes* > L. *myces*, fungus. Stromata with apothecia, fleshy-gelatinous. In the interior of the leaves of willow causing leaf fall.

Cryptomyces maximus (Fr.) Rehm: fleshy-gelatinous apothecial stromata, on branches of *Salix*, x 1.5.

Cryptomycina Höhn.

Inc. sed., Inc. sed., Inc. sed., Sordariomycetes, Pezizomycotina, Ascomycota, Fungi

Höhnel, F. v. 1917. Über die Gattung *Cryptomyces*. *Ann. mycol.* 15(5):322.

From Gr. *kryptós*, hidden, covered + *mýkes*, fungus (L. *myces*) + L. suf. -*ina*, which indicates possession or similarity. Apothecia covered or protected by a stroma formed by tissues of fungus and host. Parasitic on the aquiline fern, *Pteridium aquilinum*.

Cryptomycina pteridis (Rebent.) Höhn.: apothecia protected by a stroma; this species parasitizes the aquiline fern, x 25.

Cryptoniesslia Scheuer

Niessliaceae, Hypocreales, Hypocreomycetidae, Sordariomycetes, Pezizomycotina, Ascomycota, Fungi

Scheuer, C. 1993. *Cryptoniesslia setulosa* gen. et sp. nov. *Mycol. Res.* 97(5):543-546.

From L. *cryptos*, hidden < Gr. *kryptós* + genus *Niesslia* Auersw., referring to the immersed ascomata. Perithecia solitary, immersed, blackish-brown, not translucent, pyriform, bearing short setae around erumpent apex. Peridium composed of several layers of cells; setae blackish-brown, unicellular, thick-walled, branched or unbranched, arising around perithecial apex, rooted deeply in peridium, ostiole densely lined with periphyses. Interascal filaments extremely inconspicuous in mature perithecia, consisting of irregularly inflated, thin-walled cells; apices of interascal filaments not seen. Asci unitunicate, with a conspicuously flattened but scarcely thickened wall at apex; apical ring minute, not staining in Lugol's iodine solution. Ascospores hyaline, 1-septate. Collected on dead leaves of *Carex arenaria* in Devon, England.

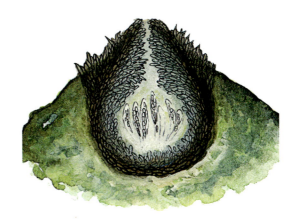

Cryptoniesslia setulosa Scheuer: vertical section of a solitary perithecium, immersed, x 270.

Cryptothecia

Cryptothecia Stirt.
Arthoniaceae, Arthoniales, Arthoniomycetidae, Arthoniomycetes, Pezizomycotina, Ascomycota, Fungi
Stirton, J. 1876. Descriptions of recently discovered foreign lichens. *Proc. Roy. Phil. Soc. Glasgow 10*:164.
From Gr. *kryptós*, concealed, hidden, or from *krýpte*, crypt + *théke*, dim. *thékion*, case. Fructifications sunken in crustous thalli and enveloped in a wall formed by thallus, carbonaceous pseudothecia lacking an excipulum. Ascostroma hysterothecia. Asci cylindrical, bitunicate, peculiar in having a thick external wall and a relatively thin internal wall, intermixed with pseudoparaphyses; in some species aggregated in fertile areas differentiated from thallus. Ascospores septate, reduced number within each ascus. Epiphytic.

Cudonia Fr.
Cudoniaceae, Rhytismatales, Leotiomycetidae, Leotiomycetes, Pezizomycotina, Ascomycota, Fungi
Fries, E. M. 1849. *Summa veg. Scand., Sectio Post.* (Stockholm):348.
From L. *cudo*, *cudonis* (from Gr. *kódion*, untanned skin), morion or helmet of skin, for the shape of the apothecia and for their coriaceous consistency. Apothecia capitate, clearly differentiated into a pileus and stipe; pileus globose, almost always involuted, i.e., doubled inward. Asci clavate, 8-spored, Ascospores acicular or clavate, at times partioned transversally, or without septa, with numerous granulations aligned longitudinally. On soil with fallen leaves, principally in forests of beech and other Fagaceae.

Cunninghamella echinulata (Thaxt.) Thaxt. ex Blakeslee: echinulate, unispored sporangiola that are borne on globose or claviform vesicles, x 550.

Cudonia confusa Bres.: capitate apothecia, clearly differentiated into a pileus and stipe, on soil, x 20.

Cunninghamella Matr.
Cunninghamellaceae, Mucorales, Inc. sed., Inc. sed., Mucoromycotina, Zygomycota, Fungi
Matruchot, M. L. 1903. Une mucorinée purement conidianne, *Cunninghamella africana. Ann. Mycol.* 1(1):45-60.
Named in honor of the Scottish researcher Dr. *D. D. Cunningham* + L. dim. suf. *-ella*. Sporangiola unispored, pedicellate, borne on globose or clavate vesicles or ampullae. Saprobic in soil.

Curvularia Boedijn (syn. Malustela Bat. & J. A. Lima)
Pleosporaceae, Pleosporales, Pleosporomycetidae, Dothideomycetes, Pezizomycotina, Ascomycota, Fungi
Boedijn, K. B. 1933. Über einige phragmosporen Dematiazeen. *Bull. Jard. bot. Buitenz, 3 Sér.* 13(1):120-134.
From L. *curvus*, bent + suf. *-aria*, connection or possesion, in reference to the conidia, which are curved, dark, and multicellular, with the end cells less pigmented. Colonies floccose, brownish-white, not zonate. Mycelial hyphae branched, septate, brown. If present synnemata erect, covered by a layer of fertile hyphae. Conidiophores acropleurogenous. Conidia ellipsoid or obovoid, phragmosporous, brown, smooth, formed from pores in conidiophores, geniculate, generally with a protuberant hilum on base. Parasitic or saprobic on grasses, although also in soil and air.

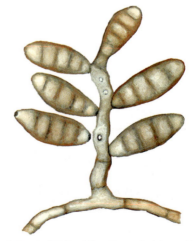

Curvularia lunata (Wakker) Boedijn: geniculate conidiophore, with curved, dark, and multicellular conidia, x 1,000.

Cylindrocarpostylus

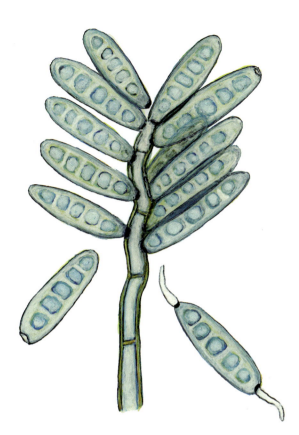

Curvularia hawaiiensis (Bugnic. ex M. B. Ellis) Manamgoda, et al. (syn. Bipolaris hawaiiensis (Bugnic. ex M. B. Ellis) J. Y. Uchida & Aragaki): geniculate conidiophore, with fusoid conidia; there is a conidium with a germ tube on each end, x 1,100.

Cyathicula De Not.
Helotiaceae, Helotiales, Leotiomycetidae, Leotiomycetes, Pezizomycotina, Ascomycota, Fungi
De Notaris, G. 1863. *Comm. Soc. crittog. Ital.* 1(fasc. 5):381.
From Gr. *kyáthos*, cup > L. *cyathos* + dim. suf. *-ula*. Apothecia scattered, stipitate, at first closed, subglobose, expanding, margin with sharp teeth, outside smooth. Asci cylindric-clavate, 8-spored. Ascospores fusoid, elongated, 1-septate. Paraphyses filiform. On plant material.

Cyathicula coronata (Bull.) Rehm: stipitate apothecia, with sharp teeth in the margin, on plant detritus, x 4.

Cyathus Haller
Agaricaceae, Agaricales, Agaricomycetidae, Agaricomycetes, Agaricomycotina, Basidiomycota, Fungi
Haller, V. A. von. 1768. *Hist. stirp. Helv.* 3:236.
From Gr. *kyáthos*, goblet, for the shape of the fructification. Fructifications cyathiform, infundibuliform or campanulate. Peridium of three layers with dark-colored, lenticular peridioles, united to internal wall by means of a complex funiculum. Spores inside peridioles, hyaline, subglobose, oval or ellipsoidal, smooth. On wood and other plant remains, dung, fertilized soils, or soils in forests, pastures and arid or semiarid regions.

Cyathus striatus (Huds.) Willd.: infundibuliform basidioma, with gray-colored peridioles, on wood, x 5.

Cyberlindnera Minter
Inc. sed., Saccharomycetales, Saccharomycetidae, Saccharomycetes, Saccharomycotina, Ascomycota, Fungi
Minter, D. W. 2009. *Cyberlindnera*, a replacement name for *Lindnera* Kurtzman et al., nom. illegit. *Mycotaxon* 110:473.
Replaced *Lindnera* in honor of the German mycologist *Paul Lindner* + L. fem. des. *-a-*, for his early work on yeasts. Asci globose to ellipsoid, unconjugated or arising from conjugation between a cell and its bud or between independent cells, some heterothallic, deliquescent or persistent, forming one to four ascospores. Ascospores hat-shaped, spherical, or spherical with an equatorial ledge. Cell division by multilateral budding on a narrow base, budded cells spherical, ovoid or elongate. Sometimes pseudohyphae and true hyphae. Glucose fermented, some species ferment other sugars as well and nitrate. Where determined, predominant ubiquinone CoQ-7; diazonium blue B reaction negative.

Cylindrocarpostylus R. Kirschner & Oberw.
Inc. sed., Inc. sed., Inc. sed., Inc. sed., Pezizomycotina, Ascomycota, Fungi

Cylindrochytridium

Kirschner, R., F. Oberwinkler. 1999. *Cylindrocarpostylus*, a new genus based on a hyphomycete rediscovered from bark beetle galleries. *Mycol. Res.* 103(9):1152-1156.
From genus *Cylindrocarpon* Wollenw. + Gr. *stýlos*, pillar, column > L. *stilus*, referring to the formation of the conidiogenous head on the tip of a long stipe. Conidiophores mononematous, macronematous, straight, penicillate, hyaline, with rough-walled stipes and metulae. Penicillia apical on stipes, composed of series of metulae and conidiogenous cells, without sterile extensions. Conidiogenous cells phialidic cylindrical. Conidia septate, curved, hyaline, aggregating in a slimy mass. Occurring in bark beetle galleries in *Picea abies* and *Pinus sylvestris* in Germany.

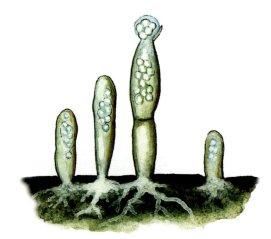

Cylindrochytridium johnstonii Karling: cylindrical zoosporangia, on decaying vegetable debris, Virginia, U.S.A., x 500.

Cylindrocarpostylus gregarious (Bres.) R. Kirschner & Oberw.: conidiophore with phialides and conidia, x 250.

Cylindrochytridium Karling
Chytridiaceae, Chytridiales, Chytridiomycetidae, Chytridiomycetes, Inc. sed., Chytridiomycota, Fungi
Karling, J. S. 1941. *Cylindrochytridium johnstonii* gen. nov. et sp. nov., and *Nowakowskiella profusum* sp. nov. *Bull. Torrey bot. Club* 68:381-387.
From Gr. *kýlindros*, cylinder + *chytrídion*, dim. of *chytrís*, pot, earthen jar, flower vase + L. dim. suf. -*ium*. Zoosporangium shaped like cylindrical receptacle, pedicellate or sessile. Saprobic on plant remains.

Cylindrosporium Grev.—see Pyrenopeziza Fuckel
Inc. sed., Helotiales, Leotiomycetidae, Leotiomycetes, Pezizomycotina, Ascomycota, Fungi
Greville, R. K. 1822. *Scott. crypt. fl.* (Edinburg) 1:27.
From Gr. *kylíndros*, cylinder + *sporá*, spore + L. dim. suf. -*ium*, referring to the cylindrical spores.

Cyphelium Ach.
Caliciaceae, Teloschistales, Lecanoromycetidae, Lecanoromycetes, Pezizomycotina, Ascomycota, Fungi
Acharius, E. 1815. *K. Vetensk-Acad. Nya Handl.* 3:261.
From NL. *cyphella* < Gr. *kýphella*, the concavities of the ears + L. suf. -*ium*, which indicates characteristic of. Thallus crustous, smooth or rugose, granulose, verrucose or areolate, rudimentary or well developed. Apothecia on upper surface of thallus. On rocks, soil or wood. Photobionts *Protococcus*, *Pleurococcus* and *Trentepohlia*.

Cystoderma Fayod
Agaricaceae, Agaricales, Agaricomycetidae, Agaricomycetes, Agaricomycotina, Basidiomycota, Fungi
Fayod, M.V. 1889. Prodrome d'une histoire naturelle des Agaricinés. *Annls Sci. Nat., Bot. sér.* 7, 9:181-411.
From the Gr. *kýstis* > L. *cystis*, bladder, pouch + *dérma*, skin, referring to the presence of cyst-shaped cellular elements in the epicutis of the fruitbody. Pileus and stipe covered with a velar layer or epicutis, which consists entirely or predominantly of sphaerocysts forming an epithelium intermixed with elongated elements. Hymenophore lamellate; lamellae adnexed, adnate or even short decurrent, in age sometimes free. Spore print white, sometimes pale cream; spores hyaline, smooth, ellipsoid to subglobose or ventricose to subrhomboid or oblong to subcylindric, amyloid or inamyloid. Among and on moss, on the ground in woods.

Cystopage Drechsler
Zoopagaceae, Zoopagales, Inc. sed., Inc. sed., Zoopagomycotina, Zygomycota, Fungi
Drechsler, C. 1941. Four phycomycetes destructive to nematodes and rhizopods. *Mycologia* 33(3):251.
From Gr. *kýstis*, bladder, vesicle, cell + *páge*, loop, snare. Chlamydospores in hyphae form traps to capture and kill amoebae and nematodes in soil. It invades the hosts by means of haustoria.

Cyttaria

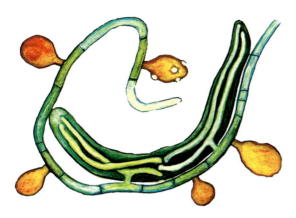

Cystopage intercalaris Drechsler: nematode attacked by the hyphae of this fungus; outside the host, chlamydospores are produced, x 150. Described in Virginia, U.S.A.

Cystotheca Berk. & M. A. Curtis
Erysiphaceae, Erysiphales, Leotiomycetidae, Leotiomycetes, Pezizomycotina, Ascomycota, Fungi
Berkeley, M. J., M. A. Curtis. 1860. Characters of new fungi, collected in the North Pacific Exploring Expedition by Charles Wright. *Proc. Amer. Acad. Arts & Sci.* 4:111-130.
From Gr. *kýstis*, bladder, pouch > L. *cysto* + Gr. *théke*, case, box > L. *theca*, for the pouch-like ascomata each with a single ascus. Ascomata cleistothecioid, dark brown, formed on surface of primary mycelium and surrounded by dense growth of secondary hyphae. Mycelium superficial, hypophyllous, forming extensive colonies on which conidia produced. Subglobose haustoria formed in host epidermal cells. Cells of secondary mycelium setose, flexuous or sickle-shaped, thick-walled, pale or dark brown, forming a dense turf on primary mycelium. Ascomata globose to subglobose in top view, somewhat flattened in side view, glabrous or with occasional scattered, short, hyphal appendages around base. Ascomatal wall composed of two distinct layers. Ascomata containing a single ascus. Asci subglobose to ellipsoidal, with a short stalk, relatively thick-walled, thinner at apex, 6-8-spored. Ascospores 1-celled, hyaline, smooth, oval to ellipsoid. Asexual morph oidium-like. On living leaves of oaks (*Quercus* sp.) and other Fagaceae.

Cytospora Ehrenb. (syn. **Valsa** Fr.)
Valsaceae, Diaporthales, Sordariomycetidae, Sordariomycetes, Pezizomycotina, Ascomycota, Fungi
Ehrenberg, C. G. 1818. *Sylvae mycologicae Berolinenses*:1-32.
From the Gr. *kýtos*, hollow receptacle, now often utilized to refer to a cell + *sporá*, spore, seed. Conidiomata that usually contain either labyrinthine chambers or clusters of pycnidia, having conidiophores and allantoid hyaline conidia; in moist conditions, the conidia exude from the fruiting bodies in gelatinous matrices, usually as yellow, orange, red or pallid tendrils. Stromata immersed, prosenchymatous, valsoid, black, erumpent. Perithecia clustered or in a circle in stroma, subglobose, with long ostiolar necks converging through disc. Ascomatal wall dark, outer cells isodiametric, thick-walled; inner cells hyaline, thin-walled, flattened. Asci unitunicate, cylindric to ellipsoid or clavate, with a non-amyloid refractive apical ring, lying free in perithecium at maturity, 4-8-spored. Ascospores hyaline, 1-celled, allantoid to subcylindric. Asexual morph pycnidial in stromata. Important plant pathogens, cause cankers and dieback on many genera of hard woods and coniferous trees, but rarely on herbaceous plants.

Cystotheca tjibodensis (Gäum.) Katum.: sagittal section of a chasmathecium, with a single ascus and ascospores, on leaf of *Castanea argentea*, Java, x 340.

Cytospora eugeniae (Nutman & F. M. Roberts) G. C. Adams & Rossman (syn. **Valsa eugeniae** Nutman & F. M. Roberts): ostiolate perithecia, derived from black, immersed, valsoid ascostromata, emerging through bark cracks of hosts (*Eucalyptus grandis*, *Eugenia caryophyllus*, *Tibouchina heteromalla*), described in Australia, Tanzania & Malaysia, x 4; unitunicate, ellipsoid ascus, with eight, 1-celled, allantoid ascospores, x 2,000.

Cyttaria Berk. (syn. **Cyttariella** Palm)
Cyttariaceae, Cyttariales, Leotiomycetidae, Leotiomycetes, Pezizomycotina, Ascomycota, Fungi
Berkeley, M. J. 1842. On an edible fungus from Tierra del Fuego. *Trans. Linn. Soc. London* 19:40.
From Gr. *kýttarion*, this from *kýtos*, cavity or breast, from which is derived the L. dim. word *cellula*, cell.

Cyttariella

Fructifications compound, round, with numerous apothecia that open by a wide orifice at maturity, fleshy, large. Apothecia formed beneath a cortical layer of fructification; at first spherical, at maturity opening independently with numerous slightly sunken cavities. Asexual morph pycnidial cyttariella-like. Parasitic on southern hemisphere beech in the genus *Nothofagus*. Fructifications edible.

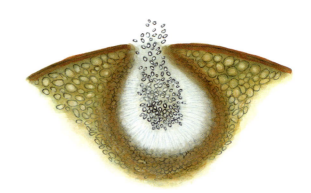

Cyttaria hookeri Berk. (syn. **Cyttariella deformans** (E. Bommer & M. Rousseau) Palm): longitudinal section of an immersed pycnidium, with conidia, x 200.

Cyttaria gunnii Berk.: compound fructifications, with numerous apothecia, on stems of *Nothofagus cunninghamii* (myrtle beech), Tasmania, x 1.

Cyttariella Palm—see **Cyttaria** Berk.
Cyttariaceae, Cyttariales, Leotiomycetidae, Leotiomycetes, Pezizomycotina, Ascomycota, Fungi
Palmer, J. E. 1932. *Annls. Mycol.* 30(5/6):418.
From genus *Cyttaria* Berk. + L. dim. suf. *-ella*.

Dinemasporium sasae

D

Dacampiosphaeria D. Hawksw.—see **Pyrenidium** Nyl.

Dacampiaceae, Inc. sed., Inc. sed., Dothideomycetes, Pezizomycotina, Ascomycota, Fungi

Hawksworth, D. L. 1980. Notes on some fungi occurring on *Peltigera*, with a key to accepted species. *Trans. Br. mycol. Soc.* 74(2):363-386.

From genus *Dacampia* A. Massal. + genus *Sphaeria* Haller.

Dacrymyces Nees

Dacrymycetaceae, Dacrymycetales, Inc. sed., Dacrymycetes, Agaricomycotina, Basidiomycota, Fungi

Nees,1816. *Syst. Pilze* (Würxburg):89.

From Gr. *dácryon*, tear, drop, distillation + *mýkes* (L. *myces*), fungus, for the often lenticular shape, resembling a tear. Fructifications generally small, less than 20 mm, with a gelatinous consistency, light colored, pulvinate, tuberculate, lenticular or discoid. On wood, in tropical and temperate regions.

Dacrymyces capitatus Schwein. (syn. **Dacrymyces ellisii** Coker): bifurcate basidia with young basidiospores, and secondary basidiospores, x 1. Collected on wood, Great Britain.

Dacryopinax G. W. Martin

Dacrymycetaceae, Dacrymycetales, Inc. sed., Dacrymycetes, Agaricomycotina, Basidiomycota, Fungi

Martin, G. W. 1948. New or noteworthy tropical fungi. IV. *Lloydia* 11:116.

From Gr. *dácryon*, tear, drop, distillation + *pínax*, plate, writing tablet, picture, drawing, painting. Fructifications yellow or orange, with droplets on surface, gelatinous-ceraceous or cartilaginous, pileate-stipitate, with the shape of a spoon, spatula, fan or petaloid. On tree bark or more frequently on dead wood. A common species in temperate forests is *D. spathularia* (Schwein.) G. W. Martin.

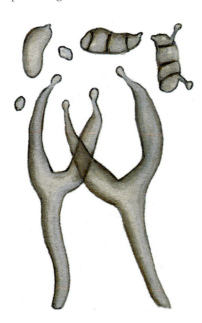

Dacrymyces chrysospermus Berk. & M. A. Curtis: cerebriform basidiomata, on dead wood, x 500.

Dacryopinax elegans (Berk. & M. A. Curtis) G. W. Martin: spathulate basidiomata, on dead wood, x 4.

Dactylomyces

Dactuliophora C. L. Leakey
Inc. sed., Pleosporales, Pleosporomycetidae, Dothideomycetes, Pezizomycotina, Ascomycota, Fungi
Leakey, C. L. A. 1964. *Dactuliophora*, a new genus of mycelia sterilia from tropical Africa. *Trans. Br. mycol. Soc.* 47(3):341-350.
From Gr. *dáktylos* > L. *dactylos*, finger + Gr. *phóros*, bearing, for the finger-like hyphae on the sclerotia. Reproduction solely by caducous sclerotia. Mycelium generally immersed, hyaline, diffuse in tissues of leaf spots, aggregated irregularly in epidermal or deeper leaf tissues forming plectenchymic masses from which sclerotiophores and sclerotia develop. Sclerotiophore remains after disjunction of sclerotium, superficial or occasionally more or less immersed structure, external continuation of immersed aggregation. Sclerotia glabrous, hispidulous, hispid or sparsely setose, spherical, subspherical, ellipsoidal, pyriform or rostrate, wholly or partly composed of dematiaceous cells on outside, hyaline, undifferentiated within, separating from sclerotiophore by fracture of many thin-walled cells joining base of mature sclerotium to center of sclerotiophore. Pathogenic on leaves of leguminous plants in Africa.

Dactylaria Sacc.
Inc. sed., Helotiales, Leotiomycetidae, Leotiomycetes, Pezizomycotina, Ascomycota, Fungi
Saccardo, P. A. 1880. Conspectus generum fungorum Italie inferiorum. *Michelia* 2(6):20.
From genus *Dactylium* Nees < Gr. *dáctylos*, finger + L. dim. suf. *-ium* + dim. suf. *-aria*, which indicates connection, in reference to the finger-like shape of the conidia. Interestingly, the name could also have been derived from Gr. *dactýlios*, ring + L. dim. suf. *-aria*, connection with, for the presence of rings. Nematode-trapping fungi with rings, constrictive or not. Conidia oblong, one or pluri-septate. In soil.

Dactylella Grove
Orbiliaceae, Orbiliales, Orbiliomycetidae, Orbiliomycetes, Pezizomycotina, Ascomycota, Fungi
Grove, W. B. 1884. New or noteworthy fungi. Part 1. *J. Bot.*, London, 22:199.
From genus *Dactylium* Nees < Gr. *dáctylos*, finger + L. dim. suf. *-ium* + dim. suf. *-ella*; it is a genus similar to the genus *Dactylaria* Sacc. but its digiform, septate conidia are similar to those of *Arthrobotrys* Corda. *Dactylella*, like *Dactylaria*, develops constrictive rings for trapping nematodes in the soil.

Dactylellina M. Morelet
Orbiliaceae, Orbiliales, Orbiliomycetidae, Orbiliomycetes, Pezizomycotina, Ascomycota, Fungi
Morelet, M. 1968. De aliquibus in mycologia novitatibus (5e note). *Bulletin de la Société des Sciences Naturelles et d'Archéologie de Toulon et du Var.* 178:6.
From Gr. *dáctylos*, finger + suf. *-ina*, derived from, denoting likeness. Vegetative hyphae hyaline, septate. Conidiophores hyaline, erect, septate, usually unbranched, gradually tapering upward, bearing a single conidium hyaline, mostly broadly spindle-shape, rounded distally, narrowing proximally, base truncate, 2-5-septate. Trapping nematodes by means of adhesive mesh like, two-dimensional networks. On soil.

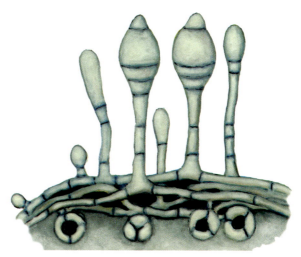

Dactylellina cionopaga (Drechsler) Ying Yang & Xing Z. Liu (syn. **Dactylella cionopaga** Drechsler): conidiophores with digitate conidia, and constrictive rings for capturing soil nematodes, x 500. Described in Oregon, U.S.A.

Dactylomyces Sopp.—see **Thermoascus** Miehe
Trichocomaceae, Eurotiales, Eurotiomycetidae, Eurotiomycetes, Pezizomycotina, Ascomycota, Fungi
Sopp, O. J. O. 1912. Monographie der Pilzgruppe *Penicillium* mit besonderer Berücksichtigung der in Norwegen gefundenen Arten. *Skr. VidenskSelsk. Christiana, Kl. 1, Math.-Natur.* 11:35-42.

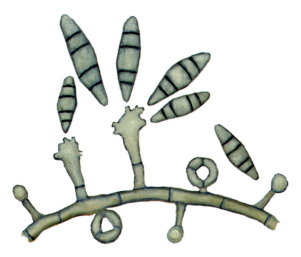

Dactylaria candida (Nees) Sacc.: conidiophores with multiseptate conidia, and constrictive rings for capturing soil nematodes, x 500.

Daedalea

From Gr. *dáktylos*, finger + *mýkes* > L. *myces*, fungus, referring to the finger-like branches of the conidiophores, on which are produced chains of conidia.

Daedalea Pers.
Fomitopsidaceae, Polyporales, Inc. sed., Agaricomycetes, Agaricomycotina, Basidiomycota, Fungi
Persoon, C. H. 1801. *Syn. meth. fung.* (Göttingen) 2:500.
From Gr. *daidáleos*, well carved, labyrinth, in reference to hymenophore composed of combs arranged as labyrinths. Fructifications annual or perennial, effuse-reflexed or pileate, often concentrically zonate or imbricate, color variable, white to dark, with labyrinthine pores. On dead wood, generally of deciduous leaved trees, in temperate or tropical forests. *Daedalea quercina* (L.) Fr., with thick, hard, woody or suberose, pale brown, sessile or substipitate fructifications, grows on trunks, stumps or oak posts.

Daedalea quercina (L.) Pers.: basidioma with labyrinthiform hymenophore, on tree trunk, x 1.

Daldinia Ces. & De Not.
Xylariaceae, Xylariales, Xylariomycetidae, Sordariomycetes, Pezizomycotina, Ascomycota, Fungi
De Notaris, G. 1863. Proposte di alcune rettificazioni al profilo dei discomiceti. *Comment. Soc. crittog. Ital.* 1 (*fasc.* 4):197.
Dedicated to the Swiss Friar *Agostino Daldini*, collaborator in the Erbario crittogamico italiano. Stromata large, black, hemispherical, pulvinate, 3-5 cm diam, in longitudinal section with endostroma having concentric layers in which perithecia develop. On dead branches of trees and shrubs.

Dasturella Mundk. & Khesw.
Phakopsoraceae, Pucciniales, Inc. sed., Pucciniomycetes, Pucciniomycotina, Basidiomycota, Fungi
Mundkur, B. B., K. F. Kheswalla. 1943. *Dasturella*-A new genus of Uredinales. *Mycologia* 35(2):201-206.
Named in honor of the Indian mycologist *J. F. Dastur* + L. dim. suf. *-ella*. Pycnia and aecia unknown. Uredinia minute, at first subepidermal, later erumpent, paraphysate, without peridia. Urediniospores sessile. Telia subepidermal, erumpent. Teliospores fascicled into large, sessile heads without cysts, 3-6- to rarely 7-celled. On leaves of *Dendrocalamus* spp. in India.

Daldinia concentrica (Bolton) Ces. & De Not.: pulvinate stromata, one of these with a sagittal section to show the embedded perithecia, on dead wood, x 1.

Davisomycella Darker
Rhytismataceae, Rhytismatales, Leotiomycetidae, Leotiomycetes, Pezizomycotina, Ascomycota, Fungi
Darker, G. D. 1967. A revision of the genera of the Hypodermataceae. *Can. J. Bot.* 45:1399-1444.
Named in honor of *J. J. Davis* + connective *o* + L. *myces*, fungus + dim. suf. *-ella*, who originally described the fungus. Ascomata apothecial, black, short-elliptical to oblong or elongate, subepidermal. Upper wall simple, parenchymatous, subhymenium thin, hyaline. Hymenium somewhat cupulate. Paraphyses numerous, filiform, simple. Asci clavate or saccate. Ascospores clavate, attenuated at base. Pycnidia small, flat, concolorous with needle. On pine needles, causing lesions.

Debaryomyces Lodder & Kreger-van Rij
Saccharomycetaceae, Saccharomycetales, Saccharomycetidae, Saccharomycetes, Saccharomycotina, Ascomycota, Fungi
Lodder, J., N. J. W. Kreger-van Rij. 1984. *The Yeasts. A Taxonomic Study*. 3rd edn. North Holland Publ. Co., Amsterdam, pp. 130, 145.
Named in honor of the German mycologist *H. A. de Bary* (1831-1888)+L. suf. *-myces*, fungus < Gr. *mýkes*. Ascospores warted, resulting from conjugation of a mother cell and a bud. Asexually multilateral budding and pseudomycelium. Yeast that produces slow and weak fermentations. In soil, cheeses, fermented foods and exudates of plants.

Decapitatus Redhead & Seifert—see Mycena (Pers.) Roussel

Mycenaceae, Agaricales, Agaricomycetidae, Agaricomycetes, Agaricomycotina, Basidiomycota, Fungi
Redhead, S.A., et al. 2000. *Rhacophyllus and Zorovaemyces - teleomorphs or anamorphs?*. *Taxon* 49:789-798.
From Gr. *dekátos*, the tenth + L. *capitatus*, having a head. Have deep yellow hemispherical heads (modified pileus), which are dehiscent (fall off), and act as propagules. This heads are characterized by masses of conidia on a synnema. Is a plant pathogen producing leaf brown and subcircular spots on coffee plants, primarily in Latin America. Can grow on all parts of the host, including leaves, stems and fruits.

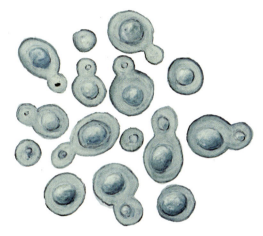

Debaryomyces hansenii (Zopf) Lodder & Kreger-van-Rij: yeast budding cells, and asci with a single ascospore, x 3,000.

Delitschia Auersw.
Delitschiaceae, Pleosporales, Pleosporomycetidae, Dothideomycetes, Pezizomycotina, Ascomycota, Fungi
Auerswald, B. 1866. *Delitschia* Awd., nov. gen., e grege *Sphaeriacearum simplicium* et affinitate *Sordariarum* et *Amphisphaeriarum*. *Hedwigia* 5:49.
Dedicated to the German geographer *Delitsch* + L. des. *-ia*. Ascomata uniloculate, perithecioid pseudothecia, scattered or gregarious, partially or entirely immersed; pseudothecia dark brown to black, smooth or roughened, globose to subglobose or obpyriform; ostiolar neck papillate to long cylindrical. Ascomatal wall pseudoparenchymatous, membranous to coriaceous. Centrum containing pseudoparaphyses. Asci bitunicate, cylindric to clavate, long- or short-stalked, 4- to many-spored. Ascospores dark brown to almost black, with an elongated germ slit, oval to broadly fusoid, 2-celled, often constricted at septum, septum median, transverse or oblique, sometimes separating into part-spores, surrounded by a gelatinous sheath. Coprophilous.

Dematiocladium Allegr., et al.
Nectriaceae, Hypocreales, Hypocreomycetidae, Sordariomycetes, Pezizomycotina, Ascomycota, Fungi
Crous, P. W., et al. 2005. *Dematiocladium celtidis* gen. sp. nov. (Nectriaceae, Hypocreales), a new genus from *Celtis* leaf litter in Argentina. *Mycol. Res.* 109(7):833-840.
From *Dematium* Pers., genus with pigmented hyphae + L. *clados* < Gr. *kládos*, branch, referring to the similarity to the genus *Cylindrocladium*. Setae unbranched, straight to flexuous, arising from pseudoparenchymatous cells in a basal stroma. Conidiophores consisting of a stipe and a penicillate arrangement of fertile branches. Stipe septate, hyaline, smooth, brown at base, arising from a basal stroma, frequently terminating in a swollen, globose apical cell that gives rise to 1-6 primary branches. Conidiogenous branches hyaline, smooth, 0-2-septate; terminal branches producing 1-6 phialides. Phialides elongate-doliiform to reniform or subcylindrical, straight to slightly curved, aseptate, apex with minute periclinal thickening and inconspicuous collarette. Conidia cylindrical, rounded at both ends, straight, hyaline, 1(-2)- septate, held in parallel clusters by hyaline slime. Chlamydospores globose, thick-walled, medium brown, arranged in intercalary mycelial chains. Collected on leaf litter of *Celtis* in Argentina.

Dendrophlebia Dhingra & Priyanka
Meruliaceae, Polyporales, Inc. sed., Agaricomycetes, Agaricomycotina, Basidiomycota, Fungi
Dhingra, G. S., Priyanka. 2011. *Dendrophlebia* (Agaricomycetes), a new corticioid genus from India. *Mycotaxon* 116:157-160.
From Gr. *dendrós*, tree + genus *Phlebia* Fr., for the presence of tree-like dendrohyphidia in addition to the characters of the genus *Phlebia*. Basidiocarps resupinate, closely adnate, effused, ceraceous; hymenial surface smooth, continuous, yellowish to brownish, turning dark-ruby with 3% KOH; margins not differentiated. Hyphal system monomitic; generative hyphae branched, septate, clamped; hyphae often agglutinated, penetrating deep into substratum, covered by yellowish-brown crystalline matter. Dendrohyphidia present. Cystidia thin-walled, hyphoid. Basidia clavate to subclavate, 4-sterigmate, with a basal clamp. Basidiospores ellipsoid, smooth, with thickened walls, inamyloid, acyanophilous. Collected on a decaying angiospermous stump in West Kameng, Bombila, Arunachal Pradesh, India.

Dendrophoma Sacc.—see **Dinemasporium** Lév.
Inc. sed., Xylariales, Xylariomycetidae, Sordariomycetes, Pezizomycotina, Ascomycota, Fungi
Saccardo, P. A. 1880. Conspectus generum fungorum italiae inferiorum nempe ad Sphaeropsideas, Melanconieas et Hyphomyceteas pertinentium, systemate sporologico dispositoru. *Michelia* 2(no. 6):4.
From Gr. *déndron*, tree + genus *Phoma* Sacc.

Dendrosporomyces

Dendrosporomyces Nawawi, et al.
Inc. sed., Inc. sed., Inc. sed., Agaricomycetes, Agaricomycotina, Basidiomycota, Fungi
Nawawi, A., et al. 1977. *Dendrosporomyces prolifer* gen. et sp. nov., a Basidiomycete with branched conidia. *Trans. Br. mycol. Soc.* 68(1):59-63.
From Gr. *dendrós*, tree + *sporá*, seed > L. *spora*, spore + *myces*, fungus, for the much-branched conidia. Mycelium of hyaline, branched, binucleate hyphae with dolipore septa. Conidiophores semi-macronematous, mononematous, erect, unbranched, hyaline, producing conidia terminally. Conidia holoblastic, branched, hyaline, consisting of a main axis bearing laterals and secondary branches in succession. Isolated from foam in the Gombak River, Selangor, Malaysia.

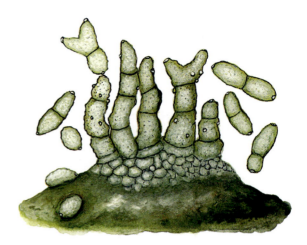

Denticularia fici Deighton: conidiophores arising from a stroma, and fusiform conidia; causes leaf spots in *Ficus exasperata*, x 800.

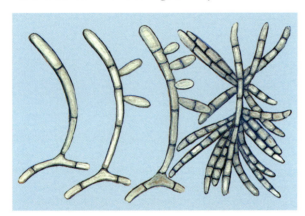

Dendrosporomyces prolifer Nawawi, J. Webster & R. A. Davey: development of conidiophores and conidia, showing the mature pluriseptate conidia of this aquatic fungus, x 120.

Dendrostilbella Höhn.—see **Claussenomyces** Kirschst.
Tympanidaceae, Helotiales, Leotiomycetidae, Leotiomycetes, Pezizomycotina, Ascomycota, Fungi
Höhnel, F. X. R. von, 1905. *Österr. Bot. Z.* 55(1):22.
From Gr. *déndron*, tree + genus *Stilbella* Lindau.

Denticularia Deighton
Inc. sed., Inc. sed., Inc. sed., Inc. sed., Pezizomycotina, Ascomycota, Fungi
Deighton, F. C. 1972. Four leaf-spotting Hyphomycetes from Africa. *Trans. Br. mycol. Soc.* 59(3):419-427.
From L. *denticulatus*, with small teeth + L. suf. *-aria*, meaning possession. Mycelium immersed. Conidiophores arising from stromata, densely crowded, brown, mostly simple, smooth, thin-walled, continuous or few septate, sympodial, polyblastic, denticulate, not cicatrized, denticles short, subcylindric with a truncate unthickened apex. Conidia pale brown, more or less fusiform, catenulate, with hila and unthickened scars, thin-walled, smooth or minutely rough-walled, continuous or 1-septate. Parasitic causing leaf spots on *Ficus exasperata* and *F. asperifolia* in Sierra Leone.

Dentinum Gray—see **Hydnum** L.
Hydnaceae, Cantharellales, Inc. sed., Agaricomycetes, Agaricomycotina, Basidiomycota, Fungi
Gray, S. F. 1821. *Nat. Arr. Brit. Pl.* (London) 1:650.
From L. *dentatus*, dentate, from *dens*, *dentis*, tooth + L. suf. *-inum*, which indicates possession or similarity, since it has a dentate hymenium.

Dermatella P. Karst.—see **Pezicula** Tul. & C. Tul.
Dermateaceae, Helotiales, Leotiomycetidae, Leotiomycetes, Pezizomycotina, Ascomycota, Fungi
Karsten, P. 1871. *Bidr. Känn. Finl. Nat. Folk* 19:16.
From genus *Dermea* Fr. + L. dim. suf. *-ella*.

Dermateopsis Nannf.
Dermateaceae, Helotiales, Leotiomycetidae, Leotiomycetes, Pezizomycotina, Ascomycota, Fungi
Nannfeldt, J. A. 1932. Studien über die Morphologie und Systematik der nicht-lichenisierten inoperculaten Discomyceten. *Nova Acta R. Soc. Scient. upsal.*, Ser. 4, 8(2):1-368.
From genus *Dermatea* Fr. + Gr. suf. *-ópsis*, appearance, i.e., similar to this genus. Apothecia erumpent, thickly gregarious or closely crowded, often cespitose, when young bilaterally compressed, hysteriform, opening with an irregular aperture leaving margin often notched or occasionally toothed, externally yellowish. Hymenium darker. Asci broadly clavate, inoperculate, 8-spored. Ascospores long-fusoid, straight or curved. Paraphyses filiform. On branches of oaks.

Dermatocarpon Eschw.
Verrucariaceae, Verrucariales, Inc. sed., Eurotiomycetes, Pezizomycotina, Ascomycota, Fungi
Eschweiler, F. G. 1824. Systema Lichenum, *Genera Exhibens rite distincta, Pluribus Novis Adaucta*.1-26.
From Gr. *dérma*, genit. *dérmatos*, skin, leather + *karpós*,

fruit. Foliose thallus with numerous fructifications, membranous or coriaceous, adnate or umbilicate, areolate, reticulate, rugose or squamulose, in layers delimited by upper and lower cortices. Fructifications minute or small perithecia, immersed one to several in each squamule or areola; at times only black ostioles visible on surface of thallus. Grows on moist or inundated rocks, soil and trees.

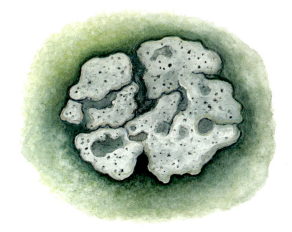

Dermatocarpon luridum (Dill. ex With.) J. R. Laundon: squamulose thallus of this lichen, with embedded perithecia, on moist rock, x 3.

Dermatodothella Viégas
Inc. sed., Inc. sed., Inc. sed., Dothideomycetes, Pezizomycotina, Ascomycota, Fungi
Viégas, A. P. 1944. Alguns fungos de Brasil II. Ascomicetos. *Bragantia* 4(1-6):1-392.
From Gr. *dérmatos*, skin, leather + *dothien*, abcess, pustule + L. dim. suf. *-ella*, referring to the clypeus around the ostiole. Hyphae subcuticular, erupting to form a black stroma. Stromata amphigenous, bearing ascigerous locules. Locules in groups, smooth, papillate, with a central pore surrounded by a clypeus. Asci clavate, with a thick wall. Ascospores fusoid, straight or slightly curved, 5-7-septate, brown, smooth. Collected on leaves of *Mikania* sp. in São Paulo State, Brazil.

Dermatodothella multiseptata Viégas: sagittal section of an erumpent black stroma, with embedded perithecia, and clypeus around the ostiole, on leaves of *Mikania* sp., in Sao Paulo State, Brazil, x 350.

Dermea Fr.
Dermateaceae, Helotiales, Leotiomycetidae, Leotiomycetes, Pezizomycotina, Ascomycota, Fungi
Fries, E. M. 1825. *Syst. orb. veg.* (Lundae) *1*:114.
From Gr. *dérma*, skin + des. *-ia*, for the erumpent ascomata. Apothecia occurring singly, or more often in cespitose clusters, often on a stromatic base, tubercular or discoid, rarely scutellate, usually dark-colored, coriaceous to subcarbonaceous. Asci usually broadly clavate, inoperculate, 8-spored. Ascospores comparatively large, occasionally minute, simple or becoming several septate with age, septation often erratic even in same species. Paraphyses colored, tips agglutinated into a dark brown or blackish epithecium. On woody hosts.

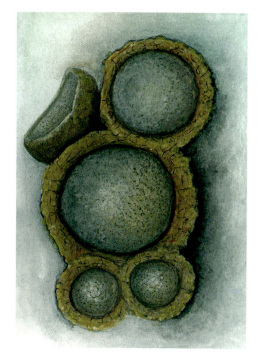

Dermea cerasi (Pers.) Fr.: cespitose and erumpent apothecia, on dry branches of *Prunus cerasus*, Germany, x 15.

Descalsia A. Roldán & Honrubia
Inc. sed., Inc. sed., Inc. sed., Inc. sed., Pezizomycotina, Ascomycota, Fungi
Roldán, A., M. Honrubia. 1989. *Descalsia*, a new aquatic hyphomycete anamorph genus. *Mycol. Res.* 92(4):494-497.
Named in honor of the Spanish mycologist *Enrique Descals* + L. des. *-ia*. Conidiophores micronematous, intercalary. Conidiogenous cells integrated or discrete, solitary, apical or lateral, with one or more loci, typically sympodial. Conidia holoblastic, branched, solitary, apical; branches paired, subacrogenous, synchronous. Conidial secession schizolytic. Detached conidia with branches lateral, nearly opposite or perpendicular to main axis, multiseptate. Isolated from stream foam in Riópar, Albacete, Spain.

Diademospora

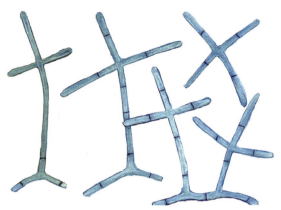

Descalsia cruciata A. Roldán & Honrubia: holoblastic, cross-shaped conidia, of this aquatic fungus, x 600.

Diademospora B. E. Söderstr. & Bååth
Inc. sed., Inc. sed., Inc. sed., Inc. sed., Pezizomycotina, Ascomycota, Fungi
Söderström B., E. Bååth. 1979. *Diademospora ramigera* gen. et sp. nov. from coniferous forest soil. *Trans. Br. mycol. Soc.* 72(2):340-342.

From Gr. *diadema*, crown, + L. *spora* < Gr. *sporá*, spore, for the arrangement of the conidia on the conidiophores. Colonies slow-growing, floccose, at first white, later becoming reddish-brown with brown reverse. Blastoconidia pleurogenous, formed singly on creeping hyphae; conidia septate, forked, consisting of one or two slightly pigmented basal cells, darker at septa, an apical cell and mostly one subterminal branch which is almost parallel to main axis; end cell and branch hyaline, provided with four subterminal and one terminal protuberance. Dacryoid hyaline microconidia in old colonies, chlamydospores absent. Isolated from coniferous forest soils in Sweden.

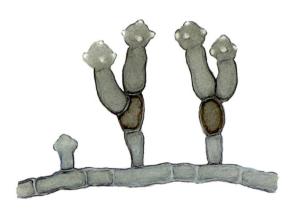

Diademospora ramigera B. E. Söderstr. & Baath: conidiophores with slightly pigmented basal cells, and conidia, whose terminal cells have protuberances, x 600.

Diaporthe Nitschke (syn. **Phomopsis** Sacc. & Roum.)
Diaporthaceae, Diaporthales, Sordariomycetidae, Sordariomycetes, Pezizomycotina, Ascomycota, Fungi
Nitschke, T. R. J. 1870. *Pyrenomyc. Germ.* 2:240.

From Gr. *diaporthéo*, to ruin, destroy, from *diá*, through + *porthéo*, devastate, in relation to the melanosis and damage caused by these fungi. Perithecia dark, generally embedded in a stroma. Conidia often of two types. Alpha globose, oval or elliptical, non-septate; beta conidia elongated, at times bent like a staff. Asexual morphs are phomopsis-like. Parasitic and saprobic on leaves and woody tissues.

Diaporthe impulsa (Cooke & Peck) Sacc.: saprobic stromata on dead wood of *Sorbus aucuparia*; at the left, sagittal section of a stroma with embedded perithecia; at the right, ostioles of the perithecia, x 10.

Diaporthe neoviticola Udayanga, et al. (syn. **Phomopsis viticola** (Sacc.) Sacc.): pycnidia with conidial cirrhus, on host tissues, x 10; alfa and beta conidia, x 400.

Diatrype Fr.
Diaporthaceae, Diaporthales, Sordariomycetidae, Sordariomycetes, Pezizomycotina, Ascomycota, Fungi
Fries, E. M. 1849. *Summa veg. Scand.*, Sectio Post. (Stockholm):384.

From Gr. *diatrypáo*, perforate, go through, pierce, from *diá*, through, + *trýpa*, *trýpes*, opening, hole. Stroma with erumpent ectostroma and endostroma through which extend long, ostiolar perithecial necks. Ascospores allantoid. On the dead wood or bark of branches and trunks of trees.

Diatrypella (Ces. & De Not.) De Not.
Diatrypaceae, Xylariales, Xylariomycetidae, Sordariomycetes, Pezizomycotina, Ascomycota, Fungi

Dictyodochium

De Notaris, G. 1863. *Sferiacei italici*. R. I. de Sordo-Muti. Genove, p. 29.

From genus *Diatrype* Fr. + L. dim. suf. *-ella*, from Gr. *diá*, through + *trýpa*, *trýpes*, opening, hole. Eustromata with ostiolate perithecia erumpent through ectostroma, embedded in endostroma. Ascospores allantoid. On the dead bark of trees.

Diatrype disciformis (Hoffm.) Fr.: front view of the pulvinate, erumpent stromata, showing the ostioles of the perithecia, on host bark, x 15.

Diatrypella quercina (Pers.) Cooke: sagittal section of the erumpent stroma on dead bark, with embedded perithecia in the lignicolous endostroma, x 40.

Dibotryon Theiss. & Syd.—see **Apiosporina** Höhn.
Venturiaceae, Venturiales, Pleosporomycetidae, Dothideomycetes, Pezizomycotina, Ascomycota, Fungi
Theissen, F., H. Sydow. 1915. Die Dothideales. *Annls. mycol.* 13(5/6):663.
From Gr. *dís*, twice + *bótrys*, genit. *bótryos*, raceme, for the ascostromata that occur aggregated or in racemes.

Dichomitus D. A. Reid
Polyporaceae, Polyporales, Inc. sed., Agaricomycetes, Agaricomycotina, Basidiomycota, Fungi
Reid, D. A. 1962. Notes on fungi which have been referred to the Thelephoraceae sensu lato. *Persoonia* 2(2):151.
From Gr. *díchoos*, double + *mitōs*, thread, referring to the branched hyphae of the context. Basidiocarps annual to perennial, resupinate to effused-reflexed. Pileus surface, white to blackish, hyphal system dimitic with arboriform skeletal binding hyphae. Basidiospores cylindrical to oblong-ellipsoid, smooth, inamyloid, thin-walled. Widespread, causing white rot on conifers and hardwoods.

Dicranophora J. Schröt.
Mucoraceae, Mucorales, Inc. sed., Inc. sed., Mucoromycotina, Zygomycota, Fungi
Schröter, J. 1886. Über die auf Hutpilzen vorkommeden Mucorineen. *Jber. schles. Ges. vaterl. Kultur* 64:183-185.
From Gr. *díkranos*, with two heads or horns + *phóros*, bearer < *phéro*, to carry. Sporangiophores branch dichotomously, with branches terminating in sporangiola. Parasitic on fleshy boletaceous basidiomycetes.

Dicranophora fulva J. Schröt.: sporangiophore that branches dichotomously, with terminal sporangiola, x 300.

Dictyochaetopsis Aramb. & Cabello
Chaetosphaeriaceae, Chaetosphaeriales, Sordariomycetidae, Sordariomycetes, Pezizomycotina, Ascomycota, Fungi
Arambarri, A., M. Cabello. 1990. Considerations about *Dictyochaeta*, *Codinaeopsis* and a new genus, *Dictyochaetopsis*. *Mycotaxon* 38:11-14.
From Gr. *díktyon*, a net + *chaité* or *chaetá*, a bristle + *-ópsis*, aspecto, appearance. Macronematous, brown and setiform conidiophores, with fertile or sterile apex, which bear phialidic, mostly discrete conidiogenous cells with distinct flared collarettes, and by hyaline, fusoid to cylindrical slimy conidia with or without setulae. Reported from litter or submerged plant debris.

Dictyodochium Sivan.—see **Gibbera** Fr.
Venturiaceae, Venturiales, Pleosporomycetidae, Dothideomycetes, Pezizomycotina, Ascomycota, Fungi
Sivanesan, A. 1984. *Acantharia*, *Gibbera* and their anamorphs. *Trans. Br. mycol. Soc.* 82(3):507-529.
From Gr. *díktyon*, net + *doché* > L. *dochium*, receptacle, container, for the dictyoseptate conidia.

Dictyonema

Dictyochaetopsis gonytrichodes (Shearer & J. L. Crane) Whitton, et al. (syn. **Codinaeopsis gonytrichoides** (Shearer & J. L. Crane) Morgan-Jones): conidiophore with phialides, and free conidia with a single setula at each end, isolated from dead stems of *Rubus*, x 400.

Dictyonema C. Agardth ex Kunth
Inc. sed., Inc. sed., Agaricomycetidae, Agaricomycetes, Agaricomycotina, Basidiomycota, Fungi
Agardth, C. A. 1822. *In*: C. S. Kunth, *Syn. pl.* (Paris) 1:1.
From Gr. *díktyon*, net + *néma*, filament. Thallus of loosely interwoven hyphae, semicircular, united to substrate on one side or forms an irregular layer. Hymenium, with basidia and cystidia, ventral part of thallus as a membranous pileus. Grows on trees.

Dictyonema thelephora (Spreng.) Zahlbr.: thallus of this basidiolichen, growing on the bark tree, x 1.

Dictyophora Desv.—see **Phallus** Junius ex L.
Phallaceae, Phallales, Phallomycetidae, Agaricomycetes, Agaricomycotina, Basidiomycota, Fungi
Desvaux, N. A. 1809. Observations sur quelques genres à établir dans la famille des champignons. *J. Bot.* (Morot) 2:92.
From Gr. *díktyon*, a net + suf. *-phóros*, < *phéro*, bear, support, for the presence of a reticulate indusium on the fructification. Immature fructification ovoid or subglobose, white or gray. At maturity pileus campanulate, reticulate, with a perforate apex; receptaculum fusiform or cylindrical, white, hollow; indusium suspended from lower part of pileus, free at base; gleba mucilaginous, fetid, on exterior surface. Spores elliptical, smooth. On highly organic soils and mulch in tropical, subtropical or temperate forests.

Dictyostomiopelta Viégas
Micropeltidaceae, Microthyriales, Inc. sed., Dothideomycetes, Pezizomycotina, Ascomycota, Fungi
Viégas, A. P. 1944. Alguns fungos de Brasil II. Ascomicetos. *Bragantia* 4(1-6):1-392.
From Gr. *díktyon*, a net + genus *Stomiopeltella* Theiss., i.e., a stomiopeltella-like fungus with dictyospores. Ascomata black thyriothecia, punctiform, with distinct margin, dimidiate-scutiform, circular to subcircular, lighter in color at margin. Upper wall composed of dark brown, intricately arranged hyphae. Asci globose-clavate, short-pedicellate or sessile, with thickened apex, 8-spored. Ascospores hyaline, oblong, muriform, constricted at the septa, smooth. Collected on *Manihot utilissima* in São Paulo State, Brazil.

Dicyma Boulanger—see **Ascotricha** Berk.
Xylariaceae, Xylariales, Xylariomycetidae, Sordariomycetes, Pezizomycotina, Ascomycota, Fungi
Boulanger, E. 1897. *Rev. gén. Bot.* 9:18.
From Gr. *dís*, twice + *kýma* > L. *cyma*, green cabbage, referring to the bouquet or inflorescence of conidiophores.

Didymella Sacc.
Didymellaceae, Pleosporales, Pleosporomycetidae, Dothideomycetes, Pezizomycotina, Ascomycota, Fungi
Saccardo, P. A. 1880. Fungi Gallici. Series II. *Michelia* 2(6):39-149.
From Gr. *dídymos*, double + L. dim. suf. *-ella*, referring to the two-celled ascospores. Ascomata uniloculate, perithecioid pseudothecia, scattered, immersed or erumpent; pseudothecia dark brown to black, globose to subglobose, ostiolate. Ascomatal wall pseudoparenchymatous, sometimes thickened around ostiole to form a clypeus. Centrum containing filamentous, often branched, pseudoparaphyses. Asci bitunicate, parallel, clavate to cylindrical, short-stalked, 8-spored. Ascospores 2-celled, septum median, oval to fusiform, hyaline. Asexual morphs are phoma-like. On leaves and stems of angiosperms conifers. *Didymella arachidicola* (Chochrjakov) Tomilin causes web blotch disease of peanut (*Arachis* spp.).

Didymosphaeria Fuckel
Didymosphaeriaceae, Pleosporales, Pleosporomycetidae, Dothideomycetes, Pezizomycotina, Ascomycota, Fungi

Fuckel, K. W. G. L. 1870. Symbolae mycologicae. Beiträge zur Kenntniss der Rheinischen Pilze. *Jb. nassau. Ver. Naturk.* 23-24:1-459.

From Gr. *dídymos*, double + genus *Sphaeria* Haller, i.e., a *Sphaeria* with 2-celled ascospores. Ascomata uniloculate, perithecioid pseudothecia, scattered or aggregated in groups, immersed, subepidermal, sometimes with slight clypeus. Pseudothecia dark brown, subglobose, with an ostiolar papilla. Ascomatal wall composed of small angular cells. Centrum with branched, filiform, trabeculate pseudoparaphyses. Asci bitunicate, cylindrical to cylindric-clavate, subclavate or broadly obovoid, short-stalked or subsessile, 1-8-spored. Ascospores 2-celled, septum median, smooth or minutely verruculose, olivaceous to dark brown. Asexual morphs pycnidial and periconia-like. On stems and leaves of herbaceous plants and on seeds.

Didymella fagi C. Z. Wei & Y. Harada: longitudinal section of a perithecioid pseudothecium, with asci, on plant tissues, x 250; one bitunicate ascus with bicellular ascospores, on inoculated leaves of *Fagus crenata*, Japan, x 1,000.

Didymosphaeria oblitescens (Berk. & Broome) Fuckel: stromata with pseudothecia, on the branch of *Rubus fruticosus*, x 10; sagittal section of a pseudothecium, x 20; one bitunicate ascus with eight ascospores, x 1,000; free bicellular ascospores, x 1,500.

Digitodesmium P. M. Kirk

Inc. sed., Pleosporales, Pleosporomycetidae, Dothideomycetes, Pezizomycotina, Ascomycota, Fungi

Kirk, P. M. 1981. New or interesting microfungi II. Dematiaceous Hyphomycetes from Esher Common, Surrey. *Trans. Br. mycol. Soc.* 77(2):279-297.

From L. *digitus*, finger + Gr. *desmos*, bundle + L. dim. suf. -*ium*, for the digitate conidia united at base. Mycelium mostly immersed in substratum. Conidiophores semi-macronematous, mononematous, fasciculate, composed of pale brown, smooth, septate, moniliform hyphae. Conidiogenous cells holoblastic, monoblastic, integrated, terminal determinate, globose to doliiform, minutely cicatrized. Conidia acrogenous, solitary, seceding schizolytically, euseptate, dry, digitate, apex of arms occasionally with a hyaline gelatinous cap. Collected on dead wood of *Quercus roboris* in Esher Common, Surrey, U.K.

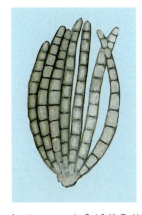

Digitodesmium heptasporum L. Cai & K. D. Hyde: digitate conidium with seven multiseptate arms, on wood submerged in small stream, in Yunnan, China, x 1,000.

Dimargaris Tiegh.

Dimargaritaceae, Dimargaritales, Inc. sed., Inc. sed., Kickxellomycotina, Zygomycota, Fungi

van Tieghem, P. 1875. Nouvelles recherches sur les mucorinées. *Annls. Sci. Nat., Bot., sér. 6*, 1:154.

From Gr. *dís*, two, double + *margarítes*, pearl. Shiny heads forming bispored sporangiola (merosporangia). Mycoparasitic on mucoraceous fungi.

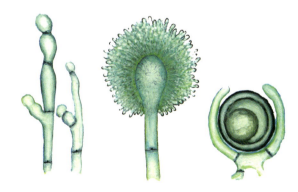

Dimargaris cristalligena Tiegh.: vesiculate sporangiophore with merosporangia, x 20; at the left, merosporangia with bispored sporangiola, x 130; at the right, zygosporangium, x 400. Collected on dung, France.

Dimorphospora

Dimorphospora Tubaki (syn. **Fluminispora** Ingold)
Helotiaceae, Helotiales, Leotiomycetidae, Leotiomycetes, Pezizomycotina, Ascomycota, Fungi
Tubaki, K. 1958. Studies on the Japanese Hyphomycetes. V. Leaf and stem group with a discussion of the classification of Hyphomycetes and their perfect stages. J. Hattori bot. Lab. 20:156.
From Gr. *di-* < *dis*, pref. meaning two, double + *morphé*, form, shape + *sporá*, spore. Mycelium aquatic, septate, branched, hyaline; aerial conidiophores penicillately branched, forming flask-shaped phialides with conspicuous collarettes, through which phialospores emerge; conidia aggregating in heads at the apex of the phialides, not in mucus, with 1-celled, hyaline, globose; submerged conidiophores similarly branched but producing buds (blastospores) at apices; aquatic, saprophytic on dead leaves.Mycelium submersed, septate, branched, hyaline. Conidiophores hyaline, branched. Conidia 1-celled, hyaline, oval or subreniform, arising from first terminal, afterwards near arising near the other. Isolated from submerged decaying leaves of deciduous trees from streams in UK.

Dimorphospora foliicola Tubaki (syn.: **Fluminispora ovalis** Ingold): conidiophores, one with macroconidia, and the other with microconidia developed in pure culture, emerging from phialides, x 1,000. Isolated from submerged decaying leaves of deciduous trees from streams in England.

Dinemasporium Lév. (syn. **Dendrophoma** Sacc.)
Inc. sed., Xylariales, Xylariomycetidae, Sordariomycetes, Pezizomycotina, Ascomycota, Fungi
Léveille 1846. Annls Sci. Nat., Bot., sér. 3, 5:274.
From Gr. *di-* < *dis*, pref. meaning two, double + *néma*, thread + *sporídion*, dim. of *sporá*, spore. Pycnidia black, cup-shaped, superficial, with long dark setae; these are closed at first and open at maturity exposing a slimy conidial mass; conidiophores rod-shaped, mostly simple; phialidic conidiogenous cells; conidia hyaline, 1-celled, elongate or allantoid, with a slender appendage at each end. Occur on various herbaceous and woody substrata. Conidiophores long, frequently branched; conidia unicellular, hyaline.

Dinemasporium sasae A. Hashim., Sat. Hatak. & Kaz.: setose conidiomata growing on the surface of dead host plant (*Sasa kurilensis*, in Japan), x 125; longitudinal section of a conidioma, showing phialides and mature unicellular conidia, x 400; phialides giving rise to hyaline, smooth and gutulate conidia, with an appendage (sometimes double) in each pole, x 1,700.

Diplocarpon F. A. Wolf (syns. **Entomosporium** Lév., **Marssonina** Magnus)
Dermateaceae, Helotiales, Leotiomycetidae, Leotiomycetes, Pezizomycotina, Ascomycota, Fungi
Wolf, F. A. 1912. The perfect stage of *Actinonema rosae*. Bot. Gaz. 54:231.
From Gr. *diplóos*, double + *karpóo*, to produce fruit, perceive fruit, profit, or from *karpós*, fruit, for having two types of fructifications. Apothecia erumpent, obconic or turbinate, black. Acervuli without setae, entomosporium-like, with appendaged conidia that appear microscopically similar to "insects" or "flame". Important pathogens of cultivated plants causing leaf spots known as scorching in apple, pear, rose and strawberry.

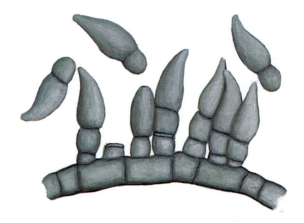

Diplocarpon fragariae (Lib.) Rossman (syn. **Marssonina fragariae** (Lib.) Kleb.): conidiogenous cells (phialides), with flame-shaped, bicelled conidia, x 1,250. Causes leaf scorch of strawberry.

Dipodascopsis

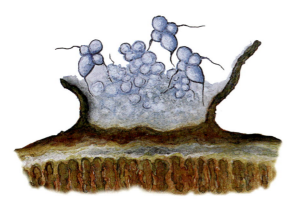

Diplocarpon mespili (Sorauer) B. Sutton (syn. **Entomosporium mespili** (DC.) Sacc.): conidia that resemble the shape of an insect, with the apical cell and each of the lateral cells has a long appendage, causing leaf blight of hawthorn, x 1,000.

Diplocarpon rosae F. A. Wolf: longitudinal sections of acervulus with conidia (asexual phase), and apothecium with asci and ascospores (sexual phase), on rose leaf, x 100.

Diplodia Fr.

Botryosphaeriaceae, Botryosphaeriales, Inc. sed., Dothideomycetes, Pezizomycotina, Ascomycota, Fungi

Montagne, J. F. C. 1834. Notice sur les plantas cryptogames récemment decouvertes en France contenant aussi d'indication précise des localités de quelques espèces les plus rares de la flore francaise. *Annls. Sci. Nat., Bot.,* sér. 2, 1:295-307.

From Gr. *diplóos* or *díploos*, double < *diplóe*, something double or divided into two parts < *diplóo*, to double + L. suf. -*ia*, characteristic of. Pycnidia producing bicellular conidia. Sexual morphs botryosphaeria-like. Parasitic on plants.

Diplodia africana Damm & Crous: sagittal section of pycnidium, with bicellular conidia, on plant host, x 200. Isolated from wood, close to pruning wound of *Prunus persica*, in Western Cape Province, South Africa.

Diplolaeviopsis Giralt & D. Hawksw.

Inc. sed., Inc. sed., Inc. sed., Inc. sed., Pezizomycotina, Ascomycota, Fungi

Giralt, M., D. L. Hawksworth. 1991. *Diplolaeviopsis ranula*, a new genus and species of lichenicolous coelomycetes growing on the *Lecanora strobilina* group in Spain. *Mycol. Res.* 95(6):759-761.

From Gr. *diplóos*, double + L. *laevis*, smooth + Gr. -*ópsis*, appearance, similar to the two-celled conidia in genus *Lichenodiplis* Dyko & D. Hawksw. Conidiomata pycnidial arising singly, uniloculate, immersed, becoming erumpent at maturity, dark brown to black, ostiolate, walls composed of olivaceous brown pseudoparenchymatous cells. Conidiophores absent or subglobose to subcylindrical, simple or sparsely branched, more or less hyaline. Conidiogenous cells subcylindrical, occasionally proliferating enteroblastically, hyaline. Conidia holoblastic, arising singly, dry, elongate-soleiform to tadpole-shaped, hyaline, 1(-2)-septate, constricted at septum, smooth. Collected on *Lecanora* Ach. in Terragona, Spain.

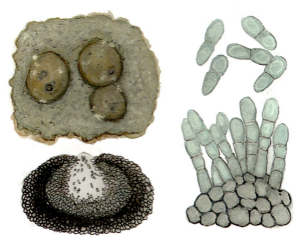

Diplolaeviopsis symmictae Diederich & Coppins: pycnidial conidiomata on apothecia in thallus of *Lecanora symmicta*, Arizona, U.S.A., x 50; sagittal section of pycnidium, x 175; conidiophores with bicellular conidia, x 650.

Dipodascopsis L. R. Batra & Millner

Lipomycetaceae, Saccharomycetales, Saccharomycetidae, Saccharomycetes, Saccharomycotina, Ascomycota, Fungi

Batra, L. R., P. C. Millner. 1978. *In:* C. V. Subramanian (ed.), 1978. *Taxonomy of Fungi* (Proc. Int. Symp. Madras, 1973) 1:209.

From genus *Dipodascus* Lagerh. < Gr. *dís*, two + *poús*, genit. *podós*, foot + *askós* > L. *ascus*, wine-skin, ascus + suf. -*opsis* < Gr. -*ópsis*, appearance, aspect, i.e., similar to *Dipodascus*. Hyphae and gametangia uninucleate, instead of multinucleate in *Dipodascus*. Reproducing asexually by blastospores instead of multinucleate oidia. Saprobic on exudates of higher plants.

Dipodascus

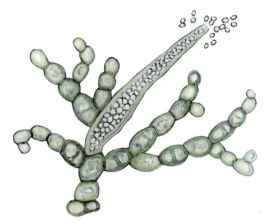

Dipodascopsis uninucleata (Biggs) L. R. Batra & Millner: pseudomycelium with budding cells, and the multisporate ascus, x 1,000.

Dipodascus Lagerh.
Dipodascaceae, Saccharomycetales, Saccharomycetidae, Saccharomycetes, Saccharomycotina, Ascomycota, Fungi
Lagertheim, G. 1892. Eine neue geschlechliche Hemiasce, *Dipodascus albidus*. *Jb. wiss. Bot.* 24:549-565.
From *dís*, two + *poús*, genit. *podós*, foot + *askós* > L. *ascus*, wine-skin, ascus. Mature ascus with two suspensors, one on each side, representing gametogamia that previously fused to form zygote, cylindrical, tapered toward apex. Ascospores ovoid to ellipsoid. Saprobic on exudates of plants.

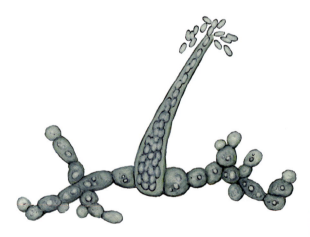

Dipodascus geniculatus de Hoog, et al.: pseudomycelium with blastospores, and an ascus with numerous ascospores, x 1,000.

Disciotis Boud.
Morchellaceae, Pezizales, Pezizomycetidae, Pezizomycetes, Pezizomycotina, Ascomycota, Fungi
Boudier, J. L. E. 1885. Nouvelle classification des discomycetes charnus. *Bull. Soc. mycol. Fr.* 1:100.
From Gr. *dískios*, very dark (from *dís*, a lot + *skiá*, dark) + *otós*, ear + L. suf. *-is*, which indicates narrow connection. Apothecia large, up to 15 cm long, dark brown, completely extended or even reflexed, lobulated, exterior surface whitish or yellowish, in center a short, wide peduncle, occasionally rudimentary. Hymenial surface often veined or circumvolutioned. Asci broadly elliptic. Paraphyses wide, septate. Grows in forests and river banks on humid soils.

Disciotis venosa (Pers.) Arnould: lobulate and large apothecium, on humid soil, x 15.

Disciseda Czern.
Agaricaceae, Agaricales, Agaricomycetidae, Agaricomycetes, Agaricomycotina, Basidiomycota, Fungi
Czernajew, B.M. 1845. Nouveaux cryptogames de l'Ukraine et quelques mots sur la flore de ce pays. *Bull. Soc. Imp. nat. Moscou* 18(2):153.
From L. *discínde*, *discíndere*, to cut, to tear, in reference to the exoperidium which, at maturity, is fissured circumscissily at the equator. Basidiocarp depressed globose or obovate. Peridium composed of two layers, exoperidium and endoperidium. Exoperidium either membranous or formed from hyphae intermixed with sand particles and vegetable debris, breaking away except for a small discoid or cupulate basal portion; endoperidium membranous or papyraceous, dehiscing by a small apical mouth. Gleba pulverulent, capillitium of short simple or branched hyphae. Spores globose, verrucose, with long, short or rudimentary pedicels. Solitary or in small groups in the ground in North and South America, Europe, Asia, Africa, Australia and New Zealand.

Disciseda candida (Schwein.) Lloyd: basidiomata, showing ostiole in the endoperidium, and sand case at the base, on soil, x 1.

Dothidea

Discostromopsis H. J. Swart—see **Seimatosporium** Corda

Discosiaceae, Amphisphaeriales, Xylariomycetidae, Sordariomycetes, Pezizomycotina, Ascomycota, Fungi
Swart, H. J. 1979. Australian leaf inhabiting fungi. X. *Seimatosporium* species on *Callistemon*, *Melaleuca* and *Leptospermum*. Trans. Br. mycol. Soc. 73(2):213-221.
Named for the similarity to genus *Discostroma* Clem. + Gr. suf. *-ópsis*, aspect, appearance.

Discula Sacc.

Gnomoniaceae, Diaporthales, Sordariomycetidae, Sordariomycetes, Pezizomycotina, Ascomycota, Fungi
Saccardo, P. A. 1884. *Syll. fung.* (Abellini) 3:674.
From L. *discus*, disk + dim. suf. *-ula*, in reference to the circular acervuli. Acervuli in epidermis. Causing dogwood anthracnose, leaf spot, twig dieback, or bract necrosis of dogwood in temperate regions.

Discula destructiva Redlin: acervuli with conidiogenus cells, and conidia, parasitizing twigs and leaves of *Cornus florida*, Maryland, U.S.A., x 700.

Dispira Tiegh.

Dimargaritaceae, Dimargaritales, Inc. sed., Inc. sed., Kickxellomycotina, Zygomycota, Fungi
van Tieghem, P. 1875. Nouvelles recherches sur les mucorinées. *Annls Sci. Nat., Bot., sér.* 6,1:160.
From Gr. *dís*, two, double + L. *spira*, spiral, a coil. Sporangiophores coiled or recurved, forming double spirals bearing sporangiola. Mycoparasitic on mucoraceous fungi.

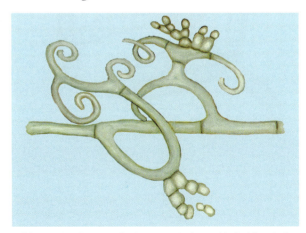

Dispira simplex R. K. Benj.: sporangiophore with double spirals branches, and sporangiola, x 1,100. Isolated in California, U.S.A.

Dolabra C. Booth & W. P. Ting

Inc. sed., Inc. sed., Inc. sed., Dothideomycetes, Pezizomycotina, Ascomycota, Fungi
C. Booth, W. P. Ting. 1964. *Dolabra nepheliae* gen. nov., sp. nov., associated with canker of *Nephelium lappaceum*. Trans. Br. mycol. Soc. 47(2):235-237.
From L. *dolabra*, hatchet, for the shape of the ascomata in vertical view. Ascostromata superficial, singular, either pyriform or dolabriform, with a black, outer crust. Locules with abundant pseudoparaphyses, ostiole in upper part, sterile stipe in lower part. Asci bitunicate, cylindrical. Ascospores scolecosporous, hyaline, septate. Collected on *Nephelium lappaceum* in Selangor, Malaysia.

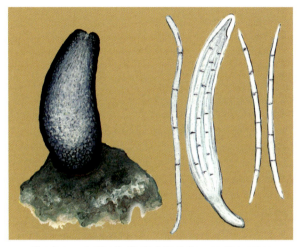

Dolabra nepheliae C. Booth & W. P. Ting: sagittal section of the hatchet ascostroma, x 200; a bitunicate ascus with ascospores, x 750, many septate ascospores, x 750.

Doratomyces Corda

Microascaceae, Microascales, Hypocreomycetidae, Sordariomycetes, Pezizomycotina, Ascomycota, Fungi
Corda, A. J. C. 1829. In: Sturm, *Deutschl. Fl., 3 Abt.* (Pilze Deutschl.) 2:65.
From Gr. *dorátion*, dart, or < *dóry*, *dóras*, genit. *dóratos*, lance, arrow + *mýkes* > L. *myces*, fungus. Synnemata composed of fascicles of long conidiophores that terminate in a lance-shaped head. Saprobic on vegetable remains, feathers, dung and soil.

Dothidea Fr.

Dothideaceae, Dothideales, Dothideomycetidae, Dothideomycetes, Pezizomycotina, Ascomycota, Fungi
Fries, E. M. 1818. *Observ. mycol.* 2:347.
From Gr. *dothién*, genit. *dothiénos*, abscess, boil, small tumor. Stromata pulvinate, erumpent on a broad base, with perithecioid, ostiolate locules, multilocular, parenchymatous. Asci short, cylindrical, fasciculate; originating from hemispherical layer in base of locule. Ascospores brown, 1-septate. On stems of plants.

Dothidella

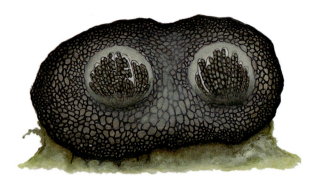

Dothidea sambuci (Pers.) Fr.: erumpent ascostroma, with two perithecioid locules, asci and ascospores, on host tissues, x 200. On branches of *Sambucus nigra* and *S. ebulus*, Germany.

Dothidella Speg.
Polystomellaceae, Inc. sed., Inc. sed., Dothideomycetes, Pezizomycotina, Ascomycota, Fungi
Spegazzini, C. 1880. Fungi argentini. *Anal. Soc. cient. argent.* 9(4):9.
From the genus *Dothidea* Fr. + L. dim. suf. *-ella*. The first comes from the Gr. *dothién*, genit. *dothíenos*, from *dothíon*, genit. *dothíonos*, small tumor, tubercle, abscess, furuncle. Pseudothecia hemiperithecioid, dimidiate, shield-shaped, developing subcuticularly with multiple ascigerous locules. Asci obclavate or cylindrical. Ascospores 1-septate. Causes leaf spots of plants.

Dothidella ulmi (C.-J. Duval) G. Winter: leaf of *Ulmus procera*, showing groups of pseudothecia, causing leaf spots, x 1.

Drechmeria W. Gams & H. B. Jansson
Clavicipitaceae, Hypocreales, Hypocreomycetidae, Sordariomycetes, Pezizomycotina, Ascomycota, Fungi
Gams, W., Jansson, H.B. 1985. The nematode parasite *Meria coniospora* Drechsler in pure culture and its classification. *Mycotaxon.* 22(1):33-38
Named to honor the American mycologist C. *Drechsler*, and Welsh professor E. *Mer* + L. suf. *-ia*, meaning belonging. Mycelium hyaline, branched; conidiophores simple, elongate, septate; conidia hyaline, 1-celled, produced singly or in clusters on lateral or apical sterigmata. Is an obligate fungal pathogen that infects nematodes via the adhesion of specialized spores to the host cuticle.

Drechmeria coniospora (Drechsler) W. Gams & H.-B. Jansson: conidiophore and obclavate, hyaline conidia, x 500.

Drechslera S. Ito
Pleosporaceae, Pleosporales, Pleosporomycetidae, Dothideomycetes, Pezizomycotina, Ascomycota, Fungi
Ito, S. 1930. On some new ascigerous stages of species of *Helminthosporium* parasitic on cereals. *Proc. Imp. Acad. Japan* 6:352-355.
Named in honor of the American mycologist *Charles Drechsler* + L. fem. ending *-a*. Conidiophores sympodial, dematiaceous. Conidia cylindrical, distoseptate, porospores that germinate from any or all cells. Parasitic and saprobic on grasses.

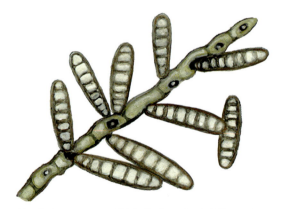

Drechslera avenacea (M. A. Curtis ex Cooke) Shoemaker: conidiophore with distoseptate conidia, x 250.

Drepanopeziza (Kleb.) Jaap (syn. Gloeosporium Desm. & Mont.)
Dermateaceae, Helotiales, Leotiomycetidae, Leotiomycetes, Pezizomycotina, Ascomycota, Fungi
Desmazières, J. B. H., J. F. C. Montagne. 1849. *Annls Sci. Nat., Bot., sér.* 3 12:295.

From. Gr. *drepané* and *drapánon*, a sickle + genus *Peziza* Dill. ex Fr. < Gr. *péziz* or *pézikes*, an ancient name to designate certain sessile fungi that emit a cloud of spores when mature. Apothecia black with irregular margin. Asci, clavate, unitunicate, apex bluing in iodine. Ascospores ellipsoidal to spherical, aseptate, biguttulate, hyaline. Acervulus mucilaginous in which conidia embedded. Asexual morph gloeosporium-like. Causes anthracnose, leaf spot, and dieback of woody plants in temperate regions.

Ductifera Lloyd
Inc. sed., Auriculariales, Inc. sed., Agaricomycetes, Agaricomycotina, Basidiomycota, Fungi
Lloyd, C. G. 1917. Rare or interesting fungi received from correspondants. *Mycol. Notes (Cincinnati)* 5:711.
From L. *ductus*, duct, brush-stroke, stroke + *fero*, to bear. Fructification surface with vesicles or irregular reliefs, brain-like convolutions, gelatinous-cartilaginous, firm, generally small, opaque mass, flat, pulvinate or cerebriform, white or grayish-yellow. On wood in humid places in tropical and temperate forests. The type species, *D. millei* Lloyd, is similar to species of *Tremella* (Dill.) Fr., but its structure is different.

Duddingtonia R. C. Cooke
Orbiliaceae, Orbiliales, Orbiliomycetidae, Orbiliomycetes, Pezizomycotina, Ascomycota, Fungi
Cooke, R. C. 1969. Two nematode-trapping hyphomycetes, *Duddingtonia flagrans* gen. et comb. nov., and *Monacrosporium mutabilis* sp. nov. *Trans. Br. mycol. Soc.* 53(2):315-319.
Named in honor of the British mycologist *C. L. Duddington* + L. des. *-ia*, for his studies of nematode-trapping fungi. Colonies effused, almost smooth, with few aerial hyphae, white to pale pink. Mycelium of branched, septate, smooth-walled hyphae. Conidiophores arise singly, erect, straight, unbranched. Conidia arise singly as blown-out ends of conidiophore; after first conidium produced, a new growing point formed to one side and slightly below it, then a second conidium forms. On nematode-infested agar, adhesive three-dimensional hyphal networks formed in which nematodes are captured. Isolated from soil from Rivelin Dam, Yorks, UK.

Duplicaria Fuckel
Rhytismataceae, Rhytismatales, Leotiomycetidae, Leotiomycetes, Pezizomycotina, Ascomycota, Fungi
Fuckel, K. W. G. L. 1870. Symbolae mycologicae. Beiträge zur Kenntniss der Rheinischen Pilze. *Jb. nassau. Ver. Naturk.* 23-24:1-459.
From L. *duplex*, twofold > *duplicaria*, double, in reference to the bifusoid ascospores. Ascomata apothecia, black, elliptical, subcuticular, opening irregularly. Hymenium cupulate at first, then becoming flat; subhymenial layer hyaline, grading into a thin basal layer of large brownish cells in contact with brown, discolored epidermal cells with dark brown hyphae. Paraphyses simple, fusiform. Asci clavate-saccate, unitunicate. Ascospores bifusiform, 1-celled, lower part tapering toward base. Pycnidia small, black, melasmia-like. Causing lesions on needles of conifers.

Durandiella Seaver
Tympanidaceae, Phacidiales, Leotiomycetidae, Leotiomycetes, Pezizomycotina, Ascomycota, Fungi
Seaver, F. J. 1932. The Genera of Fungi. *Mycologia* 24(2):248-263.
Named in honor of the American botanist *E. J. Durand* + connective *-i-* + L. dim. suf. *-ella*. Apothecia erumpent, separate or cespitose, sessile to substipitate, coriaceous, black, glabrous. Hymenium usually flat or nearly so. Asci cylindric-clavate, 8-spored. Ascospores subfiliform to filiform, hyaline, septate, variously curved. Paraphyses hyaline, filiform, tips forming an epithecium. Conidial bodies cylindric or cylindric-subulate to irregularly rounded, or ovoid to subglobose. Conidia subfiliform to filiform, hyaline, septate, variously curved. On woody stems.

Durandiella gallica M. Morelet: erumpent apothecia, on bark of *Abies alba*, France, x 10.

Durella Tul. & C. Tul. (syn. Plasia Sherwood)
Helotiaceae, Helotiales, Leotiomycetidae, Leotiomycetes, Pezizomycotina, Ascomycota, Fungi
Tulasne, L. R., C. Tulasne. 1865. *Select. fung. carpol.* (Paris) 3:177.
The etymology could not be determined. Apothecia greenish-black, erumpent, sessile. Asci clavate, inoperculate, iodine-negative, 8-spored. Ascospores cylindrical-ellipsoidal or ellipsoidal-fusiform, smooth, hyaline. Conidiomata pycnidial, discoid or short-hysteriiform, cupulate, pycnidial wall of interwoven hyphal gelatinous paraplectenchyma. Conidiogenous cells lining base of pycnidium attached directly to pycnidial wall or in short chains, enteroblastic, mono- to polyphialidic. Conidia large, cylindrical, hyaline, transversely septate, not appendaged. Gregarious on stems of herbaceous plants or dead wood, in England.

Dussiella

Dussiella Pat. (syn. **Echinodothis** G. F. Atk.)
Clavicipitaceae, Hypocreales, Hypocreomycetidae, Sordariomycetes, Pezizomycotina, Ascomycota, Fungi
Patouillard, M. N. 1890. *Dussiella*, Nouveau genre de'Hypocréacées. *Bull. Soc. mycol. Fr.* 6(1):107.
Named in honor of the Swish botanist R. P. *Antonie Düss.* + L. dim. suf. *-ella*. Stroma subfleshy or corky, light colored, pulvinate to subglobose or irregular in form, often constricted at base, sometimes entirely surrounding host, consisting of several layers. Perithecia superficial, scattered, subcylindrical, sessile, giving an echinulate appearance to stroma. Asci cylindrical, 8-spored. Ascospores linear, septate, at length separating at septa into short segments. Associated with the perennial bamboo grass *Arundinaria tecta*, in subtropical and tropical regions. *D. tuberiformis* is one of only a few fungi that are capable of infecting insects (scale insects: Coccidae, Homoptera) and plants.

Dwibahubeeja N. Srivast., et al.
Inc. sed., Inc. sed., Inc. sed., Inc. sed., Pezizomycotina, Ascomycota, Fungi
Srivastana, N., et al. 1995. New hyphopodiate hyphomycetes from North-Eastern Uttar Pradesh, India. *Mycol. Res.* 99(4):395-396.
From Skt. *dwi*, two + *bahu*, arms + *beeja*, having propagules, for the bifurcate conidia. Colonies epiphyllous, brownish-black. Mycelium branched, septate, smooth-walled, brown, hyphopodiate. Stromata and setae lacking. Conidiophores micronematous, mononematous, with conidiogenous cells borne directly on hyphae. Conidiogenous cells monoblastic, integrated, determinate. Conidia solitary, dry, acrogenous, brown, smooth-walled, bifurcate, with arms joined at base. Collected on living leaves of *Calamus tenuis* in Barhani, Uttar Pradesh, India

Dussiella tuberiformis (Berk. & Ravenel) Pat. ex Sacc. (syn. **Echinodothis tuberiformis** (Berk. & Ravenel) G. F. Atk.: corky and echinulate stroma, with superficial and scattered perithecia, on stem of *Arundinaria tecta*, x 1.

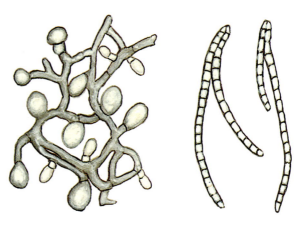

Dwibahubeeja indica N.Srivast. et al.:mycelium with capitate hyphopodia, x 700; and young and mature conidia, x 1,000.

E

Entomophthora muscae

E

Echinella Massee—see **Pirottaea** Sacc.
Dermateaceae, Helotiales, Leotiomycetidae, Leotiomycetes, Pezizomycotina, Ascomycota, Fungi
Massee, G. E. 1895. *Brit. Fung.-Fl.* (London) 4:304.
From Gr. *echínos*, hedgehog + L. dim. suf. *-ella*, referring to the hairs that cover the ascomata.

Echinocatena R. Campb. & B. Sutton
Inc. sed., Inc. sed., Inc. sed., Inc. sed., Pezizomycotina, Ascomycota, Fungi
Campbell, R., B. C. Sutton. 1977. Conidial ontogeny in *Echinocatena arthrinioides* gen. et sp. nov. (Deuteromycotina: Hyphomycetes). *Trans. Br. mycol. Soc.* 69:125-131.
From Gr. *echínos*, hedgehog, spiny + L. *catena*, a chain, referring to the chains of echinulate condiogenous cells. Conidiophores formed from superficial mycelium, micronematous, mononematous, unbranched, straight, pale brown, sparsely echinulate or smooth. Conidiogenous cells arising in simple or branched acropetal chains from apex of conidiophore, separated by prominent dark brown, thick septa, pale brown, echinulate, cylindrical to doliiform, polyblastic, integrated, indeterminate, distal part fertile with up to 7 conidiogenous loci. Conidia solitary, dry, spherical, brown, thick-walled, aseptate, echinulate.

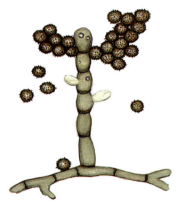

Echinocatena arthrinioides R. Campb. & B. Sutton: conidiophore with chains of echinulate conidia, on dead leaf, Rajasthan, India, x 1,250.

Echinodontium Ellis & Everh.
Echinodontiaceae, Russulales, Inc. sed., Agaricomycetes, Agaricomycotina, Basidiomycota, Fungi
Ellis, J. B., B. M. Everhardt. 1900. New species of fungi from various localities with notes on some published species. *Bull. Torrey bot. Club* 27(2):49.
From Gr. *echínos*, hedgehog + *odoús, odóntos*, tooth + L. dim. suf. *-ium*, which indicates connection or similarity, for the dentate hymenophore similar to the spines of a hedgehog. Fructifications perennial, resupinate to ungulate, i.e., similar to horse hoof, with a smooth, tuberculate or dentate hymenophore. In *E. tinctorium* hymenophore of thick, long, hard spines that are at times united laterally like a labyrinth. The Indians of the western United States extract a dye from this fungus as suggested by the specific name. Causes a severe brown rot of heart wood of living conifers, principally firs.

Echinodontium tinctorium (Ellis & Everh.) Ellis & Everh.: ungulate basidioma, with dentate hymenophore, on the bark of conifer tree, x 0.125. Collected on logs of hemlock (*Tsuga* spp.), Alaska, U.S.A.

Echinodothis G. F. Atk.—see **Dussiella** Pat.
Clavicipitaceae, Hypocreales, Hypocreomycetidae, Sordariomycetes, Pezizomycotina, Ascomycota, Fungi

Eleutherascus

Atkinson, G. F. 1894. Steps toward a revision of the linosporous species of North America graminicolous Hypocreaceae. *Bull. Torrey bot. Club* 27(2):222-225.
From Gr. *echínos*, hedeghog + *dothien*, abcess, segments.

Echinotrema Park.-Rhodes—see **Trechispora** P. Karst.
Hydnodontaceae, Trechisporales, Inc. sed., Agaricomycetes, Agaricomycotina, Basidiomycota, Fungi
Parker-Rhodes, A. F. 1955. The Basidiomycetes of Skokholm Island XIII. *Echinotrema clanculare* gen. et sp. nov. *Trans. Br. mycol. Soc.* 38:366-368.
From Gr. *echínos*, hedgehog, spiny + *tréma*, hole, referring to echinulate spores and habitat of fungus.

Eidamella Matr. & Dassonv.
Myxotrichaceae, Inc. sed., Inc. sed., Leotiomycetes, Pezizomycotina, Ascomycota, Fungi
Matruchot, L., C. Dassonville. 1901. *Eidamella spinosa*, dermatophyte préduisant des périthèces. *Bull. Soc. mycol. Fr.* 17:123-132.
Named in honor of *E. Eidam* + L. dim. suf. *-ella*. Ascomata lacking a definite peridium, i.e., a gymnothecium, with uncinate peridial appendages. Saprobic on keratinous remains of animals.

Eidamella spinosa Matr. & Dassonv.: gymnothecium with peridial appendages, x 300.

Elaphomyces Nees
Elaphomycetaceae, Eurotiales, Eurotiomycetidae, Eurotiomycetes, Pezizomycotina, Ascomycota, Fungi
Nees von Esenbeck, C. G. D. 1820. Gesch. merkw. Pilze 4:LXIX.
From Gr. *élaphos*, stag, deer + *mýkes* > L. *myces*, fungus, because apparently deer and other herbivorous animals eat these fungi. Ascomata globose, semifleshy, enclosed, fertile throughout due to absence of a sterile base in interior; surrounded by a well-defined, more or less thick peridium. Asci evanescent. Ascospores dark brown or black, in a powdery mass with hyphae of capillitial trama. Mycorrhizal with pine and oak, frequently parasitized by *Tolypocladium* W. Gams.

Elaphomyces granulatus Fr.: semifleshy ascocarp, on soil, x 1; it is frequently parasitized by **Tolypocladium** W. Gams.

Elegantimyces Goh, et al.
Inc. sed., Pleosporales, Pleosporomycetidae, Dothideomycetes, Pezizomycotina, Ascomycota, Fungi
Goh, T. K., et al. 1998. *Elegantimyces sporidesmiopsis* gen. et sp. nov. on submerged wood from Hong Kong. *Mycol. Res.* 102:239-242.
From L. *elegantis*, elegant + *myces* < Gr. *mykés*, genit. *mykétos*, fungus, an elegant fungus. Colonies on natural substratum black, hairy. Mycelium immersed. Stromata not developed. Setae and hyphopodia absent. Conidiophores macronematous, mononematous, cylindrical, euseptate, smooth, thick-walled, brown, straight, simple or branched. Conidiogenous cells integrated, terminal, cylindrical, monoblastic, determinate. Conidia acrogenous, holoblastic, solitary, dry, dematiaceous, euseptate, thick-walled, smooth, obclavate, asetulate. Conidial secession rhexolytic. On submerged wood in the Lam Tsuen River, Hong Kong.

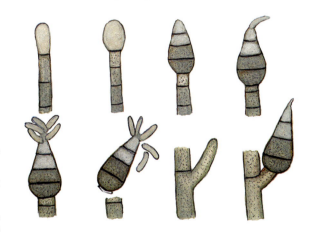

Elegantimyces sporidesmiopsis Goh, et al.: conidiogenesis to form primary conidia and conidia similar to the genus **Idriella** P. E. Nelson & S. Wilh., x 250.

Eleutherascus Arx
Ascodesmidaceae, Pezizales, Pezizomycetidae, Pezizomycetes, Pezizomycotina, Ascomycota, Fungi

Ellisiopsis

Arx, J. A. von. 1971. On *Arachniotus* and related genera of the Gymnoascaceae. *Persoonia* 6:371-380.
From Gr. *eléutheros*, free + *askós*, bag, sac > ML. *ascus*, wine skin, ascus. Ascomata lacking. Fungus producing abundant mycelium in culture; hyphae hyaline, septate, bearing curved or coiled ascogonia. Asci formed singly or in small clusters, broadly clavate, obovoid, or subspherical, thin-walled, usually 8-spored. Ascospores globose, 1-celled, brown, with a thick wall covered with spines. In soil.

Eleutherascus peruvianus L. H. Huang: subspherical and octosporate asci; isolated from soil in Perú, x 400; and one spiny ascospore viewed with the scanning electron microscope, x 3,000.

Ellisiopsis Bat.—see **Beltraniella** Subram.
Amphisphaeriaceae, Amphisphaeriales, Xylariomycetidae, Sordariomycetes, Pezizomycotina, Ascomycota, Fungi
Batista, A. C. 1956. Systematic revision of the genera *Ellisiella* Sacc. and *Ellisiellina* Camara, and the new genus *Ellisiopsis*. *Anais Soc. Biol. Pernambuco* 14(1/2):16-25.
Named in honor of the American mycologist *Job B. Ellis* + Gr. suf. *-ópsis*, aspect, appearance.

Elotespora R. F. Castañeda & Heredia
Inc. sed., Inc. sed., Inc. sed., Inc. sed., Pezizomycotina, Ascomycota, Fungi
Castañeda Ruiz, R. F., et al. 2010. *Elotespora*, an enigmatic anamorphic fungus from Tabasco, Mexico. *Mycotaxon* 111:197-203.
From Nahuatl *elote*, ear of corn + L. *spora*, spore, referring to the shape of the conidia. Conidiomata on natural substrate superficial or somewhat immersed, stromatic, unilocular, scattered, cyathiform, concave, cupulate to irregular, brown, dark brown to black. Conidiophores absent. Conidiogenous cells not observed. Conidiogenesis obscure, probably holoblastic, monoblastic, discrete, and conidial secession probably rhexolytic. Conidia solitary, muriform, brown or dark brown, fusiform, oval, ellipsoid, cylindrical to obovate, smooth or verruculose, dry. On decaying wood in Tabasco, Mexico.

Elotespora mexicana R. F. Castañeda & Heredia: stromatic conidiomata, somewhat immersed on the natural substrate (decaying wood), x 630.

Elsinoë Racib. (syn. **Sphaceloma** de Bary)
Elsinoaceae, Myrangiales, Dothideomycetidae, Dothideomycetes, Pezizomycotina, Ascomycota, Fungi
Raciborski, M. 1900. Botanischen Institut zu Buitenzorg, Batavia, Jakarta. *Parasit. Alg. Pilze Java's* (Jakarta) 1:14.
Derived from the name of the heroine homonima of the drama of the poet Krasinski. Ascostromata with numerous uniascal locules; ascospores dictyosporous. Acervuli becoming coalescent. Conidial morph sphaceloma-like. Plant pathogens that cause anthracnose or scab of crop plants such as avocado, citrus, grape, raspberry, and others.

Elsinoë fawcettii Bitanc. & Jenkins: anthracnose or scab produced by this phytopathogenic fungus on a fruit of avocado, on which produces ascostromata with numerous uniascal locules, x 0.5.

Enantioptera

Elsinoë rosarum Jenkins & Bitanc (syn. **Sphaceloma rosarum** (Pass.) Jenkins): inflammation of the host leaf parasitized with the acervuli, x 0.7; an acervulus with numerous, hyaline conidia, x 350.

Elytroderma Darker
Rhytismataceae, Rhytismatales, Leotiomycetidae, Leotiomycetes, Pezizomycotina, Ascomycota, Fungi
Darker, G. D. 1932. The Hypodermataceae of conifers. *Contr. Arnold Arbor.* 1:62.
From Gr. *elytron*, cover, sheath + *dérma*, skin, for the sunken ascomata. Apothecia deeply immersed, erumpent at maturity, dark, short-elliptical to linear, opening by a longitudinal slit. Hymenium concave. Paraphyses simple, filiform. Asci fusiform to clavate, 8-spored. Ascospores long, fusiform, 2-celled, surrounded by a gelatinous sheath. Pycnidia simple. Conidiophore layer applanate, with minute bacillar conidia. On conifer needles, causing lesions.

Elytroderma deformans (Weir) Darker: erumpent apothecia, opening by a longitudinal slit, on conifer needles, x 5.

Emericella Berk.—see Aspergillus P. Micheli
Aspergillaceae, Eurotiales, Eurotiomycetidae, Eurotiomycetes, Pezizomycotina, Ascomycota, Fungi
Berkeley, M. J. 1857. *Intr. crypt. bot.* (London):340.
Dedicated to the English cleric *Emeric Streatefeld Berkeley* + L. dim. suf. *-ella*.

Emericellopsis J. F. H. Beyma
Inc. sed., Hypocreales, Hypocreomycetidae, Sordariomycetes, Pezizomycotina, Ascomycota, Fungi
Beyma, F. H. van. 1940. Beschreibung einiger neuer Pilzarten aus dem Centraalbureau voor Schimmelcultures, Baarn (Nederland), VI. Mitteilung. *Antonie van Leeuwenhoek* 6:263-290.
From the genus *Emericella* Berk. + L. suf. *-opsis*, from Gr. *-ópsis*, appearance, aspect. Similar to *Emericella* Berk. but with cleistothecial ascomata. Ascospores 1-celled, subglobose to ellipsoid, pale brown, with hyaline, longitudinal, wing-like appendages. Asexual morph acremonium-like. In soil.

Emericellopsis minima Stolk: cleistothecial ascomata, and asci with 1-celled, pale brown, ascospores, on soil, Mozambique, x 200.

Emmonsiella Kwon-Chung—see Ajellomyces
McDonough & A. L. Lewis)
Ajellomycetaceae, Onygenales, Eurotiomycetidae, Eurotiomycetes, Pezizomycotina, Ascomycota, Fungi
McDonough, E. S., Lewis, A. L. 1968. The ascigerous stage of *Blastomyces dermatitidis*. *Mycologia*. 60(1):76-83.
Dedicated to the American mycologist *C. W. Emmons* + L. dim. suf. *-ella*.

Empusa Cohn—see Entomophthora Fresen.
Entomophthoraceae, Entomophthorales, Inc. sed., Inc. sed., Entomophthoromycotina, Zygomycota, Fungi
Cohn, F. 1855. *Empusa muscae* und die Krankheit der Stubenfliegen. *Hedwigia* 1:60.
From Gr. *émpousa*, ghost, phantom, for the appearance of the insects attacked by this fungus, or from Gr. *empýes*, relating to suppuration, pus, from *empýo*, to break out, referring to the appearance of dead insect parasitized by this fungus, destroyed from within.

Enantioptera Descals
Inc. sed., Inc. sed., Inc. sed., Inc. sed., Pezizomycotina, Ascomycota, Fungi
Descals E., J. Webster. 1983. Four new staurosporous Hyphomycetes from mountain streams. *Trans. Br. mycol. Soc.* 80:67-75.
From Gr. *enantios*, opposite + *pterón*, wing, for the arrangement of the cells on the conidia. Conidiophores micronematous. Conidiogenous cells single, short, lateral, proliferations sympodial. Conidia acrogenous,

Enchylium

staurosporous, main body elongated, lateral branches opposite. Isolated from foam from a stream in Allt Rad an Luig, Caledonia, Scotland.

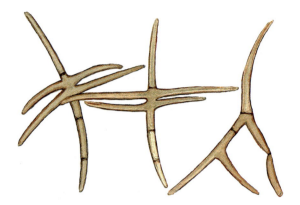

Enantioptera tetra-alata Descals: staurosporous conidia of this aquatic fungus, with lateral branches; on right, one can see a secondary conidium, x 60.

Enchylium (Ach.) Gray

Collemataceae, Peltigerales, Lecanoromycetidae, Lecanoromycetes, Pezizomycotina, Ascomycota, Fungi

Gray, 1821. *Nat. Arr. Brit. Pl.* (London) 1:396.

From Gr. *enchylos*, juicy + L. dim. suf. *-ium*. Thallus forming rosette-like cushions, foliose, deeply lobed, rather thick. Lobes numerous, crowded, contiguous, often ascending at the centre of the thallus, often channelled towards the apices with raised margins, swollen, somewhat contorted. Upper surface dark olive-green to black. Apothecia numerous, terminal and on raised margins, crowded and frequently covering the thallus except the outermost parts of lobes, mostly appearing stalked. Disc flat to convex, with a rather thin, smooth thalline exciple. Ascospores fusiform with acute apices, septate

Enchylium polycarpon (Hoffm.) Otálora, et al. (syn. **Collema polycarpon** Hoffm.): gelatinous thallus with apothecia, on bark, among mosses, x 2.

Endochytrium Sparrow

Endochytriaceae, Chytridiales, Chytridiomycetidae, Chytridiomycetes, Inc. sed., Chytridiomycota, Fungi

Sparrow, F. K., Jr. 1933. Observations on operculate chytridiaceous fungi collected in the vicinity of Ithaca, N. Y. *Amer. J. Bot.* 20:63-77.

From Gr. *éndon*, inside + *chytrís*, pot, receptacle + L. dim. suf. *-ium*. Thallus endobiotic, i.e., develops within cells of algae, either as a parasite or a saprobe in decomposing plant tissues.

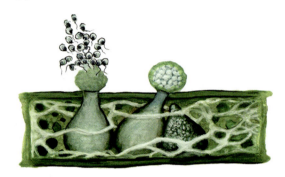

Endochytrium ramosum Sparrow: endobiotic, parasitic sporangia inside the cell of an alga, discharging uniflagellate zoospores, x 420.

Endocochlus Drechsler

Cochlonemataceae, Zoopagales, Inc. sed., Inc. sed., Zoopagomycotina, Zygomycota, Fungi

Drechsler, C. 1935. Some conidial phycomycetes destructive to terricolous amoebae. *Mycologia* 27:14.

From Gr. *éndon*, inside, + *kóchlos*, snail. Vegetative hypha endozoic, thick, spiraled, develops in amoeba invaded by fungus; when host cell dies, coiled hypha produces sporangiophores and zygophores that emerge from amoeba forming unispored merosporangia and zygospores.

Endocochlus asteroides Drechsler: endozoic, spiraled, vegetative hypha within the cell of an amoeba; when the host cell dies, the coiled hypha produces sporangiospores (with unispored merosporangia) and zygophores, x 460.

Endodothiora Petr.

Dothioraceae, Dothideales, Dothideomycetidae, Dothideomycetes, Pezizomycotina, Ascomycota, Fungi

Petrak, F. 1929. Mykologische Notizen. *Annls. mycol.* 27(5/6):345.

From Gr. *éndon*, inside + *dothíon*, genit. *dothíonos*, or *dothién*, *dothiénos*, abcess, boil, small tumor, tubercle + *-or*, state or quality, in reference to the endoparasite nature of *Dothiora* Fr. Stromata with immersed pseudothecia. Asci form a compact palisade exposed when upper part of pseudothecium breaks becoming black, mature ascomata appearing apothecioid.

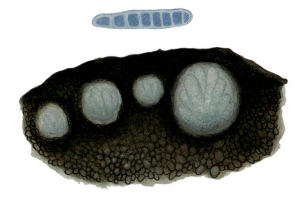

Endodothiora sydowiana Petr.: pseudothecia immersed in the ascostroma of the fungus host (**Dothidea puccinioides** (DC.) Fr., x 140; free ascospore, x 830.

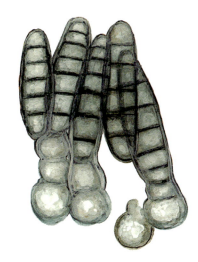

Endogenospora aspectabilis R. F. Castañeda, et al.: vase-shaped conidiogenous cells, with multiseptate conidia, x 1,400.

Endogenospora R. F. Castañeda, et al.

Inc. sed., Inc. sed., Inc. sed., Inc. sed., Pezizomycotina, Ascomycota, Fungi

Castañeda Ruiz, R. F., et al. 2010. *Endogenospora*, a new genus of anamorphic fungi from Venezuela. *Mycotaxon* 112:75-82.

From Gr. *éndon*, within, inside + *génos*, to engender, be produced + *sporá*, spore, seed > NL. *spora*, spore, because conidia arise from inner, deep-seated layers of conidiogenous cells. Colonies on natural substrate effuse, brown or black. Conidiophores mostly absent, reduced to conidiogenous cells. Conidiogenous cells endogenous-holoblastic, vase-shaped, clavate, subcylindrical or elongated infundibuliform, brown or dark, determinate or with enteroblastic percurrent proliferations, thick-walled, internal and deep, located at inflated base. Conidial secession schizolytic. Conidia solitary, clavate to cylindrical, enteroblastic, multiseptate, smooth or verrucose, brown to dark brown, accumulating in dry masses. On decaying branch of plant in Venezuela.

Endogone Link

Endogonaceae, Endogonales, Inc. sed., Inc. sed., Mucoromycotina, Zygomycota, Fungi

Link, H. F. 1809. Observationes in ordine plantarum naturales. *Mag. Gesell. naturf. Freunde, Berlin* 3(1-2):33.

From Gr. *éndon*, inside + *gónos*, the engendered, the offspring, referring to the spores or reproductive structures. Sporangiospores, zygospores and chlamydospores produced in interior of hypogeous sporocarps. Forms mycorrhizae on plants.

Endogone pisiformis Link: sporangia produced in the interior of the hypogeous sporocarps of this mycorrhizogenous fungus, x 250. Collected in Germany.

Endomyces Reess

Endomycetaceae, Saccharomycetales, Saccharomycetidae, Saccharomycetes, Saccharomycotina, Ascomycota, Fungi

Reess, M. 1870. *Bot. Unters. Alkoholgärungspilze*:77.

From Gr. *éndon*, inside + *mýkes* > L. *myces*, fungus, referring to the production of spores. Asci with ascospores, 1-8 per ascus. Mycelium producing geotrichum-like arthrospores. Ubiquitous mold saprobic or parasitic on plants and animals.

Endomycopsis Stell.-Dekk.—see **Saccharomycopsis** Schiønning

Saccharomycopsidaceae, Saccharomycetales, Saccharomycetidae, Saccharomycetes, Saccharomycotina, Ascomycota, Fungi

Stelling-Dekker, N. M. 1931. *Die sporogenen Hefen. Verhandelingen Koninklijke Nederlandse Akademie van Wetenschappen Afdeling Natuurkunde.* 28:1-547.

From genus *Endomyces* Reess < Gr. *éndon*, inside + *mýkes* > L. *myces*, fungus + suf. *-opsis* < Gr. *-ópsis*, aspect, appearance, i.e., similar to *Endomyces* Reess.

Endoptychum

Endoptychum Czern.—see **Chlorophyllum**
Agaricaceae, Agaricales, Agaricomycetidae, Agaricomycetes, Agaricomycotina, Basidiomycota, Fungi
Czerniaiev, B. M. 1845. Nouveaux cryptogames de l'Ucraine. *Bull. Soc. Imp. nat. Moscou* 18(2, III):132-157.
From Gr. *éndon*, inside + *ptyché*, *ptychós*, fold, from this *ptychódes*, folded, rough, with superimposed gills.

Endothia Fr.—see **Cryphonectria** (Sacc.) Sacc. & D. Sacc.
Cryphonectriaceae, Diaporthales, Sordariomycetidae, Sordariomycetes, Pezizomycotina, Ascomycota, Fungi
Fries, E. M. 1849. *Summa veg. Scand.*, Section Post. (Stockholm):385.
From Gr. *éndothen*, *éndothi*, from the interior, inwardly, from *éndon*, inside.

Endothiella Sacc.—see **Cryphonectria** (Sacc.) Sacc. & D. Sacc.
Cryphonectriaceae, Diaporthales, Sordariomycetidae, Sordariomycetes, Pezizomycotina, Ascomycota, Fungi
Saccardo, P. A. 1906. Notae mycologicae. *Annls mycol.* 4:273.
From genus *Endothia* Fr. < Gr. *éndothen*, *éndothi*, from the interior + L. dim. suf. *-ella*.

Enridescalsia R. F. Castañeda & Guarro
Inc. sed., Inc. sed., Inc. sed., Inc. sed., Pezizomycotina, Ascomycota, Fungi
Castañeda Ruiz, R. F., et al. 1998. Enridescalsia, a new genus of conidial fungi from submerged leaves in Cuba. *Mycol. Res.* 102:42-44.
Named in honor of the Spanish mycologist *Enrique Descals* + L. des. *-ia*. Colonies spreading, brown. Mycelium superficial, immersed. Conidiophores differentiated, mononematous, septate, brown to olivaceous. Conidiogenous cells polytretic, synchronous, terminal, with small but conspicuous pores. Conidia cylindrical, fusiform to obclavate, septate, brown, smooth or verrucose, not in chains.

Enterographa Fée
Roccellaceae, Arthoniales, Arthoniomycetidae, Arthoniomycetes, Pezizomycotina, Ascomycota, Fungi
Fée, A. L. A. 1824. *Essai Crypt. Exot.* (Paris): xxxii, cx,57.
From Gr. *énteron*, intestine + *graphé*, drawing, painting. Thallus well developed, subcartilaginous, greenish-glaucescent. Apothecia lirelliform, elongated, narrow, immersed, with a pale to dark exterior and a pale interior, similar to *Graphis* Ach. Ascospores hyaline, fusiform, elliptical, dactyloid or acicular, quadrilocular or plurilocular. Grows on tree bark.

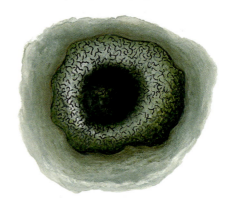

Enterographa elaborata (Lyell ex Leight.) Coppins & P. James: immersed, lirelliform apothecium of the thallus of this lichen, on tree bark, x 1.

Entoloma (Fr.) P. Kumm. (syn. **Rhodophyllus** Quél.)
Entolomataceae, Agaricales, Agaricomycetidae, Agaricomycetes, Agaricomycotina, Basidiomycota, Fungi
Kummer, P. 1871. *Führ. Pilzk.* (Zerbst):23.
From Gr. *entós*, from this the pref. *ento-*, inside, in the interior, within + *lōma*, fringe, border, margin, bank, ragged edge. Pileus lacking a veil on incurved or involuted margin, medium to large, lacking a volva and ring on stipe, generally with a tricholomatoid appearance, gills pink at maturity due to accumulation of spores. Terricolous, probably mycorrhizal; frequent in forests of oak and other broad-leaved trees. Some edible species, but others toxic.

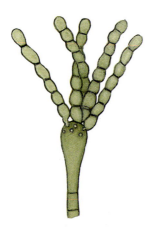

Enridescalsia speciosa R. F. Castañeda & Guarro: polytretic conidiogenous cell with septate conidia, x 900. Isolated from submerged decaying leaves of *Terminalia catappa*, Cuba.

Entoloma hochstetteri (Reichardt) G. Stev.: basidioma on soil, x 0.8.

Entyloma

Entomophthora Fresen. (syn. **Empusa** Cohn)
Entomophthoraceae, Entomophthorales, Inc. sed., Inc. sed., Entomophthoromycotina, Zygomycota, Fungi
Fresenius, G. 1856. Notiz, Insekten-Pilze betreffend. Bot. Ztg. *14*:882-883.
From Gr. *éntomos*, insect + *phthorá*, corruption, spot, death, destruction, because this fungus is a parasite of insects. Phototrophic sporangiophores violently discharge unispored sporangiola. May also infect humans.

white, becoming empty as its contents are transferred to developing spore. Walls of vesicular stalk expand to accomodate spore, forming a clear outer membrane tightly appressed to the spore. Spore wall membrane continuous, except for funnel-shaped portion which extends into the mother vesicle and is closed by a thickened plug.

Entorrhiza globoidea Vánky: sori of teliospores on the stems of *Isolepsis cernua* var. *setiformis*, x 2; teliospores seen in light microscopy, x 400. Collected on roots of *Scirpus fluitans*, Western Australia.

Entomophthora muscae (Cohn) Fresen.: house fly invaded with this parasitic fungus, x 5; magnification of the fly abdomen showing the sporangiophores with sporangiola, x 40.

Entomosporium Lév.—see **Diplocarpon** Wolf
Inc. sed., Helotiales, Leotiomycetidae, Leotiomycetes, Pezizomycotina, Ascomycota, Fungi
Léveillé, D. M. 1856. Description d'un nouveau genre de champignons (*Entomosporium*). *Bull. Soc. bot. Fr. 3*:31.
From Gr. *éntomos*, *éntomon*, cut, and by association, insect + *sporá*, spore + L. dim. suf. *-ium*.

Entorrhiza C. A. Weber
Entorrhizaceae, Entorrhizales, Entorrhizomycetidae, Entorrhizomycetes, Inc. sed., Basidiomycota, Fungi
Weber, C. A. 1884. *Bot. Ztg 42*:378.
From Gr. *entós*, inside + *rhíza*, root. Teliospores in sori. Spores single, i.e., not united in groups; on germination each one produces a simple promycelium with apical verticils of basidiospores. Causes galls in roots of plants mainly in families Cyperaceae and Juncaceae.

Entrophospora infrequens (I.R. Hall) R.N. Ames & R.W. Schneid: spore development from the saccule of this mycorrhizic fungus; this saccule becomes empty because its content fill out the mature spore, x 350.

Entrophospora R.N. Ames & R.W. Schneid. (syn. **Sacculospora** Oehl, et al.)
Acaulosporaceae, Diversisporales, Inc. sed., Glomeromycetes, Glomeromycotina, Glomeromycota, Fungi
Ames, R.N.; Schneider, R.W. 1979. *Entrophospora*, a new genus in the Endogonaceae. *Mycotaxon. 8*(2):347-352.
From Gr. *entós*, within, inside + *trophós*, that which nourishes, which serves as food + *sporá*, spore, referring to the azygospores produced singly in soil within the stalk of the mother vesicle. These have a thin walled, dense

Entyloma de Bary
Entylomataceae, Entylomatales, Exobasidiomycetidae, Exobasidiomycetes, Ustilaginomycotina, Basidiomycota, Fungi
De Bary, H. A. 1874. Beiträge zur Morphologie und Physiologie der Pilze. *Bot. Ztg. 32* (7):101.
From Gr. *entylóo*, to swell up, become swollen + suf. *-oma*, which denotes a pathological condition, a tumor. Tumor intercellular in plant tissues. Sori remain enclosed in plant after maturity, more or less agglutinated in

Eocronartium

lesions in lower part of stem or leaves forming spots. Pathogenic on grasses, Asteraceae, Chenopodiaceae, Fabaceae, and Papaveraceae.

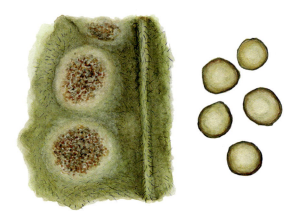

Entyloma australe Speg.: sori of teliospores on the leaf of *Physalis peruviana*, x 5; teliospores, x 1,400.

Eocronartium G. F. Atk.

Eocronartiaceae, Platygloeales, Inc. sed., Pucciniomycetes, Pucciniomycotina, Basidiomycota, Fungi
Atkinson, G. F. 1902. Preliminary note on two new genera of Basidiomycetes. *J. Mycol.* 8:107.
From Gr. *eós*, primitive, early + *Cronartium* Fr., for the supposed phylogenetic antiquity of the genus, thus related to primitive Uredinales and auriculariod fungi with simple septa. Fructifications with heterobasidia differentiated into probasidia and transversely septate metabasidia. *Eocronartium muscicola* parasitizes mosses in temperate regions.

Eocronartium muscicola (Pers.) Fitzp.: parasitic basidiomata of the moss *Eurhynchium corralense*, x 4.

Ephebe Fr.

Lichinaceae, Lichinales, Inc. sed., Lichinomycetes, Pezizomycotina, Ascomycota, Fungi
Fries, E. M. 1825. *Syst. orb. veg.* (Lundae) 1:256.
From Gr. *éphebos*, ephebe, youth who has reached puberty, adolescent, perhaps for the short, fruticose thallus that resembles the pubescence that develops in young men upon reaching puberty. Thallus dark brown or black, tangled or woolly, of intertwined filaments surrounding host alga, *Stigonema*. Apothecia inconspicuous, immersed in photobiont or becoming semi-superficial. On moist rocks.

Ephebe lanata (L.) Vain.: fruticose, woolly thallus on rock, x 3.

Ephemeroascus Emden—see Coniochaeta

Coniochaetaceae, Coniochaetales, Sordariomycetidae, Sordariomycetes, Pezizomycotina, Ascomycota, Fungi
Emden, J. H. van. 1973. *Ephemeroascus* gen. nov. (Eurotiales) from soil. *Trans. Br. mycol. Soc.* 61(3):599-601.
From Gr. *ephémeros*, living only one day, temporary + *askós*, wine skin > ML. *ascus*, referring to deliquescent ascus wall.

Epichloë (Fr.) Tul. & C. Tul. (syn. Neotyphodium Glenn, et al.)

Clavicipitaceae, Hypocreales, Hypocreomycetidae, Sordariomycetes, Pezizomycotina, Ascomycota, Fungi
Tulasne, L. R., C. Tulasne. 1865. *Select. fung. carpol.* (Paris) 3:24.
From Gr. *epí*, on + *chloë*, grass, for the habit of the fungus. Mycelial colonies white to yellowish, cottony, with abundant aerial hyphae, bearing phialides. Phialides solitary, straight, slender, tapering slightly toward apex, bearing apical conidia. Conidia oblong, ellipsoidal, lunate or fusiform, 1-celled, hyaline, smooth. Stromata effuse, superficial, often encircling stem, initially white, byssoid, becoming fleshy, brightly colored. Ascomata ostiolate perithecia, immersed in stromata, often with free ostiolar necks, obpyriform. Ascomatal wall thin, composed of elongated, thin-walled cells. Centrum containing lateral paraphyses. Asci unitunicate, long cylindrical, with a thickened, non-amyloid apex, 8-spored. Ascospores hyaline, filiform, multiseptate, not fragmenting. Obligate endophytic parasites of grasses.

Epicoccum Link

Didymellaceae, Pleosporales, Pleosporomycetidae, Dothideomycetes, Pezizomycotina, Ascomycota, Fungi
Link, J. H. F. 1815. *Mag. Gesell. naturf. Freunde, Berlin* 7:32.

Eremothecium

From Gr. *epí*, on + L. *coccus* < Gr. *kókkos*, coconut, grain, sphere. Fructifications rudimentary stromata covered by conidiophores. Conidiophores short, in compact groups or sporodochia. Common in soil and on vegetable remains.

Epichloë typhina (Pers.) Tul. & C. Tul.: stroma with immersed perithecia, on grass stem, x 2.

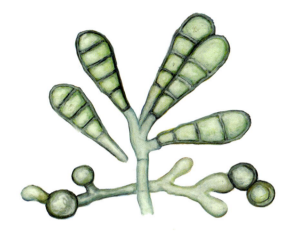

Epidermophyton floccosum (Harz) Langeron & Miloch: conidiophore with multiseptate macroconidia, and unicellular microconidia, x 2,000.

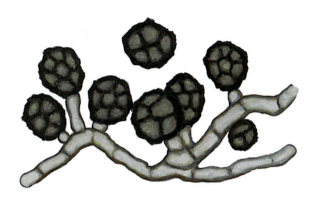

Epicoccum nigrum Link: conidiophores with multicellular and dark conidia, x 600.

Epidermophyton Sabour.

Arthrodermataceae, Onygenales, Eurotiomycetidae, Eurotiomycetes, Pezizomycotina, Ascomycota, Fungi
Sabouraud, R. 1907. *Arch. Méd. exp. Anat. path.* 19:754.
From Gr. *epidermís*, membrane or pellicle that covers the skin < *epí*, on + *dérma*, skin + *phytón*, plant. Macroconidia clavate, septate. Microconidia and sexual morph unknown. Causes tinea or cutaneous mycosis known as dermatophytoses in humans infecting skin and nails; also grows as saprobe on keratinous substrates.

Eremascus Eidam

Eremascaceae, Coryneliales, Eurotiomycetidae, Eurotiomycetes, Pezizomycotina, Ascomycota, Fungi
Eidam, E. 1883. Beitrag zur Kenntniss der Gymnoasceen. *Beitr. Biol. Pfl.* 3:385.
From Gr. *éremos, eremás*, solitary, abandoned + *askós* > L. *ascus*, wine-skin, ascus. Two uninucleate gametangia form a zygote separated from gametangia by septa, enlarges and diploid nucleus divides three times, the first two by meiosis, resulting in eight haploid nuclei that become ascospores in each ascus. Asexual morph arthrospores or blastospores. Saprobic.

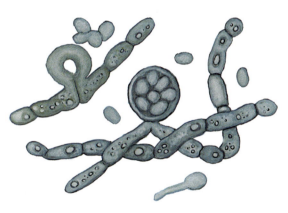

Eremascus albus Eidam: gametangial fusion to form an ascus with ascospores, and mycelium that reproduces by arthrospores and blastospores, x 1,300.

Eremothecium Borzi (syns. **Ashbya** Guilliern., **Nematospora** Peglion, **Spermophthora** S. F. Ashby & W. Nowell)

Eremotheciaceae, Saccharomycetales, Saccharomycetidae, Saccharomycetes, Saccharomycotina, Ascomycota, Fungi
Borzi, A. 1888. *Eremothecium Cymbalariae*, nuovo ascomicete. *Nuovo G. bot. ital.* 20:452-456.
From Gr. *éremos*, solitary + NL. *thecium* < Gr. *théke*, box; here, ascus. Yeast with multilateral budding; well-developed pseudomycelium; sparse, septate true mycelium with fusiform intercalary cells. Cells germinate and behave asexually, or fuse giving rise to a diploid mycelium that forms oval asci and intercalary asci. Asci originate singly at apices of multinucleate hyphae, lageniform or nearly so, each with 8-32 or more ascospores, arranged in two groups, one at each end of ascus, without bud-

Erysiphe

ding cells. Ascospores acicular or fusiform, often with a non-motile, filiform or flagelliform appendage. Capable of fermenting sugars. Parasitic on plants, especially capsules of cotton inoculated by insect punctures. Used by industry to produce vitamin B_2 (riboflavin).

might turn aside the rust of the mature wheat or cereals. Chasmothecia or cleistotheical perithecia with peridial mycelioid, simple, branched, or uncinate appendages, small globose, sometimes with apically dichotomous peridial appendages. Asexual morph oidium-like or acrosporium-like.

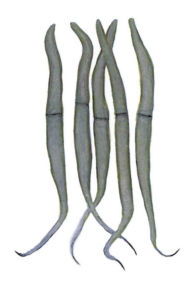

Eremothecium coryli (Peglion) Kurtzman (syn. **Nematospora coryli** Peglion): acicular, one-septate ascospores, x 2,800.

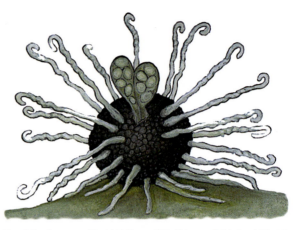

Erysiphe flexuosa (Peck) U. Braun & S. Takam.: cleistothecioid perithecium with peridial mycelioid appendages, showing two emerging asci with ascospores, x 375. Collected in New York, U.S.A.

Eremothecium gossypii (S. F. Ashby & W. Nowell) Kurtzman: acicular ascospores, x 1,500.

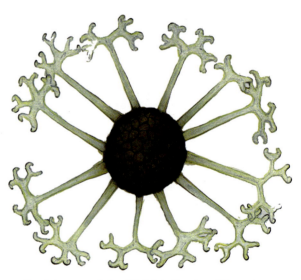

Erysiphe grossulariae (Wallr.) de Bary (syn. **Microsphaera grossulariae** (Wallr.) Sacc.): globose, cleistothecioid perithecium with apically dichotomous appendages, x 250.

Erysiphe R. Hedw. ex DC. (syns. **Microsphaera** Lév., **Uncinula** Lév.)
Erysiphaceae, Erysiphales, Leotiomycetidae, Leotiomycetes, Pezizomycotina, Ascomycota, Fungi
Lamarck, J. B. de, A. P. De Candolle. 1805. *Fl. franç.*, Edn 3 (Paris) 2:272.
From Gr. *erysibáo*, to infest with rust or mildew > *erysíbe* or *erysíphe*, which in L. means *robigo* or *rubigo*, which according to Pliny signifies a crust or moldy layer that grew on stones, but according to others it also can mean mildew, filth, blight, rust. It is also related to *Robigus*, the God that the ancient Romans venerated so that he

Erythricium J. Erikss. & Hjortstam (syn. **Marchandiobasidium** Diederich & Schultheis)
Corticiaceae, Corticiales, Inc. sed., Agaricomycetes, Agaricomycotina, Basidiomycota, Fungi
Eriksson, J, K. Hjortstam. 1970. *Erythricium*, a new genus of Corticiaceae (Basidiomycetes). *Svensk bot. Tidskr.* 64(2):165.
From Gr. pref. *erythrós*, red, reddish + L. dim, suf. *-ium*. Characterized by resupinate, rose-colored basidiomata, lack of clamps, subhymenium hyphae with short cells,

and basidiospores which are relatively large and have firm cyanophilous walls. All species are strikingly similar in their morphology, differing mainly in spore shape and size, habitat preferences and distribution. Is known from many countries with tropical and temperate climates. Causes a canker and die-back disease in a very wide host range, and many other tree crops and forests trees.

Esdipatilia Phadke
Inc. sed., Inc. sed., Inc. sed., Inc. sed., Pezizomycotina, Ascomycota, Fungi
Phadke, C. H. 1981. *Esdipatilia*, a new genus of Hyphomycetes. *Trans. Br. mycol. Soc.* 77:642-644.
Named in honor of the Indian botanist *S. D. Patil* + L. des. *-ia*. Synnemata erect, pale straw-colored, terminating in a mucilagenous head of spores. Individual conidiophores macronematous, erect, septate, filiform, simple or branched, hyaline to subhyaline, each producing a single conidium. Conidiogenous cells monoblastic, integrated, terminal, determinate or percurrent. Conidia solitary, slimy, hyaline to subhyaline, multiseptate, typically bent, truncate at base, tapering towards ends, central cell of conidium enlarged. Terminal part of conidium attenuated forming a long acute beak, which forms a wide angle with central cell. On decaying wood of *Gardenia resinifera* at the Kanakeshwar forest of Konkan, Maharashtra, India.

Esdipatilia indica Phadke: synnema with mucilaginous conidial head, x 140.

Esteya J. Y. Liou, et al.
Inc. sed., Inc. sed., Inc. sed., Inc. sed., Pezizomycotina, Ascomycota, Fungi
Liou, J. Y., et al. 1999. *Esteya*, a new nematophagous genus from Taiwan, attacking the pinewood nematode (*Bursaphelenchus xylophilus*). *Mycol. Res.*103:243-248.
Named in honor of the Canadian mycologist *R. H. Estey* + L. des. *-a*, for his work on nematophagous fungi. Colonies on potato dextrose agar grey, greyish-green to dark green. Hyphae branched, septate, hyaline, subhyaline to greyish-green, smooth, roughened. Conidiophores, conidiogenous cells, and conidia of two types. First type: conidiophores macronematous, mononematous, simple, erect, broadly ampulliform, subhyaline to greyish-green, smooth, roughened. Conidiogenous cells integrated, phialidic, rarely percurrent. Conidia solitary, 1-celled, asymmetrically ellipsoidal, lunate, concave, ends moderately apiculate, hyaline, smooth-walled, containing an endospore-like structure, adhesive. Second type: conidiophores macronematous, mononematous, simple or branched, erect, cylindrical, subulate, septate, hyaline, subhyaline to greyish-green, smooth roughened. Conidiogenous cells integrated, phialidic, enteroblastic, terminal or intercalary, indeterminate. Conidia solitary, slimy, 1-celled, bacilloid, cylindrical, hyaline, smooth, non-adhesive, often aggregated at apex forming a false head. Nematodes parasitized by adhesive, lunate conidia. Isolated from infected pinewood nematodes, *Bursaphelenchus xylophilus*, extracted from a wilted Japanese black pine tree, *Pinus thunbergii*, in Yanmingshan, Taipei City, Taiwan.

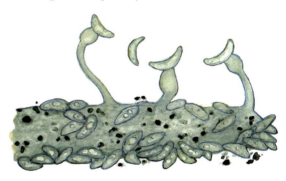

Esteya vermicola J. Y. Liou, et al.: conidiophores and lunate, hyaline conidia, parasitizing the anterior part of a pinewood nematode, *Bursaphelenchus xylophilus*, x 1,100.

Etheirophora Kohlm. & Volkm.-Kohlm.
Etheirophoraceae, Inc. sed., Hypocreomycetidae, Sordariomycetes, Pezizomycotina, Ascomycota, Fungi
Kohlmeyer, J., B. Volkmann-Kohlmeyer. 1989. Hawaiian marine fungi, including two new genera of Ascomycotina. *Mycol. Res.* 92:410-421.
From Gr. *étheiro*, hair, mane + *phóros*, bearing, referring to the hair-like appendages of the ascospores. Ascomata depressed, immersed, ostiolate, periphysate, papillate, clypeate, coriaceous, light-colored, paraphysate. Asci 8-spored, cylindrical to oblong-ventricose, pedunculate, persistent, maturing successively on ascogenous tissue at bottom of locule. Ascospores ellipsoidal, 1-septate, hyaline, with filamentous appendages at one or both ends. On submerged wood in Hawaii, U.S.A.

Eupenicillium

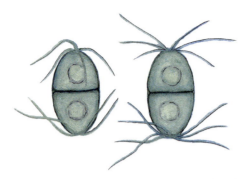

Etheirophora bijubata Kohlm. & Volkm.-Kohlm.: ellipsoidal, 1-septate, hyaline ascospores, with filamentous appendages at one or both ends, x 700.

Eupenicillium F. Ludw.—see **Penicillium** Link
Aspergillaceae, Eurotiales, Eurotiomycetidae, Eurotiomycetes, Pezizomycotina, Ascomycota, Fungi
Ludwing, F. 1892. *Lehrb. Niederen Kryptog.* (Stuttgart):256-263.
From Gr. *eú*, true + the genus *Penicillium* Link, this from L. *penicillus*, dim. of *peniculus*, brush (dim. of *penis*, plume, brush).

Eurotium Link—see **Aspergillus** P. Micheli ex Haller
Aspergillaceae, Eurotiales, Eurotiomycetidae, Eurotiomycetes, Pezizomycotina, Ascomycota, Fungi
Link, H. F. 1809. Observationes in ordines plantarum naturales. *Mag. Gesell. naturf. Freunde,* Berlin 3:1-42.
From Gr. *eurós*, genit. *eurõtos*, mold, trash, decay, decomposition + L. dim. suf. *-ium*.

Euryancale Drechsler
Cochlonemataceae, Zoopagales, Inc. sed., Inc. sed., Zoopagomycotina, Zygomycota, Fungi
Drechsler, C. 1939. Five new Zoopagaceae destructive to rhizopods and nematodes. *Mycologia* 31:410.
From Gr. *eurýs*, broad, wide + *ankále*, bent arm. Sporangiophores with ends swollen, thick, curved or bent, terminating in an apiculus with unispored merosporangia. Sporangiophores arising from external hyphae developing from internal hyphae. Destructive endoparasite of nematodes in soil.

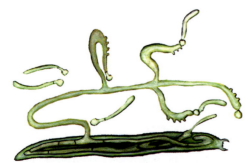

Euryancale sacciospora Drechsler: sporangiophores curved like arms, with unispored merosporangia, emerging from the internal hyphae of parasitized soil nematode, x 1,000.

Eutypa Tul. & C. Tul.
Diatrypaceae, Xylariales, Xylariomycetidae, Sordariomycetes, Pezizomycotina, Ascomycota, Fungi
Tulasne, L. R., C. Tulasne. 1863. *Select. fung. carpol.* (Paris) 2:52.
From Gr. *eú*, true + *typé*, *týpos*, mark, the last, also scar, injury, from *typóo*, to form, be prominent, model, mold, in reference to the stromata. Pseudostromata erumpent in bark, breaking to expose ectostroma with ostioles of perithecia in endostroma. Ascospores allantoid. Saprobic on wood.

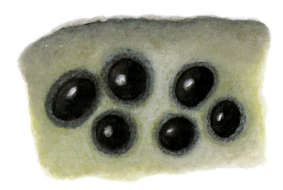

Eutypa flavovirens (Pers.) Tul. & C. Tul.: perithecial pseudostromata that erupt in the bark of the host plant, x 5.

Eutypella (Nitschke) Sacc.
Diatrypaceae, Xylariales, Xylariomycetidae, Sordariomycetes, Pezizomycotina, Ascomycota, Fungi
Saccardo, P. A. 1875. Conspectus generum pyrenomycetum italicorum systemate carpologico dispoitorum. *Atti Soc. Veneto-Trent. Sci. Nat., Padova,* Sér. 44:80.
From the genus *Eutypa* Tul. & C. Tul. (this from the Gr. *eú*, true + *typóo*, to sculpt, carve, engrave, from *typé*, *týpos*, bruise, injury, mark) + L. dim. suf. *-ella*. Pseudostroma with perithecia immersed in endostroma, black perithecia with ostiolar necks projecting from ectostroma. Ascospores allantoid. On dead bark or wood.

Evernia Ach.
Parmeliaceae, Lecanorales, Lecanoromycetidae, Lecanoromycetes, Pezizomycotina, Ascomycota, Fungi
Acharius, E. 1809. *In:* Luyken, Tent. *Hist. Lich.*:90.
From Gr. *evernís*, what grows or sprouts well, branched, flowering + L. suf. *-ia*, which indicates quality or state of something. Thallus highly branched, fruticose, often tangled. Apothecia large, terminal or marginal, with a concave disc different in color from thallus. Photobiont *Protococcus*. On trees and old wood, rarely on rocks.

Excipulariopsis P. M. Kirk & Spooner (syn. **Kentingia** Sivan. & W. H. Hsieh)
Parmulariaceae, Asterinales, Dothideomycetidae, Dothideomycetes, Pezizomycotina, Ascomycota, Fungi
Spooner, B. M., P. M. Kirk. 1982. Taxonomic notes on

Exobasidium

Excipularia and *Scolecosporium*. *Trans. Br. mycol. Soc.* 78(2):251.

From L. *excipulum*, receptacle + Gr. *-ópsis*, aspecto, view, appearance. Mycelium mostly immersed. Conidiomata superficial, pulvinate, dark brown to black, with a basal aggregation of thick-walled, dark brown cells. Setae peripheral, arising directly from cells of the basal stroma, subulate, dark brown, septate, thick-walled, smooth. Conidiophores short, cylindrical, pale brown, unbranched. Conidiogenous cells holoblastic, monoblastic, integrated, terminal, determinate. Conidia acrogenous, solitary, dry, broadly fusoid, truncate at the base, multiseptate, smooth, thick-walled, dark brown, with hyaline to very pale brown terminal cells. Ascomata discoid to pulvinate, multilocular, setose, superficial. Pseudoparaphyses filamentous, septate, hyaline. Asci bitunicate, obovoid to broadly clavate, 8-spored. Ascospores hyaline to brown, fusiform, straight to slightly curved, 1-3-septate, with smaller end cells and larger central cells, smooth or slightly verrucose. On woody substrates.

Fries, E. M. 1822. *Syst. mycol.* (Lundae) 2(1):220.
From L. *exidio*, from *exsudo*, to exude, to distill, to sweat, give off a liquid, this from *sudo*, to sweat + pref. *ex-*, which expresses the idea of something exterior or of interior origin + dim. des. *-idia*, for the viscous exudate that covers the hymenial surface of the fructification. Fructifications gelatinous, smooth, papillate or granulose, dark brown, reddish or whitish, generally small, up to 2 cm in diam, discoidal, infundibuliform or irregular. Several fructifications on same substrate, small erumpent pustules soon converge, forming large gelatinous masses up to 20 cm long, especially in type species, *E. glandulosa* Fr. Commonly on living or dead wood of conifer or broad-leaved trees.

Evernia prunastri (L.) Ach.: fruticose thallus, with apothecia, on old wood, x 1.

Exidia saccharina Fr.: gelatinous, dark brown basidiomata, on dead wood of trees, x 1.

Excipulariopsis narsapurensis (Subram.) Spooner & P. M. Kirk (syn. **Kentingia corticola** Sivan. & W. H. Hsieh): sagittal section of a multilocular, setose ascoma, with asci and fusiform ascospores, on bark in Taiwan, x 450.

Exidia Fr.

Auriculariaceae, Auriculariales, Inc. sed., Agaricomycetes, Agaricomycotina, Basidiomycota, Fungi

Exobasidium Woronin

Exobasidiaceae, Exobasidiales, Exobasidiomycetidae, Exobasidiomycetes, Ustilaginomycotina, Basidiomycota, Fungi

Woronin, M. S. 1867. *Verh. Naturf. Ges. Freiburg* 4(4):397.

From Gr. pref. *exo-*, derived from *éxo*, outside, on the exterior, external + L. *basidium*, from Gr. *basídion*, dim. of *básis*, base, basidium. Basidia on thin hymenial layer free on host producing basidiospores that germinate by budding as in yeasts, giving rise to blastospores, at least on culture media; may also form germ tubes directly. Both blastospores and germ tubes develop into mycelium. Basidia cylindrical or claviform, each producing two to eight basidiospores on sterigmata. Parasitic on wild and cultivated vascular plants.

Exophiala

Exobasidium japonicum Shirai: basidia with basidiospores, on azalea tissues, x 1,000.

Exophiala J. W. Carmich.
Herpotrichiellaceae, Chaetothyriales, Chaetothyriomycetidae, Eurotiomycetes, Pezizomycotina, Ascomycota, Fungi
Carmichael, J. W. 1966. Cerebral mycetoma of trout due to a *Phialophora*-like fungus. *Sabouraudia* 5:120-123.
From Gr. *éxo*, outside + *phiále*, glass, bottle, referring to the conidiogenous cells. Conidiogenus cells phialidic with an indistinct collarette, or, in some species, capable of forming percurrent or annellidic phialides; conidia subhyaline, unicellular, with a slightly truncate hilum. Budding, yeast-like cells may revert to a mycelial phase. Saprobic on decomposing wood, in soil, some parasites of humans causing black tinea, subcutaneous abscesses or chromomycosis.

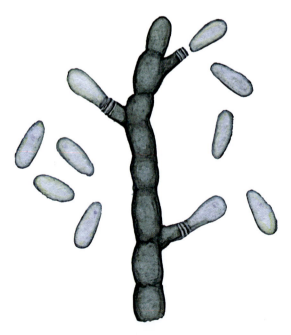

Exophiala jeanselmei (Langeron) McGinnis & A. A. Padhye: conidiogenous cell with annellidic phialides, and subhyaline, unicellular conidia, x 3,000.

Exosporiella P. Karst. (syn. **Anomalemma** Sivan.)
Melanommataceae, Pleosporales, Pleosporomycetidae, Dothideomycetes, Pezizomycotina, Ascomycota, Fungi
Karsten, P. 1892. *Finlands mögelsvampar* (Hyphomycetes fennici):160.
From Gr. *exó*, outside + *sporá*, spore + L. dim. suf. *-ella*. Intertwined hyphae mainly superficial, around the pseudothecia, composed of conidiophores cylindrical or lageniform, septate and branched at the base, conidiogenous cells proliferate enteroblastically, with percurrent proliferations and with a succession or sequential conidia formed at higher levels. Conidia brown, usually with 4-transverse septa, slightly constricted at septa, initially hyaline, pale yellowish brown to brown at maturity, with central two cells dark brown and terminal cells light brown, slightly rounded at apex, rounded to abruptly attenuated towards a more or less conspicuous tapering base, caudate, smooth-walled. Pseudothecia gregarious, superficial on a prosenchymatous stroma, ostiolate; peridium pseudoparenchymatous. Asci clavate, bitunicate, 8-spored. Ascospores 1-3 septate, brown, constricted at mid septum, smooth, fusiform. Pseudoparaphyses filiform, hyaline. On wood and bark, or ectoparasite on fruiting structures of other fungi, as *Corticium* Pers.

Exosporiella fungorum (Fr.) P. Karst.: gregarious ascomata on the surface of wood and bark, x 50; vertical section of an ascoma showing the peridium with peripheral conidiogenous cells and conidia of the asexual stage (*Epochnium fungorum*, the basionym), x 160; developing ascus with ascospores, x 700; 4-celled released ascospore, x 2,500; fusiform 4-celled conidia, slightly constricted at the septa, with central, brown cells, with the basal and terminal cells more clear, x 1,250.

Exserohilum K. J. Leonard & Suggs (syn. **Setosphaeria** K. J. Leonard & Suggs)
Pleosporaceae, Pleosporales, Pleosporomycetidae, Dothideomycetes, Pezizomycotina, Ascomycota, Fungi
Leonard, K. J., E. G. Suggs. 1974. *Setosphaeria prolata*, the ascigerous state of *Exserohilum prolatum*. *Mycologia* 66:281-297.
From L. *exsertus*, protruding + *hilum*, trifle, scar of a bean seed, for the protruding hilum on the base of the conidium. Ascomata uniloculate, perithecioid pseu-

Exserohilum

dothecium, superficial, erumpent or immersed in host tissues; pseudothecia dark brown to black, ostiolate, with or without a neck, with short, brown setae around ostiole and on upper part of pseudothecium. Ascomatal wall of angular to globose pseudoparenchymatous cells, coriaceous or carbonaceous. Centrum with filiform, septate, branched pseudoparaphyses. Asci bitunicate, thick-walled, cylindrical or clavate, straight or curved, stalked, 1-8-spored. Ascospores hyaline, fusoid to oblong, several septate, constricted at septa, smooth, surrounded by a thin gelatinous sheath. Conidiophores cylindrical, simple, olivaceous-brown, geniculate. Conidia porogenous, acrogenous, subcylindrical to fusoid or broadly obclavate-rostrate, pseudoseptate, olivaceous to brown, with a distinct hilum protruding from basal cell. Germination by bipolar germ tubes. Mostly parasitic on grasses or other monocotyledons, but some species also found in soil and other substrates.

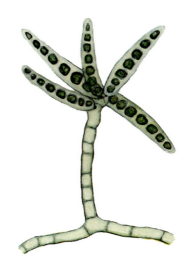

Exserohilum rostratum (Drechsler) K. J. Leonard & Suggs: geniculate conidiophore with pseudoseptate conidia, x 800.

F

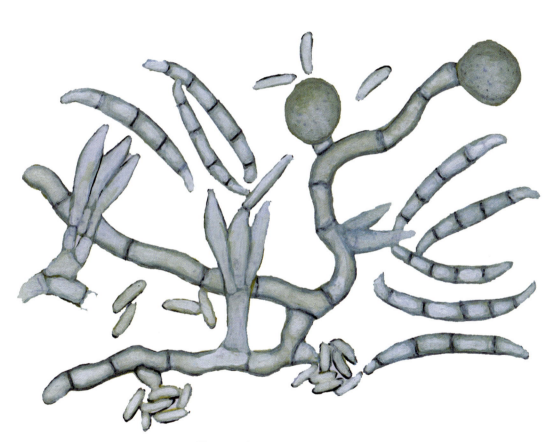

Fusarium oxysporum

F

Falciformispora K. D. Hyde

Inc. sed., Pleosporales, Pleosporomycetidae, Dothideomycetes, Pezizomycotina, Ascomycota, Fungi

Hyde, K. D. 1992. Intertidal mangrove fungi from the west coast of Mexico, including one new genus and two new species. *Mycol. Res.* 96:25-30.

From L. *falciformis*, shaped like a scythe + *spora*, spore < Gr. *sporá*, spore, seed, in reference to the ascospore appendage. Ascomata subglobose or ovoid, erumpent, becoming superficial on substratum, black, coriaceous, ostiolate, solitary or gregarious. Peridium composed of an inner layer of elongate hyaline cells and an outer layer of thick-walled angular or rounded brown cells. Pseudoparaphyses cellular and numerous. Asci 8-spored, bitunicate, thick-walled, nonamyloid, pedunculate, with an ocular chamber, with fissitunicate dehiscence, arising from ascogenous tissue at base of ascoma. Ascospores fusiform, hyaline, 2-3-seriate, 6-(7)-8-septate, slightly constricted at septa, surrounded by a thin mucilaginous sheath with a single scythe-like appendage. Collected on submersed mangrove wood at Boca de Pascuales, Colima, Mexico.

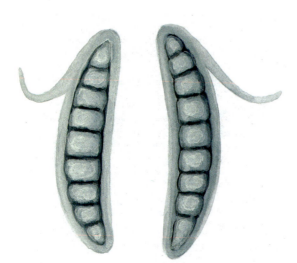

Falciformispora lignatilis K. D. Hyde: fusiform, hyaline, multiseptate ascospores, surrounded by a mucilaginous sheath, with a single scythe-like appendage, x 1,200.

Farysia Racib.

Farysiaceae, Ustilaginales, Ustilaginomycetidae, Ustilaginomycetes, Ustilaginomycotina, Basidiomycota, Fungi

Raciborski, M. 1909. Parasitische und epiphytische Pilze Javas. *Bull. int. Acad. Sci. Lett. Cracovie, Cl. sci. math. nat. Sér. B, sci. nat.* 3:354.

The etymology could not be determined. Fructifications composed of masses of spores called sori, crossed by numerous bundles of sterile hyphae, similar to elaters, and also by tissue of host. Spores teliospores, solitary, i.e., not united in groups or balls of spores or in pairs as in other smuts. *Farysia merrilli* (Henn.) Syd. and *F. javanica* Racib. cause smut or blight on plants of the genus *Carex* and other Cyperaceae.

Farysia butleri (Syd. & P. Syd.) Syd. & P. Syd.: sterile hyphae, similar to elaters, and teliospores that cause smut blight on plants of *Carex indica*, x 700.

Favolaschia (Pat.) Pat.

Mycenaceae, Agaricales, Agaricomycetidae, Agaricomycetes, Agaricomycotina, Basidiomycota, Fungi

Patouillard, N.T., G. de Lagerheim. 1895. Champignons de l'Équateur (Pugillus IV). *Bulletin de l'Herbier Boissier.* 3:53-74.

From. L. *favus* > NL. *favosus*, like a honey comb + genus *Laschia* < named in honor to German mycologist W. G. *Lasch*. Basidomata sessile or with lateral, eccentric or central pseudostipe of often extremely variable size, depending on the position of the fructifications on the substratum; pileus up to 10 mm diam, yellow, orange, salmon orange, pink, lateritius or pigmentless. Hymenophore poroid; in young carpophores the dissepiments between the originally round pores are mostly still thick in relation to the pore diameter, but the tube-walls become relatively thinner as the fungus matures; the outermost peripheral pores are the youngest and often the smallest. The pileus-trama is more or less gelatinized, mostly with the exception of a ramose-filamentose subhymenium and an interwoven infraepicuticular zone. Dermatocystidia are of two basic kinds, gloeocystidia and acanthocysts. Basidia are provided with a basal clamp, variable in size and shape. Spores thin-walled, one-celled, amyloid, smooth, ellipsoid, ovoid, cylindrical, globose. Grows typically on dead plant material, wood, herbaceous stems, woody roots or fruits. Ocurring in the warm-temperate to subtropical and tropical zone.

Favolaschia calocera R. Heim: ventral view of basidiome showing the hymenophore compound of the yellowish pores arranged in honey comb appearance, x 5. Grows on wood of coniferous forest.

Favolus P. Beauv.
Polyporaceae, Polyporales, Inc. sed., Agaricomycetes, Agaricomycotina, Basidiomycota, Fungi
Beauvois, P. 1805. *Flore Oware Benin* 1:1.
From L. *favus*, honeycomb + dim. suf. *-olus*, like a small honeycomb. Fruitbody annual, dimidiate, flabelliform to spathulate, rarely with a centrally placed stipe, soft and thin when fresh, brittle when dry. Pileus glabrous to finely tomentose, white to straw-colored but may darken with age. Pores hexagonal to radially elongate towards stipe, semi-lamellate. Context very thin, except close to base. Spores cylindrical to slightly navicular with tapering ends, smooth, thin-walled. On angiosperms. In wet rain forests, pantropical in Africa, Asia, Brazil and the Philippines.

Favostroma B. Sutton & E. M. Davison
Inc. sed., Inc. sed., Inc. sed., Inc. sed., Pezizomycotina, Ascomycota, Fungi
Sutton, B. C., E. M. Davison. 1983. Three stromatic Coelomycetes from Western Australia. *Trans. Br. mycol. Soc.* 81(2):291-301.
From L. *favus*, honeycomb + Gr. *strōma*, cushion, bed, for the many locules in the stroma. Conidiomata eustromatic, olivaceous brown, pulvinate, irregularly multilocular and convoluted but with locules filled with textura intricata; dehiscence irregular. Conidiophores hyaline, septate, irregularly branched, formed from walls of locules and conidiomata. Conidiogenous cells integrated, determinate, smooth, formed as lateral branches on conidiophores. Conidia holoblastic, solitary, aseptate, smooth, spherical, ovoid or obpyriform, thick-walled, truncate at base and with a marginal frill. Collected on *Eucalyptus calophylla*, Curara Block, Dwellingup, Western Australia.

Favostroma crypticum B. Sutton & E. M. Davison: eustromatic conidiomata with the locules full of conidia, x 150.

Felisbertia Viégas
Dermateaceae, Helotiales, Leotiomycetidae, Leotiomycetes, Pezizomycotina, Ascomycota, Fungi
Viégas, A. P. 1944. Alguns fungos de Brasil II. Ascomicetos. *Bragantia* 4:1-392.
Named in honor of the Brazilian botanist/entomologist *Felisberto C. Camargo* + L. des. *-ia*. Ascomata apothecial, immersed in host tissues, becoming erumpent through epidermis at maturity, pulvinate, dark brown, lacking an ectal excipulum. Asci clavate, unitunicate, thick-walled, with a truncate apex, 8-spored. Ascospores oblong, with rounded ends, hyaline, becoming 3-celled. An erumpent conidial state also formed. Described from *Melastoma* in Brazil.

Filobasidiella Kwon-Chung—see **Cryptococcus** Vuill.
Tremellaceae, Tremellales, Inc. sed., Tremellomycetes, Agaricomycotina, Basidiomycota, Fungi
Kwon-Chung, J. G. 1975. A new genus, *Filobasidiella*, the perfect state of *Cryptococcus neoformans*. *Mycologia* 67(6):1198.

Filobasidium

From *Filobasidium* L. Olive + L. dim. suf. *-ella*, similar, resembling, for its similarity to this genus, whose name derives from *filum*, thread, string + *basidium* < Gr. *basídion* < *básis*, base + dim. suf. *-ídion*, basidium, for the cylindrical and elongate shape of the basidium.

Filobasidium L. S. Olive
Filobasidiaceae, Filobasidiales, Inc. sed., Tremellomycetes, Agaricomycotina, Basidiomycota, Fungi
Olive, L. S. 1968. An unusual new heterobasidiomycete with tilletia-like basidia. *J. Elisha Mitchell scient. Soc.* 84:261.
From L. *filum*, thread, string, filament, fiber + *basidium* < Gr. *basídion* < *básis*, base + dim. suf. *-ídion*, basidium, for the shape of the basidium, which is long, cylindrical, and tapered in the upper part. Sexual reproduction heterothallic bipolar, dikaryotic mycelium result from conjugation of haploid, yeast-like cells that originate by budding. Dikaryotic mycelium with dolipore septa and clamp connections similar to those in *Tremella* Dill. ex L. Basidium long, cylindrical, tapered in upper part, slightly wider at apex, each basidium with five to nine basidiospores. Basidiospores sessile, elliptical, fusiform, oval or reniform. Asexual reproduction by budding of spherical, oval, elliptical or elongate cells, which may produce either a pseudomycelium or a true mycelium.

Filosporella Nawawi
Inc. sed., Inc. sed., Inc. sed., Inc. sed., Pezizomycotina, Ascomycota, Fungi
Nawawi, A. 1976. *Filosporella* gen. nov. an aquatic hyphomycete. *Trans. Br. mycol. Soc.* 67:175.
From L. *filum*, thread + *spora*, spore + dim. suf. *-ella*, for the shape of the long, filamentous spore. Mycelium gray to dark brown, septate, branched, superficial or immersed. Conidiophores erect, septate, branched at apex. Conidia holoblastic, hyaline, terminal, filiform, septate. Collected on submerged decaying leaves of *Pandanus helicopus* from Tasek Bera, an inland lake in Pahang, Malaysia.

Flabellocladia Nawawi
Inc. sed., Inc. sed., Inc. sed., Inc. sed., Pezizomycotina, Ascomycota, Fungi
Nawawi, A. 1985. Another aquatic hyphomycete genus from foam. *Trans. Br. mycol. Soc.* 85:174.
From L. *flabellum*, a small fan + Gr. *kládos* > L. *clados*, branch, for the arrangement of the cells of the conidia. Mycelium of branched, septate, hyaline hyphae. Conidiophores mostly lateral, micronematous, mononematous, simple. Conidiogenous cells integrated, determinate, monoblastic. Conidia solitary, hyaline, staurosporous, multiseptate; main axis elongate; 3-bractate, similar to a crown. Isolated from stream foam in Cameron Highlands, Malaysia.

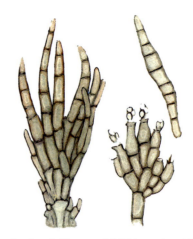

Filosporella pinguis Marvanová & Bärl.: erect conidiophores, branched at apex, with filiform and septate conidia; note the percurrent proliferations of the conidiophores and the microconidial state, x 400. Isolated from foam of stream, New Brunswick, Canada.

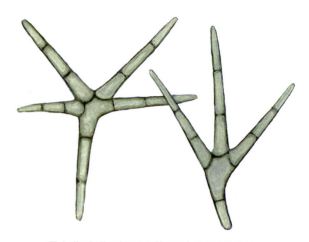

Flabellocladia gigantea Nawawi: staurosporous, hyaline, multiseptate conidia, tetra- and pentaradiate, x 600.

Flammulina P. Karst.
Physalacriaceae, Agaricales, Agaricomycetidae, Agaricomycetes, Agaricomycotina, Basidiomycota, Fungi
Karsten, P. A. 1891. Symbolae ad mycologiam Fennicam. XXX. *Meddn. Soc. Fauna Flora fenn.* 18:62.
From genus *Flammula* Fr., which is derived from L. *flammeus*, flame, bright, of the color of a flame or of fire + des. *-ina*, which indicates similarity or relation to. Fruiting bodies generally brown or bright orange, with both pileus and stipe pigmented, gills almost always yellowish. Principally on dead wood and rarely on roots. *Flammulina velutipes* (Curt. ex Fr.) Sing., type species, has a viscous, smooth, yellow to reddish-orange pileus with yellow or yellowish-white gills. In groups united at base, rarely solitary, on trunks, in conifer and oak forests.

Fluminispora Ingold—see **Dimorphospora** Tubaki
Helotiaceae, Helotiales, Leotiomycetidae, Leotiomycetes, Pezizomycotina, Ascomycota, Fungi

Ingold, C. T. 1958. New aquatic hyphomycetes: *Lemonniera*, *Brachycladia*, *Anguillospora crassa* and *Fluminispora ovalis*. *Trans. Br. mycol. Soc.* 41:365-372.

From L. *flumin*, river + *spora*, spore, for the habitat of the fungus.

Flammulina velutipes (Curtis) Singer: basidioma with the pileus and stipe yellow to reddish-orange, and yellowish-white gills, on conifer trunk, on dead or decaying wood of *Salix* or *Ulmus*, Great Britain, x 1.

Fomes (Fr.) Fr. (syn. Ungulina Pat.)

Polyporaceae, Polyporales, Inc. sed., Agaricomycetes, Agaricomycotina, Basidiomycota, Fungi

Fries, E. M. 1849. *Summa veg. Scand.*, Sectio Post. (Stockholm):319 (adnot.), 321.

From L. *fomes*, tinder, combustible, for the woody appearance of the majority of the species, which have been utilized as fuel and to produce fire by friction with other hard objects. Basidiocarps woody, hard, rigid, resinoid, sessile, ungulate with a trimitic hyphal system, dark or bright colored, with several layers of tubes, attaining large size. Hymenophore of stratified tubes, cystidia lacking, trama light or dark. Spores white. Causes white rot on dead wood. Some species transferred to other genera, e.g., *Heterobasidion annosum* Bref. (syn. *Fomes annosus* (Fr.) Cooke) causes white rot of conifers.

Fomes fomentarius (L.) Fr.: ungulate, dark colored, basidiocarp, on conifer wood, x 0.5.

Fomitiporia Murrill

Hymenochaetaceae, Hymenochaetales, Inc. sed., Agaricomycetes, Agaricomycotina, Basidiomycota, Fungi

Murrill, W. A. 1907. Agaricales: Polyporaceae-Agaricaceae. *N. Amer. Fl. (New York)* 9(1):7.

From genus *Fomes* (Fr.) Fr. + *Poria* Pers., referring to the similarity to both genera. Hymenophore resupinate, epixylous, perennial, inseparable, rigid, context thin, brown, tubes brown, stratified, usually thick walled and entire. Spores globose to ellipsoidal, smooth, hyaline. Hyphae usually brown, cystidia sometimes present. Temperate and tropical species generally on dead trunks.

Fomitopsis P. Karst.

Fomitopsidaceae, Polyporales, Inc. sed., Agaricomycetes, Agaricomycotina, Basidiomycota, Fungi

Karsten, P. A. 1881. Symbolae ad mycologiam fennicam. VIII. *Meddn Soc. Fauna Flora fenn.* 6:9.

From genus *Fomes* (Fr.) Fr. + Gr. suf. *-ópsis*, similar to the genus *Fomes*. Basidiocarps perennial or rarely annual, sessile to effused-reflexed, tough to woody, pore surface and context white to tan or pinkish, pores mostly small, regular. Hyphal system dimitic or trimitic, generative hyphae with clamps. Basidia clavate, 4-sterigmate, with basal clamps. Basidiospores subglobose to cylindric, hyaline, smooth, negative to Melzer's reagent. Causes brown cubical rot of living or dead conifers and hard woods. In subtropical and boreal forests including high elevations.

Fonsecaea Negroni

Herpotrichiellaceae, Chaetothyriales, Chaetothyriomycetidae, Eurotiomycetes, Pezizomycotina, Ascomycota, Fungi

Negroni, P. 1936. Estudio micológico del primer caso argentino de cromomicosis. *Revista Inst. Bacteriol. 'Dr. Carlos G. Malbrán'* 7:424.

Dedicated to O. da Fonseca, Brasilian physician + L. *-ea*. Predominance of one to three types of sporulation: phialidic, acroblastogenous and rachiform. *Fonsecaea pedrosoi*, type species, causes chromomycosis in humans and other animals commonly found in tropical and sub-tropical regions where it grows as a soil saprotroph.

Fonsecaea pedrosoi (Brumpt) Negroni: three types of sporulation: phialidic, acroblastogenous and rachiform conidia, on different conidiophores, x 650.

Funneliformis

Funneliformis C. Walker & A. Schüßler
Glomeraceae, Glomerales, Inc. sed., Glomeromycetes, Glomeromycotina, Glomeromycota, Fungi
Schüßler, A., C. Walker. 2010. *The Glomeromycota*, A species list with new families and new genera (Gloucester):13.

From E. *funnel*, cone-shaped utensil < ME. *funel* < L. *infundere*, to pour in + connective -i- + -*formis*, shape, for the shape of the subtending hyphae. Spores formed singly or sometimes in sporocarps with a few to several spores per sporocarp, with mono- to multiple-layered spore wall. Subtending hyphae, concolorous with spore wall color or slightly lighter; species-specific, generally funnel-shaped to cylindrical. Wall differentiation and pigmentation over long distances from spore base then mycelium becoming hyaline. Pore regularly closed by a conspicuous septum that species-specifically arises from structural wall layer, from an additional adherent innermost, (semi-) flexible lamina, or from both but lacking introverted wall thickening. Typical vesicular-arbuscular mycorrhiza, with mycorrhizal structures that stain blue to dark blue in trypan blue. In soil or rarely roots.

Funneliformis mosseae (T. H. Nicolson & Gerd.) C. Walker & A. Schüssler: funnel-shaped spores with conspicuous subtending hyphae, x 60. Collected in Great Britain.

Fusarium Link (syns. **Bidenticula** Deighton, **Gibberella** Sacc., **Neocosmospora** E. F. Sm., **Pycnofusarium** Punith.)
Nectriaceae, Hypocreales, Hypocreomycetidae, Sordariomycetes, Pezizomycotina, Ascomycota, Fungi
Link, J. H. F. 1809. Observationes in ordines plantarum naturales. *Mag. Gesell. naturf. Freunde, Berlin* 3(1-2):10.
From L. *fusus*, spindle + suf. -*arium*, container, in reference to the spores in the shape of a spindle. Leaf spots large, extending rapidly and soon occupying most of the leaf, brown, broadly zonate, soon becoming covered on both surfaces with greenish or greenish-white floccose masses of conidiophores and conidia. Primary mycelium internal, hyphae dilute olivaceous, septate, branched, deeper olivaceous in substomatal cavities, becoming dark brown on emergence and giving rise to superficial secondary mycelium from which arise conidiophores. Secondary mycelium abundant, superficial, composed of repent or arcuate hyphae, pale olivaceous, smooth, branched, septate, interwoven forming a loose network over surface of leaf spot. Perithecia globose or convex, violaceous or bluish, more or less blackish, soft, ostiolate, superficial, or within a fleshy stroma, singly or in groups. Acervuli foliicolous, subepidermal, basal stroma rudimentary, of a few layers of pseudoparenchymatous hyaline cells. Conidiophores arising as lateral branches from secondary mycelial hyphae, abundant, erect, alternately branched at a wide angle, intermingled with each other, moderate olivaceous below, paler above and almost hyaline in distal part, straight below, slightly flexuous towards apex, distantly septate, smooth, sympodular, polyblastic, denticulate, wide above often narrowing towards apex. Conidia develop on conidiogenous cells or phialides at times grouped into sporodochia. Conidia abundant, almost hyaline, fusiform, slightly curved, smooth, thin-walled, gradually tapered towards both ends, with a truncate unthickened hilum, 3-7-septate, mostly 3-septate, not constricted. Sexual morph gibberella-like. Common in soil and on plant matter. Cause rot of corn cobs and heads of barley, wheat, rye and rice, and root rot of corn and rice. On living but senescent leaves. Some species produce potent toxins, such as zearalenone, which can cause confusion and even death in humans and domestic animals.

Fusarium graminearum Schwabe (syn. **Gibberella zeae** (Schwein.) Petch): exterior view of blackish perithecium; at right, a section of another perithecium, to see the hymenium of this phytopathogenic fungus on corn cobs, x 500.

Fusarium neocosmosporiellum O'Donnell & Geiser (syn. **Neocosmospora vasinfecta** E. F. Sm.): brown perithecium, on soil, x 400; asci with eight unicellular ascospores, x 600.

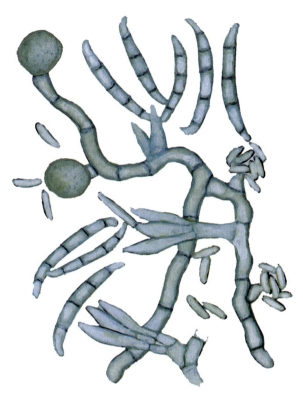

Fusarium oxysporum Schltdl.: conidiophore with phialides that produce macroconidia, microconidia and chlamydospores, x 1,000.

Fuscocerrena Ryvarden

Polyporaceae, Polyporales, Inc. sed., Agaricomycetes, Agaricomycotina, Basidiomycota, Fungi

Ryvarden, L. 1982. *Fuscocerrena*, a new genus in the Polyporaceae. *Trans. Br. mycol. Soc.* 79(2):279.

From L. *fuscus*, brown + genus *Cerrena* Gray. Fruitbody effuso-reflexed to resupinate; pileus dark brown, hymenophore poroid, sinuous to hydnoid, greenish-white, farinose when fresh, later dark brown; hyphal system dimitic; generative hyphae with clamps; skeletal hyphae pale brown; dendrohyphidia present in hymenium; cystidia none; spores hyaline, cylindrical and nonamyloid. On dead deciduous wood.

Fusicladium Bonord

Venturiaceae, Venturiales, Pleosporomycetidae, Dothideomycetes, Pezizomycotina, Ascomycota, Fungi

Bonorden, H. F. 1851. *Handb. Allgem. mykol.* (Suttgart):80.

From L. *fusus*, spindle + Gr. *kládos*, branch + L. dim. suf. *-ium*. Mycelium internal, subcuticular, intraepidermal to intercellular, sometimes substomatal. Stromata pseudostromatic, pigmented. Conidiophores solitary or fasciculate, arising from internal hyphae or stromata, erumpent, occasionally emerging through stomata, sometimes forming well-developed sporodochial conidiomata; conidiophores often reduced to conidiogenous cells or composed of several cells, erect, cylindrical, pyriform, subclavate, narrowly obclavate, geniculate-sinuous, unbranched, pale olivaceous to dark brown, smooth to verruculose. Conidia solitary or catenate, in simple or branched acropetal chains, ellipsoid-ovoid, obovoid, fusiform, obclavate-subcylindrical, straight to curved, 0-3(-4)-septate, subhyaline to medium brown, but mostly olivaceous, smooth to verruculose, ends pointed or rounded to truncate. Pseudothecia are formed in the spring, and the asci produce two-celled ascospores. On leaves, fruits and twigs, causing leaf spots, scab diseases, necroses and deformations.

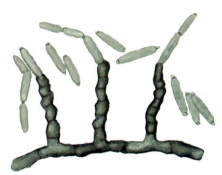

Fusicladium paraamoenum Crous et al.: dark brown conidiophores with conidiogenous cells provided with several denticles and short chains of conidia, 1-3 septate, subcylindrical and guttulate, x 1,000. Isolated from the host plant (*Eucalyptus regnans*) in Victoria, Australia.

G

Gautieria morchelliformis

G

Gaeumannomyces Arx & D. L. Olivier
Magnaporthaceae, Magnaporthales, Sordariomycetidae, Sordariomycetes, Pezizomycotina, Ascomycota, Fungi
Arx, J. A. von, D. L. Olivier. 1952. The taxonomy of *Ophiobolus graminis* Sacc. *Trans. Br. mycol. Soc.* 35:29-32.
Named in honor of the Swiss mycologist *Ernst Gäumann* + L. *myces* < Gr. *mykés*, fungus. Mycelium superficial, brown, forming a thin mat on surface of host tissues; hyphae with simple or lobed, dark brown hyphopodia. Ascomata ostiolate perithecia, solitary, immersed, black, globose to broadly oval; ostiolar neck short, broad, oblique. Ascomatal wall composed of flattened cells. Centrum with filamentous paraphyses. Asci unitunicate, with an apical ring, elongate-clavate, 8-spored. Ascospores filiform to narrowly clavate, tapering toward ends, straight or curved, several-septate, hyaline or yellowish. Asexual morph phialophora-like. On roots, stems and leaves of grasses and sedges. *Gaeumannomyces graminis* (Sacc.) Arx & Olivier causes take-all disease of wheat (*Triticum* spp.) and other cereals.

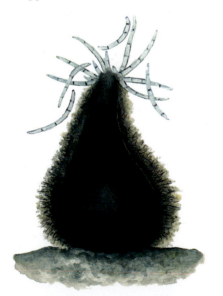

Gaeumannomyces graminis (Sacc.) Arx & D. L. Olivier: ostiolate perithecium, with filiform, several-septate ascospores, on host tissues of wheat, x 180.

Galactinia (Cooke) Boud.—see **Peziza** Dill. ex Fr.)
Pezizaceae, Pezizales, Pezizomycetidae, Pezizomycetes, Pezizomycotina, Ascomycota, Fungi
Boudier, J. L. E. 1885. Nouvelle classification des discomycètes charnus. *Bull. Soc. mycol. Fr.* 1:101.
From Gr. *gála*, genit. *gálactos*, milk + L. suf. *-inia*, which indicates similarity or possession, for the milky juice contained in the apothecia.

Galactomyces Redhead & Malloch—see **Dipodascus** Lagerh.)
Dipodascaceae, Saccharomycetales, Saccharomycetidae, Saccharomycetes, Saccharomycotina, Ascomycota, Fungi
Redhead, S. A., D. W. Malloch. 1977. The Endomycetaceae: new concepts, new taxa. *Can. J. Bot.* 55(13):1708.
From Gr. *gálaktos* > L. *galactos*, milky + *myces*, fungus.

Galerina Earle
Hymenogastraceae, Agaricales, Agaricomycetidae, Agaricomycetes, Agaricomycotina, Basidiomycota, Fungi
Earle, F. S. 1901. The genera of North American gill fungi. *Bull. New York Bot. Gard.* 5:423.
From genus *Galera* Fr., derived from L. *galerus* or *galerum*, cap, pileus, covering for the head similar to a helmet + L. suf. *-ina*, pertaining to something. Pileus generally small, reddish-brown with orange or yellow tones; gills brilliantly colored due to reddish-brown spores. In beds of moss and among grasses, on sandy soils, also lignicolous or humicolous. Some species are toxic due to the presence of amatoxins.

Galiella Nannf. & Korf
Sarcosomataceae, Pezizales, Pezizomycetidae, Pezizomycetes, Pezizomycotina, Ascomycota, Fungi
Korf, R. P. 1957. Two bulgarioid genera: *Galiella* and *Plectania*. *Mycologia* 49:107-111.
From L. *galea*, helmet > *galeola*, helmet-shaped glass + dim. suf. *-ella*, for the shape of the apothecium.

Gautieria

Galerina marginata (Batsch) Kühner: reddish-brown basidiomata, on soil, x 1.5.

Ganoderma P. Karst.
Ganodermataceae, Polyporales, Inc. sed., Agaricomycetes, Agaricomycotina, Basidiomycota, Fungi
Karsten, P. A. 1881. Enumeratio boletinearum et polyporearum. Fennicum, systemate novo dispositarium. *Revue mycol.*, Toulouse 3(9):17.
From Gr. *gános*, bright, from *ganóo*, to polish, to shine, to gleam + *dérma*, skin, cutis, for the laquered surface of the pileus. Basidiocarps annual or perennial, stipitate or sessile, woody or suberose, generally dark colored on upper part of pileus, light on hymenium, with one or several layers of tubes; occasionally very large, up to 60 cm long. Grows on living or dead trees, causing a white rot, although some species are mycorrhizal with epiphytic orchids. *Ganoderma applanatum* (Pers. ex Wallr.) Pat. is one of the most common species on living and dead trees in subtropical regions; it has a sessile fructification in the shape of a semicircular shelf, whereas *G. lucidum* (Leys. ex Fr.) P. Karst. has basidiocarps with a long lateral stipe.

Gasteroagaricoides D. A. Reid
Psathyrellaceae, Agaricales, Agaricomycetidae, Agaricomycetes, Agaricomycotina, Basidiomycota, Fungi
Reid, D. A. 1986. New or interesting records of Australasian Basidiomycetes: VI. *Trans. Br. mycol. Soc.* 86 (3):429-440.
From Gr. *gasterós*, belly + *agarikón*, mushroom + NL. suf. *-oides*, likeness, for the gastroid character of the basidioma. Sporophores agaricoid, convex, with persistently and strongly incurved margin, with hymenophore more or less exposed, cinnamon-brown, densely asperate with small granular warts. Hymenophore formed of recognizable, intricately fused lamellae; lamellae very thin, paler than cap. Stipe broader toward fusoid base, pallid with a few recurved fibrillose scales below. Warts on pileus formed of thin-walled, globose, ovate or fusoid elements initially in chains. Hyphae often with slightly thickened walls and minute clamp-connections. Hymenophore consisting of basidia separated by large globose cells, brachycystidia, pseudoparaphyses, but lacking cystidia. Spore print not formed. Spores elongate-elliptic, symmetrical, smooth, hyaline or subhyaline with a distinct wall but lacking a germ pore. Collected in debris between rotting logs, Norfolk Island.

Gautieria Vittad.
Gomphaceae, Gomphales, Phallomycetidae, Agaricomycetes, Agaricomycotina, Basidiomycota, Fungi
Vittadini, C. 1831. *Monogr. Tuberac.* (Milano):25.
Named in honor of the Italian mycologist *Giuseppe Gautieri* + L. des. *-a*. Fructifications subglobose, tuberiform or pyriform, with a prominent basal rhizomorph. Peridium either persistent or evanescent at maturity. Gleba of tramal plates anastomosed to form labyrinthiform or cellular cavities lined with a definite palisade hymenium; columella simple or more often dendroid, traversing gleba; sterile base usually present. Spores elliptical or obovate, longitudinally ribbed; basidia bearing 1-4 brownish spores. Growing superficially or partly submerged in soils rich in humus or vegetable debris. In North and South America, Europe, Asia, Africa, Australia, Tasmania and New Zealand.

Ganoderma lucidum (Curtis) P. Karst.: basidioma with laquered surface of the pileus, and lateral stipe, on a rotten hazel stump, Great Britain, x 0.3.

Gautieria morchelliformis Vittad.: vertical section of a subglobose basidiome, showing the labyrinthiformic tramal plaques (gleba) that produce the basidiospores, x 0.5. This fungus is hypogeous in soil rich of humus, among mosses.

Geastrum

Geastrum Pers.
Geastraceae, Geastrales, Phallomycetidae, Agaricomycetes, Agaricomycotina, Basidiomycota, Fungi
Persoon, C. H. 1794. *Neues Mag. Bot.* 1:85.
From Gr. *gé*, earth + *astér*, *ástron*, star, for the dehiscence of the exoperidium in the shape of a star when the fructification matures. Fructifications globose or subglobose, acuminate at apex when young; composed of an exoperidium of three layers; an endoperidium that delimits gleba at maturity; layers of exoperidium split together from apex to base in several segments (rays) during dehiscence. Gleba sessile or pedicellate, membranous or papyraceous, opens by an apical pore, composed of capillitium and spores; spores globose, pigmented, rugose. Grows in pine and oak forests and in semiarid zones such as pastureland and scrub land.

Geastrum violaceum Rick: star-shaped, mature basidioma, showing the sporiferous sac, on soil, x 2.

Gelasinospora Dowding—see **Neurospora** Shear & B.O. Dodge
Sordariaceae, Sordariales, Sordariomycetidae, Sordariomycetes, Pezizomycotina, Ascomycota, Fungi
Dowding, E. E. 1933. *Gelasinospora*, a new genus of Pyrenomycetes with pitted spores. *Canadian Journal of Research, Section C* 9:294.
From Gr. *gelasínoi*, *gelasínon*, the wrinkles formed on laughing, from *gelao*, *geláso*, to laugh + *sporá*, spore.

Gelatinopsis Rambold & Triebel
Helicogoniaceae, Helotiales, Leotiomycetidae, Leotiomycetes, Pezizomycotina, Ascomycota, Fungi
Rambold, G., D. Triebel. 1990. *Gelatinopsis*, *Geltingia* and *Phaeopyxis*: three helotialean genera with lichenicolous species. *Notes R. bot. Gdn Edinb.* 46(3):375.
From. L. *gelare* < *gele*, to freeze, to congeal + Gr. *-ópsis*, aspect, appearance.
Ascomata apothecioid, sessile on a thin subiculum, reduced, small, gelatinous, aggregated or dispersed, immersed to errumpent, but then broadly attached to the substratum, roundish to elongate, pale to dark brown, gray or blackish, immarginate or indistinctly marginate, Disc exposed, flat or slightly convex. Excipulum composed of laterally of hyphoid cells, immersed in a more or less abundant, hyaline or pale colored gel, basally of isodiametric cells. Paraphyses simple or branched, septate, apically in some species swollen, without pigments. Asci clavate, thin-walled, 8-spored. Ascospores hyaline, 0-1(-3) septate, smooth, without a distinct perispore, subsphaerial, ellipsoid to fusiform, straight or slightly curved, with obtuse ends. Pycnidia with phialidic conidiogenous cells and conidia rod-shaped, straight, usually with truncate ends. On basidiomata and ascomata of fungi, or thalli of lichens.

Gelineostroma H. J. Swart
Inc. sed., Rhytismatales, Leotiomycetidae, Leotiomycetes, Pezizomycotina, Ascomycota, Fungi
Swart, H. J. 1988. Australian leaf-inhabiting fungi XXVII. Two Ascomycetes on *Athrotaxis* from Tasmania. *Trans. Br. mycol. Soc.* 90:445-449.
From L. *gelineus*, jelly-like + Gr. *strōma*, mattress, bed, referring to the gelatinized outer walls of the stroma. Stromata on abaxial surface of needles, circular to elliptical, partly confluent when close together, intraepidermal, when mature rupturing outer epidermal wall and opening with a short slit. Stroma center containing a hymenium surrounded by pale brown vertical hyphae. Basal layer of stroma thin, lacking structural detail. Apical layer of stroma hyphal at margin, absent elsewhere. Asci 8-spored, at first much shorter than the paraphyses, saccate, with a firm wall and without recognizable apical structures; mature asci elongating and releasing pores through a circular or somewhat irregular opening. Ascospores oval to elliptical, unicellular, hyaline while maturing, finally turning dark brown; in young spores outer wall layer gelatinous. Paraphyses with a narrow lumen and a much-swollen wall. Vegetative mycelium destroying host mesophyll. On *Athrotaxis selaginoides*, Cradle Mt National Park, Tasmania, Australia.

Gelineostroma swartii P. R. Johnst.: vertical section of the stroma, on *Arthotaxis cupressoides*, Tasmania, x 30; ascus with young ascospores, x 380; two mature uniseptate ascospores, with the outer wall layer gelatinous, x 480.

Genistella

Genabea Tul. & C. Tul.
Pyronemataceae, Pezizales, Pezizomycetidae, Pezizomycetes, Pezizomycotina, Ascomycota, Fungi
Tulasne, L. R., C. Tulasne. 1845. *G. bot. ital.* 1(7-8):60.
From L. *Genabum*, city (now Orleans) in the Roman Province of Gallia Lugdunensis (now Lyon), France, the place where it was collected for the first time. Hypogeous ascocarps lacking a basal mycelial layer, very lobed and folded, with various openings. Sterile zones in hymenium isolated by pseudoparenchyma. Ascospores ellipsoid or globose. Collected in Canada, Europe and Australia (Tasmania).

Genabea cerebriformis (Harkn.) Trappe: cerebriform ascocarp, among mosses, x 4.

Genea Vittad. (syn. **Myrmecocystis** Harkn.)
Pyronemataceae, Pezizales, Pezizomycetidae, Pezizomycetes, Pezizomycotina, Ascomycota, Fungi
Vittadini, C. 1831. *Monogr. Tuberac.* (Milano):27.
Dedicated to the Italian cleric J. Gené, with the addition of the letter *a*, because the name ends in a vowel. Ascocarps hypogeous, globose or lobed, folded, fleshy, waxy, hollow, with or without a basal mycelial layer, a single or several apical openings, with a well defined hymenial layer, which is covered by a secondary covering lined by fusion of apices of paraphyses. Asci persistent. Ascospores ellipsoidal or globose with a sculptured wall. Occurs near surface of humus or beneath mosses, often associated with ant nests, in fagaceous forests.

Genea thaxteri Gilkey: globose, lobate, fleshy ascocarp, on soil, x 8.

Geniculifera Rifai
Orbiliaceae, Orbiliales, Orbiliomycetidae, Orbiliomycetes, Pezizomycotina, Ascomycota, Fungi
Rifai, M. A. 1975. *Geniculifera* Rifai nom. nov. *Mycotaxon* 2:214-216.
From L. *geniculum*, knee + *fero*, to bear, for the geniculations on the conidiophore. Colonies fast growing, pale pink to whitish. Hyphae branched, hyaline, septate, smooth-walled. Conidiophores hyaline, smooth, septate, at first straight, then becoming geniculate or flexuous by repeated subapical growth. Conidiogenous cells terminal, integrated, polyblastic, sympodial and geniculate, with denticles of various length. Conidia solitary, acrogenous at first, later acropleurogenous, obpyriform to obovoid-turbinate or broadly fusoid-ellipsoidal, 1- to many septate, hyaline singly, pinkish-white en masse. Parasitic on nematodes in soil.

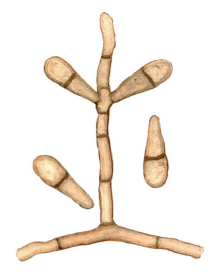

Geniculifera effusa (Jarow.) Oorschot: flexuous conidiophore, with obovoid-turbinate, one-septate conidia, x 700. Isolated in Poland.

Geniculosporium Chesters & Greenh.—see **Nemania** Gray
Xylariaceae, Xylariales, Xylariomycetidae, Sordariomycetes, Pezizomycotina, Ascomycota, Fungi
Chesters C. G. C., G. N. Greenhalgh. 1964. *Geniculosporium serpens* gen. et sp. nov., the imperfect state of *Hypoxylon serpens*. *Trans. Br. mycol. Soc.* 47:393-401.
From L. *genu*, dim. *geniculum*, knee + *spora*, spore + dim. suf. *-ium*, for the knee-like bends in the conidiogenous apical region of the conidiophore.

Genistella L. Léger & Gauthier—see **Legeriomyces** Pouzar
Legeriomycetaceae, Harpellales, Inc. sed., Inc. sed., Kickxellomycotina, Zygomycota, Fungi
Léger, L., M. Gauthier. 1932. Endomycetes nouveaux des larves aquatiques d'insectes. *C. r. hebd. Séanc. Acad. Sci., Paris* 194(26):2264.
From L. *genista*, a name applied to various shrubby plants, in particular to the broom < Celtic *gen*, shrub + dim. suf. *-ella*, for the shape of the branched thallus.

Genistellospora

Genistellospora Lichtw.
Legeriomycetaceae, Harpellales, Inc. sed., Inc. sed., Kickxellomycotina, Zygomycota, Fungi
Lichtwardt, R. W. 1972. Undescribed genera and species of Harpellales (Trichomycetes) from the guts of aquatic insects. *Mycologia* 64(1):167-197.
From genus *Genistella* L. Léger & Gauthier < L. *genista*, a name applied to various shrubby plants, in particular to the broom + dim. suf. -*ella* + Gr. *sporá*, spore, referring to the spores. Trichospores produced on branched thallus of this endocommensal of dipteran insect larvae.

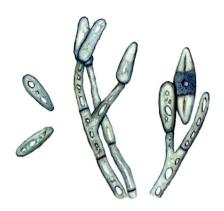

Genistellospora tropicalis Ríos-Velásquez, et al.: branched thallus, with spores and zygospore, x 400.

Geoglossum Pers. (syns. **Corynetes** Hazsl., **Gloeoglossum** E. J. Durand)
Geoglossaceae, Geoglossales, Inc. sed., Geoglossomycetes, Pezizomycotina, Ascomycota, Fungi
Persoon, C. H. 1794. Neuer Versuch einer systematischen Einteilung der Schwämme. *Neues Mag. Bot.* 1:116.
From Gr. pref. *geo-*, derived from *gé*, *géa*, earth, soil + *glõssa*, tongue, for the flat, elongate shape of the ascocarp, resembling a tongue, combined with its terricolous habitat, i.e., "earth tongue", with the neutral L. ending -*um*. Apothecia clavate, fleshy, viscid-gelatinous, erect, stipitate, black, brownish-black, or purplish-black, without a delimitation between pedicel and upper part bearing hymenium. Asci clavate-cylindric, opening by a pore, pore blue in iodine, 8-spored. Ascospores brown or fuliginous, cylindric or curved, ends rounded, transversally 3-15 multiseptate. Paraphyses numerous, septate, thickened, or coiled and brown above, not confined to hymenium but continuing down stem to base. Grows in shady, humid forests on decaying wood or soil rich in humus.

Georgefischeria Thirum. & Naras.
Georgefischeriaceae, Georgefischeriales, Exobasidiomycetidae, Exobasidiomycetes, Ustilaginomycotina, Basidiomycota, Fungi
Narasimhan, M. J., et al. 1963. *Georgefischeria*, a new genus of the Ustilaginales. *Mycologia* 55:30-34.
Named in honor of the American mycologist *George W. Fischer* + L. des. -*ia*. Mycelium permeating host tissues, forming internal sori of interwoven hyphae in wide lysigenous cavities that are delimited by marginal mycelium. Hyphal cells of sori converted into a mass of individual spores that are immersed in a mucous matrix. Spores spherical, 1-celled, subhyaline to pale cinnamon-brown, liberated by disintegration of surrounding host tissues. Spore germination by formation of promycelium that produces a whorl of 2-4 spores at apex. Collected on leaves, shoots and fruits of *Rivea hypocrateriformis* in Poona, India.

Geoglossum glabrum Pers.: elongate ascocarps, resembling a tongue, in grass, Germany, x 1.

Georgefischeria riveae Thirum. & Naras: internal sorus, with cinnamon-brown teliospores, in witches-broom of *Rivea hypocrateriformis*, x 80.

Geotrichum Link (syn. **Galactomyces** Redhead & Malloch)
Dipodascaceae, Saccharomycetales, Saccharomycetidae, Saccharomycetes, Saccharomycotina, Ascomycota, Fungi
Link, J. H. F. 1809. Observationes in ordines plantarum naturales. *Mag. Gesell. naturf. Freunde, Berlin* 3(1-2):17.
From Gr. *geo* < *gé*, *géa*, earth, soil + *thríx*, genit. *trichós*, hair + L. suf. -*um*, referring to the hyphae of this common soil mold. Colonies white with hyphae forming arthrospores (holoarthric conidia). Mycelium abundant, of broad, hyaline hyphae. Ascomata lacking. Asci formed from gametangia. Gametangia globose to subclavate, arising in pairs around a septum, with each

pair forming from adjacent cells that fuse to form an ascus. Asci formed singly, oval to globose or ellipsoidal, short-stalked, nonproliferating, containing a single ascospore, not rupturing at maturity. Ascospores 1-celled, oval to globose or ellipsoidal, pale yellow-brown, with a prominent oil globule, wall roughened with a thin, smooth exosporium and a hyaline equatorial furrow. Isolated from soil, fruits, foods, and sometimes from animal tissues; also a saprobe in lacteous and amylaceous foods. Causes oral, intestinal, bronchial or pulmonary geotrichosis in humans and higher animals.

Geotrichum candidum Link (syn. **Dipodascus geotrichum** (E.E. Butler & L.J. Petersen) Arx): giant colony on agar, x 1. Isolated in Puerto Rico.

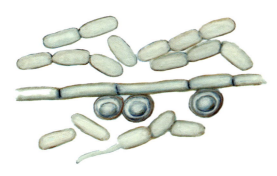

Geotrichum candidum Link (syn. **Dipodascus geotrichum** (E.E. Butler & L.J. Petersen) Arx): arthrospores and asci with one ascospore, x 950. Isolated in Puerto Rico.

Gerdemannia C. Walker et al.—see **Pacispora** Sieverd. & Oehl

Pacisporaceae, Diversisporales, Inc. sed., Glomeromycetes, Glomeromycotina, Glomeromycota, Fungi
Walker, C., et al. 2004. *Gerdemannia* gen. nov, a genus separated from *Glomus* and Gerdemanniaceae fam. nov., a new family in the Glomeromycota. *Mycol. Res.* 108:707-718.
Named in honor of the American mycologist *James W. Gerdemann*, for his extensive studies of glomoid fungi.

Gerronema Singer

Marasmiaceae, Agaricales, Agaricomycetidae, Agaricomycetes, Agaricomycotina, Basidiomycota, Fungi
Singer, R. 1951. New genera of fungi V. *Mycologia* 43:599.
Probably from Gr. *gérrhon*, anything made of wickerwork + *néma*, thread, referring to the context of the pileus tissue or of the hymenophore trama. Pileus and stipe white or pigmented, generally xantinic or carotenoid (yellow or orange), hyphae with or without fibulae; sometimes slightly thickened; lamellae decurrent, veil none, context white or concolorous with surface of pileus, usually unchanging but sometimes becoming reddish chestnut with age and on drying. Spores hyaline to yellowish, inamyloid, with a thin homogeneous wall. On decayed wood and on mosses or directly on soil, often lichenized with algae of the genus *Coccomyxa*; wide distribution in tropical, subtropical, temperate zones and Antarctica.

Gibbera Fr.

Venturiaceae, Venturiales, Pleosporomycetidae, Dothideomycetes, Pezizomycotina, Ascomycota, Fungi
Fries, E.M. 1825. *Syst. orb. veg.* (Lundae) 1: 110.
From. L. *gibbus*, bent, hunched > *gibber*, a hunch, hump + *-a*, fem. ending for Latin first declension. Ascomata perithecial, erumpent through the host cuticle and immersed rather than superficial mycelium, sometimes encircling it, black, composed of large dark brown angular cells. The ostiolar region hardly papillate, black, the entire exposed. Surface covered with short setae. Asci cylindrical, thick-walled and fissitunicate, the apex rounded with a fairly well-defined ocular chamber, 8-spored. Ascospores arranged obliquely uniseriately, fusiform-ellipsoidal, 1-septate, distinctly narrower at the slightly supramedian septum, initially greenish but becoming golden brown at maturity, fairly thick-walled, smooth or slightly roughened, without a perispore or gelatinous sheath or appendages. Apparently causing a disease on leaves and stems of plants, as *Vaccinum* species.

Gibbera prinsepiae (Chona, Munjal & J.N. Kapoor) E. Müll.: at left, vertical section of a stroma showing various ascostromate and sporodochia (one of these with spermogonium); at right, an sporodochium with conidiogenous cells and conidia and dark brown setae, x 650. The asexual stage is *Dictyodochium prinsepiae*.

Gibberella

Gibberella Sacc.—see **Fusarium** Link
Nectriaceae, Hypocreales, Hypocreomycetidae, Sordariomycetes, Pezizomycotina, Ascomycota, Fungi
Saccardo, P. A. 1877. *Michelia* 1(1):43.
From L. *gibber*, giboso, hunched, vaulted + dim. suf. *-ella*.

Gigaspora Gerd. & Trappe
Gigasporaceae, Diversisporales, Inc. sed., Glomeromycetes, Glomeromycotina, Glomeromycota, Fungi
Gerdemann, J. W., J. M. Trappe. 1974. The Endogonaceae in the Pacific Northwest. *Mycol. Mem.* 5:25.
From Gr. *gígas*, giant + *sporá*, spore, for the large size of the azygospores. Spores produced in a bulbous suspensor cell, larger than those of other endogonaceous fungi. Forms endotrophic mycorrhizae on poinsettia (*Euphorbia pulcherrima*).

Gigaspora margarita W. N. Becker & I. R. Hall: azygospore with a suspensor cell, x 600. Isolated in Illinois, U.S.A.

Gilbertella Hesselt.
Choanephoraceae, Mucorales, Inc. sed., Mucoromycetes, Mucoromycotina, Zygomycota, Fungi
Hesseltine, C.W. 1960. *Gilbertella* gen. nov. (Mucorales). *Bull. Torrey bot. Club.* 87(1):21-30
Dedicated to E. M. Gilbert + L. dim. suf. *-ella*. Sporangia many-spored, with a persistent wall that opens by circumscissile dehiscence along a suture. Sporangiospores smooth-walled, with long, slender polar appendages. Lacking sporangiola. Causes dry rot of peach and other fruits.

Glaphyriopsis B. Sutton & Pascoe
Inc. sed., Inc. sed., Inc. sed., Inc. sed., Pezizomycotina, Ascomycota, Fungi
Sutton, B. C., I. G. Pascoe. 1987. Some cupulate coelomycetes from native Australian plants. *Trans. Br. mycol. Soc.* 88(2):169.
From Gr. *glaphyrós*, hollow + *ópsis*, aspect, for the appearance of the conidiomata. Mycelium branched, septate, pale brown, anastomosing, superficial. Conidiomata covering apices of host hairs but sometimes arising laterally, eustromatic, shallow cupulate to tubular, brown, extending down hairs and originating from superficial mycelium, crenulate to irregular. Conidiophores absent. Conidiogenous cells formed from inner cells of conidiomatal wall, discrete, hyaline, ampulliform to doliiform or lageniform, producing a succession of conidia at same level by percurrent enteroblastic proliferation; collarette absent, periclinal thickening with cytoplasmic channel. Conidia holoblastic, hyaline or pale brown, smooth, 0-3 euseptate, fusiform.

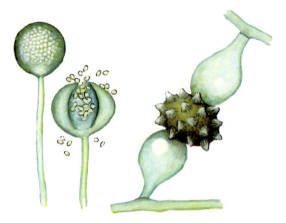

Gilbertella persicaria (E. D. Eddy) Hesselt.: many-spored sporangia, zygophores with a zygospore, on decaying fruit of *Prunus persica*, New York, U.S.A., x 300.

Glaphyriopsis brunnea B. Sutton & Pascoe: conidioma covering the apex of hair, on the stem and leaf hairs of *Acacia dealbata*, New South Wales, Australia, x 40; conidiogenous cells and fusiform, with 2- or 3-euseptate conidia, x 2,000.

Glaziella Berk.
Glaziellaceae, Pezizales, Pezizomycetidae, Pezizomycetes, Pezizomycotina, Ascomycota, Fungi

Berkeley, M. J. 1880. Symbolae ad floram Braziliae centralis cognoscendam. *In: Glaziou, Vidensk. Meddel. Dansk Naturhist. Foren. Kjøbenhavn* 80:31.

Named in honor of *A. Glaziou* + L. dim. suf. *-ella*. Large hollow sporocarps with thin walls in which chlamydospores originate. Known only from tropical soils.

Glaziella vesiculosa Berk.: large and hollow sporocarp, on the ground, Rio de Janeiro, Brazil, x 3.

Gliocladium Corda—see **Sphaerostilbella** (Henn.) Sacc. & D. Sacc.

Hypocreaceae, Hypocreales, Hypocreomycetidae, Sordariomycetes, Pezizomycotina, Ascomycota, Fungi

Corda, A. C. J. 1840. *Icon. fung.* (Prague) 4:30.

From Gr. *gloiós*, gum and other glutinous substances + *kládos*, branch + L. dim. suf. *-ium*, due to the fact that the conidiophores branch to form penicillia.

Gloeoglossum E. J. Durand—see **Geoglossum** Pers.

Geoglossaceae, Geoglossales, Inc. sed., Geoglossomycetes, Pezizomycotina, Ascomycota, Fungi

Durand, E. J. 1908. The Geoglossaceae of North America. *Annls mycol.* 6(5):418.

From Gr. *gloiós*, *gloeó*, viscid + *glóssa*, tongue.

Gloeophyllum P. Karst.—see **Osmoporus** Singer

Gloeophyllaceae, Gloeophyllales, Inc. sed., Agaricomycetes, Agaricomycotina, Basidiomycota, Fungi

Karsten, P. A. 1882. Enumeratio fungorum et myxomycetum in Lapponia orientali aestate. *Bidr. Känn. Finl. Nat. Folk* 37:79.

From Gr. *gloiós* > L. *gloeo*, resin, gluten, gum, from this the pref. *gleo-* or *gloeo-*, which signifies resinous, glutinous, sticky + *phýllon* > L. *phyllum*, leaf, gill.

Gloeosporium Desm. & Mont.—see **Drepanopeziza** (Kleb.) Jaap

Dermateaceae, Helotiales, Leotiomycetidae, Leotiomycetes, Pezizomycotina, Ascomycota, Fungi

Desmazières, J. B. H., J. F. C. Montagne. 1849. *Annls Sci. Nat., Bot., sér.* 3 12:295.

From Gr. *gloiós*, gum, glutinous substance + *sporá*, spore + L. dim. suf. *-ium*.

Gloeotinia M. Wilson, et al.

Inc. sed., Helotiales, Leotiomycetidae, Leotiomycetes, Pezizomycotina, Ascomycota, Fungi

Wilson, M., et al. 1954. *Gloeotinia* - a new genus of the Sclerotiniaceae. *Trans. Br. mycol. Soc.* 37(1):31.

From Gr. *gloiós*, gum and other glutinous substances, from *gloiódes*, glutinous, viscous + L. des. *-inia*, derived from the Gr. suf. *-ínos*, which denotes material, nature or source of something. Apothecia pedicellate, dark, cupulate to discoid, develops from a stroma in infected tissues of host. Asci clavate or cylindrical, with an apical pore. Conidia cylindrical, unicellular, suballantoid. On seeds of the grasses *Lolium* and *Secale*.

Gloeotinia granigena (Quél.) T. Schumach.: cupulate apothecia growing from a stroma, in the infected tissues of seed of *Lolium multiflorum*, x 12.

Glomerella Spauld. & H. Schrenk—see **Colletotrichum** Corda

Glomerellaceae, Glomerellales, Hypocreomycetidae, Sordariomycetes, Pezizomycotina, Ascomycota, Fungi

Spaulding, P., H. von Schrenk. 1903. The bitter-rot fungus. *Science* 17:750-751.

From L. *glomero*, to roll up, pile up, to group + dim. suf. *-ella*.

Glomus Tul. & C. Tul. (syns. **Simiglomus** Sieverd., et al., **Sclerocystis** Berk. & Broome)

Glomeraceae, Glomerales, Inc. sed., Glomeromycetes, Glomeromycotina, Glomeromycota, Fungi

Tulasne, L. R., C. Tulasne. 1845. Fungi nonnulli hypogaei, novi v. minus cogniti act. *G. bot. ital.* 1(2):63.

From L. *glomus*, ball, globe, ball of yarn, for the shape of the intracellular vesicles. Sporocarps rounded, cottony. Spores formed singly or in very loose, small clusters. Spores with a mono-to-multiple layered spore wall. Wall of the subtending hypha conspicuously continuous and concolorous with the spore wall, or slightly lighter

Glonium

in color than the spore wall. Subtending hyphae are funnel-shaped to cylindrical. Wall at spore attachment not with introverted wall thickening. Pore at spore base open but several septa in hyphae some distance from spore base can separate spore contents from mycelial contents. Walls of subtending hyphae thickened over very long distances from the spore base. Forming typical vesicular-arbuscular mycorrhiza, with mycorrhizal structures that stain blue to dark blue in trypan blue. Endomycorrhizal fungus develops in roots of plants.

Glomus glomerulatum Sieverd.: intracellular vesicles of this endomycorrhizal fungus within the root cell of plant host, from pot culture on *Pueraria phaseoloides*, Colombia, x 350.

Glonium Muhl.

Gloniaceae, Mytilinidiales, Pleosporomycetidae, Dothideomycetes, Pezizomycotina, Ascomycota, Fungi
Muhlenberg, G. H. E. 1813. *Cat. Pl. Amer. Sept.*:101.
From Gr. *klón*, genit. *klonós*, branch, bud + dim. *klónion*, with the des. *-ion* > L. dim. *-ium*. Ascostromata frequently elongated, linear, branched, radial or fastigiate, wider than high, black, carbonaceous, partially immersed at base, superficial or emergent, or situated on subiculum. Asci cylindrical, bitunicate intermixed with pseudoparaphyses. Ascospores uniseptate, generally hyaline. Develops as a saprobe on wood and tree branches.

Glonium stellatum Muhl. ex Fr.: part of the carbonaceous, radial, and fastigiate perithecial ascostroma, on rotten wood, U.S.A., x 20.

Glutinoagger Sivan. & Watling

Inc. sed., Inc. sed., Inc. sed., Agaricomycetes, Agaricomycotina, Basidiomycota, Fungi
Sivanesan, A., R. Watling. 1980. A new mitosporic Basidiomycete. *Trans. Br. mycol. Soc.* 74(2):424.
From L. *gluten*, genit. *glutinis*, glue + *agger*, material in a mound. Conidiomata effused or sporodochial, pale to ochraceous-buff, formed of conidiogenous cells fused into textura oblita at base. Conidiophores hyaline, branched, septate. Conidiogenous cells sympodially and percurrently proliferating, integrated with clamp connections. Conidia holoblastic, pyriform, ellipsoid or lacryform with a persistent hyphal appendage, enclosed in thick mucilaginous sheath. Collected on *Trabutia cocoicola* growing on *Cocos nucifera* in the Seychelles.

Glutinoagger fibulatus Sivan & Watling: sporodochial conidioma, x 40; conidiogenous cells, and lacrymorphic conidia with a persistent hyphal appendage, x 600; on leaves and stems of *Cocos nucifera*.

Gnomonia Ces. & De Not.

Gnomoniaceae, Diaporthales, Sordariomycetidae, Sordariomycetes, Pezizomycotina, Ascomycota, Fungi
Cesati, V., De Notaris, G. 1863. Schema di classificazione degle sferiacei italici aschigeri piu' o meno appartenenti al genere *Sphaeria* nell'antico significato attribuitoglide Persono. *Comm. Soc. crittog. Ital.* 1(4):177-420.
From L. *gnomon*, *gnomonis*, iron needle of a sundial, in reference to the long neck of the perithecium. Perithecia develop in host tissues with long neck that extends to surface. Conidia produced in acervuli. Grows on hardwoods.

Gnomonia fimbriata (Pers.) Fuckel (syn. **Asteroma carpini** (Lib.) B. Sutton): groups of conidiomata growing in the upper leaf of the host, x 0.8. On the right, there is a longitudinal section of conidioma, with conidia, x 280.

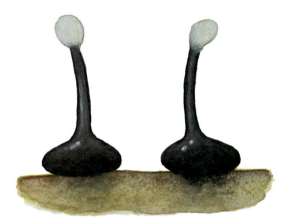

Gnomonia gnomon (Tode) J. Schröt.: long neck perithecia, on host leaves, x 70.

Gnomoniopsis Berl.
Gnomoniaceae, Diaporthales, Sordariomycetidae, Sordariomycetes, Pezizomycotina, Ascomycota, Fungi
Berlese, A. N. 1893. *Icon. fung.* (Abellini) *1*(3):93.
From L. *gnomon, gnomonis*, iron needle of a sundial, in reference to the long neck of the perithecium. Pycnidia brightly colored, with a fleshy or waxy texture, globose, errumpent or superficial, more or less papillate, conspicuous, ostiolate, perithecioid. In bark and leaves of plants producing leaf spots.

Gnomoniopsis comari (P. Karst.) Sogonov (syn. **Zythia fragariae** Laib.): brightly colored, erumpent pycnidia in the leaf spots produced by this phytopathogenic fungus, x 0.75; vertical section of a pycnidium showing the conidia in the interior, and the expelling cirrhus at the ostiole, x 150.

Godroniopsis Diehl & E. K. Cash
Helotiaceae, Helotiales, Leotiomycetidae, Leotiomycetes, Pezizomycotina, Ascomycota, Fungi
Diehl, W. W., E. K. Cash. 1929. The taxonomy of *Peziza quernea. Mycologia* 21:243.
From genus *Godronia* Moug. & Lév. + Gr. suf. -*ópsis*, appearance, i.e., similar to this genus. Apothecia sessile or subsessile, seated on a black, blister-like subiculum, erumpent through outer bark, at first closed, later cup-shaped or subdiscoid, corky-leathery. Asci clavate, inoperculate, 8-spored. Ascospores simple, hyaline, fusiform. Paraphyses filiform, slightly enlarged above, forming an epithecium. Occurring on various species of oak.

Godroniopsis quernea (Schwein.) Diehl & E. K. Cash: sessile apothecia seated on a black, erumpent subiculum, x 15; vertical section of a cup-shaped apothecium, showing the ascospores, on oak bark, x 30.

Gomphidius Fr.
Gomphidiaceae, Boletales, Agaricomycetidae, Agaricomycetes, Agaricomycotina, Basidiomycota, Fungi
Fries, E.M. 1836. *Fl. Scan.*:339.
From Gr. *gómphos*, nail + suf. -*eídos*, similar to; with the shape of a nail. Pileus glabrous, with context white, frequently viscid to glutinous or more rarely dry, small to rather large, sometimes reddened by autoxidation; hymenophore lamellate; lamellae rather thick and decurrent; stipe frequently yellow at base; gomphidic and atromentinic acids present. Specialized and obligatory ectomycorrhizal, with practical use in forestry, especially in reforestation projects. Present in north temperate areas. All species are edible.

Gomphus Pers.
Gomphaceae, Gomphales, Phallomycetidae, Agaricomycetes, Agaricomycotina, Basidiomycota, Fungi
Persoon, C. H. 1797. *Tent. disp. meth. fung.* (Lipsiae):74.
From Gr. *gómphos*, nail, in the shape of a wedge or pin, club, for the shape of the fructification. Fructification fleshy, claviform, infundibuliform or bugle-shaped; hymenium covers anastomosed veins or folds arranged radially and decurrent toward base of stipe, confluent with pileus; spores white or pale ochraceous. *Gomphus floccosus* (Schw.) Sing. (= *Cantharellus floccosus* Schw.) is one of the most esteemed edible species in the genus; it has a pileus that is convex-truncate at first, later bugle or infundibuliform, yellow or reddish-orange. It grows on the duff layer in fir forests or in mixed conifer and oak forests.

Gonapodya A. Fisch.
Gonapodyaceae, Monoblepharidales, Inc. sed., Monoblepharidomycetes, Inc. sed., Chytridiomycota, Fungi

Gongronella

Fischer, A. 1892. Phycomycetes. Die Pilze Deutschlands, Österreichs und der Schweiz. *In*: Winter, *Rabenh. Krypt.-Fl.*, Edn 2 (Leipzig) *1*(4):382.

From Gr. *gónos*, *góne*, the engendered, offspring + *poús*, genit. *podós*, foot. Thallus produces asexual zoosporangia and sexual gametangia. Saprobic on branches and fruits submerged in water.

Gomphus clavatus (Pers.) Gray: nail-shaped fructifications, in grass in woodland, Germany, x 0.5.

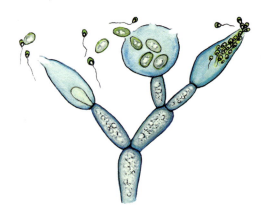

Gonapodya polymorpha Thaxt.: thallus producing asexual (zoosporangia) and sexual (gametangia) reproductive structures, x 950.

Gongronella Ribaldi

Cunninghamellaceae, Mucorales, Inc. sed., Inc. sed., Mucoromycotina, Zygomycota, Fungi

Ribaldi, M. 1952. Sopra un interessante zigomicete terricolo: *Gongronella urceolifera* n. gen. et nov. spec. *Riv. Biol.* 44:164.

From Gr. *gongróne*, tumor of the throat + L. dim. suf. *-ella*, in reference to the globose apophysis. Sporangium formed beneath apophysis, partially separated by a constriction, thus appearing like a small tumor located between sporangiophore and sporangium. Saprobic in soil.

Grandigallia M. E. Barr, et al.

Inc. sed., Inc. sed., Inc. sed., Dothideomycetes, Pezizomycotina, Ascomycota, Fungi

Barr, M. E., et al. 1987. A spectacular loculoascomycete from Venezuela. *Mycotaxon* 29:195-198.

From L. *grandis*, large + *galla*, gallnut > ME. *galla*, gall + L. *-ia*, quality or state of being. Large black, rough galls formed on branches of tree host consisting of stromatic tissues intermixed with cells of host. Ascomata obpyriform on surface of stroma with bases immersed. Asci bitunicate with reddish-brown dictyospores. Occurs on branches of *Polylepis sericea* in the Venezuelan Andes above 3,000 m.

Gongronella butleri (Lendn.) Peyronel & Dal Vesco: sporangiophores with apophysate sporangia of this mucoraceous fungus, saprobe in soil, x 700.

Grandigallia dictyospora M. E. Barr, et al.: vertical section of an ascostroma, intermixed with the cells of the host branches (*Polylepis sericea*), showing the asci and ascospores inside the locule, x 60; ascus with young ascospores, x 250; free ascospores, x 350.

Graphiola Poit.

Graphiolaceae, Exobasidiales, Exobasidiomycetidae, Exobasidiomycetes, Ustilaginomycotina, Basidiomycota, Fungi

Poiteau, M. A. 1824. Description du *Graphiola*, nouveau genre des plantes parasite de la famille des champignons. *Ann. Sci. Nat. (Paris)* 3:473.

From Gr. *graphé*, drawing, painting (from *grápho*, to scratch, to record, to paint) + L. dim. suf. *-ola*. Fructifications erumpent on upper, rarely lower, surface of leaves forming spots or scars. Basidiocarps cupuliform at first, then subglobose or saculiform, with a flexible peridium that partially encloses hyphal tissue where basidia are produced. Basidia arranged in chains of spherical fertile cells on which are formed the bicellular basidiospores at maturity. Spores emerge through an opening or ostiole of basidiocarp, attached to white, flexuous filaments that function as elaters or dispersal elements. Basidiospores germinate by budding (yeast phase) on surface of stomata producing infection hyphae that penetrate through stomata. *Graphiola phoenicis* (Moug. & Fr.) Poit. parasitizes palms of the genus *Phoenix*, in particular *Ph. canariensis* (Canary Island palm) and *Ph. dactylifera* (date palm).

Graphiola phoenicis (Moug. & Fr.) Poit.: erumpent basidioma on the surface of the leaf of bamboo (*Bambusa*), showing the group of basidiospores, x 30.

Graphis Adans.
Graphidaceae, Ostropales, Ostropomycetidae, Lecanoromycetes, Pezizomycotina, Ascomycota, Fungi
Adanson, M. 1763. *Fam. Pl.* 2:11.
From Gr. *graphé*, drawing, painting representation, writing, or from *graphís*, burin, stiletto, awl for engraving or the engraving itself < *grápho*, to draw, to scratch, scribble. Apothecia linear or curved, sometimes branched, known as lirellae, black, resembling scratches or writings on tree bark. Thallus crustous, slender or smooth, light gray.

Graphium Corda—see **Ophiostoma** Syd. & P. Syd.
Microascaceae, Microascales, Hypocreomycetidae, Sordariomycetes, Pezizomycotina, Ascomycota, Fungi
Corda, A. J. C. 1837. *Icon. fung.* (Prague) 1:18.
From L. *graphis*, paint brush < *graphium* < Gr. *graphíon*, instrument for writing or drawing, stylus with which the ancients wrote on waxed tablets.

Grifola Gray
Meripilaceae, Polyporales, Inc. sed., Agaricomycetes, Agaricomycotina, Basidiomycota, Fungi

Gray, S.F. 1821. *Nat. Arr. Brit. Pl.* (London) 1:643.
From *grifole*, the Italian name of the mushroom. Basidiocarps annual, stipitate, stipe simple or branched, giving rise to numerous petaloid pilei; upper surface gray to brownish, finely tomentose to glabrous; pore surface white to cream colored, pores angular; context white to pale buff; tubes decurrent on stipe; hyphal system dimitic, generative hyphae with clamps. Basidiospores ovoid to ellipsoid. Produces white rot of hardwoods and conifers. Some species are edible, frequent in North America.

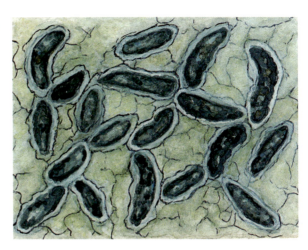

Graphis scripta (L.) Ach.: crustous thallus, with elongate apothecia (lirellae), on tree bark, Europe, x 30.

Guanomyces M. C. González, et al.
Chaetomiaceae, Sordariales, Sordariomycetidae, Sordariomycetes, Pezizomycotina, Ascomycota, Fungi
González, M.C., et al. 2000. *Guanomyces* a new genus of Ascomycetes from Mexico. *Mycologia* 92:1138-1148.
From E. *guano*, sea bird droppings < Am. Sp. *huano* < Quechua, *wanu*, dung for fuel + L. *myces* < Gk. *mýkes*, fungus. Ascomata gregarious, superficial, ovoidal, ostiolate, with long neck, membranous, brown to pale brown, opaque, pseudoparenchymatous, covered with glandular hairs. Ascomatal neck thin-walled, subcylindrical, erect to slightly curved, pale brown; lateral hairs arising singly, subcylindrical, straight, septate, translucent, pale brown, glandular, hyaline at tip, unbranched. Asci clavate, stipitate, rounded at top, 8-spored, lacking an apical pore, non-amyloid, thin-walled, unitunicate, deliquescing, maturing successively, paraphysate. Ascospores ellipsoidal, 1-celled, hyaline, without germ pores or slits, non-amyloid, minutely aculeate, extruded in liquid droplet. Chlamydospores spherical or obovate, sessile or short stipitate, lateral, dark brown, thick-walled, smooth, coriaceous.

Guedea Rambelli & Bartoli
Inc. sed., Inc. sed., Inc. sed., Inc. sed., Pezizomycotina, Ascomycota, Fungi

Guepinia

Rambelli, A., A. Bartoli. 1978. *Guedea*, a new genus of dematiaceous Hyphomycetes. *Trans. Br. mycol. Soc.* 71:340-342.

Named in honor of Ivory Coast Science Minister *J. Lourognon-Guede*. Conidiophores erect, dark brown, simple or repeatedly branched, acroauxic, indeterminate. Conidiogenous cells intercalary, integrated, holoblastic, denticulate. Conidia oval, dark brown, transversely septate, pedunculate. Collected on fallen bark in Tal, Ivory Coast.

Guanomyces polythrix M. C. González, et al.: long neck perithecium covered with glandular hairs, with ascospores extruded in liquid droplets, on agar culture, x 20. Isolated from bat guano, Morelos, Mexico.

Guepinia Fr. (syn. **Phlogiotis** Quél.)

Inc. sed., Auriculariales, Inc. sed., Agaricomycetes, Agaricomycotina, Basidiomycota, Fungi
Quélet, L., 1886. *Enchir. fung.* (Paris):202.
Named in honor of the French mycologist and botanist *J. P. Guépin* + L. *-inia*, belonging to, like.
Basidiomata gelatinous-elastic, erect, substipitate, spatuliform or infundibuliform, in this case furrowed on one side as in shape of an ear, pinkish-orange or reddish, at times with a paler base, 3-12 cm high. Among mosses and rotted wood with fallen leaves in conifer forests, in humus, or in humid grasslands.

Guepiniopsis Pat.

Dacrymycetaceae, Dacrymycetales, Inc. sed., Dacrymycetes, Agaricomycotina, Basidiomycota, Fungi
Patouillard, N.T. 1883. *Tab. analyt. Fung.* (Paris) 1(1):27.
From genus *Guepinia* Fr. + Gr. suf. *-ópsis*, i.e., similar to, or with the aspect of the genus *Guepinia*. Basidiocarps cupuliform, coriaceous, bent, gelatinous, attenuated in stipe. Hymenium in a layer at inner surface of cup. Basidia at first cylindric, developing at apex of hyphae in which protoplasm accumulates and a wall formed to isolate basidia near apex of hyphae. Spores hyaline, navicular.

Guepinia helvelloides (DC.) Fr. (syn. **Phlogiotis helvelloides** (DC.) G. W. Martin): spathuliform, pinkish-orange basidiomata, on soil in conifer forest, x 1.

Guignardia Viala & Ravaz—see **Phyllosticta** Pers.

Phyllostictaceae, Botryosphaeriales, Inc. sed., Dothideomycetes, Pezizomycotina, Ascomycota, Fungi
Viala, P., L. E. Ravaz. 1892. *Bull. Soc. mycol. Fr.* 8(2):63.
Dedicated to the French botanist L. Guignard + L. suf. *-ia*, which denotes pertaining to.

Gymnoascus Baran.

Gymnoascaceae, Onygenales, Eurotiomycetidae, Eurotiomycetes, Pezizomycotina, Ascomycota, Fungi
Baranetzky, J. 1872. Entwicklungsgeschichte des *Gymnoascus reessii*. *Bot. Ztg.* 30:158.
From Gr. *gymnós*, naked, exposed + *askós* > L. *ascus*, wine-skin, sac, ascus. Gymnothecium with undefined appendages having free, spiny tips. Asci more or less protected by a loose network of hyphae. Saprobic in soil.

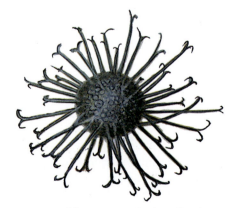

Gymnoascus reessii Baran.: gymnothecium with spine-shaped tips, x 350. Isolated from dung of sheep and horse, Germany.

Gymnoeurotium Malloch & Cain—see **Aspergillus** P. Micheli
Aspergillaceae, Eurotiales, Eurotiomycetidae, Eurotiomycetes, Pezizomycotina, Ascomycota, Fungi
Malloch, D., R. F. Cain. 1972. The Trichocomataceae: Ascomycetes with *Aspergillus*, *Paecilomyces*, and *Penicillium* imperfect states. *Can. J. Bot.* 50(12):2619.
From Gr. *gymnós*, naked + genus *Eurotium* Link.

Gymnogaster J. W. Cribb
Agaricaceae, Agaricales, Agaricomycetidae, Agaricomycetes, Agaricomycotina, Basidiomycota, Fungi
Cribb, J.W. 1956. The Gasteromycetes of Queensland. II. Secotiaceae. *Pap. Dept. Bot. (formerly Biol.) Univ. Qd.* 3:109.
From Gr. *gymnós*, naked, uncovered + *gastér*, *gastéros*, stomach, belly, because it is a naked gasteromycete. Fruitbody centrally stipitate, globose depressed, without peridium, columella percurrent through gleba to apex. Gleba pale brown, but on sectioning turns blue, later changing to dark brown. Basidia 4-spored, sterigmate. Spores elliptic or ovate, brown. Color changes similar to those observed in many boletes. Collected in a jungle of Queensland, Australia.

Gymnopaxillus E. Horak
Serpulaceae, Boletales, Agaricomycetidae, Agaricomycetes, Agaricomycotina, Basidiomycota, Fungi
Horak, E., M. Moser. 1966. Fungi Austroamericani. 8. Über neue Gastroboletaceae aus Patagonien. *Singeromyces* Moser, *Paxillogaster* Horak und *Gymnopaxillus* Horak. In Horak & Moser, *Nova Hedwigia* 10:335.
From Gr. *gymnós*, naked, nude + genus *Paxillus* Fr., i.e., a naked *Paxillus*, without a peridium and with the gleba exposed. Basidiocarps epigeous, rarely subhypogeous. Pileus ovate-conical to irregularly subglobose, without peridium; gleba exposed, cerebriform or irregularly subglobose, undulate cavernose, similar to that of *Gautieria*.Vittad. Stipe more or less cylindric, generally attenuate at base, with upper part (columella) ramose, subfistulose; without veil. Spores bilaterally symmetric, inequilaterally ellipsoid or subfusoid, cystidia absent. In soil of *Nothofagus* forests of South America.

Gymnopilus P. Karst.
Hymenogastraceae, Agaricales, Agaricomycetidae, Agaricomycetes, Agaricomycotina, Basidiomycota, Fungi
Karsten, P. A. 1879. Rysslands, Finlands och den Skandinaviska halföns Hattsvampar. Förra Delen: Skifsvampar. *Bidr. Känn. Finl. Nat. Folk* 32:XXI.
From Gr. *gymnós*, bare, nude, naked + *pilós*, hat, head, i.e., with the head uncovered; naked head or hat. Pileus usually brightly colored: yellow, fulvous, red, blue, lilac, green; some pigments incrusting hyphal wall; viscid or hygrophanous or dry, glabrous, fibrillose, squamulose or squarrose; lamellae narrow to broad, rusty in dried mature carpophores, adnexed to decurrent; context often bitter; stipe well developed, pigmented, frequently with a prominent annulus; hyphae with clamp connections. Spore print ferruginous-fulvous, brown or orange; spores ellipsoid, distinctly warty, with double wall, without germ pore. On coniferous wood or frondose wood. Some species endotrophically mycorrhizal with orchids; but not ectotrophically mycorrhizal. Almost cosmopolitan, except in the Antarctic continent.

Gymnopilus aeruginosus (Peck) Singer: gregarious basidiocarps on dead wood, x 0.5. Toxic and hallucinogenous.

Gymnopus (Pers.) Gray
Omphalotaceae, Agaricales, Agaricomycetidae, Agaricomycetes, Agaricomycotina, Basidiomycota, Fungi
Gray, S. F. 1821. *Nat. Arr. Brit. Pl.* (London) 1:627.
From Gr. *gymnós*, naked, nude + *poús*, genit. *podós*, foot. Basidiocarps relatively small, and range from browns to white in color, collybioid, rarely tricholomatoid or marasmioid with a pileus convex, plano-convex to applanate or slightly concave, dry or slightly viscid, glabrous or innately radially fibrillose, lamellae free, emarginate or adnate, crowded to sometimes fairly distant, a stipe central, non-insititious, and a spore print white. Basidiospores are ellipsoid to oblong, rarely subglobose to globose or lacrimoid, thin-walled, hyaline, non-amyloid; cheilocystidia often present, cylindrical, flexuous, clavate or irregularly coralloid; pleurocystidia usually absent or in some species well-developed; a pileipellis in the form of a cutis or ixocutis of radially arranged cylindrical hyphae, or interwoven, made up of irregular coralloid terminal elements; hyphae never amyloid or dextrinoid, and clamp connections mostly present. Most species are found in leaf and woody litter, and act as descomposers. The genus has a widespread, and cosmopolitan distribution.

Gymnosporangium R. Hedw. ex DC.
Pucciniaceae, Pucciniales, Inc. sed., Pucciniomycetes, Pucciniomycotina, Basidiomycota, Fungi
Lamarck, J. de, A. P. De Candolle. 1805. *In:* Lamarck &

Gyoerffyella

de Candolle, *Fl. franç.,* Edn. 3 (Paris) 2:216.
From Gr. *gymnós*, naked + L. *sporangium*, sporangium (this from Gr. *sporá*, spore, seed + *angeíon*, glass, or receptacle). Telia exposed when wet, gelatinous, on members of Cupressaceae. Aecia generally on members Rosaceae. Causes the rust diseases of junipers or cedars, and apple, pear and blackthorn.

Gymnopus iocephalus (Berk. & M. A. Curtis) Halling (syn. **Collybia iocephala** (Berk. & M. A. Curtis) Singer): mature, pinkish basidiomata, on soil, x 1.

Gymnosporangium juniperi-virginianae Schwein.: telia in a gall, causing the rust disease on juniper, x 10.

Gyoerffyella Kol (syn. **Ingoldia** R. H. Petersen)
Inc. sed., Inc. sed., Inc. sed., Inc. sed., Pezizomycotina, Ascomycota, Fungi
Kol, 1928. *Folia Cryptog.* 1:618.
Name in honor of the Hungarian bryologist *Istvan Györffy* + L. dim. suf. *-ella*. Fungus submerged aquatic with septate, branching, hyaline mycelium. Conidia hyaline, septate, consisting of a curved, attenuated axis, two attenuated lateral appendages, and a single attenuated secondary appendage. On submerged rotting leaves.

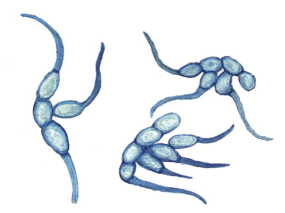

Gyoerffyella craginiformis (R. H. Petersen) Marvanová (syn. **Ingoldia craginiformis** R. H. Petersen): aquatic fungus with hyaline, septate conidia (aleuriospores), with a central axis, with three or four cells, and lateral cells, with or without appendages, x 600.

Gyrodon Opat.
Paxillaceae, Boletales, Agaricomycetidae, Agaricomycetes, Agaricomycotina, Basidiomycota, Fungi
Opatowski, G. 1836. De Familia fungorum Boletoideorum. *Arch. Naturgesch.* 2(1):5.
From Gr. *gyrós*, round + *odón*, genit. *odóntos*, tooth, rounded and toothed. Pileus viscid when wet; hymenophore of irregular arranged (gyrose) or boletinoid tubes; stipe central or eccentric, without a veil; context either changing color or unchanging on exposure; spore print brown to olive; spores short ellipsoid to subreniform, smooth, brownish; hyphae with clamp connections. On wood or ground in woods, often forming mycorrhizae with trees such as *Alnus*, *Fraxinus*, and *Quercus*.

Gyromitra Fr.
Discinaceae, Pezizales, Pezizomycetidae, Pezizomycetes, Pezizomycotina, Ascomycota, Fungi
Fries, E. M. 1849. *Summa veg. Scand.,* Sectio Post. (Stockholm):346.
From Gr. *gyrós*, rounded, curved + L. *mitra*, cap, headdress, due to the cerebriform pileus of the ascoma. Ascomata large apothecia, generally lobulate with brainlike convolutions, sometimes saddle-shaped, similar to fructifications of *Helvella* L. *Helvella* includes edible or mildly toxic species, whereas *Gyromitra* includes species that on occasion are toxic or even fatal. Grows on humid soils in coniferous and mixed woods.

Gyrophora Ach.—see **Umbilicaria** Hoffm.
Umbilicariaceae, Umbilicariales, Umbilicariomycetidae, Lecanoromycetes, Pezizomycotina, Ascomycota, Fungi
Acharius, E. 1803. *Methodus*, Sectio prior (Stockholmiæ):100.

From Gr. *gyrós*, round, rounded off, made in a circular shape + suf. *-phóros*, bearer, from *phéro* or *phoréo*, to carry, have, to be endowed with, possess.

Gyromitra esculenta (Pers.) Fr.: apothecium with cerebriform pileus, on soil, x 1.

Gyrophragmium Mont.
Agaricaceae, Agaricales, Agaricomycetidae, Agaricomycetes, Agaricomycotina, Basidiomycota, Fungi
Montagne, C. 1843. Considérations générales sur la tribu des Podaxinées et fondation du nouveau genre *Gyrophragmium*, appartenant à cette tribu. *Annls Sci. Nat.*, Bot., sér. 2 20:77.
From Gr. *gyrós*, round, a circle + *phragmós*, fragment, hedge, wall, septum, fence + L. dim. suf. *-ium*; the name refers to the radially arranged lamellae of the gleba. Fruitbody at first turbinate, agaricoid at maturity with fibrose central stipe which grows through gleba as a columella. Gleba lamellate without capillitium, spores small, smooth, without germ pore. In arid sandy soil of North Africa, South Europe, Central Asia and America, especially Argentina and Mexico.

Gyroporus Quél.
Gyroporaceae, Boletales, Agaricomycetidae, Agaricomycetes, Agaricomycotina, Basidiomycota, Fungi
Quélet, L. 1886. *Enchir. fung.* (Paris):161.
From Gr. *gyrós*, round + *póros* > L. *porus*, pore, referring to the round pores of the basidiocarp. Pileus glabrous to coarsely fibrous-subsquamose, non-viscid. Hymenophore tubulose with small to medium sized or large pores, depressed around stipe; stipe hollow or solid, surface glabrous, fibrous or subfurfuraceus, not reticulate, without veil; context white or whitish, unchanging, not bluing; hyphae of carpophore with clamp connections. Spore print yellow; spores subhyaline to yellowish, ellipsoid or elongate. On the ground under conifers and frondose trees normally forming ectomycorrhizae. Both in temperate and tropical regions, widely scattered but uncommon or absent in the southern temperate zone. All species are edible and highly esteemed by mycophagists.

Gyroporus castaneus (Bull.: Fr.) Quél.: basidiome boletoid, chestnut brown, dry, cap up to 10 cm, stipe hollow or cavernose, brittle, more or less smooth. Tubes at first white, then cream to straw. Broadleaf or coniferous forests, mycorrhizal mostly with oaks (*Quercus*), but also with beech (*Fagus*), sweet chestnut (*Castanea*) or birch (*Betula*), x 1.

H

Herpothallon rubrocinctum

H

Haematomma A. Massal.—see **Ophioparma** Norman
Haematommataceae, Lecanorales, Lecanoromycetidae, Lecanoromycetes, Pezizomycotina, Ascomycota, Fungi
Massalongo, A. B. 1852. *Ric. auton. lich. crost.* (Verona):32.
From Gr. pref. *haíma-*, genit. *haímatos-*, blood + *omma*, aspect, eye, for the apothecia similar to blood-red eyes.

Hallenbergia Dhingra & Priyanka
Inc. sed., Inc. sed., Inc. sed., Inc. sed., Inc. sed., Basidiomycota, Fungi
Dhingra, G. S., G. Priyanka. 2011. *Hallenbergia* (Agaricomycetes), a new corticioid genus. *Mycotaxon* 118:289. Named in honor of the Swedish mycologist *Nils Hallenberg* + L. des. *-ia*. Basidiocarp resupinate, adnate, effused, thin, ceraceous; hymenial surface smooth, farinose, continuous, some cracks developing on drying; margins not well differentiated. Hyphal system monomitic; generative hyphae thin-walled, septate, clamped; basal hyphae irregularly branched and interwoven into a dense texture; subhymenial hyphae short-celled, compactly packed and appearing like a cellular tissue. Cystidia absent. Basidia subclavate to suburniform, 4-sterigmata. Basidiospores ellipsoid to ovoid or subglobose, smooth, with thickened walls, cyanophilous, inamyloid. On decaying angiospermous twigs in Nawephu, Thimphu, Bhutan.

Hallenbergia singularis Dhingra & Priyanka: sagittal section of a resupinate basidiocarp, showing basidia, ellipsoidal basidiospores, and generative hyphae, x 240.

Halonectria E. B. G. Jones
Bionectriaceae, Hypocreales, Hypocreomycetidae, Sordariomycetes, Pezizomycotina, Ascomycota, Fungi
Jones, E. B. G. 1965. *Halonectria milfordensis* gen. et sp. nov., a marine pyrenomycete on submerged wood. *Trans. Br. mycol. Soc.* 48:287.
From Gr. *háls*, genit. *halós*, the sea + genus *Nectria* (Fr) Fr., i.e., a marine fungus resembling *Nectria*. Perithecia solitary, immersed or partly immersed in substratum, brightly colored, membranous, with long necks. Asci unitunicate, clavate, 8-spored, deliquescing at maturity. Ascospores hyaline, fusiform, sometimes lunate or falcate, unicellular, guttulate. On submerged Scots pine test blocks in South Wales, UK.

Halonectria milfordensis E. B. G. Jones: immersed brightly colored perithecia, in Scots pine test blocks (*Pinus sylvestris*), Great Britain, x 90, with fusiform, unicellular ascospores, x 800.

Halophiobolus Linder—see **Lulworthia** G. K. Sutherl.
Lulworthiaceae, Lulworthiales, Inc. sed., Sordariomycetes, Pezizomycotina, Ascomycota, Fungi
Linder, D. L. 1944. I. Classification of the marine fungi. pp. 401-433, *In*: Barghoorn, E. S., D. L. Linder, Marine fungi: their taxonomy and biology. *Farlowia* 1:395-467.
From Gr. *háls*, genit. *halós*, the sea + genus *Ophiobolus* Riess, i.e., a marine *Ophiobolus*.

Halosarpheia Kohlm. & E. Kohlm.
Halosphaeriaceae, Microascales, Hypocreomycetidae, Sordariomycetes, Pezizomycotina, Ascomycota, Fungi
Kohlmeyer, J., E. Kohlmeyer. 1977. Bermuda marine fungi. *Trans. Br. mycol. Soc.* 68:208.

Name formed as an anagram of the closely related genus *Halosphaeria* Linder. Ascocarps solitary or gregarious, obpyriform to subglobose, immersed or partly immersed, ostiolate, papillate, coriaceous, brown to black. Pseudoparenchyma of large, thin-walled cells filling venter of young ascocarps, breaking up to form catenophyses. Asci 8-spored, clavate, pedunculate, unitunicate, thick-walled below apex, thin-walled in peduncle, persistent, developing at base of venter; mature asci breaking off at base from ascogenous tissue. Ascospores ellipsoidal, 1-septate, hyaline, with apical appendages; appendages exosporic remnants, at first cap-like, stiff, homogeneous, attached to apices, at maturity becoming soft and transforming into fibers. An obligately marine species. On loose submerged wood.

Halorosellinia Whalley, et al.
Xylariaceae, Xylariales, Xylariomycetidae, Sordariomycetes, Pezizomycotina, Ascomycota, Fungi
Whalley, A. J. S., et al. 2000. *Halorosellinia* gen. nov. to accommodate *Hypoxylon oceanicum*, a common mangrove species. *Mycol. Res.* 104:368.

From Gr. *halós*, the sea + genus *Rosellinia* De Not., for the habitat of the fungus and similarity to the latter genus. Pseudostromata poorly developed. Ascomata immersed in pseudostroma, subglobose or hemispherical, soft to leathery, black, ostiole papillate. Asci 8-spored, cylindrical, unitunicate, with relatively large apical apparatus, dark blue in Melzer's reagent, tapering with a distinct apical rim. Ascospores unicellular, dark brown, ellipsoidal, prominent, straight germ slit on ventral side, smooth-walled.

Halosphaeria Linder
Halosphaeriaceae, Microascales, Hypocreomycetidae, Sordariomycetes, Pezizomycotina, Ascomycota, Fungi
Barghoorn, E. S., D. H. Linder. 1944. Marine fungi: their taxonomy and biology. *Farrowia* 1:412.

From Gr. *háls*, genit. *halós*, the sea, salt + genus *Sphaeria* Haller, i.e., a marine *Sphaeria*-like fungus. Ascomata ostiolate perithecia, solitary or gregarious, superficial or partially immersed in substrate. Perithecium globose or subglobose to ovoid or ellipsoidal, hyaline, cream-colored, brown or black, papillate or with a long ostiolar neck. Wall coriaceous or membranous, thick. Centrum of young ascomata filled with large, thin-walled pseudoparenchymatous cells that later form chains of vesiculose cells resembling paraphyses, later deliquescing at maturity. Asci unitunicate, undifferentiated at apex, clavate or subfusiform, short-stipitate, 8-spored, deliquescing early. Ascospores ellipsoidal, sometimes cylindrical, with rounded ends or rhomboidal, hyaline, 2-celled, appendaged; appendages apical and/or lateral, attached to ascospore wall, variable in shape, subcylindric, obclavate or subglobose, persistent or gelatinizing. On dead wood, stems and other cellulosic materials in marine habitats.

Halosphaeriopsis T.W. Johnson
Halosphaeriaceae, Microascales, Hypocreomycetidae, Sordariomycetes, Pezizomycotina, Ascomycota, Fungi
Johnson Jr, T.W. 1958. Some lignicolous marine fungi from the North Carolina coast. *J. Elisha Mitchell scient. Soc.* 74:42-48.

From Gr. *hals*, genit. *halós*, the sea + *sphaíra*, ball, sphere + *-ópsis*, aspect, appearance. Ascomata mostly immersed, subglobose to ellipsoidal, 300-330 μm diameter ostiolate, subcarbonose, black, solitary or gregarious. Asci unitunicate, mostly clavate, without an apical ring or thickening, often early deliquescing. Ascospores hyaline, 1-septate at maturity, with bipolar and central appendages; the polar appendages cup-like, small, equatorial appendages lunate. Marine fungi, on wood flotant or plant substrates.

Halorosellinia oceanica (S. Schatz) Whalley, et al.: sagittal section of a pseudostroma, showing two ascocarps releasing ascospores, x 20; ascus with eight ascospores, x 370; free dark brown ascospores, with germ slit, x 650. On *Rhizophora mangle*, in Florida, U.S.A.

Halosphaeriopsis mediosetigera (Cribb & J. W. Cribb) T. W. Johnson (syn. **Halosphaeria mediosetigera** Cribb & J. W. Cribb): ellipsoidal, 2-celled ascospore, with apical and lateral appendages attached to ascospore wall, x 2,400.

Hanliniomyces

Hanliniomyces Raja & Shearer
Inc. sed., Inc. sed., Sordariomycetidae, Sordariomycetes, Pezizomycotina, Ascomycota, Fungi
Raja, H. A., C. A. Shearer. 2008. Freshwater ascomycetes: new and noteworthy species from aquatic habitats in Florida. *Mycologia* 100(3):467-489.
Named in honor of the American mycologist *Richard T. Hanlin* for his outstanding contributions to ascomycete systematics + Gr. *mýkes* > L. *myces*, fungus. Ascomata scattered, partially immersed to superficial, black, ostiolate, with a cylindrical, periphysate neck; venter globose to subglobose, membranous. Peridium composed of pseudoparenchymatic cells, of textura angularis in surface view. Paraphyses wide at the base, tapering toward the apex, septate, slightly constricted at the septa. Asci unitunicate, narrowly fusoid, with or without a stalk, with a refractive, nonamyloid apical apparatus, containing eight uniseriate to overlapping biseriate ascospores. Ascospores ellipsoidal, aseptate and hyaline when young, becoming 3-septate and brown with age, surrounded by a gelatinous sheath.

Hanliniomyces hyaloapicalis Raja & Shearer: vertical section of an ascoma, globose, membranaceous, black, ostiolate, necked, x 150; unitunicate ascus, x 330, with 8 ascospores, ellipsoidal, aseptate, multigutulate, x 500; the mature ascospores are dark brown, 3-septate, surrounded with a mucous sheath, x 570. Found in Big Cypress National Reserve, Florida, U.S.A.

Hanseniaspora Zikes (syn. **Kloeckera** Janke)
Saccharomycodaceae, Saccharomycetales, Saccharomycetidae, Saccharomycetes, Saccharomycotina, Ascomycota, Fungi
Zikes, H. 1911. Zur Nomenklaturfrage der apiculatus Hefe. *Centbl. Bakt. ParasitKde*, Abt. II 30:148.
Dedicated to the Danish mycologist *E. Chr. Hansen* (1842-1909) + Gr. *sporá*, spore. Yeast characterized by apiculate cells with bipolar budding, usually without pseudomycelium, with hat-shaped, helmet-shaped or saturnoid ascospores; ferments vigorously. Asexual morph kloeckera-like. Isolated principally from soft fruits, such as grapes, sap of plants, soil, and fruit flies.

Hansenula Syd. & P. Syd. (syn. **Wickerhamomyces** Kurtzman, et al.)
Pichiaceae, Saccharomycetales, Saccharomycetidae, Saccharomycetes, Saccharomycotina, Ascomycota, Fungi
Sydow, H., P. Sydow. 1919. Mykologische Mitteilungen. *Annls mycol.* 17(1):44.
Dedicated to the Danish mycologist *E. Chr. Hansen* (1842-1909) + L. dim. suf. *-ula*. Asci globose to ellipsoid, unconjugated or arise from conjugation between a cell and its bud or between independent cells. Some species heterothallic. Asci deliquescent or persistent and form one to four ascospores that are hat-shaped or spherical with an equatorial ledge. Cell division by multilateral budding on a narrow base and budded cells spherical, ovoid or elongate. Pseudohyphae and true hyphae formed by some species. Glucose is fermented by most species and some species ferment other sugars as well. A variety of sugars, polyols and other carbon sources are assimilated by most species, but not methanol and hexadecane. Nitrate is utilized by some species. Where determined, the predominant ubiquinone is CoQ-7. The diazonium blue B reaction is negative. The genus is phylogenetically circumscribed from analysis of LSU and SSU rRNA and EF-1a gene sequences.

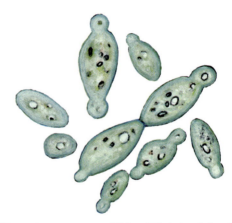

Hanseniaspora uvarum (Niehaus) Shohata, et al. ex M.T. Sm. (syn. **Kloeckera apiculata** (Reess) Janke): apiculate yeast cells, with bipolar germination, x 1,600.

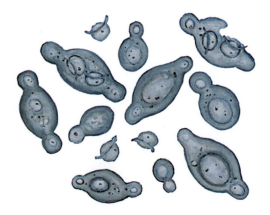

Hanseniaspora valbyensis Klöcker: yeast cells with bipolar budding, and asci with saturnoid ascospores, x 2,500.

Harpezomyces

Hansfordiopeltopsis M.L. Farr
Inc. sed., Inc. sed., Inc. sed., Inc. sed., Pezizomycotina, Ascomycota, Fungi
Farr, M. L. 1986. Amazonian foliicolous fungi. II. Deuteromycotina. *Mycologia* 78(2):269-286.
Named in honor of the English mycologist *C. G. Hansford* + connective *-io-* + Gr. *pélte*, small shield + *ópsis*, aspect. Free mycelium lacking. Pycnidial conidiomata superficial, dimidiate-scutate, orbicular, greenish-black, membranous, smooth, with margins pelliculose and hyaline. Upper wall prosenchymatous-meandering, not radiate, with ostiole central or eccentric. Hymenium inverted. Basal wall gelatinous, hyaline, thin. Conidiophores lacking. Conidia sessile, ellipsoidal or bacillar, 1-celled, hyaline. On leaves of *Coelocaryon* sp. in Kivu, Belgian Congo.

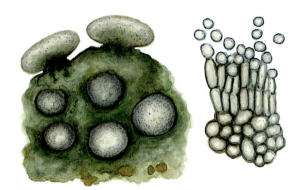

Haradamyces foliicola Masuya, et al.: sclerotium-like propagules in leaf of *Cornus florida*, Honshu, Japan, x 120; microconidiophores with microconidia, originated from terminal cells of a propagule, x 800.

Harpella L. Léger & Duboscq
Harpellaceae, Harpellales, Inc. sed., Inc. sed., Kickxellomycotina, Zygomycota, Fungi
Léger, L., O. Duboscq. 1929. *Harpella melusinae* n. g., n. sp. Entophyte eccriniforme, parasite des larvas de simulie. *C. r. hebd. Séanc. Acad. Sci., Paris* 188,88:951-954.
From L. *harpe*, scimitar, curved sabre, sickle, scythe < Gr. *hárpe* + dim. suf. *-ella*, because thallus has the shape of a sabre or sickle. Lives as an endocommensal in the middle intestine of dipteran simulid larvae.

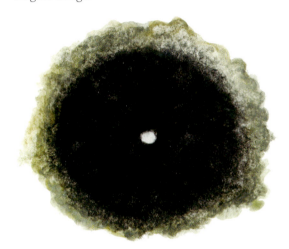

Hansfordiopeltopsis amazonensis M.L. Farr: front view of a black pycnothyrium on the surface of a host plant, showing the clear ostiole in the center, x 350.

Haradamyces Masuya, et al.
Inc. sed., Helotiales, Leotiomycetidae, Leotiomycetes, Pezizomycotina, Ascomycota, Fungi
Masuya, H., et al. 2009. *Haradamyces foliicola* anam. gen. et sp. nov., a cause of zonate leaf blight disease in *Cornus florida* in Japan. *Mycol. Res.* 113:173-181.
Named in honor of the Japanese mycologist *Yukio Harada* + L. *myces* < Gr. *mýkes*, fungus, for his contributions to knowledge of sclerotiniaceous fungi in Japan. Propagules multicellular, discoid, white to cream, becoming grey to dark brown with age, comprised of two elements: an inner layer of repeatedly centrifugally branching chains of cells and an outer layer of tightly appressed terminal cells, with 1-2 knob-like outgrowths. Each propagule arising individually from a hyaline, multihyphal, sclerotium-like structure in leaf tissue, seceding by simultaneous schizolytic secession at a number of septa. Terminal cells producing elongated hyphae on detached propagules at maturity, or rarely developing into microconidiophores on plants.

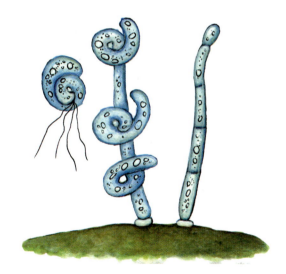

Harpella melusinae L. Léger & Duboscq: thalli, one of them with sickle-shaped trichospores; adhered to the intestine of diptera simulid larvae (*Melusina ornata*), France, x 85.

Harpezomyces Malloch & Cain—see **Aspergillus** P. Micheli ex Link
Trichocomaceae, Eurotiales, Eurotiomycetidae, Eurotiomycetes, Pezizomycotina, Ascomycota, Fungi
Malloch, D., R. F. Cain. 1972. The Trichocomataceae: Ascomycetes with *Aspergillus*, *Paecilomyces* and *Penicillium* imperfect states. *Can. J. Bot.* 50:2613-2628.
From Gr. *harpeza*, a thorn hedge + *mýkes* > L. *myces*, fungus, referring to the tomentose ascocarps.

Harposporium

Harposporium Lohde
Clavicipitaceae, Hypocreales, Hypocreomycetidae, Sordariomycetes, Pezizomycotina, Ascomycota, Fungi
Lohde, G. 1874. Über einige neue parasitische Pilze. *Tagbl. Versamml. Ges. Deutsch. Naturf.* 47:203-206.
From L. *harpe* < Gr. *hárpe*, scimitar, sickle, scythe + *spóros*, spore + L. dim. suf. *-ium*. Endoparasitic. Conidia small and sickle-shaped, with a sharp tip that facilitates their sticking in esophagus of nematodes that ingest them; once inside conidia germinate and the resulting hyphae invade it, forming a large number of chlamydospores in interior of animal, and conidiophores with phialides that erupt through cuticle to continue dissemination.

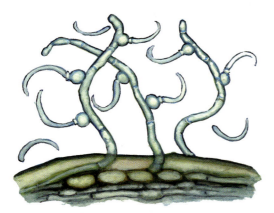

Harposporium harposporiferum (Samuels) Spatafora & Kepler (syn. **Harposporium anguillulae** Lohde): endoparasitic fungus inside a soil nematode, with hyphae and chlamydospores, and emerging conidiophores with sickle-shaped conidia, x 160.

Harrya Halling, et al.
Boletaceae, Boletales, Agaricomycetidae, Agaricomycetes, Agaricomycotina, Basidiomycota, Fungi
Halling, R.E., et al. 2012. Affinities of the *Boletus chromapes* group to *Royoungia* and the description of two new genera, *Harrya* and *Australopilus*. *Aust. Syst. Bot.* 25(6):418-431.
Named in honor of *Harry D. Thiers*, American boletologist, teacher and mentor. Fruitbodies epigeous. Pileus rose pink to brownish-pink or pinkish-grey. Hymenophore tubulose, white, then vinaceous-pink. Stipe white above, yellow at base, with fine pink scabers on surface. Spores pinkish- to reddish-brown, smooth. On litter, soil in forests associated with conifers, Betulaceae and *Quercus* in North America, Costa Rica and China.

Hebeloma (Fr.) P. Kumm. (syn. **Anamika** K. A. Thomas, et al.)
Hymenogastraceae, Agaricales, Agaricomycetidae, Agaricomycetes, Agaricomycotina, Basidiomycota, Fungi
Kummer, P. 1871. *Führ. Pilzk.* (Zerbst):22.
From Gr. *hébe*, youth, young, puberty, pubescence + *lõma*, stripe, border, margin, bank, fringe, in reference to the presence of an internal fibrous veil, like a curtain. Basidiomata with collybioid or naucorioid habit, with remains of veil on incurved margin of pileus in young specimens. Pileus viscous, conico-convex or convex, glabrous, white to buff or pale brown. Lamellae adnate, yellowish, pale orange, greyish-orange or dull brown with border commonly whitish, dentate. Stipe continuous, almost equal or enlarged towards both ends. Spore deposit brown. Basidiospores sublimoniform to amygdaliform, thick-walled, with a conspicuous callus but without germ-pore, with cavernous type of epitunica ornamentation. Basidia 4-spored, rarely 2-spored, clavate with a median constriction. Cheilo- and pleurocystidia present, versiform, clavate, pedicellate-clavate, uteriform or fusoid, frequently with a subcapitate, mucronate or rostrate apex. Caulocystidia either in small clusters or scattered, versiform, similar to cheilocystidia. Hymenophoral trama regular. Pileipellis and epicutis of thin-walled cylindrical hyphae. Hypocutis distinct, subcellular, of swollen, encrusted hyphae. Clamp connections present. Terricolous, some specific, such as abandoned mole or wasp nests. Often obligately mycorrhizal. Common in coniferous forests, but also in other types of forests and meadows.

Hebeloma mesophaeum (Pers.) Quél.: basidioma, with the silky pubescence in the pileus margin, terricolous, x 0.5.

Helicobasidium Pat.
Helicobasidiaceae, Helicobasidiales, Inc. sed., Pucciniomycetes, Pucciniomycotina, Basidiomycota, Fungi
Patouillard, N. T. 1885. Note sur un genre nouveau d' hyménomycètes. *Bull. Soc. bot. Fr.* 32:171.
From Gr. *hélix*, genit. *hélikos*, spiral, turn, helix, in a spiral, curved + L. *basidium*, derived from the Gr. *basídion*, dim. of *básis*, base. Fructifications membranous, resupinate, smooth hymenium; unicellular, hyaline, reniform spores. *Helicobasidium purpureum* Pat. is the sexual morph of synonym "*Rhizoctonia*" *crocorum* Fr. Causes violaceous root rot of plants, including saffron (*Crocus*), carnation (*Dianthus*), potato (*Solanum*), carrot (*Daucus*), asparagus (*Asparagus*), celery (*Apium*), young trees and various herbs.

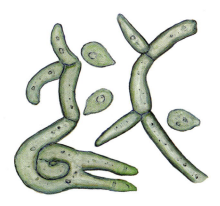

Helicobasidium corticioides Bandoni: helicoid basidia, with unicellular basidiospores, x 1,000. Collected in Colorado, U.S.A.

Helicocentralis Sri-indr et al.

Inc. sed., Inc. sed., Inc. sed., Leotiomycetes, Pezizomycotina, Ascomycota, Fungi

Sri-indrasutdhi, et al. 2015. *Helicocentralis hyalinus* gen. et sp. nov., an aero-aquatic helicosporous fungus (Leotiomycetes, Ascomycota) in Thailand. *Mycol. Progr.* 14(10/81):1-12.

From Gr. *hélix*, genit. *hélikos*, spiral + L. *centrum*, center + suf. *alis*, referring to the spiral or helical shape of conidia, and to the origin center in conidial rotation. Colonies on natural substrate effuse, dispersed, hairy and colourless. Mycelium partly superficial and partly immersed, hyaline, composed of branched, septate, smooth, and creeping hyphae. Conidiophores short, micronematous or semi-macronematous, mononematous, solitary or sometimes forming 3-4 in a group, arising laterally from the creeping mycelium, unbranched or rarely branched, tapering at the apex, septate, smooth, and hyaline. Conidiogenous cells monoblastic, integrated, terminal, determinate, conical to cylindrical. Conidia holoblastic, hyaline, solitary, dry, acrogenous, simple, circinate, and centrifugally coiled to clockwise or counterclockwise. Conidial filament coiled in three-dimensional plane, non-hygroscopic, hyaline, septate, and smooth walled.

Helicocentralis hyalina Sri-indr et al.: septate hyphae with acrogenous conidiophores, holoblastic, and hyaline conidia rolled centrifugaly (excentric), helicoidal or circinate, 2-3 turns (clockwise or anticlockwise), x 1,000. This aero-aquatic fungus was isolated in Thailand.

Helicogermslita Lodha & D. Hawksw.

Xylariaceae, Xylariales, Xylariomycetidae, Sordariomycetes, Pezizomycotina, Ascomycota, Fungi

Hawksworth, D. L., B. C. Lodha. 1983. *Helicogermslita*, a new stromatic xylariaceous genus with a spiral germ slit from India. *Trans. Br. mycol. Soc.* 81:91-96.

From Gr. *helix*, genit. *hélikos*, twisted, wound + L. *germen*, bud, offshoot + ME. *slitte* > E. *slit*, a narrow cut + euphonic *-a-*, for the helical germ slits on the ascospores. Ascomata stromatic, single or united in small groups under a common clypeus-like structure, perithecioid, subglobose, black, carbonaceous, ostiolate, periphysate. Paraphyses filiform, persistent. Asci elongate-cylindrical, unitunicate, non-amyloid, lacking an apical ring. Ascospores ellipsoid, simple, brown, with a helical germ slit running along its length, lacking a gelatinous sheath. Asexual morph unknown. Collected on angiospermous twigs in India.

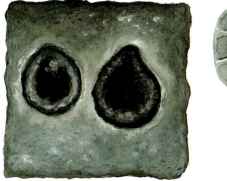

Helicogermslita celastri (S. B. Kale & S. V. S. Kale) Lodha & D. Hawksw.: stroma with two carbonaceous perithecia, on branches of angiospermous plants, Maharashtra, India, x 4; ellipsoidal ascospores with helical germ slit running along the length of the spore, x 875.

Helicogoosia Hol.-Jech.

Inc. sed., Inc. sed., Inc. sed., Inc. sed., Pezizomycotina, Ascomycota, Fungi

Holubová-Jechová, V. 1991. *Helicogoosia*, a new genus of lignicolous hyphomycetes. *Mycotaxon* 41:445-450.

From Gr. *hélikos* > L. *helicos*, twisted, wound, for the shape of the spore + *goos*, in honor of American mycologist *Roger D. Goos* + des. *-ia*, for his contribution to the study of helicosporous fungi. Colonies cottony or tomentose, effuse, mycelium superficial, composed of smooth, pale brown hyphae. Condiophores arising laterally and singly on hyphae, semimacronematous, mononematous, usually unbranched, straight, pale brown, septate or aseptate, thick-walled, smooth. Conidiogenous cells monoblastic, integrated, terminal or lateral, lanceolate, smooth. Conidia holoblastic, acrogenous, solitary, dry, simple, tightly coiled in excentric direction, septate, brown, seceding schizolytically.

Helicosingula

Helicogoosia paradoxa Hol.-Jech.: conidiophores with lanceolate conidiogenous cells originated from diminute pores in the conidiophores; circinate conidia, 5-12 septate, light brown when inmature, and darker and larger when mature, x 1,000. Described in dead bark of *Pinus sylvestris*.

Helicosingula P. S van Wyk, et al.
Inc. sed., Inc. sed., Inc. sed., Inc. sed., Pezizomycotina, Ascomycota, Fungi
Van Wyk, P. S., et al. 1985. *Helicosingula*, a new genus of dematiaceous Hyphomycetes on *Leucadendron tinctum* in South Africa. *Trans. Br. mycol. Soc.* 85:183-187.
From Gr. *hélix*, genit. *hélikos*, twisted, wound + L. *singularis*, solitary, for the coiled, solitary conidia. Colonies on leaves circular, radiating, with a sooty black appearance. Mycelium on leaf surface of dark brown, smooth hyphae branched, intertwined and anastomosed to form a dense, crust-like network; superficial mycelium attached to subcuticular mycelium by narrow, cylindrical, brown hyphae that penetrate cuticle. Subcuticular layer of hyphae at first hyaline, becoming dark brown, thick-walled, stroma-like. Conidiophores indistinguishable from superficial vegetative hyphae. Condiogenous cells borne laterally on vegetative hyphae, discrete, globose, at first hyaline, later brown, giving rise to a single holoblastic terminal conidium, remaining as a cup-shaped cell after conidial secession. Conidia arise solitarily as blown-out ends of apex of conidiogenous cell, dry, dark brown, smooth, thick-walled, multiseptate, not constricted at septa, helicoid, tightly coiled in three dimensions to resemble dictyospores. On living leaves of *Leucadendron tinctum* in Stellenbosch, Cape Province, Republic of South Africa.

Helicostylum Corda
Mucoraceae, Mucorales, Inc. sed., Inc. sed., Mucoromycotina, Zygomycota, Fungi

Corda, A. C. J. 1842. *Icon. fung.* (Prague) 5:18, 55.
From Gr. *hélix*, genit. *hélikos*, helix + *stýlos*, column. Verticillate branches, curved like a helicoid that bear sporangiola. Saprobe in soil and dung.

Helicosingula leucadendri P. S. van Wyk, et al.: conidiogenous cells on tissues of the leaf host, giving rise to holoblastic terminal conidia, x 620.

Heliscella Marvanová
Inc. sed., Inc. sed., Inc. sed., Inc. sed., Pezizomycotina, Ascomycota, Fungi
Marvanová, L. 1980. New or noteworthy aquatic hyphomycetes. *Clavatospora*, *Heliscella*, *Nawawia* and *Heliscina*. *Trans. Br. mycol. Soc.* 75:221-231.
From genus *Heliscus* Sacc. + L. dim. suf. *-ella*. Hyphae hyaline, becoming brown. Conidiophores hyaline, simple or branched. Phialides ampulliform, often curved, with distinct collarettes, formed either on the mycelium or on short conidiophores. Conidia enteroblastic, clavate, with a short apical denticle or tetraradiate, minute. On submersed wood in aquatic habitats.

Heliscella stellata (Ingold & V. J. Cox) Marvanová: tetraradiate conidia of this aquatic fungus, x 2,250. Isolated in Great Britain.

Heliscina Marvanová
Inc. sed., Inc. sed., Inc. sed., Inc. sed., Pezizomycotina, Ascomycota, Fungi
Marvanová, L. 1980. New or noteworthy aquatic hyphomycetes. *Clavatospora*, *Heliscella*, *Nawawia* and

Heliscina. Trans. Br. mycol. Soc. 75:221-231.

From genus *Heliscus* Sacc. + L. des. *-ina*. Hyphae hyaline, septate, branched. Conidiophores simple or sparsely branched, bearing apical conidiogenous cells. Conidiogenous cells ampulliform, forming an apical blastic conidium, after which a new conidiogenous cell proliferates through narrow conidial scar to form another conidium, forming short chains. Conidia campanulate to clavate, apically denticulate. Isolated from debris in rivers in the Hrubý Jeseník Mountains, Bela pod Pradedern, Moravia.

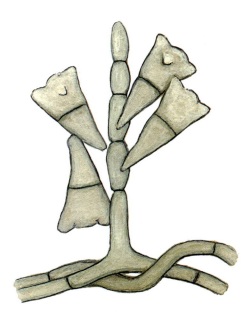

Heliscina campanulata Marvanová: conidiophore with ampulliform conidiogenous cells, and clavate, bicellular conidia, x 1,600. Collected on rotting leaves of *Fagus sylvatica* in mountain stream, former Czechoslovakia.

Helminthocarpon Fée

Graphidaceae, Ostropales, Ostropomycetidae, Lecanoromycetes, Pezizomycotina, Ascomycota, Fungi

Fée, A. L. A. 1837. *Essai Crypt. Exot.,* Suppl. Révis (Paris) 2:156.

From Gr. *hélmis*, genit. *hélminthos*, worm + *karpós*, fruit. Apothecia frequently elongated, straight, curved or flexuous forming on surface of whitish, generally smooth, crustous thallus. Apothecia covered partially or totally by a thin, dark-brown thalloid excipulum. Grows on tree bark.

Helminthosporium Link

Massarinaceae, Pleosporales, Pleosporomycetidae, Dothideomycetes, Pezizomycotina, Ascomycota, Fungi
Link, J. H. F. 1809. Observationes in ordines plantarum naturales. *Mag. Gesell. naturf. Freunde, Berlin* 3(1-2):10.
From Gr. *hélmis*, genit. *hélminthos*, worm + *spóros*, spore + L. dim. suf. *-ium*. Conidia long, generally obclavate, pseudoseptate, frequently with a basal scar (hilum), dark brown or black, formed through pores in pigmented, erect, often fasciculate conidiophores. Saprobic on wood. Many species placed in *Helminthosporium* are placed in *Bipolaris* Shoemaker, *Curvularia* Boedijin and *Drechslera* S. Ito.

Helminthocarpon scriptellum J. Steiner: crustous thallus with elongated apothecia (lirellae), on tree bark, x 5.

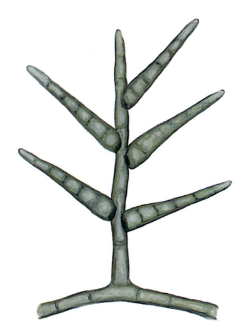

Helminthosporium solani Durieu & Mont.: conidiophore of this fungus, saprobic, or pathogen of grasses, with obclavate, pseudoseptate, dark brown conidia (porospores), x 800.

Helotiella Sacc.

Inc. sed., Helotiales, Leotiomycetidae, Leotiomycetes, Pezizomycotina, Ascomycota, Fungi

Helvella

Saccardo, P. A. 1884. *Botan. Zbl.* 18:213-220.
From genus *Helotium* Sacc. + L. dim. suf. *-ella*. Apothecia minute, sessile or subsessile, usually bright colored, or occasionally dark, clothed with hairs. Asci cylindric or subcylindric, inoperculate, usually 8-spored. Ascospores ellipsoid or fusoid, becoming 1-septate. Paraphyses filiform. On dead plant material.

Helvella L. (syns. **Macropodia** Fuckel, **Paxina** Kuntze)
Helvellaceae, Pezizales, Pezizomycetidae, Pezizomycetes, Pezizomycotina, Ascomycota, Fungi
Linnaeus, C. 1753. *Species plantarum* 2:1180.
From L. *helvella*, grass or small vegetable; an old name applied ambiguously to a variety of small cabbage, which has a certain similarity to the genus described here, because of the presence in both of folds and spaces or gaps. Ascomata apothecial, differentiated into a pileus and stipe; pileus of a thin sheet shaped like a saddle, miter, or concave plate, sometimes subglobose or cerebriform, commonly with edges folded, sometimes with whitish veins that project in relief from outer surface, on occasion united with stipe, with a white, grayish, pale brown, dark brown, reddish or yellowish surface or short, inconspicuous or sessile; outer surface bears hymenium with asci and paraphyses. Stipe cylindrical, slender or thick, small or long, hollow or solid with a smooth surface, lacunose or fluted. Asci cylindrical, operculate, 8-spored. Ascospores unicellular, hyaline, elliptical or fusiform. In humid soils, among mosses, branches and fallen leaves in coniferous and mixed forests with oak and madrone.

Helvella lacunosa Afzel.: cerebriform ascoma, with the saddle-shaped pileus and lacunose stipe, on soil, among mosses, x 1.

Hemicarpenteles A. K. Sarbhoy & Elphick—see **Penicillium** Link
Trichocomaceae, Eurotiales, Eurotiomycetidae, Eurotiomycetes, Pezizomycotina, Ascomycota, Fungi

Sarbhoy, A. K., J. J. Elphick. 1968. *Hemicarpenteles paradoxus* gen. & sp. nov.: The perfect state of *Aspergillus paradoxus. Trans. Br. mycol. Soc.* 51:155-157.
From Gr. *hémi*, half + genus *Carpenteles* Langeron.

Hemileia Berk. & Broome
Inc. sed., Pucciniales, Inc. sed., Pucciniomycetes, Pucciniomycotina, Basidiomycota, Fungi
Broome, C. E. 1869. *Gard. Chron.,* London:1157.
From Gr. *hémi-*, half + *leíos*, smooth. Rusts with unknown spermogonia and aecia. Urediniospores generally asymmetrical, with a smooth concave side and a warted dorsal convex side. Teliospores germinate without dormancy, producing an external basidium. Parasitic on tropical plants of economic importance. One of the most aggressive species is *H. vastatrix* Berk. & Broome, the cause of coffee rust.

Hemileia vastatrix Berk. & Broome: this fungus causes coffee rust, characterized by the asymmetrical urediniospores, with a smooth concave side and a dorsal convex and warted side, emerging from a stoma on leaf of *Coffea*, x 1,600. Collected in Sri Lanka.

Henicospora P. M. Kirk & B. Sutton
Inc. sed., Inc. sed., Inc. sed., Inc. sed., Pezizomycotina, Ascomycota, Fungi
Kirk, P. M., B. C. Sutton. 1980. *Henicospora* gen. nov. (Hyphomycetes). *Trans. Br. mycol. Soc.* 75:249-253.
From Gr. *hénikos*, single > L. *henicos* + *spóra*, spore, referring to the single spore on the condiogenous cell. Colonies effuse, minutely hairy. Mycelium partly superficial, partly immersed in substratum, composed of smooth, thin-walled, hyaline, branched, septate hyphae. Conidiophores micronematous or semi-macronematous, mononematous, simple, arising singly, terminally and laterally from hyphae. Conidiogenous cells integrated or discrete, terminal, monoblastic, determinate. Conidia acrogenous, solitary, dry, cylindrical, distoseptate, pale olivaceous-brown, seceding by fracture of basal cell of conidium or wall of conidiogenous cell. On dead leaves and stems.

Herpomyces

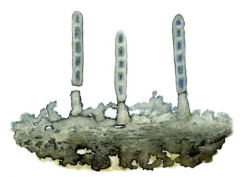

Henicospora minor P. M. Kirk & B. Sutton: superficial, hyaline mycelium, on decomposing stem of *Rubus fruticosus*, United Kingdom; conidiogenous cells with cylindrical, distoseptate conidia, x 1,100.

Henningsomyces Kuntze

Marasmiaceae, Agaricales, Agaricomycetidae, Agaricomycetes, Agaricomycotina, Basidiomycota, Fungi

Kuntze, O. 1898. *Revis. gen. pl. (Leipzig)* 3(2):437-576.

Named in honor of the German mycologist *Paul Christoph Hennings* + connective -o- + L. *myces* < Gr. *mýkes*, fungus. Basidiomes tiny, tubular, 1-1.5 mm high x 10 µm wide; usually in large, crowded groups, but also singly. Tubes white, of smooth, interwoven hyphae, often appearing glossy when fresh. Outer surface of tubes covered with short, finely branched marginal hairs (dendrohyphidia), especially around mouth of tubes. Inner surface of tubes lined with basidia that produce hyaline, globose to subglobose basidiospores.

Henningsomyces candidus (Pers.) Kuntze: tubular basidiomes, of interwoven, glossy, white hyphae; the interior of the basidiomes produce basidiospores, x 1.2. On dead wood, in Idaho, U.S.A.

Hericium Pers.

Hericiaceae, Russulales, Inc. sed., Agaricomycetes, Agaricomycotina, Basidiomycota, Fungi

Persoon, C. H. 1794. *Baier. Reise* 1:109.

From L. *ericius, hericius*, hedgehog + L. suf. *-ium*, which indicates connection or similarity, for the shape of the basidioma, similar to a hedgehog. Fructifications fleshy-leathery, clavarioid or hydnoid, with branches that support hymenophore, covered with cylindrical spines or teeth of different length, at times projecting horizontally but, for the most part, geotropic. Fructifications of *Hericium coralloides* (Scop. ex Fr.) Gray [=*H. flagellum* (Scop.) Pers.] at first white, later yellowish, composed of branches from whose tips hang long teeth of hymenophore. Includes edible, lignicolous fungi that grow on dead trunks of fir and others conifers. *Hericium erinaceus* (Bull. ex Fr.) Pers. grows in cracks or injuries on trunks of living oaks as well as on trunks of walnut, apple and other broad-leaved trees.

Hericium erinaceus (Bull.) Pers.: lignicolous, hydnoid basidioma, with geotropic cylindrical spines or teeth that bear the hymenophore, on wood, x 0.2.

Herpomyces Thaxt.

Herpomycetaceae, Laboulbeniales, Laboulbeniomycetidae, Laboulbeniomycetes, Pezizomycotina, Ascomycota, Fungi

Thaxter, R. 1903. Notes on the Laboulbeniaceae with descriptions of new species. *Proc. Amer. Acad. Arts & Sci.* 38(2):11.

From Gr. *hérpo*, that which meanders or crawls + *mýkes* (L. *myces*), fungus, for the shape of the thallus of this fungus, which lives adhered to the exoskeleton of cockroaches, into which it inserts its haustoria, the organs of attachment and nutrient absorption. Feminine thalli with one or more perithecia and masculine thalli with two or many spermogonia.

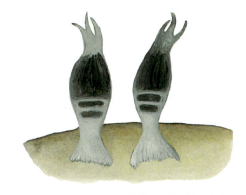

Herpomyces paranaensis Thaxt.: feminine thalli, with perithecium, adhered to the exoskeleton of cockroaches, x 290.

Herpothallon

Herpothallon Tobler
Arthoniaceae, Arthoniales, Arthoniomycetidae, Arthoniomycetes, Pezizomycotina, Ascomycota, Fungi
Tobler 1937, *Flora* 131:446.
From Gr. *hérpo*, that which meanders or crawls + *thallós*, thallus. The genus is characterized by the byssoid prothallus and hypothallus, more or less felty heteromerous thallus with felty pseudoisidia, pustules, soredia-like granules, or minute granules, pigmented hypothallus; confluentic acid and chiodectonic acid are common thalline compounds, and *Trentepohlia* as photobiont. It is growing abundantly on bark in the high humid, shady understories of tropical rainforests.

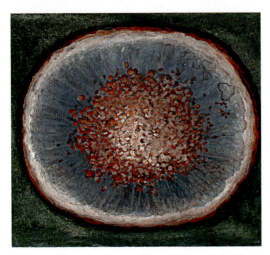

Herpothallon rubrocinctum (Ehrenb.) Aptroot, et al. (syn. **Cryptothecia rubrocincta** (Ehrenb.) G. Thor): carbonaceous, with reddish pseudothecia, sunken in a crustous thallus of this lichen, x 250.

Hertella Henssen
Placynthiaceae, Peltigerales, Lecanoromycetidae, Lecanoromycetes, Pezizomycotina, Ascomycota, Fungi
Henssen, A. 1985. *Hertella*, a new lichen genus in the Peltigerales from the Southern Hemisphere. *Mycotaxon* 22:381-397.
Named in honor of German botanist H. Hertel + L. des. *-ella*, who discovered the type specimen. Thallus small, olive, filamentous, fastened by a gelatinous attachment base. Apothecia adnate, olive or brown, with a pale proper margin. Apothecial development hemiangiocarpous, with large-celled ascogonia. Hymenium strongly gelatinous, staining dark blue in iodine, epithecium and hypothecium pale brown. Proper excipulum cupulate, thick. Paraphyses septate. Asci cylindrical to clavate, with amyloid cap or ring, 8-spored. Spores simple, hyaline, ellipsoid or slightly curved. Pycnidia adnate, forming rod-shaped conidia. On Prince Edward Island.

Heterobasidion Bref. (syn. **Spiniger** Stalpers)
Bondarzewiaceae, Russulales, Inc. sed., Agaricomycetes, Agaricomycotina, Basidiomycota, Fungi
Brefeld, O. 1888. *Unters. Gesammtgeb. Mykol.* (Leipzig) 8:154.
From Gr. *héteros*, another type, different + *basídion*, basidium; with basidia different from those of the genus *Fomes* Fr., in particular to those of *F. annosum* Bref., which is the type species. Fructifications of a dimitic hyphal system with dextrinoid skeletal hyphae predominant; cuticle on pileus and asperulate spores. Annual to perennial fructifications, sessile, resupinate or pileate, hard and woody; pileus generally some tone of brown color, finely tomentose, with a light colored hymenium. Conidia with conical spines or denticles on surface of terminal ampullae of conidiophores, on which the dacryoid conidia arise synchronously. On living or dead wood, predominantly of conifers. An important pathogen, in particular *H. annosum* Bref., previously known as *Fomes annosus* (Fr.) Cooke, causes white rot of wood and the destruction of conifer roots. Its asexual morph is represented by its synonym *Spiniger meineckellus* (Olson) Stalpers.

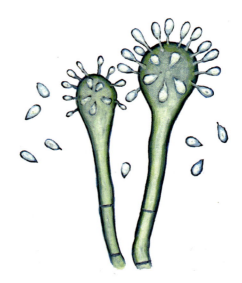

Heterobasidion annosum (Fr.) Bref. (syn. **Spiniger meineckellus** (A. J. Olson) Stalpers): conidiophores, with conical spines or denticles on the surface of terminal ampullae, and on which dacryoid conidia arise synchronously (asexual morph), x 500. Isolated in California, U.S.A.

Hexacladium D. L. Olivier
Inc. sed., Inc. sed., Inc. sed., Inc. sed., Pezizomycotina, Ascomycota, Fungi
Olivier, D. L. 1983. Phyllosphere fungi which capture wind-borne pollen grains. II. *Hexacladium corynephorum* gen. et sp. nov. *Trans. Br. mycol. Soc.* 80:237-245.
From Gr. *hex*, six + *kládos*, dim. *kladion*, branch > L. *cladium*, for the six arms of the conidium. Fungus with hyaline septate hyphae bearing clavate lateral branches that capture anemophilous pollen grains. Conidia holoblastic, single, hyaline, six-branched with terminal clavate swellings. Conidiogenous cells integrated, in-

Histoplasma

tercalary in long and short hyphae, producing conidia laterally; usually with determinate, sympodial proliferation. On leaves of *Alberta magna* in the National Botanic Gardens, Kirstenbosch, near Cape Town, South Africa.

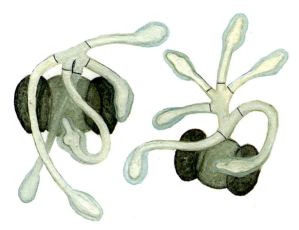

Hexacladium corynephorum D. L. Olivier: hexaseptate conidia, with the clavate and mucilaginous swellings, which capture anemophilous pollen grains, x 1,000.

Hexagonia Fr.

Polyporaceae, Polyporales, Inc. sed., Agaricomycetes, Agaricomycotina, Basidiomycota, Fungi

Fries, E. M. 1838. *Epicrisis systematis mycologici, seu synopsis hymenomycetum. Hort. Veron. Pl. Nov.*:35.

From Gr. *hexágonos*, from *héx*, six, from which is derived the pref. hexa- + *gonía*, angle, in reference to the hexagonal pores of the hymenophore. Basidiomata pileate annual or perennial, of medium size, developing in shape of a semicircular shelf, principally in tropical regions. Hymenophore with pores more or less large (0.3-2 mm diam), regular, similar to a bee honeycomb. Destroyers of dead wood of deciduous trees; e.g., *H. tenuis* Fr., whose fructification has a surface with alternating gray and light or dark brown concentric zones, whereas the pores are light in color.

Hexagonia tenuis (Fr.) Fr.: basidiomata showing the hexagonal pores of hymenophore, on dead wood, x 0.5.

Hiospira R. T. Moore

Inc. sed., Inc. sed., Inc. sed., Dothideomycetes, Pezizomycotina, Ascomycota, Fungi

Moore, R. T. 1962. *Hiospira*, a new genus of the Helicosporae. *Trans. Br. mycol. Soc.* 45:143-146.

From L. *hio*, open + *spira*, coiled, referring to the shape of the spores. Somatic mycelium micronemic, smooth. Conidiophores fuscous, erect or ascending, sterile above, several basal cells narrowed, smooth, remainder with surface markings same as on conidia. Conidia fuscous, pleurogenous, single or producing secondary conidiophores with conidia, but not catenate, tightly coiled when dry, relaxing when moistened, tending to be borne at a downward angle to conidiophore when singlemarkings forming a surface reticulum composed of more or less hexagonal areolae; filaments thick, forming up to 10 gyres, basal initial cells straight to curved, the proximal one tapering to conidiophore peg. Collected on plants in Njala, Sierra Leone.

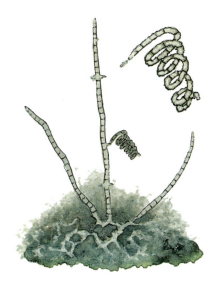

Hiospira hendrickxii (Hansf.) R. T. Moore: conidiophores with pleurogenous, tightly coiled conidia, x 300. On plants, in Congo. New Caledonia, New Zealand, Western Samoa.

Histoplasma Darling (syns. Ajellomyces McDonough & A. L. Lew., Emmonsiella Kwon-Chung)

Ajellomycetaceae, Onygenales, Eurotiomycetidae, Eurotiomycetes, Pezizomycotina, Ascomycota, Fungi

Darling, S. T. A. 1906. A protozoon general infection producing pseudotubercles in the lungs and focal necroses in the liver, spleen and lymph nodes. *J. Amer. Med. Assoc.* 46:1283-1285.

From Gr. *hístos*, tissue, referring to the organic tissues + *plásma*, plasma, primordial fluid material of living beings. Dimorphic. Sexual morph emmonsiella-like, with buff, gray or reddish cleistothecia with peridium of compact, coiled hyphae. Saprobic phase in soil, mycelial, with macro- and microconidia. Initially confused with

Hormisciella

a protozoan, this organism is an intracellular parasite causing histoplasmosis, a systemic mycosis of reticuloendothelial system (lymphatics, lungs, spleen, liver, etc.) in yeast phase.

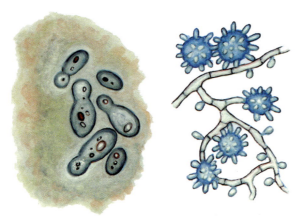

Histoplasma capsulatum Darling: dimorphic fungus producing histoplasmosis; at left, intracellular parasite, in its yeast phase, x 900; at right, saprobic phase, with the mycelium producing macro- and microconidia, x 90. Isolated from a human patient in Panama.

Histoplasma capsulatum Darling: gymnothecium with spiral, radiating hyphae, x 200.

Hormisciella Bat.—see **Antennatula** Fr. ex F. Strauss
Euantennariaceae, Inc. sed., Inc. sed., Dothideomycetes, Pezizomycotina, Ascomycota, Fungi
Batista, A. C. 1956. Novos fungus Demateaceae. *Anais da Sociedade de Biologia de Pernambuco* 14:98-110.
From Gr. *hórmos*, chain + connective -*ci*- + L. dim. suf. -*ella*, for the catenulate hyphal cells.

Hormodendrum Bonord.—see **Cladosporium** Link
Cladosporiaceae, Capnodiales, Dothideomycetidae, Dothideomycetes, Pezizomycotina, Ascomycota, Fungi
Bonorden, H. F. 1851. *Handb. Allgem. mykol.* (Stuttgart):76-77.

From Gr. *hórmos*, collar, chain + *déndrion*, shrub, in reference to the conidiophores.

Hughesiella Bat. & A. F. Vital—see **Chalara** (Corda) Rabenh.
Inc. sed., Inc. sed., Inc. sed., Inc. sed., Pezizomycotina, Ascomycota, Fungi
Batista, A. C., A. F. Vital. 1956. *Hughesiella* - novo e curioso gênero de fungos Dematiaceae. *Anais Soc. Biol. Pernambuco* 14 (1-2):141-144.
Named in honor of the Welsh mycologist *Stanley J. Hughes* + connective -*i*- + L. dim. suf. -*ella*.

Humicola Traaen
Chaetomiaceae, Sordariales, Sordariomycetidae, Sordariomycetes, Pezizomycotina, Ascomycota, Fungi
Traaen, A. E. 1914. Untersuchungen über Bodenpilze aus Norwegen. *Nytt Mag. Natur.* 52:31-32.
From L. *humi*, humus, earth, soil + *cola*, inhabitant, i.e., living in soil. Aleuriospores large, dark, unicellular, more or less spherical and borne directly on vegetative hyphae or on short, lateral conidiophores, individually or in short chains.

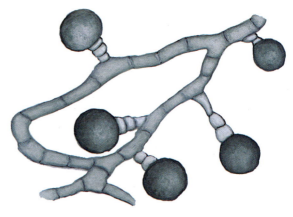

Humicola fuscoatra Traaen: large, black, unicellular aleuriospores, x 900.

Humidicutis (Singer) Singer
Hygrophoraceae, Agaricales, Agaricomycetidae, Agaricomycete, Agaricomycotina, Basidiomycota, Fungi
Singer, R. 1958. Fungi Mexicani, series secunda - Agaricales. *Sydowia* 12(1-6): 221-243.
From L. *humidis*, moist + *cutis*, skin, referring to the moist caps. Pileus convex, convex-umbonate or conic, margin rarely and not deeply splitting; surface subhygrophanous, moist, rarely viscid, colors usually bright; lamellae thick, sinuate or broadly adnate, often with a decurrent tooth; odor absent or disagreeable; carotenoid pigments usually present; pileipellis hyphae parallel, prostrate, cylindric; basidiospores hyaline, thin-walled, inamyloid, ellipsoid or broadly ellipsoid, not constricted; lamellae trama subregular or regular, rarely tapered,

with right-angled septa; clamp connections absent in context and pellis, but toruloid clamps present at the base of basidia and or basidioles.

Hyalocladium moubasheri Moustafa: conidiophore with hyaline, acrogenous dictyospores; isolated from air samples, on agar, x 450.

Humidicutis brunneovinacea Garibay-Orijel R.: basidiomes growing among mosses, in *Cupressus* and *Quercus* forest, in Santa María Huitepec, Cempoaltepetl, Oaxaca, Mexico, x 1.

Humphreya Steyaert

Ganodermataceae, Polyporales, Inc. sed., Agaricomycetes, Agaricomycotina, Basidiomycota, Fungi

Steyaert, R.L 1972. Species of *Ganoderma* and related genera mainly of the Bogor and Leiden Herbaria. *Persoonia* 7(1):55-118.

Dedicated to the memory of the American pathologist *C. J. Humphrey*, pioneer of the anatomical study of the genus *Ganoderma* + L. des. -*a*. Basidioma convex funnel-shaped, centrally or laterally stipitate, greyish-brown. Context and tube layer concolorous, honey-red. Cutis sharply defined; melanoid substance similar to that of *Ganoderma*, but of different composition. Basidiospores two-walled, the epi- and endosporium separated by reticulate or disjoined red cristae.

Hyalocladium Moustafa

Inc. sed., Inc. sed., Inc. sed., Inc. sed., Pezizomycotina, Ascomycota, Fungi

Moustafa, A. F. 1976. *Hyalocladium*, a new Hyphomycete genus from Kuwait. *Trans. Br. mycol. Soc.* 67:537-539.

From Gr. *hyálinos*, glassy, colorless + *kládos*, branch > L. *clados*. Mycelium superficial, composed of hyaline, pale pink, septate, smooth hyphae; conidiophores micronematous or semimacronematous, arising singly, terminally and laterally on hyphae, flexuous, sometimes nodose but typically geniculate. Conidia dictyospores, monoblastic, acrogenous, hyaline. Isolated from air samples in Kuwait.

Hyaloseta A. W. Ramaley

Niessliaceae, Hypocreales, Hypocreomycetidae, Sordariomycetes, Pezizomycotina, Ascomycota, Fungi

Ramaley, A. W. 2001. *Hyaloseta nolinae*, its anamorph *Monocillium noliniae*, and *Niesslia agavacearum*, Hypocreales, from leaves of Agavaceae. *Mycotaxon* 79:267-274.

From Gr. *hýlos*, glass, genit. *hyálinos*, glossy > L. + *seta*, saeta, stiff hair, bristle. Ascomata superficial and solitary or aggregated on a hyaline subiculum on substrate; dark, globose perithecia with a periphysate ostiole, but without a papilla. Ascomatal peridium soft, thin, bearing hyaline setae thick-walled, with swollen, ellipsoid tips. Asci unitunicate, 8-spored, with 1-septate, hyaline ascospores. Asexual morph monocillium-like. On dead leaves of *Nolina micrantha* in New Mexico.

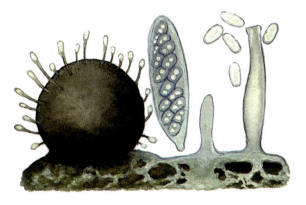

Hyaloseta nolinae A. W. Ramaley: globose perithecium, covered by hyaline setae, with swollen ellipsoidal tips, x 300; unitunicate ascus with one-septate ascospores, x 980; phialide with conidia, x 1,200.

Hyalotiella

Hyalotiella Papendorf
Bartaliniaceae, Amphisphaeriales, Xylariomycetidae, Sordariomycetes, Pezizomycotina, Ascomycota, Fungi
Papendorf, M. C. 1967. Two new genera of soil fungi from South Africa. *Trans. Br. mycol. Soc.* 50:69-75.
From genus *Hyalotia* Guba + L. dim. suf. *-ella*, for the similarity to this genus. Fruiting pustule pycnidium-like, dark brown to black, irregular, globose or lobate, finally discoid; conidiophores elongate, cylindrical, simple or branched; conidia aleuriospores, cylindric, at first continuous, then 3-septate, rarely 4-septate, subhyaline, apical cell hyaline, narrow-conical, pointed, crested with two to four simple or rarely branched setulae, basal cell rounded or truncate at base, without supporting pedicel. Isolated from soil of an *Acacia karroo* community near Johannesburg, South Africa.

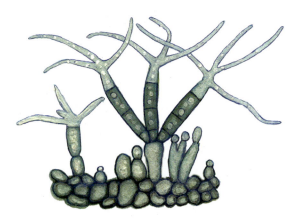

Hyalotiella transvalensis Papendorf: short conidiophores, with cylindrical, multiseptate aleuriospores, crested with two to four branched setulae, x 550.

Hydnangium Wallr.
Hydnangiaceae, Agaricales, Agaricomycetidae, Agaricomycetes, Agaricomycotina, Basidiomycota, Fungi
In: Dietrich, A. 1839. *Fl. Regn. Boruss.* 7:465.
From Gr. *hýdnum*, tuber + *angeîos*, glass, vessel, in reference to the form of the carpophore. Hypogeous or semihypogeous globose or pyriform fructification attached by a radicate rhizomorph. Peridium simple of woven gelatinized hyphae. Gleba with open cavities lined with a true hymenium. Spores globose or subglobose, echinulate, slightly tinted, thin-walled. Basidia clavate, commonly 2-4-spored. Although a gasteromycete closely related to the mushroom *Laccaria* Berk. & Broome. Growing in soil rich in humus or in leaf mold of the forest floor. Known from Europe, Japan, Australia (Tasmania), New Zealand.

Hydnopolyporus D. A. Reid
Meripilaceae, Polyporales, Inc. sed., Agaricomycetes, Agaricomycotina, Basidiomycota, Fungi
Reid, D. A. 1962. Notes on fungi which have been referred to the Thelephoraceae sensu lato. *Persoonia* 2(2):151.
From the genus *Hydnum* L. + *Polyporus* P. Micheli ex Adans. Sporophores white, lignicolous, discrete or caespitose, often forming small rosettes of numerous, irregular flabelliform pilei. Fruitbodies usually consist of a large number of narrow flattened segments united into a short stipe. Hymenial surface bearing isolated warts, spines or ridges or becoming tardily poroid. Spores smooth, thin-walled, hyaline, non-amyloid, broadly elliptical to ovoid. Known from the southern United States, West Indies, Bolivia, Venezuela, Brazil and Argentina.

Hydnum L. (syn. **Dentinum** Gray)
Hydnaceae, Cantharellales, Inc. sed., Agaricomycetes, Agaricomycotina, Basidiomycota, Fungi
Linnaeus, C. 1753. *Sp. pl.* 2:1178.
From Gr. *hýdnon*, an ancient name applied to a type of edible fungi in the time of Theophrastus; this name was given to the truffles (*Tuber* Mich. ex Fr.), which are unrelated to this genus. Probably it is derived from *hýdor*, water, in reference to the fungus as "the damp one". Pileus repand, pale yellow, reddish, pubescent, pruinose, smooth; hymenophore dentate, conspicuous, in lower part upper portion of stipe. Terricolous in temperate forests, edible and mycorrhizal. *Hydnum* was previously applied to all dentate species, but is now restricted to species related to *H. repandum* L. ex Fr. (synonyms: *Sarcodon repandum* Quél., *Dentinum repandum* S. F. Gray). *Hydnum repandum* and the closely related *H. rufescens* Fr. are both edible.

Hydnum rufescens Pers.: basidioma, with stipe, and pileus with dentate hymenophore, on soil, x 1.

Hygrocybe (Fr.) P. Kumm.
Hygrophoraceae, Agaricales, Agaricomycetidae, Agaricomycetes, Agaricomycotina, Basidiomycota, Fungi
Kummer, P. 1871. *Führ. Pilzk.* (Zwickau):26.
From Gr. *hygrós*, moist, wet + *kybe*, head, pileus, be-

cause in several of the species the pileus appears to be moist. Pileus viscous or dry, generally brightly colored, small to medium. Common in temperate forests, rare in the tropics. Often in open places, terricolous, less frequently on wood, probably not mycorrhizal. Several species are edible.

Hygrocybe punicea (Fr.) P. Kumm.: brightly colored, terricolous basidiomata, x 1.

Hygrophoropsis (J. Schröt.) Maire ex Martin-Sans
Hygrophoropsidaceae, Boletales, Agaricomycetidae, Agaricomycetes, Agaricomycotina, Basidiomycota, Fungi
Martin-Sans, 1929. *L'Empoisonnem. Champ.*:99.
From genus *Hygrophorus* Fr. + L. suf. *-opsis*, similar to > Gr. *-ópsis*. Pileus somewhat tomentose with involute margin when young; hymenophore lamellate; lamellae decurrent, narrow, usually repeatedly forked; spore print white to yellowish; spores subhyaline, subglobose to elliptic or cylindric; stipe, when present, fleshy, central or eccentric, without veil; context fleshy. On the ground, among moss and on wood; facultatively ectomycorrhizal in temperate and tropical regions. Some species are edible after boiling but generally toxic when consumed raw.

Hygrophorus Fr. (syn. **Camarophyllus** (Fr.) P. Kumm.)
Hygrophoraceae, Agaricales, Agaricomycetidae, Agaricomycetes, Agaricomycotina, Basidiomycota, Fungi
Fries, E. M. 1836. *Fl. Scan.*:339.
From Gr. *hygrós*, moist + Gr. suf. *-phóros* (L. *-phorus*), from *phéro*, to take, to indicate that it has dampness, referring to the hygrophanous character of the pileus. Pileus mostly clitocyboid, appearing waxy, viscous or glutinous, principally on gills, generally bright colored, sometimes with violet colors, occasionally with an olivaceus hue. Gills more or less distant, rounded, adnate, sinuous (adnexed) or decurrent, i.e., adherent to the stipe and at times with a sinus near it, or extended below from point of insertion with stipe. Spores white in mass, hyaline, smooth with homogenous, thin wall, inamyloid. Terricolous and/or mycorrhizal. Common in coniferous and broad-leaved forests as well as in humid grasslands.

Hygrophorus hypothejus (Fr.) Fr.: hygrophanous, brightly colored basidioma, with decurrent lamellae between the pileus and the stipe, on soil, x 0.5.

Hymenochaete Lév. (syn. **Stipitochaete** Ryvarden)
Hymenochaetaceae, Hymenochaetales, Inc. sed., Agaricomycetes, Agaricomycotina, Basidiomycota, Fungi
Léveille, J.H. 1846. Description des champignons. *Annls Sci. Nat.*, Bot., sér. 3, 5:150.
From L. *hymenium*, hymenium < Gr. *hyménio*, membrane + *chaité*, long flowing hair, mane, a bristle. Leathery or almost woody, in many cases perennial basidiomes have a more or less smooth hymenophore; they are brown, but this colour turns almost black when treated with alkali (xanthochroic reaction). The basidiomes are effused (resupinate), or with stereum-like thin pilei, in some rare cases with a primitive stipe. In the hymenium, there are numerous very thick-walled dark brown cystidia-like setae (20-200 µm long); all hyphae are without clamps. Basidiospores are hyaline, thin-walled and comparatively small. All species cause white fibrous or pocket rot of dead wood.

Hymenochaete rubiginosa (Dicks.) Lév.: basidioma on dead wood, x 1; hymenial setae and basidiospores, x 300.

Hymenogaster

Hymenogaster Vittad.
Hymenogastraceae, Agaricales, Agaricomycetidae, Agaricomycetes, Agaricomycotina, Basidiomycota, Fungi
Vittadini, C. 1831. *Monogr. Tuberac.* (Milano):30.
From Gr. *hymén*, genit. *hyménos*, membrane, sheet + *gastér*, stomach. Fructifications hypogeous, globose, subglobose or tuberiform, having a radicular or rhizomorphic base. Peridium of one or two layers. Gills anastomosed and delimited in fertile cavities of gleba in interior of fructifications. Spores pigmented, verrucose or with a gelatinous layer. Grows under surface of soil in conifer and oak forests.

Hymenogaster sulcatus R. Hesse: globose basidioma; at right, longitudinal section of this basidioma to show the fertile cavities of the gleba, x 4.

Hymenoscyphus Gray (syns. **Ciboriella** Seaver, **Phaeohelotium** Kanouse)
Helotiaceae, Helotiales, Leotiomycetidae, Leotiomycetes, Pezizomycotina, Ascomycota, Fungi
Gray, S.F. 1821. *Natural Arrangement of British Plants* (London) 1:673.
From L. *hymenium*, hymenium < Gr. *hyménio*, membrane + *scýphos*, a cup. Fruitbody consisting of a tiny shallow cup and a conspicuous stalk; cup up to 4 mm wide with an entire margin; inner surface smooth, pale yellow; outer surface smooth, colored like the cup or paler; stalk up to 15 mm long, up to 1 mm thick, smooth, shiny when wet, colored like the cup. Asci 8-spored, amyloid pore; paraphyses filiform, sometimes forked, with slightly enlarged tips. Scattered or in groups on acorn cups and beechnut husks.

Hyphodermella J. Erikss. & Ryvarden
Phanerochaetaceae, Polyporales, Inc. sed., Agaricomycetes, Agaricomycotina, Basidiomycota, Fungi
Eriksson, J., L. Ryvarden. 1976. *Corticiaceae of Northern Eur.* (Oslo) 4:579.
From Gr. *hýphos*, filament, tissue + *dérma*, skin + L. dim. suf. *-ella*, i.e., a little *Hyphoderma* Wallr. Fruitbodies resupinate, effuse, thickening with age, crustose. Hymenium ceraceous, at first whitish, with age darkening to ochraceous, hymenophore irregularly odontoid, with small spines, scattered in young hymenium, but numerous, irregularly crowded in older fructifications; with more or less incrusted projecting cystidioid hyphae, readily visible under lens and giving a characteristic appearance; margins mostly determinate, hyphal system monomitic; basidia clavate with four sterigmata; spores ellipsoid, smooth nonamyloid. A monotypic genus.

Hyphodiscosia Lodha & K. R. C. Reddy
Inc. sed., Inc. sed., Inc. sed., Inc. sed., Pezizomycotina, Ascomycota, Fungi
Lodha, B. C., K. R. Chandra Reddy. 1974. *Hyphodiscosia* gen. nov. from India. *Trans. Br. mycol. Soc.* 62:418-421.
From Gr. *hyphé*, web, hypha + genus *Discosia* Lib., for the resemblance to this genus. Conidiophores brown, simple, septate, producing conidia in clusters on swollen apices from denticles. Conidia producing blastospores, solitary, pale pink, 2-celled, subcylindrical, dorsiventral, truncate at base, obtuse or rounded above, with a single, hyaline, mucoid, hair-like, lateral appendage arising from each cell on ventral side. On bark of dead wood from Jaipur, Rajasthan, India.

Hypholoma (Fr.) P. Kumm.
Strophariaceae, Agaricales, Agaricomycetidae, Agaricomycetes, Agaricomycotina, Basidiomycota, Fungi
Kummer, P. 1871. Anleitung zum methodischen, leichten und sichern Bestimmen der in Deutschland vorkommenden Pilze: mit Ausnahme der Schimmel- und allzu winzigen Schleim- und Kern-Pilzchen. *Führ. Pilzk.* (Zerbst):21.
From Gr. *hýphos*, filament, tissue + *lóma*, margin, fringe; with filaments in the margin of the pileus. Basidiocarps agaricoid, pileus at first hemispheric, at maturity flat-convex, yellow-red or brown. Stipe fibrous, cylindric or curved; internal veil arachnoid, disappearing with age; lamellae gray colored. Spores oval to elliptic, smooth. Growing in dead wood of different trees. Some species have a bitter taste, not edible.

Hypholoma fasciculare (Huds.) P. Kumm. (syn. **Naematoloma fasciculare** (Huds.) P. Karst.): gregarious basidiocarps, with yellowish stipes and reddish pileus, on dead wood, x 0.15.

Hyphomyces C.H. Bridges & C.W. Emmons
Inc. sed., Inc. sed., Inc. sed., Inc. sed., Pezizomycotina, Ascomycota, Fungi
Bridges, C. H., C. W. Emmons.1961. A phycomycosis of horses caused by *Hyphomyces destruens*. *Journal of the American Veterinary Medical Association.* 138(11):579-589.

From Gr. *hýpho*, *hyphé*, spider web, hypha + *mýkes* > L. *myces*, fungus. One of the causal agents of equine phycomycosis. Affects the skin and/or nasal mucosa of horses in tropical and subtropical parts of the world. It has been describes in the U.S.A., Australia, Brazil, India and Colombia, and is known under various local names such as leeches, swamp cancer, and rhinophycomycosis. Is usually seen as an area or areas of exuberant granulation (most often on the limbs between the hoof and knee or hock) but may be found on the ventral abdomen, neck, lips and skin surrounding the nostrils. Purulent exudates may drain from openings on the surface of the lesion. The animal may traumatize or destroy part of the lesion by biting or licking, apparently as a result of severe pruritus. Lesions of the nostril may cause inspiratory dyspnea; those on the lips may impair prehension. The fungus can be seen as branching, occasionally septate hyphae which are scattered throughout the larger foci of necrosis and in the centers of the smaller ones. The hyphae frequently are numerous in the necrotic blood vessels. No type of sporulation has been observed.

Hyphomyces destruens C. H. Bridges & C. W. Emmons: coenocytic, nonsporulating hyphae that causes granulomatose illness in tissues of horses (*Equus*), x 550.

Hypnotheca Tommerup
Inc. sed., Inc. sed., Inc. sed., Inc. sed., Pezizomycotina, Ascomycota, Fungi
Tommerup, I. C. 1970. *Hypnotheca graminis* gen. et sp. nov., perfect state of *Monochaetiella themedae*. *Trans. Br. mycol. Soc.* 55:463-475.

From Gr. *hypnós*, sleep + *théke*, box > L. *theca*, case, referring to the long dormant period required for the asci to mature. Pseudothecia basin-shaped, solitary or gregarious, subepidermal or erumpent, wall yellow, one cell thick, bearing a fascicle of aparaphysate asci. Asci cylindrical, sessile, bitunicate, thickened at apex, 8-spored. Ascospores 1-celled, uniseriate, subglobose, hyaline, thin-walled. Paraphyses lacking. The asexual morph is monochaetiellopsis-like. On leaves of *Heteropogonis contortus* in Queensland, Australia.

Hypocrea Fr.—see **Trichoderma** Pers.
Hypocreaceae, Hypocreales, Hypocreomycetidae, Sordariomycetes, Pezizomycotina, Ascomycota, Fungi
Fries, E. M. 1825. *Syst. orb. veg.* (Lundae) 1:104.
From Gr. *hypó*, under + *kréas*, fleshy.

Hypoderma De Not.
Rhytismataceae, Rhytismatales, Leotiomycetidae, Leotiomycetes, Pezizomycotina, Ascomycota, Fungi
De Notaris, G. J. 1847. Prime linee di una nuova disposizione de' Pirenomiceti Isterini. *Fl. franç.*, Edn. 3 (Paris) 2:13.

From Gr. *hypó*, beneath, under + *dérma*, skin, in reference to the ascomata immersed in host tissues. Apothecia sunken in substrate, dark, dull or shining black, mostly short elliptical, opening by a narrow, erumpent longitudinal slit. Hymenium flat or occasionally concave. Paraphyses simple, filiform, sometimes lacking. Asci clavate to cylindric, unitunicate, 8-spored. Ascospores bacillar to fusiform, hyaline, 1-celled, surrounded by a thick gelatinous sheath. Pycnidia simple, applanate, with minute bacillar or clavate conidia. On needles of conifers, causing lesions and needle cast.

Hypodermella Tubeuf
Rhytismataceae, Rhytismatales, Leotiomycetidae, Leotiomycetes, Pezizomycotina, Ascomycota, Fungi
Tubeuf, K. v, 1895. *Botan. Zbl.* 61:48.
From genus *Hypoderma* De Not. + L. dim. suf. *-ella*. Apothecia with a covering of its own tissue, beneath cuticle or epidermis of leaves of conifers. Causes premature loss of needles.

Hypodermella laricis Tubeuf: apothecia on yellowish needles of fir tree (*Larix*), x 0.3.

Hypogymnia (Nyl.) Nyl.
Parmeliaceae, Lecanorales, Lecanoromycetidae, Lecanoromycetes, Pezizomycotina, Ascomycota, Fungi
Nylander, W. 1896. *Lich. Envir. Paris*: 39:139.

Hypomyces

From Gr. *hypó*, underneath + *gymnós*, naked, uncovered, for the foliose thallus with the lower surface bare. Thallus wiith long, narrow, inflated, hollow lobules, black on lower part, whitish or greenish-gray on upper part where apothecia or soredia develop. Grows principally on conifers and oaks, or on mosses on rocks.

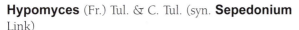

Hypogymnia inactiva (Krog) Ohlsson: foliose thallus, with hollow lobules, black on the lower part, and greenish-gray in the upper part, with apothecia, on conifer twig, x 3.

Hypomyces (Fr.) Tul. & C. Tul. (syn. **Sepedonium** Link)

Hypocreaceae, Hypocreales, Hypocreomycetidae, Sordariomycetes, Pezizomycotina, Ascomycota, Fungi
Tulasne, L. R. 1860. *Annls Sci. Nat.*, Bot., sér. 4 *13*:11.
From Gr. *hypó*, under, inferior + *mýkes* > L. *myces*, fungus, for developing principally on the hymenophore of basidiomycetes, principally of the genus *Russula* Pers. and *Lactarius* Pers. Conidiophores produce aleuriospores in clusters; large, rough-walled aleuriospores, hyaline or light-colored. Parasitizes and rots the fruiting bodies of agarics and boletes, on which they cover and deform its surface. Also, saprobic in soil. Some species, such as *H. lactifluorum* (Schw.) Tul. are edible and highly regarded, although there are cases in which the host is a toxic species.

Hypomyces lactifluorum (Schwein.) Tul. & C. Tul.: red *Russula* basidiocarp, parasitized by the mycelium of this fungus, on soil, x 1.

Hypomyces chrysospermus Tul. & C. Tul. (syn. **Sepedonium chrysospermum** (Bull.) Fr.): agaricous basidioma parasitized by the mycelium of the sepedonium-like fungus, x 0.5; this species forms conidiophores and rough-walled, light-colored aleuriospores, on the fructification host, x 500.

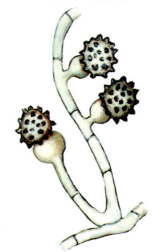

Hypomyces perniciosus Magnus (syn. **Mycogone perniciosa** (Magnus) Delacr.): mycoparasitic fungus growing on fructifications of Agaricales, x 1; bicellular aleuriospores with the upper cell larger and with a tuberculate wall, x 430.

Hysterographium

Hypotrachyna (Vain.) Hale
Parmeliaceae, Lecanorales, Lecanoromycetidae, Lecanoromycetes, Pezizomycotina, Ascomycota, Fungi
Hale, M.E. 1974. Delimitation of the lichen genus *Hypotrachyna* (Vainio) Hale. *Phytologia*. 28(4):340-342
From Gr. *hypó*, under, inferior + *trachýs*, rough + L. dim. suf. *-ina*. Is a segregate of the collective genus *Parmelia*, characterized by narrow, apically truncate lobes, a black lower surface, and dichotomously branched rhizines. Is extremely homogeneous in overall lobe configuration. Soredia originate in soralia and erupt as a powder; pustules coarse, inflated, extremely fragile, isidialike structures; isidia fingerlike, cylindrical, densely filled with medullary hyphae; lobules abundantly or distinctively produced; maculae are mottled, light to dark patches in the cortex. Apothecia uniformly small, adnate, and imperforate; ascospores one-celled. Pycnidia with microconidia are often produced by apothecia-bearing species. This genus is primarily of higher elevations throughout its range in the tropics, concentrated en the New world. Corticolous or saxicolous.

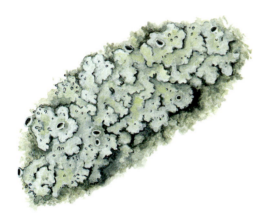

Hypotrachyna imbricatula (Zahlbr.) Hale (syn. **Parmelia imbricatula** Zahlbr.): folious thallus, on tree bark, x 1.

Hypoxylon Bull. (syn. **Sphaeria** Haller)
Xylariaceae, Xylariales, Xylariomycetidae, Sordariomycetes, Pezizomycotina, Ascomycota, Fungi
Bulliard, J. B. P. 1791. *Hist. Champ. France* (Paris) 1:168.
From Gr. *hypó-*, beneath, inferior + Gr. *xýlon*, wood, for its hard consistency, similar to that of wood. Stromata carbonaceous, pulvinate to applanate, perithecia formed in internal layer of stroma (endostroma), ostiolar necks extend through external layer (ectostroma), on which produced a layer of conidia. On trunks and branches of living trees, which it can parasitize, generally weakly, or continue growing as a saprobe after the host has died.

Hysterium Pers.
Hysteriaceae, Hysteriales, Pleosporomycetidae, Dothideomycetes, Pezizomycotina, Ascomycota, Fungi
Persoon, C. H. 1797. *Tent. disp. meth. fung.* (Lipsiae) 5.
From Gr. *hystéra*, genit. *hystericós*, womb, long and cleft, groove, crack + L. suf. *-ium*, in reference to the hysterothecium. Ascostroma elongate, opens at maturity by a long fissure that follows a line of dehiscence, hymenium completely exposed. Pseudothecia free, carbonaceous, with thick walls. Asci cylindrical. Ascospores multiseptate (phragmospores). On old stumps and tree bark.

Hypoxylon haematostroma Mont.: perithecial stromata, developed in the endostroma, and their ostiolar necks protrude in the ectostroma, on host trunk, x 7. Collected in Cuba.

Hysterium pulicare Pers.: mature, black, hysterothecium, showing the long fissure of dehiscence, on old tree bark of *Quercus*, Germany, x 15.

Hysterographium Corda
Inc. sed., Inc. sed., Pleosporomycetidae, Dothideomycetes, Pezizomycotina, Ascomycota, Fungi
Corda, A. C. J. 1842. *Icon. fung.* (Prague) 5:34, 77.
From Gr. *hystéra*, genit. *hystericós*, womb, long and cleft, groove, crack + *grápho*, to draw, represent, write > *graphé*, drawing, painting inscription + L. suf. *-ium*. Fructification an elongated hysterothecium, dehiscent by means of a furrow that leaves the hymenium exposed, which looks like a linear inscription on the substrate. Hysterothecia carbonaceous with thick walls, asci cylindrical; ascospores dictyosporous. Saprobic on bark of woody plants.

Hysterographium fraxini (Pers.) De Not.: carbonaceous, black, elongated hysterothecia, dehiscent by means of a furrow that leaves the hymenium exposed, on bark of *Fraxinus*, x 25.

Irenopsis tortuosa

I

Ileodictyon Tul. & C. Tul.
Phallaceae, Phallales, Phallomycetidae, Agaricomycetes, Agaricomycotina, Basidiomycota, Fungi
Tulasne, R., C. Tulasne. 1844. Choix de plantes de la Nouvelle-Zélande. *In*: Raoul, *Annls Sci. Nat.*, Bot., sér. 3 2:114.
From Gr. *íleon*, gut + *díktyon*, net, referring to the form of the fructification. Fructifications a clathriform net or globose basket, with holes bordered by hollow arms similar to lower part of small intestine called the ileon. Immature fruit body sub-hypogeous, peridium white, rupturing irregularly at apex. Gleba olivaceous-brown, mucilaginous, covering inner surface of receptacle. Receptacle sessile, more or less globose but becoming flaccid, of a hollow sphere of anastomosing arms forming isodiametric polygonal meshes; arms tubular almost 1 cm in diameter, detached from peridial volva. Odor foetid. Spores elliptical-cylindrical, subhyaline, smooth. On disturbed soil or at the edge of woodland clearings. Common in Australia (Tasmania) and New Zealand. Introduced into England but rare in that country.

Ingoldia R. H. Petersen—see **Gyoerffyella** Kol
Inc. sed., Inc. sed., Inc. sed., Inc. sed., Pezizomycotina, Ascomycota, Fungi
Petersen, R. H. 1962. Aquatic hyphomycete from North American. I. Aleuriospore (Part I), and key to the genera. *Mycologia* 54:117-151.
Dedicated to the British mycologist *C. T. Ingold* + L. des. *-a*, for his work on aquatic fungi.

Ingoldiella D. E. Shaw
Hydnaceae, Cantharellales, Inc. sed., Agaricomycetes, Agaricomycotina, Basidiomycota, Fungi
Shaw, D. E. 1972. *Ingoldiella hamata* gen. et sp. nov. a fungus with clamp connexions from a stream in North Queensland. *Trans. Br. mycol. Soc.* 59(2):255-259.
From genus *Ingoldia* R. H. Petersen + L. dim. suf. *-ella*, Hyphae hyaline, branched, septate, with clamps. Conidia arising terminally from hyphal branches, hyaline, septate, with clamps, of a main arm or stem and 2-3 lateral arms or branches produced from basal cells of main arm; lateral arms spreading at right angles, mostly 4-septate, main and lateral arms narrowed towards apex and terminating in paired hooks. On submerged decaying leaves from a stream in North Queensland, Australia.

Inocybe (Fr.) Fr.
Inocybaceae, Agaricales, Agaricomycetidae, Agaricomycetes, Agaricomycotina, Basidiomycota, Fungi
Fries, E. M. 1863. *Monogr. Hymenomyc. Suec.* (Upsaliae) 2 (2):346.
From Gr. *ís*, *inós*, fiber, nerve + *kýbe*, head, pileus, due to the presence of small fibers on the pileus. Pileus conical to umbonate, dry, with small fibers on upper part, arranged radially, often with rimose or lacerate appearance, even forming scales. Gills grayish, brownish or rusty-brown; spores similar in color. On soil or decomposing wood, possibly mycorrhizal. Some species have been described as edible but others contain toxic alkaloids that can cause severe poisoning or intoxication.

Inocybe rimosa (Bull.) P. Kumm.: dark colored basidiocarp, with the umbonate, rimose pileus, on soil, x 1.5.

Inonotus P. Karst.
Hymenochaetaceae, Hymenochaetales, Inc. sed., Agaricomycetes, Agaricomycotina, Basidiomycota, Fungi
Karsten, P. A. 1879. Symbolae ad mycologiam Fennicam. VI. *Meddn Soc. Fauna Flora fenn.* 5:39.
From Gr. *inós*, fiber + *oús, otós*, ear, a fibrous ear. Fruitbody annual, resupinate to pileate, mostly broadly attached, flexible to woody, hard when dry, solitary to imbricate, yellowish to dark brown, smooth, tomentose to hispid, hymenophore poroid, frequently with yellow tint, pores small, rarely large and irregular, context cinnamon-brown, shiny to dull, pale yellowish to pale brown or bay. Spores ellipsoid to globose, rarely cylindrical, smooth, hyaline to yellowish-brown. On dead or living deciduous or coniferous trees. Cosmopolitan.

Insolibasidium Oberw. & Bandoni
Platygloeaceae, Platygloeales, Inc. sed., Pucciniomycetes, Pucciniomycotina, Basidiomycota, Fungi
Oberwinkler, F., R. Bandoni. 1984. *Herpobasidium* and allied genera. *Trans. Br. mycol. Soc.* 83:639-658.
From L. *insolitus*, unusual, uncommon + *basidium*, small base, basidium, referring to the uncommon characteristics of being an auricularioid parasite. Hyphae hyaline, thin-walled, without clamps, inter- and intracellular, fertile hyphae giving rise to terminal basidia that protrude through stomata, then proliferating to form a cluster of basidia. Probasidial cystidia lacking. Basidia cylindric, curved, 4-celled. Sterigmata hypha-like to cornute. Basidiospores hyaline, thin-walled, short-cylindric, slightly curved, non-amyloid, apiculus oblique, germinating by germ tube or repetition. Asexual morph present. Parasitic in living leaves of dicotyledonous plants.

Intraornatospora B.T. Goto, et al.
Inc. sed., Diversisporales, Inc. sed., Glomeromycetes, Glomeromycotina, Glomeromycota, Fungi
Goto, B. T., et al. 2012. Intraornatosporaceae (Gigasporales), a new family with two new genera and two new species. *Mycotaxon* 119:117-132.
From L. *intra*, inside + *ornata*, adorned, ornamented + *spora*, spore < Gr. *sporá*, seed, spore, referring to the position of the spore ornamentation. Spores on bulbous sporogenous cells that arise terminally on mycelial hyphae. Spores with two walls: outer wall generally triple-layered, continuous with wall of sporogenous cell; inner surface of outer wall with tuberculate or spiny projections; inner wall hyaline, 2-3 layered, forms *de novo*. Spores formed singly in soils, rarely in roots.

Irenopsis F. Stevens
Meliolaceae, Meliolales, Inc. sed., Sordariomycetes, Pezizomycotina, Ascomycota, Fungi
Stevens, F. L. 1927. The Meliolineae. I. *Annls. Mycol.* 25:411.

From genus *Irene* (a woman's name) + L. suf. *-opsis*, from Gr. *-ópsis*, aspect, appearance. Ascomata with setae, not mycelium. Conidia lacking. Black mildew or sooty mold, Parasitic on living leaves of tropical plants.

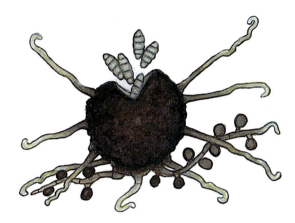

Irenopsis tortuosa (G. Winter) F. Stevens: perithecium releasing ascospores, covered with setae, and basal mycelium with capitate hyphopodia, causing black mildew of tropical plants, x 110.

Irpex Fr.
Meruliaceae, Polyporales, Inc. sed., Agaricomycetes, Agaricomycotina, Basidiomycota, Fungi
Fries, E. M. 1825. *Syst. orb. veg.* (Lundae) 1:81.
From L. *irpex*, dentate, from *irpicinus*, provided with teeth, like a rake, for the lacerate walls of the pores, which resemble teeth or flattened spines. Basidiocarps annual, generally white to pale buff, resupinate, effuse-reflexed, with a thin context, hymenophore of flattened teeth, irregular, more or less united at base forming teeth, pores, or reticulate and irregular gills. Causing a white rot on dead wood of conifers and hardwoods. *Irpex lacteus* Fr. is an important destroyer of wood of broad-leaved trees and conifers; it is characterized by white, crustaceous, reflexed, effuse fructifications, with hymenophore of dentate tubular or laciniate and labyrinthiform spines.

Irpex lacteus (Fr.) Fr.: resupinate basidiocarp, pale buff, with the hymenophore composed of flattened teeth, on dead wood, x 3.5.

Isia

Isia D. Hawksw. & Manohar.
Inc. sed., Sordariales, Sordariomycetidae, Sordariomycetes, Pezizomycotina, Ascomycota, Fungi
Hawksworth, D. L., C. Manoharachary. 1978. *Isia*, a new genus in the Sordariaceae *sensu lato* for *Thielavia neocaledoniensis*. *Trans. Br. mycol. Soc.* 71(2):332-335.
Name based on the acronym '*ISI*', for the name of the scanning electron microscope used in its characterization + L. des. *-a*. Ascomata arising singly, dispersed, ostiolate, black, subglobose to obpyriform; peridium pseudoparenchymatous, textura angularis, dark brown, neck elongate, delicate, thin-walled. Paraphyses absent. Asci subcylindrical to elongate-clavate, unitunicate, with a distinct apical annular ring, 8-spored. Ascospores ellipsoid to fusiform, golden brown, simple, walls verruculose, of compacted convex swollen verrucae. On leaves of *Pandanus* spp. from Bihar, India.

thin-walled, rounded at apex, 8-spored. Ascospores consisting of two fusoid cells connected by narrow isthmus, surrounded by a gelatinous sheath. Pycnidia epiphyllous, intraepidermal, concolorous with leaf at maturity. Causing lesions needles on conifers.

Isthmiella quadrispora Ziller: needles of *Abies lasiocarpa*, in British Columbia, Canada, attacked by needle cast or needle blight by this fungus, x 1. Vertical section of an ascoma, showing its subcuticular development, with asci with four ascospores, x 70.

Itersonilia Derx

Cystofilobasidiaceae, Cystofilobasidiales, Inc. sed., Tremellomycetes, Agaricomycotina, Basidiomycota, Fungi
Derx, H. G. 1948. *Itersonilia*, nouveau genre de sporobolomycètes a mycélium bouclé. *Bull. bot. Gdns Buitenz.* 17(4):465-472.
The etymology could not be determined. Dimorphic producing dikaryotic mycelium with clamps and a yeast-like phase with blastospores forming ballistospores. Causes seedling blight, root canker, leaf spot/necrosis, dieback of Apiaceae such as dill.

Isia neocaledoniensis (C. Moreau) D. Hawksw. & Manohar.: vertical section of a black perithecium, carbonaceous, globose-obpyriform, x 160, immersed in the leaf of *Pandanus fascicularis* in India. Mature ascus, ellipsoidal-fusiform, golden-brown, with verruculose wall, x 1,850.

Isthmiella Darker

Rhytismataceae, Rhytismatales, Leotiomycetidae, Leotiomycetes, Pezizomycotina, Ascomycota, Fungi
Darker, G. D. 1967. A revision of the genera of the Hypodermataceae. *Can. J. Bot.* 45:1399-1444.
From Gr. *ísthmos*, narrow passage or connection + L. dim. suf. *-ella*, referring to the shape of the ascospores. Ascomata linear apothecial, hypophyllous, dark brown, immersed in host tissues, opening by a longitudinal fissure; covering layer of dark pseudoparenchyma tissue; basal layer plectenchymatous, hyaline. Asci clavate,

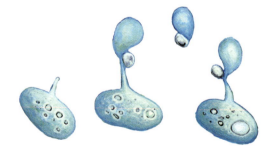

Itersonilia perplexans Derx: vegetative yeast cells forming ballistospores, x 2,300. Isolated in Netherlands.

J

Jerainum triquetrum

J

Jerainum Nawawi & Kuthub.

Inc. sed., Inc. sed., Inc. sed., Inc. sed., Pezizomycotina, Ascomycota, Fungi

Nawawi, A., A. J. Kuthubutheen. 1992. *Jerainum triquetrum* gen. et sp. nov., a new hyphomycete with muriform and appendaged conidia. *Mycotaxon* 45:409-415.

From Gunung *Jerai*, Kedah, Malaysia, the locality in which the fungus was isolated. Colonies effuse, brown, mostly inconspicuous. Mycelium partly superficial, partly immersed in the substratum, consisting of smooth, brown, branched, septate hyphae. Conidiophores macronematous, mononematous, solitary or in groups, erect, straight to flexuous, septate, arising from the superficial hyphae. Conidiogenous cells incorporated in the conidiophores, terminal, monoblastic and proliferating percurrently. Conidia acrogenous, solitary, dry, brown, smooth, ellipsoidal to obconical, muriform, with appendages at the crown and at the base formed during and after conidial secession, which is schizolytic. On decaying twigs collected from a small freshwater stream at Gunung Jerai, Malaysia.

Jerainum triquetrum Nawawi & Kuthub.: conidia at various stages arranged in developmental series as observed on decaying twigs: typical conidia, a conidium in side view, and a conidium view from above, x 1,500.

Joergensenia Passo, et al.

Inc. sed., Lecanorales, Lecanoromycetidae, Lecanoromycetes, Pezizomycotina, Ascomycota, Fungi

Passo, A., et al. 2008. *Joergensenia*, a new genus to accommodate *Psoroma cephalodinum* (lichenized Ascomycota). *Mycol. Res.* 112:1465-1474.

Named in honor of the Danish botanist *Peter M. Joergensen* + L. des. -*ia*. Thallus heteromerous, squamulose to small-foliose, loose or closely attached to substratum, orbicular, corticolous. Prothallus present, thin, black or rarely absent. Squamules incised to sublobulate, whitish-green, medulla white, main photobiont green. Cephalodia conspicuous, growing over thallus, photobiont filamentous, clustered in glomeruli (*Scytonema*). Apothecia lecanorin, laminar, thalline exciple crenulate, persistent, concolorous with thallus. Hymenium I+ blue. Asci clavate, apex with internal amyloid cap-shaped plug. Ascospores simple, hyaline, ovoid to ellipsoid.

Joergensenia cephalodina (Zahlbr.) Passo, et al.: general view of squamulose to foliose thallus, with apothecia and cephalodia, on tree bark, x 3. Collected in Juan Fernández Islands, Chile.

Jola Möller

Eocronartiaceae, Platygloeales, Inc. sed., Pucciniomycetes, Pucciniomycotina, Basidiomycota, Fungi

Möller, A. 1895. *Bot. Mitt. Trop.* 8:22.

Related to the names *Jóle*, *Yole*, *Yola* and *Yolanda*, which are derived from the Gr. name *Ióle*. Fructifications gelatinous, soft, white, irregular, small patches when turgid, on drying a thick layer like a scab. Basidial apparatus composed of elongate-clavate protobasidia and cylindrical metabasidia transversely septate at maturity; sterigmata short, cylindrical; basidiospores germinate by repetition. Parasitic of capsules of mosses; e.g., *J. hookerianum* Möller in neotropical American regions.

Ascospores hyaline, ellipsoid, symmetrical to asymmetrical, straight to curved, bicellular, with a central or slightly eccentric septum, constricted at the septum, smooth, with or without blunt or pointed appendages.

Jola javensis Pat.: gelatinous fructifications (basidial apparatuses), parasitic on mosses (*Sematophyllum swartzii*), x 5. Collected in Java.

Juglanconis juglandina (Kunze) Voglmayr & Jaklitsch: subepidermal acervulus, with phialides producing a conidial dark mass, on walnut tree (*Juglans regia*), Germany, x 750.

Jubispora B. Sutton & H. J. Swart

Inc. sed., Inc. sed., Inc. sed., Inc. sed., Pezizomycotina, Ascomycota, Fungi

Sutton, B. C., H. J. Swart. 1986. Australian leaf-inhabiting fungi XXIII. *Colletogloeum* species and similar fungi on *Acacia. Trans. Br. mycol. Soc.* 87(1):93-102.

From L. *juba*, mane + Gr. *sporá* > L. *spora*, spore, for the persistent mucilaginous appendage down one side of the conidium. Conidiomata epidermal, flat, black, circular to ellipsoid, shining, eustromatic, consisting of a basal layer and a thinner upper layer of dark brown, thick-walled textura angularis; dehiscence by rupture of overlying fungal tissue, upper epidermal wall and cuticle. Conidiophores absent. Conidiogenous cells discrete, dark brown, verrucose, lageniform to cylindrical, with several enteroblastic percurrent proliferations. Conidia holoblastic, more or less straight, gradually tapered to an obtuse to subacute apex and a narrow truncate base, verrucose especially at base, brown, 3-euseptate, formed in a mucilaginous sheath, which persists along one side of conidium. On pods of *Acacia* spp. in South Africa and Zambia.

Juglanconis Voglmayr & Jaklitsch

Juglanconidaceae, Diaporthales, Diaporthomycetidae, Sordariomycetes, Pezizomycotina, Ascomycota, Fungi

Voglmayr, H., et al. 2017. *Juglanconis* gen. nov. on Juglandaceae, and the new family Juglanconidaceae (Diaporthales). *Persoonia* 38:142.

Referring to its occurrence on Juglandaceae, a family of the order Fagales. Pseudostromata consisting of an inconspicuous, erumpent, light to dark coloured ectostromatic disc causing a more or less pustulate bark surface. Perithecia inconspicuous at the bark level with long lateral ostioles Asci oblong or fusoid, 8-spores.

Jumillera J. D. Rogers, et al.

Xylariaceae, Xylariales, Xylariomycetidae, Sordariomycetes, Pezizomycotina, Ascomycota, Fungi

Rogers, J. D., et al. 1997. *Jumillera* and *Whalleya*, new genera segregated from *Biscogniauxia*. *Mycotaxon* 64:39-50.

Named in honor of the American mycologist *Julian H. Miller* + des. *-a*, for his numerous studies on the xylariaceous fungi. Stromata applanate or effused-pulvinate, solitary or confluent, lacking KOH-extractable pigments. Outer dehiscing layer dark brown, thin; inner exposed surface white, gray, greenish or black. Perithecia globose to depressed-globose, monostichous. Ostioles depressed, appearing punctate. Asci 8-spored, cylindrical, stipitate, persistent, with discoid apical ring, amyloid. Ascospores brown to dark brown, unicellular, ellipsoid, nearly equilateral or inaequilateral, with rounded ends and straight germ slit, smooth. Asexual morph libertella-like with a geniculosporium-like synanamorph.

Jumillera mexicana J. D. Rogers, et al.: perithecial stroma on decorticated wood, San Luis Potosí, Mexico, x 1, dark brown ascospores with straight germ slit, x 1,500.

K

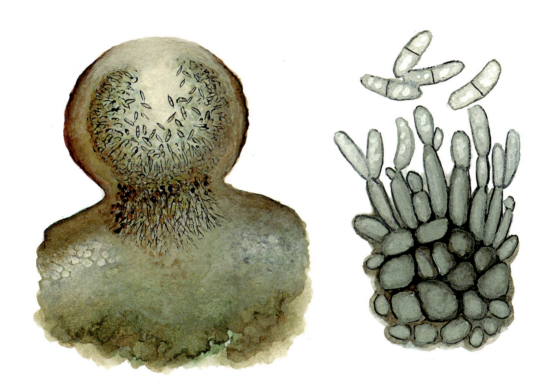

Karsteniomyces peltigerae

K

Kalamarospora G. Delgado
Inc. sed., Inc. sed., Inc. sed., Inc. sed., Pezizomycotina, Ascomycota, Fungi
Delgado, G. 2010. South Florida microfungi: *Kalamarospora multiflagellata* gen. et sp. nov. (hyphomycetes), with additional new records from U.S.A. *Mycotaxon* 114:231-246.
From Gr. *kalamári*, squid + *spóros*, seed, in reference to the squid-like shape of the conidia. Colonies on natural substratum effuse, hairy. Mycelium predominantly immersed in substrate, composed of branched, septate, smooth, pale brown to brown hyphae. Stromata lacking. Conidiophores macronematous, mononematous, single or in groups, simple, erect, straight or slightly flexuous, mostly transversely striate, cylindrical, septate, dark brown or blackish-brown, regenerating percurrently. Conidiogenous cells monoblastic, integrated, terminal, cylindrical, pale brown to brown, transversely striate, percurrent. Conidial secession rhexolytic. Conidia acrogenous, solitary, obclavate or ellipsoidal, pale brown to brown, smooth, internally filled with a visible mass of subhyaline, septate filaments protruding apically or subapically as multiple long, filiform, subhyaline or hyaline, dichotomously branched appendages. Sexual morph unknown.

Kallichroma Kohlm. & Volkm.-Kohlm.
Bionectriaceae, Hypocreales, Hypocreomycetidae, Sordariomycetes, Pezizomycotina, Ascomycota, Fungi
Kohlmeyer, J., B. Volkmann-Kohlmeyer. 1993. Observations on *Hydronectria* and *Kallichroma* gen. nov. *Mycol. Res.* 97(6):753-761.
From Gr. *kalós*, beautiful + *chrōma*, color, for the brightly colored ascomata. Ascomata subglobose to ellipsoidal, at first immersed, then erumpent, ostiolate, periphysate, epapillate, indistinctly clypeate, fleshy-leathery, orange-yellowish, gregarious or frequently confluent. Peridium hyaline, three-layered. Hamathecium composed of apical paraphyses, filaments septate, simple. Asci eight-spored, subcylindrical to clavate, short pedunculate, thin-walled at maturity, persistent, without apical apparatus, IKI negative. Ascospores biseriate, ellipsoidal, 1-septate in middle, hyaline, longitudinally striate by thin ribs or smooth with or without an early dissolving mucilaginous sheath. Saprobic on wood in marine habitats.

Kalamarospora multiflagellata G. Delgado: conidiophores, conidiogenous cells (without filiform appendages), and rhexolytic conidia, with up to 12 filiform appendages, x 600. Found in dead leaves of *Sabal palmetto* (Areaceae) in Florida, U.S.A.

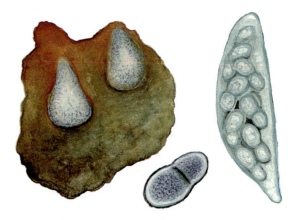

Kallichroma asperum Abdel-Wahab, et al.: sagittal section of confluent, orange-yellowish ascomata, on decomposing wood of mangrove tree (*Avicenia marina*), South Arabia, x 50; ascus with young ascospores, x 720; mature ascospore, ellipsoidal, 1-septate in the middle, x 1,000.

Kickxella

Kamatia V. G. Rao & Subhedar—see **Pseudodictyosporium** Matsush.

Dictyosporiaceae, Pleosporales, Pleosporomycetidae, Dothideomycetes, Pezizomycotina, Ascomycota, Fungi

Rao, V. G., A. W. Subhedar. 1976. *Kamatia*—A new genus of Hyphomycetes. *Trans. Br. mycol. Soc.* 66(3):539-541.

Named in honor of the Indian mycologist *M. N. Kamat* + L. des. *-ia*.

Karsteniomyces D. Hawksw.

Inc. sed., Inc. sed., Inc. sed., Inc. sed., Pezizomycotina, Ascomycota, Fungi

Hawksworth, D. L. 1980. Notes on some fungi occurring on *Peltigera*, with a key to accepted species. *Trans. Br. mycol. Soc.* 74(2):363-386.

Named in honor of the Finnish mycologist *P. Karsten* + connective *-io-* + L. suf. *-myces*, fungus. Conidiomata single, uniloculate, dispersed to loosely aggregated, subglobose, superficial, nectrioid, pale orange to dark red, ostiole irregular, with a thick, pseudoscleренchymatous wall. Conidiophores cylindrical, branched. Conidiogenous cells acrogenous to pleurogenous, cylindrical, hyaline. Conidia holoblastic, elongate-ellipsoid, hyaline, 1-septate, smooth, guttulate.

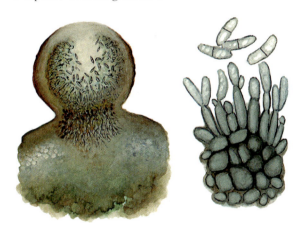

Karsteniomyces peltigerae (P. Karst.) D. Hawksw.: longitudinal section of a conidioma, parasitizing the thallus of *Peltigera canina*, x 135; conidiogenous cells, with hyaline, one-septate conidia, x 1,400.

Kendrickomyces B. Sutton, et al.

Inc. sed., Inc. sed., Inc. sed., Inc. sed., Pezizomycotina, Ascomycota, Fungi

Sutton, B. C., et al. 1976. *Kendrickomyces* gen. nov. and *Waydora* nom. nov., two unusual stromatic Coelomycetes. *Trans. Br. mycol. Soc.* 67(2):243-249.

Named in honor of the Canadian mycologist *W. Bryce Kendrick* + L. *myces*, fungus. Mycelium immersed, branched, septate, hyaline to pale brown. Fructifications immersed, globose to collabent, stromatic, at maturity furfuraceous; peripheral wall of dark brown, sclerotioid pseudoparenchyma; inner tissue of two distinct layers. Locules initiated in upper layer, aggregated into 1 or 2 separate groups within stroma, vertically elongated, regularly arranged, each with an ostiolar channel merging into a single communal ostiole. Ostiole papillate, surrounded by large dark brown pseudoparenchyma. Conidiophores septate, irregularly branched, hyaline, smooth, formed from inner cells of locular walls. Conidiogenous cells enteroblastic, phialidic, integrated, determinate, hyaline, tapered. Conidia hyaline, aseptate, smooth, ± guttulate, falcate. Collected on dead branches of *Terminalia belerica* in Teenai Ghat, Goa, India.

Kentingia Sivan. & W. H. Hsieh—see **Excipulariopsis** P.M. Kirk & Spooner

Parmulariaceae, Asterinales, Dothideomycetidae, Dothideomycetes, Pezizomycotina, Ascomycota, Fungi

Sivanesan, A., W. H. Hsieh. 1989. *Kentingia* and *Setocampanula*, two new ascomycete genera. *Mycol. Res.* 93(1):83-90.

From *Kenting*, the type location in Taiwan + L. des *-ia*.

Kernia Nieuwl.

Microascaceae, Microascales, Hypocreomycetidae, Sordariomycetes, Pezizomycotina, Ascomycota, Fungi

Nieuwland, J. A. 1916. Critical notes on new and old genera of plants—VIII. *Am. Midl. Nat.* 4:379-386.

Probably named in honor of the American mycologist *Frank D. Kern* + L. des. *-ia*. Ascomata cleistothecia, globose to ovoid, ellipsoidal or irregular, dark brown to black, glabrous or with hyphoid hairs, especially when young, some species with appendages at maturity. Appendages occurring in one or more fascicles consisting of several hairs arising from ascomatal wall, long, slender, unbranched, brown, with slightly thickened walls, straight to flexuous, usually with circinate tips. Wall of ascomata several layers thick. Asci irregularly arranged in centrum, globose to ovoid, 8-spored, with walls evanescent at maturity. Ascospores 1-celled, ellipsoidal, pale brown, dextrinoid when young, reddish-brown en mass at maturity, containing a prominent droplet, with a germ pore at each end, smooth. Asexual morphs graphium-like and scopulariopsis-like. On dung, decaying vegetable matter, wood, and in soil.

Kickxella Coem.

Kickxellaceae, Kickxellales, Inc. sed., Kickxomycetes, Kickxomycotina, Zygomycota, Fungi

Coemans, E. 1862. Notice sur un champignon nouveau: *Kickxella alabastrina* Cms. *Bull. Soc. R. Bot. Belg.* 1:155-159.

Named in honor of the Belgian botanist *J. Kickx* + L. dim. suf. *-ella*. Sporangiophores borne on sporocladia arranged in terminal verticils. Pseudophialides formed on upper surface of sporocladia, which give rise to unispored sporangiola. Saprobic on mouse, horse and zebra dung.

Kiehlia

Kickxella alabastrina Coem.: sporangiophore with the terminal verticil of sporocladia, which give rise to unispored sporangiola, x 800. Isolated in Belgium.

Kiehlia Viégas
Parmulariaceae, Asterinales, Dothideomycetidae, Dothideomycetes, Pezizomycotina, Ascomycota, Fungi
Viégas, A. P. 1944. Alguns fungos de Brasil II. Ascomicetos. *Bragantia* 4(1-6):1-392.
Dedicated to the Brazilian mycologist/phytopathologist *Jorge Kiehl* + L. des. -ia. Mycelium subepidermal in host, amphigenous, forming lesions bordered by a distinct yellow margin, erupting through epidermis. Stromata elongate, black, opening by a longitudinal slit. Locules globose to elongate, 1-3 per stroma, containing pseudoparaphyses. Asci clavate, 8-spored, J-. Ascospores ovoid to clavate, hyaline, 1-celled, smooth. Pycnidia globose to elongate, immersed in similar stromata, opening by an apical fissure. Conidiophores hyaline, simple. Conidia hyaline, 1-celled, oblong to fusiform or plano-convex, with a filiform appendage at each end, smooth. On an unidentified grass in São Paulo State, Brazil.

Kimbropezia Korf & W. Y. Zhuang
Pezizaceae, Pezizales, Pezizomycetidae, Pezizomycetes, Pezizomycotina, Ascomycota, Fungi
Korf, R. P., W-Y. Zhuang. 1991. *Kimbropezia* and *Pfistera*, two new genera with bizarre ascus apices (Pezizales). *Mycotaxon* 40:269-279.
Named in honor of the American mycologist *James W. Kimbrough*, for his studies on pezizaceous fungi + Gr. *pézis* or *pézikes*, an ancient name to designate certain sessile fungi that are earth flowers that emit a cloud of spores when mature; from this, L. *pezicae* or *pezise*, and *peziza*. Apothecia discoid, sessile, margin slightly to strongly enrolled. Disc pale brown to brown. Excipulum 4-layered. Subhymenium hyphae with cyanophilic rings of thickened wall material at septa. Hymenium with subcylindrical paraphyses with thin walls. Asci subcylindrical, thick-walled with strongly dextrinoid contents when young, becoming thin-walled, eguttulate, uniseriate when mature. On soil in Tenerife, Canary Islands.

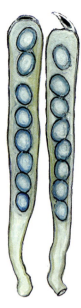

Kimbropezia campestris Korf & W. Y. Zhuang: operculate asci with cyanophilic ascospores, x 200. On soil, among grasses and weeds.

Kirramyces J. Walker, et al.—see Teratosphaeria Syd. & P. Syd.
Mycosphaerellaceae, Mycosphaerellales, Dothideomycetidae, Dothideomycetes, Pezizomycotina, Ascomycota, Fungi
Walker, J., et al. 1992. *Phaeoseptoria eucalypti* and similar fungi on *Eucalyptus*, with description of *Kirramyces* gen. nov. (Coelomycetes). *Mycol. Res.* 96(11):911-924.
From Australian aborigine word *kirra*, leaf + L. *myces*, fungus.

Kloeckera Janke—see Hanseniaspora Zikes
Saccharomycodaceae, Saccharomycetales, Saccharomycetidae, Saccharomycetes, Saccharomycotina, Ascomycota, Fungi
Janke, A. 1928. Über die Formgattung *Kloeckera* Janke. *Zentralbl. Bakteriol. Parasitenk., Abt. II,* 59:311.
Dedicated to the mycologist *A. Klöcker* (1862-1923) + L. des. -a.

Kluyveromyces Van der Walt
Saccharomycetaceae, Saccharomycetales, Saccharomycetidae, Saccharomycetes, Saccharomycotina, Ascomycota, Fungi
Van der Walt, J. P. 1956. *Kluyveromyces*—A new yeast genus of the Endomycetales. *Antonie van Leeuwenhoek* 22:265-272.
Named in honor of the Dutch microbiologist *A. J. Kluyver* + L. suf. *-myces* < Gr. *mýkes*, fungus. Yeast that reproduces

asexually by budding to form pseudomycelium, true mycelium lacking. Asci mostly 1-4-spored, up to 16 or more. Ascospores generally reniform or crescentiform, also round or oval. Ferments vigorously, some species synthesize a red pigment close to or identical with pulcherrimina. In diverse habitats, such as estuarine mud, sea water, soils, fermented dairy products, fleshy basidiocarps and others.

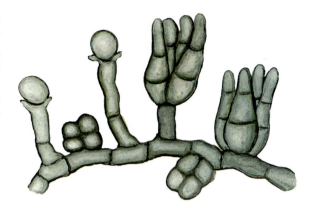

Kodonospora tetracolumnaris K. Ando: hypha with conidiogenous cells and two-septate conidia, and inverted bell-shaped conidia, of four obclavate, multiseptate columns cells, x 1,450.
On dead leaves of *Pandanus boninensis*, Japan.

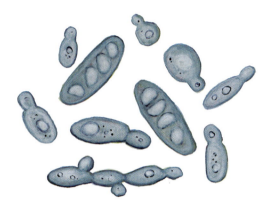

Kluyveromyces marxianus (E. C. Hansen) Van der Walt: budding vegetative yeast cells, asci with four reniform ascospores, x 2,100. Isolated from beer wort.

Kmetiopsis Bat. & Peres

Inc. sed., Inc. sed., Inc. sed., Inc. sed., Pezizomycotina, Ascomycota, Fungi

Batista, A. C, G. E. P. Peres. 1960. *Kmetiopsis*—Um novo gênero de fungos Tuberculariaceae. *Publicações Inst. Micol. Recife* 245:1-9.

From genus *Kmetia* + Gr. suf. *-ópsis*, aspect, appearance, for the apparent similarity to this genus. Superficial mycelium absent. Intramatrical mycelium subhyaline, forming a pseudostroma that erupts to form a superficial sporodochium. Sporodochia oblong, pulvinate, minute, gelatinous. Phialides grouped in fascicles, hyaline or pigmented. Conidia 1-celled, acicular, hyaline, acrogenous, aggregated in mucilaginous heads. Collected on leaves of Hymenaeae sp. in São Lourenço da Mata, Pernambuco, Brazil.

Kodonospora K. Ando

Inc. sed., Inc. sed., Inc. sed., Inc. sed., Pezizomycotina, Ascomycota, Fungi

Ando, K. 1993. *Kodonospora*, a new staurosporous hyphomycete genus from Japan. *Mycol. Res.* 97(4):506-508.

From Gr. *kódon*, a bell + L. *spora*, spore, for the shape of the spores. Mycelium composed of branched, septate hyphae. Conidiophores absent, conidia-bearing hyphae undifferentiated. Conidiogenous cells enteroblastic, integrated in hyphae. Conidia inverted bell-shaped, brownish, composed of four columns of cells, columns obclavate, multiseptate. Collected on fallen leaves of *Pandanus boninensis* on Kitaiou Island, Japan.

Kordyana Racib.

Brachybasidiaceae, Exobasidiales, Exobasidiomycetidae, Exobasidiomycetes, Ustilaginomycotina, Basidiomycota, Fungi

Raciborski, M. 1900. *Parasit. Alg. Pilze Java's* (Jakarta) 2:35.

The etymology could not be determined. Pileus reduced to a loose group of cylindrical basidia with oblong or fusoid basidiospores. Parasitic on Commelinaceae.

Korfia J. Reid & Cain

Hemiphacidiaceae, Helotiales, Leotiomycetidae, Leotiomycetes, Pezizomycotina, Ascomycota, Fungi

Reid, J., R. F. Cain. 1963. A new genus of the Hemiphacidiaceae. *Mycologia* 55(6):781-785.

Named in honor of the American mycologist *Richard P. Korf* + L. des. *-ia*. Apothecia hypophyllous, subepidermal, simple, exposed by circumscissile removal of epidermis, elliptical to orbicular, light in color when moist, dark when dry. Excipulum poorly developed. Paraphyses filiform, septate, branched, recurved at tips, with a yellowish mucilaginous coating, agglutinating at tips forming an epithecium. Asci broadly clavate or subfusoid, narrowing to a short stipe, broadly rounded at apex, inoperculate, 8-spored, J+. Ascospores hyaline, 1-celled, ellipsoid, obtuse at ends, straight or curved, sometimes becoming brown and several-celled with age. On conifer needles.

Korfiella D. C. Pant & V. P. Tewari

Sarcosomataceae, Pezizales, Pezizomycetidae, Pezizomycetes, Pezizomycotina, Ascomycota, Fungi

Pant, D. C., V. P. Tewari. 1970. *Korfiella*, a new genus of Sarcoscyphaceae. *Trans. Br. mycol. Soc.* 54(3):492-495.

From the genus *Korfia* J. Reid & Cain + L. dim. suf. *-ella*. Apothecia gregarious or scattered, leathery, split on one side to base, subsessile or rarely with a short stem-like base. Outer surface "Argus Brown" when fresh,

Kregervanrija

hymenium brownish-black. Excipulum two layered: medullary excipulum made up of textura intricata, of brown hyphae running parallel to outside; ectal excipulum dark chestnut-brown, textura angularis. Asci not bluing in iodine, suboperculate, long-cylindrical; ascospores hyaline, oval to elliptical, uni- to biguttulate; paraphyses simple or branched, septate, anastomosing, hyaline. Collected on a rotting stump of a tree in Nainital, U.P., India.

similarities to this genus. Stromata circular or elliptical, superficial, black, carbonaceous, sessile, with the shape of an inverted dinner plate, flat on top, smooth, with slightly protruding ostiolar necks. Perithecia globose, immersed in stroma in a single layer, with a thin, carbonaceous wall. Asci unitunicate, elongate, 8-spored. Ascospores plano-convex, brown, 1-celled, smooth. Collected on stems of *Guadua* sp. in São Paulo State, Brazil.

Korfia tsugae Reid & Cain: vertical section of an apothecium, elliptical to orbicular (x 350), with asci containing 8-septate ascospores of brown color, and sometimes with several cells at maturity, ellipsoidal, straight or curved; the paraphyses are filiform, septate, and upon maturing yellowish, sticky, agglutinated at their apex to form an epithecium. It is found in coniferous foliage in Ottawa and Toronto, Canada.

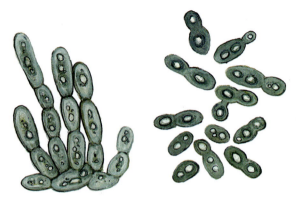

Kregervanrija pseudodelftensis Kurtzman: pseudomycelium, vegetative yeast cells, asci with ascospores, x 3,400. Isolated from fallen fruit of apple, Ohio, U.S.A.

Kregervanrija Kurtzman
Pichiaceae, Saccharomycetales, Saccharomycetidae, Saccharomycetes, Saccharomycotina, Ascomycota, Fungi
Kurtzman, C. P. 2006. New species and new combinations in the yeast genera *Kregervanrija* gen. nov., *Saturnispora* and *Candida*. FEMS Yeast Res. 6(2):28-297.
Named in honor of the Dutch zymologist *Nelly Jeanne Wilhelmina Kreger-van Rij* + L. des. -*a*-, for her many contributions to yeast systematics. Budding multilaterally on a narrow base. Cells globose, ovoid or elongate. Pseudohyphae, if formed, not well differentiated, true hyphae not produced. Asci with one to four ascospores, unconjugated, or arising from conjugations between independent cells or between a cell and its bud. Asci persistent or deliquescent. Ascospores hat-shaped or spherical, with or without a subequatorial ring. Some species are homothallic whereas *Kregervanrija fluxuum* may be heterothallic.

Kretzschmariella Viégas
Xylariaceae, Xylariales, Xylariomycetidae, Sordariomycetes, Pezizomycotina, Ascomycota, Fungi
Viégas, A. P. 1944. Alguns fungos de Brasil II. Ascomicetos. Bragantia 4(1-6):1-392.
From genus *Kretzschmaria* Fr. + dim. suf. -*ella*, for

Kuehneromyces Singer & A. H. Sm.—see **Pholiota** (Fr.) P. Kumm.
Strophariaceae, Agaricales, Agaricomycetidae, Agaricomycetes, Agaricomycotina, Basidiomycota, Fungi
Singer, R. 1946. Two new species in the Agaricales. Mycologia 38(5):504.
Dedicated to the French mycologist *R. Kühner* + Gr. *mýkes* > L. *myces*, fungus.

Kuehniella G. F. Orr—see **Amauroascus** J. Schröt.
Onygenaceae, Onygenales, Eurotiomycetidae, Eurotiomycetes, Pezizomycotina, Ascomycota, Fungi
Orr, G. F. 1976. *Kuehniella*, a new genus of the Gymnoascaceae. Mycotaxon 4:171-178.
Named in honor of the American mycologist *Harold H. Kuehn* + L. dim. suf. -*ella*, for his contributions to taxonomy of the gymnoascaceous fungi.

Kurtzmaniella Lachance & Starmer
Saccharomycetaceae, Saccharomycetales, Saccharomycetidae, Saccharomycetes, Saccharomycotina, Ascomycota, Fungi
Lachance, M.-A., W. T. Starmer. 2008. *Kurtzmaniella* gen. nov. and description of the heterothallic, haplontic yeast species *Kurtzmaniella cleridarum* sp. nov., the teleomorph of *Candida cleridarum*. Int. J. Syst. Evol. Microbiol. 58:520-524.
Named in honor of the American mycologist *Cletus P. Kurtzman* + L. dim. suf. -*ella*, in recognition of his major contributions to yeast systematics. Vegetative reproduction by multilateral budding. Vegetative cells ovoid to

ellipsoidal. Asci slowly evanescent, conjugated, usually forming two flattened, hat-shaped ascospores with a conspicuous basal ledge. Rudimentary pseudomycelium formed. Fermentative. Collected from flowers of cacti and associated beetles of the genus *Carpophilus* in the southwest United States.

Kusanoopsis F. Stevens & Weedon—see **Uleomyces** Henn.
Cookellaceae, Inc. sed., Inc. sed., Dothideomycetes, Pezizomycotina, Ascomycota, Fungi
Stevens, F. L., A. G. Weedon. 1923. Three new myriangiaceous fungi from South America. *Mycologia* 15(5):197-206.
From genus *Kusanoa* Henn. + Gr. -*ópsis*, for the similarity to this genus.

Kweilingia Teng (syn. **Tunicopsora** Suj. Singh & P. C. Pandey)
Phakopsoraceae, Pucciniales, Inc. sed., Pucciniomycetes, Pucciniomycotina, Basidiomycota, Fungi
Teng, S.C. 1940. Supplement to higher fungi of China. *Sinensia*, Shanghai 11:124.
From *Kweilin*, a city of Zhuang, China, the country of origin + L. -*ia*, suf. for terminating the scientific genera. The specimen type (*K. bambusae*) was collected in Yangshuo, on bamboo. Referring to the layer of intertwined hyphae that covers the telia. Uredia subepidermal, erumpent, linear, aparaphysate; uredospores sessile, subglobose, ovoid to pyriform, sparsely echinulate. Telia in flat, black sheaths, sporogenous tissues on pseudoparenchymatous hyphal layer with radiating black hyphae in lower regions, teliospore chains not laterally or vertically adherent, teliospores pale brown to brown, single-celled, with two germ pores; promycelium external, four-celled. Pycnia and aecia not observed. On leaf sheaths of bamboo in tropical region.

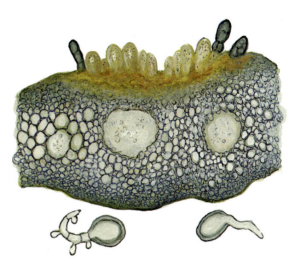

Kweilingia bagchii (Suj. Singh & P. C. Pandey) Buriticá (syn. **Tunicopsora bagchii** Suj. Singh & P. C. Pandey): vertical section of a bamboo leaf sheath (*Dendrocalamus strictus*), Uttar Pradesh, India, showing the orange-yellowish uredinium with scantily echinulate urediniospores, and some dark, with transversal and longitudinal septate teliospores, at the periphery of the uredinium, x 13; below, some germinating teliospores, x 30.

L

Lycoperdon nigrescens

L

Laboulbenia Mont. & C. P. Robin
Laboulbeniaceae, Laboulbeniales, Laboulbeniomycetidae, Laboulbeniomycetes, Pezizomycotina, Ascomycota, Fungi
Robin, C.P. 1853. *Histoire naturelle des végétaux parasites qui croissent sur l'homme et sur les animaux vivants*:1-702. Named in honor of the French entomologist A. Laboulbène + L. suf. -*ia*, which denotes pertaining to. Perithecia with receptacle generally of five cells, an external appendage and an internal appendage, often abundantly branched. Obligate ectoparasites of different orders of insects and a few acarids.

Laboulbenia formicarum Thaxt.: thallus of this obligate ectoparasite on ants, with receptacle, perithecium and appendages, x 550.

Laccaria Berk. & Broome
Hydnangiaceae, Agaricales, Agaricomycetidae, Agaricomycetes, Agaricomycotina, Basidiomycota, Fungi
Berkeley, M. J., Ch. E. Broome. 1883. Notices of British fungi. *Amer. Mag. nat. Hist.* Ser. 5 12:370.
From L. *laccaria*, relative to dyed skin, from *laccatus*, varnished with shellac, polished, from the Persian *laka* and the Arabic *lakk*, or from the Hindu *lakh*, shellac + L. suf. -*aria*, which indicates connection or possession, perhaps referring to the color and texture of the surface of the pileus of the type species, *L. laccata* (Scop. ex Fr.) Berk. & Br. Basidiocarps small to medium, some large; pileus convex-plane, not viscous; gills thick, slightly waxy, violaceous to purple; stipe usually rigid and fibrous. Terrestrial, many considered edible and mycorrhizal.

Laccaria laccata (Scop.) Cooke: mature basidiocarps, on forest soil, x 1.

Laccariopsis Vizzini
Physalacriaceae, Agaricales, Agaricomycetidae, Agaricomycetes, Agaricomycotina, Basidiomycota, Fungi
Vizzini, A., et al. 2012. *Laccariopsis*, a new genus for *Hydropus mediterraneus* (Basidiomycota, Agaricales). *Mycotaxon* 121:393-403.
From genus *Laccaria* Berk. & Broome + Gr. -*ópsis*, for the resemblance of the basidiomata to this genus. Basidiomata agaricoid, lamellae thick, stipe deeply rooting with long pseudorhiza, veil absent, spore-print whitish; spores large, thin- to thick-walled, smooth, inamyloid; basidia large, thin- to thick-walled (sclerobasidia); cheilo- and pleurocystidia abundant, thin- to thick-walled; pileipellis strongly gelatinized, of a loose hymeniderm with slender, cryptic, thin to moderately thick-walled pileocystidia; stipitipellis with caulocystidia, localized only at apex, thin to slightly thick-walled; stipititrama monomitic; clamp-connections present.

Lachnella Fr.
Lachnaceae, Helotiales, Leotiomycetidae, Leotiomycetes, Pezizomycotina, Ascomycota, Fungi
Fries, E. M. 1836. *Fl. Scan.*:343.
From Gr. *lachné*, woolly + L. dim. suf. *-ella*, for the hairy ascomata. Apothecia sessile or stipitate, externally densely clothed with hairs. Hairs usually flexuose, smooth, or more often delicately roughened. Asci cylindric to clavate, usually 8-spored. Ascospores ellipsoid to fusoid, simple, or rarely pseudoseptate. Paraphyses filiform to lanceolate. On bark and wood of conifers.

Lachnellula P. Karst.
Hyaloscyphaceae, Helotiales, Leotiomycetidae, Leotiomycetes, Pezizomycotina, Ascomycota, Fungi
Karsten, P. A. 1884. Symbolae ad mycologiam Fenniacam. *Meddn Soc. Fauna Flora fenn.* 11:138.
From genus *Lachnella* Boud. + L. dim. suf. *-ula*. Apothecia stipitate or subsessile, bright-colored, yellowish or orange, externally whitish, densely clothed with white hairs or, in one species, slightly rufous. Asci cylindric or clavate, inoperculate, 8-spored. Ascospores globose, hyaline. Paraphyses filiform or subclavate. On bark of conifers.

Lachnum Retz.
Lachnaceae, Helotiales, Leotiomycetidae, Leotiomycetes, Pezizomycotina, Ascomycota, Fungi
Retzius, A. J. 1795. *K. svenska Vetensk-Akad. Handl.*, ser. 1:329.
From Gr. *lachné*, *lachnos*, woolly hair, for the ascomata clothed in hairs. Apothecia scattered or gregarious, usually stipitate, with disc plane or concave, smooth, white to yellow, ochre, orange, grayish or brownish. Receptacle cupulate or patellate, white or pigmented, clothed with hairs. Hairs cylindrical or tapered, straight or curved, sometimes wider at tips, thin- or thick-walled, septate, hyaline or pigmented, smooth or granulated, occasionally bearing incrustations. Ectal excipulum hyaline or pigmented. Medullary excipulum composed of interwoven hyphae. Asci unitunicate, cylindric to cylindric-clavate, inoperculate, with amyloid pore in conical apex, 8-spored. Ascospores hyaline, fusoid to filiform, occasionally ellipsoid, 1-several celled, smooth. Paraphyses filiform to lanceolate. On the bark and wood of conifers.

Lacrymaria Pat.
Psathyrellaceae, Agaricales, Agaricomycetidae, Agaricomycetes, Agaricomycotina, Basidiomycota, Fungi
Patouillard, N. 1887. *Hyménomyc. Eur.*, 1-166.
From L. *lacrima*, tear, something related to tears + des. *-ia*. Pileus usually large; stipe often relatively short and fleshy; veil sometimes annular, with fibrils on surface of pileus, sometimes distinctly pigmented. Spores smooth, verruculose or verrucose, cystidia and metuloids often with amyloid contents, lamellae frequently spotted at maturity by dark spores. Many authors include this genus as a subgenus of *Psathyrella* Pat.

Lachnum imbecille P. Karst.: sessile ascomata with the hymenium orange-reddish when young, with the excipulum surrounded by hyaline septate hairs, x 20. Grows on dead leaves of *Eriophorum angustifolium* in Alto de la Casa del Puerto (Tineo-Asturias), Spain.

Lactarius Pers.
Russulaceae, Russulales, Inc. sed., Agaricomycetes, Agaricomycotina, Basidiomycota, Fungi
Persoon, C. H. 1797. *Tent. disp. meth. fung.* (Lipsiae):63.
From L. *lactarius*, lactic, this from *lac*, *lactis*, milk + suf. *-arius*, which indicates connection or possession, for the secretion of latex when the fructification is cut or separated. Fructifications small to medium, sometimes large, fleshy. Hymenophore of thick or thin gills, generally decurrent, i.e., descendent on upper part of stipe. Pileus including gills and stipe with white latex that frequently changes color, or latex of other colors, e.g., yellow, red, orange or blue. Grows in conifer and oak forests in temperate, subtropical and tropical forests. Many species are also edible and of great culinary value, although some species are toxic.

Lactarius indigo (Schwein.) Fr.: blue basidioma with decurrent lamellae, and white latex secretion, on soil, x 0.5.

Laetiporus Murrill
Fomitopsidaceae, Polyporales, Inc. sed., Agaricomycetes, Agaricomycotina, Basidiomycota, Fungi

Laetisaria

Murrill, W. A. 1904. The Polyporaceae of North America. *Bull. Torrey bot. Club* 31(11):607.

From L. *laetus*, bright, beautiful, abundant + *porus*, pore < Gr. *póros*, for having brightly colored pores. Basidiocarps annual, yellow, orange to pinkish-brown, with generally small pores, shelf-shaped, white flesh semibland, spongy, with a penetrating aroma. Leaves in subtropical forests and oak groves.

Laetiporus portentosus (Berk.) Rajchenb.: lignicolous basidioma, with poroid hymenophore, x 0.5. Collected in Western Australia.

Laetisaria Burds.
Corticiaceae, Corticiales, Inc. sed., Agaricomycetes, Agaricomycotina, Basidiomycota, Fungi

Burdsall, H. H., Jr. 1979. Laetisaria (Aphyllophorales, Corticiaceae), a new genus for the teleomorph of *Isaria fuciformis*. *Trans. Br. mycol. Soc.* 72:419-422.

Name formed from genus *Laeticorticium* Donk. + genus *Isaria* Pers. Basidiocarps effused, pellicular, smooth, with broad hyphae, up to 10 μm diam, septate, lacking clamp connections; probasidia sphaeropedunculate; metabasidium cylindrical to clavate, four sterigmate; hyphidia simple; basidiospores ovoid, hyaline, thin-walled, non-amyloid, acyanophilous. On grasses.

Lagenulopsis Fitzp.
Coryneliaceae, Coryneliales, Eurotiomycetidae, Eurotiomycetes, Pezizomycotina, Ascomycota, Fungi

Fitzpatrick, H. M. 1942. Revisionary studies in the Coryneliaceae. *Mycologia* 34:487.

From L. *lagenula*, small bottle < Gr. *lágenos*, bottle, flask, jar + suf. *-opsis* < Gr. *-ópsis*, appearance. Ascocarps erect, often lobed, with a perforation after dehiscence, through which clavate, pedicellate asci freed; ascospores globose. Parasitic on bark of tropical conifers.

Lambertella Höhn.
Rutstroemiaceae, Helotiales, Leotiomycetidae, Leotiomycetes, Pezizomycotina, Ascomycota, Fungi

Höhnel, F. von. 1918. *Lambertella* n. g. v. H. Fragmente zur Mykologie XXI. Sitzungsberichte Akademie Wissenschaften Wien. *Sber. Akad. Wiss. Wien*, Math.-naturw. Kl., Abt. 1, 127:375.

Named for the discoverer of the fungus, Lambert + L. dim. suf. *-ella*. Apothecia on a dark stromatic base consisting of a single layer of cells and loosely interwoven hyphae, stipitate, gregarious or scattered, brown, or yellowish-brown, fleshy, becoming coriaceous or corneous on drying. Hymenium slightly darker. Stem relatively stout, variable in length, or occasionally wanting, hirsute or furfuraceous. Asci cylindric to clavate, attenuated below, rounded or truncate at tip, 8-spored. Ascospores usually 1-seriate, or occasionally becoming partially 2-seriate, simple, broadly ellipsoid, ovoid or lunate, usually unequal sided, smooth or rough, golden-brown or olivaceous when mature. Paraphyses often branched, hyaline, slightly enlarged above. Spermatia usually present. No conidial state observed. On decaying fruits and leaves.

Lagenulopsis bispora Fitzp.: stroma with long necked perithecia, on leaf of *Podocarpus milanjianus*, x 40.

Lambertella palmeri Raitv. & R. Galán: stipitate, cupulate, white apothecia, growing on the upper surface of *Quercus ilex* subsp. *ballota* in La Rioja, Spain, x 20; a young ascus with 8-guttulate ascospores, x 700; and three liberated greenish ascospores, elipsoidal-fusiform shaped, with a horn-like appendage in each pole ("Napoleonic hat"), two ascospores germinating, x 1,200.

Lasiobertia Sivan.
Xylariaceae, Xylariales, Xylariomycetidae, Sordariomycetes, Pezizomycotina, Ascomycota, Fungi
Sivanesan, A. 1978. *Lasiobertia africana* gen. et sp. nov. and a new variety of *Bertia moriformis*. *Trans. Br. mycol. Soc.* 70 (3):383-387.
From Gr. *lasios*, hairy, woolly + genus *Bertia* De Not. Colonies dense, effused, velvety. Ascocarps uniloculate, nonstiolate, tuberculate, single or aggregated, dark brown to black. Asci eight spored, unitunicate, stalked, with an amyloid apical ring. Ascospores fusoid, one septate, hyaline. Paraphyses numerous, filiform, hyaline. On a palm in Bunsu, Ghana.

Lasiosphaeriella Sivan.
Inc. sed., Inc. sed., Sordariomycetidae, Sordariomycetes, Pezizomycotina, Ascomycota, Fungi
Sivanesan, A. 1975. *Lasiosphaeriella*, a new genus of the family Lasiosphaeriaceae. *Trans. Br. mycol. Soc.* 64 (3):441-445.
From genus *Lasiosphaeria* Ces. & De Not. + L. dim. suf. *-ella*, for the similarity to this genus. Stromata dull white when young, later dark brown to black, erumpent to superficial, with one to many locules, aggregated, with ostiole lined with periphyses. Asci unitunicate, long-stalked, without any apical apparatus. Ascospores hyaline, one-celled at first, later becoming brown, many septate after discharge, surrounded by a mucilaginous sheath, germinating by numerous phialidic germ tubes. Paraphyses numerous. On decaying wood in Uganda.

Lasiosphaeriopsis D. Hawksw. & Sivan.
Nitschkiaceae, Coronophorales, Hypocreomycetidae, Sordariomycetes, Pezizomycotina, Ascomycota, Fungi
Hawksworth, D. L. 1980. Notes on some fungi occurring on *Peltigera*, with a key to accepted species. *Trans. Br. mycol. Soc.* 74:363-386.
From genus *Lasiosphaeria* Ces. & De Not. + Gr. suf. *-ópsis*, aspect, appearance. Stromata superficial, carbonaceous, black, 1-multiloculate, aggregated, short-stalked, irregularly obovoid, coarsely warted, ostiolate; ostiole with periphyses; walls thick, many-layered, composed of dark brown, pseudoparenchymatous cells with perforations 'Munk pores', cells elongated in stipe. Paraphyses absent. Asci elongate-clavate, long-stalked, unitunicate, without any discernible apical apparatus, 2- to 4-spored. Ascospores broadly fusiform, 3- to 4-septate, brown, but with terminal cells subhyaline, smooth-walled, without a gelatinous sheath.

Lawalreea Diederich
Inc. sed., Inc. sed., Inc. sed., Inc. sed., Pezizomycotina, Ascomycota, Fungi
Diederich, P. 1990. New or interesting lichenicolous fungi 1. Species from Luxembourg. *Mycotaxon* 37:297-330.
Named in honor of Professor A. Lawalrée, of Brussels, for his assistance and encouragement. Conidiomata pycnidial, immersed, sometimes erumpent, subsphaerical to ellipsoid, non-ostiolate, opening irregularly at apex. Conidiophores absent. Conidiogenous cells lining pycnidial cavity, enteroblastic, with short collarette, discrete, determinate, ampulliform to ellipsoid, hyaline, smooth-walled. Conidia simple, hyaline, smooth-walled, falciform, widest in upper part, with rounded apex, weakly truncate base. Lichenicolous on *Lecanora persimilis* (Th. Fr.) Nyl. in Luxembourg.

Lasiosphaeriopsis salisburyi D. Hawksw. & Sivan: lichenicolous fungus. Vertical section of stromata, originating in the algal layer of the host (**Peltigera rufescens** L.), x 150; peridium compound of pseudoparenchymatous cells with 'Munk pore', x 600; ascus 4-spored, x 240; free ascospores, x 350. Described in the British Isles.

Lawrynomyces Karasiński
Inc. sed., Hymenochaetales, Inc. sed., Agaricomycetes, Agaricomycotina, Basidiomycota, Fungi
Karasiński, D. 2013. *Lawrynomyces*, a new genus of corticitoid fungi in the Hymenochaetales. *Acta Mycologica*, Warrzawa 48(1):5-11.
Named in honor of Polish mycologist *Maria Ławrynowicz* + L. *-myces*, fungus. Basidiomata resupinate, effused, adnate, thin. Hymenophore even, margin indeterminate, without rhizomorphs. Hyphal system monomitic. Hyphae with simple septa, thin to thick-walled, hyaline. Cystidia present. Hyphidia sometimes present. Basidia suburniform to subcylindrical, constricted, more or less pedunculate, basally without clamps, with (2)4 prominent sterigmata. Basidiospores relatively large, broadly ellipsoid to subglobose, with slightly thickened walls and distinct apiculus, smooth, inamyloid, indextrinoid, acyanophilous. On decaying coniferous wood.

Lazulinospora Burds. & M. J. Larsen—see **Amauroderma** Murrill
Ganodermataceae, Polyporales, Inc. sed., Agaricomycetes, Agaricomycotina, Basidiomycota, Fungi
Burdsall, H. H., Jr., M. J. Larsen. 1974. *Lazulinospora*, a new genus of Corticiaceae, and a note on *Tomentella atrocyanea*. *Mycologia* 66(1):96-100.
From ML. *lazulum*, genit. *lazuli*, blue + connective *-no-* + L. *spora* < Gr. *sporá*, because the spores turn blue in KOH.

Lecanicillium

Lecanicillium W. Gams & Zare
Cordycipitaceae, Hypocreales, Hypocreomycetidae, Sordariomycetes, Pezizomycotina, Ascomycota, Fungi
Gams, W., R. Zare. 2000. A revision of *Verticillium* sect. Prostrata. III. Generic classification. *Nova Hedwigia* 72:47-55.

From Gr. *lekáne*, dim. of *lekánion*, dish, pan, kettle, plate, bowl + suf. dim. L. *-ium*. Conidiophores commonly arising from aerial, sometimes submerged, hyphae, either erect and differentiated from the subtending hyphae by thicker walls, or prostrate and little differentiated from the subtending hyphae. Conidiogenous cells discrete aculeate phialides and/or aphanophialides, usually verticillate, sometimes also solitary. Conidia adhering in more or less globose slimy head or in fascicles, which may be inserted transversely on the tip of the phialide, in a few cases forming chains, short- to long-ellipsoidal to fusiform or falcate with more or less pointed ends; sometimes the first formed conidia are longer (macroconidia) than subsequent ones (microconidia). Species mostly entomogenous or fungicolous.

Lecanora argentata (Ach.) Röhl.: crustaceous thallus, with lecanorine apothecia, x 15.

Leccinum Gray
Boletaceae, Boletales, Agaricomycetidae, Agaricomycetes, Agaricomycotina, Basidiomycota, Fungi
Gray, S. F. 1821. *Nat. Arr. Brit. Pl.*:646.

From L. *lecio*, oak (*Quercus*), in reference to the habitat of the genus. Pileus viscid or dry, glabrous, tomentose or granulose; hymenophore yellow, yellowish or whitish; tubes very long in comparison with diameter of context and radius of pileus, but drastically shortened around stipe, almost free, pores small (less than 1 mm diam). Spores olivaceous to ferruginous, fusoid-cylindric or fusoid-oblong, smooth, usually hyaline, small to medium sized; stipe easily breaking, with rough asperulation, often with a network. Ectomycorrhizal, frequently with Ericaceae, Fagales, and Salicales. Widely distributed from the arctic regions to the subtropics.

Lecidea Ach.
Lecideaceae, Lecideales, Lecanoromycetidae, Lecanoromycetes, Pezizomycotina, Ascomycota, Fungi
Acharius, E. 1803. *Methodus*, Sectio prior (Stockholmiæ):xxx, 32.

From Gr. *lékos*, plate, small shield + suf. *-ídes*, which indicates similarity, for the lecideiform or lecideine apothecia, i.e., orbicular and without a thalline margin. Black apothecia; spores simple, hyaline. Photobiont *Protococcus*. On rocks and wood, at times on soil or mosses.

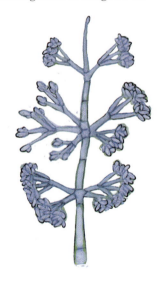

Lecanicillium lecanii (Zimm.) Zare & W. Gams (syn. **Verticillium lecanii** (Zimm.) Viégas): this saprobic fungus in soil has a verticillate arrangement of the phialides along the principal axis, with unicellular, hyaline, phialospores embedded in mucilage, forming gloeoid balls on the apices of the phialides, x 1,000.

Lecanora Ach.
Lecanoraceae, Lecanorales, Lecanoromycetidae, Lecanoromycetes, Pezizomycotina, Ascomycota, Fungi
Acharius, E. 1809. In Luyken, J. A. *Tent. Hist. Lich.*:90.

From Gr. *lekáne*, dim. *lekánion*, plate, pot + suf. *-or*, beautiful, colorful, in reference to the apothecia. Thallus crustous, rarely subfoliose or foliose, attached to substrate by hyphal rhizoids. Apothecia small to large, with disc flat or convex, rounded off, bordered by a thalline margin of same color as thallus (lecanorine apothecium). Photobiont *Protococcus*. On rocks, soil, mosses and trees.

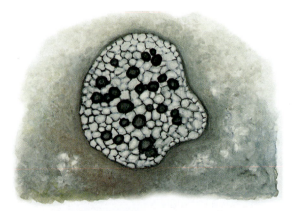

Lecidea tessellata Flörke: crustaceous thallus on soil, with black, lecideiform apothecia, x 7.

Legeriomyces Pouzar (syn. **Genistella** L. Léger & Gauthier)

Legeriomycetaceae, Harpellales, Inc. sed., Inc. sed., Kickxellomycotina, Zygomycota, Fungi

Pouzar, Z. 1972. *Genistella* Léger et Gauthier vs. *Genistella* Ortega: a nomenclatural note. *Folia geobot. phytotax.* 7:319.

Named in honor to the French protozoologist *L. Léger* + Gr. *mýkes* > L. *myces*, fungus. Live in the intestine of ephemeropteran insect larvae. Has long obpyriform trichospores, bearing two long appendages and biconical zygospores, that upon detachment display a collar and a single appendage.

Legeriomyces hungaricus J.K. Misra, et al.: branched thalli with trichospores, borne from a basal structure (holdfast), one free trichospore with two filiform appendages, and one zygospore with one central appendage, x 100. This fungus lives as commensal adhered to the hind gut of nympha of *Baetis rhodani* (Ephemeroptera, Baetidae), in Hungary.

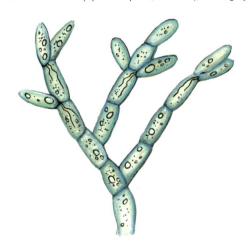

Legeriomyces ramosus (L. Léger & M. Gauthier) Pouzar (syn. *Genistella ramosa* L. Léger & M. Gauthier): branched thallus, which produces trichospores, x 400.

Lembosia Lév. (syn. **Morenoella** Speg.)

Asterinaceae, Asterinales, Dothideomycetidae, Dothideomycetes, Pezizomycotina, Ascomycota, Fungi

Léveillé, J.-H. 1845. Description des champignons de l'herbier du Museum de Paris. *Annls Sci. Nat.*, Bot., Sér 3, 3:58.

From Gr. *lêmbos*, a small boat + L. des. *-ia*, for the shape of the ascomata. Mycelial hyphae superficial, thick-walled, dark brown, with frequent septa and lateral hyphopodia much-branched, forming dark colonies in lesions on host surface. Ascostroma linear, oblong to oval, sometimes branched, Y-shaped, covered by a black aggregation of hyphae or shield. Shield composed of laterally united hyphae that radiate outward from the center, with a lighter central area that opens by a longitudinal slit at maturity to expose the asci. Ascomata elongate, dimidiate pseudothecia. Centrum containing branched, filamentous pseudoparaphyses. Asci bitunicate, ellisoid, oblong, clavate or subglobose, thickened at apex, 4-8-spored. Ascospores 2-celled, broadly cylindrical to ovoid, septum median or submedian, smooth, hyaline, becoming brown at maturity. On living leaves of vascular plants, especially in the tropics.

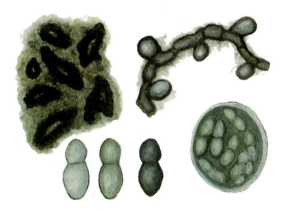

Lembosia albersii Henn.: thyriothecia on the surface of the host plant (*Roupala elegans*), x 10; superficial hyphae with appressoria, x 300; ascus with 8 ascospores, x 350; young and mature ascospores, x 800.

Lentinus Fr.

Polyporaceae, Polyporales, Inc. sed., Agaricomycetes, Agaricomycotina, Basidiomycota, Fungi

Fries, E. M. 1825. *Syst. orb. veg.* (Lundae)*1*:77.

From L. *Lentinus*, a proper name of antiquity, from *lentus*, strong, firm, flexible + suf. *-inus*, which indicates material, consistency or relation. Basidiomata fleshy-hard, coriaceous, or very firm, but flexible, hirsute. Gills adnate or decurrent gills dentate, aserrate or frayed borders. Stipe sessile, subsessile or with a central, eccentric or lateral, with a more or less defined, frequently fugaceous, ring. Spores white. Lignicolous, of medium size, solitary or cespitose. Several species are edible and capable of being cultivated.

Lenzites

Lentinus arcularius (Batsch) Zmitr. (syn. **Polyporus arcularius** (Batsch) Fr.): mushroom-shaped fructification, the pileus with many pores on the lower part of the hymenophore, on tree branch, x 1.

Lentinus crinitus (L.) Fr.: lignicolous, hirsute basidioma, with central stipe, x 1.

Lenzites Fr.
Polyporaceae, Polyporales, Inc. sed., Agaricomycetes, Agaricomycotina, Basidiomycota, Fungi
Fries, E. M. 1836. *Fl. Scan.* (Uppsala):339.
Dedicated to the German botanist *Harold Othmar Lenz* (1799-1870) + Gr. suf. *-ites*, which denotes pertaining to. Basidiomata annual, pileate, semicircular, white to grayish with age, coriaceous or suberose, with the hymenium laminar to daedaloid or labyrinthiform. On dead wood of deciduous trees, causing a white rot. *Lenzites betulina* (L.) Fr. grows on wood of birch and other Betulaceae.

Leotia Pers.
Leotiaceae, Helotiales, Leotiomycetidae, Leotiomycetes, Pezizomycotina, Ascomycota, Fungi
Persoon, C. H. 1801. *Syst. mycol.* 2(1):25.
From Gr. *leiótes*, smoothness, brillance, for the appearance of the surface of the ascoma, which is quite lubricous, smooth, and brilliant. Ascomata fleshy, more or less gelatinous, erect, with a pedicel from two to eight cm high. Asci claviform. Ascospores hyaline, fusiform, three to seven septa. Terricolous in coniferous or deciduous forests. The most common species is *L. lubrica* Pers. ex Fr.

Leotia lubrica (Scop.) Pers.: brilliantly colored, gelatinous ascomata, on coniferous forest, Slovenia, x 1.

Lepidopterella Shearer & J. L. Crane
Argynnaceae, Inc. sed., Inc. sed., Dothideomycetes, Pezizomycotina, Ascomycota, Fungi
Shearer, C. A., J. L. Crane. 1980. Taxonomy of two cleistothecial Ascomycetes with papilionaceous ascospores. *Trans. Br. mycol. Soc.* 75:193-200.
From insect order *Lepidoptera* + L. dim. suf. *-ella*, for the shape of the ascospores. Ascocarps solitary, superficial to slightly immersed, globose to subglobose, black, nonostiolate. Peridium pseudoparenchymatous, several layered, brown to black. Asci bitunicate, globose to subglobose, stalked, thick-walled, lacking a subapical chamber, 8-spored. Ectoascus rigid, rupturing by an apical slit. Ascospores papilionaceous, 2-celled, brown. Isolated from an unidentified submerged twig, Elvira Cypress Swamp, Johnson County, Illinois, U.S.A.

Lepidopterella tangerina Raja & Shearer: vertical section of ascoma (cleistothecial), globose-subglobose, yellowish-brown; asci with 8 young ascospores, x 200; fissitunicate ascus with the extended endoascus, x 700; and reniform ascospores, 1-septate, tangerine-yellow, smooth walled, x 800.

Lepiota (Pers.) Gray
Agaricaceae, Agaricales, Agaricomycetidae, Agaricomycetes, Agaricomycotina, Basidiomycota, Fungi

Gray, S. F. 1821. *Nat. Arr. Brit. Pl.* (London) *1*:601.
From Gr. *lepís*, scale + L. suf. *-otus*, *-ota*, which indicates possession or similarity, i.e., scaly, for the presence of scales on the pileus. Basidiocarps of variable size, pileus characteristically ornamented, fine tomentum to scales formed by cracking of pileus, but center generally smooth. Gills free, white. Spores white. Stipe lacks a volva at base, but with a membranous ring on upper part. Terrestrial, common in coniferous and subtropical forests as well as tropical jungles. Some are edible but not recommended due to the difficulty of distinguishing edible from poisonous species.

Lepiota grandispora Murrill: scaly basidioma, common on forest soil, x 0.5.

Lepista (Fr.) W. G. Sm.
Tricholomataceae, Agaricales, Agaricomycetidae, Agaricomycetes, Agaricomycotina, Basidiomycota, Fungi
Smith, W. G. 1870. *Clavis agaricorum*: an analytical key to the British agaricini. *J. Bot. Lond.* 8:248.
From L. *lepista*, from Gr. *lepasís*, genit. *lepasídos*, or *lepasté*, wine glass or large beverage glass, or a small, wide glass like a shell, which held water in the temples, referring to the shape of the pileus of some of the species. Basidiocarp medium size. Pileus smooth or with small fibers, dry or subviscous, generally thick, fleshy; stipe uniform or somewhat bulbous, smooth or fibrillose-scaly. Spore print white, buff, yellowish or somewhat pink. Similar to *Clitocybe* Kummer, but with the gills less decurrent, purple to purplish-blue, pinkish-orange, reddish-white, or buff-pinkish with age. On soil in conifer and oak forests. Some species are edible.

Lepraria Ach.
Stereocaulaceae, Lecanorales, Lecanoromycetidae, Lecanoromycetes, Pezizomycotina, Ascomycota, Fungi
Acharius, E. 1803. *Methodus*, Sectio prior (Stockholmiæ):3.
From Gr. *lépra*, rough, and *leprós* scaly, leprose + L *-aria*, which indicates connection or possession. Thallus sorediate, granulose-pulvinate, forms fragmented crusts. Apothecia and sexual structures lacking. Photobiont is *Protococcus*. At base of trees, humus, soil and protected rocks.

Lepista nuda (Bull.) Cooke: fleshy, purplish basidiomata, on oak forest soil in France, x 0.5.

Lepraria lobificans Nyl.: leprose and scaly thallus, with soredia, on dead wood, x 0.5.

Leprieuria Læssøe, et al.
Xylariaceae, Xylariales, Xylariomycetidae, Sordariomycetes, Pezizomycotina, Ascomycota, Fungi
Læssøe, T., et al. 1989. *Camillea*, *Jongiella* and light-spored species of *Hypoxylon*. *Mycol. Res.* 93:121-155.
Dedicated to C. *Leprieur*, who collected Xylariaceae in French Guiana + L. des *-ia*. Stromata erect, cylindrical, black, fragile, straight or slightly curved. Perithecia immersed in apex of stroma beneath depressed apical disc with visible ostioles. Asci 8-spored, deliquescent. Ascospores reniform, blackish-brown, smooth, with a straight ventral germ slit. Asexual morph geniculosporium-like. On dead wood in the tropics, especially in the Amazon Region.

Leptodiscella

Leptodiscella Papendorf
Inc. sed., Inc. sed., Inc. sed., Inc. sed., Pezizomycotina, Ascomycota, Fungi
Papendorf, M. C. 1969. *Leptodiscella africana* gen. et comb. nov. *Trans. Br. mycol. Soc.* 53(1):145-147.
From genus *Leptodiscus* Gerd. + L. dim. suf. *-ella*.
Colonies without sclerotia. Hyphae frequently septate, hyaline, often funiculose. Conidiophores uni- or multicellular, simple or branched, often aggregated, forming pseudostromatic structures resembling acervuli or sporodochia. Conidia single or in groups on hyphae or cells of conidiophores, cylindrical with rounded ends, smooth, thin-walled, hyaline or faintly colored, medially 1-septate, bearing a single filamentous setula sublaterally at each end.

Leptogium (Ach.) Gray
Collemataceae, Peltigerales, Lecanoromycetidae, Lecanoromycetes, Pezizomycotina, Ascomycota, Fungi
Gray, S. F. 1821. *Nat. Arr. Brit. Pl.* (London) 1:400.
From Gr. *leptós*, slender, delicate < *leptógeios*, *leptógaios* or *leptógeos*, of light soil, sterile + L. dim. suf. *-ium*.
Thin thallus, of simple structure, with a delicate trama, gelatinous consistency due to presence of blue-green *Nostoc* photobiont with a viscous sheath. On soil, rocks, mosses, old wood, and trees.

Leptogium phyllocarpum var. **phyllocarpum** (Pers.) Mont.: blue-green, gelatinous thallus, with red apothecia, x 1.

Leptographium Lagerb. & Melin (syn. **Verticicladiella** S. Hughes)
Ophiostomataceae, Ophiostomatales, Sordariomycetidae, Sordariomycetes, Pezizomycotina, Ascomycota, Fungi
Lagerberg, T. et al. 1927. Biological and practical researches into blueing in pine and spruce. *Svensk Skogsvårdsförening Tidskr.* 25:257.
From Gr. *leptós*, slender, fine, delicate + L. *graphium* < Gr. *graphíon*, stylus with which the ancients wrote on wax-coated tablets. Conidiophores arranged on slender synnemata, with verticillate branches, thick, darkly pigmented, with penicillate heads of series of metulae and conidiogenous cells. Conidia aseptate, hyaline, ovoid, aggregated in mucilage. Causes blue stain of lumber. Saprobic on rotted wood and other vegetable detritus in the soil.

Leptographium procerum (W. B. Kendr.) M. J. Wingf.: slender synnemata, with the shape of a stylus; it develops as a saprobe on rotted wood, x 90; synnema with conidiogenous cells and unicellular conidia, x 700. Collected in Québec, Canada.

Leptopeltis Höhn.
Leptopeltidaceae, Inc. sed., Inc. sed., Dothideomycetes, Pezizomycotina, Ascomycota, Fungi
Höhnel, F. von. 1917. *Ber. dt. bot. Ges.* 35:422.
From Gr. *leptós*, tenuis, thin + *pélte*, shield. Pseudothecia apothecioid, subcuticular or rarely intraepidermal, often fused to form black ribs, with a generally radial scutellum. Asci ovoid or clavate at maturity. Ascospores 1-septate, hyaline. On leaves and branches of plants.

Leptopeltis aquilina (Fr.) Petr.: sagittal section of an apothecioid, subcuticular pseudothecium, with radial scutellum, asci and uniseptate ascospores, on dead leaf of aquiline fern (*Pteridium aquilinum*), x 375.

Leptosphaeria Ces. & De Not.
Leptosphaeriaceae, Pleosporales, Pleosporomycetidae, Dothideomycetes, Pezizomycotina, Ascomycota, Fungi
Cesati, V., G. De Notaris. 1863. Sferiacei italici. *Comm. Soc. crittog. Ital.* 1(fasc. 4):227.
From Gr. *leptós*, slender, thin, fine + *sphaíra* (L. *sphaera*), sphere, in reference to the characteristics of the pseudothecium. Fructification perithecioid delimited by layers of thick-walled cells, especially at base. Ascospores generally fusiform, triseptate or multiseptate,

hyaline, brown or yellow. Asexual morph pycnidial, phoma-like. On stems of herbaceous plants, causing scab, rots or cankers.

Leptosphaeria acuta (Fuckel) P. Karst.: perithecioid fructifications on the herbaceous plant, with multiseptate ascospores, x 600.

Leptosphaerulina McAlpine

Didymellaceae, Pleosporales, Pleosporomycetidae, Dothideomycetes, Pezizomycotina, Ascomycota, Fungi
McAlpine, D. 1902. *Fungus Diseases of stone-fruit trees in Australia*, Melbourne:103.
From Gr. *leptós*, slender, delicate, fine + *sphaíra*, sphere > L. *sphaera* and dim. *sphaerula* + Gr. *-ina*, which indicates similarity or possession. Pseudothecia small, thin-walled, easily ruptured, perithecioid, immersed. Asci clavate, few. Ascospores hyaline with transverse and longitudinal septa (dictyospores). On leaves and stems. Causing a leaf spot disease of peanut.

Leptosphaerulina trifolii (Rostr.) Petr.: immersed perithecioid pseudothecia, on grass tissues, releasing asci with hyaline ascospores (dictyospores), x 450.

Leptostromites Poinar

Fossil Fungi.
Poinar, G., Jr. 2003. Coelomycetes in Dominican and Mexican amber. *Mycol. Res.* 107(1):117-122.
From genus *Leptostroma* Fr. + *-ites* < Gr. des. suf. for a fossil. Pycnidia subsuperficial, separate, flattened, elongate-ovoid, only upper portion well developed; pale tan in incident light; dark in transmitted light; mostly opening by a cleft, sometimes by irregular fractures. On dicotyledonous leaf in an amber mine, Dominican Republic.

Leptothyrites Poinar

Fossil Fungi
Poinar, G. Jr. 2003. Coelomycetes in Dominican and Mexican amber. *Mycol. Res.* 107(1):117-122.
From genus *Leptothyrium* Kunze + *-ites* < Gr. des. suf. for a fossil. Pycnidia superficial, shield-shaped, without ostiole, opening by fragmentation. Conidia minute, hyaline, 1-celled, round to ovoid. On a monocotyledonous leaf in an amber mine, Dominican Republic.

Leptotrochila P. Karst.

Dermateaceae, Helotiales, Leotiomycetidae, Leotiomycetes, Pezizomycotina, Ascomycota, Fungi
Karsten, P. 1871. *Bidr. Känn. Finl. Nat. Folk* 19:245.
From Gr. *leptós*, small + genus *Trochila* Fr., i.e., a small *Trochila*. Ascomata apothecial, formed in a small stroma immersed in host tissues, becoming erumpent at maturity, solitary or in groups, saucer-shaped to flat, with a foot-like hypostroma or seated on a pseudostroma with indefinite margins, exterior dark brown to black, glabrous. Excipulum pseudoparenchymatous, well-developed. Hymenium yellowish, composed of asci interspersed with paraphyses. Paraphyses filiform, septate, slightly longer than asci. Asci unitunicate, inoperculate, clavate to cylindric-clavate, with a short stipe, apical pore usually bluing in iodine, 8-spored. Ascospores clavate, ellipsoidal, oblong or oval, 1-2-celled, hyaline, smooth. Asexual morph sporonema-like. Causing lesions on living or overwintered leaves and stems of herbaceous plants.

Letharia (Th. Fr.) Zahlbr.

Parmeliaceae, Lecanorales, Lecanoromycetidae, Lecanoromycetes, Pezizomycotina, Ascomycota, Fungi
Zahlbruckner, A. 1892. *In*: O. Kuntze, Revisio generum plantarum mit Bezug auf einige Flechtengattungen. *Hedwigia* 31:36.
From Gr. *léthos* or *léthi*, forgetfulness < *letháno*, to cause to forget + L. suf. *-aria*, which denotes a thing that is similar to or connected to something, probably in relation to certain popular beliefs. Thallus conspicuous, fruticose, golden, lemon-green or greenish-yellow, irregularly rounded branches, with a rugose surface and internal medullary bands. Apothecia a brown disc with a lobulate border. On conifer trunks and branches.

Leucoagaricus Locq. ex Singer

Agaricaceae, Agaricales, Agaricomycetidae, Agaricomycetes, Agaricomycotina, Basidiomycota, Fungi
Singer, R. 1948. Diagnoses fungorum novorum agaricalium. *Sydowia* 2(1-6):35.

Leucocoprinus

From Gr. *leukós*, white, clear + *Agaricus* L., a white mushroom. Similar in appearance to *Lepiota* (Pers.) Gray, but differs in microscopic characters, principally by having thick-walled, metachromatic spores. Terricolous, coprophilous or lignicolous on dead wood. Some are edible, whereas others, such as *L. naucinus* (Fr.) Sing., are toxic.

Leucoagaricus nympharum (Kalchbr.) Bon.: mature basidiocarp, with scaly pileus, on dead wood, x 0.5.

Leucocoprinus Pat.
Agaricaceae, Agaricales, Agaricomycetidae, Agaricomycetes, Agaricomycotina, Basidiomycota, Fungi
Patouillard, N.T. 1888. Quelques points de la classification des Agaricinées. *J. Bot.*, Paris 2:16.
From Gr. *leukós*, white + genus *Coprinus* Pers., because it is similar to this genus, but with white spores. Pileus radially split, sulcate-pectinate, at least marginal half; lamellae thin, soft, sometimes subdeliquescent. Spore print white. Spores with germ pores, smooth or rugose. Cystidia absent or few on sides of lamellae; cheilocystidia usually numerous; fibulae generally absent. Stipe usually with annulus movable at least in age. On soil or on various hosts in warmer part of American continent. Common in the tropics of both hemispheres. Few species are toxic.

Leucopaxillus Boursier
Tricholomataceae, Agaricales, Agaricomycetidae, Agaricomycetes, Agaricomycotina, Basidiomycota, Fungi
Boursier, J. 1925. *Leucopaxillus* nov. gen. *Bull. trimest. Soc. mycol.* Fr. 41(3); em. Singer, *Rev. Mycol.* 4:69, 1939.
From Gr. *leukós*, white + genus *Paxillus* Fr., similar to this genus, but the spore print is white. Hyphae inamyloid and with numerous fibulae. Pileus opaque, non hygrophanous with initially involute margin; lamellae decurrent, adnexed to adnate. Stipe central, rarely excentric, fleshy to somewhat tough. Veil none. Context unchanging on bruising. Spores hyaline, rough to warty, amyloid, small to medium, short elliptic, ovoid or subglobose. On humus and debris in boreal to subtropical regions in both hemispheres. Some species are facultative ectotrophic mycorrhizal formers. Some species edible, others produce clitocybin with a potential application against tuberculosis.

Leveillula G. Arnaud (syn. Oidiopsis Scalia)
Erysiphaceae, Erysiphales, Leotiomycetidae, Leotiomycetes, Pezizomycotina, Ascomycota, Fungi
Arnaud, G. 1921. Étude sur les champignons parasites. 4. Tribu des erysiphées Tulasne. *Ann. Épiphyt.* 7:82-108.
Dedicated to the French mycologist M. J. H. Léveillé + L. dim. suf. *-ula*. Ascocarps cleistothecioid (chasmothecia) with peridial mycelioid appendages. Conidia ellipsoid, catenulate, ovulariopsis-like. Obligate parasite of vascular plants causing powdery mildew or oidium.

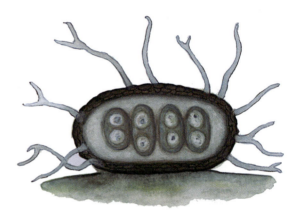

Leveillula taurica (Lév.) G. Arnaud (syn. **Oidiopsis taurica** (Lév.) E. S. Salmon): sagittal section of a cleistothecioid ascocarp, with peridial mycelioid appendages; obligate parasite on leaves of *Zygophyllum fabago*, in Krym, Russia, x 800.

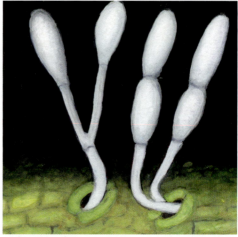

Leveillula taurica (Lév.) G. Arnaud: conidiophores emerging from the stomas of the host (leaves of *Zygophyllum fabago*, Krym, Russia), with catenulate and ellipsoidal conidia, x 400.

Lichina C. Agardh
Lichinaceae, Lichinales, Inc. sed., Lichinomycetes, Pezizomycotina, Ascomycota, Fungi

Agardh, C. 1817. *Syn. Alg. Scand.* :xii, 9.
From L. *lichen* < Gr. *leichén*, liquen + L. suf. *-ina*, which signifies possession or similarity, because at first it was thought to be an alga that was similar to the lichens. Thallus fruticulose, very small, composed of slender, erect branches grouped in racemes. Apothecia inconspicuous, more or less immersed in tips of branches. Photobiont *Calothrix*. On marine rocks.

Lichina pygmaea (Lightf.) C. Agardh: fruticose thallus, with branches grouped in racemes, and inconspicuous apothecia, on marine rock, Great Britain, x 10.

Limacella Earle

Amanitaceae, Agaricales, Agaricomycetidae, Agaricomycetes, Agaricomycotina, Basidiomycota, Fungi
Earle, F. S. 1909. The genera of North American gill fungi. *Bull. New York Bot. Gard.* 5:447.
From L. *limax, limaxis*, slug (a viscous mollusk) + L. dim. suf. *-ella*, in reference to the viscid surface of the pileus. Pileus without a volva; lamellae free or nearly so; lamellulae not truncate, cystidia none, subhymenium cellular. Stipe dry or viscid, with a membranous cortinoid or gelatinous to fleshy annulus; hyphae fibulate. Spore print white; spores small or medium, smooth to finely roughened, hyaline, ovoid to globose or broad ellipsoid, inamyloid. On the ground or rotten wood. Some species may be facultative ectotrophic mycorrhizal formers. Some species are edible but have no economic importance.

Limacinia Neger

Inc. sed., Inc. sed., Inc. sed., Inc. sed., Inc. sed., Fungi
Neger, F. W. 1895. *Centbl. Bakt. ParasitKde*, Abt. I, 1:538.
From L. *limax, limacis*, slug, viscous mud, similar to slime + L. suf. *-inia*, which signifies similarity or pertinence to, for the dark color and the viscous consistency, of the fungus. Pseudothecia smooth or with hyphal growth, at times pedicellate, or on agglutinated synnemata. On sugary exudates of insect parasites of plants.

Linderia G. Cunn.—see Clathrus P. Micheli ex L.

Phallaceae, Phallales, Phallomycetidae, Agaricomycetes, Agaricomycotina, Basidiomycota, Fungi
Cunningham, G. H. 1931. The Gasteromycetes from Australasia, IX. The Phallales, Part II. Families Clathraceae and Claustulaceae. *Proc. Linn. Soc. N.S.W.* 56(3):192.
Dedicated to the American mycologist *David H. Linder* (1899-1946) + L. *-ia*, denotes belonging to.

Linderina Raper & Fennell

Kickxellaceae, Kickxellales, Inc. sed., Inc. sed., Kickxellomycotina, Zygomycota, Fungi
Raper, K. B., D. I. Fennell. 1952. Two noteworthy fungi from Liberian soil. *Am. J. Bot.* 39:79-86.
Dedicated to the American mycologist *David H. Linder* + L. dim. suf. *-ina*. Sporocladia lenticular, globose on apex of sporangiophore. Sporocladia with pseudophialides on upper surface, with a single unispored sporangiolum (merosporangium) on each pseudophialide. In soil.

Linderina pennispora Raper & Fennell: globose sporocladium, with pseudophialides and unispored sporangiola, x 650. Isolated from soil, Liberia.

Linkosia A. Hern.-Gut. & B. Sutton

Inc. sed., Inc. sed., Inc. sed., Inc. sed., Pezizomycotina, Ascomycota, Fungi
Hernández-Gutiérrez, A., B. C. Sutton. 1997. *Imimyces* and *Linkosia*, two new genera segregated from *Sporidesmium sensu lato*, and redescription of *Polydesmus*. *Mycol. Res.* 101(2):201-209.
Named in honor of the German botanist *Johann Heinrich Friedrich Link* + Gr. des. *-osia*, who first described the genus *Sporidesmium* Link. Colonies effuse, hypophyllous, brown. Mycelium superficial, of septate, brown to dark brown hyphae. Conidiophores a single conidiogenous cell, solitary, short, simple, ampulliform, truncate at apex, brown to dark brown. Conidia holoblastic, straight or slightly curved, narrowly obclaviform to obclaviform-rostrate, conico-truncate at base, disto-

Linospora

septate, constricted at septa, pale brown, brown at constrictions, subhyaline at apex, smooth. On dead leaves of *Coccothrinax* sp. in Camagüay, Cuba.

Linospora Fuckel
Gnomoniaceae, Diaporthales, Sordariomycetidae, Sordariomycetes, Pezizomycotina, Ascomycota, Fungi
Fuckel, L. 1870. Symbolae mycologicae. Beiträge zur Kenntniss der Rheinischen Pilze. *Jb. nassau. Ver. Naturk.* 23-24:1-459.
From Gr. *línon*, line, thread + *sporá*, spore, seed, for the thread-like ascospores. Ascomata ostiolate perithecia, formed in host tissues beneath blackened clypeus and surrounded by pseudostroma delimited by blackened region; perithecia single, lying horizontal in leaf, ostiole lateral, curved, extending to leaf surface. Ascomatal wall brown. Asci unitunicate, elongate to cylindrical, 4-8-spored. Ascospores hyaline, filiform or cylindrical, 1-2-several-celled, maturing after overwintering. In living leaves of deciduous trees.

Lipomyces Lodder & Kreger-van Rij
Lipomycetaceae, Saccharomycetales, Saccharomycetidae, Saccharomycetes, Saccharomycotina, Ascomycota, Fungi
Lodder, J., N. J. W. Kreger-van Rij. 1952. *Yeasts, a taxonomic study*, [Edn 1] (Amsterdam):669-670.
From Gr. *lípos*, fat + *mýkes* > L. *myces*, fungus. Vegetative cells often with a large lipid globule, sometimes also a capsule. Culture with a mucose appearance and consistency. Asci formed in various manners. Ascospores oval, brown or amber, smooth or rough, 4-16 or more per ascus. Incapable of fermenting. Isolated from soil.

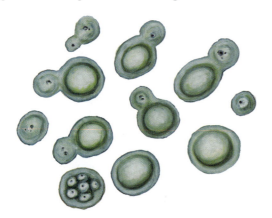

Lipomyces starkeyi Lodder & Kreger-van Rij: vegetative yeast cells, with large lipid globule, an ascus with ascospores, x 2,000. Isolated from soil, Netherlands.

Lirula Darker
Rhytismataceae, Rhytismatales, Leotiomycetidae, Leotiomycetes, Pezizomycotina, Ascomycota, Fungi
Darker, G. D. 1967. A revision of the genera of the Hypodermataceae. *Can. J. Bot.* 45:1399-1444.
From L. *lira*, ridge, furrow + dim. suf. *-ula*, in reference to the appearance of the ascomata when open. Ascomata linear apothecia, hypophyllous, dark brown, immersed, opening by a longitudinal fissure, covering layer of dark pseudoparenchyma tissue; basal layer plectenchymatous, hyaline. Paraphyses simple, filiform, inconspicuous at maturity. Asci clavate, unitunicate, 8-spored, rounded at apex, tapering gradually toward base. Ascospores clavate, hyaline, 1-celled, with a gelatinous sheath. Pycnidia continuous, in groove of upper surface of needle, dark brown. On needles of conifers, causing lesions.

Lobaria (Schreb.) Hoffm.
Lobariaceae, Peltigerales, Lecanoromycetidae, Lecanoromycetes, Pezizomycotina, Ascomycota, Fungi
Hoffman, G. F. 1796. *Deutschl. Fl.*, Zweiter Theil (Erlangen):138.
From NL. *lobus* < Gr. *lobós*, lobe, cluster, the lower end of the ear + L. suf. *-aria*, which indicates connection or possession. Thallus foliose, lobate, with cephalodia, lower surface lacking cyphellae and pseudocyphellae. Photobiont cyanophycea and/or chlorophycea, principally *Protococcus*. On rocks, mosses and moist wood.

Lobaria pulmonaria (L.) Hoffm.: foliose, lobate thallus with apothecia, on old trees in shady woods of *Fagus* and *Quercus*, Europe, x 6.

Lobatopedis P. M. Kirk
Inc. sed., Inc. sed., Inc. sed., Inc. sed., Pezizomycotina, Ascomycota, Fungi
Kirk, P. M. 1979. A new dematiaceous Hyphomycete from leaf litter. *Trans. Br. mycol. Soc.* 73(1):75-79.
From NL. *lobatus*, lobed + L. *pedis*, foot, base, referring to the lobed basal cells of the conidiophores. Colonies amphigenous, effuse, hairy, dark brown to black. Mycelium immersed, composed of pale to mid-brown septate hyphae. Conidiophores macromanetous, mononematous, arising singly from dark brown, radially lobed basal cells, erect to recumbent, straight or flexuous, simple or with one to several short or long acropleurogenous branches, mid to dark brown with prominent septa. Conidiogenous cells integrated, terminal, monoblastic,

determinate. Conidia holoblastic, acrogenous, solitary or catenate, dry, mid to dark brown with prominent septa. On decaying leaves of deciduous trees in England.

Loboa Cif. et al.
Ajellomycetaceae, Onygenales, Eurotiomycetidae, Eurotiomycetes, Pezizomycotina, Ascomycota, Fungi
Ciferri, R. et al. 1956. Taxonomy of Jorge Lobo's disease fungus. Publicações *Inst. Micol. Recife.* 53:1-21.
Name dedicated to the Brazilian physician *Jorge Lobo*, who discovered the disease lobomycosis in an Indian patient in the Amazonas, Brazil. An uncultivated fungal pathogen, which cause a chronic cutaneous and subcutaneous fungal granulomatous disease, known as Lobo's disease, lobomycosis and keloidal blastomicosis, characterized by the development of nodular keloidal and sometimes verrucoid and ulcerated lesions, in which there are large numbers of spherical to lemon-shaped cells, conspicuous and easily seen, particularly on the pinnae, face, upper and lower limbs, and with no involvement of mucous membranes. The fungus reproduces by budding, usually at the end of the cell, and the bud may remain attached to the parent cell by a tubular connection; the tubular attachment of the bud may be persistent, and bead-like chains of cells are commonly seen. The cell is multinucleate, and may bud at one or several points.

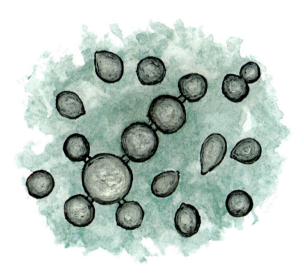

Loboa loboi (Fonseca & Leao) Cif., et al.: yeast cells in human tissues, Brazil, x 1,600.

Loculohypoxylon M. E. Barr
Teichosporaceae, Pleosporales, Pleosporomycetidae, Dothideomycetes, Pezizomycotina, Ascomycota, Fungi
Barr, M. E. 1976. *Hypoxylon grandineum*: a loculoascomycete. *Mycotaxon* 3(3):325-329.
From L. *loculus*, a little place, dim. of *locus*, a place + genus *Hypoxylon* Adans., i.e., an *Hypoxylon* with locules.

Ascomata pseudothecia, small, immersed or erumpent in host periderm. Asci bitunicate, oblong or clavate with four or eight ascospores. Ascospores broadly elliptical, pale to dark brown, with a long, faint germ slit. On dead trunks of *Quercus* spp.

Lolia Abdel-Aziz & Abdel-Wahab
Inc. sed., Inc. sed., Inc. sed., Inc. sed., Pezizomycotina, Ascomycota, Fungi
Abdel-Aziz, F. A., M. A. Abdel-Wahab. 2010. *Lolia aquatica* gen. et sp. nov. (Lindgomycetaceae, Pleosporales), a new coelomycete from freshwater habitats in Egypt. *Mycotaxon* 114:33-42.
From Arabic *loli*, pearl, in reference to the color of the conidiomata. Conidiomata acervular, superficial, pearl white, single or aggregated. Peridium forming textura intricata, hyaline, embedded in gel. Conidiogenesis holoblastic. Conidia unicellular, clavate, cylindrical, ellipsoidal, hyaline, smooth, thin-walled, with basal and apical cellular, tapering, attenuating appendages. On decayed stem of *Phragmites australis* in El Balyana city, Sohag, Egypt.

Lomentospora Hennebert & B. G. Desai
Microascaceae, Microascales, Hypocreomycetidae, Sordariomycetes, Pezizomycotina, Ascomycota, Fungi
Hennebert, G. L., B. G. Desai. 1974. *Lomentospora prolificans*, a new hyphomycete from greenhouse soil. *Mycotaxon* 1(1):45-50.
From E. *loment*, pod constricted between the seeds < L. *lomentum*, bean pod + *spora*, spore, for the appearance of the conidia on the conidiophore rachis. Colonies slow-growing, grayish-brown to dark brown when sporulating. Conidiophores forming conidiogenous cells successively and basipetally on hyphae or on short hyphal branches. Conidiogenous cells sympodial, flask-shaped, narrower at tip, hyaline, producing an apical conidium, then proliferating acropetally to form a new conidium by swelling and budding close to previous conidium, forming a long, delicate, flexuous rachis with conidial attachment scars. Conidia holoblastic, numerous, borne singly, successively at apex of conidigenous cell, ovoid to elliptical, truncate at base, 1-celled, smooth, dry, pale brown. Isolated from soil from mixed forest litter, Heverlee, Belgium.

Lophium Fr.
Mytilinidiaceae, Mytilinidiales, Inc. sed., Dothideomycetes, Pezizomycotina, Ascomycota, Fungi
Fries, E. M. 1818. *Observ. mycol.* (Havniae) (2):345.
From Gr. *lóphe*, genit. *lóphos*, hill, crest, end, top, small mountain + L. dim. suf. *-ium*, for the shape of the hysterothecia, which appear as small elevations, sometimes in the shape of a mollusk shell or wedge, on wood and tree branches. Ascostromata hysterothecia opening by

Lophodermella

a longitudinal furrow, black, carbonaceous, partially immersed at base, superficial or on a subiculum. Asci bitunicate, cylindrical, intermixed with pseudoparaphyses. Ascospores filiform.

Lophium mytilinum (Pers.) Fr.: black, carbonaceous hysterothecia, with the shape of mollusk shell, on wood of tree branch, x 20.

Lophodermella Höhn.
Rhytismataceae, Rhytismatales, Leotiomycetidae, Leotiomycetes, Pezizomycotina, Ascomycota, Fungi
Höhnel, F. von. 1917. System der Phacidiales. *Ber. dt. bot. Ges.* 35(3):416-422.
From genus *Lophodermium* Chevall.+ L. dim. suf. *-ella*. Ascomata apothecia, subhypodermal, short to elongate, circular to elliptical, concolorous with needle surface. Hymenium cupulate; subhymenium thin. Paraphyses simple, filiform. Asci clavate to subcylindric, unitunicate. Ascospores clavate to subbiseriate to fasciculate. Pycnidia small, flask-like.

Lophodermium Chevall.
Rhytismataceae, Rhytismatales, Leotiomycetidae, Leotiomycetes, Pezizomycotina, Ascomycota, Fungi
Chevallier, F. F. 1826. *Fl. gén. env.* (Paris) 1:435.
From Gr. *lóphos*, crest, long hair + *dérma*, skin + *-ium*, L. dim. suf., referring to the filiform or acicular ascospores. Apothecia subcuticular or intraepidermal, similar to hysterothecia, with a longitudinal line of dehiscence. Asci with elongate ascospores grouped in the shape of a raceme or crest that occupies almost all the length of the ascus. Parasitic, causing needle fall in pines.

Lophodermium pinastri (Schrad.) Chevall.: subcuticular apothecia, on pine needle, x 15.

Lophomerum Ouell. & Magasi
Rhytismataceae, Rhytismatales, Leotiomycetidae, Leotiomycetes, Pezizomycotina, Ascomycota, Fungi
Ouellette, G. B., L. P. Magasi. 1966. *Lophomerum*, a new genus of Hypodermataceae. *Mycologia* 58(2):275-280.
From Gr. *lóphos*, crest, mane + *méros*, part, in reference to strands of hyphae that extend beyond the edge of the ruptured epidermis in mature apothecia. Apothecia amphigenous, scattered or crowded, frequently confluent, elliptical, shining black, subcuticular, opening by a longitudinal slot. Basal layer thick, plectenchymatous; covering layer black, pseudoparenchymatous. Hymenium flat, hyaline. Paraphyses filiform, septate, recurved or sometimes branched at maturity. Asci clavate to subcylindric, unitunicate, 8-spored, pore J-. Ascospores hyaline, filiform, phragmosporous at maturity, with a gelatinous sheath. Pycnidia simple, flat; conidia bacillar. On needles of *Abies* and *Picea*, causing lesions and needle cast.

Lophotrichus R. K. Benj.
Microascaceae, Microascales, Hypocreomycetidae, Sordariomycetes, Pezizomycotina, Ascomycota, Fungi
Benjamin, R. K. 1949. Two species representing a new genus of the Chaetomiaceae. *Mycologia* 41(3):347.
From Gr. *lóphos*, tuft, crest, comb + *thríx*, genit. *trichós*, hair. Perithecia with a crest of hairs on apical part of neck. Cellulolytic saprobe on dung, straw and other organic substrates.

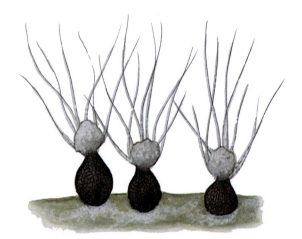

Lophotrichus bartlettii (Massee & E. S. Salmon) Malloch & Cain: cellulolytic perithecia, with hairs in the apical part of the ostiolar neck, x 60.

Lulworthia G. K. Sutherl. (syn. **Halophiobolus** Linder)
Lulworthiaceae, Lulworthiales, Inc. sed., Sordariomycetes, Pezizomycotina, Ascomycota, Fungi
Sutherland, G. K. 1915. Additional notes on marine pyrenomycetes. *Trans. Brit. Mycol. Soc.* 5:259.
This fungus was found on a seaweed known as the bladder wrack of *Lulworth*, a place on the coast of Dorset,

U. K. + -*ia*, ending of generic name. Perithecia semi-immersed, becoming superficial, subglobose to globose, black, coriaceous, solitary, ostiole periphysate; paraphyses lacking. Asci 8-spored, cylindrical, unitunicate, thin-walled, deliquescing early. Ascospores fasciculate, filiform, hyaline, unicellular, smooth-walled, with tapering end chambers, provided with mucilaginous appendages. On various aquatic substrata, usually in marine habitats.Perithecia solitary or gregarious, flask-shaped to elongate-ellipsoid or subcylindrical, membranous to subcarbonaceous, black, fuscous or light colored, often dark above, lighter below; ostiole conical to elongate, cylindrical, straight or undulate, deep fuscous or black. Paraphyses absent. Asci elongate-clavate, 8-spored. Ascospores hyaline, elongate, slender, cylindrical or filamentous, mostly non-septate at first, only becoming septate on germination, at each end a conspicuous hyaline, conoid, or somewhat inflated appendage. On immersed test blocks.

Lulworthia purpurea (I.M. Wilson) T.W. Johnson (syn. **Halophiobolus purpureus** I. M. Wilson): subcarbonaceous, black perithecium, immersed in test blocks, with cylindrical and undulate ostiole, x 200; two filamentous ascospores, each end with inflated appendage, x 570. On wood, Great Britain.

Lutypha Khurana, et al.
Typhulaceae, Agaricales, Agaricomycetidae, Agaricomycetes, Agaricomycotina, Basidiomycota, Fungi
Khurana I. P. S., et al. 1977. *Lutypha sclerotiophila* gen. et sp. nov. (Aphyllophorales) from India. *Trans. Br. mycol. Soc.* 68(3):478-483.
Name formed as an anagram of *Typhula* (Pers.) Fr. Fruitbodies white, branched or simple, with or without a sclerotium, appearing pulverulent or villous, with sparse to profuse branching, irregular or often monaxial, branches cylindrical to slightly flattened, erect or prostrate on plant debris, ultimate branches filiform, terete, with acute sterile tips. Sclerotium yellowish-brown, exposed. Fruitbodies arising exogenously from sclerotium; if without sclerotium, then fruitbodies arising from an effused snow-white mycelium. Hyphal system monomitic; subhymenial hyphae not inflated, lacking clamps; hyphae of branches inflated and with clamps generally in whorls, long-celled; hyphae in stipitate sterile base mostly sclerified, slightly gelatinous, relatively less inflated than those of branches. Hymenium amphigenous, not thickening, generally as loose hypochnoid granules, sometimes continuous in parts of fruitbody. Branches producing loose arachnoid hyphae with hypochnoid basidial clusters along their length. Basidia 4-spored, broadly cylindric to suburniform. Spores subhyaline, thin-walled, non-amyloid, acyanophilous. Sclerotium with a medulla and outer rind. Rind 3-5 cells thick, yellowish-brown, thick-walled. Internal medullary hyphae appearing cellular in section, cells thick-walled, slightly gelatinous. Collected in subtropical forest of Sattal, Nainital, Uttar Pradesh, India.

Lycoperdon Pers.
Agaricaceae, Agaricales, Agaricomycetidae, Agaricomycetes, Agaricomycotina, Basidiomycota, Fungi
Persoon, C. H. 1794. *Nov. pl. gen.* (Florentiae):217.
From Gr. *lýkos*, wolf + *pérdo*, to make wind, for the way the spores leave the mature fructification. Fruitbodies globose, subglobose or pyriform with a mass of spores intermixed with fibers (gleba); internal basal part (subgleba) alveolate sterile tissue. Spores globose, subglobose, or ellipsoidal, pedicellate or nonpedicellate, echinulate, or occasionally, smooth. On decayed wood in grasslands or forests of conifers and oaks.

Lycoperdon nigrescens Pers.: pyriform fructification, on soil among mosses, x 1.

Lylea Morgan-Jones
Inc. sed., Inc. sed., Inc. sed., Inc. sed., Pezizomycotina, Ascomycota, Fungi
Morgan-Jones, G. 1975. Notes on hyphomycetes. VIII. *Lylea*, a new genus. *Mycotaxon* 3(1):129-132.
Named in honor of the plant pathologist *James Albert Lyle*, of Auburn University, Alabama + L. des. -*a*. Colonies effuse, thin, olive-brown. Mycelium partly superficial, partly immersed, composed of branched, septate, pale brown, smooth-walled or minutely verruculose hyphae.

Lyophyllum

Conidiophores micronematous or semi-micronematous, inconspicuous, formed on short, cylindrical hyphal branches. Conidiogenous cells monoblastic, integrated, determinate. Conidia catenate, dry, acrogenous, formed in short, often branched, acropetal chains, simple, mid to dark brown, smooth, 1-7, primary 4-septate. On bark of *Pinus taeda* in Auburn, Alabama, U.S.A.

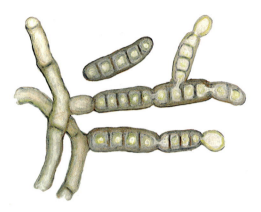

Lyophyllum loricatum (Fr.) Kühner: mature, cespitose basidiocarps, on soil, x 1.

Lylea catenulata Morgan-Jones: conidiophores, with monoblastic conidiogenous cells, and acrogenous, multiseptate conidia, x 375. Collected on bark of *Pinus taeda*.

Lyophyllum P. Karst.
Lyophyllaceae, Agaricales, Agaricomycetidae, Agaricomycetes, Agaricomycotina, Basidiomycota, Fungi
Karsten, P. A. 1881. *Acta Soc. Fauna Flora fenn.* 2(no. 1):29:3.
From Gr. *lýo*, to release, dissolve, separate + *phýllon*, leaf, gill, for having gills only adhered to the stipe but generally not decurrent. Pileus yellowish-brown, grayish, brown, blackish or grayish-white, gills and stipe whitish to yellowish. Basidiocarps small to medium, cespitose, forming large groups, stipe somewhat eccentric. Spores white. Terrestrial in pine and oak forests. All species are edible, of considerable culinary value, but can be confused with toxic species of *Clitocybe* (Fr.) Staude and *Entoloma* (Fr.) Kummer.

Lysurus Fr.
Phallaceae, Phallales, Phallomycetidae, Agaricomycetes, Agaricomycotina, Basidiomycota, Fungi
Fries, E.M. 1823. *Syst. mycol.* (Lundae) 2(2):279, 285.
From Gr. *lýa*, free + *ourós*, tail, with reference to the free arms of the peridium at maturity. Fruitbodies subhypogeous, globose to obovoid, attached by a white mycelial cord, comprising an outer peridium which ruptures irregularly containing an unexpanded receptacle. Endoperidium white, membranous with meridional grooves, retained as a persisting volva. Gleba yellowish-brown, with a foetid odor, covering inner surface of receptacle arms. Receptacle with a cylindric white sterile stipe, surmounted by short, conical tapering arms, initially apically fused but soon free, remaining attached to peridial volva. Spores subhyaline, elliptic to cylindric, smooth. On rotting hay or in rich highly manured soil. Common in Australia and New Zealand, rare elsewhere.

M

Marasmius haematocephalus

M

Macalpinomyces Langdon & Full.
Ustilaginaceae, Ustilaginales, Ustilaginomycetidae, Ustilaginomycetes, Ustilaginomycotina, Basidiomycota, Fungi
Langdon, R. F. N., R. A. Fullerton. 1977. *Macalpinomyces*, a new genus of smut fungi. *Trans. Br. mycol. Soc.* 68(1):27-30.
Named in honor of the Australian mycologist *Daniel McAlpine*, for his significant contributions to knowledge of smut fungi + L. *myces*, fungus. Sori in ovaries, dark. Peridium composed of hyphae associated inter- and intracellularly with cells of pericarp. Spores mostly irregularly polygonoid, sometimes subglobose, spore walls cinnamon to fulvous, smooth or minutely granular. Giant cells mixed with spores, globose, subglobose or ellipsoid, smooth, buff to ochraceous. On *Eriachne* sp. in northern Australia.

Macalpinomyces eriachnes Langdon & Full.: spores, globose or subglobose, but mostly angular, 11-14 μm in greatest dimension, x 730, mixed with giant cells, globose, subglobose or ellipsoid, up to 35 μm in longest dimension, with wall up to 5 μm thick, smooth, buff to ochraceous, multiguttulate with lipid globules, x 1,000. At left, the spike of *Eriachne* sp., x 2, attacked by the smut fungus which produce the teliospores and giant cells.

Macbridella Seaver—see **Byssosphaeria** Cooke
Melanommataceae, Pleosporales, Pleosporomycetidae, Dothideomycetes, Pezizomycotina, Ascomycota, Fungi
Seaver, F. J. 1909. The Hypocreales of North America-II. *Mycologia* 1(5):177-207.
Named in honor of the American mycologist *T. H. Macbride* + L. dim. suf. *-ella*.

Macrohyporia I. Johans. & Ryvarden
Polyporaceae, Polyporales, Inc. sed., Agaricomycetes, Agaricomycotina, Basidiomycota, Fungi
Johansen I., L. Ryvarden. 1979. Studies in the Aphyllophorales of Africa VII. Some new genera and species in the Polyporaceae. *Trans. Br. mycol. Soc.* 77(2):189-199.
From Gr. *makrós* > L. *macros*, large + *hy(phé)* hypha + L. *por(us)*, pore + des. *-ia*, an apparent reference to the large hyphae around the pores. Fruitbodies annual to perennial, resupinate, in small patches to widely effused, brittle to hard when dry, pore surface cream to ochraceous or pale brown, dull, pores 1-5 per mm or larger, context thin. Hyphal system monomitic-dimitic generative hyphae thin-walled, simple-septate, in trama and context up to 20 μm wide, binding hyphae or strongly branched thick-walled generative hyphae dominating in trama and context, inamyloid to weakly amyloid, of large diameter; spores hyaline to pale yellow, thin to weakly thick-walled, subglobose to ellipsoid, non-amyloid. Rarely with sclerotium. On charred wood in Torvoomba, Queensland, Australia.

Macrolepiota Singer
Agaricaceae, Agaricales, Agaricomycetidae, Agaricomycetes, Agaricomycotina, Basidiomycota, Fungi
Singer, R. 1948. New and interesting species of Basidiomycetes. *Pap. Mich. Acad. Sci.* 32:141.
From Gr. *makrós*, big, large + *Lepiota* (Pers.) Gray, due to its similarity to this genus. Pileus generally dark-colored, brown or reddish in young specimens, cuticle cracks at maturity forming distinct scales on light background; center characteristically smooth, dark-colored, up to 30 cm or more high. On soil or remains of wood. Edible.

Macropodia Fuckel—see **Helvella** L.
Helvellaceae, Pezizales, Pezizomycetidae, Pezizomycetes, Pezizomycotina, Ascomycota, Fungi
Fuckel, L. 1870. Symbolae mycologicae. Beiträge der rheinischen Pilze. *Jb. nassau. Ver. Naturk.* 23-24:331.
From Gr. *makrós*, big, large + L. suf. *-podo, -poda, -podia*, derived from Gr. *poús*, genit. *podós*, foot, for the large base that supports the apothecium.

Maculatifrondis K. D. Hyde

Phyllachoraceae, Phyllachorales, Inc. sed., Sordariomycetes, Pezizomycotina, Ascomycota, Fungi

Hyde, K. D., et al. 1996. Fungi associated with leaf spots of palms. *Maculatifrondis aequatoriensis* gen. et sp. nov., with a *Cyclodomus* anamorph, and *Myelosperma parasitica* sp. nov. Mycol. Res. 100(12):1509-1514.

From L. *maculatus*, spotted + *frondis*, leaf, referring to the leaf spots characteristic of the fungus. Ascomata immersed within host palisade and mesophyll cells, ellipsoidal, neck central and emerging through a clypeus. Paraphyses not seen. Asci 8-spored, clavate-saccate, thin-walled, unitunicate, lacking apical structures, early deliquescing. Ascospores 3-4-seriate, oblong-ellipsoidal, hyaline, unicellular, wall roughened, surrounded by a thin persistent mucilaginous sheath. On a palm leaf in Ecuador.

Maculatipalma J. Fröhl. & K. D. Hyde

Valsaceae, Diaporthales, Sordariomycetidae, Sordariomycetes, Pezizomycotina, Ascomycota, Fungi

Fröhlich, J., K. D. Hyde. 1993. *Maculatipalma fronsicola* gen. et sp. nov. causing leaf spots on palm species in north Queensland with descriptions of related genera: *Apioplagiostoma* and *Plagiostoma*. Mycological Research 99(6):727-734.

From L. *maculatus*, spotted + *palma*, palm tree, for the habit of the fungus. Ascomata immersed, cylindrical, with an eccentric, periphysate ostiole curving upwards piercing host surface. Peridium comprising flattened to cuboidal, brown walled cells. Paraphyses few, hypha-like, filiform, septate, tapering distally, dissolving at maturation. Asci 8-spored, broadly cylindrical, unitunicate, apex truncate, with a J-, refractive, subapical ring. Ascospores uniseriate, occasionally biseriate in ascus center, ellipsoidal or obpyriform, hyaline, unequally bicelled.

Maculatipalma frondicola J. Fröhl. & K. D. Hyde: ascomata immersed in a living leaf of palm tree (*Linospadix microcarya*), Queensland, Australia, x 0.66; sagittal section of an ascoma, x 360; bitunicate ascus with eight hyaline, bicelled ascospores, x 500.

Madurella Brumpt

Inc. sed., Sordariales, Sordariomycetidae, Sordariomycetes, Pezizomycotina, Ascomycota, Fungi

Brumpt, M. E. 1905. Sur le mycétome á grains noirs, maladie produite par une mucédinée du genre *Madurella* n. g. Compt.-Rend. Séances Mém. Soc. Biol. 58:997-999.

A name derived from the *Madura* region of India + L. dim. suf. *-ella*, because the majority of the first cases recorded of mycetoma were from this region of India, with the names "Madura foot", maduramycosis or maduromycosis. Phialospores, aleuriospores and sclerotia. In soil. Causing granulomatose, localized infection, usually of the hands and feet, whose abscesses discharge pus with characteristic granules.

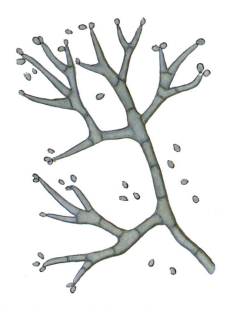

Madurella mycetomatis (Laveran) Brumpt: conidiophore with phialides and phialospores, x 1,000.

Magnohelicospora R. F. Castañeda, et al.

Inc. sed., Inc. sed., Inc. sed., Inc. sed., Pezizomycotina, Ascomycota, Fungi

Castañeda-Ruiz, R. F., et al. 2012. Two new microfungi from Portugal: *Magnohelicospora iberica* gen. & sp. nov. and *Phaeodactylium stadleri* sp. nov. Mycotaxon 121:171-179.

From L. *magnus*, great + Gr. *hélix*, genit. *hélikos*, twisted, coiled + *sporá*, spore, in reference to the large, coiled spores. Colonies on natural substratum effuse, hairy, brown to dark brown. Mycelium superficial, immersed. Conidiophores macronematous, mononematous, erect, smooth or verruculose, brown to dark brown. Conidiogenous cells polyblastic, integrated, sympodial, flattened, indeterminate. Conidial secession schizolytic. Conidia solitary, tightly coiled in three dimensions, doliiform to conical, multi-euseptate, acrogenous to acropleurogenous, brown to dark brown or olivaceous, smooth or verrucose. On decayed leaves in Portugal.

Malassezia

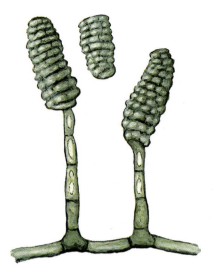

Magnohelicospora iberica R. F. Castañeda, et al.: conidiophore, conidiogenous cells with tightly coiled conidia, x 600.

Malassezia Baill. (syn. **Pityrosporum** Sabour.)
Malasseziaceae, Malasseziales, Inc. sed., Malasseziomycetes, Ustilaginomycotina, Basidiomycota, Fungi
Baillon, E. H. 1889. *Traité Bot. Méd. Crypt.*:234.
Dedicated to the French doctor *L. Malassez* + L. suf. *-ia*, which denotes belonging to. Anascosporogenic yeast, nonfermentative, with ellipsoidal or cylindrical cells, monopolar germination. Inhabits skin of humans and some higher animals causing tinea or pityriasis versicolor and other similar infections.

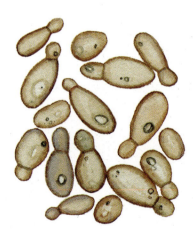

Malassezia furfur (C. P. Robin) Baill.: yeast cells with monopolar germination, on skin of human patient, France, x 1,800.

Malustela Bat. & J. A. Lima— see **Curvularia** Boedijn
Pleosporaceae, Pleosporales, Pleosporomycetidae, Dothideomycetes, Pezizomycotina, Ascomycota, Fungi
Batista, A. C., et al. 1960. Dois novos deuteromycetes de air atmosferico. *Publicações Inst. Micol. Recife* 263:1-18.
From L. *malus*, mast, pole + *tela*, cloth, something woven, in reference to the erect synnemata covered by a layer of fertile hyphae.

Manoharachariella Bagyan., et al.
Inc. sed., Inc. sed., Inc. sed., Dothideomycetes, Pezizomycotina, Ascomycota, Fungi
Bagyanarayana, G., et al. 2009. *Manoharachariella*, a new dematiaceous hyphomycetous genus from India. *Mycotaxon* 109:301-305.
Named in honor of the Indian mycologist *C. Manoharachary* + connective *-i-* + L. dim. suf. *-ella*. Colonies effuse, thin, pale brown to mid-brown, hairy, mycelium immersed, hyphae brown, smooth. Stroma none, setae and hyphopodia absent. Conidiophores macronematous, mononematous, sparsely branched, branches loosely fasciculate, arising laterally and apically from immersed mycelium, erect, usually flexuous, septate, septa few, hyaline to subhyaline. Conidiogenous cells acroauxic, monoblastic, integrated, cylindrical, hyaline to subhyaline. Conidia solitary, dry, acrogenous, acropleurogenous, simple, doliiform, obpyriform, ellipsoidal, apiculate, smooth, dictyoseptate, tiered, mid-brown to dark-brown to blackish-brown, apical and basal tiers hyaline to subhyaline. On unidentified twigs, Darakonda, Andhra Pradesh, India.

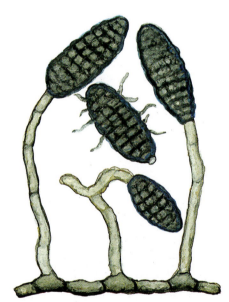

Manoharachariella indica Rajeshk. & S. K. Singh: conidiogenous cells with blackish-brown, ellipsoidal, dictyoseptate conidia, x 300. Isolated from decayed leaves.

Marasmius Fr.
Marasmiaceae, Agaricales, Agaricomycetidae, Agaricomycetes, Agaricomycotina, Basidiomycota, Fungi
Fries, E. M., 1836. *Fl. Scan.*:339.
From Gr. *marasmós*, withered, from *maraíno*, withered, to dry + L. suf. *-ius*, which indicates connection, similarity or something characteristic, referring to the characteristic that these fungi have of drying, instead of decaying. Fruitbodies small, whitish or reddish, with little context in pileus, distant gills. Stipe usually

Margaretbarromyces

central, thin, cartilaginous, rigid like a wire, drying rapidly and reviving again when rehydrated. Spores white. Lignicolous or humicolous.

Marasmius haematocephalus (Mont.) Fr.: lignicolous, reddish basidiomata, with distant gills, x 2. On fallen leaves, Brazil.

Maravalia Arthur (syn. **Angusia** G. F. Laundon)
Chaconiaceae, Pucciniales, Inc. sed., Pucciniomycetes, Pucciniomycotina, Basidiomycota, Fungi
Arthur, J.C. 1922. New species of Uredineae—XV. *Bot. Gaz.* 73:60.
The type specimen was found on leaves of *Pithecellobium latifolium* in *Maraval* Valley, a northern suburb of Port of Spain, Republic of Trinidad and Tobago + L. *-ia*, suf. for terminating the scientific genera. Spermogonia subcuticular or subepidermal with flat hymenia. Aecia usually aggregate around spermogonia, erumpent, spores pedicellate, verrucose, echinulate, or covered by spines on mound-like projections. Uredinia similar to aecia, but more scattered; spores pedicellate, similar to aeciospores in the same species. Basidiosori subepidermal in origin, soon erumpent, often replacing uredinia and/or aecia, pale, often compact and waxy; probasidia pedicellate, one-celled, thin-walled, rarely slightly thickened at subapical portions, mostly hyaline, rarely pale yellowish-brown, formed successively on laterally free basidiogenous cells; pedicels hyaline, persistent, sometimes hygroscopic; metabasidia formed by continuous apical elongation of probasidia.

Marchandiobasidium Diederich & Schultheis—see
Erythricium J. Erikss. & Horstam
Corticiaceae, Corticiales, Inc. sed., Agaricomycetes, Agaricomycotina, Basidiomycota, Fungi
Diederich, P., et al. 2003. *Marchandiobasidium aurantiacum* gen. sp. nov., the teleomorph of *Marchandiomyces aurantiacus* (Basidiomycota, Ceratobasidiales). *Mycol. Res.* 107(5):523-527.
Named after the genus *Marchandiomyces*, of which it is the sexual morph.

Marchandiomyces Diederich & D. Hawksw.
Corticiaceae, Corticiales, Inc. sed., Agaricomycetes, Agaricomycotina, Basidiomycota, Fungi
Diederich, P. 1990. New or interesting lichenicolous fungi 1. Species from Luxembourg. *Mycotaxon* 37:297-330.
Named in honor of *Louis Marchand* + L. *myces*, fungus, who was the first mycologist and lichenologist in Luxembourg. Colonies forming subspherical sclerotia, sometimes developing into convex sporodochia, composed of hyaline cells. Conidiophores semi-macronematous, erect, branched at base, hyaline, thick, smooth walled. Conidigenous cells monoblastic integrated, terminal, simple, obpyriform to obovoid, distinctly truncate at base, hyaline, with a thick, finely verrucose wall. Parasitic on lichen thalli.

Margaretbarromyces Mindell, et al.
Inc. sed., Pleosporales, Pleosporomycetidae, Dothideomycetes, Pezizomycotina, Ascomycota, Fungi
Mindell, R. A., et al. 2007. *Margaretbarromyces dictyosporus* gen. sp. nov.: a permineralized corticolous ascomycete from the Eocene of Vancouver Island, British Columbia. *Mycol. Res.* 111(6):681.
Named in honor of the American mycologist *Margaret E. Barr* + L. *myces*, fungus, for her numerous contributions to the taxonomy of the loculate fungi. Ascomata uniloculate, solitary, erumpent through periderm, pyriform, tapering to ostiole. Ascostromata pseudoparenchymatous, two layered, inner layer stratified, compressed, wide; outer layer irregular, thickened around ostiole. Peridial hyphae septate, highly branched. Asci elongate, attachment basal. Ascospores uniseriate-biseriate, multicellular, unconstricted, up to 24 times asymmetrically dictyoseptate. On fossilized rocks on Vancouver Island, British Columbia, Canada.

Margaretbarromyces dictyosporus Mindell, et al.: sagittal section of an ascoma of this fossil fungus, on fragment of a peridem plant, from Eocene, x 100.

Marssonina

Marssonina Magnus—see **Diplocarpon** F. A. Wolf
Dermateaceae, Helotiales, Leotiomycetidae, Leotiomycetes, Pezizomycotina, Ascomycota, Fungi
Magnus, P. 1906. Notwendige Umänderung des Namens der Pilzgattung *Marssonia* Fisch. *Hedwigia* 45:88-91.
Like the genus *Marssonia* Fisch., this was also dedicated to the German botanist *Theodor Friedrich Marsson* + Gr. suf. *-ina*, which indicates similarity or possession.

Martensella Coem.
Kickxellaceae, Kickxellales, Inc. sed., Inc. sed., Kickxellomycotina, Zygomycota, Fungi
Coemans, E. 1863. Quelques Hyphomycètes nouveaux. 1. *Mortierella polycephala* et *Martensella pectinata*. *Bull. Acad. R. Sci. Belg., Cl. Sci.*, sér. 2 15:536-544.
Dedicated to the French mycologist *Martens* + L. dim. suf. *-ella*. Sporocladia septate forming pseudophialides on lower surface. Unispored sporangiola (merosporangia) borne singly on each pseudophialide. Parasitic on basidiocarps of *Corticium* Pers.

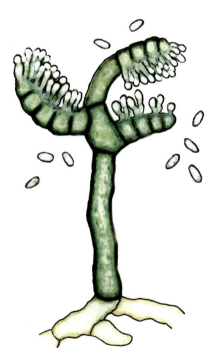

Martensella corticii Thaxt. ex Linder: septate sporocladia with pseudophialides that produce unispored sporangiola, x 550.

Martininia Dumont & Korf
Sclerotiniaceae, Helotiales, Leotiomycetidae, Leotiomycetes, Pezizomycotina, Ascomycota, Fungi
Dumont, K. P., R. P. Korf. 1970. Nomenclatural notes. VI. A new name, *Martininia*, to replace *Martinia* (Sclerotiniaceae). *Mycologia* 62:608.
Named for the discoverer of the fungus, the American mycologist *George W. Martin* + L. des *-ia*. Apothecia arising singly or several from minute, hemispherical sclerotia on surface of substratum, thin, membranous, shallow cup-shaped, stipitate. Hymenium olivaceous to olive-brown when spores mature. Stem long, slender, hair-like. Asci 8-spored. Ascospores simple, ellipsoid, olive-brown. On tree bark and wood in Panama.

Masoniella G. Sm.—see **Microascus** Zukal
Microascaceae, Microascales, Hypocreomycetidae, Sordariomycetes, Pezizomycotina, Ascomycota, Fungi
Smith, G. 1952. *Masoniella* Nom. Nov. *Trans. Br. mycol. Soc.* 35(3):237.
Named in honor of the English mycologist *E. W. Mason* + L. dim. suf. *-ella*, for his work on Hyphomycetes.

Massospora Peck
Entomophthoraceae, Entomophthorales, Inc. sed., Inc. sed., Entomophthoromycotina, Zygomycota, Fungi
Peck, C. H. 1878. *Massospora cicadina* n. g. et sp. *Ann. Rep. N.Y. St. Mus. nat. Hist.* 31:44.
From Gr. *másso*, to crush, grind + *sporá*, spore. Spores not forcibly discharged from sporangiophores. Parasitizes and destroys all of the interior of the body of the host insects (grasshoppers and locusts), which on breaking release powdery masses of spores mixed with the remains of the destroyed tissues.

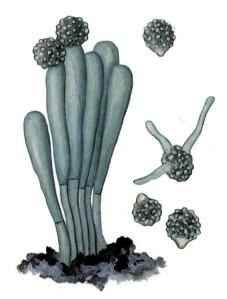

Massospora cicadina Peck: conidiophores with tuberculate conidia, x 1,200.

Mastigobasidium Golubev
Leucosporidiaceae, Leucosporidiales, Inc. sed., Microbotryomycetes, Pucciniomycotina, Basidiomycota, Fungi
Golubev, W. I. 1999. *Mastigobasidium*, a new teleomorphic genus for the perfect state of ballistosporous yeast *Bensingtonia intermedia*. *Int. J. Syst. Bacteriol.* 49 (3):1301-1305.

Megalonectria

From Gr. *mástix*, genit. *mástigos*, whip + L. *basidium*, small base. Colonies whitish to cream-colored. Cells oval, reproducing by budding and ballistospores. Cell wall ultrastructure of basidiomycetous type. Pseudomycelium and hyphae without clamp connections. True mycelium with clamp connections and teliospores arise after conjugation of a compatible mating pair. Teliospore germinates to produce several long aseptate hyphae on apices with clavate, usually curved, phragmometabasidia. Basidiospores lateral, terminal, ellipsoidal or bacilliform, single or in clusters, sessile or on pegs, germinate by budding. Urease and D-glucuronate-positive. Fermentation absent. Major ubiquinone system: Q-9. Xylose absent in extracellular polysaccharides. Nitrate not assimilated.

Mazosia A. Massal.
Roccellaceae, Arthoniales, Arthoniomycetidae, Arthoniomycetes, Pezizomycotina, Ascomycota, Fungi
Massalongo, A. B. 1854. *Geneac. lich.* (Verona) 9.
From Gr. *mazós*, breast, teat, wet nurse + L. suf. *-ia*, which denotes quality or state of a thing, probably for the appearance of the surface of the thallus. Thallus crustous, with round or lirelliform pseudothecia with an excipular margin. Spores hyaline, with transverse septa. On trees and rocks in tropical regions.

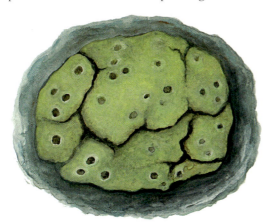

Mazosia ocellata (Nyl.) R. C. Harris: crustous thallus with lirelliform pseudothecia, on bark tree, x 1.

Megacapitula J. L. Chen & Tzean
Inc. sed., Inc. sed., Inc. sed., Inc. sed., Pezizomycotina, Ascomycota, Fungi
Chen, J. L., S. S. Tzean. 1993. *Megacapitula villosa* gen. et. sp. nov. from Taiwan. *Mycol. Res.* 97(3):347-350.
From Gr. *mégas*, great + L. *capitulum*, head, with reference to the large conidia. Mycelium of branched, septate, smooth or roughened, hyaline or pigmented hyphae. Conidiophores micronematous, semimacronematous, mononematous, simple or branched, brown, smooth or roughened. Conidiogenous cells integrated, terminal, lateral or rarely intercalary, determinate. Conidia holoblastic, solitary, ovoid, obclavate, ellipsoidal or obpyriform, muriform, pigmented, with densely packed, branched or unbranched, hair-like appendages at apex. On decayed petioles of fallen broadleaf trees in Chaiyi, Taiwan.

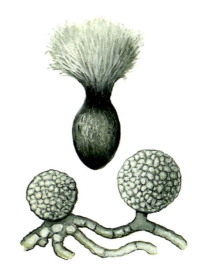

Megacapitula villosa J. L. Chen & Tzean: conidiophore, with young, pigmented conidia, and mature conidium with hair-like apical appendages, x 130.

Megachytrium Sparrow
Cladochytriaceae, Chytridiales, Chytridiomycetidae, Chytridiomycetes, Inc. sed., Chytridiomycota, Fungi
Sparrow, F. K., Jr. 1931. A note on a new chytridiaceous fungus parasitic in *Elodea*. *Occas. Pap. Boston Soc. Nat. Hist.* 8:9-10.
From Gr. *mégas*, large + *chytrís*, pot, earthen jar, flower vase, receptacle + L. dim. suf. *-ium*. Thallus polycentric, well developed. On leaves of *Elodea*.

Megachytrium westonii Sparrow: polycentric thallus (with many zoosporangia), developed in tissues of the host plant, x 600.

Megalonectria Speg.—see **Nectria** (Fr.) Fr.
Nectriaceae, Hypocreales, Hypocreomycetidae, Sordariomycetes, Pezizomycotina, Ascomycota, Fungi
Spagazzini, C. 1881. Fungi Argentini. *Anal. Soc. cient.*

Megasporoporia

argent. 12(5):208-227.
From Gr. *mégas*, fem. *mégale*, great, large + genus *Nectria* (Fr.) Fr.

Megasporoporia Ryvarden & J.E. Wright
Polyporaceae, Polyporales, Inc. sed., Agaricomycetes, Agaricomycotina, Basidiomycota, Fungi
Ryvarden, L., et al. 1982. *Megasporoporia*, a new genus of resupinate polypores. *Mycotaxon* 16(1):172-182.
From Gr. *méga*, large + *spóro*, spore + *poría*, pores, with large spores and pores. Fruitbodies resupinate, pores large, angular to round, pore surface cream, greyish to pale brown or cinnamon, context usually thin, white to cream or pale brown, generative hyphae with clamps, skeletal hyphae thick-walled, dextrinoid. Spores cylindrical, thin-walled, large. Crystals usually present in subhymenium and context. On wood, producing white rot. Generally in the tropics as a pantropical genus in America, Africa and Asia.

Melampsora Castagne
Melampsoraceae, Pucciniales, Inc. sed., Pucciniomycetes, Pucciniomycotina, Basidiomycota, Fungi
Castagne, J. L. M. 1843. Observations sur Quelques Plantes Acotylédonées de la Famille des Urédinées et dans les sous-tribus des Némasporées et des Aecidinées recueillies dans le départment des Bouches-du-Rhône. *Observ. Uréd.* 2:1-24.
From NL. *melampsora* < Gr. *mélas*, black + *psóra*, itch, scab, for the crust-like layer of colored teliospores. Pycnia subepidermal or subcuticular, without paraphyses. Aecia when present foliicolous, with rudimentary peridium or none, orange-yellow when fresh. Aeciospores globoid or ellipsoid, with verrucose walls. Uredinia when present erumpent, pulverulent, with a delicate, evanescent peridium. Urediniospores borne singly on pedicels, globoid or ellipsoid, interspersed with capitate paraphyses more slender at periphery of sorus. Telia subcuticular or subepidermal, forming crusts of a single layer of spores. Teliospores 1-celled, adhering laterally, with colored walls.

Melampsorella J. Schröt.
Pucciniastraceae, Pucciniales, Inc. sed., Pucciniomycetes, Pucciniomycotina, Basidiomycota, Fungi
Schröter, J. 1874. *Melampsorella*, eine neue Uredineen-Gattung. *Hedwigia* 13:81-85.
From genus *Melampsora* Castagne + L. dim. suf. *-ella*. Pycnia inconspicuous, without paraphyses. Aecia hypophyllous, conspicuous, bullate or cylindric, dehiscent by irregular rupture. Aeciospores ellipsoid, with orange-yellow contents, thin, verrucose wall. Uredinia hypophyllous, somewhat bullate, opening by central pore, peridium delicate, scarcely noticeable. Urediniospores ellipsoid or obovate having bright yellow contents and thin, sparsely echinulate walls. Telia hypophyllous, forming large whitish or pinkish areas. Teliospores loosely grouped within epidermal cells, globoid, usually 1-celled, with thin, hyaline walls.

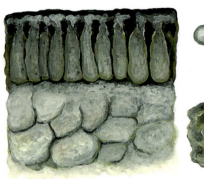

Melampsora caprearum Thüm.: sagittal section of subcuticular telium, forming a black crust of a single layer of teliospores, in willow leaf, x 450, and mature teliospore producing the basidium with four basidiospores, x 1,200.

Melanconium Link
Melanconidaceae, Diaporthales, Sordariomycetidae, Sordariomycetes, Pezizomycotina, Ascomycota, Fungi
Link, J. H. F. 1809. Observationes in ordines plantarum naturales. *Mag. Gesell. naturf. Freunde, Berlin* 3(1-2):9.
From Gr. *mélas*, genit. *mélanos*, black + *kónis*, dust + L. dim. suf. *-ium*, referring to the fine black or brown spores. Acervuli lenticular, subepidermal or peridermic, separate or confluent, often with extended conidial masses. Sexual morph melanconis-like. Pathogenic on woody plants.

Melanconium ershadii Kunze: surface view of conidiomata on the branch of *Pterocarya rhoifolia* in Iran, x 5; conidiophores with conidiogenous cells (annellidic) giving rise to young and mature conidia, obpyriform, x 300.

Melanelixia O. Blanco, et al.
Parmeliaceae, Lecanorales, Lecanoromycetidae, Lecanoromycetes, Pezizomycotina, Ascomycota, Fungi
Blanco, O., et al. 2004. *Melanelixia* and *Melanohalea*, two new genera segregated from *Melanelia* (Parmeliaceae)

based on molecular and morphological data. *Mycol. Res.* 108 (8):873-884.

From genus *Melanelia* Essl. + dedicated in honor of the Australian lichenologist *John A. Elix* for his immense contributions to lichen systematics and chemistry, especially in Parmeliaceae. Thallus foliose, loosely to moderately adnate; lobes plane to concave, flat, short, apices rounded, eciliate; upper surface olive-green to dark brown, smooth to rugose, maculate or not, lacking pseudocyphellae, with or without soredia, isidia and cortical hairs; upper cortex paraplectenchymatous, covered by an epicortex with either dispersed pores or fenestrations; cell walls containing isolichenan; medulla white to pale yellow or occasionally orange in lower parts; lower cortex flat, smooth, dark brown to black; rhizines simple with white tips. Ascomata apothecial, laminal, sessile to subpedicellate, disc imperforate, concave, becoming plane with age, pale to dark brown, amphithecium commonly maculate and with an abundantly fenestrated or pored epicortex. Asci elongate, clavate, lecanora-type, apically thickened, without an internal apical beak, 8-spored. Ascospores ellipsoid to ovoid, hyaline, thin-walled, simple. Conidiomata pycnidial, immersed, laminal. Conidia cylindrical to fusiform, simple, hyaline.

Melanoderma B. K. Cui & Y. C. Dai

Polyporaceae, Polyporales, Inc. sed., Agaricomycetes, Agaricomycotina, Basidiomycota, Fungi

Cui, B.-K., et al. 2011. *Melanoderma microcarpum* gen. et sp. nov. (Basidiomycota) from China. *Mycotaxon* 116:295-302.

From Gr. *mélano*, black + *dérma* skin; referring to the black crust on the pileal surface. Basidiocarps perennial, pileate to effused-reflexed, pileus circular to irregularly formed. Pileal surface black when fresh, color unchanged when dry, concentrically zonate, glabrous; margin obtuse. Pore surface white when fresh, cream buff when dry; pores circular, dissepiments thick, entire. Context cream-buff, woody hard. Tubes cream-buff, woody hard, stratified. Hyphal system dimitic; generative hyphae bearing clamp connections; skeletal hyphae dextrinoid, CB+; tissue unchanged in KOH. Generative hyphae clamped, scanty, hyaline, thin-walled, unbranched; skeletal hyphae dominant, thick-walled to subsolid, usually branched, strongly interwoven. Cystidia clavate to ventricose, hyaline, thin-walled, usually apically encrusted; cystidioles clavate, hyaline, thin-walled. Basidia clavate, with four sterigmata, basal clamp connections; basidioles similar in shape to basidia, but slightly smaller. Rhomboid crystals frequently present in trama and hymenium. Basidiospores cylindrical, hyaline, thin-walled, smooth, IKI-, CB-. On fallen angiosperm trunk in Mangshan Forest Park, Hunan Province, Yizhang County, China.

Melanohalea O. Blanco, et al.

Parmeliaceae, Lecanorales, Lecanoromycetidae, Lecanoromycetes, Pezizomycotina, Ascomycota, Fungi

Blanco, O., et al. 2004. *Melanelixia* and *Melanohalea*, two new genera segregated from *Melanelia* (Parmeliaceae) based on molecular and morphological data. *Mycol. Res.* 108:873- 884.

From genus *Melanelia* Essl., and in honor of the American lichenologist *Mason E. Hale* for his contributions to our knowledge of the Parmeliaceae. Thallus foliose, loosely to moderately adnate; lobes plane to concave, flat, short, apices rounded, eciliate; upper surface olive-green to dark brown, smooth to rugose, emaculate, commonly pseudocyphellate on warts or on tips of isidia, with or without soredia and isidia; upper cortex paraplectenchymatous, epicortex not pored; cell walls containing isolichenan; medulla white; lower surface flat, smooth, pale brown to black; rhizines simple. Ascomata apothecial, laminal, sessile to subpedicillate; disc imperforate, concave, becoming convex with age, brown, amphithecium with pseudocyphellate papillae, without maculae. Asci elongate, clavate, lecanora-type, apically thickened, without an internal apical beak, 8-32 spored. Ascospores globose to ovoid or ellipsoid, thin-walled, hyaline. Conidiomata pycnidial, immersed, laminal. Conidia cylindrical to fusiform, simple, hyaline.

Melanoleuca Pat.

Tricholomataceae, Agaricales, Agaricomycetidae, Agaricomycetes, Agaricomycotina, Basidiomycota, Fungi

Patouillard, N. T. 1897. *Catalogue Raisonné des Plantes Cellulaires de la Tunisie* (Paris):22.

From Gr. *mélas*, genit. *mélanos*, black + *leukós*, white, referring to the white color of the gills and the generally dark color of the pileus and stipe. Basidiomata solitary or cespitose, small to medium, opaque, gills and spores white. Stipe semicartilaginous, slender, solid, firm, longitudinally striate. Terricolous, common in forests and grasslands.

Melanoleuca melaleuca (Pers.) Murrill: terricolous basidiomata, with dark-colored stipe and pileus, and white gills, x 0.5.

Melanophoma

Melanophoma Papendorf & J. W. du Toit
Inc. sed., Inc. sed., Inc. sed., Inc. sed., Pezizomycotina, Ascomycota, Fungi
Papendorf, M. C., J. W. du Toit. 1967. *Melanophoma*: a new genus of the Sphaeropsidales. *Trans. Br. mycol. Soc.* 50(3):503-506.
From Gr. *mélano*, black + genus *Phoma* Fr., referring to the pigmented conidia. Pycnidia solitary, globose to flask-shaped, membranous, pale brown, ostiole simply impressed or papillate; phialides globose to broadly flask-shaped with short or inconspicuous neck, hyaline; spores exogenous, single, unicellular, ovoid, broad-elliptical or subglobose, smooth or asperulate, thick-walled, pigmented, exosporium hyaline, membranous. Isolated from litter and surface soil of a mixed *Acacia karroo* community in Potchefstroom, South Africa.

Melanoporia Murrill
Polyporaceae, Polyporales, Inc. sed., Agaricomycetes, Agaricomycotina, Basidiomycota, Fungi
Murrill, W. A. 1907. Polyporaceae, Part 1. *N. Amer. Fl.* (New York) 9(1):14.
From Gr. *mélas*, genit. *mélanos*, black + *porós* > L. *porus*, pore + L. des. *-ia*, which indicates possession, referring to the dark-colored surface of the pores. Fruitbodies resupinate, perennial, purplish-black to fuliginous; pores small, hyphal system dimitic, generative hyphae afibulate, skeletal hyphae pale to dark brown; cystidia none. Spores ellipsoid, smooth. Causing brown rot in North America.

Melanospora Corda
Ceratostomataceae, Melanosporales, Hypocreomycetidae, Sordariomycetes, Pezizomycotina, Ascomycota, Fungi
Corda, A. C. J. 1837. *Icon. fung.* (Prague) 1:24.
From Gr. *mélas*, genit. *mélanos*, black + *sporá*, spore. Perithecia ostiolate, globose to subglobose, superficial, immersed, or erumpent on host, golden-yellow or blackish, covered with hyaline hairs. Perithecial wall of somewhat flattened, angular, pseudoparenchymatous cells. Ostiolar neck cylindrical, with or without a crown of hyaline setae. Centrum of thin-walled pseudoparenchymatous cells. Asci clavate to cylindrical-fusiform, thin-walled, undifferentiated at apex, evanescent, 8-spored. Ascospores 1-celled, ellipsoid to ellipsoid-fusiform, brown, with faintly pitted walls and two small, terminal germ pores. At maturity ascospores extruded through ostiole, collecting in a cluster. A phialidic state has been reported. Common in soil, on bark; also isolated from seedlings with a *Fusarium* sp., on which it may be parasitic.

Meliola Fr.
Meliolaceae, Meliolales, Inc. sed., Sordariomycetes, Pezizomycotina, Ascomycota, Fungi
Fries, E. M. 1825. *Syst. orb. veg.* (Lundae) 1:111.
From Gr. *mélon*, apple, melon, quince, or other similar fruit + L. dim. suf. *-ola*, *-iola*, for the appearance of the fructifications. Mycelium black. Ascomata sphaeroidal, non-ostiolate with setae. Conidia are lacking. Black mildew or sooty mold on sugary exudate of insects on living leaves of trees and shrubs (e.g., citrus).

Melanospora zamiae Corda: yellowish perithecium, with semi-translucid wall, containing dark brown ascospores, on soil, x 190.

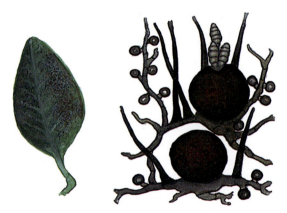

Meliola marthomaensis Jac. Thomas, et al.: non-ostiolate ascomata on leaf of *Hymenodictyon obovatum*, x 1, discharging ascospores, with dark setae and mucronate hyphopodia of this sooty mold on host plant, x 200.

Meloderma Darker
Rhytismataceae, Rhytismatales, Leotiomycetidae, Leotiomycetes, Pezizomycotina, Ascomycota, Fungi
Darker, G. D. 1967. A revision of the genera of the Hypodermataceae. *Can. J. Bot.* 45:1399-1444.
From Gr. *mélas*, black + *dérma*, skin, referring to the black ascomata. Ascomata apothecia, elliptical, often in a row, black, usually surrounded by a grayish zone, subepidermal, subcuticular near opening. Outer cells of opening hyaline, with a row of swollen cells.

Covering layer dark, pseudoparenchymatous; basal layer plectenchymatous. Hymenium flat. Paraphyses initially simple, filiform, later becoming hooked, swollen, branched at tip. Asci subcylindrical, unitunicate. Ascospores bacilliform, subclavate to fusiform, with a gelatinous sheath. Pycnidia simple, flat.

Melogramma Fr.
Melogrammataceae, Xylariales, Xylariomycetidae, Sordariomycetes, Pezizomycotina, Ascomycota, Fungi
Fries, E.M. 1849. Summa vegetabilium Scandinaviae. 2:259-572.
From Gr. *mélas*, black, dark + *grámma*, a mark, line, writing. Stroma depressed-pulvinate, cinereous then black, the surface mammillate by the numerous subjacent perithecia. Perithecia small, ovoid, papillate, lying in a single layer. Asci cylindric, with a short stalk, octosporous, paraphysate; spores fusiform, slightly curved, 3-septate, yellow-brown. On dead branches.

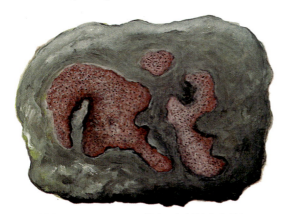

Melogramma gyrosum (Schwein.) Tul. & C. Tul.: perithecial ascomata, on eucalyptus wood, x 500.

Meria Vuill.—see **Rhabdocline** Syd.
Hemiphacidiaceae, Helotiales, Leotiomycetidae, Leotiomycetes, Pezizomycotina, Ascomycota, Fungi
Vuillemin, P. 1896. Les hypostomaceés, nouvelle famille des champignons parasites. *C. r. hebd. Séanc. Acad. Sci., Paris* 122:546.
Dedicated to the Welsh Professor *E. Mer* + L. suf. *-ia*, which denotes pertaining to.

Meripilus P. Karst.
Meripilaceae, Polyporales, Inc. sed., Agaricomycetes, Agaricomycotina, Basidiomycota, Fungi
Karsten, P. A. 1882. Rysslands, Finlands och den Skandinaviska halföns Hattsvampar. *Bidr. Känn. Finl. Nat. Folk* 37:33.
From Gr. *meríso*, to divide, to split + *pílos*, hat > L. *pilus*. Fructifications annual, large, pileate, of numerous imbricate fan-shaped to spathulate, pilei arising from a common base or short stipe. Pileus brown with radial lines, smooth, with concentric zones; pore surface white, darkening when touched or dried, pores entire, small, tubes white, short; hyphal system monomitic; generative hyphae thin- to thick-walled, smooth, hyaline, simple septate; spores subglobose, smooth, thin-walled. On deciduous wood or close to stumps causing a white rot, cosmopolitan.

Meruliporia Murrill
Coniophoraceae, Boletales, Agaricomycetidae, Agaricomycetes, Agaricomycotina, Basidiomycota, Fungi
Murrill, W. A. 1942. Florida resupinate polypores. *Mycologia* 34(5):596.
From genus *Merulius* Fr. + L. *poria*, in reference to the merulioid hymenophore. Basidiocarps annual, resupinate, with mycelial mats and rhizomorphs, pore surface pale buff to grey when fresh, darkening to nearly black on drying, merulioid; tube layer darkening on drying; clamp connections present, hyphal system monomitic. Basidia narrowly clavate; basidiospores broadly ellipsoid, brown at maturity, dextrinoid. Causing brown rot in structural timber.

Merulius Fr.—see **Phlebia** Fr.
Meruliaceae, Polyporales, Inc. sed., Agaricomycetes, Agaricomycotina, Basidiomycota, Fungi
Fries, E. M. 1821. *Syst. mycol.* (Lundae) 1:326.
From L. *merula*, blackbird, thrush, European bird of black or brown color + suf. *-ius*, which indicates connection or similarity.

Metabourdotia L. S. Olive
Inc. sed., Auriculariales, Inc. sed., Agaricomycetes, Agaricomycotina, Basidiomycota, Fungi
Olive, L. S. 1957. Two new genera of Ceratobasidiaceae and their phylogenetic significance. *Am. J. Bot.* 44 (5):429-435.
From Gr. *metá*, afterwards, from a different position + genus *Bourdotia* Bres. & Torrend, for the apparent similarity with this genus. Fructifications resupinate, waxy-pruinose with gleocystidia. Basidia claviform, wide, septate in apical part, with incomplete septa in basal part; basidiospores germinate repetitively.

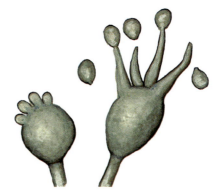

Metabourdotia tahitiensis L. S. Olive: claviform basidia with basidiospores, x 2,000. Collected in Society Is., French Poly-

Metarhizium

Metarhizium Sorokin
Clavicipitaceae, Hypocreales, Hypocreomycetidae, Sordariomycetes, Pezizomycotina, Ascomycota, Fungi
Sorokin, N. V. 1879. *Veg. Parasitenk. Mensch Tieren* 2:268-291.
Prob. < L. *meta*, a conical or pyramidal column at the ends of the Roman circus, pillar + Gr. *rhíza*, root + L. dim. suf. *-ium*. Conidiophores simple or aggregated, profusely branching in a penicillate manner, terminating in phialides producing catenulate conidia that adhere in aggregated, bright green, olive-brown or purple columns. Causing green muscardine disease of insects, used in biological control.

Metarhizium anisopliae (Metschn.) Sorokin: penicillate conidiophores, with phialides producing bright green-colored conidia, on agar, x 375.

Metschnikowia T. Kamienski (syns. **Chlamydozyma** Wick., **Monosporella** Keilin)
Metschnikowiaceae, Saccharomycetales, Saccharomycetidae, Saccharomycetes, Saccharomycotina, Ascomycota, Fungi
Kamienski, T. 1889. Notice préliminaire sur l'espèce de *Metschnikowia* (*Monospora* Metsch.). *Trudy S. Petersb. Obschch. Est. Otd. Bot.* 30 (1):363-364.
Dedicated to Russian microbiologist *E. Metschnikoff* + L. suf. *-ia*, which denotes pertaining to. Yeast with multilateral budding, pseudomycelium rudimentary or absent. Asci clavate, spheropedunculate or ellipsoid-pedunculate, with 1-2 acicular ascospores. Ascospores without flagelliform appendages. Chlamydospores with a thick, double wall, which can function as asci. Undergoes protosexuality in which haploid cells form from diploid cells in culture without producing ascospores. Capable of fermenting various sugars. Some species are parasites of invertebrates, others are free-living in aquatic and terrestrial habitats.

Meyerozyma Kurtzman & M. Suzuki
Debaryomycetaceae, Saccharomycetales, Saccharomycetidae, Saccharomycetes, Saccharomycotina, Ascomycota, Fungi
Kurtzman, C. P., M. Suzuki. 2010. Phylogenctic analysis of ascomycete yeasts that form coenzyme Q-9 and the proposal of the new genera *Babjeviella*, *Meyerozyma*, *Millerozyma*, *Priceomyces*, and *Scheffersomyces*. *Mycoscience* 51(1):2-14.
Named in honor of the American mycologist *Sally A. Meyer* + connective *-o-* + Gr. *zýme*, yeast, for her pioneering studies in the molecular relatedness of *Candida* species. Cell division by multilateral budding on a narrow base. Cells ovoid to elongate. Pseudohyphae often abundant, true hyphae are absent. Asci usually unconjugated, evanescent or persistent, with 1-4 ascospores. Ascospores hat-shaped or ovoid, possibly ornamented with a narrow ledge. Heterothallic or possibly homothallic. Sugars are fermented and many common hexoses, disaccharides, polyols, and organic acids are assimilated, but nitrate is not utilized as a sole source of nitrogen. Where known, the mol% G + C content of the nuclear DNA is 45-50%, and the major ubiquinone is CoQ-9.

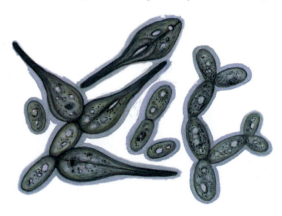

Metschnikowia vanudenii Gim.-Jurado, et al.: yeast cells with multilateral budding, pseudomycelium and acicular ascospores, x 1,200.

Microascus Zukal (syns. **Masoniella** G. Sm., **Scopulariopsis** Bainier)
Microascaceae, Microascales, Hypocreomycetidae, Sordariomycetes, Pezizomycotina, Ascomycota, Fungi
Zukal, H. 1885. Über einige neue Plize, Myxomyceten, und Bakterien. *Verh. zool.-bot. Ges. Wien* 35:333-342.
From Gr. *mikrós*, small + *askós* (L. *ascus*), wine-skin, sac, ascus. Ascomata ostiolate; asci small. Asexual morph scopulariopsis-like. Common in soil on decomposing organic matter, seeds, and dung; causes onychomycosis in humans.

Microglossum Gillet
Geoglossaceae, Geoglossales, Inc. sed., Geoglossomycetes, Pezizomycotina, Ascomycota, Fungi
Gillet, C. C. 1879. *Champignons de France, Discom.* (1):25.
From Gr. *mikrós*, small > L. *micros*, + Gr. *glossa*, tongue, for the size and shape of the ascomata. Ascomata bright-

Microsporum

colored, usually yellow, brown or green, fleshy, erect, stipitate, clavate, ascigerous only in upper portion. Asci clavate-cylindric, 8-spored. Ascospores 2-seriate, hyaline, smooth, ellipsoid, fusiform or cylindric, becoming 3-many-septate. Paraphyses present. On decayed plant materials.

From Gr. *mikrós*, small + *pélte*, shield. Pseudothecia hemiperithecioid, small, hyaline, inconspicuous shield on surface of host leaves and branches, of hyphal reticulum, sinuous, lobate, brown or hyaline. Asci cylindrical or obclavte. Ascospores multiseptate, fusoid or cylindrical.

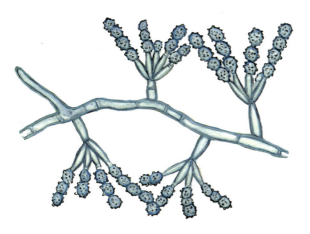

Microascus brevicaulis S. P. Abbott (syn. **Scopulariopsis brevicaulis** Bainier): conidiophores, with the appearance of a small brush or broom, with annellid phialides which produce catenulate, verrucose and basal truncate conidia, x 650.

Micropeltis symploci W. H. Hsieh, et al.: hemiperithecioid pseudothecium on a host leaf (*Symplocos glauca*), Taiwan, and asci with multiseptate, fusoid ascospores, x 160.

Microsphaera Lév.—see **Erysiphe** R. Hedw. Ex DC.
Erysiphaceae, Erysiphales, Leotiomycetidae, Leotiomycetes, Pezizomycotina, Ascomycota, Fungi
Léveillé, J. H. 1851. Organisation et disposition méthodique des espèces qui composent le genre *Erysiphe*. *Annls Sci. Nat.*, Bot., sér. 3 15:381.
From Gr. *mikrós*, small + *sphaíra* (L. *sphaera*), sphere, referring to the small, globose, cleistothecioid perithecia.

Microsporum Gruby (syn. **Nannizzia** Stockdale)
Arthrodermataceae, Onygenales, Eurotiomycetidae, Eurotiomycetes, Pezizomycotina, Ascomycota, Fungi
Gruby, D. 1843. Recherches sur la nature, la siège et le développment du porrigo decalvans ou phytoalopécie. *C. r. hebd. Séanc. Acad. Sci.*, Paris 17:301.
From Gr. *mikrós*, small + *sporón*, spore, i. e., with small spores. Cleistothecia gymnothecial, with a few spiral appendages and hyphae with short, thick-walled cells, spiny at the ends. Producing macro- and microconidia. Keratinophilic saprobe of soil, causing tineas in animals and humans.

Microascus trigonosporus C. W. Emmons & B. O. Dodge: ostiolate ascoma with triangular ascospores, on soil, x 250.

Micromyces P. A. Dang.
Synchytriaceae, Chytridiales, Chytridiomycetidae, Chytridiomycetes, Inc. sed., Chytridiomycota, Fungi
Dangeard, P. A. 1889. Mémoire sur les Chytridinées. *Botaniste* 1:39-74.
From Gr. *mikrós*, small + *mýkes* > L. *myces*, fungus. Thallus microscopic, endoparasitic on algae (*Spirogyra*, *Closterium* and others).

Micropeltis Mont.
Micropeltidaceae, Microthyriales, Inc. sed., Dothideomycetes, Pezizomycotina, Ascomycota, Fungi
Montagne, C., 1842. *In*: Sagra, *Annls Sci. Nat.*, Bot., sér. 2 17:122.

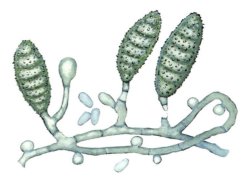

Microsporum gypseum (E. Bodin) Guiart & Grigoraki: conidiogenous cells with multiseptate, hyaline macroconidia, which separate by rhexolysis, and microconidia, x 750.

Microthyrium

Microthyrium Desm.
Microthyriaceae, Microthyriales, Inc. sed., Dothideomycetes, Pezizomycotina, Ascomycota, Fungi
Desmazières, J. B. H. J. 1841. *Annls. Sci. Nat., Bot., sér. 2, 15*:137.
From Gr. *mikrós*, small + *thyreós*, shield + dim. suf. *-ium*. Superficial and internal mycelium. Pseudothecium hemiperithecioid thyrothecium, superficial or subcuticular, dimidiate, flat, with a small ostiole in center. Ascospores hyaline, uniseptate. On leaves and dead branches, principally in tropical and subtropical regions.

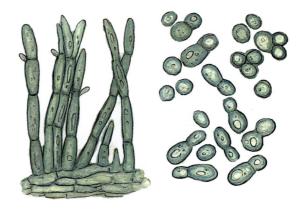

Millerozyma phetchabunensis Tammaw., et al.: pseudohyphae, budding yeast cells, and asci with 1-4 ascospores, x 3,000. Isolated from soil of evergreen forest, Thailand.

Microthyrium microscopicum Desm.: superficial, dark thyriothecium, with radially arranged plates, and a small ostiole in the center, x 375. On upper surface of dry leaves of *Fagus*, *Castanea* and *Quercus*, France.

Millerozyma Kurtzman & M. Suzuki
Debaryomycetaceae, Saccharomycetales, Saccharomycetidae, Saccharomycetes, Saccharomycotina, Ascomycota, Fungi
Kurtzman, C. P., M. Suzuki. 2010. Phylogenetic analysis of ascomycete yeasts that form coenzyme Q-9 and the proposal of the new genera *Babjeviella*, *Meyerozyma*, *Millerozyma*, *Priceomyces* and *Scheffersomyces*. *Mycoscience 51*(1):2-14.
Named in honor of the American zymologist *Martin W. Miller* + Gr. *zýme*, yeast, for his work on yeast taxonomy and ecology. Cell division by multilateral budding on a narrow base, budded cells spherical to elongate. Pseudohyphae formed, without true hyphae. Asci form following conjugation between a cell and its bud, or less commonly, between independent cells. Asci evanescent or persistent, with 1-4 ascospores. Ascospores hat-shaped with a narrow brim or spherical to ovoid with a subequatorial ledge. Homothallic. Glucose fermented; galactose and trehalose sometimes fermented. Many sugars, polyols, and organic acids assimilated, but nitrate not utilized. The major ubiquinone formed is CoQ-9. The mol% G + C content of nuclear DNA ranges from 39 to 46%, and diazonium blue B test is negative.

Mintera Inácio & P. F. Cannon
Parmulariaceae, Asterinales, Dothideomycetidae, Dothideomycetes, Pezizomycotina, Ascomycota, Fungi
Inácio, C. A., P. F. Cannon. 2003. *Viegasella* and *Mintera*, two new genera of Parmulariaceae (Ascomycota), with notes on the species referred to *Schneepia*. *Mycol. Res. 107*(1):82-92.
Named in honor of the British mycologist *David W. Minter* + L. suf. *-a-*. Colonies superficial on upper surface of host leaves. Ascostromata black, at first circular, then becoming stellate. External mycelium brown, branching dichotomously. Ascomatal locules round, ellipsoidal or elongate. Interascal tissue of cellular pseudoparaphyses. Asci clavate to broadly clavate, thick-walled, not bluing in iodine, 2- or 8-spored, with rostrate dehiscence. Ascospores cylindric-ellipsoidal to ellipsoidal, becoming pale brown, verrucose, 1-septate. On leaves in Paraguay and Spain.

Mintera reticulata (Starbäck) Inácio & P. F. Cannon: stellate, dark green ascostromata, on the surface of unidentified plant leaf, x 30.

Minutoexcipula V. Atienza & D. Hawksw.
Inc. sed., Inc. sed., Inc. sed., Inc. sed., Pezizomycotina, Ascomycota, Fungi

Atienza, V., D. L. Hawksworth. 1994. *Minutoexcipula tuckerae* gen. et sp. nov., a new lichenicolous deuteromycete on *Pertusaria texana* in the United States. *Mycol. Res.* 98(5):587-592.

From L. *minutus*, minute + *excipilum*, exciple, for the presence of an exciple of small, thickened cells. Conidiomata sporodochia-like, superficial, convex, arising from upper cortex of host, single, dark brown to black, stromatic and pseudoparenchymatic centrally, with a delimiting exciple-like rim of brown, unevenly thickened polyhedral to swollen cells. Conidiophores macronematous, short-cylindrical, branched, septate, smooth, hyaline to pale brown. Conidiogenous cells integrated, terminal, percurrently proliferating enteroblastically, annellate, cylindrical, subhyaline, smooth. Conidia arising singly, dry, holoblastic, ellipsoid, rounded at apex and truncated at base (0-)1-septate, pale to dark brown, smooth, without appendages or a gelatinous sheath.

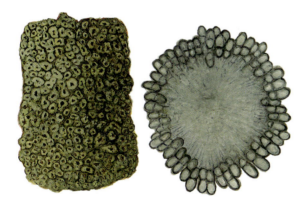

Minutoexcipula tephromelae V. Atienza, et al.: sporodochia (dark spots) parasitizing the lichen thallus of *Tephromela atra* var. *atra*, Spain, x 3; sporodochium with one-septate conidia, x 1,000.

Minutoexcipula tuckerae V. Atienza & D. Hawksw.: sagittal section of a sporodochium, showing the parenchymatous cells of excipulum, annellate conidiogenous cells, and conidia, x 700.

Mitrula Fr.

Sclerotiniaceae, Helotiales, Leotiomycetidae, Leotiomycetes, Pezizomycotina, Ascomycota, Fungi

Fries, E. M. 1821. *Syst. mycol.* (Lundae) *1*:463, 491.

From Gr. *mítra*, headdress, miter + L. dim. suf. *-ula*, i.e., a small headdress or miter, for the shape of the apothecia. Apothecia fleshy, erect, with a short pedicel; upper part bearing hymenium elliptical, pyriform or subglobose, well delimited from pedicel, brown, yellow or orangish. Asci opening by pore, claviform, with eight ascospores. Ascospores hyaline, unicellular, rarely with a septum when mature. On soils rich in humus among pine needles or on dead wood.

Mitrula elegans Berk.: mitriform, stipitate, yellow apothecia, with the fertile portion in the apex, on soil, x 3.

Mixia C. L. Kramer

Mixiaceae, Mixiales, Inc. sed., Mixiomycetes, Pucciniomycotina, Basidiomycota, Fungi

Kramer, C. L. 1959. A new genus in the Protomycetaceae. *Mycologia* 50(6):924.

Dedicated to A. J. Mix + L. suf. *-ia*, which denotes pertaining to. Mycelium multinucleate, within cell walls of host cells. Basidiomata lacking, with a sporiferous sac having a columella-like structure. Chlamydospores thin-walled, germinating to form sporiferous sacs, separated from pedicellar cells by septum. On ferns (*Osmunda*).

Mixia osmundae (Nishida) C. L. Kramer: mature sporiferous sacs, with a central column and spores, on the pedicellar cells, that emerge from chlamydospores in epidermal cells, of fern host (*Osmunda regalis* var. *japonica*), Japan, x 550.

Moana

Moana Kohlm. & Volkm.-Kohlm.
Halosphaeriaceae, Microascales, Hypocreomycetidae, Sordariomycetes, Pezizomycotina, Ascomycota, Fungi
Kohlmeyer, J., B. Volkmann-Kohlmeyer. 1989. Hawaiian marine fungi, including two new genera of Ascomycotina. *Mycol. Res.* 92(4):410-421.
From Hawaiin *moana*, ocean, referring to the habitat. Ascomata subglobose, immersed, ostiolate, periphysate, papillate, coriaceous, cream-colored, single. Peridium thin, textura angularis. Catenophyses developing from thin-walled pseudoparenchyma. Asci 8-spored, clavate, pedunculate, rounded at top, without a pore, non-amyloid, thin-walled, unitunicate, more or less persistent, maturing successively on an ascogenous tissue at bottom of locule. Ascospores subglobose, one-celled, hyaline, with a single, top-shaped appendage that unwinds in water to produce a long ribbon. On intertidal wood in Hawaii, U.S.A.

and smooth; they can vary in shape, from cylindrical, to fusoid, ventricose, or almost ovoid. Pycnidial-acervular conidiomata, with one to several locules per stroma; slender flask-shaped to cylindrical phialides arranged in a compact hymenium, with or without paraphyses, brightly coloured and slimy conidial masses, and fusoid, unicellular conidia. Aschersonia-like anamorphs. Have parasitic association with scale insects and white flies; common in the Neotropics.

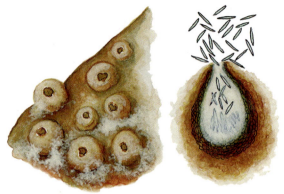

Moelleriella libera (Syd. & P. Syd.) P. Chaverri & M. Liu (syn. **Aschersonia goldiana** Sacc. & Ellis): stromatic conidiomata, x 8; sagittal section of conidioma, and conidia, x 420. Growing in soft scale insects on leaf of *Vitex*, from Pará, Brazil.

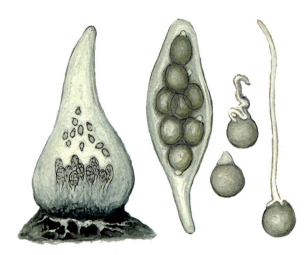

Moana turbinulata Kohlm. & Volkm.-Kohlm.: sagittal section of the ascoma, on submerged wood, Hawaii, U.S.A., with asci and ascospores, x 115; pedunculate, unitunicate ascus with eight ascospores, x 650; subglobose ascospores, with the single, top-shaped appendage that unwinds in water to produces a long ribbon, x 720.

Moelleriella Bres. (syn. **Aschersonia** Mont.)
Clavicipitaceae, Hypocreales, Hypocreomycetidae, Sordariomycetes, Pezizomycotina, Ascomycota, Fungi
Bresadola, G. 1897. *Boll. Soc. bot. ital.*: 292.
Dedicated to the German mycologist *Alfred Möller* + L. suf. dim. *-ella*. Stromata are highly variable in size, colour and shapes; from flat or effuse with loose hyphal tissue, to somewhat pulvinate or knob-shaped, to large, globose and hard stromatal tissue. White and orange are the most common colours. Perithecia can be formed in well-separated or gregarious tubercles, or they can be completely embedded in the stroma; are generally elongated to subglobose, sometimes globose. Asci cylindrical, and all species have filiform, multiseptate ascospores that disarticulate at maturity inside the ascus. The part-spores are always hyaline

Moellerodiscus Henn. (syn. **Ciboriopsis** Dennis)
Sclerotiniaceae, Helotiales, Leotiomycetidae, Leotiomycetes, Pezizomycotina, Ascomycota, Fungi
Hennings, P. 1902. Fungi blumenaeviensis. II. a cl. Alfr. Möller lecti. *Hedwigia* 41:1-33.
Dedicated to the German mycologist *Alfred Möller* + L. *discus* < Gr. *dískos*, round plate, a thin circular object. Apothecia substipitate to long stipitate, produced directly from host tissue with or without association to black line stroma, and directly from blackened, stromatized host tissue; receptacle concave to convex; stipe long and thread-like or short and nearly absent. Asci 8-spored, cylindric to cylindric clavate. Ascospores ellipsoid, fusoid, ovoid, obovoid, hyaline, 0-3 guttulate, smooth. Paraphyses hyaline or pigmented, cylindric-filiform, septate, branched. On dead leaves, twigs and fruits.

Mollisiella (W. Phillips) Massee—see **Unguiculariopsis** Rehm
Dermateaceae, Helotiales, Leotiomycetidae, Leotiomycetes, Pezizomycotina, Ascomycota, Fungi
Massee, G. E. 1895. *Brit. Fung.-Fl.* (London) 4:221.
From genus *Mollisia* (Fr.) P. Karst. + L. dim. suf. *-ella*.

Monascus Tiegh.
Monascaceae, Eurotiales, Eurotiomycetidae, Eurotiomycetes, Pezizomycotina, Ascomycota, Fungi
Tieghem, P. van. 1884. *Monascus*, genre nouveau de

l'ordre des ascomycetes. *Bull. Soc. bot. Fr. 31*:226-231. From Gr. *mónos*, alone, only + *askós* > L. *ascus*, wine-skin, sac, ascus, i.e., an ascus. Cleistothecia with several asci. In soil. Used in China and other Oriental countries to color foods such as red rice or angkak.

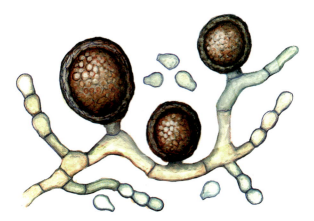

Monascus purpureus Went: purplish cleistothecia, with few asci, and catenulate, hyaline conidia, x 400.

Monilia Bonord.—see Monilinia Honey

Sclerotiniaceae, Helotiales, Leotiomycetidae, Leotiomycetes, Pezizomycotina, Ascomycota, Fungi

Bonordin, 1851. *Handb. Allgem. mykol.* (Stuttgart) 7. From L. *monile*, collar + suf. *-ia*, which denotes quality of or state of a being, because the conidia that are formed by budding remain united in acropetal chains, like a beaded necklace.

Monilinia Honey

Sclerotiniaceae, Helotiales, Leotiomycetidae, Leotiomycetes, Pezizomycotina, Ascomycota, Fungi

Honey, E. E. 1928. The monilioid species of *Sclerotinia*. *Mycologia 20*(4):153.

From L. *monile*, *monilis*, necklace + L. des. *-inia*, derived from Gr. suf. *-ínos*, which denotes material, nature or source of something. Apothecia pedicellate from stromatic tissue. Conidiophores with conidia in chains, monilia-like. In tissues of mummified fruits, causes rot in peach and other trees with drupaceous fruits.

Monilinia fructicola (G. Winter) Honey: pedicellate apothecia, emerging from stromatic tissue developed in apple fruit, x 1.

Monoblepharis

Monoblepharella Sparrow

Gonapodyaceae, Monoblepharidales, Inc. sed., Monoblepharidomycetes, Inc. sed., Chytridiomycota, Fungi

Sparrow, F. K., Jr. 1940. Phycomycetes recovered from soil samples collected by W. R. Taylor on the Allan Hancock 1939 Expedition. Allan Hancock Pacific Expeditions. *Allan Hancock Found. Publ., Occass. Pap., Ser. 1 3*(6):101-112.

From genus *Monoblepharis* Cornu < Gr. *mónos*, alone, only + *blepharís*, cilium, eyelash + L. dim. suf. *-ella*. Male gamete or antherozoid with a single flagellum on its posterior end. Zoospores uniflagellate. Saprobic on seeds and branches in soil and fresh water.

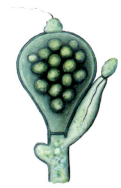

Monoblepharella taylori (Sparrow) Sparrow: oogonium with many oospheres, antheridium and antherozoids, x 350.

Monoblepharis Cornu

Monoblepharidaceae, Monoblepharidales, Inc. sed., Monoblepharidomycetes, Inc. sed., Chytridiomycota, Fungi

Cornu, M. 1871. Note sur deux genres nouveaux de la famille des Saprolégniées. *Bull. Soc. bot. Fr. 18*:58-59.

From Gr. *mónos*, alone, only + *blepharís*, cilium, eyelash, i.e., with a single cilium or flagellum. Males gametes or antherozoids uniflagellate. Zoospores uniflagellate. Saprobic on plant and animal detritus submerged in water.

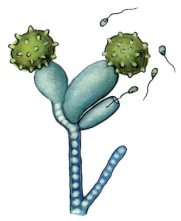

Monoblepharis polymorpha Cornu: two ellipsoidal oogonia, each one with an apical oospore; at the base of right oogonium there is an antheridium liberating antherozoids, x 375.

Monochaetia

Monochaetia (Sacc.) Allesch.
Amphisphaeriaceae, Xylariales, Xylariomycetidae, Sordariomycetes, Pezizomycotina, Ascomycota, Fungi
Allescher, A. 1902. *In*: G. L. Rabenhorst, *Krypt.-Fl.,* Edn 2 (Leipzig) *1*(7):661-662.
From Gr. *mónos*, one, only + *chaíte*, long hair, seta + L. suf. *-ia*, characteristic of. Conidia with filiform, simple or branched, apical appendage, also with a short, unbranched, basal appendage. Pathogenic producing spots on leaves, fruits and branches of plants, e.g., chestnuts and oaks.

Monosporascus cannonballus Pollack & Uecker: ostiolate, dark brown perithecium, releasing a mass of asci, x 15; stipitate ascus with a single, jet-black ascospore, x 950. Collected in Arizona, U.S.A.

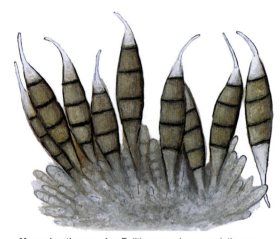

Monochaetia quercina Politis: acervulum on oak tissues, with multiseptate, brown conidia, each one with an apical and basal hyaline appendage, x 1,100. Collected in Greece.

Monosporascus Pollack & Uecker (syn. **Bitrimonospora** Sivan., et al.)
Diatrypaceae, Xylariales, Xylariomycetidae, Sordariomycetes, Pezizomycotina, Ascomycota, Fungi
Pollack, F. G., F. A. Uecker. 1974. *Monosporascus cannonballus*, an unusual ascomycete in cantaloupe roots. *Mycologia* 66(2):346-349.
From Gr. *mónos*, single + *sporá*, spore, seed + *askós*, bag > ML. *ascus*, wine skin, ascus, for the usually single-spored ascus. Ascomata perithecia, solitary, immersed, becoming partially erumpent at maturity, black. Perithecia subglobose, pale brown to black, with papillate to cylindrical ostiolar neck. Wall of perithecium several cells thick. Young centrum containing slender, septate, paraphyses. Asci unitunicate, clavate to pyriform, stipitate, thick-walled when young. Mature asci monosporous, sometimes bi- or trisporous, appearing thickened at apex. Ascospores globose, 1-celled, smooth, brown to black, opaque, shiny, with thick outer wall. Ascospores freed upon dissolution of ascus walls aggregating at apex of perithecium in a droplet of liquid. Parasitic on stems and roots of herbaceous plants, especially Cucurbitaceae, in warm, temperate and tropical regions. *Monosporascus cannonballus* Pollack & Uecker, causes root rot/vine decline of cantaloupe (*Cucumis melo* L.).

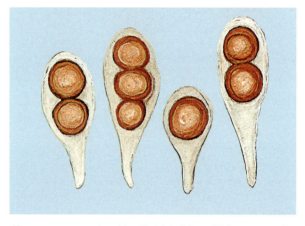

Monosporascus eutypoides (Petr.) Arx) (syn. **Bitrimonospora indica** Sivan., et al.): one, two or three ascospores per ascus, x 500.

Monosporella Keilin—see **Metschnikowia** T. Kamienski
Metschnikowiaceae, Saccharomycetales, Saccharomycetidae, Saccharomycetes, Saccharomycotina, Ascomycota, Fungi
Keilin, D. 1920. On a new Saccharomycete, *Monosporella unicuspidata* g. n. nom., n. sp., parasitic in the body cavity of a dipterous larva (*Dasyhelea obscura* Winnertz). *Parasitology* 12:83-91.
From Gr. *mónos*, one, only + *spóros*, spore + L. dim. suf. *-ella*, because the asci frequently have a single acicular ascospore.

Monosporonella Oberw. & Ryvarden—see **Oliveonia** Donk
Oliveoniaceae, Auriculariales, Inc. sed., Agaricomycetes, Agaricomycotina, Basidiomycota, Fungi
Oberwinkler, F., L. Ryvarden. 1991. *Monosporonella*, a new genus in the Tulasnellaceae, Basidiomycetes. *Mycol. Res.* 95(3):377-379.
From Gr. *mónos* > L. *monos*, one, single + *spora*, spore + suf. *-nella*, for the genus *Tulasnella* J. Schröt.

Montagnea Fr.

Agaricaceae, Agaricales, Agaricomycetidae, Agaricomycetes, Agaricomycotina, Basidiomycota, Fungi
Fries, E. M. 1836. *Fl. Scan.*:339.
Dedicated to the French doctor and mycologist *Jean Pierre Camille Montagne* (1784-1866) + L. des. *-a*. Basidiocarps epigeous, pedunculate, i.e., differentiated into pileus and stipe, volvate. Pileus with a lamelloid gleba with black or dark brown spores. In sandy soils, frequently near beaches.

Montagnea arenaria (DC.) Zeller: epigeous basidiocarp, differentiated into a stipe and dark brown pileus, the latter with a lamelloid gleba, on sandy soil, x 0.5.

Morchella Dill. ex Pers.

Morchellaceae, Pezizales, Pezizomycetidae, Pezizomycetes, Pezizomycotina, Ascomycota, Fungi
Persoon, C. H. 1794. Neuer Versuch einer systematischen Einteilung der Schwämme. *Neues Mag. Bot.* 1:116.
From G. *Morchel*, which signifies morel (from OG. *Morhil*), for the shape of the ascomata resembling a small beehive + L. dim. suf. *-ella*. Fructifications conical or almost spherical, with an alveolate pileus, i.e., with shallow depressions delimited by walls or sterile ribs. Hymenium with asci lining depressions of alveoli and cylindrical, multiseptate paraphyses. Ascospores elliptical, wide. On soils of gardens and grasslands in coniferous forests. Most of species are edible.

Morenoella Speg.—see Lembosia Lév.

Asterinaceae, Asterinales, Dothideomycetidae, Dothideomycetes, Pezizomycotina, Ascomycota, Fungi
Spegazzini, C. L. 1885. Fungi Guaranitici. *Anal. Soc. cient. argent.* 19(6):241-265.
Named in honor of the Argentinian *Francisco Moreno* + L. dim. suf. *-ella*, for his tireless pursuit of nature.

Mortierella Coem.

Mortierellaceae, Mortierellales, Inc. sed., Inc. sed., Mortierellomycotina, Zygomycota, Fungi

Coemans, E. 1863. Quelques hyphomycetes nouveaux. 1. *Mortierella polycephala* et *Martensella pectinata*. *Bull. Acad. R. Sci. Belg.*, Cl. Sci., sér. 2 15:536.
Dedicated to *E. L. Mortier* + L. dim. suf. *-ella*. Sporangia and sporangiola with columella reduced or absent, chlamydospores stylospores. Common saprobe in soil.

Morchella esculenta (L.) Pers.: conical ascoma in ancient forest, Europe, with alveolate, brown pileus, on soil, x 0.5.

Moserella Pöder & Scheuer

Sclerotiniaceae, Helotiales, Leotiomycetidae, Leotiomycetes, Pezizomycotina, Ascomycota, Fungi
Pöder, R., C. Scheuer. 1994. *Moserella radicicola* gen. et sp. nov., a new hypogeous species of Leotiales on ectomycorrhizas of *Picea abies*. *Mycol. Res.* 98(11):1334-1338.
Named in honor of the Austrian mycologist *Meinhard Moser* + L. dim. suf. *-ella*. Apothecia hypogeous, stipitate, arising from clumps of interwoven host rootlets, initially obconical, then distinctly capitate, white, rarely yellowish. Stipe sometimes branched, bearing 2-3 apothecia. Paraphyses hyaline, filiform, with clavate tips. Asci 8-spored, unitunicate, narrrowly cylindric-clavate, with apical ring bluing in iodine. Ascospores uniseriate, unicellular, ellipsoid or ovoid, with rounded ends, hyaline. Ectomycorrhizae on rootlets of *Picea abies*, in the Kristberg mountain region, Vorarlberg, Austria.

Moserella radicicola Pöder & Scheuer: two apothecia emerging from a mycorrhizal rootlet of *Picea abies*, Austria, x 5; white superficial hyphae belong to secondary infectionby **Cenococcum geophilum**. Apothecium showing the hymenium, with asci and ascospores, x 220;

Mucidula

Mucidula Pat.
Physalacriaceae, Agaricales, Agaricomycetidae, Agaricomycetes, Agaricomycotina, Basidiomycota, Fungi
Patouillard, N. 1887. *Les Hyménomycètes d'Europe*:1-166.
From L. *mucidus*, mold, moldy < *mucus*, mucus + Gr. *doulós*, slave, servant. Basidiomata armillarioid, white to pale-brown; pileus shallowly convex with inrolled margin, randomly wrinkled, always with transparent glutinous surface so as to appear as porcelain, usually darker over disc, somewhat lighter toward margin. Lamellae adnate with significant decurrent tooth, subdistant, white to pale buff, without conspicuous pleurocystidial necks. Stipe longitudinally lined, silky-striate, white or cream above annulus when fresh, downward increasingly gray to gray brown, more or less equal, expanded at base into an obpyriform or pad-like base on bark; annulus tightly tomentose. Globose basidiospores. On dead trunks. Was first described in Germany.

Mucidula mucida (Schrad.) Pat. (syn. **Oudemansiella mucida** (Schrad.) Höhn.): whitish basidiomata, on buried wood, x 1.

Muciturbo P. H. B. Talbot—see **Ruhlandiella** Henn.
Pezizaceae, Pezizales, Pezizomycetidae, Pezizomycetes, Pezizomycotina, Ascomycota, Fungi
Warcup, J. H., P. H. B. Talbot. 1989. *Muciturbo*: a new genus of hypogeous ectomycorrhizal Ascomycetes. *Mycol. Res.* 92(1):95-100.
From L. *mucus*, mucus + *turbo*, a top, for the turbinate ascomata that become mucilaginous at maturity.

Mucobasispora Moustafa & Abdul-Wahid
Inc. sed., Inc. sed., Inc. sed., Inc. sed., Pezizomycotina, Ascomycota, Fungi
Moustafa, A. F., O. A. Abdul-Wahid. 1990. *Mucobasispora*, a new dematiaceous hyphomycete genus from Egyptian soil. *Mycol. Res.* 94(1):131-135.
From L. *mucus*, mucus + Gr. *básis*, base + *spóra*, seed, spore. Colonies grey to dark black, restricted, velvety, with fasciculate surface. Aerial mycelium subhyaline to brown, smooth to finely verrucose. Conidiophores micronematous to semimacronematous. Conidiogenous cells discrete to verticillate, ampulliform, with swollen venters, narrow, cylindrical necks which shorten during spore formation. Conidia mostly obovate, olivaceous-brown, in long chains, each conidium surrounded by a subhyaline, relatively wide, mucoid membrane, which collapses, gradually giving rise to a conspicuous reticulate conidial surface in mature conidia.

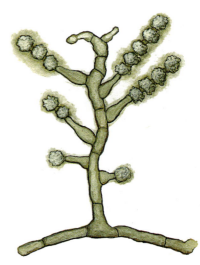

Mucobasispora tarikii Moustafa & Abdul-Wahid: conidiophore, conidiogenous cells, and long chains of conidia, surrounded by mucoid membrane, x 600. Isolated from sandy soil, Egypt.

Mucor Fresen. (syn. **Zygorhynchus** Vuill.)
Mucoraceae, Mucorales, Inc. sed., Inc. sed., Mucoromycotina, Zygomycota, Fungi
Fresenius, G. 1850. *Mucor mucedo. Beitr. Mykol.* 1:7-10.
From L. *mucor*, mold < *muceo*, to become moldy, to spoil. Common in soil, dung, decomposing plants, foods, etc. Saprobic with some pathogens of humans and higher animals. Zygospores heteromorphic, i.e., different in shape and size, one erect, small; the other curved, thicker.

Mucor hiemalis Wehmer: sporangiophores with sporangia, each one showing the columella and spores, x 350.

Mutinus

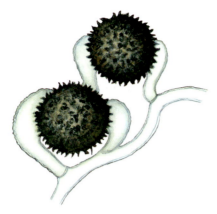

Mucor moelleri (Vuill.) Lendn. (syn. **Zygorhynchus moelleri** Vuill.): saprobic mucor in soil, characterized by the unequal suspensors of the tuberculate zygospore, x 200.

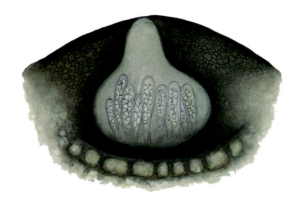

Munkiella caa-guazu Speg.: sagittal section of immersed, subcuticular ascostroma (pseudothecium), on host tissues (Malpighiaceae), with a small ostiole in the center of the scutellum, showing bitunicate asci and bicellular ascospores, x 550.

Muelleromyces Kamat & Anahosur
Phyllachoraceae, Phyllachorales, Inc. sed., Sordariomycetes, Pezizomycotina, Ascomycota, Fungi
Anahosur, K. H. 1968. *Muelleromyces*, a new member of the Sphaeriales (Ascomycetes). *Experientia* 24:849-850.
Named in honor of the Swiss mycologist *Emil Müller* + L. *myces*, fungus < Gr. *mýkes*. Ascomata black perithecia, immersed, with a prominent erumpent ostiolar neck surrounded by a dark clypeus. Asci unitunicate, with a thickened apex, 8-spored, with base gelatinizing at maturity. Ascospores dark brown, thick-walled, unequally 2-celled. On living leaves of *Eugenia jambolanae* in Mysore, India.

Munkiella Speg.
Polystomellaceae, Inc. sed., Inc. sed., Dothideomycetes, Pezizomycotina, Ascomycota, Fungi
Spegazzini, C. 1885. Fungi guaranitici pugillus. *Anal. Soc. cient. argent.* 19(6):248.
Dedicated to the Danish mycologist *A. Munk* + L. dim. suf. *-ella*. Pseudothecia subcuticular, hemiperithecioid with a small round ostiole in center of shield, multiloculate. Asci cylindrical or clavate. Ascospores hyaline, uniseptate. On leaves and branches of trees.

Muricopeltis Viégas
Micropeltidaceae, Microthyriales, Inc. sed., Dothideomycetes, Pezizomycotina, Ascomycota, Fungi
Viégas, A. P. 1944. Alguns fungos de Brasil II. Ascomicetos. *Bragantia* 4:1-392.
From L. *murus*, wall + connective *-co-* + *pélte*, light shield < Gr. *pélte*, shield. Ascomata superficial, punctiform, black, upper wall shield-shaped, hypophyllous, aggregated, broadly conical, with a central pore, margin fimbriate, lighter. Wall formed from brown, interwoven, septate, branched hyphae. Paraphyses lacking. Asci globose, 8-spored. Ascospores clavate, brown, smooth, muriform. On living leaves of *Piper* sp. in São Paulo State, Brazil.

Murrilloporus Ryvarden—see **Cristelloporia** I. Johans. & Ryvarden
Bondarzewiaceae, Russulales, Inc. sed., Agaricomycetes, Agaricomycotina, Basidiomycota, Fungi
Ryvarden, L. 1985. Type studies in the Polyporaceae 17. Species described by W. A. Murrill. *Mycotaxon* 23:169-198.
Named in honor of the American mycologist *William A. Murrill* + *-o-* connective + L. *porus*, pore, in recognition of his work on the poroid fungi.

Muscodor Worapong, et al.
Xylariaceae, Xylariales, Xylariomycetidae, Sordariomycetes, Pezizomycotina, Ascomycota, Fungi
Worapong, J., et al. 2001. *Muscodor albus* anam. gen. et sp. nov., an endophyte from *Cinnamomum zeylanicum*. *Mycotaxon* 79:67-79.
From Gr. *móskos, móschos* > ML. *moschus* > LL. *muscus*, musk + L. *odorus*, odor, for the strong, musky odor given off by cultures of the fungus. In culture mycelium of intertwined, hyaline hyphae. Described largely on basis of molecular characteristics. Cultures produce volatile compounds with a strong odor. Isolated as endophytes from bark and leaves of tropical plants. The type species was isolated from bark of cinnamon (*Cinnamomum zeylandicum*) in Honduras.

Mutinus Fr.
Phallaceae, Phallales, Phallomycetidae, Agaricomycetes, Agaricomycotina, Basidiomycota, Fungi
Fries, E. M. 1849. *Summa veg. Scand.*, Section Post. (Stockholm):434.
From L. *Mutinus*, a name also given to the god *Priapus* (L. *Priapus* < Gr. *Príapos*), son of Dionysus (Baco) and Aphrodite (Venus), symbol of creative energy of genesis desire. Fructifications with receptaculum lacking a pileus, gleba supported by subapical part of column. Spores elliptical, smooth. In the soil or on rotted wood in tropical and temperate regions.

Mycena

Mutinus caninus (Huds.) Fr.: mature basidiocarp, showing the mucilaginous, dark brown to black gleba, supported by the subapical part of the foot, on soil, x 0.8.

Mycena citricolor (Berk. & Curt.) Sacc. (syns. **Decapitatus flavidus** (Cooke) Redhead & Seifert, **Stilbella flavidum** (Cooke) Henn.): brightly colored synnemata with mucoid heads of cause of the disease of coffee leaves called "rooster eye", x 70.

Mycena (Pers.) Roussel (syn. **Decapitatus** Redhead & Seifert)
Mycenaceae, Agaricales, Agaricomycetidae, Agaricomycetes, Agaricomycotina, Basidiomycota, Fungi
Roussel, H. F. A. de. 1806. *Fl. Calvados*, Edn. 2:64.
From Gr. *mýkes* (L. *myces*), fungus + *kenós*, empty, opening, for having a hollow stipe in many species. Fruiting bodies generally small, fragile. In humus, on dead or living wood or on pine cones. Some species, such as *M. citricolor* (Berk. et Curt.) Sacc., are parasites of plants in tropical regions. Several species are luminescent.

Mycenastrum Desv.
Agaricaceae, Agaricales, Agaricomycetidae, Agaricomycetes, Agaricomycotina, Basidiomycota, Fungi
Desvaux, N.A. 1842. Sur le genre *Mycenastrum* du group des Lycoperdales. *Annls Sci. Nat.*, Bot., sér. 2(17):147.
From Gr. *mýkes*, fungus, mushroom + *-ástron*, star; star fungus. Basidiocarps globose, subglobose or pyriform to obovate, at maturity large, area of attachment broad fibrous, mycelioid. Peridium covered at first by a thick whitish coat, becoming cracked, collapsing to form grayish patches that fall away, exposing corky, nearly smooth surface, deep brown to deep purplish-brown. Outer region about 2 mm thick, opens at apex by irregular fissures forming segments that curve back to give appearance of a giant member of *Geastrum* Pers. Gleba white when young then yellowish to olive-brown, finally purplish-brown. Spores globose, 10-13 µm diam, warted, reticular, covered by a thin hyaline envelope. Capillitium branched, with thorn-like spines, dark rusty brown in iodine. Widely distributed.

Mycena lux-coeli Corner: luminescent, greenish basidiocarps, x 2. Collected in Japan.

Mycoëmilia Kurihara, et al.
Kickxellaceae, Kickxellales, Inc. sed., Inc. sed., Kickxellomycotina, Zygomycota, Fungi
Kurihara, Y., et al. 2004. Two novel kickellalean fungi, *Mycoëmilia scoparia* gen. sp. nov. and *Ramicandelaber brevisporus* sp. nov. *Mycol. Res.* 108(10):1143-1152.
From Gr. *myco-*, combining form of *mýkes*, fungus + *Emilia sonchifolia* (Asteraceae), for the resemblance of the fungus to this flower. Vegetative hyphae septate. Sporophores erect, septate, branched or unbranched, producing one to several fertile parts; apex of sporophore slightly enlarged, bearing (4-)6(-8) aggregated sporocladia. Sporocladia lageniform, with densely aggregated spores at tip. Spores hyaline, fusiform, deciduous, covered by a liquid droplet at maturity. Zygospores spherical, thick-walled, containing a globule.

Mycosphaerella

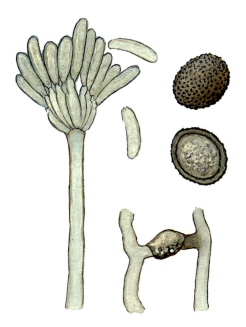

Mycoëmilia scoparia Kurihara, et al.: erect sporangiophore, with lageniform sporocladia and fusiform spores, x 1,700; young and mature zygospores, x 500. On soil, under a shrub containing dead bodies of *Armadillidium vulgare*, Hokkaido, Japan.

Mycogone Link

Hypocreaceae, Hypocreales, Hypocreomycetidae, Sordariomycetes, Pezizomycotina, Ascomycota, Fungi
Link, J. H. F. 1809. Observationes in ordines plantarum naturales. *Mag. Gesell. naturf. Freunde, Berlin* 3(1-2):18.
From Gr. *mýkes*, fungus + *góne*, descendents, prole, referring to the fact that this fungus grows on other fungi, since it is a parasite of the fructifications of Agaricales, including cultivated mushrooms. Aleuriospores bicellular aleuriospores with upper cell larger, with a tuberculate wall. Asexual morph hypomyces-like. Rarely in soil.

Myconymphaea Kurihara, et al.

Kickxellaceae, Kickxellales, Inc. sed., Kickxellomycetes, Kickxellomycotina, Zygomycota, Fungi
Kurihara, Y., et al. 2001. A new genus *Myconymphaea* (Kickxellales) with peculiar septal plugs. *Mycol. Res.*105: 1397-1402.
From Gr. *myco-*, combining form of *mýkes*, fungus + water lily genus *Nymphaea*, for the resemblance of the fungus to this flower. Sporangiophores erect, septate, branched or unbranched, asperulate, producing one to several fertile parts; apical end of sporangiophores enlarged, forming a subglobose to oblate vesicle bearing sporocladia massed on upper hemisphere. Sporocladia cylindrical, of one, sometimes two, cell(s), bearing pseudophialides. Pseudophialides flask-shaped, with long necks, bearing a single sporangiole. Sporangiola monosporic, hyaline, cylindrical and aciculate. Sporangiospores cylindrical and aciculate. Isolated from a dead soil insect, *Metriocampa* sp.

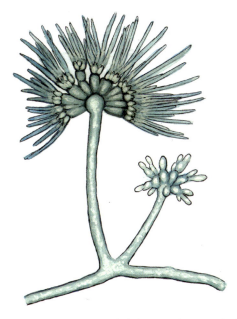

Myconymphaea yatsukahoi Kurihara, et al.: sporangiophores with sporocladia, pseudophialides and spores, x 1,000. Isolated from a dead body of *Diplura* insect (*Metriocampa* sp.), Japan.

Mycosphaerella Johanson—see Ramularia Unger

Mycosphaerellaceae, Capnodiales, Dothideomycetidae, Dothideomycetes, Pezizomycotina, Ascomycota, Fungi
Johanson, C. J. 1884. *Öfvers. K. Svensk. Vetensk.-Akad. Förhandl.* 41(9):163.
From Gr. *mýkes*, genit. *mýketos*, fungus, from this the pref. *mýko-*, equivalent to *mýketo-*, + *sphaíra* (L. *sphaera*), sphere + L. dim. suf. *-ella*, for the shape of the ascocarps.

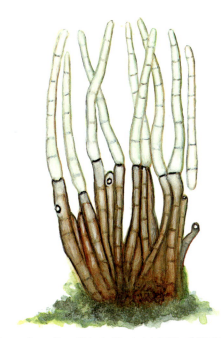

Mycosphaerella coffeicola (Cooke) J. A.Stev. & Wellman (syn. **Cercospora coffeicola** Berk. & Cooke): geniculate conidiophores with filamentous, septate conidia, on coffee tissues, x 1,000.

Mycosylva

Mycosylva M. C. Tulloch
Inc. sed., Inc. sed., Inc. sed., Inc. sed., Pezizomycotina, Ascomycota, Fungi
Tulloch, M. 1973. A new synnematous Hyphomycete. *Trans. Br. mycol. Soc.* 60(1):155-157.
From Gr. *mykés* > L. *myces*, fungus + L. *sylva*, a woods, for the collection site of the fungus. Colony at first raised, restricted, later fasciculate, finally with distinct synnemata. Synnemata of compressed septate branched hyphae, dark-walled in stalk, hyaline, contorted at apex spreading out to form expanded head. Conidia holoblastic, formed laterally from contorted hyphae, borne on a short narrow neck, delimited by a septum. Additional conidia in short acropetal chains with narrow connectives between conidia. On animal droppings in Warwickshire, England.

Mycotypha africana R. O. Novak & Backus: cylindrical vesicles that bear unispored sporangiola ("conidia"), x 275. Isolated in Zimbawe.

Mycosylva clarkii M. C. Tulloch: synnemata, x 15; conidiogenous cells, and short acropetal chains of conidia, x 1,100. On animal droppings in Great Britain.

Mycotypha Fenner
Mycotyphaceae, Mucorales, Inc. sed., Inc. sed., Mucoromycotina, Zygomycota, Fungi
Fenner, E. A. 1932. *Mycotypha microspora*, a new genus of the Mucoraceae. *Mycologia* 24(2):187-198.
From Gr. *mýkes*, fungus + genus *Typha*, a plant whose cylindrical inflorescences resemble the cylindrical vesicles that bear the unispored sporangiola ("conidia") of this saprobic fungus. On soil and dung.

Myriangium Mont. & Berk.
Myriangiaceae, Myriangiales, Dothideomycetidae, Dothideomycetes, Pezizomycotina, Ascomycota, Fungi
Berkeley, M. J. 1845. Decades of fungi. *London J. Bot.* 4:72.
From Gr. *myríos*, innumerable + *angeíon*, vessel, receptacle. Ascostromata with differentiated fertile layer, locules uniascal. Ascospores dictyosporous. Parasitic on scale insects on surface of leaves and stems of living plants, principally in tropical or subtropical regions. Some species parasitize monocotyledonous plants, such as cultivated bamboo.

Myriangium duriaei Mont. & Berk.: sagittal section of ascostroma, on insect flake (*Chrysomphalus aonidium*), showing numerous uniascal locules with dictyosporous ascospores, x 100; free dictyosporous ascospore, x 1,000. On bark, including that of *Morus alba*,

Myrioconium Syd. & P. Syd. (syn. **Botryophialophora** Linder)
Sclerotiniaceae, Helotiales, Leotiomycetidae, Leotiomycetes, Pezizomycotina, Ascomycota, Fungi
Sydow, H., P. Sydow. 1912. Sydow, Mycotheca germanica Fasc. XXII-XXIII (No. 1051-150. *Annls mycol.* 10(5):448.
From. Gr. *myríos*, numberless + NL. *conidium*, a propagative body of fungi < Gr. *kónis*, dim. *kónidion*, dust.
Sterile mycelium white or light-colored, cottony in culture. Conidiophores phialides, usually three or more from an enlarged globose or subglobose basal cell produced pleurogenously on main hyphae or their branches; occasionally conidiophores produced singly

either pleurogenous or acrogenous at ends of short side branches. Conidia hyaline or light-colored, tending to aggregate in globose clusters. Isolated from driftwood.

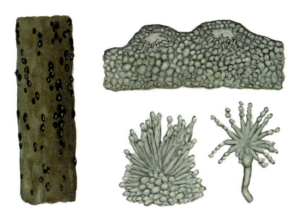

Myrioconium scirpi Syd. & P. Syd.: acervuli on branch of *Scirpi lacustris*, x 5; vertical section of two acervuli with conidiogenous cells inside, x 50; young and mature conidiogenous cells giving rise to hyaline and catenulate conidia; at right, mature conidia penicillate arranged, x 350.

Myriogenospora G. F. Atk.

Clavicipitaceae, Hypocreales, Hypocreomycetidae, Sordariomycetes, Pezizomycotina, Ascomycota, Fungi
Atkinson, G. F. 1894. Steps toward a revision of the linosporous species of North American graminicolous Hypocreaceae. *Bull. Torrey bot. Club* 21(5):222-225.
From Gr. *myrías*, 10,000 > *myríos*, myriad, numberless + *génos*, to be produced + *sporá*, spore, seed, for the numerous part-spores that are formed. Mycelium superficial on host tissues, forming two narrow, linear black stromata parallel to long axis of leaf. Ascomata formed in a single row in stroma; ascomata ostiolate perithecia, immersed in stroma, with slightly protruding ostiolar papilla. Ascomatal wall thin, composed of elongated prosenchyma cells. Centrum containing lateral paraphyses with filamentous apices and vesiculose basal cells. Asci unitunicate, formed in a basal cluster, elongate-fusoid, with a dome-shaped apical cap, 8-spored. Ascospores hyaline, filamentous, septate, separating into fusoid part-spores. On grasses including sugar cane in both temperate and tropical regions. Mycelium in leaf apex entraps tips of young leaves, thus as leaves elongate at basal meristem, a series of loops is formed producing a condition known as "tangle-top disease", caused by *Myriogenospora atramentosa* (Berk. & Curt.) Diehl.

Myriostoma Desv.

Geastraceae, Geastrales, Phallomycetidae, Agaricomycetes, Agaricomycotina, Basidiomycota, Fungi
Desvaux, N. A., 1809. Observations sur quelques genres à établir dans la famille des champignons. *J. Bot.*, Paris 2:103.

From Gr. *myríos*, ten thousand, uncountable, numerous + *stóma*, mouth, due to the dehiscence of the endoperidium by means of numerous pores. Fructifications subglobose or pyriform, hypogeous when young, stellate, epigeous when mature, due to shape resulting from splitting of exoperidium. Endoperidium supported at base of exoperidium by small, slender columns. Gleba dark, of branched filaments (capillitium) with spherical, echinulate spores. In subarid pastures, rarely in subtropical forests.

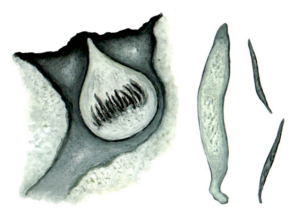

Myriogenospora atramentosa (Berk. & M. A. Curtis) Diehl: sagittal section of a stroma, with an immersed perithecium, on the leaf of *Eragrostis hirsuta*, x 60; ascus with part-spores, x 320, free part-spores, x 900.

Myriostoma coliforme (Dicks.) Corda: stellate, mature basidiocarp, showing the exoperidium and endoperidium (sporiferous sac), with many ostioles, on soil, Great Britain, x 0.5.

Myrmecocystis Harkn.—see Genea Vittad.

Pyronemataceae, Pezizales, Pezizomycetidae, Pezizomycetes, Pezizomycotina, Ascomycota, Fungi
Harkness, H. W. 1899. Californian hypogaeous fungi. *Proc. Calif. Acad. Sci.*, Ser. 3, Bot. 1:269.
From Gr. *mýrmex*, genit. *mýrmekos*, ant + *kýstis*, bladder, vesicle, cell, because it lives buried in the soil, in relation to ant nests.

Myrotheciastrum

Myrotheciastrum Abbas & B. Sutton
Inc. sed., Inc. sed., Inc. sed., Inc. sed., Pezizomycotina, Ascomycota, Fungi
Abbas, S. Q., B. C. Sutton. 1988. *Myrotheciastrum* gen. nov., an addition to the Coelomycetes. *Trans. Br. mycol. Soc.* 91(2):352-357.
From genus *Myrothecium* Tode + L. suf. *-astrum*, incomplete resemblance, i.e., similar to *Myrothecium* but not identical. Conidiomata eustromatic, black, solitary to aggregated, superficial to semiimmersed, cupulate to convoluted, unilocular, rarely multilocular, mainly of a central, semi-immersed disk-like region and a marginal excipulum; wall of many cells, generally differentiated into two layers. Paraphyses hyaline, smooth, septate, branched, produced from phialides, mainly confined to peripheral area. Conidiophores mostly absent; if present, then only 1- to 2-septate, cylindrical, hyaline, branched in upper part. Conidiogenous cells cylindrical, hyaline, occasionally dark brown to black, discrete or integrated, with enteroblastic proliferations variably spaced; collarettes and periclinal thickening prominent, cytoplasmic channels broad, funnel-shaped, apical area becoming dark brown to black. Conidia holoblastic, aseptate, ovoid, fusoid, ellipsoid to globose, greenish-black to black, apex obtuse, base obtuse to truncate, with a hyaline, tubular, empty appendage demarcated from conidial base by a septum. Sometimes conidia, conidiogenous cells and conidiophores enclosed in a mucilaginous sheath. On dead branches of *Salvadora oleoides*, Bawalpur, Pakistan.

Myrotheciastrum salvadorae Abbas & B. Sutton: sagittal section of a black, unilocular conidioma, on dead branches of *Salvadora oleoides*, showing paraphyses and conidiogenous cells with conidia, x 120; annellate conidiogenous cells with aseptate conidia, x 900.

Myrothecium Tode
Inc. sed., Hypocreales, Hypocreomycetidae, Sordariomycetes, Pezizomycotina, Ascomycota, Fungi
Tode, H.J. 1790. *Fung. mecklenb. sel.* (Lüneburg) 1:1-47.
From Gr. *myrías*, genit. *myriados*, the number 10,000; akin to *myríos*, myriad, numberless, infinite, immense + NL. *thecium* < Gr. *thekíon* < *théke*, a case of something. Sporodochia brightly colored, generally greenish, covered by a fertile layer of irregularly branched conidiophores. Conidia hyaline with membranous appendage on apical part. In soil and plant rhizosphere.

Mytilinidion Sacc.
Mytilinidiaceae, Mytilinidiales, Inc. sed., Dothideomycetes, Pezizomycotina, Ascomycota, Fungi
Saccardo, P. A. 1875. Conspectus generum pyrenomycetum italicorum systemate carpologico dispositorum. *Atti Soc. Ven-Trent. Sci. Nat.* 4:99.
From L. *mytilus*, *mitylus* (from Gr. *mítylos*), clam + Gr. dim. suf. *-ídion*, for the clam-shaped hysterothecia. Pseudothecia with longitudinal dehiscence, small, elongate, black, carbonaceous. Asci bitunicate, cylindrical. Ascospores multiseptate. Saprobic on wood and tree branches.

Mytilinidion thujarum (Cooke & Peck) M. L. Lohman: carbonaceous, black, clamp-shaped hysterothecia, on *Thuja occidentalis*, x 50.

Myxothyriopsis Bat. & A. F. Vital
Inc. sed., Inc. sed., Inc. sed., Inc. sed., Pezizomycotina, Ascomycota, Fungi
Batista, A. C., et al. 1956. *Myxothyriopsis*, e novas species de *Leptothyrium*. *Anais da Sociedade de Biologia de Pernambuco* 14:89-97.
From genus *Myxothyrium* Bubák & Kabát + Gr. suf. *-ópsis*, aspect, appearance, for the mucous material surrounding the conidia. Mycelium epiphyllous, pelliculose, thin, composed of septate, brown, flexuous, sparsely branched hyphae. Secondary hyphae subhyaline, reticulate, with indistinct septa, branched, lacking hyphopodia and setae. Pycnidia subcuticular, orbicular, uniloculate, membranous, brown, semitranslucent, without a pore. Conidia aggregated in radiating fascicles, covered with a mucous substance, fusoid, cylindrical to oblong, 1-celled, hyaline. On living leaves of *Astronium* sp. in Recife, Brazil.

Myxotrichum

Myxotrichum Kunze
Myxotrichaceae, Inc. sed., Inc. sed., Leotiomycetes, Pezizomycotina, Ascomycota, Fungi

Kunze, G., J. C. Schmidt. 1823. Einige neue order verkannten Pilzgattungen und Arten. X. *Myxotrichum*. Mykologische Hefte (Leipzig) 2:108-110.

From Gr. *mýxa*, slime + *thríx*, genit. *trichós*, hair + L. suf. *-um*. Ascocarps more or less spherical, dark-colored, often appearing bramble-like. Peridial hyphae branched, thick-walled, septate, smooth, usually ending partly in short, septate spines and partly in elongate appendages. Appendages more or less elongate, septate, rigid, straight or curved, simple or branched, apices straight, curved, bent, or uncinate, lateral branchlets straight or deflexed, long or short, dark or hyaline. Centrum at first white, then yellowish at maturity. Asci more or less globose and pedicellate, 8-spored, evanescent. Ascospores hyaline or light-colored, somewhat fusiform, lenticular, elliptical, cymbiform or navicular, occasionally with a hyaline rim round the longitudinal axis, smooth or delicately striate. Asexual stages, when present, represented by intercalary, more or less oblong or cylindrical arthro-aleuriospores or by spherical or clavate, terminal aleuriospores. Chlamydospores occasionally present. Widespread on paper, cardboard, straw, rotting wood, dung and soil.

Myxotrichum stellatum Udagawa & Uchiy.: surface view of an ascoma with peridial hyphae, x 200; young ascus, yellowish-brown, with 8 striate, fusiform ascospores, with 5-6 ridges running longitudinally, stellate or fluted in section, x 3,000. Isolated from forestal soil in Japan.

N

Nectria cinnabarina

N

Nadsonia Syd.
Saccharomycodaceae, Saccharomycetales, Saccharomycetidae, Saccharomycetes, Saccharomycotina, Ascomycota, Fungi
Sydow, H. 1912. Referate und kritische Besprechungen. *Annls mycol.* 10(3):347-348.
Dedicated to the Ucranian microbiologist G. A. *Nadson* + L. suf. *-ia*, which denotes pertaining to. Cultures with abundant sporulation dark brown. Asci forming by pedogamy with 2-4 ascospores per ascus. Ascospores large, round, with an oil globule and long, slender protuberances in external wall, yellowish-brown. Capable of fermenting some sugars. Isolated from viscous exudates of trees such as *Betula* and *Carpinus*.

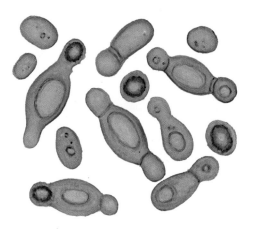

Nadsonia fulvescens (Nadson & Konok.) Syd.: yeast cells that form asci by pedogamy, ascospores with oil globule, and vegetative cells that reproduce by bud-fission, x 1,250.

Nannizzia Stockdale—see **Microsporum** Gruby
Arthrodermataceae, Onygenales, Eurotiomycetidae, Eurotiomycetes, Pezizomycotina, Ascomycota, Fungi
Stockdale, P. M. 1961. *Nannizzia incurvata* gen. nov., sp. nov., a perfect state of *Microsporum gypseum* (Bodin) Guiart et Grigorakis. *Sabouraudia* 1:41-48.
Dedicated to the Italian mycologist A. *Nannizzi* + L. suf. *-ia*, which denotes pertaining to.

Naothyrsium Bat.
Inc. sed., Inc. sed., Inc. sed., Inc. sed., Pezizomycotina, Ascomycota, Fungi
Batista, A. C., J. L. Bezerra. 1960. *Naothyrsium* - novo gênero de Rhizothyriaceae, associado a *Palawania brosimi* n. sp. *Publicações Inst. Micol. Recife* 250:1-15.
From Gr. *náos*, temple + *thýrsos*, stalk + L. des. *-ium*, with reference to the upper wall of the conidioma supported by vertical columns. Free mycelium absent. Pycnostromata superficial, dimidiate-scutate, membranous-carbonaceous, brownish-black. Upper wall of radiate hyphae, supported by several vertical columns connected to a hypostroma, smooth, opening by an irregular stellate dehiscence. Conidiophores short, conoid or cylindrical, hyaline, 1-celled, formed on underside of pycnostroma. Conidia clavate, pyriform or cordiform, 1-septate, smooth, brown. On leaves of *Brosimum discolor* in Recife, Brazil.

Naranus Ts. Watan.
Inc. sed., Inc. sed., Inc. sed., Inc. sed., Pezizomycotina, Ascomycota, Fungi
Watanabe, T. 1995. *Naranus cryptomeriae* gen. et sp. nov. from Japanese cedar seed. *Mycol. Res.* 99(7):806-808.
From *Nara* + connective *-n-* + L. suf. *-us*, ending for geographic names, for the collection site Nara City, Japan. Mycelium of branched, brown, septate hyphae. Conidiophores lacking. Conidiogenous cells blastic, integrated. Conidia globose or subglobose, brown, muriform, initially catenate, then aggregated into spherical or subspherical masses developed on aerial hyphae with age. Isolated from seeds of *Cryptomeria japonica*.

Natantiella Réblová
Inc. sed., Inc. sed., Inc. sed., Sordariomycetes, Pezizomycotina, Ascomycota, Fungi
Réblová, M., V. Štìpánek. 2009. New fungal genera, *Tectonidula* gen. nov. for *Calosphaeria*-like fungi with holoblastic-denticulate conidiogenesis and *Natantiella* gen. nov. for three species segregated from *Ceratostomella*. *Mycol. Res.* 113(9):991-1002.

From L. *natantis*, floating + dim. suf. *-ella*, referring to the asci floating freely in centrum at maturity. Perithecia nonstromatic, gregarious or solitary, dark brown to black, venter globose to subglobose, immersed; neck elongated, emerging above substratum, straight or slightly curved, ostiole periphysate. Perithecial wall leathery, two-layered. Paraphyses septate, hyaline, tapering towards tip, longer than asci. Asci unitunicate, clavate, 8-spored, truncate to broadly rounded at apex, long-stipitate, when young arising in large bundles from hymenium, floating freely at maturity. Ascospores oblong to ellipsoidal, hyaline, aseptate. Asexual morph unknown.

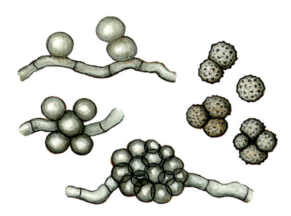

Naranus cryptomeriae Ts. Watan.: conidial formation from aerial hyphae, mature, muriform conidia, with echinulate wall, x 1,300. Isolated from seeds of Japanese cedar, *Cryptomeria japonica*, Japan.

Nautosphaeria E. B. G. Jones

Halosphaeriaceae, Microascales, Hypocreomycetidae, Sordariomycetes, Pezizomycotina, Ascomycota, Fungi

Jones, E. B. G. 1964. *Nautosphaeria cristaminuta* gen. et sp. nov., a marine pyrenomycete on submerged wood. *Trans. Br. mycol. Soc.* 47(1):97-101.

From Gr. *nautés*, sailor, sea + genus *Sphaeria* Haller, for habitat of the fungus. Perithecia solitary, immersed in substratum, hyaline or cream, membranous, neck long. Asci unitunicate, 8-spored, clavate, deliquescing. Ascospores grey or fuscous, unicellular, ellipsoid, apically and laterally appendiculate. On beech test blocks submerged in the Irish Sea.

Nawawia Marvanová

Inc. sed., Inc. sed., Inc. sed., Inc. sed., Pezizomycotina, Ascomycota, Fungi

Marvanová, L. 1980. New or noteworthy aquatic hyphomycetes. *Clavatospora*, *Heliscella*, *Nawawia* and *Heliscina*. *Trans. Br. mycol. Soc.* 75(2):221-231.

Named for the Malay mycologist A. Nawawi + L. des. *-a*, who first discovered the fungus. Hyphae brown, superficial or immersed, septate, forming minute stromata. Conidiophores rigid, erect, septate, thick-walled, brown, lighter toward apex, forming terminal phialides. Conidia enteroblastic, hyaline, tetraradiate or pyramidal, 1-celled, bearing long apical appendages.

Nautosphaeria cristaminuta E. B. G. Jones: sagittal section of a perithecium, immersed on submerged test block of *Fagus sylvatica*, Irish Sea, x 130; two free ellipsoidal ascospores, apically and laterally appendiculate, x 950.

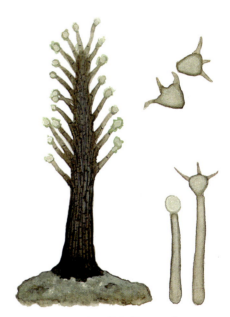

Nawawia dendroidea K. D. Hyde, et al.: synnema with conidiophores forming terminal phialides, x 90; pyramidal, 1-celled conidia, bearing long apical appendages, x 700. In dead submerged stems of *Phragmites*, KwaZulu-Natal, South Africa.

Nectria (Fr.) Fr. (syns. **Creonectria** Seaver, **Megalonectria** Speg., **Scoleconectria** Seaver, **Sphaerostilbe** Tul. & C. Tul., **Tubercularia** Tode)

Nectriaceae, Hypocreales, Hypocreomycetidae, Sordariomycetes, Pezizomycotina, Ascomycota, Fungi

Fries, E. M. 1849. *Summa veg. Scand.*, Sectio Post. (Stockholm):387.

From L. *necto*, unite, interweave, connect + suf. *-ia*, characteristic of, because the perithecia are intimately united in the stromata. Stromata fleshy, tubercular or depressed, occasionally of a slender stalk with subglobose

Nectriopsis

to globose head, red, yellow, brown or occasionally black in age. Perithecia superficial or surrounding stroma, orange or red, subglobose to globose, with ostiole often depressed with age, smooth, verrucose or furfuraceous, fleshy, superficial on or surrounding stroma, scattered or in dense clusters. Asci clavate or cylindrical, 2-8-spored with spores occasionally accompanied by numerous minute spore-like bodies. Ascospores elliptical to fusoid or subfiliform, straight or curved, 1- to many septate or muriform, hyaline. Paraphyses present or not evident. Sporodochia sessile, cushion-like or tuberculate with compact layers of erect conidiophores that produce phialospores. On bark of trees, causing necrosis and canker known as coral spot in fruit trees (apple, plum, peach) and shade trees (maple, chestnut, mulberry, willow).

Nectria cinnabarina (Tode) Fr.: mature, reddish perithecia, on wood, Germany, x 400.

Nectria cinnabarina (Tode) Fr. (syn. **Tubercularia vulgaris** Tode): tuberculate sporodochia on deciduous woody plant, parasite causing the disease called coral spot, x 4; conidiophores with unicellular, hyaline phialospores, x 1,000.

Nectriopsis Maire

Bionectriaceae, Hypocreales, Hypocreomycetidae, Sordariomycetes, Pezizomycotina, Ascomycota, Fungi
Maire, R. 1911. Remarques sur queleues hypocréacées. *Annls mycol.* 9(4):315-325.
From genus *Nectria* (Fr.) Fr. + Gr. suf. -ópsis, aspect, appearance, for the general similarity to *Nectria*. Ascomata ostiolate perithecia, superficial or immersed in mycelium, but not stromatic, solitary, gregarious or scattered. Perithecia obpyriform to oval or subglobose, with an ostiolar papilla, glabrous or with hairs on upper portion of perithecium, pale yellow to white. Walls thin, of several layers of flattened cells, ostiolar papilla of conical palisade of slender, parallel hyphae or of small pseudoparenchymatous cells. Ostiole lined with slender periphyses. Centrum of young perithecium with apical paraphyses. Asci unitunicate, clavate to cylindrical, apex undifferentiated or with a small, non-amyloid ring, short-stipitate, 8-spored. Ascospores hyaline, 2-celled, ellipsoidal to oblong, constricted or not at septum, smooth, spinulose or striate. Asexual morph acremonium-like. On other fungi, lichens or myxomycetes, occasionally on plants.

Nemania Gray (syn. Geniculosporium Chesters & Greenh.)

Xylariaceae, Xylariales, Xylariomycetidae, Sordariomycetes, Pezizomycotina, Ascomycota, Fungi
Gray, S. F. 1821. *A natural arrangement of British plants.* 1:1-824.
From Gr. néma, genit. némators, thread + -ia, ending of nouns denoting quality of or state of being. Stromata effused-pulvinate, pulvinate, discoid, occasionally applanate, superficial or partially embedded in host tissue, gregarious, solitary or confluent, with sloped or steep margins, unipartite; surface whitish, gray or grayish-brown when young, at length dark-brown, blackish brown, or black; mature stromata plane or with inconspicuous to conspicuous perithecial mounds; carbonaceous tissue immediatelly beneath surface; tissue between the perithecia either carbonaceus and persistent of whitish, soft, becoming grayish-brown, and often disintegrated with age, composed of fungal tissue, host tissue, or both. Perithecia spherical, obovoid, monostichous, with carbonaceous stromatal material surrounding individual perithecia when aged in some taxa; ostioles higher than stromatal surface and with openings slightly papillate, papillate, or coarsely papillate, with or without encircling disc. Asci 8-spored, cylindrical, short- or long-stipitate, persistent, with apical ring urn-shaped, inverted hat-shaped, cuneate or, less frequently, discoid, amyloid or dextrinoid in Melzer's reagent. Ascospores pale brown, light brown, brown, dark brown, or blackish-brown, unicellular in mature ascospores, ellipsoidal, cylindrical, or fusoid, inequilateral, slightly inequilateral, or nearly equilateral, with acute, narrowly rounded, or broadly rounded ends, with a straight, conspicuous or inconspicuous germ slit of spore length so much less than spore-length.

Nematospora Peglion—see Eremothecium Borzi

Eremotheciaceae, Saccharomycetales, Saccharomycetidae, Saccharomycetes, Saccharomycotina, Ascomycota, Fungi
Peglion, V. 1897. Sopra un nuovo blastomicete, parassita

del fruto del Noccivela. *Atti R. Acad. Lincei,* Rendiconti Cl. Sci. Fis., sér. 5 6:216-278.

From Gr. *néma,* genit. *nématos,* thread, filament + *sporá,* spore, in reference to the fusiform ascospores with a non motile, filiform or flagelliform appendage.

Nemania serpens (Pers.) Gray (syn. **Geniculosporium serpens** Chesters & Greenh.): branched conidiophores, with one-celled conidia, x 1,000.

Neocarpenteles Udagawa & Uchiy.

Aspergillaceae, Eurotiales, Eurotiomycetidae, Eurotiomycetes, Pezizomycotina, Ascomycota, Fungi

Udagawa, S., S. Uchiyama, 2002. *Neocarpenteles*: a new ascomycete genus to accommodate *Hemicarpenteles acanthosporus. Mycoscience* 43(1):3-6.

From Gr. *néos,* new, recent + genus *Carpenteles* Langeron. Colonies producing coiled branches of hyphae that develop into the ascoma-bearing stromata. Stromata superficial, more or less globose, yellowish-brown to grayish-brown, tomentose, hard and sclerotioid, non-ostiolate, uniloculate, containing asci but lacking an inner differentiated ascomatal peridium; asci irregularly disposed, maturing outward from the center of the stroma, globose to subglobose, 8-spored, evanescent; ascospores one-celled, hyaline, lenticular, variously sculptured, with two equatorial crests; aspergillus, asexual state, characterized by phialides borne directly on the vesicle (uniseriate aspergilla).

Neocosmospora E. F. Sm.—see **Fusarium** Link

Nectriaceae, Hypocreales, Hypocreomycetidae, Sordariomycetes, Pezizomycotina, Ascomycota, Fungi

Smith, E. F. 1899. *U.S.D.A. Div. Veg. Pathol. Bull.* 17:45.

From Gr. pref. *néo-,* from *néos,* recent, new + the genus *Cosmospora* Rabenh., this from Gr. *kósmos,* world, ornate, beauty + *sporá,* spore.

Neofusicoccum Crous, et al.

Botryosphaeriaceae, Botryosphaeriales, Inc. sed., Dothideomycetes, Pezizomycotina, Ascomycota, Fungi

Crous, P. W. et al. 2006. Phylogenetic lineages in the Botryosphaeriaceae. *Stud. Mycol.* 55:235-254.

From Gr. *néos,* new, recent + L. *fusus,* spindle + *coccus* < Gr. *kókkos,* grain, coccus, a berry. Ascomata forming botryose clusters, each comprising many ascomata, erumpent through the bark, globose, with a short conical papilla, dark brown to black, smooth, thick-walled. Asci clavate, 8-spored, bitunicate. Ascospores broadly elipsoidal to fusoid, hyaline, smooth, aseptate, occasionally becoming 1-septate. Conidiomata globose and non-papillate, entire locule lined with conidiogenous cells. Conidiogenous cells holoblastic, hyaline, subcylindrical, proliferating percurrently to form 1-2 annellations, or proliferating at the same level to form periclinal thickenings. Conidia ellipsoidal with apex round and base flat, unicellular, hyaline, old conidia becoming 1-2 septate hyaline, or light brown with the middle cell darker than the terminal cells. Dichomera asexual form. Conidia subglobose to obpyriform, brown, apex obtuse, base truncate, 1-3 transverse septa, 1-2 longitudinal septa and 1-2 oblique septa. This genus comprising numerous species found on a wide range of plant hosts of agricultural, forestry, ecological and economic importance. Infected plants can exhibit a multiplicity of disease symptoms, such as fruit rots, leaf spots, seedling damping-off, and collar rot, cankers, blight of shoots and seedlings, gummosis, blue-stain of the sapwood, dieback and tree death.

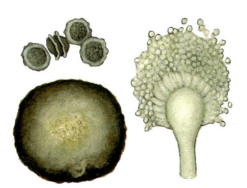

Neocarpenteles acanthosporum (Udagawa & Takada) Udagawa & Uchiy. (syn. **Hemicarpenteles acanthosporum** Udagawa & Takada): sclerotium-like cleistothecium, x 45; ascospores with two equatorial crests, x 1,350; conidiophore with phialides and conidia aspergillus-like, x 80. Isolated in North Solomons.

Neofusicoccum ribis (Slippers, et al.) Crous, et al. (syn. **Botryosphaeria ribis** Grossenb. & Duggar): vertical section of pseudothecia on dead wood, with asci, and ascospores, x 70.

Neolysurus

Neolysurus O. K. Mill., et al.
Phallaceae, Phallales, Phallomycetidae, Agaricomycetes, Agaricomycotina, Basidiomycota, Fungi
Miller, O. K., et al. 1991. *Neolysurus*: a new genus in the Clathraceae. *Mycol. Res.* 95(10):1230-1234.
From Gr. *néo*, new + genus *Lysurus* Fr. Basidiomata columnar, with a thick, hollow, spongy stalk, a volval sac surrounding base. Five hollow columns arise from top of stalk forming an arch that supports cushion-shaped gleba. Gleba olive green, divided into a fine tubular mesh of interwoven, filamentous, branched hyphae. Basidiospores cylindric, thin-walled, olive gray en mass. Odor of basidioma offensive. Collected in La Selva Biological Station, Puerto Viejo, Costa Rica.

Neophaeosphaeria M. P. S. Câmara, et al.
Leptosphaeriaceae, Pleosporales, Pleosporomycetidae, Dothideomycetes, Pezizomycotina, Ascomycota, Fungi
Câmara, M., et al. 2003. *Neophaeosphaeria* and *Phaeosphaeriopsis*, segregates of *Paraphaeosphaeria*. *Mycol. Res.* 107(5):515-522.
From Gr. *néo*, new + genus *Phaeosphaeria* I. Miyake, for the similarity to this genus. Ascomata immersed, subepidermal, usually erumpent at maturity, globose to sphaeroid to pyriform, often papillate, solitary or gregarious in a stroma, often surrounded by septate, brown hyphae extending into host tissues. Asci bitunicate, cylindric. Ascospores cylindric, broadly rounded at apex, tapering to a narrowly rounded base, 3- or 4-septate, first septum submedian, often constricted, brown, punctate to verrucose. Asexual state coniothyrium-like. Conidiomata pseudoparenchymatous, sometimes stromatic. Conidiogenous cells lining entire locule, conidiogenesis holoblastic, proliferating percurrently, usually resulting in conspicuous annellations. Conidia globose, ovoid or ellipsoid, aseptate, yellowish-brown often becoming brown at maturity, verrucose to punctate.

Neosartorya Malloch & Cain—see **Aspergillus** P. Micheli
Trichocomaceae, Eurotiales, Eurotiomycetidae, Eurotiomycetes, Pezizomycotina, Ascomycota, Fungi
Malloch, D., R. F. Cain. 1973. The Trichocomataceae: Ascomycetes with *Aspergillus*, *Paecilomyces* and *Penicillium* imperfect states. *Can. J. Bot.* 50(12):2620.
From Gr. *néos*, new + genus *Sartorya* Vuill. (named in honor of *A. Sartory*).

Neotyphodium Glenn, et al.—see **Epichloë** (Fr.) Tul. & C. Tul.
Clavicipitaceae, Hypocreales, Hypocreomycetidae, Sordariomycetes, Pezizomycotina, Ascomycota, Fungi
Glenn, A. E., et al. 1996. Molecular phylogeny of *Acremonium* and its taxonomic implications. *Mycologia* 88 (3):369-383.
From Gr. *néos*, new + genus *Typhodium* Link, in recognition of morphological and molecular differences from the latter genus.

Nephroma Ach.
Nephromataceae, Peltigerales, Lecanoromycetidae, Lecanoromycetes, Pezizomycotina, Ascomycota, Fungi
Acharius, E. 1809. In Luyken, *Tent. Hist. Lich.*:92.
From Gr. *nephrós*, kidney + suf. -*oma*, which refers to a pathological state or to generalization of a concept, for the kidney-shaped lobules of the thalli of some species, especially when they form apothecia. Thallus foliose, small or large, irregularly lobulate. Apothecia marginal or submarginal, at times large, on inferior surface of lobules of thallus, with disc brown or reddish-brown. Photobiont *Nostoc* or *Palmella*. Terricolous, rupicolous or arboricolous.

Nephroma resupinatum (L.) Ach.: kidney-shape lobules of the foliose, terricolous thallus, with brown apothecia, x 2.

Neurospora Shear & B. O. Dodge (syn. **Gelasinospora** Dowding)
Sordariaceae, Sordariales, Sordariomycetidae, Sordariomycetes, Pezizomycotina, Ascomycota, Fungi
Shear, C. L., B. O. Dodge. 1927. Life histories and heterothallism of the red bread-mold fungi of the *Monilia sithophila* group. *J. Agric. Res.*, Washington 34:1025.
From Gr. *neũron*, nerve + *sporá*, spore, because of the spore ornamentation, which consists of furrows and ribs on ascospores arranged in manner of nerves. Perithecia conicial or pyriform, brown or black. Asci cylindrical, opening by apical pore, stipitate, with four or eight ascospores. Ascospores hyaline at first, then brown or black, with a punctate epispore. Asexual morph produces masses of pinkish conidia on substrates rich in organic substances. Causing pink bread mold. Saprobic in soil with abundant organic matter, used in genetic experiments.

Nia R. T. Moore & Meyers
Niaceae, Agaricales, Agaricomycetidae, Agaricomycetes, Agaricomycotina, Basidiomycota, Fungi

Moore, R.T., S.P. Meyers. 1959. Thalassiomycetes I. Principles of delimitation of the marine mycota with the description of a new aquatically adapted Deuteromycete genus. *Mycologia* 51(6):871-876.

From *Ni*, Andean Indian personification of the sea + L. des. *-a*. Fructification sclerotial, producing conidia abundantly over surface of leathery cortex; center gelatinous. Conidia spherical to oval, with one apical and several basal bristles. Isolated on beech wood shavings submerged for several months in a bay of Florida, U.S.A.

Neurospora cerealis (Dowding) Dania García, et al. (syn. **Gelasinospora cerealis** Dowding): pyriform, black perithecium, on soil; x 100, ascus with eight ascospores, x 200; two free ascospores with punctate epispore, x 450.

Neurospora sitophila Shear & B. O. Dodge: furrows and ribs on the ascospore wall, x 1,000.

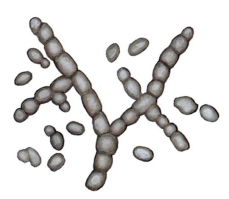

Neurospora sitophila Shear & B. O. Dodge (syn. **Monilia sitophila** (Mont.) Sacc.): conidiophores with catenulate, acropetal conidia, x 400.

Nidula V. S. White

Agaricaceae, Agaricales, Agaricomycetidae, Agaricomycetes, Agaricomycotina, Basidiomycota, Fungi

White, V. S. 1902. The Nidulariaceae of North America. *Bull. Torrey bot. Club* 29:271.

From L. *nidus*, nest + L. dim. suf. *-ula*, because the fructification resembles a small bird's nest. Fructification with peridioles lacking funiculum, with a viscous surface to adhere to objects that they contact. Peridium persistent, epihragm on upper part of fructification that breaks at dehiscence exposing peridioles. Lignicolous, rarely terrestrial.

Nidula niveotomentosa (Henn.) Lloyd: fructifications that resemble small bird's nest, showing the brown peridioles, on wood, x 2.5.

Nidularia Fr.

Agaricaceae, Agaricales, Agaricomycetidae, Agaricomycetes, Agaricomycotina, Basidiomycota, Fungi

Fries, E. M. 1818. *Symb. gasteromyc.* (Lund) 1:2.

From L. *nidula*, small nest + suf. *-aria*, which indicates receptacle, case. Fructifications globose, containing sporiferous peridioles without a funiculum, peridioles with mucilage, thin wall fragments at maturity, lacking epihragm. Lignicolous, terricolous, or fimicolous; frequently hygrophilic, develops among mosses.

Nidularia deformis (Willd.) Fr.: terricolous, sphaerical fructification, that contains sporiferous, brown peridioles, x 5.

Nigrospora Zimm.

Inc. sed., Inc. sed., Inc. sed., Sordariomycetes, Pezizomycotina, Ascomycota, Fungi

Nitschkia

Zimmerman, A. 1902. *Centbl. Bakt. ParasitKde,* Abt. I 8:220.

From L. *nigro,* to be black < *niger,* black + Gr. *sporá,* spore, because its conidia are black, aseptate, smooth, ovoid to subglobose, somewhat flattened in horizontal axis, borne on short, hyaline pedicels, at a more or less upright angle with respect to vegetative hyphae. Common as a parasite of cereals, rarely a soil saprobe. Common species, *N. oryzae* Hudson, sexual morph khuskia-like.

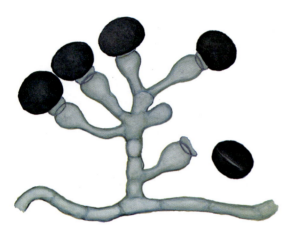

Nigrospora oryzae (Berk. & Broome) Petch: conidiophore with hyaline conidiogenous cells, bearing black conidia, with germ slit, x 1,000. Isolated in Sri Lanka.

Nitschkia G. H. Otth ex P. Karst. (syn. **Scortechiniellopsis** Sivan.)

Nitschkiaceae, Coronophorales, Hypocreomycetidae, Sordariomycetes, Pezizomycotina, Ascomycota, Fungi

Karsten, P. A. 1873. *Bidr. Känn. Finl. Nat. Folk* 23:13.

Named in honor of *Th. Nitschke* + L. suf. *-ia,* which denotes pertaining to. Ascomata superficial, scattered to densely crowded, black, on a dense subiculum of brown, septate, branched hyphae, black with a metallic iridescence, turbinate to clavate, ornamented with short, tooth-like, often furcate spines. The ascomata lack a definite ostiole but are provided with a "Quellkörper" of mucilaginous cells at the region below the apex. Asci clavate, long stalked, unitunicate, eight spored, evanescent. Ascospores brown, unicellular to one septate, allantoid. Paraphyses filiform and hyaline. Saprobic on dead wood.

Nitschkia cupularis (Pers.) P. Karst.: gregarious, erumpent perithecia, on a stroma, in branch of *Acer,* x 2.

Nowakowskiella J. Schröt.

Nowakowskiellaceae, Chytridiales, Chytridiomycetidae, Chytridiomycetes, Inc. sed., Chytridiomycota, Fungi

Schröter, J. 1893. Phycomycetae. *In:* A. Engler and K. Prantl, *Nat. Pflanzenfam.,* Teil. I (Leipzig) 1(1):63-141.

From genus *Nowakowskia* Borzi, dedicated to the naturalist L. Nowakowski + L. dim. suf. *-ella.* Polycentric with extensive, branched rhizoidal system. Zoosporangia operculate. Saprobic on decomposing plants.

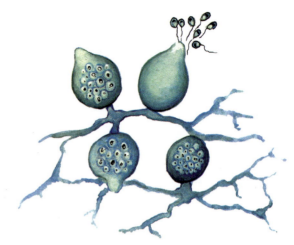

Nowakowskiella atkinsii Sparrow: polycentric thallus, with rhizoidal system, many operculate zoosporangia, one of these discharging uniflagellate zoospores, x 500. Isolated in Cuba.

Nummularia Tul. & C. Tul.—see **Biscogniauxia** Kuntze

Xylariaceae, Xylariales, Xylariomycetidae, Sordariomycetes, Pezizomycotina, Ascomycota, Fungi

Tulasne, L. R., C. Tulasne. 1863. *Select. fung. carpol.* (Paris) 2:42.

From L. *nummulus,* small coin + suf. *-aria,* which indicates connection or possesion, referring to the shape of the stromata, which are crustous, rounded or discoidal, similar to the head of a nail or small coin.

Nypaella K. D. Hyde & B. Sutton

Inc. sed., Inc. sed., Inc. sed., Inc. sed., Pezizomycotina, Ascomycota, Fungi

Hyde, K. D., B. C. Sutton. 1992. *Nypaella frondicola* gen. et sp. nov., *Plectophomella nypae* sp. nov., and *Pleurophomopsis nypae* sp. nov. (Coelomycetes) from intertidal fronds of *Nypa fruticans. Mycol. Res.* 96(3):210-214.

From host genus *Nypa* + L. dim. suf. *-ella.* Conidiomata pycnidial, formed on subiculum, superficial, separate or aggregated, globose to pyriform, apricot or pale brown, with a central ostiole. Conidiophores formed from base of conidioma, hyaline, branched, septate, filiform. Conidiogenous cells phialidic, determinate, cylindrical, hyaline, smooth, proliferating enteroblastically. Conidia holoblastic, hyaline, aseptate, ellipsoid, smooth.

O

Ophiostoma ulmi

O

Obstipispora R. C. Sinclair & Morgan-Jones
Inc. sed., Inc. sed., Inc. sed., Inc. sed., Pezizomycotina, Ascomycota, Fungi
Sinclair, R. C., G. Morgan-Jones. 1979. Notes on Hyphomycetes. XXIX. *Obstipispora chewaclensis* gen. et sp. nov. *Mycotaxon* 8(1):152-156.
From L. *obstipus*, bent, inclined to one side + *spora*, spore, for arrangement of spores. Mycelium of branched, septate, hyaline, smooth hyphae, mostly immersed. Conidiophores micronematous, mononematous, hyaline, filiform, erect, straight or flexuous, smooth, septate. Conidiogenous cells monoblastic, integrated, terminal. Conidia solitary, acrogenous, hyaline, smooth, multiseptate, cylindrical, tapering gradually towards ends, abruptly bent towards middle, obtuse at apex, subtruncate at base. On leaves of *Populus deltoides* submerged in stream in Lee County, Alabama, U.S.A.

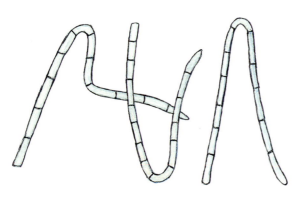

Obstipispora chewaclensis R. C. Sinclair & Morgan-Jones: multiseptate, cylindrical conidia, x 700. On leaves of *Populus deltoides*, submerged in a stream, in Alaska, U.S.A.

Ochrolechia A. Massal.
Ochrolechiaceae, Pertusariales, Ostropomycetidae, Lecanoromycetes, Pezizomycotina, Ascomycota, Fungi
Massalongo, A. B. 1852. *Ric. auton. lich. crost.* (Verona):30.
From Gr. pref. *ocro-*, derived from Gr. *ochrós*, ochre, yellow + *léchos*, *lécheos*, bed. Thallus crustose, thick, grayish or whitish. Apothecia with disc more or less concave or flat, frequently brown or yellowish, pink or brick color, with a thick border, similar to thallus in color and structure, at times rugose, verrucose or flexuous. Photobiont *Pleurococcus*. On trees and rocks.

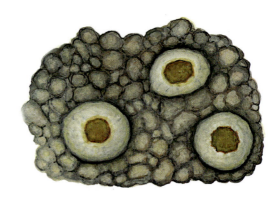

Ochrolechia parella (L.) A. Massal.: crustose thallus, with yellowish apothecia, x 8.

Ocotomyces H. C. Evans & Minter (syn. **Uyucamyces** H. C. Evans & Minter)
Inc. sed., Rhytismatales, Leotiomycetidae, Leotiomycetes, Pezizomycotina, Ascomycota, Fungi
Evans, H. C., D. W. Minter. 1985. Two remarkable new fungi on pine from Central America. *Trans. Br. mycol. Soc.* 84(1):57-78.
From Am. Sp. *ocote* < Nahuatl *ocotl*, Central American pine tree + L. *-myces*, fungus. Colonies scattered, of clusters containing between 10 and 30 fruitbodies, erumpent, markedly blackened by stroma. Stromata immersed, of brown-walled cells. Fruitbodies apothecioid, cupulate, erumpent, superficial, black, shiny, opening by irregular splits in upper surface. Asci cylindrical to saccate, thick-walled when young, becoming thin-walled in lower part when mature, 8-spored, J-. Ascospores fusiform to biapiculate, hyaline, 1-celled, smooth. Sporophores of two types: first type hyaline, thin-walled, smooth, septate, of cells dichotomously branched at an acute angle, apical cell of each branch slightly swollen in its upper half, producing a single spore. Spores allantoid,

hyaline, thin-walled, smooth, guttulate, aseptate, with an apical and basal beak; second type spores hyaline, thin-walled, aseptate, smooth, filiform, curved, produced from swollen cells lower down sporophores of first spore type. On living leaves, branches and stems of *Pinus maximinoi* on Cerro Uyuca, Honduras.

Oedocephalum glomerulosum (Bull.) Sacc.: conidiophores with apical ampulla, on which are borne botryoblastospores, on agar culture, x 350.

Ocotomyces parasiticus H. C. Evans & Minter: sagittal section of an ascoma, on living leaves of *Pinus maximinoi*, Honduras, x 70; an ascus, x 250, with eight, fusiform ascospores, x 270.

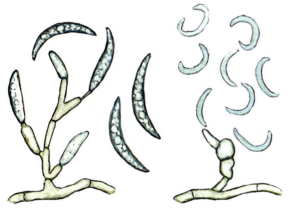

Ocotomyces parasiticus H. C. Evans & Minter (syn. **Uyucamyces parasiticus** H. C. Evans & Minter): two different sporophores, the first is hyaline, composed of cells dichotomously branched in an acute angle, with allantoid, aseptate conidia, x 175; the second sporophore is filiform with conidia probably functioning as spermatia, x 1,700. On living, fallen leaves of *Pinus maximinoi*, in Cerro Uyuca, Honduras.

Oedocephalum Preuss

Inc. sed., Pezizales, Pezizomycetidae, Pezizomycetes, Pezizomycotina, Ascomycota, Fungi
Preuss, G. T. 1851. Übersicht untersuchter Pilze, besonders aus der Umgegend von Hoyerwerda. *Linnaea* 24:131-132.
From Gr. *oídos*, tumor, swelling < *oidéo*, to swell up + *kephalé*, genit. *kephalós*, head + dim. suf. *-um*, i.e., a small swollen head. Conidiophore with apical ampulla bearing botryoblastospores more or less simultaneously. Soil fungus.

Oidiopsis Scalia—see **Leveillula** G. Arnaud

Erysiphaceae, Erysiphales, Leotiomycetidae, Leotiomycetes, Pezizomycotina, Ascomycota, Fungi
Scalia, G. 1902. *Agric. Calabro-Siculo* 27:393-397.
From genus *Oidium* < Gr. *oón*, egg + L. dim. suf. *-ium* + L. suf. *-opsis* < Gr. *-ópsis*, aspect, appearance, i.e., similar to *Oidium* Link, the asexual morph of *Erysiphe* Hedw. f. ex DC. emend. Lév.

Oidium Link—see **Blumeria** Golovin ex Speer

Erysiphaceae, Erysiphales, Leotiomycetidae, Leotiomycetes, Pezizomycotina, Ascomycota, Fungi
Link, J. H. F. 1824. *In*: Willdenow, *Sp. pl.*, Edn 4, 6 (1):121.
From Gr. *ooídion* < *oón*, egg + suf. *-ídion* < L. dim. suf. *-idium*, in reference to the mature conidia, which are more or less egg-shaped.

Okeanomyces K. L. Pang & E. B. G. Jones (syn. **Remispora** Linder):

Halosphaeriaceae, Microascales, Hypocreomycetidae, Sordariomycetes, Pezizomycotina, Ascomycota, Fungi
Pang, K. L., et al. 2004. *Okeanomyces*, a new genus to accommodate *Halosphaeria cucullata* (Halosphaeriales, Ascomycota). *Bot. J. Linn. Soc.* 146(2):223-229.
From Gr. *okeanos*, ocean + *mykés*, fungus, in reference to the habitat of the fungus. Ascomata subglobose or ellipsoidal immersed or semi-immersed, brown or black, coriaceous or subcarbonaceous. Periphyses absent. Asci clavate, with short peduncles, thin-walled, early deliquescing. Catenophyses present. Ascospores cylindrical, thin-walled, with or without a single cap-like appendage. On intertidal and drifting wood, test panels, mangrove roots and pneumatophores, and seedlings of *Rhizophora mangle*, submerged branches of *Pluchea fosbergii*, endocarp of *Cocos nucifera*; sometimes in the calcareous lining of teredinid tubes.

Oligoporus

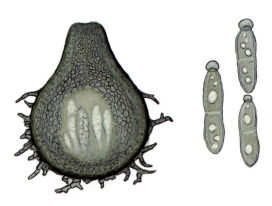

Okeanomyces cucullatus (Kohlm.) K. L. Pang & E. B. G. Jones (syn. **Remispora cucullata** Kohlm.): vertical section of a perithecium, showing its pseudoparenchyma, utricular cells, and asci with ascospores, x 300; three liberated, one-septate ascospores with vacuoles and appendanges (caps) in the apical end, x 1,000. On immersed test blocks, Atlantic Ocean, North Carolina, U.S.A.

Oligoporus Bref.—see Postia Fr.
Fomitopsidaceae, Polyporales, Inc. sed., Agaricomycetes, Agaricomycotina, Basidiomycota, Fungi
Brefeld, O. 1888. Basidiomyceten III. Autobasidiomyceten. *Unters. Gesammtgeb. Mykol.* (Leipzig) 8:114.
From Gr. *olígos*, few, small + *póros* > L. *porus*; with few or small pores.

Oliveonia Donk (syn. Monosporonella Oberw. & Ryvarden)
Oliveoniaceae, Auriculariales, Inc. sed., Agaricomycetes, Agaricomycotina, Basidiomycota, Fungi
Donk, M. A. 1958. Notes on resupinate Hymenomycetes V. *Fungus.* 28(1-4):16-36.
Dedicated to the American mycologist L. S. Olive + Gr. *on*, genit. *ontós*, a being + L. suf. *-ia*, ending of nouns denoting quality of or state of being. Basidiomata thin and waxy, usually grey. Hymenium dense, at least when young, comprising one or more layers of basidia, with scattered cystidia, arising from narrow, agglutinated, subhymenial hyphae. Hyphae binucleate, thin-walled, often agglutinate, with or without clamp-connexions; subicular hyphae sometimes wider, with slightly thickened walls. Cystidia, if present, tubular, obtuse, thin-walled, hyaline, projecting with age. Basidia globose to widely clavate, narrowly stalked, pleural in young or thin basidiomes, 4-sterigmata. Basidiospores subglobose to oblong or citriform, producing secondary spores by replication. Widespread, known from north temperate and tropical zones. Presumably saprobic on rotten wood, or associated with termite nests.

Olpidium (A. Braun) J. Schröt.
Olpidiaceae, Olpidiales, Chytridiomycetidae, Chytridiomycetes, Inc. sed., Chytridiomycota, Fungi
Schröter, J. 1868. Dir Pilze Schlesiens:180. *In*: F. Cohn, *Flora Europaea algarum aquae dulcis et submarinae* 3:1-814.
From Gr. *olpís*, oil-jar, any small vessel + dim. des. *-idion* > L. dim. suf. *-idium*, for the shape of the sporangium. Sporangium endobiotic, predominantly spherical or ellipsoidal, with a discharge tube for zoospores. Parasitic on freshwater algae such as *Zygnema*, phanerogamic plants, and microscopic aquatic animals.

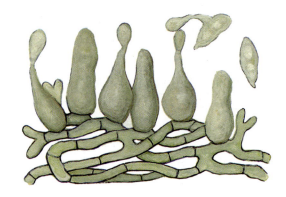

Oliveonia termitophila (Oberw. & Ryvarden) P. Roberts (syn. **Monosporonella termitophila** Oberw. & Ryvarden): efibulate hyphae, cylindrical cystidia, and subglobose basidia with a single basidiospore, x 460. Isolated from termite runs, Zambia.

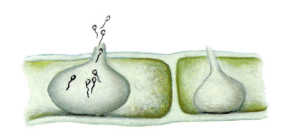

Olpidium endogenum (A. Braun) J. Schröt.: endobiotic, sphaerical sporangia, inside the thallus of freshwater alga (*Zygnema*), discharging uniflagellate zoospores, x 450.

Ombrophila Fr.
Helotiaceae, Helotiales, Leotiomycetidae, Leotiomycetes, Pezizomycotina, Ascomycota, Fungi
Fries, E. M. 1849. *Summa veg. Scand.,* Sectio Post. (Stockholm):357.
From Gr. *ómbros*, rain + *phílos*, having an affinity for, i.e., wet places. Apothecia fleshy to gelatinous, stipitate, stem often reduced in length, pallid to purplish. Asci clavate to cylindric, usually 8-spored. Ascospores elongate-ellipsoid to fusoid, often unequal-sized, hyaline, simple. Paraphyses variable. On leaves and wood in water or wet places.

Omphalina Quél.
Tricholomataceae, Agaricales, Agaricomycetidae, Agaricomycetes, Agaricomycotina, Basidiomycota, Fungi
Quélet, L. 1886. *Enchir. fung.* (Paris):42.

Onychophora

Quélet, L., 1886. *Enchiridion fungorum in Europa et praesertim in Gallia vigentium* (Paris):42.

From Gr. *omphalós*, navel + L. suf. *-ina*, which denotes similarity, for the umbilicate or omphaloid pileus. It is one of the few fungi in the order Agaricales that can become lichenized. The pileus generally is pigmented and brightly colored (yellow, green, purple) or opaque (gray or almost black), and more or less hygrophanous, smooth or scaly near the center. It has decurrent gills and a fleshy to subcartilaginous stipe that is hollow in mature specimens. It grows on sandy or rocky soil, dead wood or among mosses.

Omphalodium Meyen & Flot.
Parmeliaceae, Lecanorales, Lecanoromycetidae, Lecanoromycetes, Pezizomycotina, Ascomycota, Fungi
Meyen, F. J. F., J. Ch. Flotow. 1843. Observationes botanicas. *Nova Acta Phys.-Med. Acad. Caes. Leop.-Carol. Nat. Cur.* Suppl. 1 *119*:223.

From L. *omphalodium* < Gr. *omphalós*, navel + L. suf. *-odium* < Gr. *-oeídes*, similarity, i.e., similar to a navel. Thallus foliose with a central depression where united to substrate, coriaceous, yellowish-green, upper surface with dark brown or black apothecia. On rocks and boulders in high areas of the mountains.

Omphalora T. H. Nash & Hafellner
Parmeliaceae, Lecanorales, Lecanoromycetidae, Lecanoromycetes, Pezizomycotina, Ascomycota, Fungi
Nash, et al. 1990. *Omphalora*, a new genus in the Parmeliaceae. *Lichenologist* 22(4):355-365.

From Gr. *omphalós*, navel, umbilicus; the center of a small central projection + *óra*, care, concern. Thallus umbilicate (foliose), attached only by the umbilicus, otherwise free and raised from substrate, often markedly folded, leathery; upper surface yellow to yellowish-green or dull greenish-yellow, dull to somewhat shiny, usually with a network of ridges that are rounded and not very prominent, with conspicuous papillae developing in circular rows; pseudocyphellae punctiform; lower surface dark blue-green to black. Apothecia very common and conspicuous; thalline margin concolorous with thallus, predominantly smooth; disc red-brown, plane to slightly convex; asci clavate, 8-spored; ascospores simple, ellipsoid, hyaline, 1-celled. Pycnidia partially immersed, surrounded by blue pigments; conidia fusiform to weakly bifusiform. On exposed acidic rocks and cliff faces. Described from Arizona, U.S.A.

Omphalotus Fayod
Omphalotaceae, Agaricales, Agaricomycetidae, Agaricomycetes, Agaricomycotina, Basidiomycota, Fungi
Fayod, V. 1889. Padrome d' une histoire naturelle des agaricinées. *Annls Sci. Nat., Bot.*, sér. 7, 9:338.

From Gr. *omphalós*, navel + L. suf. *-otus*, which indicates possession or relation, referring to the generally depressed shape of the pileus. Fruiting in large numbers, pleurotoid or clitocyboid, generally of bright colors. Hymenophore frequently luminescent when fresh. On fallen conifer and broad-leaved wood. Majority of species toxic, due to muscarine.

Omphalora arizonica (Tuck. ex Willey) T. H. Nash & Hafellner (syn. **Omphalodium arizonicum** (Tuck. ex Willey) Tuck.): foliose, yellowish-green thallus, with dark brown or black apothecia, x 0.5.

Omphalotus nidiformis (Berk.) O. K. Mill.: pleurotoid, greenish, luminescent basidioma, x 1. Collected in Western Australia.

Onychophora W. Gams, et al.
Inc. sed., Inc. sed., Inc. sed., Inc. sed., Pezizomycotina, Ascomycota, Fungi
Gams, W., et al. 1984. *Onychophora*, a new genus of phialidic Hyphomycetes from dung. *Trans. Br. mycol. Soc.* 82(1):174-177.

From Gr. *ónyx*, genit. *ónychos*, talon, claw + *phóros*, bearing, for the hook-shaped phialides. Conidiophores densely crowded along hyphal tufts, arising as lateral branches from vegetative hyphae, usually consisting of 2-3 cells, distal one a hooked phialide, subterminal ones producing lateral phialides with more or less curved necks. Phialides flask-shaped, forming small clusters

Opegrapha

of conidia. Conidia obovoidal to ellipsoidal, hyaline, smooth-walled. Isolated from rabbit pellets at Devon, British Isles.

Opegrapha Ach.

Opegraphaceae, Arthoniales, Arthoniomycetidae, Arthoniomycetes, Pezizomycotina, Ascomycota, Fungi
Acharius, E. 1809. *K. Vetensk-Acad. Nya Handl. 30*:97.
From Gr. *opés*, hole, opening, crevice + *graphé*, drawing, engraving < *grápho*, to draw, scratch, depressed in surface appearing cracked or with drawings, dark when excipulum opens. Apothecia linear, fusiform or ellipsoid, occasionally branched or circular, generally black. On trees and rocks.

Opegrapha vulgata (Ach.) Ach.: crustose thallus on tree bark, with linear apothecia (lirellae), x 8.

Ophiociliomyces Bat. & I. H. Lima

Meliolaceae, Meliolales, Inc. sed., Sordariomycetes, Pezizomycotina, Ascomycota, Fungi
Batista, A. C., I. H. Lima. 1955. *Ophiociliomyces*, um novo gênero dos Parodiopsidaceae. *Anais Soc. Biol. Pernambuco* 13(2):29-37.
From Gr. *ophís*, serpent > genit. *opheós*, snake + NL. *cilium*, hair or hair-like structure + L. *myces* < Gr. *mykés*, fungus, in reference to the filiform ascospores. Superficial mycelium effuse, olivaceous, branched at right angles, without hyphopodia. Internal mycelium subhyaline, subepidermal in mesophyll. Pseudothecia sparse, superficial, globose, ostiolate, seated on a subiculum, wall prosenchymatous, brownish-black. Pseudoparaphyses filiform, hyaline, simple or branched. Asci ellipsoidal, numerous, with a thick wall, bitunicate, 8-spored. Ascospores filiform-cylindric, parallel in ascus, phragmosporous, with each terminal cell bearing a single short appendage. On living leaves of *Bauhinia radcliana* in Recife, Brazil.

Ophiocordyceps Petch (syn. Paraisaria Samson & B. L. Brady)

Ophiocordycipitaceae, Hypocreales, Hypocreomycetidae, Sordariomycetes, Pezizomycotina, Ascomycota, Fungi
Petch, T. 1931. Notes on entomogenous fungi. *Trans. Br. mycol. Soc.*16(1):55-75.
From Gr. *ophís*, a snake, a serpent, *ophíon*, name of a fabulous animal + *kordýle*, swelling, tumor + NL. *ceps*, head, a headdress. Stroma solitary, cylindrical, red, orange or black, with roughened surface due to prominent perithecial necks. Perithecia partially erumpent, flask-shaped; asci 8-spored, hyaline, cylindrical, usually with a gradually thickened apex; ascospores fusoid, hyaline, thin-walled, multiseptate, and do not divide into part-spores at the septa, arranged more or less in two overlapping bundles. This entomopathogenic genus is commonly collected on Hymenoptera adults (worker ants), or on Lepidoptera larvae.

Ophiodothella (Henn.) Höhn.

Phyllachoraceae, Phyllachorales, Inc. sed., Sordariomycetes, Pezizomycotina, Ascomycota, Fungi
Höhnel, F. X. R. von. 1910. *Sber. Akad. Wiss. Wien, Math.-naturw. Kl., Abt. 1* 119:940.
From Gr. *ophís*, genit. *opheós*, serpent, snake + *dothién*, abscess + L. dim. suf. *-ella*, for the long, filiform spores. Ascomata ostiolate perithecia, immersed in host tissue, single or clustered, with erumpent ostiolar neck surrounded by a black clypeus; perithecium brown, broadly obpyriform to laterally ovoid, with a single ostiolar neck, but sometimes ostioles amphigenous; ostiole lined with periphyses. Perithecial wall several layers thick. Centrum containing filamentous, branched paraphyses. Asci unitunicate, arranged at sides of perithecium, ellipsoidal to oval, short-stalked, with a thickened apex, 8-spored. Ascospores filiform, 1-celled, lying parallel in ascus or coiled around one another, hyaline. Asexual morph an acervular coelomycete. Parasitic in living leaves of flowering plants, causing diseases, ascomata sometimes maturing in overwintered leaves.

Ophioparma Norman (syn. Haematomma A. Massal.)

Haematommataceae, Lecanorales, Lecanoromycetidae, Lecanoromycetes, Pezizomycotina, Ascomycota, Fungi
Norman, J. M. 1852. Conatus praemissus redactionis novae generum nonnullorum Lichenum in organis fructificationes vel sporis fundatae. *Nytt Magazin for Naturvidenskapene* 7:213-252.
From Gr. *ophís*, a snake, a serpent + L. *parma*, a small round shield. Thallus crustose, corticate, sorediate. Apothecia sessile, disc round to irregular, blood red to chestnut brown, proper margin thick, concolorous with the disc; hymenium pigmented orange to umber throughout; hypothecium hyaline in the upper parts but pigmented rose to buff in the lower parts. Paraphyses rarely branched and anastomosing, somewhat thickened

at the tip. Asci with an I+ blue tholus, but without an ocular chamber or axial mass. Ascospores hyaline, thin-walled, transversely 3-7-septate. Pycnidia immersed; ostiole region pigmented dark green. On rocks and trees.

Ophioparma lapponica (Räsänen) Hafellner & R. W. Rogers (syn. **Haematomma lapponicum** Räsänen): crustose thallus, with red apothecia, on quartzile, Finland, x 1.

Ophiostoma Syd. & P. Syd. (syns. **Graphium** Corda, **Pesotum** J. L. Crane & Schokn.)

Ophiostomataceae, Ophiostomatales, Sordariomycetidae, Sordariomycetes, Pezizomycotina, Ascomycota, Fungi

Sydow, H., P. Sydow. 1919. Mykologische Mitteilungen. *Annls mycol.* 17:43.

From Gr. *ophís*, genit. *opheós*, snake, serpent > L. *ophio* + Gr. *stóma*, mouth > L. *stoma*, for the long, slender ostiolar neck. Ascomata perithecia, solitary, scattered, superficial or partially immersed, dark brown to black. Perithecia minute, ostiolate or nonostiolate; subglobose to globose, bearing short hairs on exterior; with external hairs, long, slender ostiolar neck with a fringe of bristle-like hairs around ostiole. Wall pseudoparenchymatous. Asci globose, oval or subclavate, formed in chains, 8- spored, with thin walls that deliquesce early. Ascospores 1-celled, hyaline, often appearing yellowish en mass, variable in shape, frequently asymmetrical, oval, ellipsoidal, reniform, or hat-shaped, sometimes appearing sheathed due to presence of hyaline outer wall layer. Asexual morphs occasionally with synnematal conidiophores, 2-5 mm tall, dark-colored, with a mucilaginous head, sporothrix-like. Conidia on phialides at tips of conidiophores. On woody plants as saprobes or parasites, occasionally on herbaceous plants and in soil; dispersal often by insects. *Ophiostoma ulmi* (Buisman) Nannf. is the cause of Dutch elm disease.

Orbilia Fr.

Orbiliaceae, Orbiliales, Orbiliomycetidae, Orbiliomycetes, Pezizomycotina, Ascomycota, Fungi

Fries, E. M. 1836. Corpus Florarum provincialium suecicae I. *Fl. Scan.*:1-349.

From L. *orbis*, circle, ring + suf. *-ia*, possibly referring to the round apothecia. Apothecia sessile, membranous, bright-colored, white, yellow or red, typically smooth, subcorneous when dry. Asci cylindric to clavate, inoperculate, usually 8-spored. Ascospores bacilliform or ellipsoid to subglobose. Paraphyses filiform, with apices often clavate to subglobose. On dead plant materials.

Ophiostoma ulmi (Buisman) Nannf. (syn. **Graphium ulmi** M. B. Schwarz): synnemata with conidiophores and conidial mucilaginous heads, on rotten wood, x 40.

Orbimyces Linder

Inc. sed., Inc. sed., Inc. sed., Inc. sed., Pezizomycotina, Ascomycota, Fungi

Barghoorn, E. S., D. H. Linder. 1944. Marine fungi: their taxonomy and biology. *Farlowia* 1:404-405.

From L. *orbis*, round, globe + *myces* < Gr. *mýkes*, fungus, in reference to the shape of the conidia. Conidia of a globose principal cell, with a small basal prominence and 4-5 septate apical appendages arranged radially, giving conidia a star-like appearance, contributes to flotability in water. Saprobe on organic remains in marine water.

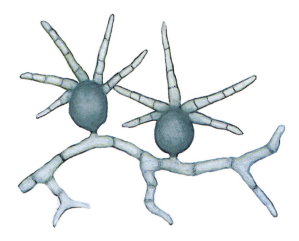

Orbimyces spectabilis Linder: staurospores, with five-septate apical appendages, arranged radially, x 1,200.

Orbispora

Orbispora Oehl, et al.—see **Scutellospora** C. Walker & F. C. Sanders

Gigasporaceae, Diversisporales, Inc. sed., Glomeromycetes, Glomeromycotina, Glomeromycota, Fungi

Oehl, F., et al. 2011. *Orbispora* gen. nov., ancestral in the Scutellosporaceae (Glomeromycetes). *Mycotaxon* 116:161-169.

From L. *orbis*, circle, orb + *spora*, spore, referring to the mono-lobed, coiled, orb-like germination shield of the spores.

Orchesellaria Manier ex Manier & Lichtw.

Asellariaceae, Asellariales, Inc. sed., Inc. sed., Kickxellomycotina, Zygomycota, Fungi

Manier, J.-F. 1958. *Orchesellaria lattesi* n. g., n. sp., Trichomycete rameux Asellariidae commensal d'un apterygote collembole *Orchesella villosa* L. *Annls Sci. Nat.,* Zool. sér. 20, *11*:131-139.

A name based on the genus of the collembolid, *Orchesella*, that serves as its host + L. suf. -*aria*, which indicates connection. Thallus branched. Arthrospores, attached to proctodeum of terrestrial collembolid insects.

Orchesellaria mauguioi Manier: branched thallus, with arthrospores, x 1,100; this trichomycete lives adhered to the proctodeum of terrestrial collembolid insects (*Isotomurus palustris*), in France.

Ornatispora K. D. Hyde, et al.—see **Stachybotrys** Corda

Inc. sed., Inc. sed., Inc. sed., Sordariomycetes, Pezizomycotina, Ascomycota, Fungi

Hyde, K. D., et al. 1999. *Byssosphaeria, Chaetosphaeria, Niesslia* and *Ornatispora* gen. nov., from palms. *Mycol. Res. 103*(11):1423-1439.

From L. *ornatus*, adorned, ornamented + *spora*, spore, for the ornamented spores.

Osmoporus Singer (syn. **Gloeophyllum** P. Karst.)

Gloeophyllaceae, Gloeophyllales, Inc. sed., Agaricomycetes, Agaricomycotina, Basidiomycota, Fungi

Singer, R. 1944. Notes on taxonomy and nomenclature of the polypores. *Mycologia* 36(*1*):65-69.

From Gr. *osmós*, to push, action of going through, to put into + *póros* > L. *porus*, pore, passage. Basidiomata annual or perennial, resupinate to pileate, leathery to woody, gills frequently confluent, anastomosed and daedaloid or labyrinthiform; hymenophore laminar. Lignicolous on dead wood of different angiosperms in tropical regions or on conifers in temperate forests, causing brown rot.

Osmoporus mexicanus (Mont.) Y. C. Dai & S. H. He (syn. **Gloeophyllum mexicanum** (Mont.) Ryvarden): labyrinthiform hymenophore of the basidioma, on dead wood, x 1.

Ostracoderma Fr.

Pezizaceae, Pezizales, Pezizomycetidae, Pezizomycetes, Pezizomycotina, Ascomycota, Fungi

Fries, E. M., 1825. *Systema orbis vegetabilis* (Lundae), 1:150.

From < Gr. *óstrakon*, head, covered with mud, soil composed of ground brick and lime + *dérma*, skin, cortex. Apparently there was confusion in considering this genus as the conidial phase of *Peziza ostracoderma*, which in reality belongs to the genus *Chromelosporium* Corda of the hyphomycetes. The name *Ostracoderma*, which belongs to the myxomycetes, is now considered a synonym of *Dictydiaethalium* Rost., and its name refers to the characteristics of the peridium of its fructifications (aethalia).

Ostracoderma minutum (R. Heim) Hennebert: conidiomata, globose, white on their velvety surface, x 1, with filamentous conidiophores that produce yellowish conidia. This pezizaceous fungus is associated with hazel plantation, in soil with decomposing vegetable detritus.

Ovadendron

Ostropa Fr.
Stictidaceae, Ostropales, Ostropomycetidae, Lecanoromycetes, Pezizomycotina, Ascomycota, Fungi
Fries, E. M. 1825. *Syst. orb. veg.* (Lundae) *1*:109.
From Gr. *óstreon*, oyster + *opé*, hole, opening, crevice, crack. Ascomata perithecioid with a prominent beak that opens by a fissure, partially or totally immersed. Asci cylindrical with thickened apex and narrow pore. Ascospores filiform, hyaline, multiseptate, violently discharged. Lignicolous and foliicolous species.

Otidea (Pers.) Bonord.
Pyronemataceae, Pezizales, Pezizomycetidae, Pezizomycetes, Pezizomycotina, Ascomycota, Fungi
Bonorden, H. F. 1851. *Handb. Allgem. mykol.* (Stuttgart):205.
From Gr. *oús*, *otós*, ear + L. suf. *-ideus*, *-idea*, derived from Gr. *eídos*, aspect, shape, similarity, and from *eído*, to appear or to look like, in reference to the shape of the apothecia. Apothecia sessile or with a wide, poorly defined pedicel, smooth, yellow, orange or grayish-yellow, cupuliform or ear-shaped due to slit in one side. On soil in coniferous forests.

Otidea leporina (Batsch) Fuckel:
ear-shaped, greyish-yellow apothecia, on forest soil, x 0.6.

Otwaya G. W. Beaton
Hyaloscyphaceae, Helotiales, Leotiomycetidae, Leotiomycetes, Pezizomycotina, Ascomycota, Fungi
Beaton, G., G. Weste. 1978. Four inoperculate Discomycetes from Victoria, Australia. *Trans. Br. mycol. Soc.* 71(2):215-221.
Named after the type location, *Otway Range* + L. des. *-a*. Apothecia superficial, scattered or aggregated on a green pigmented substrate, black, disk depressed, pale jade green when fresh, drying darker; receptacle cup-shaped on a tapering stalk or almost sessile, appearing darker than disk, black at base of stalk from an irregular covering of dark green pigment, with a thin margin to disk, marginal and upper basal area tomentose with cylindrical or slightly tapering, obtuse, usually septate, smooth, hyaline, rather thick-walled, slightly stiff hairs originating as continuations of hyphal strands of outer excipular layers. Asci cylindrical with a long, slightly tapering stalk, 8-spored, pore strongly bluing in Melzer's reagent before and after treatment with 2½% KOH. Ascospores ellipsoidal, hyaline, uniseriate, with one or two oil drops, covered with scattered, irregular, small warts that stain deeply in cotton blue. Paraphyses cylindrical with thickened tips, unbranched, basally 1-2-septate, same average length as asci. Hymenial hairs associated with paraphyses, distributed rather uniformly throughout the hymenium, filamentous, apparently continuous and same length as paraphyses. On decaying stumps and branches of *Nothofagus cunninghamii* in Melba Gully, Otway Range, Victoria, Australia.

Oudemansiella Speg.
Physalacriaceae, Agaricales, Agaricomycetidae, Agaricomycetes, Agaricomycotina, Basidiomycota, Fungi
Spegazzini, C. 1881. Fungi guaranici. *Anal. Soc. cient. argent.* 12(1):24.
Dedicated to the Dutch botanist C. A. J. A. Oudemans (1825-1906) + L. dim. suf. *-ella*. Stipe prolonged into a root-like base (pseudorrhiza). Apparently growing on buried wood or tree roots, also terricolous. Most species considered edible. Cancerstatic or antitumor properties attributed to *O. radicata* (Relh. ex Fr) Sing.

Ovadendron Sigler & J. W. Carmich.
Inc. sed., Inc. sed., Inc. sed., Inc. sed., Pezizomycotina, Ascomycota, Fungi
Sigler, L., J. W. Carmichael. 1976. Taxonomy of *Malbranchea* and some other hyphomycetes with arthroconidia. *Mycotaxon* 4(2):349-488.
From L. *ovum*, egg + Gr. *déndron*, tree, referring to the swollen conidia formed on upright conidiophore branches. Vegetative hyphae hyaline, septate. Fertile branches narrow, regularly septate in more or less basipetal succession. Alternate arthroconidia broader than fertile hyphae, developing in long chains. Mature conidia released by lysis of intervening segments, smooth, hyaline, barrel-shaped. On keratinous substrates.

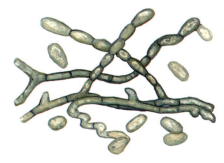

Ovadendron sulphureo-ochraceum (J. F. H. Beyma) Sigler & J. W. Carmich.: vegetative hyphae, with alternate, barrel-shaped arthroconidia, x 1,800.

Oviascoma

Oviascoma Y. J. Yao & Spooner
Pyronemataceae, Pezizales, Pezizomycetidae, Pezizomycetes, Pezizomycotina, Ascomycota, Fungi
Yao, Y.-J., B. M. Spooner. 1996. *Oviascoma*, a new genus of Otideaceae. *Mycol. Res.* 100(1):102-104.
From L. *ovum*, egg + *ascus*, ascus + *-oma*, tissue, i.e., tissue bearing asci, for the egg-shaped ascomata. Apothecia solitary or gregarious; whitish when fresh, brownish-orange to brown after drying; cylindric at first, becoming ovoid or obovoid to almost globose. Disc strongly convex, smooth. Receptacle emarginate, deeply cupulate, outer surface glabrous, attached to substratum on a narrow base. Excipular cells thin-walled, broadly ellipsoid to subglobose, marginal cells slightly elongate or clavate. Asci operculate, cylindric. Ascospores unicellular, hyaline, globose to subglobose, ornamented with spines. Paraphyses filiform, septate, flexuous. In soil in coniferous forests.

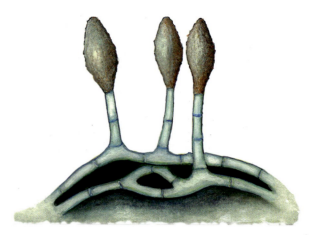

Ovulariopsis ellipsospora G. J. M. Gorter: conidiophores with ovoid-shaped conidia, x 300.

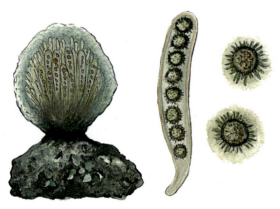

Oviascoma paludosum (Dennis) Y. J. Yao & Spooner: sagittal section of an egg-shaped ascoma, on rotten leaf of *Carex*, Great Britain, showing the hymenium, x 90; ascus with eight ornamented ascospores, x 140; two free, spiny ascospores, x 300.

Ovulariopsis Pat. & Har.
Erysiphaceae, Erysiphales, Leotiomycetidae, Leotiomycetes, Pezizomycotina, Ascomycota, Fungi
Patouillard, N., P. Hariot. 1900. *J. Bot.*, Paris 14:245.
From L. *ovularia* < *ovulum*, dim. of *ovum*, egg + L. suf. *-aris*, belonging to + suf. *-opsis* < Gr. *-ópsis*, appearance, in reference to the presence of small conidia with an ovoid shape. Chasmothecia with appendages with bulbous base and pointed at apex. Conidiophores producing a single conidium per day. Sexual morph phyllactinia-like, *Phyllactina* Lev. Powdery mildew on living leaves.

Ovulinia F. A. Weiss
Sclerotiniaceae, Helotiales, Leotiomycetidae, Leotiomycetes, Pezizomycotina, Ascomycota, Fungi
Weiss, F. A. 1940. *Ovulinia*, a new generic segregate from *Sclerotinia*. *Phytopathology* 30:236-244.
From L. *ovulium*, small egg + des. *-inia* < Gr. *-ínos*, nature of something, an apparent reference to the conidia. Sclerotia irregularly discoid to shallow cup-shaped, thin, black, formed within host tissue. Microconidia minute, globose, produced in chains on short, fusoid hyphae forming tufts on surface of host. Conidia large, obovoid, hyaline, produced singly on short, simple or branched conidiophores, forming a thin mat on surface of host. Apothecia arising singly or in groups from sclerotium. Asci slender, cylindric or subcylindric, inoperculate, 8-spored. Ascospores ellipsoid. Paraphyses mostly simple. Causing blight on flowers of *Azalea* and *Rhododendron*.

Ozonium Link—see **Coprinellus** P. Karst.
Psathyrellaceae, Agaricales, Agaricomycetidae, Agaricomycetes, Agaricomycotina, Basidiomycota, Fungi
Link, J. H. F. 1809. Observationes in ordines plantarum naturales. *Mag. Ges. Naturf. Freunde* 3:21.
From Gr. *ozómenos*, branched, nodose < *ózos*, branch, shoot, knot of a tree + L. dim. suf. *-ium*; or < Gr. *othónion*, piece of cloth, handkerchief, in reference to its hyphae, which are decumbent and branched, at first aggregated in fascicles, later ascendent and penicillate, without spores.

P

Phyllactinia guttata

P

Pacispora Sieverd. & Oehl (syn. **Gerdemannia** C. Walker et al.)

Pacisporaceae, Diversisporales, Inc. sed., Glomeromycetes, Glomeromycotina, Glomeromycota, Fungi

Sieverding C., F. Oehl, *In*: F. Oehl & C. Sieverding. 2004. *Pacispora*, a new vesicular arbuscular mycorrhizal fungal genus in the Glomeromycetes. *J. Appl. Bot.*, Angew. Bot. 78:74.

From L. *paci*, peace + Gr. *sporá*, spore, dedicated to the peace in the world. Form single spores terminally on hyphae in the soil. Spores globose, subglobose to ellipsoid, have a three-layered outer wall and an inner wall which usually is three layered. The middle layer of the inner wall reacts to Melzer's reagent. The hyphal attachment is cylindric or often slightly constricted at the base of the spore. Germination of spores is from the inner wall directly through the outer wall; one or several germ tubes are formed. Formation of sporocarps is not known. Forms vesicular arbuscular mycorrhiza.

Bainier, G. 1907. Mycothèque de l'Ecole de Pharmacie. XI. *Bull. Soc. mycol. Fr.* 23(1):26-27.

From Gr. *poikílos*, varied, diverse + *mýkes* > L. *myces*, fungus, for the varied appearance and color of its colonies. Ascomata with peridium of fine, soft, loosely interwoven hyphae. Conidiophores pedicillate with verticillate branching or penicillate heads; conidiogenous cells phialidic. Conidia ovoid or fusoid, catenulate. Saprobic in soil, some species opportunistic pathogens of animals. Sexual morph byssochlamys-like.

Paecilomyces niveus Stolk & Samson (syns. **Byssochlamys lagunculariae** (C. Ram) Samson, et al., **Byssochlamys nivea** Westling): asci with ascospores, intermixed with the mycelium, x 150.

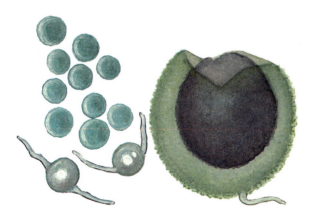

Pacispora scintillans (S. L. Rose & Trappe) C. Walker, et al. (syn. **Gerdemannia scintillans** (S. L. Rose & Trappe) C. Walker, et al.): spores, x 300; germinated spores, x 350; reaction to the Melzer's solution of endospore wall, x 500. Collected under *Cercocarpus ledifolius*, Oregon, U.S.A.

Paecilomyces Bainier (syn. **Byssochlamys** Westling)

Trichocomaceae, Eurotiales, Eurotiomycetidae, Eurotiomycetes, Pezizomycotina, Ascomycota, Fungi

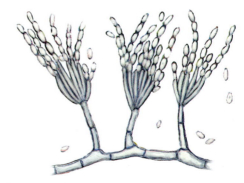

Paecilomyces variotii Bainier: conidiophores with penicillate heads of phialides, and fusoid, catenulate conidia, x 250.

Parachionomyces

Panaeolus (Fr.) Quél. (syn. **Anellaria** P. Karst.)
Inc. sed., Agaricales, Agaricomycetidae, Agaricomycetes, Agaricomycotina, Basidiomycota, Fungi
Quélet, L. 1872. Les champignons du Jura et les Voges. *Mém. Soc. Émul. Montbéliard*, Sér. 2, 5:151.
From Gr. *panaíolos*, resplendent, of various colors, variegated, from the pref. *pan-*, derived from *pan*, all + *aiólos*, mottled, streaked, variegated. Frucitifications solitary or caespitose, white or with a slight ochraceous hue on disc, often with sterile projecting margin, campanulate. Stipe with or without traces of veil, either solid or annulate or both, not pigmented, central, sometimes viscid. Gills. variegated. Basidia normally clavate, 4-spored; spore print black; spores under microscope deep purplish-fuscous to black, smooth, always large with a brown germ pore, ellipsoid to lemon-shaped. On soil in meadows and gardens; on dung and manured soil. Some species, such as *P. sphinctrinus* (Fr.) Quél., are hallucinogenic; the majority are toxic.

Verma, R. K., Kamal. 1986. *Paraaoria himalayana* gen. et sp. nov., a foliicolous Coelomycete on citrus from Nepal. *Trans. Br. mycol. Soc.* 87(4):645-647.
From Gr. *pará*, near, beside + genus *Aoria* Cif., i.e., similar to *Aoria*. Tar spot lesions epiphyllous, circular, black, affected area of leaf becoming thicker. Mycelium immersed, branched, septate, either hyaline or dark brown. Conidiomata eustromatic, shield-shaped, subcuticular, black, multilocular, sometimes unilocular. Ostiole absent, dehiscence by irregular fissures or breakdown of upper wall. Conidiophores pale brown, septate, straight, branched at base or unbranched, formed from upper cells of lower wall. Conidiogenous cells indeterminate, integrated, filiform, hyaline, smooth, with 1-4 percurrent enteroblastic proliferations, collarettes flared, periclinal thickening and channel present. Conidia holoblastic, pale brown, aseptate, verruculose, straight, apex obtuse, base truncate with a marginal frill, guttulate. On living leaves of *Citrus* sp., in Kathmandu Valley, Nepal.

Panaeolus papilionaceus (Bull.) Quél.: variegated basidiomes, on herbivore dung, x 1. Collected in Morocco.

Papulaspora immersa Hotson: microsclerotia (papulaspores), from an isolate on agar culture, x 300.

Papulaspora Preuss
Inc. sed., Inc. sed., Inc. sed., Sordariomycetes, Pezizomycotina, Ascomycota, Fungi
Preuss, G. T. 1851. Übersicht untersuchter Pilze besonders aus der Umgegend von Hoyerswerda. *Linnaea* 24:112.
From L. *papula*, protuberance, pustule + L. *spora*, spore. Consisting of microsclerotia or thallic propagules, known as papulaspores. Saprobic in soil. One species causes a disease of commercially cultivated mushrooms.

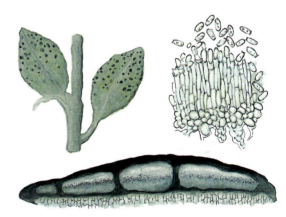

Paraaoria himalayana R. K. Verma & Kamal: habit of epiphyllous, black conidiomata on the leaves of *Citrus* sp., x 0.5; conidiomata full of conidia, x 200; conidiophores with conidia, truncate at base, x 1,500.

Paraaoria R. K. Verma & Kamal
Inc. sed., Inc. sed., Inc. sed., Inc. sed., Pezizomycotina, Ascomycota, Fungi

Parachionomyces Thaung
Inc. sed., Inc. sed., Inc. sed., Inc. sed., Pezizomycotina, Ascomycota, Fungi

Paracoccidioides

Thaung, M. M. 1979. Two new microfungi from Burma. *Trans. Br. mycol. Soc.* 71(2):333-337.

From Gr. *pará*, beside, near + genus *Chionomyces* Deighton & Piroz., for the resemblance to this genus. Mycelium superficial, hyperparasitic, composed of hyaline, apparently aseptate or hardly septate, smooth, repent hyphae that bear conidiophores as lateral and terminal branches. Conidiophores macronematous, mononematous, hyaline, simple or branched, straight or flexuous, ascending or erect, septate, geniculate, thin-walled, smooth, sympodial, cicatricized; old apices eventually displaced to one side, persistent. Conidial scars truncate-obtuse, small but prominent, thickened. Conidiogenous cells polyblastic, integrated, terminal, sympodial, more or less cylindric or inflated or truncate-triangular, cicatricized. Conidia hyaline, simple, fusiform, navicular or obclavate, attenuated towards apex, often rostrate, with a truncate or obtuse thickened hilum, three or more septate, smooth. On colonies of *Acroconidiellina arecae* M. B. Ellis on living leaves of *Areca catechu*, near Sinthawt, Yezin (Pyinrnana), Burma.

Paracoccidioides F. P. Almeida

Ajellomycetaceae, Onygenales, Eurotiomycetidae, Eurotiomycetes, Pezizomycotina, Ascomycota, Fungi

Almeida, F. P. 1930. Différences entre l'agent étiologique du granulome coccidioïdique des Etats-Unis et celui du Brésil. Nouveau genre pour le champignon brésilien. *C. r. hebd. Séanc. Mém. Soc. Biol.* 106:315-316.

From Gr. *pará*, at the side, almost + genus *Coccidioides* Stiles < L. *coccidium* < Gr. *kokkíon*, little ball + L. suf. *-oides* < Gr. *-oeídes*, similar to, i.e., it is similar to *Coccidioides* (the causal agent of coccidioidomycosis). Dimorphic fungus characterized by its multipolar budding in parasitic phase and by aleuriospores in saprobic phase. Causes paracoccidioidomycosis or South American blastomycosis in humans, a pulmonary mycosis that becomes systemic.

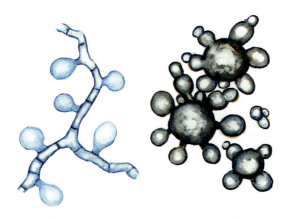

Paracoccidioides brasiliensis (Splend.) F. P. Almeida: conidiophores with aleuriospores in the saprobic phase, and multipolar budding of cells in the parasitic phase, x 1,000. Isolated from dead human patient, Brazil.

Paradentiscutata B.T. Goto, et al.

Inc. sed., Inc. sed., Inc. sed., Glomeromycetes, Glomeromycotina, Glomeromycota, Fungi

Goto, B.T., et al. 2012. Intraornatosporaceae (Gigasporales), a new family with two new genera and two new species. *Mycotaxon* 119:117-132.

From L. *para*, near, equal + *dentatus*, dentate + *scutatus*, with a shield, referring to the similarities with the germination shields of spores of Dentiscutataceae. Sporocarps unknown. Spores formed singly on bulbous sporogenous cells, terminally on subtending hypha that arise from mycelial hyphae. Spores three-walled; outer spore wall with 3-4 layers, middle wall with an expanding outer, tuberculate ornamentation towards inner wall, and an inner wall with 2-3 layers. Germination shields yellow-brown to brown, with 4-8(-10) wave-like lobed projections forming outer surface of shield; folds separating lobes on shield; each lobe with initiation from where germ tubes arise and penetrate outer wall. Recovered from tropical rainforest fragment, Serra de Jibóia, Santa Terezinha, Bahia State, Brazil.

Paradischloridium Bhat & B. Sutton

Inc. sed., Inc. sed., Inc. sed., Inc. sed., Pezizomycotina, Ascomycota, Fungi

Bhat, D. J., B. C. Sutton. 1985. Some 'phialidic' Hyphomycetes from Ethiopia. *Trans. Br. mycol. Soc.* 84(4):723-730.

From Gr. *pará*, by the side of, near + genus *Dischloridium*, similar to this genus. Colonies effuse, dark brown, velvety. Conidiophores mononematous, erect, dark brown. Conidiogenous cells terminal, integrated, lacking a collarette, proliferating enteroblastically to produce a succession of conidia at same level. Conidia holoblastic, solitary, slimy, cylindrical, obtuse at both ends, distoseptate, smooth. On dead branches in Kaffa, Ethiopia.

Paraeutypa Subram. & Ananthap.

Diatrypaceae, Xylariales, Xylariomycetidae, Sordariomycetes, Pezizomycotina, Ascomycota, Fungi

Subramanian, C. V., D. Ananthapadmanaban. 1988. *Paraeutypa*, a new genus of the Diatrypaceae. *Trans. Br. mycol. Soc.* 90(2):327-330.

From Gr. *pará*, near, beside + genus *Eutypa* Tul. & C. Tul. Clypeus effuse, forming cushion-like masses partly enclosing perithecial necks on substrate. Perithecia solitary, immersed in host-tissue, thick-walled, ostiolate. Peridium two-layered: outer layer of thick-walled, dark brown cells; inner layer of thin-walled, hyaline cells. Asci unitunicate, thin-walled, clavate with prominent stalks, mostly attached to sub-hymenium, nonamyloid, 8-spored. Ascospores allantoid, 1-celled. Paraphyses absent. On deciduous trunks in Kakkachi, Chengeltheri, Tirunelvcli Dt., Tamil Nadu State, India.

Paraisaria Samson & B. L. Brady—see **Ophiocordyceps** Petch

Ophiocordycipitaceae, Hypocreales, Hypocreomycetidae, Sordariomycetes, Pezizomycotina, Ascomycota, Fungi

Samson, R. A., B. L. Brady. 1983. *Paraisaria*, a new gen. for *Isaria dubia*, the anamorph of *Cordyceps gracilis*. *Trans. Br. mycol. Soc.* 81:285-290.

From Gr. *pará*, near + genus *Isaria* Pers., for the similarity to this genus.

Paralepistopsis Vizzini

Tricholomataceae, Agaricales, Agaricomycetidae, Agaricomycetes, Agaricomycotina, Basidiomycota, Fungi

Vizzini, A., E. Ercole. 2012. *Paralepistopsis* gen. nov. and *Paralepista* (Basidiomycota, Agaricales). *Mycotaxon* 120:253-267.

From Gr. *pará*, near + genus *Lepista* (Fr.) W. G. Sm.+ Gr. *-ópsis*, appearance, aspect, for the similarity to this genus. Basidiomata agaricoid, veil absent, pileal surface a cutis of repent to interwoven, cylindrical hyphae; spores whitish to cream, thin-walled, smooth, inamyloid, slightly cyanophilous; clamp-connections present, no sarcodimitic texture in any part of basidioma.

Paraphysoderma Boussiba, et al.

Physodermataceae, Blastocladiales, Inc. sed., Blastocladiomycetes, Inc. sed., Chytridiomycota, Fungi

James, T. Y., et al. 2011. *Paraphysoderma sedebokerense*, gen. et sp. nov., an aplanosporic relative of *Physoderma* (Blastocladiomycota). *Mycotaxon* 118:177-180.

From Gr. *pará*, near, beside + genus *Physoderma* Wallr. Sporangium monocentric, eucarpic, epibiotic. Aplanospores few to dozens per sporangium, lacking flagella, swarming before being released, exiting through tear in sporangium, with filose pseudopodia, crawling on surface of host cell before encysting and developing endogenously. Rhizoidal system developing from single point on sporangium, forming an apophysis inside host. Resting sporangium epibiotic, darker, with thicker wall than zoosporangium, germinating to release aplanospores. Closely related to, but distinct morphologically and in host preference, from the genus *Physoderma*. Parasitic on green algae.

Parapithomyces Thaung

Inc. sed., Inc. sed., Inc. sed., Inc. sed., Pezizomycotina, Ascomycota, Fungi

Thaung, M. M. 1976. New Hyphomycetes from Burma. *Trans. Br. mycol. Soc.* 66(2):211-215.

From Gr. *pará*, beside, near + genus *Pithomyces* Berk. & Broome. Colonies effuse, olivaceous brown to blackish-brown, sparse, puncticulate to glebulate or scarcely confluent in habit. Mycelium superficial, composed of branched, anastomosing, septate, olivaceous brown or brown, repent, thin-walled, flexuous, smooth hyphae. Stromata, setae and hyphopodia absent. Conidiophores semi-macronematous or macronematous, mononematous, simple or branched, straight or flexuous, septate, concolorous with mycelium, smooth. Conidiogenous cells monoblastic or polyblastic, integrated terminal or subterminal or intercalary, sympodial, cylindrical or somewhat clavate or short, inflated, mostly denticulate; denticles conic to more or less cylindrical, truncate, occasionally divided. Conidia dry, solitary or in short acropetal chains, developing from attenuated tips of conidiophores or branches, also from apices of denticles, simple, pleurogenous or acropleurogenous, straight, olivaceous brown to dark blackish-brown, smooth or apparently minutely rugulose, ovoid, obovoid, clavate, pyriform, obpyriform, ellipsoid to broadly ellipsoid, oblong or subglobose, muriform, sometimes constricted at septa, basal cell often becoming paler and protruding, with or without relics of fracturing denticles. On living leaves of *Bridelia retusa* in the Botanical Garden, Maymyo, Burma.

Paraphysoderma sedebokerense Boussiba, et al.: flagellate zoospore and amoeboid swarmer with visible pseudopodia, x 8,300, of this pathogen of green alga cells (*Haematococcus pluvialis*, commercially grown to produce astaxanthin), x 600.

Parapleurotheciopsis P. M. Kirk

Inc. sed., Inc. sed., Inc. sed., Inc. sed., Pezizomycotina, Ascomycota, Fungi

Kirk, P. M. 1982. New or interesting microfungi IV. Dematiaceous Hyphomycetes from Devon. *Trans. Br. mycol. Soc.* 78(1):55-74.

From Gr. *pará*, near + genus *Pleurotheciopsis* B. Sutton, for the similarity to this genus. Colonies effuse, hairy, brown to dark blackish-brown, often inconspicuous. Mycelium partly superficial, partly immersed in substratum, composed of pale brown to brown, smooth, branched, septate hyphae. Conidiophores macronematous, mononematous, erect, simple, smooth, septate, straight or slightly flexuous, brown to dark brown, inflated at base to form a radially lobed basal cell. Conidiogenous cells integrated, holoblastic, monoblastic, terminal, cylindrical to lageniform, percurrent. Conidia acrogenous, dry, smooth, hyaline to pale brown, composed of a single, septate or non-septate primary ramoconidium with one or more broad denticles at apex, with or without second-

Parasitella

ary or tertiary, septate or non-septate ramoconidia, also with apical denticles, ramoconidia bearing short chains of ellipsoid to broadly fusiform, septate or non-septate conidia. On decaying leaves in Exeter, Devon, U.K.

Parasitella Bainier
Mucoraceae, Mucorales, Inc. sed., Inc. sed., Mucoromycotina, Zygomycota, Fungi
Bainier, G. 1903. Sur quelques espéces de mucorinées nouvelles ou peu connues. *Bull. Soc. mycol. Fr.* 19 (2):153-172.
From L. *parasitus* < Gr. *parásitos* < *pará*, near, at the side of + *sítos*, bread, food in general + L. dim. suf. *-ella*. The name refers to the mode of life of this fungus, which parasitizes other Mucorales.

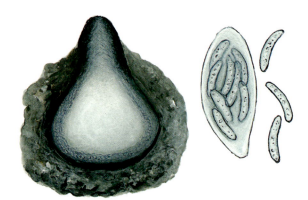

Paravalsa indica Ananthap.: sagittal section of an immersed perithecium in the tree bark, x 300; ascus with allantoid, one-celled ascospores, and free allantoid ascospores, x 1,700.

Parasitella parasitica (Bainier) Syd.: sporangiophores with sporangia, x 200.

Parasteridiella H. Maia—see **Asteridella** McAlpine
Meliolaceae, Meliolales, Inc. sed., Sordariomycetes, Pezizomycotina, Ascomycota, Fungi
Maia, H. da Silva. 1960. Fungos diversos. *Publicações Inst. Micol. Recife* 267:1-51.
From Gr. *pará*, beside, near + genus *Asteridiella* McAlpine, for the similarity to this genus.

Paravalsa Ananthap.
Valsaceae, Diaporthales, Sordariomycetidae, Sordariomycetes, Pezizomycotina, Ascomycota, Fungi
Ananthapadmanaban, D. 1990. *Paravalsa indica* gen. et sp. nov. from India. *Mycol. Res.* 94(2):275-276.
From Gr. *pará*, beside, near + genus *Valsa* Adans., for the similarity to this genus. Stroma absent. Perithecia solitary, immersed in host tissue, ostiolate, with prominent necks. Asci unitunicate, thin-walled, clavate, becoming free in perithecial cavity, non-amyloid, 8-spored. Ascospores allantoid, 1-celled. On bark in Chengeltheri, Tirunelveli, Tamil Nadu State, India.

Parberya C. A. Pearce & K. D. Hyde
Phyllachoraceae, Phyllachorales, Inc. sed., Sordariomycetes, Pezizomycotina, Ascomycota, Fungi
Pearce, C. A., K. D. Hyde. 2001. Two new gen. in the Phyllachoraceae: *Sphaerodothella* to accommodate *Sphaerodothis danthoniae*, and *Parberya* gen. novo. *Fungal Diversity* 6:83-97.
Named in honor of the Australian plant pathologist D. G. *Parbery* + L. des. *-a*, for his contributions to our understanding of the Phyllachoraceae on Poaceae. Stromata developing as suboblate to linear, black, shiny tar spots. Ascomata immersed in host tissue, beneath a black clypeus, globose to irregular with a central to eccentric ostiolar canal, lined with periphyses. Peridium of several layers of flattened, elongate, thin-walled cells. Paraphyses present. Asci 8-spored, cylindrical to clavate, unitunicate, with a nonreactive apical, disc-like apparatus. Ascospores oval to ellipsoidal, with rounded poles, aseptate, golden-brown, ornamented with short blunt spines, lacking a germ pore or germ slit. On living leaves on Mt. Kosciusko, Australia.

Parmelia Ach.
Parmeliaceae, Lecanorales, Lecanoromycetidae, Lecanoromycetes, Pezizomycotina, Ascomycota, Fungi
Acharius, E. 1803. *Methodus,* Sectio prior (Stockholmiæ): xxxiii, 153.
From L. *parma*, small round shield, buckler + suf. *-elis*, belonging to. Thallus foliose, grayish, with elongated pseudocyphellae on upper cortex; lower part black or pale, with simple or branched rhizines. Apothecia flat or saucer-shaped, generally brown, with a thin margin. Soredia and isidia also present. On rocks, conifers and deciduous trees.

Parmotrema A. Massal.
Parmeliaceae, Lecanorales, Lecanoromycetidae, Lecanoromycetes, Pezizomycotina, Ascomycota, Fungi
Massalongo, A. B. 1860. Esame comparativo di alcuni

generi di licheni. *Atti Inst. Veneto Sci. lett., ed Arti,* Sér. 3, 5:248.

From L. *parma*, small round shield, buckler + *trema*, hole, for the appearance of the apothecia. Thallus foliose with a highly modified cortex, similar to that of *Parmelia* Ach., with rhizines and pseudocyphellae. Apothecia lecanorine. Spores hyaline, unicellular. On tree bark and rocks.

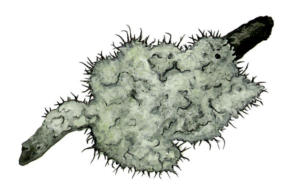

Parmotrema hypotrypum (Nyl.) Hale: folious, gray thallus with rhizines in the border, and pseudocyphellae, on tree bark, x 1.

Parmularia Nilson
Parmulariaceae, Asterinales, Dothideomycetidae, Dothideomycetes, Pezizomycotina, Ascomycota, Fungi
Nilson, B. 1907. *Flecht. Sarekgebirg.*:34.
From L. *parmula*, dim. of *parma*, buckler, small round shield, small shell + suf. *-aria*, which indicates connection or possession. Pseudothecia apothecioid, developing superficially, although connected to mycelium in interior tissues of host plant, having ascigerous loculi arranged linearly and radially from a sterile center, on which is formed a longitudinal dehiscence furrow, exposing clavate or cylindrical asci. Ascospores 1-septate.

Parmularia styracis Lév.: apothecioid, dark gray, superficial pseudothecia in the leaf of *Styrax ferrugineus*, Goiás, Brazil, x 30.

Parmulariopsella Sivan.
Parmulariaceae, Asterinales, Dothideomycetidae, Dothideomycetes, Pezizomycotina, Ascomycota, Fungi
Sivanesan, A. 1970. *Parmulariopsella burseracearum* gen. et sp. nov. and *Microcyclus placodisci* sp. nov. *Trans. Br. mycol. Soc.* 55(3):509-514.
From genus *Parmulariopsis* Petr. + L. dim. suf. *-ella*. Ascostromata epiphyllous, superficial, black, discrete, irregularly suborbicular, flattened-convex, with radiate ridges marking position of loculi, plurilocular, connected with hypostroma in epidermal cells of leaf by numerous single hyphae which penetrate cuticle. Loculi many, opening linearly along radiate ridges. Asci numerous, 8-spored, erect, ellipsoid to obovoid, rounded at apex, sessile or shortly stipitate. Paraphyses branched. Ascospores conglobate or obliquely biseriate, smooth, brown, pyriform, unequally 1-septate, with upper cell longer and broader than lower cell, constricted at septum. On living leaves of *Santiria trimera* in Njala (Kori), Sierra Leone.

Parodiellina Viégas
Parodiopsidaceae, Inc. sed., Inc. sed., Dothideomycetes, Pezizomycotina, Ascomycota, Fungi
Viégas, A. P. 1944. Alguns fungos de Brasil II. Ascomicetos. *Bragantia* 4(1-6):1-392.
From genus *Parodiella* Speg. + L. suf. *-ina*, which denotes likeness. Hyphae subhyaline, septate, intercellular, epiphyllous, subcuticular, erupting through host cuticle to form ascomata. Ascomata uniloculate pseudothecia, globose, sessile, black, rough, separate but aggregated in circular groups, with a carbonaceous wall and apical pore. Asci clavate, bitunicate, short-pedicellate, 8-spored. Ascospores ovoid, yellowish-brown, smooth, 2-celled, with a small basal cell. On leaves of *Cordia corymbosa* in São Paulo State, Brazil.

Parvomyces Santam.
Laboulbeniaceae, Laboulbeniales, Laboulbeniomycetidae, Laboulbeniomycetes, Pezizomycotina, Ascomycota, Fungi
Santamaria, S. 1995. *Parvomyces*, a new gen. of Laboulbeniales from Spain. *Mycol. Res.* 99(9):1071-1077.
From L. *parvus*, small + L. *-myces* < Gr. *-mýkes*, fungus, for the small size of the fungus. Dioecious. Male thallus of four superposed cells, uppermost functioning as a simple antheridium. Antheridium with a spinous process, i.e. persistent original spore apex, on its dorsal margin. Female thallus of a tricellular receptacle. Primary appendage unbranched, of a series of superposed cells, with a spinous process on its inferior margin. Perithecial wall cells arranged in four vertical rows. Perithecial stalk cell short. On a beetle, *Merophysia formicaria*, in Spain.

Patella F. H. Wigg.—see Scutellinia (Cooke) Lambotte
Pyronemataceae, Pezizales, Pezizomycetidae, Pezizomycetes, Pezizomycotina, Ascomycota, Fungi
Wiggers, F. H. 1780. *Prim. fl. holsat.* (Kiliae):106.

Patescospora

From L. *patella*, little plate, for the shape of the apothecia, which are discoid or cup-shaped, with a wide base.

Patescospora Abdel-Wahab & El-Shar.
Aliquandostipitaceae, Jahnulales, Inc. sed., Dothideomycetes, Pezizomycotina, Ascomycota, Fungi
Pang, K. L., et al. 2002. Jahnulales (Dothideomycetes, Ascomycota): a new order of lignicolous freshwater ascomycetes. *Mycol. Res. 106*(9):1031-1042.
From L. *patesco*, to be opened + *spora* < Gr. *sporá*, in reference to the deliquescent asci. Ascomata pseudothecia, immersed to erumpent, globose to subglobose, brown to dark brown, papillate, solitary, borne on tip of broad, erect, supporting hyphae. Asci ovoid to clavate, scattered in small locules in centrum, bitunicate, with walls dissolving at maturity, releasing spores. Ascospores ellipsoidal, hyaline, 1-septate, deeply constricted at septum, with a large mucilaginous sheath. On immersed, decaying tree branches in Egypt.

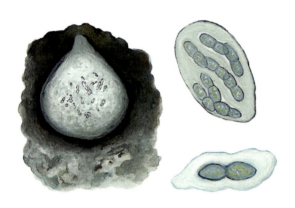

Patescospora separans Abdel-Wahab & El-Shar.: vertical section of an immersed pseudothecium in decomposing wood, x 125; ascus with ascospores, x 450; an ellipsoidal, one-septate ascospore, deeply constricted at septum, with a large mucilaginous sheath, x 400.

Paxillogaster E. Horak
Boletaceae, Boletales, Agaricomycetidae, Agaricomycetes, Agaricomycotina, Basidiomycota, Fungi
Horak, E., M. Moser. 1966. Fungi Austroamericani VIII, Über neue Gastroboletaceae aus Patagonien: *Singeromyces* Moser, *Paxillogaster* Horak und *Gymnopaxillus* Horak. *Nova Hedwigia 10*:332.
From genus *Paxillus* Fr. + Gr. *gáster*, stomach, belly. Fructifications epigeous, with lycoperdiform or pyriform habit at first, later cylindrical, of more or less irregular context. Pileus dry with loculate to sublamelliform, rarely exposed gleba. Stipe well developed, without fragments of veil. Spores bilaterally symmetrical, fusoid or ellipsoid, smooth, yellowish. Cystidia clavate, hyphae clampless. In soils of *Nothofagus* forests in South America.

Paxillus Fr.
Paxillaceae, Boletales, Agaricomycetidae, Agaricomycetes, Agaricomycotina, Basidiomycota, Fungi
Fries, E. M. 1835. Corpus Florarum provincialium suecicae I. *Fl. Scan.* 1-349.
From L. *paxillus*, little stick, pin, rack, clothes hanger, maybe in reference to the shape of the fruitbody. Pileus tomentose with involute margin when young; lamellae often anastomosing, especially near stipe, sometimes venose rugose on side, easily separable from context of pileus. Stipe central, eccentric, lateral or absent; veil present or absent; context sometimes becoming brown. Spores yellowish to brownish, smooth, with moderately thin wall, without germ pore. Widely distributed, but particularly abundant in the temperate and tropical regions of South America.

Paxina Kuntze—see **Helvella** L.
Helvellaceae, Pezizales, Pezizomycetidae, Pezizomycetes, Pezizomycotina, Ascomycota, Fungi
Kuntze, O. 1891. *Revis. gen. pl.* (Leipzig) 2:864.
From L. *patina*, disk, small glass, plate, pan + L. dim. suf. *-ina*, which denotes similarity, by the form of the apothecia.

Pedumispora K. D. Hyde & E. B. G. Jones
Diatrypaceae, Xylariales, Xylariomycetidae, Sordariomycetes, Pezizomycotina, Ascomycota, Fungi
Hyde, K. D., E. B. G. Jones. 1992. Intertidal mangrove fungi: *Pedumispora* gen. nov. (Diaporthales). *Mycol. Res.* 96(1):78-80.
From L. *pedum*, shepherd's crook + *-myces* < Gr. *-mýkes*, fungus + *sporá*, spore, in reference to the shape of the spores. Pseudostroma brown to black, covering wood surface and enclosing elements of wood, with erumpent pustules containing 1-4 immersed ascomata. Ascomata brown or black, subglobose, coriaceous, aggregated in groups, ostiolate, papillate, periphysate, with paraphyses. Asci 8-spored, fusiform, pedunculate, unitunicate, apically truncate. Ascospores filiform, multiseptate, curved, longitudinally striate, with tapering poles, one or both ends crook-like. On branches of *Rhizophora* in Thailand.

Peltigera Willd.
Peltigeraceae, Peltigerales, Lecanoromycetidae, Lecanoromycetes, Pezizomycotina, Ascomycota, Fungi
Willdenow, C. L. 1787. *Fl. berol. prodr.* (Berlin):347.
From L. *pelta* < Gr. *pélte*, small shield + L. *gero*, to bear, for the shape of the apothecia. Thallus foliose, generally large, more or less lobulate, loosely attached to substrate. Apothecia medium size to large, up to 8-9 mm diam, originating on margins of lobules; disc brown or reddish-brown, at times oblong or irregular, rolled up toward margin of thallus. Spores hyaline

Penicillium

or brownish-gray, fusiform or acicular, 3-7 septate. On soil, mosses, mossy rocks, logs and old wood.

Pedumispora rhizophorae K. D. Hyde & E. B. G. Jones: fusiform, 8-spored ascus, x 310; and filiform, multiseptate ascospores with both ends crook-like, x 350. In decomposing branches of mangrove tree (*Rhizophora apiculata*), Thailand.

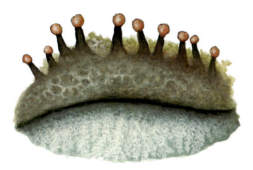

Peltigera membranacea (Ach.) Nyl.: foliose, lobulate thallus with reddish-brown apothecia, on old wood, x 1.

Penicillium Link (syns. **Carpenteles** Langeron, **Eupenicillium** F. Ludw., **Hemicarpenteles** A. K. Sarbhoy & Elphick)

Trichocomaceae, Eurotiales, Eurotiomycetidae, Eurotiomycetes, Pezizomycotina, Ascomycota, Fungi

Link, J. H. F. 1809. Observationes in ordines plantarum naturales. *Mag. Gesell. naturf. Freunde, Berlin* 3(1-2):16. From L. *penicillus*, dim. of *peniculus*, brush, dim. of *penis*, feather duster, brush, due to the shape of the conidiophore. Sexual morph cleistothecial, produced on surface of cultures, at first resembling compact sclerotia, of polyhedral, sclerenchymatous cells, globose or nearly so, covered in a thin network of branching hyphae, unilocular, producing ascogenous hyphae at center and eventually filling cleistothecium; at maturity bright-colored, surrounded by a hard peridium of several layers of thick-walled cells. Asci globose to ovoid, 8-spored. Ascospores hyaline, lenticular, with two equatorial crests. Conidiophores with a pedicel, branching pattern mono-, bi-, tri-, tetra- or poly-verticillate, i.e., with only phialides on apex, or with branches, metulae and phialides in different levels. Phialides producing catenulate conidia; conidia generally globose or ovoid, hyaline or pigmented, smooth or rough. The genus *Penicillium* is perhaps the most ubiquitous of all the fungi, with a large number of species, the majority saprobic, although some are opportunistic pathogens, several of great economic, industrial and medical importance e.g., the producers of antibiotics.

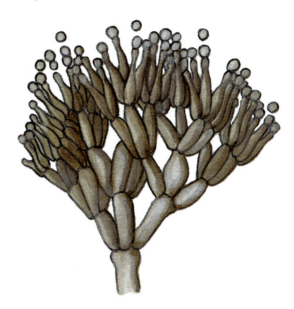

Penicillium arenicola Chalab.: conidiophore with polyverticillate branching, phialides, and globose, brownish conidia, x 1,500. Isolated in Russia.

Penicillium citrinum Thom: conidiophore with biverticillate branching, phialides, and globose, yellow conidia, x 1,500.

Peniophora

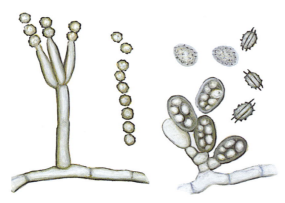

Penicillium javanicum J. F. H. Beyma (syn. **Eupenicillium javanicum** (J. F. H. Beyma) Stolk & D. B. Scott): conidiophore with phialides, x 360; and conidia, x 1,000; and asci with ascospores, x 1,250; ascospores in front and lateral views, x 1,650.

Peniophora Cooke
Peniophoraceae, Russulales, Inc. sed., Agaricomycetes, Agaricomycotina, Basidiomycota, Fungi
Cooke, M. C. 1879. On *Peniophora. Grevillea* 8:17-21.
From Gr. *peníon*, spindle, distaff, spool of a shuttle on a sewing machine or weaver + suf. *-phora*, from Gr. suf. *phóros*, from *phéro*, to carry, sustain, for the shape and disposition of the cystida in the hymenium. Similar to *Corticium* Pers., but differs in having fructifications with cystidia in hymenium. *Peniophora gigantea* (Fr.) Massee forms flat, resupinate basidiocarps, and decomposes wood, principally in pine forests. Due to its rapid growth, this species competes with *Heterobasidion annosum* Bref., cause of root rot of pines, thus inhibiting or suppressing its development. This phenomenon serves as biological control such that *P. gigantea* is cultivated in order to inoculate its spores on the stumps of trees to outcompete the pathogen.

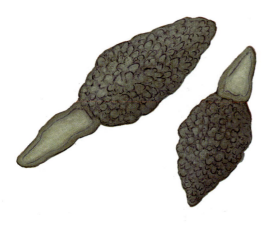

Peniophora junipericola J. Eriksś.: spindle-shaped, incrusted walled cystidia, from the hymenium, x 1,000. Collected in Sweden.

Perenniporia Murrill
Polyporaceae, Polyporales, Inc. sed., Agaricomycetes, Agaricomycotina, Basidiomycota, Fungi
Murrill, W. A. 1942. Florida resupinate polypores. *Mycologia.* 34(5):595-596.
From L. *perenis*, persistent, durable, eternal + genus *Poria* P. Browne. Basidiocarps mostly perennial, rarely annual, resupinate to pileate; pileus smooth, ochraceous to blackish in age; pore surface white to cream, pores small, isodiametric; context white to light ochraceous and fibrous to woody hard; hyphal system dimitic to trimitic; basidiospores thin- to thick-walled, gobose to ellipsoid, dropshaped to truncate, hyaline, non-dextrinoid to strongly dextrinoid, often variable within the same basidiocarp. On dead and living hardwoods and conifers. Cosmopolitan.

Perenniporia ochroleuca (Berk.) Ryvarden: resupinate, poroid, wood-inhabiting basidiome, that produce white rot, cosmopolitan, x 0.5.

Peresia H. Maia—see **Colletotrichum** Corda
Glomerellaceae, Glomerellales, Hypocreomycetidae, Sordariomycetes, Pezizomycotina, Ascomycota, Fungi
Maia, H. da Silva. 1960. Fungus diversos. *Publicações Inst. Micol. Recife* 267:1-51.
Named in honor of the Brazilian mycologist *Generosa E. P. Peres* + L. des. *-ia*.

Periconia Tode
Inc. sed., Pleosporales, Pleosporomycetidae, Dothideomycetes, Pezizomycotina, Ascomycota, Fungi
Tode, H. J. 1791. *Fung. mecklenb. sel.* (Lüneburg) 2:2.
From< Gr. *perí*, around, near + *kónis*, dust. Conidiophores simple, pigmented, producing spherical, dark, blastic conidia that are formed individually or in acropetal chains at apex. Causes leaf spots, generally associated with other fungi. Also saprobic in soil and on vegetable remains.

Pericystis Betts—see **Bettsia** Skou ex Pitt et al.
Ascosphaeraceae, Onygenales, Eurotiomycetidae, Eurotiomycetes, Pezizomycotina, Ascomycota, Fungi
Betts, A. D. 1912. A bee-hive fungus, *Pericystis alvei*, gen. et sp. nov. *Ann. Bot., Lond.* 26:795-799.

From Gr. *perí*, around, round about + *kýstis*, bladder, vesicle, cell, in reference to the sporocysts that contain spherical groups of asci.

Periconia byssoides Pers.: pigmented conidiophores, with blastic, spherical, dark conidia, on vegetable remains, x 500.

Peridiospora C. G. Wu & S. J. Lin
Endogonaceae, Endogonales, Inc. sed., Inc. sed., Mucoromycotina, Zygomycota, Fungi
Wu, C.-G., S.-J. Lin. 1997. Endogonales in Taiwan: a new genus with unizygosporic sporocarps and a hyphal mantle. *Mycotaxon* 64:179-188.
From Gr. *perí*, around + dim. suf. *-ídion* > L. *-idium* + *spora*, spore, for the hyphal enclosure of the zygospores. Zygosporocarps produced singly or in aggregates in soil or on roots, containing one zygosporangium. Zygosporangia enclosed by a hyphal peridium. Gametangia ephemeral, uniting at or near their tips with zygosporangium, budding at point of union or from one of two gametangia. Collected from National Yushan Park, Taiwan.

Peridiospora reticulata C. G. Wu & Suh J. Lin: mature zygosporangia, each one containing one zygosporangium, enclosed by hyphal peridium, x 230. Isolated from rhizosphere of native weed, Taiwan.

Peritrichospora Linder—see **Corollospora** Werderm.
Halosphaeriaceae, Microascales, Hypocreomycetidae, Sordariomycetes, Pezizomycotina, Ascomycota, Fungi
Linder, D. L. 1944. I. Classification of the marine fungi, pp. 401-433, *In*: Barghoorn, E. S. and D. L. Linder, Marine fungi: their taxonomy and biology. *Farlowia* 1:395-467.
From Gr. pref. *perí-* > NL. *peri-*, around + Gr. *trichós*, hair + *sporá*, seed, spore > L. *spora*, for the ring of cilia around the center of the ascospore.

Peroneutypa Berl.
Diatrypaceae, Xylariales, Xylariomycetidae, Sordariomycetes, Pezizomycotina, Ascomycota, Fungi
Berlese, A. 1902, *Icon. Fung.* 3:80.
From Gr. *péra*, *peré*, purse, knapsack, sack, or < *peróne*, bubble, ampoule, tassel + genus *Eutypa* Tul. & C. Tul. Stromata solitary to gregarious, immersed, becoming raised to erumpent by a long ostiolar canal, dark brown to black, glabrous, circular to irregular in shape, arranged in longitudinal or valsoid configuration, multiascoma, with conspicuous, clustered, roundish to cylindrical prominent ostioles in the centre. Ascomata perithecial, immersed in an stroma, tightly aggregated in a horizontal perspective in various rows, dark brown to black, globose to subglobose, glabrous individual ostiole with long neck. Asci unitunicate, cylindrical or clavate, sessile, with small, truncated apices; ascospores subhyaline to pale yellowish, elongate-allantoid, aseptate, smooth-walled, with small guttules. Saprobic on bark.

Peroneutypa scoparia Carmarán & A. I. Romero (syn. **Eutypella scoparia** (Schwein.) Ellis & Everh.): perithecial pseudostromata in dead bark or wood, x 50.

Persiciospora P. F. Cannon & D. Hawksw.
Nectriaceae, Hypocreales, Hypocreomycetidae, Sordariomycetes, Pezizomycotina, Ascomycota, Fungi
Cannon, P. F., D. L. Hawksworth. 1982. A re-evaluation of *Melanospora* Corda and similar Pyrenomycetes, with a revision of the British species. *J. Linn. Soc., Bot.* 84:115-160.

Pertusaria

From L. *persica*, peach + connective *-io-* + *spora*, spore, seed, because the pitted spores resemble the pit of a peach. Ascomata ostiolate perithecia, globose to subglobose, superficial, immersed or erumpent, golden-yellow or blackish, covered with hyaline hairs. Perithecial wall of flattened, angular, pseudoparenchymatous cells. Ostiolar neck cylindrical, with or without a crown of hyaline setae. Centrum of thin-walled pseudoparenchymatous cells. Asci clavate to cylindrical-fusiform, thin-walled, undifferentiated at apex, evanescent, 8-spored. Ascospores 1-celled, ellipsoid to ellipsoid-fusiform, brown, with faintly pitted walls and two small, terminal germ pores. At maturity ascospores extruded through ostiole, collecting in a cluster. On bark; also isolated from seedlings with a *Fusarium* sp., on which it may be parasitic.

Pertusaria pertusa (L.) Tuck.: crustous, verrucose, brown-reddish thallus, with apothecia, on rock, Europe, x 15.

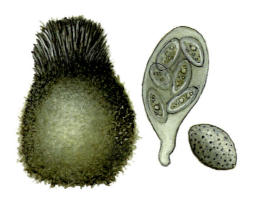

Persiciospora africana J. C. Krug: ostiolate, subglobose, blackish perithecium, covered with hyaline hairs, x 230; 8-spored ascus, x 750; one mature, brown ascospore, with faintly pitted wall, and terminal germ pore, x 1,000. On forest soil, Botswana.

Pertusaria DC.

Pertusariaceae, Pertusariales, Ostropomycetidae, Lecanoromycetes, Pezizomycotina, Ascomycota, Fungi
Lamarck, J.B. de; De Candolle, A.P. 1805. *Flore française* 2:1-600.
From L. *pertusus*, bored through, pierced, perforated + suf. *-aria*, which indicates connection or possession. Thallus crustose, sometimes with verrucose protuberances on upper surface. Apothecia immersed in depressions, opens through a pore. Photobiont *Protococcus*. On old wood, trees, rocks, soil and moss.

Pesotum J. L. Crane & Schokn.—see Ophiostoma Syd. & P. Syd. e

Ophiostomataceae, Ophiostomatales, Sordariomycetidae, Sordariomycetes, Pezizomycotina, Ascomycota, Fungi
Crane, J. L., J. D. Schoknecht. 1973. Conidiogenesis in *Ceratocystis ulmi*, *Ceratocystis piceae*, and *Graphium penicillioides*. *Am. J. Bot.* 60:346-354.
From L. *pes*, foot + des. *-otum*, which indicates possession, because of the presence of a foot at the base of the synnema.

Pestalopezia Seaver

Helotiaceae, Helotiales, Leotiomycetidae, Leotiomycetes, Pezizomycotina, Ascomycota, Fungi
Seaver, F. J. 1942. Photographs and descriptions of cup-fungi-XXXVI. *Mycologia* 34(3):298-301.
From genus *Pestalotia* De Not. + *Pezia*, from genus *Peziza* Dill. ex Fr., a *Peziza* with a pestalotia-like asexual morph. Apothecia superficial, sessile or subsessile, at first globose, becoming expanded, subdiscoid, externally pruinose or tomentose, light-colored; hymenium nearly plane, dark-colored, almost black. Asci subcylindric, inoperculate, 8-spored. Paraphyses filiform, strongly enlarged above, pale brown. On leaf spots on *Rhododendron*.

Pestalotia De Not.

Amphisphaeriaceae, Xylariales, Xylariomycetidae, Sordariomycetes, Pezizomycotina, Ascomycota, Fungi
De Notaris, G. 1841. Micromycetes italici velminus cognite. *Mém. R. Accad. Sci. Torino*, Ser. 2 3:80.
Dedicated to the Italian physician and botanist of the XIX century, *Fortunato Pestalozza*. Acervuli epidermal or subepidermal, separate or confluent. Conidia fusiform, septate with two or more apical appendages and one basal appendage that is simple or rarely branched. On vines in Europe.

Pestalotiopsis Steyaert

Pestalotiopsidaceae, Amphisphaeriales, Xylariomycetidae, Sordariomycetes, Pezizomycotina, Ascomycota, Fungi
Steyaert, R. L. 1949. Contribution à l'étude monographique de *Pestalotia* de Not. et *Monochaetia* Sacc. *Bull. Jard. bot. État Brux.* 19(3):285-354.
Name dedicated to the Italian physician and botanist *Fortunato Pestalozza* + Gr. *-ópsis*, appearance, aspect, similar to the genus *Pestalotia* De Not. Fusiform conidia formed within compact acervuli. The conidia are usu-

ally 5-celled, with 3 brown median cells and hyaline end cells, and with two or more apical appendages arising from the apical cell. Species of this genus are ubiquitous in distribution, occurring on a wide range of substrata. Many are saprobes while others are pathogenic or endophytic on living plant leaves and twigs. Some species have gained attention in recent years as they have been found to produce many important secondary metabolites. Are asexual state members of the family Amphisphaeriaceae.

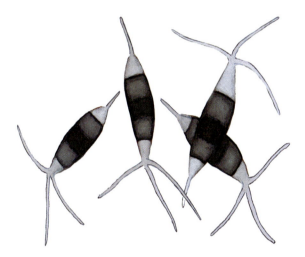

Pestalotiopsis microspora Speg.: fusiform, pigmented, septate conidia, with two or more apical, hyaline appendages, and one basal, hyaline appendage, x 1,400.

Petriella Curzi
Microascaceae, Microascales, Hypocreomycetidae, Sordariomycetes, Pezizomycotina, Ascomycota, Fungi
Curzi, M. 1930. Petriella, nuovo genere di pirenomicete. *Boll. R. Staz. Patalog. Veget. Roma* 10:380-422.
Dedicated to the Italian botanist *Lionello Petri* + L. dim. suf. *-ella*. Ascomata perithecial, nonstromatic, scattered on substrate, superficial or with base immersed, ostiolate, papillate or with a short neck, subglobose to obpyriform, black, usually with hairs or setae on upper part of perithecium and around ostiole. Perithecial wall membranous. Centrum composed of several layers of pseudoparenchymatous cells, with short, slender paraphyses. Asci unitunicate, subglobose to clavate or obovate, short-stalked, 8-spored, undifferentiated at apex, evanescent at maturity. Ascospores 1-celled, dextrinoid when young, yellowish to reddish-brown when mature, asymmetrical, elongate to ovoid, usually inequilateral, plano-convex or concavo-convex, smooth, with a germ pore at each end, extruded through ostiole in a cirrhus. Asexual morphs scedosporium-like. On plant debris, seeds, dung and in soil.

Petriellidium Malloch —see Pseudallescheria Negr. & I. Fisch.
Microascaceae, Microascales, Hypocreomycetidae, Sordariomycetes, Pezizomycotina, Ascomycota, Fungi
Malloch, D. 1970. New concepts in the Microascaceae illustrated by two new species. *Mycologia* 62(4):738-739.
From genus *Petriella* Curzi + L. dim. suf. *-idium*. This pathogen forms subspherical grains, more or less hard.

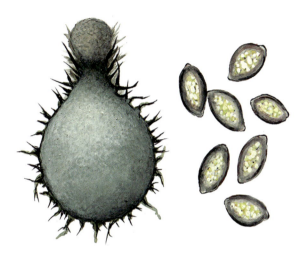

Petriella setifera (Alf. Schmidt) Curzi: ostiolate, black perithecium, with setae in the upper part, and a cirrhus with ascospores, x 280; yellowish to reddish-brown ascospores, with lipid drops and a germ pore at each end, x 1,800.

Petromyces Malloch & Cain—see Aspergillus P. Micheli ex Haller
Trichocomaceae, Eurotiales, Eurotiomycetidae, Eurotiomycetes, Pezizomycotina, Ascomycota, Fungi
Malloch, D., R. F. Cain. 1972. The Trichocomataceae: Ascomycetes with *Aspergillus*, *Paecilomyces* and *Penicillium* imperfect states. *Can. J. Bot.* 50(12):2613-2628.
From Gr. *petrós*, rock, stone + *mýkes* > L. *myces*, fungus, for the hard consistency of the stromata.

Peyritschiella Thaxt.
Laboulbeniaceae, Laboulbeniales, Laboulbeniomycetidae, Laboulbeniomycetes, Pezizomycotina, Ascomycota, Fungi
Thaxter, R. 1890. Papers read before the Academy of Arts and Sciences. Contributions from the Physical Laboratory of the Massachussets Institute of Technology II. *Proc. Amer. Acad. Arts & Sci.* 25:8.
Named in honor of the Austrian botanist *J. Peyritsch* + L. dim. suf. *-ella*. On species of coleopterans. Receptacle multicellular, bilaterally asymmetrical; subterminal column of cells forming a single compound spermogonium, superior column of cells giving rise to one or more perithecia and numerous simple, unicellular appendages.

Pezicula Tul. & C. Tul. (syn. Dermatella P. Karst.)
Dermateaceae, Helotiales, Leotiomycetidae, Leotiomycetes, Pezizomycotina, Ascomycota, Fungi

Peziza

Tualsne L. R., C. Tulasne 1865. *Select. fung. carpol.* (Paris) 3:182.

From L. *pezica*, a sessile mushroom + dim. suf. *-ula*. Apothecia usually occurring in cespitose clusters on a stromatic base, sessile or with a short thick stem-like base, light colored whitish or yellowish, rarely exceeding 1 mm in diam, tubercular or discoid, soft and fleshy; asci broad-clavate, 8-spored; spores ellipsoid, simple or becoming tardily 1-several-septate; paraphyses hyaline or subhyaline and usually free, not agglutinated and not usually forming an epithecium. The pycnospores are broad-ellipsoid and borne externally on the surface of the stroma, or in well-developed pycnidia. On trunks and dead branches.

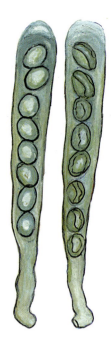

Peziza pyrophila (Korf & W. Y. Zhuang) Brumm. (syn. **Pfistera pyrophila** Korf & W. Y. Zhuang): asci showing arched apical dome, with eight ascospores; in the left ascus, the ascospores have cyanophilic perispore; in the right ascus, the mature ascospores are folded and wrinkled, x 350. On a burn site, La Palma, Canary Islands, Spain.

Peyritschiella protea Thaxt.: hermaphrodite thalli, each one with an apical perithecium, x 260. On the exoskeleton of species of coleopterans.

Peziza Dill. ex Fr. (syns. **Galactinia** (Cooke) Boud., **Pfistera** Korf & W. Y. Zhuang)
Pezizaceae, Pezizales, Pezizomycetidae, Pezizomycetes, Pezizomycotina, Ascomycota, Fungi
Fries, E. M. 1822. *Syst. mycol.* (Lundae) 2(1):40.

From Gr. *pézis* or *pézikes*, an ancient name to designate sessile fungi that are earth flowers that emit a cloud of spores when mature; from this, L. *pezicae* or *pezise*, and *peziza*. Fructifications fleshy, disk or cup-shaped apothecia with margin entire or crenulate, sessile or adhering to substrate by means of a central pedicel, commonly from 1-8 cm diam, dark brown, brown, yellowish or whitish. Receptacle often slightly lighter than disc when dry, somewhat pustulate. Ectal excipulum thick, with strongly cyanophilic walls. Medullary excipulum consisting of a loose mixture of thin-walled hyphae, ovoid to globose cells. Subhymenium hyphae with cyanophilic rings at septa. Asci 8-spored, with a thin arched apical dome at maturity, J-. Ascospores ellipsoid, smooth, contents highly refractive when mature. On humus in conifer forests, meadows and gardens fertilized with manure, and on decomposing trunks or branches.

Peziza varia (Hedw.) Alb. & Schwein.: cup-shaped, yellowish apothecia, on soil, x 1.

Pfistera Korf & W. Y. Zhuang—see **Peziza** Dill. ex Fr.
Pezizaceae, Pezizales, Pezizomycetidae, Pezizomycetes, Pezizomycotina, Ascomycota, Fungi
Korf, R. P., W. Y. Zhuang. 1991. *Kimbropezia* and *Pfistera*, two new genera with bizarre ascus apices (Pezizales). *Mycotaxon* 40:269-279.

Named in honor of the American mycologist *Donald H. Pfister* for his studies on the genus *Peziza* + L. des *-a*.

Phacidiopycnis Potebnia (syn. **Potebniamyces** Smerlis)
Cryptomycetaceae, Rhytismatales, Leotiomycetidae, Leotiomycetes, Pezizomycotina, Ascomycota, Fungi
Potebnia, A. 1912. Ein neuer Krebserreger des Apfel-

baumes *Phacidiella discolor* (Mout. & Sacc.) A. Pot., seine Morphologie und Entwicklungsgeschichte. *Zeitschr. Pflanzenkrankh.* 22:143.

From Gr. *phákos*, a lentil, lentil-shaped; in modern technical terms it often means a lens + *pyknós*, compact, dense, solid, strong, in reference to the pycniostroma full of conidia. Pycnidia partly immersed or nearly free, subglobose to more or less flattened, uni to multiloculate; ostioles or irregular apertures single or several. Conidiophores multicellular lining the pycnidial cavity, producing conidiogenous cells integrated or discrete, lageniform to irregular with conidiogenous loci single to several. Conidia hyaline, lacriform with the dehiscence end flattened or ovoid to ellipsoid without the cicatrice, smooth. The fungus causes fruit rot on apples during storage and is associated with a twig dieback and canker disease of crabapple trees and dead twigs of pear trees.

Phacidiopycnis pyri (Fuckel) Weindlm. (syn. **Potebniamyces pyri** (Berk. & Broome) Dennis): pycnidia on agar on a Petri dish, x 0.3; macro- and microconidia, x 1,000; infected pear fruit, x 0.5; biguttulate ascospores, x 800.

Phacidium Fr.
Phacidiaceae, Helotiales, Leotiomycetidae, Leotiomycetes, Pezizomycotina, Ascomycota, Fungi
Fries, E. M. 1815. *Observ. mycol.* (Havniae) 1:167.
From Gr. *phaké*, lentil, or *phakós*, from lenticular figure or wart + L. dim. suf. *-idium*. Apothecia lenticular, amphigenous, subepidermal, erumpent, dehiscent. Parasitic on undersurface of leaves of conifers causing snow blight and inducing needle fall.

Phacidium lacerum Fr.: erumpent and dehiscent apothecia, on dead, fallen needles of *Pinus sylvestris*, Sweden, x 20.

Phaeobulgaria Seaver—see Bulgaria Fr.
Bulgariaceae, Leotiales, Leotiomycetidae, Leotiomycetes, Pezizomycotina, Ascomycota, Fungi
Seaver, F. J. 1932. The Genera of Fungi. *Mycologia* 24 (2):248-263.
From L. *phaeo* < Gr. *phaiós*, dark, brown + genus *Bulgaria* Fr.

Phaeocandelabrum R.F. Castañeda, et al.
Inc. sed., Inc. sed., Inc. sed., Inc. sed., Pezizomycotina, Ascomycota, Fungi
Castañeda Ruiz, R. F., et al. 2009. *Phaeocandelabrum*, a new genus of anamorphic fungi to accommodate *Sopagraha elegans* and two new species, *Ph. callisporum* and *Ph. joseiturriagae*. *Mycotaxon* 109:221-232.
From L. *phaeo* < Gr. *phaiós*, dark, brown + genus *Candelabrum* Beverw., referring to a dark *Candelabrum*. Colonies on natural substrate effuse, epiphyllous, sometimes amphigenous, hairy, brown or black. Mycelium superficial, immersed. Conidiophores macronematous, mononematous, erect, straight, septate, smooth or verruculose, brown below to pale brown towards apex. Conidiogenous cells monoblastic, terminal, integrated, determinate or indeterminate with enteroblastic percurrent proliferations. Conidial secession rhexolytic. Conidia solitary, acrogenous, complex, multicellular, branched, irregular, pyramidal, turbinate, globose to Y-shaped, brown to dark brown with a basal frill. Synanamorph selenosporella-like, arising from the tubercles of tertiary cells. On decaying leaves in Brazil, Cuba, Mexico and Venezuela.

Phaeocollybia R. Heim
Cortinariaceae, Agaricales, Agaricomycetidae, Agaricomycetes, Agaricomycotina, Basidiomycota, Fungi
Heim, R. 1931. *Le genre Inocybe* (Paris) Encyclop. Mycol. 70.
From L. *phaeo* < Gr. *phaiós*, dark, brown + genus *Collybia* (Fr.) Staude, a dark *Collybia*. Pileus generally glabrous, viscid or glutinous, conical or campanulate, usually umbonate, later expanding; lamellae more or less free or adnexed; hyphae of context with or without clamp connections. Spores ochraceus-brown, verrucose to almost smooth. On soil and humus, frequently in coniferous woods, especially of *Pinus*. Facultative ectomycorrhizal in temperate and tropical regions.

Phaeohelotium Kanouse—see Hymenoscyphus Gray
Helotiaceae, Helotiales, Leotiomycetidae, Leotiomycetes, Pezizomycotina, Ascomycota, Fungi
Kanouse, B. B. 1935. Notes on new or unusual Michigan Discomycetes. II. *Pap. Mich. Acad. Sci.* 20:65-78.
From L. *phaeo* < Gr. *phaiós*, dark, brown + genus *Helotium* Tode, i.e., a *Helotium* with brown spores.

Phaeolus

Phaeolus (Pat.) Pat.
Fomitopsidaceae, Polyporales, Inc. sed., Agaricomycetes, Agaricomycotina, Basidiomycota, Fungi
Patouillard, N. 1900. *Essai Tax. Hyménomyc.* (Lons-le-Saunier):86.
From L. *phaeo* < Gr. *phaiós*, dark, brown + dim. suf. *-olus*, in reference to the dark color of the fruitbody. Basidocarps annual, sessile to stipitate; upper surface orange at first, becoming brown, strigose to fibrillose; pore surface orange to greenish-brown; pores daedaleoid to circular; context orange to brown, fibrous to spongy, hyphal system monomitic, basidia clavate, simple, septate at base, basidiospores ellipsoid to cylindric, hyaline, smooth. Causes brown rot of living conifers. Cosmopolitan monotypic genus.

Phaeosphaeriopsis M. P. S. Câmara, et al.
Phaeosphaeriaceae, Pleosporales, Pleosporomycetidae, Dothideomycetes, Pezizomycotina, Ascomycota, Fungi
Câmara, M. P. S., et al. 2003. *Neophaeosphaeria* and *Phaeosphaeriopsis*, segregates of *Paraphaeosphaeria*. *Mycol. Res.* 107:515-522.
From genus *Phaeosphaeria* I. Miyake + Gr. *-ópsis*, aspect, for the similarity to this genus. Ascomata immersed, subepidermal, usually erumpent at maturity, globose to subglobose to pyriform, often papillate, solitary or gregarious in a stroma, often surrounded by septate, brown hyphae extending into host tissues. Asci bitunicate, cylindric. Ascospores cylindric, broadly rounded at apex, tapering to narrowly rounded base, 4-5-septate, first septum submedian, often constricted, brown, echinulate, punctate or verrucose. Conidiomata pseudoparenchymatous, sometimes of scleroplectenchyma. Conidiogenous cells lining locule; conidiogenesis holoblastic, proliferating percurrently resulting in inconspicuous annellations. Conidia cylindrical, often truncate at ends, 0-3-septate, yellowish-brown, punctate. Sometimes only bacillar microconidia produced from simple, apparently nonproliferating phialides. Asexual morph coniothyrium-like or phaeostagonospora-like.

Phaeotremella Rea
Tremellaceae, Tremellales, Inc. sed., Tremellomycetes, Agaricomycotina, Basidiomycota, Fungi
Rea, C. 1912. New and rare British fungi. *Trans. Br. mycol. Soc.* 3:376-380.
From L. *phaeo* < Gr. *phaiós*, dark, brown + genus *Tremella* Pers. Basidocarps large, foliaceous, varying in colour from pale brown to totally black; lobes arising directly from the point of attachment, entire, rotund, even to undulate. Basidia varying in shape (ovoid to subglobose), slightly thick-walled, with hyaline to brownish content; swollen cells often present on basal lobes, broadly ellipsoid to globose, slightly to very thick-walled, producing ellipsoid or subglobose conidial cells; basidiospores broadly ellipsoid or subglobose, hyaline to brownish. On fallen logs of conifers and deciduous trees and is known to parasitize crust fungi of the genus *Stereum* Hill ex Pers.

Phaeosphaeriopsis glauco-punctata (Grev.) M. P. S. Câmara, et al.: vertical section of ascoma, in leaf of *Ruscus aculeatus*, x 175; ascus with eight, multiseptate ascospores, x 650; free, four-septate ascospores, x 500.

Phaeotremella foliacea (Pers.) Wedin, et al. (syn. **Tremella foliacea** Pers.): foliose, lobulate, brightly colored basidioma, on wood in soil, x 1.

Phaeotrichosphaeria Sivan.
Phaeosphaeriaceae, Pleosporales, Pleosporomycetidae, Dothideomycetes, Pezizomycotina, Ascomycota, Fungi
Sivanesan, A. 1983. Studies on Ascomycetes. *Trans. Br. mycol. Soc.* 81(2):313-332.
From L. *phaeo* < Gr. *phaiós*, dark, brown + genus *Trichosphaeria* Fuckel, i.e., a *Trichosphaeria* with dark spores. Perithecia superficial, solitary and scattered or aggregated in small groups, globose, subglobose or ovoid, setose, with an apical, short, papillate, ostiole lined on inside by hyaline, filiform periphyses. Perithecial wall of polygonal, brown cells, towards interior becoming less thick-walled, subhyaline to hyaline, compressed, elongated. Asci cylindrical, short-stalked, 8-spored, thin-walled, unitunicate, evanescent, with a non-amyloid indistinct apical structure. Ascospores pale brown to brown when mature, smooth, ovoid, 0-1-septate. Paraphyses filiform, hyaline. On stems of *Lawsonia alba* in Narsingpur, India.

Phaffia M. W. Mill., et al.—see **Xanthophyllomyces** Golubev
Mrakiaceae, Cystofilobasidiales, Inc. sed., Tremellomycetes, Agaricomycotina, Basidiomycota, Fungi
Miller, M. W., et al. 1976. *Phaffia*, a new yeast genus in Deuteromycotina (Blastomycetes). *Int. J. Syst. Bacteriol.* 26(2):286-291.
Dedicated to the Dutch mycologist H. J. Phaff + L. des. -ia.

Phallus Junius ex L. (syn. **Dictyophora** Desv.)
Phallaceae, Phallales, Phallomycetidae, Agaricomycetes, Agaricomycotina, Basidiomycota, Fungi
Linnaeus, C. 1753. *Sp. pl.* 2:1178.
From L. *phallus* < Gr. *phallós*, phallus, for the shape of the fructification. Receptaculum of a hollow and spongy column that supports a campanulate pileus on upper part, surface alveolate, granulose-reticulate or smooth, with a conspicuous pore at apex. Gleba viscous, fetid with elliptical, smooth spores. On rich soils, vegetable residue or rotted wood in forest clearings as well as in gardens and grasslands of temperate and subtropical regions, and sometimes in sandy soils.

Phellinus Quél.
Hymenochaetaceae, Hymenochaetales, Inc. sed., Agaricomycetes, Agaricomycotina, Basidiomycota, Fungi
Quélet, L. 1886. *Enchir. fung.* (Paris):172.
From Gr. *phellinós*, cork-like, similar to cork, referring to the consistency of the fructification. Fruitbodies pileate to resupinate, perennial, rarely annual, pileus dark brown to black, hirsute to glabrous, often sulcate, pores mostly small, tubes usually stratified, context thin, hyphal system dimitic, generative hyphae generally hyaline, spores of variable shapes, hyaline to rusty brown, sometimes dextrinoid. On dead wood. Cosmopolitan.

Phallus impudicus L.: mature basidioma, with stipe, and reticulate pileus, on soil, x 0.6. Collected in forests, Europe.

Phallus indusiatus Vent. (syn. Dictyophora indusiata (Vent.) Desv.): basidioma showing the indusium and mucilaginous gleba, on soil, x 0.5.

Phellorinia Berk.
Phelloriniaceae, Agaricales, Agaricomycetidae, Agaricomycetes, Agaricomycotina, Basidiomycota, Fungi
Berkeley, M. J. 1843. Enumeration of fungi, collected by Herr Zeyher in Uitenhage. *London J. Bot* 2:421.
From Gr. *phellós*, cork + L. des. -inia, stuff, material, nature of something, in reference to the woody or cork-like material of the stem. Peridium two-layered, supported upon a stout stem. Exoperidium continuous with exterior of stem. Endoperidium on expanded apex of stem, at dehiscence breaking irregularly at apex of endoperidium. Stem thick, woody, somewhat cork-like. Gleba with capillitium of hyphae, rarely branched, sparingly septate. Spores globose, yellowish, finely echinulate. Solitary in sandy soils of Africa, Australia, India and North America.

Phialocephala W. B. Kendr.
Vibrisseaceae, Helotiales, Leotiomycetidae, Leotiomycetes, Pezizomycotina, Ascomycota, Fungi
Kendrick, W. B. 1961. The *Leptographium* complex. *Phialocephala* gen. nov. *Can. J. Bot.* 39(5):1079-1081.
From Gr. *phiále*, cup, glass + *kephalé*, head. Gloeoid heads whitish at first, becoming dark with age, formed on ends of long, penicillate conidiophores. Conidiophores borne on aerial mycelial cords with one or several series of metulae. Phialides with an evident collarette. Conidia gloeoid. Saprobic on vegetable remains.

Phialophora Medlar
Herpotrichiellaceae, Chaetothyriales, Chaetothyriomycetidae, Eurotiomycetes, Pezizomycotina, Ascomycota, Fungi

Phialophorophoma

Medlar, E. M. 1915. A new fungus, *Phialophora verrucosa*, pathogenic for man. *Mycologia* 7(4):201-203.

From Gr. *phiále*, cup, glass + *phóros*, bearer < *phéro*, to carry, support. Phialides bottle-shaped conidiogenous cells with a collarette on apex. Conidia aseptate, spherical or ovoid, congregating in gloeoid balls. Common in soils, especially in forests, frequently isolated from wood and other cellulosic substrates. Some species are parasitic on humans and other animals, causing chromoblastomycosis.

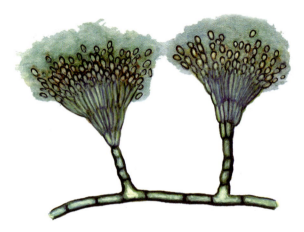

Phialocephala fortinii C. J. K. Wang & H. E. Wilcox: penicillate conidiophores, with mucoid heads composed of aggregate, unicellular conidia, x 300. Isolated from root of *Pinus sylvestris*, Finland.

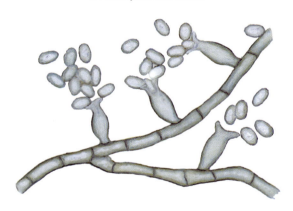

Phialophora verrucosa Medlar: bottle-shaped conidiogenous cells, with aseptate, ovoid conidia that congregate in gloeoid balls on the apex, x 1,200. Isolated from human skin lesion.

Phialophorophoma Linder

Inc. sed., Inc. sed., Inc. sed., Inc. sed., Pezizomycotina, Ascomycota, Fungi

Linder, D. L. 1944. I. Classification of the marine fungi:401-433. *In*: Barghoorn, E. S. and D. L. Linder, Marine fungi: their taxonomy and biology. *Farlowia* 1:395-467.

From genus *Phialophora* Medlar + *Phoma* Sacc., for a phoma-like fungus with endogenous conidia. Pycnidia black, subcarbonaceous, subglobose or ellipsoidal, ostiolate, immersed in substratum. Conidiophores hyaline, simple or branched below, lining cavity of pycnidium. Conidia endogenous, hyaline, ellipsoidal, 1-celled. On oak bark above low tide.

Phialotubus R. Y. Roy & Leelav.

Inc. sed., Inc. sed., Inc. sed., Inc. sed., Pezizomycotina, Ascomycota, Fungi

Roy, R. Y., K. M. Leelavathy. 1966. *Phialotubus microsporus* gen. et sp. nov., from soil. *Trans. Br. mycol. Soc.* 49 (3):495-498.

From Gr. *phiále*, bottle, jar + L. *tubus*, tube, for the shape of the phialide with a tubular apex. Conidiophores simple or highly branched with primary, secondary and tertiary branches, bearing phialides at tips; phialides narrower towards base, broad, round at distal end, distal ends prolonged into long or short, narrow, hyaline, tube-like structures bearing conidial chains; conidial chains long, with hyaline interconnections; conidia hyaline, becoming yellowish-brown at maturity, double walled, smooth, inner wall dark, outer wall hyaline. Isolated from the rhizosphere of *Dichanitium annulatum* on the campus of Banaras Hindu Unioersiiy, Varanasi, India.

Phillipsia Berk.

Sarcoscyphaceae, Pezizales, Pezizomycetidae, Pezizomycetes, Pezizomycotina, Ascomycota, Fungi

Berkeley, M. J. 1881. Australian fungi - II. *J. Linn. Soc., Bot.* 18:383-389.

Dedicated to the cleric *William Phillips* + L. suf. -*ia*, which denotes pertaining to. Apothecia sessile or with a short pedicel, discoid or slightly concave, intensely reddish-violet in interior, little lighter or whitish on exterior. Frequent in tropical forests and jungle throughout the world.

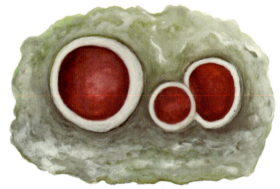

Phillipsia subpurpurea Berk. & Broome: slightly concave, reddish-violet apothecia, on tree twigs, x 1. Collected in Queensland, Australia.

Phlebia Fr. (syn. Merulius Fr.)

Meruliaceae, Polyporales, Inc. sed., Agaricomycetes, Agaricomycotina, Basidiomycota, Fungi

Fries, E. M. 1821. *Syst. mycol.* (Lundae) 1:426.

From Gr. *phléps*, *phlebós*, vein, in reference to the characteristic radial folds, in the margin of the fructification. Fructifications whitish, yellowish, reddish, brown or black, resupinate, effuse-reflexed or pileate, coriaceous-gelatinose, ceraceous, membranous or floccose with hymenium at first smooth, then reticulate, with irregular folds, veins or pores, whose edges are fertile, having small cavities instead of tubes. Pileus annual, sometimes perennial, sessile, resupinate, effuse, seldom fleshy, tough, venous-wrinkly. Hymenium concolorous with pileus; subhymenium generally thickening at maturity, with or without clamp connections; texture coriaceous. Basidia narrowly clavate. Spores hyaline to white. Lignicolous, rarely terricolous. Associated with white rot decay of angiosperms and gymnosperms.

Phlebia tremellosa (Schrad.) Nakasone & Burds.: lignicolous, floccose, reddish fructification, on tree trunks in woodland, Germany, x 1.

Phlogiotis Quél.—see **Guepinia** Fr.
Inc. sed., Auriculariales, Inc. sed., Agaricomycetes, Agaricomycotina, Basidiomycota, Fungi
Quélet, L., 1886. *Enchir. fung.* (Paris):202.
From Gr. *phlóx*, *phlogós*, flame, fire, bright reddish + *oús*, *otós*, ear, in reference to the shape and color of the fructification.

Phlyctidium (A. Braun) Rabenh—see **Rhizophydium** Schenk ex Rabenh.
Rhizophydiaceae, Rhizophydiales, Chytridiomycetidae, Chytridiomycetes, Inc. sed., Chytridiomycota, Fungi
Rabenhorst, L. 1868. *Flora Europaea algarum aquae dulcis et submarinae* 3:278.
From Gr. *phlyctís* < *phlýctaina*, tumor, scab, blister, pimple + dim suf. *-ídion* > L. dim. suf. *-ium*, i.e., a small tumor, blister or pimple, because of the appearance of the thallus.

Phlyctochytrium J. Schröt.
Chytridiaceae, Chytridiales, Chytridiomycetidae, Chytridiomycetes, Inc. sed., Chytridiomycota, Fungi
Schröter, J. 1893. Phycomycetae. *In*: A. Engler and K. Prantl, *Nat. Pflanzenfam.*, Teil. I (Leipzig) 1:63-141.
From Gr. *phlyctís* < *phlýctaina*, tumor, scab, blister, pimple + *chytrís*, kettle, receptacle + L. dim. suf. *-ium*, for the shape of the interbiotic thallus. Parasite or saprobe in freshwater algae.

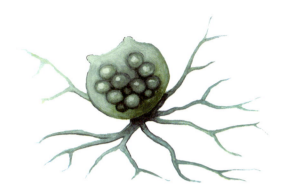

Phlyctochytrium papillatum: papillate zoosporangium, with young zoospores inside, and rhizoids at base, x 1,000.

Pholiota (Fr.) P. Kumm. (syn. **Kuehneromyces** Singer & A. H. Sm.)
Strophariaceae, Agaricales, Agaricomycetidae, Agaricomycetes, Agaricomycotina, Basidiomycota, Fungi
Kummer, P. 1871. *Führ. Pilzk.* (Zerbst):22.
From Gr. *pholís*, scale + L. suf. *-ota*, which indicates connection or possession. Pileus and/or stipes with scales, principally in *Ph. squarrosa* (Müller ex Fr.) Kumm., the type species. On living and dead wood, remains of herbaceous plants, among mosses and terricolous or humicolous. Causing brown rot, saprobic on wood, sawdust or volcanic ash, rarely on humus, in temperate mountainous regions, also in the tropics. *Pholiota destruens* (Brond.) Gill., destructive parasite of wood. Probably all species are edible, the one most consumed is *P. mutabilis* (Schaeff. ex Fr.) P. Kumm.

Pholiota mutabilis (Schaeff.) P. Kumm. (syn. **Kuehneromyces mutabilis** (Schaeff.) Singer & A. H. Sm.): mature basidiocarps, on buried wood, x 0.5.

Phoma

Pholiota squarrosoides (Peck) Sacc.: scaly, reddish-brown basidiomata, on soil, x 0.3.

Phoma Sacc.
Didymellaceae, Pleosporales, Pleosporomycetidae, Dothideomycetes, Pezizomycotina, Ascomycota, Fungi
Saccardo, P. A. 1880. Conspectus generum fungorum italiae. Inferiorum nempe ad Sphaeropsideas, Melanconieas et Hyphomyceteas pertinentium, systemate sporologico dispositoru. *Michelia* 2(6):4.
From Gr. *phós*, genit. *phodós*, red spot on the skin, or from *phóix*, pustule, blister, burn < *pho* or *pháo*, to kill + suf. *-ma*, result of an action. Causes lesions, spots and pustules on leaves and stems of host plants resulting in rotting and wilts.

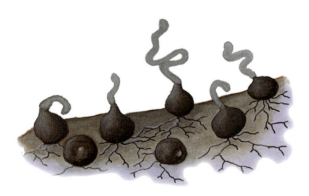

Phoma herbarum Westend.: mature, ostiolate pycnidia, showing the conidial cirrhus, on host stem, x 200.

Phomopsis (Sacc.) Bubák—see Diaporthe Nitschke
Diaporthaceae, Diaporthales, Sordariomycetidae, Sordariomycetes, Pezizomycotina, Ascomycota, Fungi
Bubák, F. 1905. *Öst. bot. Z.* 55:78.
From genus *Phoma* Sacc. + L. suf. *-opsis* < Gr. *-ópsis*, appearance, aspect, for its similarity to this phytopathogenic genus.

Phragmidium Link
Phragmidiaceae, Pucciniales, Inc. sed., Pucciniomycetes, Pucciniomycotina, Basidiomycota, Fungi
Link, J. H. F. 1816. Observationes in ordines plantarum naturales. *Mag. Gesell. naturf. Freunde, Berlin* 7:30.
From Gr. *phragmós*, partition, barrier, septum, wall + L. dim. suf. *-idium* < Gr. *-ídion*, for the septate teliospores, composed of several cells delimited by transverse septa. Rust autoecious on plants in the Rosaceae. Spermogonia without paraphyses. Aecia, when present, with or without peripheral paraphyses, aeciospores catenulate, with a warted or sometimes echinulate wall. Uredinia, when formed, generally surrounded by paraphyses. Urediniospores pedicellate with a warted or echinulate wall and indistinct, scattered pores. Telia without paraphyses. Teliospores pedicellate, of one to ten cells.

Phragmidium mucronatum (Pers.) Schltdl.: multiseptate, pedicellate, brown teliospores, x 1,200. On leaves of *Rosa*, Germany.

Phycomyces Kunze
Phycomycetaceae, Mucorales, Inc. sed., Inc. sed., Mucoromycotina, Zygomycota, Fungi
Kunze, G., J. C. Schmidt. 1823. *Mykologische Hefte* (Leipzig) 2:113.
From Gr. *phýcos*, alga, marine herb + *mýkes* > L. *myces*, fungus. Fungus looks superficially like an alga, but has mycelium, thallus and sporangiophores. In soil and dung.

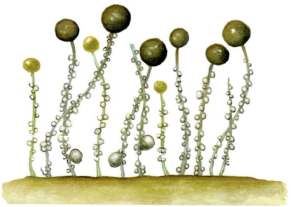

Phycomyces blakesleeanus Burgeff: roridous, green sporangiophores, with sporangia, on agar culture, x 8.

Phymatotrichopsis

Phyllachora Nitschke ex Fuckel
Phyllachoraceae, Phyllachorales, Inc. sed., Sordariomycetes, Pezizomycotina, Ascomycota, Fungi
Fuckel, K. W. G. L. 1870. *Symbolae mycologicae. Beiträge zur Kenntnis der rheinischen Pilze. Jb. nassau. Ver. Naturk.* 23-24:216.
From Gr. *phýllon*, leaf + *achór*, genit. *achōros*, dandruff, scab, scaly, wound, ulcer, sore + des. -*a*. Stromata black with perithecia. Causes lesions on leaves of host plants, e.g., on grasses and clovers, disease known as black pox or tar spot.

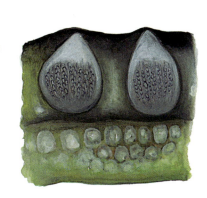

Phyllachora graminis (Pers.) Fuckel: black perithecial stromata (tar spots), x 1; vertical section of perithecia, showing the hymenium, x 60. On leaves of *Elymus* and *Lolium perenne*, Germany.

Phyllactinia Lév.
Erysiphaceae, Erysiphales, Leotiomycetidae, Leotiomycetes, Pezizomycotina, Ascomycota, Fungi
Léveillé, J. H. 1851. Organisation et disposition méthodique des espèces qui composent le genre *Erysiphe. Annls. Sci. Nat.*, Bot., sér. 3, 15:144.
From Gr. *phýllon*, leaf + *aktís*, genit. *aktínos*, radius, ray, Ascocarps cleistothecioid perithecia, chasmothecia with peridial appendages bulbous at base. sharp at apex, radially arranged on equatorial zone. Asexual morph ovulariopsis-like. Powdery mildew, obligate parasite of plants.

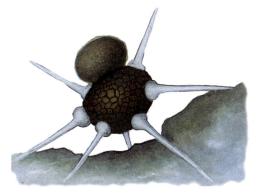

Phyllactinia guttata (Wallr.) Lév.: cleistothecioid perithecium on host leaf, with peridial appendages (bulbous at the base and sharp at the apex), radially arranged on the equatorial zone, x 200.

Phylloporia Murrill
Hymenochaetaceae, Hymenochaetales, Inc. sed., Agaricomycetes, Agaricomycotina, Basidiomycota, Fungi
Murrill, W. A. 1904. A new polyporoid genus from South America. *Torreya* 4:141-142.
From Gr. *phýllon*, leaf + *póros*, pore, i.e., a leaf with pores, because of the appearance of the basidioma. Fruitbody annual, resupinate to pileate. Pileus cinnamon to dark brown, soft, tomentose, with concentric regions, pore surface brown, pores entire, angular to round, tubes concolorous with pore surface, context light to dark brown, thin, distinctly delimited, hyphal system monomitic. Spores elliptic, pale yellowish in maturity. On dead wood, often on living bushes. Mainly tropical.

Phylloporus Quél.
Boletaceae, Boletales, Agaricomycetidae, Agaricomycetes, Agaricomycotina, Basidiomycota, Fungi
Quélet, L. 1888. *Fl. mycol. France* (Paris):409.
From Gr. *phýllon*, leaf + *póros*, pore, i.e., with lamellae and pores. Pileus tomentose to subtomentose in dry weather, more or less viscid in wet weather; hymenophore lamellate, often with venose anastomoses, or tubulose with pores generally wide, larger than 1 mm wide, often irregular and compound. Spores fusoid-subcylindric, smooth or slightly striate or rugulose. On the ground or decayed wood, obligately ectomycorrhizal with *Alnus* and *Fagus* in America, tropical Africa, South and East Asia. Almost all species are edible.

Phyllosticta Pers. (syn. **Guignardia** Viala & Ravaz)
Phyllostictaceae, Botryosphaeriales, Inc. sed., Dothideomycetes, Pezizomycotina, Ascomycota, Fungi
Persoon, C. H. 1818. *Traité champ.* (Paris):55, 147.
From Gr. *phýllon*, leaf + *stiktós*, marked, cauterized, distinguished with points < *stízo*, to mark with fire, pierce, inflame, in reference to the spots produced on the leaves. Ascomata pseudothecioid, centrum parenchymatous. Asci originating from base of locules of the pseudothecia; containing eight spores. Ascospores hyaline or pale brown. Pycnidia ostiolate, lacking periphyses. Sexual morph guignardia-like. On living or dead leaves of plants, produces leaf spots, rot or blight; cosmopolitan.

Phymatotrichopsis Hennebert
Rhizinaceae, Pezizales, Pezizomycetidae, Pezizomycetes, Pezizomycotina, Ascomycota, Fungi
Hennebert, G. L. 1973. *Botrytis* and *Botrytis*-like genera. *Persoonia* 7:199.
From genus *Phymatotrichum* Bonord. < Gr. *phýma*, genit. *phýmatos*, tumor, tubercle, excrescence + *thríx*, genit. *trichós*, hair + L. suf. -*um* + suf. -*opsis* < Gr. -*ópsis*, aspect, appearence. Rhizomorphs developing in roots, covered with rigid lateral cruciate trichome-like branches arranged at right angles to main fibers. Conidia

Phymatotrichum

botryoblastospores, borne on ampullae. *Phymatotrichopsis omnivora* attacks more than 2,000 species of hosts, including cotton, alfalfa, almost all legumes, all common fruits, and many important trees and herbs.

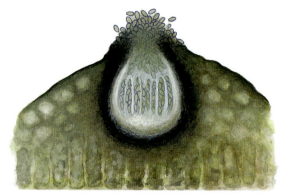

Phyllosticta ampelicida (Engelm.) Aa (syn. **Guignardia bidwellii** (Ellis) Viala & Ravaz): sagittal section of a pseudothecium, with asci and ascospores, on dried grapes of *Vitis rotundifolia*, New Jersey, U.S.A., x 300.

Phyllosticta convallariae Pers.: stem spot caused by ostiolate pycnidia, x 10. Collected in France.

Phymatotrichum Bonord.—see **Botrytis** P. Micheli ex Pers.
  Sclerotiniaceae, Helotiales, Leotiomycetidae, Leotiomycetes, Pezizomycotina, Ascomycota, Fungi
  Bonorden, H. F. 1851. *Handb. Allgem. mykol.* (Stuttgart):116.
  From Gr. *phýma*, genit. *phýmatos*, tumor, tubercle, excrescence + *thríx*, genit. *trichós*, hair + L. suf. -*um*.

Physalacria Peck
  Physalacriaceae, Agaricales, Agaricomycetidae, Agaricomycetes, Agaricomycotina, Basidiomycota, Fungi
  Peck, C. H. 1882. Fungi in wrong genera. *Bull. Torrey bot. Club* 9:1-4.
  From Gr. *physális*, bladder + *ákra*, the top. Pileus mostly a globose or irregularly inflated hollow club, with a short thin stipe, hymenial surface smooth, covering pileus, white, becoming yellowish, more or less uneven with irregular depressions or wrinkles; stipe slender, solid, straight; hyphae with clamp connections; oleocystidia present; basidia normal, basidioles fusoid, sterigmata 2 or 4; cystidioles often present; spores small to large, ellipsoid, fusoid or acicular, smooth, thin walled, inamyloid. On dead and living leaves and wood. Tropical West Africa, tropical and temperate America, tropical and Eastern Asia, Australia, Madagascar, and New Zealand.

Phymatotrichopsis omnivora (Duggar) Hennebert: at left, this phytopathogen develops rhizomorphs and sclerotia in the attacked host root, x 2; at right, the rhizomorphs are covered with rigid lateral cruciate trichome-like branches, arranged at a right angle with respect to the main fibers, x 3; in the center below, the arrangement of cells inside the sclerotium, x 40.

Physalosporopsis Bat. & H. Maia
  Inc. sed., Inc. sed., Inc. sed., Dothideomycetes, Pezizomycotina, Ascomycota, Fungi
  Batista, A. C., et al. 1955. Ascomycetidae aliquot novarum. *Anais Soc. Biol. Pernambuco* 13(2):72-86.
  From genus *Physalospora* Niessl + Gr. suf. -*ópsis*, aspect, appearance, for the similarity to this genus. Ascomata pseudothecia, uniloculate, globose or obpyriform, hypophyllous, subepidermal, immersed, with erumpent ostiolar neck, brown, with a smooth prosenchymatous wall. Paraphyses filiform, septate, branched. Asci clavate, short-pedicellate, 4-8-spored. Ascospores globose, 1-celled, hyaline, covered with a mucous substance. On living leaves of *Rhizophora mangle* in Recife, Brazil.

Physcia (Schreb.) Michx.
  Physciaceae, Teloschistales, Lecanoromycetidae, Lecanoromycetes, Pezizomycotina, Ascomycota, Fungi
  Michaux, A. 1803. *Flora Boreali-Americana* 2:326.
  From Gr. *physkíon*, bladder, blister. Thallus foliose with apothecia and soredia, giving surface a vesiculose aspect when these reproductive structures develop.

Physoderma Wallr. (syn. **Urophlyctis** J. Schröt.)
  Physodermataceae, Blastocladiales, Inc. sed., Blastocladiomycetes, Inc. sed., Chytridiomycota, Fungi
  Wallroth, K. F. W. 1833. *Fl. crypt. Germ.* (Norimbergae) 2:1-923.

From Gr. *phýsa*, blister, bubble, bladder + *dérma*, skin, i.e., a blister of the epidermis of the host plant. Epibiotic and endobiotic thallus. Resting spore with swollen, turbinate cell in the basal or posterior part, from which can originate another series of turbinate cells. Obligate parasite of aquatic and terrestrial plants.

Physcia aipolia (Ehrh. ex Humb.) Fürnr.: foliose thallus, which acquires a vesiculose aspect when the apothecia and soredia develop, on dead wood, x 2.

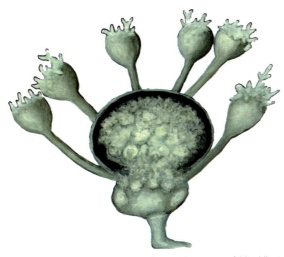

Physoderma alfalfae (Lagerh.) Karling: resting spores of this obligate pathogen, endobiotic and epibiotic, in the plant host tissues, x 1,100.

Phytocordyceps C. H. Su & H. H. Wang—see **Cordyceps** Fr.
Cordycipitaceae, Hypocreales, Hypocreomycetidae, Sordariomycetes, Pezizomycotina, Ascomycota, Fungi
Su, C.-H., H.-H. Wang. 1986. *Phytocordyceps*, a new genus of the Clavicipitaceae. *Mycotaxon* 26:337-344.
From Gr. *phytón*, plant + genus *Cordyceps* Fr.

Pichia E. C. Hansen
Pichiaceae, Saccharomycetales, Saccharomycetidae, Saccharomycetes, Saccharomycotina, Ascomycota, Fungi
Hansen, E. C. 1904. Grundlinien zur Systematik der Saccharomyceten. *Centbl. Bakt. ParasitKde,* Abt. I, 12:529-538.

Dedicated to the Italian botanist *R. E. G. Pichi-Sermolli* + L. ending *-a*. Yeast with multilateral budding and pseudomycelium, at times with true mycelium. Asci evanescent with 1-4 or more ascospores. Ascospores hat-shaped or saturnoid with an oil droplet, generally a smooth wall. Fermentation weak or absent. In liquids with ethanol forms a rugose pellicle, dry, pink or reddish. Inhabits soil, tree exudates, shed insect exoskeletons, beer and wine, fruit juices, brine and other substrates.

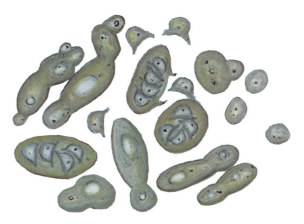

Pichia membranifaciens (E. C. Hansen) E. C. Hansen: vegetative, multilateral budding yeast cells, and asci with four hat-shaped ascospores, with an oil droplet, x 1,200. Isolated from beer wort, Germany.

Piedraia Fonseca & Leão
Piedraiaceae, Capnodiales, Dothideomycetidae, Dothideomycetes, Pezizomycotina, Ascomycota, Fungi
Fonseca, O. da, A. E. de Leão. 1928. Sobre os cogumelos da piedra brasileira. *Mem. Inst. Oswaldo Cruz* 4 (Suppl.):125.
From Sp. *piedra* < L. *petra*, stone + L. suf. *-ia*, which denotes pertaining to, due to the appearance and consistency of the ascostromata. Ascostromata black concretions with ascigerous locules. Ascospores elongated, both ends tapered, ending in a filiform extension. Causes the tropical disease called *black piedra*, on hair of humans and certain primates.

Piedraia hortae Fonseca & Leão: black concretions of an ascostroma in human hair developed by this pathogen (*black piedra*), x 25. Isolated from a human cadaver, Brazil.

Pilaira

Pilaira Tiegh.
Mucoraceae, Mucorales, Inc. sed., Inc. sed., Mucoromycotina, Zygomycota, Fungi
Tieghem, P. van. 1875. Nouvelles recherches sur les mucorinées. *Annls. Sci. Nat. Bot.*, sér. 6, 1:51.
From L. *pila*, ball, cannon ball < Gr. *pílos*, ball, shot, hat, helmet + Gr. *aíra*, hammer, in reference to the shape of the sporangium. Sporangiophore bearing sporangium on upper tip. Sporangium oblate to hemispherical, separates passively. Coprophilous.

Pilaira anomala (Ces.) J. Schröt.: sporangiophores of this coprophilic species, with oblate to hemispherical sporangia, on agar culture, x 20.

Pilobolus Tode
Pilobolaceae, Mucorales, Inc. sed., Inc. sed., Mucoromycotina, Zygomycota, Fungi
Tode, H. J. 1784. *Schr. Ges. naturf. Freunde, Berlin* 5:46.
From Gr. *pílos*, ball, shot, hat, helmet + *bállo* > L. *bolus*, to throw. Sporangium violently hurled from subsporangial vesicle formed on apex of sporangiophore. Coprophilous on cow and horses dung.

Pilobolus crystallinus (F. H. Wigg.) Tode: sporangiophore of this coprophilic species, with subsporangial vesicle and sporangium, on agar culture, x 40. Isolated from herbivore dung.

Piptocephalis de Bary
Piptocephalidaceae, Zoopagales, Inc. sed., Inc. sed., Zoopagomycotina, Zygomycota, Fungi
de Bary, A. 1865. Zur Kenntnis der Mucorineen. *Abh. senckenb. naturforsch. Ges.* 5:356.
From Gr. *pípto*, to fall, sucumb + *kephalé*, head + L. suf. *-alis*, belonging to or with reference to. Vesicles or heads bearing sporangiola, typically deciduous when mature. Mycoparasite of Mucorales.

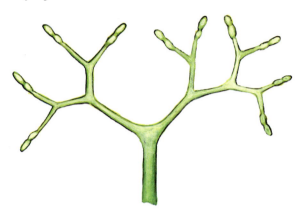

Piptocephalis lepidula (Marchal) P. Syd.: dichotomous branches of a sporangiophore, with terminal sporangiola, x 1,500. Mycoparasitic, especially of Mucorales.

Pirella Bainier
Mucoraceae, Mucorales, Inc. sed., Inc. sed., Mucoromycotina, Zygomycota, Fungi
Bainier, G. 1882. *Étud. Mucor* (Thèse, Paris):83.
From L. *pirum*, pear + dim. suf. *-ella*, for the pear-shaped sporangia of this mucor saprobic in the soil.

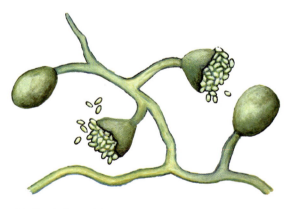

Pirella circinans Bainier: pear-shaped sporangia with spores, x 200. Saprobic in the soil.

Pirottaea Sacc. (syn. Echinella Massee)
Dermateaceae, Helotiales, Leotiomycetidae, Leotiomycetes, Pezizomycotina, Ascomycota, Fungi
Saccardo P. A. 1878. Fungi veneti. *Michelia* 1(no. 4):424.
Named in honor of the Italian botanist P. R. Pirotta + euphonic des. *-ea*. Apothecia sessile, at first closed, then expanding, becoming cup-shaped, clothed with black

or brown hairs. Asci clavate, inoperculate, 8-spored. Ascospores irregularly 2-seriate, hyaline, narrowly fusiform, becoming 3-many-septate. Paraphyses filiform or slightly clavate. On plant materials.

Piskurozyma Xin Zhan Liu, et al.
Filobasidiaceae, Filobasidiales, Inc. sed., Tremellomycetes, Agaricomycotina, Basidiomycota, Fungi
Liu X-Z, et al. *In*: X-Z Liu, et al. 2015. Towards an integrated phylogenetic classification of the Tremellomycetes. *Stud. Mycol. 81*:120.
Named in honor of the Slovenian Prof. *Jure Piškur*, for his contribution to yeast genetics, physiology and evolutionary biology. Basidiomes, if present, within the hymenium of the host. Dikaryotic hyphae with clamp connections. Haustoria not reported. In culture true hyphae occur occasionally. Pseudohyphae absent. Sexual reproduction observed for some species. Holobasidia slender with terminal sessile basidiospores. Budding cells present. Ballistoconidia may be present. Fermentation occasionally present. Phylogenetic analysis suggests close relationships with the sexual stages of species *Filobasidium capsuligenum* (Fell et al.) Rodr. Mir. and *Syzygospora sorana* Hauerslev, and two recently described cryptococci, *C. filicatus* Golubev & J. P. Samp. and *C. fildesensis* T. T. Zhang & Li Y.Yu.

Piskurozyma capsuligera (Fell et al.) Rodr. Mir. et al. (syn. **Filobasidium capsuligenum** Fell et al.) Rodr. Mir.: cylindrical basidia with five to nine sessile, elliptical basidiospores, x 750. Collected in South Africa.

Pisolithus Alb. & Schwein.
Sclerodermataceae, Boletales, Agaricomycetidae, Agaricomycetes, Agaricomycotina, Basidiomycota, Fungi
Albertini, J. B. de, L. D. Schweinitz. 1805. *Consp. fung.* (Leipzig):376.
From Gr. *písos*, pea, or *písos*, wet place, meadow + *líthos*, stone, because the fructifications resemble stones and contain globose structures or peridioles similar to peas. Fructifications pseudostipitate or sessile with thin peridium that fragments, exposing peridioles. Peridiole sectors of gleba in interior in which spores mature. Spores globose, dark brown, verrucose, forming a powdery mass when peridioles break. Partially buried in soil in pine and oak forests and in crops of eucalyptus, forming mycorrhizae with trees important in forestry.

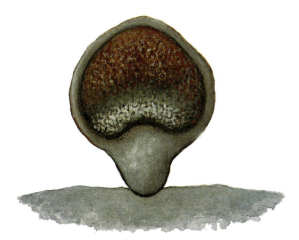

Pisolithus arhizus (Scop.) Rauschert: vertical section of a pseudostipitate basidioma, showing the peridioles with dark brown spores inside, x 1.

Pithoascus Arx
Microascaceae, Microascales, Hypocreomycetidae, Sordariomycetes, Pezizomycotina, Ascomycota, Fungi
Arx, J. A. von. 1973. Ostiolate and nonostiolate pyrenomycetes. *Proc. K. Ned. Akad. Wet.*, Ser. C, Biol. Med. Sci. 76:289-296.
From Gr. *píthos*, large jar with a wide mouth + sac < ML. *ascus*, wine skin, ascus, an apparent reference to the broad, barrel-shaped asci. Ascomata perithecia, solitary to confluent, superficial to semi-immersed, non-ostiolate, or with an inconspicuous pore, smooth, black. Wall thick, of flattened pseudoparenchyma cells. Asci doliiform to ovoid, often catenulate, 8-spored. Ascospores fusoid to navicular, 1-celled, straw-yellow, with an inconspicuous germ pore at both ends.

Pithoascus lunatus Jagielski, et al.: black perithecium, with a long ostiolar neck, x 150; 8-spored ovoid ascus, x 1,000, and free navicular, 1-celled, yellow ascospores, x 1,500. Isolated from human patients with *Tinea plantaris*, Germany.

Pityrosporum

Pityrosporum Sabour.—see **Malassezia** Baill.
Malasseziaceae, Malasseziales, Inc. sed., Malasseziomycetes, Ustilaginomycotina, Basidiomycota, Fungi
Sabouraud, R. 1904. *Maladies Chevelu; II Les Maladies Desquamatives* 2:1-716.
From Gr. *píiyron*, dandruff, tinea + *spóron*, seed, spore, because of the association of the spores of this yeast with diverse skin infections, such as tinea or pityriasis versicolor of humans and higher animals.

Placopsis (Nyl.) Linds.
Baeomycetaceae, Baeomycetales, Ostropomycetidae, Lecanoromycetes, Pezizomycotina, Ascomycota, Fungi
Lindsay, W. L. 1866. Observations on New Zealand lichens. *Trans. Linn. Soc. London* 25:536.
From Gr. *pláx*, genit. *plakós*, plaque, sheet, plate, plain > *plakódes*, tabular, foliate + L. suf. *-opsis* < Gr. *-ópsis*, appearance, aspect. Thallus crustose, lobate at margin. Cephalodia large, brown, verrucose. Soralia large capitate, up to 1 mm wide. Apothecia often lacking. On moist rocks.

Placopsis lambii Hertel & V. Wirth: crustous thallus, lobate on the margin, with large, brown cephalodia, on rock, France, x 0.5.

Plasia Sherwood—see **Durella** Tul. & C. Tul.
Helotiaceae, Helotiales, Leotiomycetidae, Leotiomycetes, Pezizomycotina, Ascomycota, Fungi
Sherwood, M. A. 1981. *Plasia*, a new genus for the macroconidial anamorph of *Durella atrocyanea*. *Trans. Br. mycol. Soc.* 77(1):196-200.
Named in honor of *Daniel Plas* + L. des. *-ia*.

Plectania Fuckel
Sarcosomataceae, Pezizales, Pezizomycetidae, Pezizomycetes, Pezizomycotina, Ascomycota, Fungi
Fuckel, K. W. G. L. 1870. Symbolae mycologicae. Beiträge zur Kenntnis der rheinischen Pilze. *Jb. nassau. Ver. Naturk.* 23-24:323.
From Gr. *plektós*, plaited, folded, twisted, interwoven, curly + L. suf. *-anus*, *-ania*, pertaining to or related to something, or a characteristic. Apothecia with outer surface downy to hirsute, at least tomentose or with interwoven hairs, disk shaped or more commonly a deep cup or funnel, margins incurved, smooth or laciniate, interior surface bright red or scarlet. On humid soils, branches or totally or partially buried wood.

Plectania melastoma (Sowerby) Fuckel: funnel-shaped, circular yellowish apothecia, with black interior, on old wood, x 7.

Plectosphaerella Kleb. (syn. **Plectosporium** M.E. Palm, et al.)
Plectosphaerellaceae, Inc. sed., Hypocreomycetidae, Sordariomycetes, Pezizomycotina, Ascomycota, Fungi
Klebahn, H. 1929. Vergilbende junge Trebgurken, ein daruf gefundenes *Cephalosporium* und dessen Schaluchfrüchte. *Phytopath. Z.* 1:31-44.
From Gr. *plektós*, plaited, twisted > L. *plectos* + genus *Sphaerella* (Fr.) Rabenh. Ascomata ostiolate perithecia, immersed in host tissues. Perithecia single, scattered, obpyriform, brown, with hyaline to subhyaline, with erumpent ostiolar neck. Ostiole lined with slender periphyses. Perithecial wall of flattened cells with pigmented walls. Centrum containing paraphyses of vesiculose basal cells, septate, constricted at septa, tapering toward apex, arising from a basal cushion of flattened, thin-walled, hyaline cells. Asci cylindrical, unitunicate, with undifferentiated apex, not bluing in iodine, short-stalked, 8-spored. Ascospores biseriate in ascus, hyaline, 2-celled, with median septum, subcylindrical with rounded ends, wall with fine warts that stain in aniline blue. Parasitic on herbaceous plants.

Pleochaeta Sacc. & Speg.
Erysiphaceae, Erysiphales, Leotiomycetidae, Leotiomycetes, Pezizomycotina, Ascomycota, Fungi
Saccardo, P. A. 1881. Fungi aliquot extra-europaei. *Michelia* 2 (no. 7):373.
From Gr. *pléon* < *pleíon*, more numerous + *chaíte*, crest, mane. Ascomata cleistothecioid perithecium or chasmothecium with rigid peridial appendages with curved or uncinate tips. Asexual morph ovulariopsis-like. Powdery mildew of *Celtis* and other plants.

Pluteus

Pleochaeta indica N. Ahmad, et al.: vertical section of a cleistothecioid perithecium, with the hymenium and numerous peridial appendages, rigid or uncinate, causing the powdery mildew of leaves of *Celtis caucasica*, Uttar Pradesh, India, x 200.

Pleospora Rabenh. ex Ces. & De Not.—see **Stemphylium** Wallr.

Pleosporaceae, Pleosporales, Pleosporomycetidae, Dothideomycetes, Pezizomycotina, Ascomycota, Fungi
Cesati, V., G. De Notaris. 1863. Sferiacei italici. *Comm. Soc. crittog. Ital.* 1 (fasc. 4):217.
From Gr. *pléos*, full + *sporá*, spore, due to the formation of numerous ascospores in the fructifications.

Pleotrichiella Sivan.

Inc. sed., Inc. sed., Inc. sed., Dothideomycetes, Pezizomycotina, Ascomycota, Fungi
Sivanesan, A. 1984. *Pleotrichiella australiensis* gen. et sp. nov. on *Lasiopetalum* from Australia. *Trans. Br. mycol. Soc.* 83(3):531-533.
From Gr. *pléos*, full + *thríx*, genit. *trichós*, hair + L. dim. suf. *-ella*, referring to the numerous setae on the ascomata. Mycelium hypophyllous, of superficial, hyaline to brown, septate, branched, smooth hyphae. Pseudothecia superficial, hypophyllous, globose, dark brown to black, setose, ostiolate. Setae simple, dark brown, thick-walled, septate, paler towards apex. Wall distinctly three layered. Asci cylindrical, pedicellate, bitunicate, 8-spored. Ascospores pale brown to brown, smooth, uniseriate, fusiform, dictyoseptate. Pseudoparaphyses filiform, hyaline, branched, septate. On leaves of *Lasiopetalum ferrugineum* in Bouddi N. P., Australia.

Pleurotus (Fr.) P. Kumm.

Pleurotaceae, Agaricales, Agaricomycetidae, Agaricomycetes, Agaricomycotina, Basidiomycota, Fungi
Kummer, P. 1871. *Führ. Pilzk.* (Zerbst):24.
From Gr. *pleurá* or *pleurón*, side, rib + L. suf. *-otus* < Gr. *otós*, ear + L. des. *-us*, in reference to the shape of the fructification and the laminar hymenophore. On living or dead wood, also on the remains of vegetable material, rarely terricolous. All species considered edible of good quality and some of them, such as *P. ostreatus* (Jacq. et Fr.) Kummer, are cultivated industrially. Medicinal properties have been attributed to other species, such as *P. tuberregium* (Fr.) Sing.

Pleurotus ostreatus (Jacq.) P. Kumm.: mature basidiomata, with the conchiform pileus and eccentric stipe, on dead wood, Austria, x 0.8.

Ploioderma Darker

Rhytismataceae, Rhytismatales, Leotiomycetidae, Leotiomycetes, Pezizomycotina, Ascomycota, Fungi
Darker, G. D. 1967. A revision of the genera of the Hypodermataceae. *Can. J. Bot.* 45:1399-1444.
From Gr. *ploíon*, small boat + *dérma*, skin, in reference to the boat-shaped ascomata. Apothecia elliptical, subepidermal, black to nearly hyaline, opening by a slit at maturity. Hymenium concave, of simple, filamentous paraphyses; subhymenium thin, hyaline. Asci unitunicate, clavate to saccate. Ascospores hyaline, short, rod-like. Asexual morph leptostroma-like. On living needles of conifers, causing needle cast disease.

Plurisperma Sivan.

Verrucariaceae, Verrucariales, Inc. sed., Eurotiomycetes, Pezizomycotina, Ascomycota, Fungi
Sivanesan, A. 1970. *Plurisperma dalbergiae* gen. et sp. nov. *Trans. Br. mycol. Soc.* 54(3):495-496.
From L. *pluri*, several + Gr. *spérma*, seed, spore, for the multispored asci. Pseudothecia solitary to aggregated, immersed in host tissue, black, with a papillate ostiole. Asci bitunicate, clavate, multi-spored. Pseudoparaphysate. Ascospores one-celled, hyaline to pale brown. On dead twigs of *Dalbergia sissoo* in Bhagat, West Pakistan.

Pluteus Fr.

Pluteaceae, Agaricales, Agaricomycetidae, Agaricomycetes, Agaricomycotina, Basidiomycota, Fungi
Fries, E. M. 1836. *Fl. Scan.*:338.
From L. *pluteus*, name of an ancient war machine with the form of a conical and unfolded hat; i.e., conical and spread out pileus. Pileus generally with cellular epicutis

Pneumatospora

or with filamentous, allantoid or cylindric, fusoid hyphae; gills free; spores ovoid or short cylindric, ellipsoid or globose, smooth; stipe central; veil reduced, fugacious or absent; context of pileus white, fleshy, inamyloid, hyphae with or without clamp connections. On dead and living plant tissue, on humus, sand and decayed wood in the forest. Cosmopolitan. Many species are edible of good culinary qualities.

Pneumatospora B. Sutton, et al.—see **Minimedusa** Weresub & P.M. LeClair
Inc. sed., Cantharellales, Inc. sed., Agaricomycetes, Agaricomycotina, Basidiomycota, Fungi
Sutton, B. C., et al. 1984. *Pneumatospora obcoronata* gen. et sp. nov. from Malaysia. *Trans. Br. mycol. Soc.* 83 (3):423-429.
From Gr. *pnéuma*, genit. *pnéumatos*, air, wind, breathing + *sporá*, seed > L. *spora*, spore, for the apparerent resemblance to pneumatophores.

Pneumocystis P. Delanoë & Delanoë
Pneumocystaceae, Pneumocystales, Pneumocystomycetidae, Pneumocystomycetes, Taphrinomycotina, Ascomycota, Fungi
Delanoë P., M. Delanoë. 1912. *C. r. hebd. Séanc. Acad. Sci., Paris* 155:660.
From Gr. *pnéuma*, genit. *pnéumatos*, air, wind, breathing + *kýstis*, bladder, vesicle, cell. Is an opportunistic agent that causes significant morbidity and mortality in immunosuppressed patients, ususally in the form of a severe pneumonia

Pneumocystis jirovecii Frenkel: trophic cells that give rise to cysts in human pulmonary alveoli, x 1,300.

Podaxis Desv.
Agaricaceae, Agaricales, Agaricomycetidae, Agaricomycetes, Agaricomycotina, Basidiomycota, Fungi
Desvaux, N. A. 1809. Observations sur quelques genres à établir dans la famille des champignons. *J. Bot. (Desvaux)* 2:84-105.
From Gr. *poús*, genit. *podós*, foot + L. *axis*, axis. Fructifications formed by an axis or foot that supports upper part of pileus, remaining in central part and apex. Pileus of a gleba with abundant capillitium. Spores dark, smooth. Common in sandy soil in arid or semiarid places.

Podaxis pistillaris (L.) Fr.: basidiomata, growing in arid, sandy soil, composed by an axis or foot that supports the pileus containing a gleba with dark brown spores, x 1.

Podophacidium Niessl
Dermateaceae, Helotiales, Leotiomycetidae, Leotiomycetes, Pezizomycotina, Ascomycota, Fungi
Niessl, G. 1868. Fungi europaei exsiccati, etc. Ed. nova. Series secunda (No.1101-1200). *In*: Rabenhorst, *Bot. Ztg.* 26:558-559.
From Gr. *podós*, foot + genus *Phacidium* Fr. Apothecia contracted at base, substipitate, obconic to turbinate, opening with a laciniate aperture. Hymenium freely exposed at maturity, bright colored, yellow-olivaceous. Asci clavate, 8-spored. Ascospores simple, hyaline. On soil.

Podoscypha Pat.
Meruliaceae, Polyporales, Inc. sed., Agaricomycetes, Agaricomycotina, Basidiomycota, Fungi
Patouillard, N. 1900. *Essai Tax. Hyménomyc.* (Lons-le-Saunier):70.
From Gr. *poús, podós*, foot, stalk + *skýphos*, cup, i.e., a cup with a stalk or foot. Basidiomata stipitate, cup-like or flabelliform, coriaceous, thin. Stipe woody or cartilaginous, simple or branched. Upper surface smooth or velutinous, striate, with crests and furrows, frequently zonate; trama homogeneous, almost without color. Hymenium thin, smooth or with longitudinal grooves; basidia short, clavate, cystidia projecting from surface of hymenium, hyaline, cylindrical or acute, thin-walled; spores ovoid, smooth. Most species in tropical regions, but some present in Europe and North America.

Podosordaria Ellis & Holw.
Xylariaceae, Xylariales, Xylariomycetidae, Sordariomycetes, Pezizomycotina, Ascomycota, Fungi
Ellis, J. B., E. W. Holway. 1897. Mexican fungi. *In*: Holway, *Bot. Gaz.* 24(1):37.
From Gr. *poús*, genit. *podós*, foot, support + genus *Sordaria* Ces. & De Not., this from L. *sordes*, to be dirty

+ suf. *-aria*, denoting connection or possession, i.e., fimicolous. Segregated from *Sordaria* Ces. & de Not. by pedicellate or stipitate perithecial stromata. On dung of various animals.

Podosordaria mexicana Ellis & Holw.: stipitate perithecial stroma, on herbivorous dung, x 10.

Podosphaera Kunze (syn. **Sphaerotheca** Lév.)
Erysiphaceae, Erysiphales, Leotiomycetidae, Leotiomycetes, Pezizomycotina, Ascomycota, Fungi
Kunze, G., J. K. Schmidt. 1823. In Kunze & Schmidt, *Mykologische Hefte* (Leipzig) 2:111-113.
From Gr. *poús*, genit. *podós*, foot + *sphaíra* (L. *sphaera*), sphere. Chasmothecial, i.e., cleistothecial perithecia globose with short, simple basal appendages, also with equatorial or apical dichotomously branched appendages. Asexual morph oidium-like. Causing powdery mildew of apple and related species.

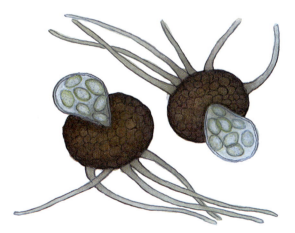

Podosphaera macularis (Wallr.) U. Braun & S. Takam.: cleistothecioid perithecia, with short and simple appendages at the base, releasing asci with ascospores, x 250. Cause of powdery mildew of apple and related species. Collected in the Maritime Territory, Russian Federation.

Podospora Ces.
Lasiosphaeriaceae, Sordariales, Sordariomycetidae, Sordariomycetes, Pezizomycotina, Ascomycota, Fungi
Cesati, V. 1856. *In*: Rabenhorst, *Klotzschii Herb. Viv. Mycol.*, Fasc.: no. 259 (vel 258).
From Gr. *poús*, genit. *podós*, foot + *sporá*, spore. Ascospores with a pedicel formed by a hyaline basal cell; also with an apical germ pore, a dark apical cell, and gelatinous appendages although these may be reduced or absent. Asci with 4, 8 or even 32 ascospores. Coprophilous and cosmopolitan.

Podospora perplexens (Cain) Cain: dark perithecium, with long ostiolar neck, of this coprophilous species, x 50; ascus with eight, dark ascospores, each one with a hyaline basal cell, x 240; one free ascospore with basal and apical hyaline appendages, x 260. On horse dung, Canada.

Polyancora Voglmayr & Yule
Xylariaceae, Xylariales, Xylariomycetidae, Sordariomycetes, Pezizomycotina, Ascomycota, Fungi
Voglmayr, H., C. M. Yule. 2006. *Polyancora globosa* gen. sp. nov., an aeroaquatic fungus from Malaysian peat swamp forests. *Mycol. Res. 110*(10):1242-1252.
From Gr. *polýs*, many + L. *ancora* < Gr. *ankýra*, anchor, for the anchor-like hooks on the cells of the conidia. Conidiophores mononematous, erect. Conidiogenous cells integrated, holoblastic, terminal. Conidia acrogenous, multicellular, globose, of three distinct elements: branching globose central cells; an outermost globose cell; and several thinner, elongated, radially oriented cylindrical cells branching several times at tip to form arched branchlets. Conidial secession schizolytic. Isolated from submerged leaves and twigs in Pahang, Malaysia.

Polypaecilum G. Sm.
Trichocomaceae, Eurotiales, Eurotiomycetidae, Eurotiomycetes, Pezizomycotina, Ascomycota, Fungi
Smith, G. 1961. *Polypaecilum* gen. nov. *Trans. Br. mycol. Soc. 44*(3):437-440.
From Gr. *polýs*, many, much + *poikílos*, varied, an apparent reference to the irregularly branched conidiophores. Colonies slow-growing; conidiophores irregularly branched with final branches producing at tip a number of sterigma-like projections without basal septa,

Polyphagus

each of which is an annellophore, producing chains of spores. Terminal branches of sporing structures termed compound annellophores. Chlamydospores abundant in some species. Isolated from the ear of a male patient in United Leeds Hospital, England.

Polyancora globosa Voglmayr & Yule: conidiophore of this aero-aquatic fungus, with aerial conidia consisting of three elements: branching globose central cells, outermost globose cells, and several radially oriented cylindrical cells to form arched branches, x 430. Conidia are borne from the aerial part of this fungus. Isolated from submerged leaves in tropical peat swamp forest, Peninsular Malaysia.

Polyphagus Nowak.
Chytridiaceae, Chytridiales, Chytridiomycetidae, Chytridiomycetes, Inc. sed., Chytridiomycota, Fungi
Nowakowski, L. 1876. Beitrag zur Kenntniss der Chytridiaceen. II. *Polyphagus euglenae*, p. 203. *In*: J. F. Cohn, *Beitr. Biol. Pfl.* 2(2):201-219.
From Gr. *polýs*, many + *phágos* < *phágomai*, to eat. Parasitizes many species of freshwater algae.

Polyphagus parasiticus Nowak.: zoosporangium of this chytrid, which can parasitize many species of freshwater algae, with its rhizoids attacking the algal filaments, x 700.

Polyporus P. Micheli ex Adans. (syn. **Poria** Pers.)
Polyporaceae, Polyporales, Inc. sed., Agaricomycetes, Agaricomycotina, Basidiomycota, Fungi

Adanson, M. 1763. *Familles de Plantes* 2:10.
From Gr. *polýs*, *polý*, much, numerous, with many + *póros* > L. *porus*, pore, for the presence of many pores on the lower part of the hymenophore. Fructifications membranous, coriaceous or suberose pileus, or resupinate, with dimitic hyphal system, pileus somewhat scaly, hymenophore of tubes whose openings are round or angular, with a white or dark context. Spores smooth, white or colored. Lignicolous, usually on wood of deciduous trees, rarely on conifers. *Polyporus arcularius* Batsch. ex Fr. is a common species on logs or branches in subtropical forests or in semiarid regions. *Polyporus vulgaris* Fr. (=*Poria vulgaris* Fr.) on dead wood or on branches of leafy trees. Previously this genus was universally accepted for all tough species with a poroid hymenium, but, based principally on microscopic differences, its species were segregated into diverse genera.

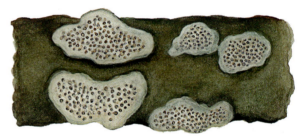

Polyporus efibulatus (A. M. Ainsw. & Ryvarden) Melo & Ryvarden: basidiomata, on dead wood, x 1.

Polystictus Fr.—see **Coltricia** Gray
Hymenochaetaceae, Hymenochaetales, Inc. sed., Agaricomycetes, Agaricomycotina, Basidiomycota, Fungi
Fries, E. M. 1851. *Nova Acta R. Soc. Scient. upsal.*, Ser. 3, 1(1):70.
From Gr. *polýstiktos*, with many punctures, from *polýs*, *polý*, many, with a lot, very + *stictós*, punctured, pitted, stigmatized, speckled. Small pores with pricks or punctate marks on surface.

Polysynnema Constant. & Seifert
Inc. sed., Inc. sed., Inc. sed., Inc. sed., Pezizomycotina, Ascomycota, Fungi
Constantinescu, O., K. A. Seifert. 1988. *Polysynnema*, a new genus of Hyphomycetes. *Trans. Brit. mycol. Soc.* 90(2):332-335.
From Gr. *polý*, many + L. *synnema* < Gr. *sýn*, together + *nêma*, filament, referring to the compound conidiomata. Conidiomata compound, of an axis with multiple fertile capituli that may be sessile or terminal on determinate synnematous stipes; white or pale brown, not changing color in 2% KOH or 85% lactic acid. Hyphae of stipe parallel, lacking clamp connexions. Marginal hyphae differentiated from hyphae of stipe. Conidiogenous cells usually integrated, proliferating sympodially, cicatrized with dark protuberant scars, pale brown, arranged in

a conidiogenous hymenium. Capituli black or dark brown. Conidia amerosporous, pale brown, with conspicuous basal scar, usually not darkly pigmented. On wood in tropical regions.

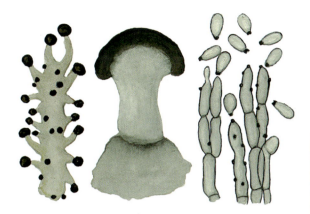

Polysynnema tropicum Constant. & Seifert: habit of conidiomatal synnema on wood, with multiple black, fertile capituli, that may be sessile or stipitate on determinate synnematous stipes, x 180; close-up of a capitulum, x 600; conidiogenous cells cicatrized with dark protuberant scars, and conidia with conspicuous basal scar, x 1,000. On wood, Hawaii, U.S.A.

Poria Pers.—see Polyporus P. Micheli ex Adans.

Polyporaceae, Polyporales, Inc. sed., Agaricomycetes, Agaricomycotina, Basidiomycota, Fungi

Persoon, C. H. 1794. Neuer Versuch einer systematischen Einteilung der Schwämme. *Neues Mag. Bot.* 1:109.

From L. *porus*, way or road, pore (Gr. *póros*) + L. suf. *-ia*, which indicates a characteristic or relation.

Porina Ach.

Porinaceae, Ostropales, Ostropomycetidae, Lecanoromycetes, Pezizomycotina, Ascomycota, Fungi

Acharius, E. 1809. *K. Vetensk-Acad. Nya Handl.* 30:158.

From L. *porus*, pore + suf. *-ina*, which denotes similarity or possession. Thallus with numerous perithecia, which open by pores, at times inconspicuous, on upper surface, crustose, thin, smooth or finely rugose, partially or totally sunken in substrate, at times fugaceous. On tree bark and rocks.

Poroconiochaeta Udagawa & Furuya—see Coniochaeta (Sacc.) Cooke

Coniochaetaceae, Coniochaetales, Sordariomycetidae, Sordariomycetes, Pezizomycotina, Ascomycota, Fungi

Udagawa, S., K. Furuya. 1979. *Poroconiochaeta*, a new genus of the Coniochaetaceae. *Trans. Mycol. Soc. Japan* 20(1):5-12.

From L. *porus*, pore + genus *Coniochaeta* (Sacc.) Cooke, in reference to the rounded pits in the ascospore wall.

Porodiscella Viégas

Xylariaceae, Xylariales, Xylariomycetidae, Sordariomycetes, Pezizomycotina, Ascomycota, Fungi

Viégas, A. P. 1944. Alguns fungos de Brasil II. Ascomicetos. *Bragantia* 4(1-6):1-392.

From genus *Porodiscus* Murrill + dim. suf. *-ella*, for similarities to this genus. Fungus forming a thick, subperidermal, reddish-brown subiculum that gives rise to ascomata borne on a short foot. Ascomata globose, brown perithecia, ostiolate, aggregated in groups. Paraphyses filiform. Asci clavate-cylindrical, 8-spored. Ascospores brown, smooth, 1-celled, plano-convex, ovoid-oblong or subfusiform. On fallen limbs in São Paulo State, Brazil.

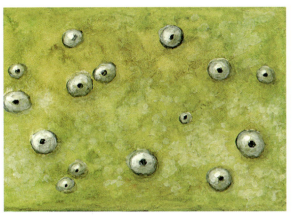

Porina epiphylla Fée: crustous, finely rugose thallus growing on tree bark, with numerous perithecia which open by pores, x 18.

Poroleprieuria M. C. González, et al.

Xylariaceae, Xylariales, Xylariomycetidae, Sordariomycetes, Pezizomycotina, Ascomycota, Fungi

González, M. C., et al. 2004. *Poroleprieuria*, a new xylariaceous genus from Mexico. *Mycologia* 96(3):675-681.

From L. *porus*, pore + genus *Leprieuria* Lassoë, et al., i.e., like *Leprieuria*, but with ascospores having germ pores. Stroma erect, cylindrical, dark brown, up to 7 mm high x 1 mm wide, erumpent through host bark, containing several ostiolate perithecia immersed in apex. Asci clavate, unitunicate, 8-spored. Ascospores reniform, pale brown, 1-celled, smooth, with an apical germ pore. On decayed bark of *Heliocarpus* sp. in Puebla, Mexico.

Poronia Willd.

Xylariaceae, Xylariales, Xylariomycetidae, Sordariomycetes, Pezizomycotina, Ascomycota, Fungi

Willdenow, C. L. 1787. *Fl. berol. prodr.*, Wilhelmi Viewegii, Berlin:400.

From L. *porus* < Gr. *póros*, way or road, pore + suf. L. *-ia*, which indicates connection, similarity or some characteristic. Ascomata with small orifices or pores in surface, which correspond to ostioles of stromatic perithecia. Stromata pale, clavate, with an apical disc that bear ostiolate perithecia. Principally on herbivore excrement.

Postia

Poroleprieuria rogersii M. C. González, et al.: lateral view of two stromata, x 4, apical view of stromatal ostioles, x 10; young ascus with eight ascospores, x 700; and liberated ascospores with germ pore, x 1,320. In decomposing bark of host (*Heliocarpus* sp.), in Puebla, Mexico.

Poronia punctata (L.) Fr.: stromata white to pale yellow, with an apical disc in which are borne the ostiolate black perithecia, on herbivore excrement, x 5.

Postia Fr. (syn. **Oligoporus** Bref.)
Fomitopsidaceae, Polyporales, Inc. sed., Agaricomycetes, Agaricomycotina, Basidiomycota, Fungi
Fries E. M. 1874. *Hymenomyc. eur.* (Upsaliae):586.
Named in honor to the Swedish naturalist *Hampus von Post* + *-ia*, ending of generic names. Irregular bracket, sometimes roughly semicircular but more often shell shaped and very occasionally in the form of a lopsided spinning top; upper surface finely velvety, uneven, white, becoming light ochre with age; margins rounded in young specimens, more acute as fruitbodies age; lower surface with white tubes and pores; watery droplets are exuded mainly from margin region and from the pores. Spores ellipsoidal to cylindrical, smooth, inamyloid. Strong fungal odour; very bitter taste. On felled trunks and large fallen branches of conifers.

Potebniamyces Smerlis—see **Phacidiopycnis**
Potebnia
Cryptomycetaceae, Rhytismatales, Leotiomycetidae, Leotiomycetes, Pezizomycotina, Ascomycota, Fungi
Smerlis, E. 1962. Taxonomy and morphology of *Potebniamyces balsamicola* sp. nov. associated with a twig and branch blight of balsam fir in Quebec. *Can. J. Bot. 40*:352.
Named in honor of the mycologist *A. Potebnia* + Gr. *mýkes* > L. *myces*, fungus.

Proboscispora S. W. Wong & K. D. Hyde
Inc. sed., Inc. sed., Inc. sed., Inc. sed., Pezizomycotina, Ascomycota, Fungi
Wong, S.-W., K. D. Hyde. 1999. *Proboscispora aquatica* gen. et sp. nov., from wood submerged in freshwater. *Mycol. Res. 103*(1):81-87.
From L. *proboscis*, proboscis + *spora*, spore, referring to the appendages on the ascospore. Ascomata immersed, ellipsoidal, pale or dark-brown, coriaceous, ostiolate, papillate, solitary or gregarious, with a short neck. Peridium of several layers of flattened cells. Paraphyses hypha-like, septate, tapering. Asci 8-spored, cylindrical, unitunicate, pedicellate, with a relatively large refractive apical ring. Ascospores uniseriate or overlapping uniseriate, ellipsoidal or fusiform, hyaline, 1-3-septate, with bipolar filamentous appendages. Appendages initially coiled or proboscis-like, subsequently unfurling in water to form long threads. On submerged wood in streams.

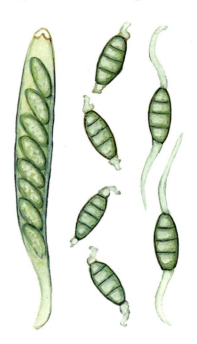

Proboscispora aquatica S. W. Wong & K. D. Hyde: uniseriate young ascus, with a refringent apex, with eight triseptate ascospores, with polar appendanges, x 550; these appendages are unfolded to give a proboscis-like appearance, x 600. In submerged wood in stream, Queensland, Australia, and Phillipines.

Prostratus Sivan., et al.
Melanconidaceae, Diaporthales, Sordariomycetidae, Sordariomycetes, Pezizomycotina, Ascomycota, Fungi

Sivanesan, A., et al. 1993. *Prostratus* a new diaporthaceous ascomycete genus on *Cyclobalanopsis* from Taiwan. *Mycol. Res.* 97(10):1179-1182.

From L. *prostratus*, prostrate < *prostenere*, to prostrate, throw down, for the horizontal ascomata. Stromata hypophyllus, scattered, immersed, subcuticular, erumpent, black. Ascomata single or in pairs, unilocular, horizontal, depressed globose to elliptical, with an erumpent periphysate ostiolar beak that bends upward. Asci cylindrical, 8-spored, rarely 4-spored, with a distinct apical structure not bluing in iodine. Ascospores ovoid to ellipsoidal, 1-septate at basal or apical end, with a large brown cell and hyaline dwarf cell. On leaves of *Cyclobalanopsis mori* in Kuanwu, Hsichu Hsien, Taiwan.

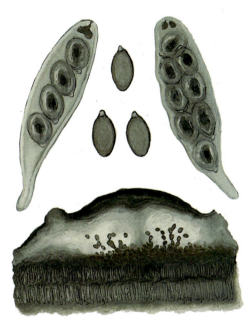

Prostratus cyclobalanopsidis Sivan., et al.: vertical section of an ascoma, with two horizontal locules, x 15; one ascus with four ascospores, and another ascus with eight ascospores, x 600; in the center free ascospores, with a dwarf cell in the apex, x 330.

Protomyces Unger

Taphrinaceae, Taphrinales, Taphrinomycetidae, Taphrinomycetes, Taphrinomycotina, Ascomycota, Fungi

Unger, F. 1833. *Exanth. Pflanzen* (Wien) 12:341.

From Gr. *prōtos*, first + *mýkes* > L. *myces*, fungus, due to its relatively simple morphology and structure, it is considered as one of the most primitive of the ascomycetes. An intracellular parasite of members of the family Apiaceae. Its sporiferous structure is interpreted as a compound ascus, known as a synascus.

Protostegiomyces Bat. & A. F. Vital

Inc. sed., Inc. sed., Inc. sed., Inc. sed., Pezizomycotina, Ascomycota, Fungi

Batista, A. C., et al. 1955. Alguns fungos hyperparasitas. *Anais Soc. Biol. Pernambuco* 13(2):94-107.

From genus *Protostegia* Cooke + L. suf. *-myces* < Gr. *-mýkés*, fungus, in apparent reference to the hyperparasitic habitat. Mycelium indistinct, pycnidia globose, initially enclosed, then cupulate, finally discoid-scutate, superficial, on a black membranous subiculum, gregarious or confluent, black. Conidiophores filiform, simple, hyaline. Conidia acrogenous, falcoid, non-septate, hyaline. Parasitic on *Lembosia byrsonimae* in Recife, Brazil.

Protomyces pachydermus Thüm.: parasitic sporiferous structures inside the host tissues; in the center, the developed compound ascus, known as synascus, x 500.

Protubera Möller

Phallogastraceae, Hysterangiales, Phallomycetidae, Agaricomycetes, Agaricomycotina, Basidiomycota, Fungi

Möller, A. 1895. Zusammen stellurg der durah die vortiegende Arbek veränderten und der Beschreibunger neuer Gattungen und Arten. Brasilischen Pilzblumen, Jena, Verlag von Gustav Fischer, Jena, Schimper, A. F. W. *Bot. Mitt. Trop.* 7:10, 145.

From L. *protubera*, to swell, to grow forth; because the carpophore is similar to a tumor or tuber. Basidiocarps rounded with a thin brown cuticle and a white, gelatinous volva with radial partition walls. Spores similar to those of Clathraceae and Hysterangiaceae. Related to *Hysterangium* Vittad.

Psathyrella (Fr.) Quél.

Psathyrellaceae, Agaricales, Agaricomycetidae, Agaricomycetes, Agaricomycotina, Basidiomycota, Fungi

Quélet, L. 1872. Les Champignons du Jura et des Vosges. *Mém. Soc. Émul. Montbéliard*, Sér. 2, 5:178.

From Gr. *psáthyros*, brittle, friable + L. dim. suf. *-ella*, in reference to the consistency of the fructification. Pileus campanulate or conical, generally hygrophanous; lamellae adnexed or adnate or sub-decurrent, rarely almost free, wedge-shaped in cross section, not deliquescent. Spore print cinnamon or purplish-fuscous or black, spores smooth or finely verruculose, generally with a truncate germ pore, slightly rhomboid or tetrahedric, small or medium sized, rarely above 20 μm, stipe central,

Pseudallescheria

radicant or not, often with short rhizomorphs; pileus context fragile, fleshy, thin. On soil, in or outside the woods, on dung, among mosses or straw, even on living plants or parasitizing other agarics. Cosmopolitan.

Pseudallescheria Negroni & I. Fisch. (syn. **Petriellidium** Malloch)

Microascaceae, Microascales, Hypocreomycetidae, Sordariomycetes, Pezizomycotina, Ascomycota, Fungi
Negroni P., I. Fischer. 1944. *Revista Inst. Bacteriol. 'Dr. Carlos G. Malbrán'* 12(no. 201):5-9.
From Gr. *pseûd*, *pseúdos*, false + genus *Allescheria* Sacc. & P. Syd. Ascocarps formed rarely. Asexual morph monosporium-like. Saprobic in soil. Causes mycetoma in humans and other vertebrates, which suppurates from granulomas and abscesses of lesions.

Pseudevernia Zopf

Parmeliaceae, Lecanorales, Lecanoromycetidae, Lecanoromycetes, Pezizomycotina, Ascomycota, Fungi
Zopf, F. W. 1903. Vergleichende Vntersuchungen über Flechten in Bezug auf ihre Stoffwechselprodukte. *Beih. Botan. Centralbl.* 14:124.
From Gr. *pseûd*, *pseúdos*, false, illegitimate, lie + genus *Evernia* Ach., from *evernís*, what grows or sprouts well. Thallus similar to that of *Evernia*, fruticose or subfruticose; lower surface furrowed, with goblet-shaped apothecia. Photobionts *Protococcus* and *Trebouxia*. Mainly on bark of conifers.

Pseudevernia furfuracea (L.) Zopf: fructicose thallus, with the lower surface furrowed, with goblet-shaped apothecia, on conifer bark, x 3.

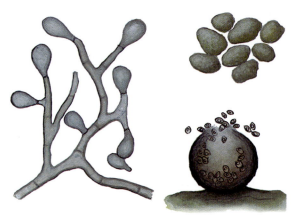

Pseudallescheria boydii (Shear) McGinnis, et al. (syn. **Petriellidium boydii** (Shear) Malloch): conidiophore with aleuriospores, x 1,000; subsphaerical, hard grains, x 1,200; ascoma with elliptical, brown ascospores, x 200.

Pseudeurotium J. F. H. Beyma

Pseudeurotiaceae, Inc. sed., Leotiomycetidae, Leotiomycetes, Pezizomycotina, Ascomycota, Fungi
Beyma, J. F. H. 1937. Beschreibung einiger neuer Pilzarten aus dem Centraalbureau voor Schimmelcultures Baarn (Holland). *Centbl. Bakt. ParasitKde*, Abt. II 96 (20-23):415.
From Gr. *pseûd*, *pseúdos*, false + genus *Eurotium* Link, i.e., a fungus similar to *Eurotium*. Ascomata cleistothecia, spherical, scattered to gregarious on substrate, brown to black. Ascomatal wall coriaceous, of a single layer of dark, thick-walled, angular to flattened cells, sometimes lined by rows of flattened, thin-walled pseudoparenchymatous cells. Asci formed laterally from ascogenous hyphae, scattered throughout interior of ascomata, globose to ellipsoidal, 8-to many-spored, evanescent at maturity. Ascospores conglobate, 1-celled, brown, globose or ellipsoidal, smooth, without germ pores. Isolated from nematode cysts, plywood, and ambrosia beetle tunnels.

Pseudocercospora Speg.

Mycosphaerellaceae, Mycosphaerellales, Dothideomycetidae, Dothideomycetes, Pezizomycotina, Ascomycota, Fungi
Spegazzini, C. 1911. Mycetes Argentinenses (Series V). *Anal. Mus. nac. B. Aires, Ser.* 3 13:329-467.
From Gr. *pseûd*, *pseúdos*, false + genus *Cercospora* Fresen. ex Fuckel. Mycelium internal and external, consisting of smooth, septate, subhyaline to brown, branched hyphae. Stroma absent to well-developed. Conidiophores arranged in loose to dense fascicles, sometimes forming distinct synnemata or sporodochia, emerging through stomata or erumpent through the cuticle, short to long, septate or continuous, pale to dark brown, smooth to finely verruculose. Conidiogenous cells integrated, terminal, occasionally intercalary, polyblastic, sympodial, or monoblastic, pale to dark brown. Conidia solitary, rarely in simple chains, subhyaline, olivaceous, pale to dark brown, usually scolecosporous, obclavate-cylindrical, filiform, acicular and transversely plurieuseptate. Ascomata pseudothecial, containing bitunicate asci with one-septate ascospores. Foliicolous, chiefly phytopathogenic, but also endophytic; commonly associated with leaf spots, but also occurring on fruit.

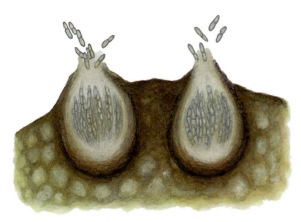

Pseudocercospora fijiensis (M. Morelet) Deighton (syn. **Mycosphaerella fijiensis** M. Morelet): sagittal section of stromatic, perithecioid pseudothecia, immersed in the tissues of the banana host (*Musa* sp., Hawaii, U.S.A.), showing asci and bicellular ascospores, x 580.

Pseudoclitocybe (Singer) Singer

Tricholomataceae, Agaricales, Agaricomycetidae, Agaricomycetes, Agaricomycotina, Basidiomycota, Fungi
Singer, R. 1956. New genera of fungi. VII. *Mycologia* 48(5):725.
From Gr. *pseûd*, *pseúdos*, false + genus *Clitocybe* (Fr.) Staude, a false *Clitocybe*. Pileus glabrous or fibrillose, more or less hygrophanous; lamellae narrow to rather broad, decurrent, but at times sinuate or adnate, occasionally forked; context not reddening when bruised; hyphae without clamp connections; subhymenium cellular or subcellular. Spores amyloid, smooth. On woody debris, rarely on dung, in temperate regions of northern hemisphere; only one species in South America. One species is edible (*P. cyathiformis*).

Pseudocyphellaria Vain.

Lobariaceae, Peltigerales, Lecanoromycetidae, Lecanoromycetes, Pezizomycotina, Ascomycota, Fungi
Vainio, E. A. 1890. Lichens du Brésil, I-II. *Acta Soc. Fauna Flora fenn.* 7 (no. 1):182.
From Gr. *pseúdos*, falsehood, lie, which indicates falseness or illegitimacy + NL. *cyphella* < Gr. *kýphella*, the concavities of the ears + L. suf. *-aria*, which indicates connection, possession. Thallus with pseudocyphellae or conspicuous pores, generally white, occasionally yellow, on lower part, similar to cyphellae, but smaller. On trees, especially oaks.

Pseudodictyosporium Matsush. (syn. **Kamatia** V. G. Rao & Subhedar)

Dictyosporiaceae, Pleosporales, Pleosporomycetidae, Dothideomycetes, Pezizomycotina, Ascomycota, Fungi
Matsush. *In*: Kobayasi, *Bull. natn. Sci. Mus.*, Tokyo, N.S. 14:473.
From Gr. *pseúdos*, false + *díktyon*, net + *sporá*, spore + NL. *-ium*, suf. added to generic names. Sporodochia

Pseudofuscophialis

superficial, punctiform to effuse, scattered, sometimes coalescing, pale brown to dark brown, with or without a mucilage covering, rarely inconspicuous. Mycelium immersed, composed of septate, branched, subhyaline to pale brown, smooth-walled hyphae. Conidiophores micronematous, aseptate, simple, hyaline to pale brown, smooth. Conidiogenous cells integrated, holoblastic, terminal, determinate, doliiform to cylindrical. Conidia acrogenous, solitary, dry, cheiroid, very pale brown, smooth-walled, euseptate or distoseptate, consisting of a truncate basal cell on which three rows of cells arise in parallel and compactly with all three rows in different planes, with or without appendages. Saprobic on submerged decayed wood in aquatic habitats.

Pseudocyphellaria aurata (Ach.) Vain.: foliose thallus, with yellow pseudocyphellae on the lower part, on tree bark, x 10.

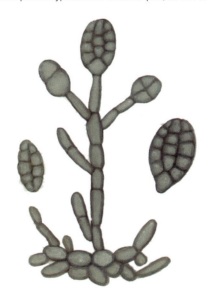

Pseudodictyosporium elegans (Tzean & J. L. Chen) R. Kirschner: conidiophores and conidia from the woody branches of *Ficus*, in Taiwan; the conidia are chiroid, light brown, with a basal black hilum, x 1,500.

Pseudofuscophialis Sivan. & H. S. Chang

Inc. sed., Inc. sed., Inc. sed., Inc. sed., Pezizomycotina, Ascomycota, Fungi

Pseudohydnum

Sivanesan, A., H. S. Chang. 1995. *Pseudofuscophialis lignicola* gen. et sp. nov. and *Chaetosphaeria capitata* sp. nov. from wood in Taiwan. *Mycol. Res.* 99(6):711-716.
From Gr. *pseúdos*, falsehood, lie > pref. *pseúdo-*, which indicates falseness or illegitimacy, i.e., false, illegitimate + genus *Fuscophialis* B. Sutton, for the similarity to this genus. Colonies effuse, sparse, mycelium superficial, of branched, septate, smooth, brown hyphae. Conidiophores macronematous, mononematous, straight, simple, rarely branched, smooth, multiseptate, brown to dark brown. Conidiogenous cells mono- to polyphialidic, integrated, rarely discrete, intercalary, terminal, sympodial. Phialides terminated by a funnel-shaped collarette. Conidial proliferation at conidiogenous locus enteroblastic. Conidia holoblastic, endogenous, acropleurogenous, simple, straight to slightly curved, smooth, apex obtuse or acute, truncate at base, transversely multiseptate, with large, pale brown central cells and smaller, subhyaline to pale brown end cells.

Pseudohydnum P. Karst.
Inc. sed., Auriculariales, Inc. sed., Agaricomycetes, Agaricomycotina, Basidiomycota, Fungi
Karsten, P. 1868. Auriculariei, Clavariei et Tremuellini, in paroecia of *Tammela crescentes*. *Not. Sällsk. Fauna et Fl. Fenn. Förh.* 9:374.
From Gr. pref. *pseúdo-*, derived from *pseúdos*, false, illegitimate + *hydnum*, for the presence of a dentate hymenium, similar to the genus *Hydnum* L. ex Fr. Fructifications brown, grayish or whitish, 2-6 cm wide, sessile, or with a short lateral stipe, gelatinous. Heterobasidia characteristic of Heterobasidiomycetes. Humicolous or lignicolous, commonly on well decomposed wood in temperate and tropical forests. In Mexico, *P. gelatinosum* (Fr). Karst. is a common species, principally in fir forests.

Pseudohydnum gelatinosum (Scop.) P. Karst.: lignicolous basidiomata on decomposed wood, with dentate hymenium, x 0.5. Collected in New Zealand.

Pseudolagarobasidium J. C. Jang & T. Chen
Phanerochaetaceae, Polyporales, Inc. sed., Agaricomycetes, Agaricomycotina, Basidiomycota, Fungi

Jang, J. C., T. Chen. 1985. *Pseudolagarobasidium leguminicola* gen. et sp. nov. on *Leucaena* in Taiwan. *Trans. Br. mycol. Soc.* 85:374-377.
From Gr. *pseúdos*, false + *lagarós*, lax, empty + L. *basidium*, small base < Gr. *básis*, base. Basidiocarps annual, resupinate, effused, adnate, soft-membranous, or often pellicular, rhizomorphs present at margin. Hymenophore odontoid, deep-violet or purplish, ochraceous to pale brown when dry, context not darkening in KOH, inamyloid, in ferric sulphate not turning green. Hyphal system monomitic. Hyphae hyaline or slightly yellowish, cylindrical, densely interwoven, branching often near clamp-connections, thin to slightly thick-walled, smooth, clamps always present. Gloeohyphae present. Gloeocystidia present, abundant, of tramal origin, hyaline, cylindrical, or clavate, subglobose when young, thin-walled, smooth, basal clamp connections present, cyanophilous. Basidia small, hyaline, clavate or suburniform, thin-walled, smooth, contents homogeneous or guttulate, a basal clamp connexion present, with (2-)4 subulate, rather slender sterigmata, up to 8 μm long. Spores hyaline, globose to subglobose, thin-walled, smooth, with small hilar appendix, inamyloid, cyanophilous. Spore print white to pale yellow. Parasitic on the roots and stems of *Leucaena leucocephala* in Taiwan.

Pseudonectria Seaver
Nectriaceae, Hypocreales, Hypocreomycetidae, Sordariomycetes, Pezizomycotina, Ascomycota, Fungi
Seaver, F. J. 1909. The Hypocreales of North America - I. *Mycologia* 1:41-76.
From Gr. *pseúdos*, false + genus *Nectria* (Fr.) Fr., for the unicellular spores. Perithecia superficial on substratum, globose to ovoid, bright-colored, yellow, red, smooth or minutely roughened, soft, membranous. Asci cylindrical, 8-spored. Ascospores elliptical or subelliptical, simple, hyaline. On dead wood.

Pseudopeziza Fuckel
Dermateaceae, Helotiales, Leotiomycetidae, Leotiomycetes, Pezizomycotina, Ascomycota, Fungi
Fuckel, K. W. G. L. 1870. *Symbolae mycologicae. Beiträge zur Kenntnis der rheinischen Pilze. Jb. nassau. Ver. Naturk.* 23-24:290.
From Gr. *pseúdos*, false, falsehood, in this case in relation to the genus *Peziza* Dill. ex Fr. Apothecia in groups. On living leaves, occasionally on small stems and peduncles. *Pseudopeziza medicaginis* (Lib.) Sacc. causes serious damage to alfalfa (*Medicago* spp.).

Pseudopileum Canter
Chytridiaceae, Chytridiales, Chytridiomycetidae, Chytridiomycetes, Inc. sed., Chytridiomycota, Fungi
Canter, H. M. 1963. Studies on British chytrids XXIII. New species on chrysophycean algae. *Trans. Br. mycol.*

Soc. 46(3):305-320.

From Gr. *pseúdos*, false + L. *pileus* < Gr. *pílos*, cap, perhaps in reference to the formation of both a cap and an endooperculum on the sporangium. Thallus monocentric, epibiotic, of a sporangium and rhizoidal system. Sporangium produced by direct enlargement of zoospore. Zoospores liberated after detachment of a lid secondarily formed within sporangium. Resting spores sexual, borne like the sporangium. Parasitic on cysts of *Mallomonas* sp. in the plankton of Elterwater, English Lake District, England.

Pseudopeziza medicaginis (Lib.) Sacc.: apothecia, in groups causing leaf spots on living leaves of *Medicago*, x 20.

Pseudopithomyces Ariyaw. & K.D. Hyde
Didymosphaeriaceae, Pleosporales, Pleosporomycetidae, Dothideomycetes, Pezizomycotina, Ascomycota, Fungi
Ariyawansa, H. A., et al. 2015. Fungal diversity notes 111-252 - taxonomic and phylogenetic contributions to fungal taxa. *Fungal Diversity* 75:27-274.
From Gr. *pseúdos*, false + *píthos*, barrel, tub, large-mouthed jar + *mýkes* > L. *myces*, fungus. Conidia dark, subglobose to oval, obovoid, oblong or doliiform. Aleuriospores with transverse and longitudinal septa that originate as lateral evaginations of conidiophores. In soil as saprobes but one species, *P. chartarum* (Berk. & M. A. Curtis) J. F. Li et al., is the causal agent of facial eczema of sheep, due to the production of the mycotoxin sporidesmin in pasture forage.

Pseudotulasnella Lowy
Tulasnellaceae, Cantharellales, Inc. sed., Agaricomycetes, Agaricomycotina, Basidiomycota, Fungi
Lowy, B. 1964. A new genus of the Tulasnellaceae. *Mycologia* 56:696-700.
From Gr. pref. *pseúdo-*, false < *pseúdos*, illegitimate, false + genus *Tullasnella* Schröt., for the similarity between these genera. Fructifications resupinate, waxy-gelatinous. Basidia partially septate, septa formed at base of sterigmata. Probasidia subglobose, cruciate-septate in apical part, incompletely septate in basal part; metabasidia clavate-capitate, with tulasnelloid sterigmata that produce basidiospores capable of germinating repetitively. *Pseudotulasnella guatemalensis* Lowy forms waxy-gelatinous, grayish fructifications that on drying remain as pruinose, whitish-gray layers.

Pseudoxenasma

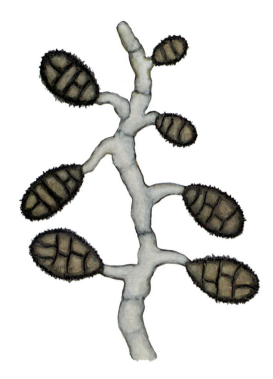

Pseudopithomyces chartarum (Berk. & M. A. Curtis) Jin F. Li, et al. (syn. **Pithomyces chartarum** (Berk. & M. A. Curtis) M. B. Ellis): conidiophore with obovoid, muriform, dark aleuriospores, x 800.

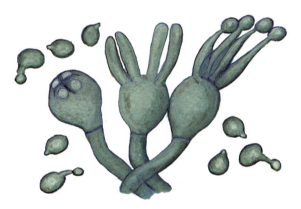

Pseudotulasnella guatemalensis Lowy: subglobose, cruciate-septate probasidia in the apical part, and clavate-stipitate metabasidia, which produce basidiospores capable of germinating repetitively, x 1,500. Collected in Guatemala.

Pseudoxenasma K. H. Larss. & Hjortstam
Russulaceae, Russulales, Inc. sed., Agaricomycetes, Agaricomycotina, Basidiomycota, Fungi
Hjortstam, K., K.-H. Larsson. 1976. *Pseudoxenasma*, a new genus of Corticiaceae (Basidiomycetes). *Mycotaxon* 4:307-311.
From Gr. *pseúdos*, false + genus *Xenasma* Donk, for the similarity of the two genera. Fruitbodies resupinate, effused, thin, hymenium more or less ceraceous, smooth. Hyphal system monomitic. Sulphocystidia present, with positive reaction to sulphovanilline. Cystidia with globose apical appendices. Basidia clavate, with four

Psilocybe

sterigmata. Spores verrucose, thick-walled, broadly ellipsoid to subglobose, with strong amyloid reaction. On dead branches of *Picea* in Sweden.

Psilocybe (Fr.) P. Kumm.
Hymenogastraceae, Agaricales, Agaricomycetidae, Agaricomycetes, Agaricomycotina, Basidiomycota, Fungi
Kummer, P. 1871. *Führer in die Pilzkunde* (Zerbst):21.
From Gr. *psilós*, smooth, nude, bald + *kýbe*, head, pileus. Pileus generally lacking ornamentation. Terricolous, among mosses, lignicolous on dead wood, coprophilous and on the remains of plants and seeds. Some species are utilized as a drug in religious rites, medicine or free consumption, due to the hallucinogenic substances, such as psilocybine and psilocine. This genus also contains toxic species that can produce serious intoxications and even death.

Photobiont *Protococcus*. On soil in arid and semiarid regions, on rocks, in mountainous forests, and on occasion growing on mosses or old wood.

Psora cerebriformis W. A. Weber: squamulose thalli, intimately attached to the substrate, on gypsum knolls, Colorado, U.S.A., x 8.

Ptechetelium Oberw. & Bandoni
Eocronartiaceae, Platygloeales, Inc. sed., Pucciniomycetes, Pucciniomycotina, Basidiomycota, Fungi
Oberwinkler, F., R. Bandoni. 1984. *Herpobasidium* and allied genera. *Trans. Br. mycol. Soc.* 83(4):645.
From Gr. *ptýche*, layer > L. *pteche* + Gr. *télos* > NL. *telium*, finished. Hyphae thin-walled, inter- and intracellular; emerging through stomata, forming small, rarely confluent patches of basidiocarps on underside of green sporophylls, whitish to cream-ochraceous or yellowish to brownish, horny when dry, soaking cartilaginous-gelatinous. Hyphae in leaves and arising from them mostly thin-walled, hyaline; those of subhymenium and hymenium thick-walled, hyaline, whitish to cream-colored; terminating in probasidia and conspicuous thick-walled, branched, light-colored dendrohyphidia with blunt terminations. Probasidia commonly stalked, globose to elongate-pyriform, thick-walled, proliferating percurrently with wall multi-layered after development of successive probasidia. Basidia cylindrical, straight to slightly curved, 4-celled when mature, sterigmata cornute. Basidiospores hyaline, thin-walled, smooth, irregularly sickle-shaped, with oblique apiculi; germinating by repetition. On living leaves of *Cyathea stuebelii* in Puyo, Napo Pastaza Prov., Ecuador.

Psilocybe cubensis (Earle) Singer: coprophilous basidiomata, with a ring in the stipe and almost smooth pileus, x 1.

Psora Hoffm.
Psoraceae, Lecanorales, Lecanoromycetidae, Lecanoromycetes, Pezizomycotina, Ascomycota, Fungi
Hoffmann, G. F. 1796. *Deutschlands Flora oder botanisches Taschenbuch,* Zweiter Theil (Erlangen):161.
From Gr. *psóra*, cutaneous disease, scaly, dry dermatosis. Thallus squamulose, crustous or subfoliose, intimately attached to substrate. Apothecia small or medium-size, with disc flat or convex, reddish-brown or black. Hymenium hyaline or brownish-gray. Spores hyaline.

Ptychoverpa Boud.—see **Verpa** Sw.
Morchellaceae, Pezizales, Pezizomycetidae, Pezizomycetes, Pezizomycotina, Ascomycota, Fungi
Boudier, J. L. E. 1907. Histoire et classification des discomycetes d'Europe. *Hist. Class. Discom. Eur.,* Paris:34.
From Gr. *ptýx, ptychós*, pleat, leather or metal sheet that covers a shield, for the consistency of the pileus, and because it is lightly folded longitudinally, + the genus *Verpa* Sw. (this from L. *verpa*, penis, phallus).

Puccinia Pers.
Pucciniaceae, Pucciniales, Inc. sed., Pucciniomycetes, Pucciniomycotina, Basidiomycota, Fungi
Persoon, C. H. 1794. Neuer Versuch einer systematischen Eintheilung der Schwämme. *Neues Mag. Bot. 1*:118.
Dedicated to *T. Puccini*, Professor in Florence. Pycnia and telia subepidermal. Teliospores pedicellate, bicellular. Autoecious or heteroecious, micro- or macrocyclic. The genus contains more than 3,000 species, many of them considered among the most destructive phytopathogenic fungi of important crops, e.g., *P. graminis* Pers., the rust of wheat and other grains, whose aecia are formed on intermediate hosts of the genus *Berberis*, especially *B. vulgaris*.

Puccinia porri (Sowerby) G. Winter: uredinia of this rust, with urediniospores, in leaf of garlic (*Allium*), x 10.

Pullularia Berkhout—see **Aureobasidium** Viala & G. Boyer
Saccotheciaceae, Dothideales, Dothideomycetidae, Dothideomycetes, Pezizomycotina, Ascomycota, Fungi
Berkhout, C. M. 1923. De schimmelgeslachten *Monilia, Oidium, Oospora en Torula*. Thesis, Utrecht:54-55.
From L. *pullulus*, dim. of *pullus*, blackish color + suf. *-aria*, which indicates connection or possession, in reference to the color of the mature colonies, which are almost black, wet and yeast-like. The name also can be derived from L. *pullulo, pullulas, pullulare*, to bud, germinate < *pullulus*, sprout, renew + dim. suf. *-aria*, in relation to its development as a saprobe on diverse substrates.

Pulveroboletus Murrill
Boletaceae, Boletales, Agaricomycetidae, Agaricomycetes, Agaricomycotina, Basidiomycota, Fungi
Murrill, W.A. 1909. The Boletaceae of North America. *Mycologia 1*(1):4-18.
From L. *pulvus, pulveris*, powder, dust + genus *Boletus* L., a powdery mushroom, because the ripe fructification has a powdery surface. Pileus with tubular hymenophore, wide to narrow pores of various colors, generally yellow, orange, red or reddish; spores fusoid, short elliptic to short cylindric. Spore print brown or olivaceous; surface of pileus often viscid when wet, gelatinous, covered by a brown to yellow pulverulence; stipe smooth, reticulate or fibrillose-pulverulent, solid or hollow, with a white or colored mycelium, context bluing or not. Mycorrhizae sometimes formed with species of *Quercus* and *Pinus* or other genera of trees; frequent in north temperate and tropical regions. Some species are edible, one is considered poisonous in the Philippines.

Pulvinella A. W. Ramaley
Inc. sed., Inc. sed., Inc. sed., Inc. sed., Pezizomycotina, Ascomycota, Fungi
Ramaley, A. W. 2001. *Pulvinella*, a new genus with prosenchymatous propagules. *Mycotaxon 79*:51-56.
From L. *pulvinus*, cushion, pillow + dim. suf. *-ella*, a small cushion or cushion-like structure. Hyphae immersed in host plant tissues, forming cushion-shaped propagules. Propagules sphaeroid, greenish clusters of prosenchymatous cells, uniform in structure, with no distinguishable rind and medulla, separating easily from mycelium. On decomposing leaves of *Nolina micrantha* in New Mexico, U.S.A.

Pustularia Fuckel—see **Tarzetta** (Cooke) Lambotte
Pyronemataceae, Pezizales, Pezizomycetidae, Pezizomycetes, Pezizomycotina, Ascomycota, Fungi
Fuckel, K. W. G. L. 1870. Symbolae mycologicae. Beiträge zur Kenntnis der rheinischen Pilze. *Jb. nassau. Ver. Naturk.* 23-24:328.
From L. *pustula*, pustule, grain + L. suf. *-aria*, which indicates possession or connection, for developing ascomata whose external surface has vesicles that resemble pustules.

Pycnofusarium Punith.—see **Fusarium** Link
Nectriaceae, Hypocreales, Hypocreomycetidae, Sordariomycetes, Pezizomycotina, Ascomycota, Fungi
Hawksworth, D. L., E. Punithalingam. 1973. New and interesting microfungi from Slapton, South Devonshire: Deuteromycotina. *Trans. Br. mycol. Soc. 61*:57-69.
From Gr. *pyknós* > L. *pycnos*, compact, dense + genus *Fusarium* Link, for the spores that resemble those of *Fusarium*.

Pycnopeziza W. L. White & Whetzel
Sclerotiniaceae, Helotiales, Leotiomycetidae, Leotiomycetes, Pezizomycotina, Ascomycota, Fungi
White, W. L., H. H. Whetzel. 1938. Pleomorphic life cycles in a new genus of the Helotiaceae. *Mycologia 30*:187-203.
From Gr. *pyknós*, dense, compact > L. *pycnos* + genus *Peziza* Dill. ex Fr., for a peziza-like fungus with a pycnidial asexual morph. Apothecia small, solitary or gregarious, short-stipitate to practically sessile, brown or brownish, at first closed, opening irregularly or by a pore, finally expanded, margin stellate or circular. Hymenium pale-brown or buff. Asci clavate or cylindric-clavate, 8-spored. Ascospores small, ellipsoid, hyaline,

Pycnoporus

simple. Paraphyses filiform, simple. Pycnidial morph acarosporium-like. On buds and flowers.

Pycnoporus P. Karst.
Polyporaceae, Polyporales, Inc. sed., Agaricomycetes, Agaricomycotina, Basidiomycota, Fungi
Karsten, P. A. 1881. Enumeratio boletinearum et polyporearum. Fennicum, systemate novo dispositarium. *Revue mycol., Toulouse* 3:18.
From Gr. *pycnós*, compact, dense, thick, very close + *póros* > L. *porus*, pore, way, road, for the appearance of the pores of the hymenophore. Basidiomata annual, usually cinnabar to reddish-orange. Lignicolous, on wood of broad-leaved trees, rarely on conifers, causing white rot. *Pycnoporus sanguineus* (= *Polyporus sanguineus* L. ex Fr.) grows in disturbed areas, on fallen or burned logs in tropical regions; it has a bright, showy, semicircular, shelf-shaped fructification that is orangish-red.

Pycnoporus sanguineus (L.) Murrill: cinnabar to reddish-orange, zonate basidioma, causing white rot of wood of broad-leaved tree, x 0.5.

Pyrenidium Nyl. (syn. **Dacampiosphaeria** D. Hawksw.)
Dacampiaceae, Inc. sed., Inc. sed., Dothideomycetes, Pezizomycotina, Ascomycota, Fungi
Nylander W. 1965. *Flora*, Regensburg 48:210.
From. Gr. *pyrén*, kernel, the pit of a fruit + *-ídion* > E. *-idium*, dim. suf. Ascomata perithecioid pseudothecia, subglobose, ovoid or obpyriform, dark brown or black, ostiolate, immersed, semiimmersed or subsessile, scattered or in smalls groups, sometimes confluent in necrotic patches or in gall-like deformations of the host thallus. Asci clavate to subcylindrical, distinctly stipitate, bitunicate, thickened with a distinct internal apical beak at the apex, I-, (2-)4- or 8-spored; ascospores brown or dark brown in the oldest, with the tips of the end cells often pale brown to subhyaline, ellipsoid to broadly fusiform, often slightly curved, rounded or obtuse at the apices, usually 3-septate, rarely 2- or 4-septate; usually with a conspicuous pore at the central part of the septa, constricted at the septa or not; smooth-walled. Conidiomata pycnidial, black, immersed, with a dark brown wall; conidia hyaline, simple, short-oblong. On thallus of lichens occurring on bark, rocks, soil, bryophytes or living leaves.

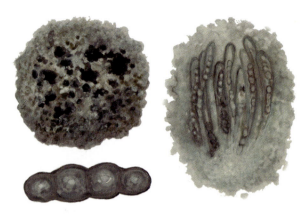

Pyrenidium baeomycearium (Linds.) Nav.-Res. & Cl. Roux: lichenicolous fungus, with the perithecia forming dark brown galls in the thallus of **Baeomyces** Pers., x 1; vertical section of a perithecium showing the asci with 8, 3-septate young ascospores, x 400; a liberated, mature ascospore, dark-brown, 3-septate, x 1,800.

Pyrenochaeta De Not.
Inc. sed., Inc. sed., Pleosporomycetidae, Dothideomycetes, Pezizomycotina, Ascomycota, Fungi
De Notaris, G. 1849. Micromycetes italici velminus cogniti. *Mém. R. Accad. Sci. Torino*, Ser. 2 10:348.
From Gr. *pyrén*, genit. *pyrénos*, nucleus, stone of a fruit, grain + *chaíte*, hair, seta. Pycnidia with abundant hairs or setae around ostiole, fewer on rest of pycnidium. Parasitic on plants and saprobic on decomposing stems and leaves.

Pyrenochaeta nobilis De Not.: pycnidium, growing as saprobe on decomposing plant remains, covered with abundant hairs or setae around the ostiole, x 290.

Pyrenopeziza Fuckel (syn. **Cylindrosporium** Grev.)
Dermateaceae, Helotiales, Leotiomycetidae, Leotiomycetes, Pezizomycotina, Ascomycota, Fungi

Fuckel, K. W. G. L. 1870. *Symbolae mycologicae. Beiträge zur Kenntnis der rheinischen Pilze. Jb. nassau. Ver. Naturk.* 23-24:293.

From Gr. *pyrén*, genit. *pyrénos*, pit of fruits, seed, nucleus + genus *Peziza* Dill. ex Fr., with which it is compared due to the formation of apothecia. Acervuli subcuticular. Spores cylindrical. Parasitic on vascular plants on which it causes leafspots. *Pyrenopeziza brassicae* Sutton & Rawlinson (= *Cylindrosporium concentricum* Grev.) causes light leaf spot of *Brassica oleracea* and related hosts.

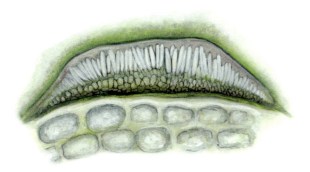

Pyrenopeziza brassicae B. Sutton & Rawl. (syn. **Cylindrosporium concentricum** Grev.): subcuticular acervulus, that produces cylindrical spores, on host tissues, x 300.

Pyrenopeziza carduorum Rehm: numerous apothecia of this parasite, causing leaf spots on dead stem of cactus (*Cirsium palustre*), Germany, x 3.

Pyrenophora Fr.

Pleosporaceae, Pleosporales, Pleosporomycetidae, Dothideomycetes, Pezizomycotina, Ascomycota, Fungi

Fries, E. M. 1849. *Summa veg. Scand.*, Sectio Post. (Stockholm):2:397.

From Gr. *pyrén*, *pyrénos*, nucleus, seed, stone of a fruit + suf. *-phóros*, bearer, from *phéro*, to carry, sustain, in reference to the central part of the pseudothecium. Ascomata pseudothecioid, with pseudoparaphyses differentiated in centrum before formation of asci. Ascospores large, hyaline or brownish-gray, oblong, triseptate or pluriseptate, with a longitudinal septum in one or all cells. Asexual morph drechslera-like. On dead herbaceous stems and grasses.

Pyrenula Ach.

Pyrenulaceae, Pyrenulales, Chaetothyriomycetidae, Eurotiomycetes, Pezizomycotina, Ascomycota, Fungi

Acharius, E. 1814. *Synopsis Methodica Lichenum.* (Lund):117.

From Gr. *pyrén*, genit. *pyrénos*, grain, nucleus, stone of a fruit + L. dim. suf. *-ula*. Thallus crustous, thin, smooth or rugose, lacking differentiation into layers. Perithecia small to medium, immersed in thallus like black granules. On bark of trees in which it is partially or totally sunken.

Pyrenophora tritici-repentis (Died.) Drechsler: vertical section of a pseudothecium in wheat leaf, x 40, showing the centrum with asci and brownish-gray, oblong, triseptate ascospores, x 750.

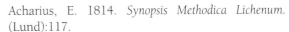

Pyrenula occidentalis (R. C. Harris) R. C. Harris: crustous, thin, smooth thallus in tree bark, with the immersed, black granules-like perithecia, x 0.8. Collected in Washington, U.S.A.

Pyronema Carus

Pyronemataceae, Pezizales, Pezizomycetidae, Pezizomycetes, Pezizomycotina, Ascomycota, Fungi

Carus, C. G. 1835. Beobachtung einer sehr eigentümlichen Schimmel vegetation (*Pyronema marianummihi*) auf Kohlenboden. *Nova Acta Phys.-Med. Acad. Caes. Leop.-Carol. Nat. Cur.* 17(1):367-375.

From Gr. *pýr, pyrós*, fire + *némo*, to occupy, inhabit, consume, eat. Apothecia discoid or lenticular, generally bright-colored, red or pink. Asci 8-spored. Ascospores hyaline, unicellular, ellipsoid. On soil and wood in burned areas, among leaves formed by carbonaceous masses; on wet, disintegrating, unburned leaves and in greenhouses where the soil has been steam sterilized.

Pyronema

Pyronema omphalodes (Bull.) Fuckel: discoid, brilliantly colored, red or pink apothecia in burned wood, x 10.

Pyronema omphalodes (Bull.) Fuckel.: cylindrical ascus, 8-spored, and 2 paraphyses, x 300; free ascospores, ellipsoidal, 1-celled, non-guttulate, smooth, x 460.

Q

Queletia mirabilis

Q

Quasiconcha M. E. Barr & M. Blackw.
Mytilinidiaceae, Mytilinidiales, Incertae sedis, Dothideomycetes, Pezizomycotina, Ascomycota, Fungi
Barr, M. E., M. Blackwell. 1980. A new genus in the Lophiaceae. *Mycologia* 72(6):1224-1227.
From Gr. *quasi*, as if + *konchç*, dim. of *konchíon*, shell > L. *concha*, mussel, shell, for the shape of the ascomata. Ascomata pseudothecial, shaped like a mussel shell standing on edge; superficial, shiny black, attached to substrate by sparse brown, distantly septate, rough-walled hyphae; peridium brittle, thin, dark reddish brown, composed of radiating rows of parallel cells, fanning out from base to sides and above to elongate opening that spans length of ascoma, two or three layers thick in side view. Asci cylindric, bitunicate, numerous, thin walled at maturity. Ascospores brown, broadly ellipsoid, symmetric, 1-septate, slightly constricted at septum; coarsely reticulate in surface view, Isolated from undigested seeds of Juniperus virginiana, Texas, U.S.A.

Queletia Fr.
Agaricaceae, Agaricales, Agaricomycetidae, Agaricomycetes, Agaricomycotina, Basidiomycota, Fungi
Fries, E. 1872. *Queletia*, novum Lycoperdaceorum genus. Accedit nova Gyromitrae species. *Öfvers. K. Svensk. Vetensk.-Akad. Förhandl.* 28(2):171.
Named in honor of the French mycologist *Lucien Quélet* + L. *-ia*, ending of generic names. Basidiomata scattered to gregarious, stipitate. Pileus globose to subglobose. Gleba ochre, abundant, with filaments of whitish capillitium. Dehiscence apical, irregular, without formation of a stoma or defined structure. Stipe cylindrical, curved, fading towards base, not twisted, with a strongly folded surface and appearing torn, ochre and brown, woody. Basidiospores yellow-ochraceous, globose to subglobose, with a prominent hilar appendix, strongly verrucose, with verrucae grouped into conical structures. Capillitium abundant. Grows on bark and wood debris, or in sandy soil, from France, England, Spain, U.S.A., and Mexico.

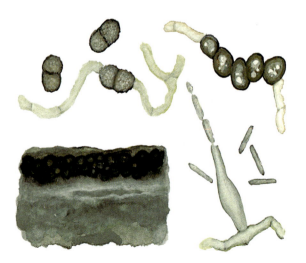

Quasiconcha reticulata M. E. Barr & M. Blackw.: black ascomata (pseudothecia) in root phloema of Allepo pine (*Pinus halapensis*), x 1.5; it was also found in dead roots of *Thuja occidentalis*. Reticulate ascospores germinating in both poles, x 650; phialides releasing aseptate conidia, x 400.

Queletia mirabilis Fr.: basidiomata gregarious on sandy soil, in Canary Islands, Spain, x 1.

R

Rhodotus palmatus

R

Radiomyces Embree
Radiomycetaceae, Mucorales, Inc. sed., Inc. sed., Mucoromycotina, Zygomycota, Fungi
Embree, R. W. 1959. *Radiomyces*, a new genus in the Mucorales. *Am. J. Bot. 46*:25-30.
From L. *radio* < *radius*, ray, radius + Gr. *mýkes* > L. *myces*, fungus, referring to the origin of the secondary vesicles that bear pedicellate sporangiola growing radially from a primary vesicle. Coprophilous on lizard and mouse dung, pathogenic to mice.

Radiomyces embreei R. K. Benj.: sporangiophore with a terminal primary vesicle, which produce secondary vesicles that support pedicellate sporangiola, arranged radially from a primary vesicle, x 350. Isolated in California, U.S.A.

Radulomycetopsis Dhingra, et al.
Inc. sed., Agaricales, Agaricomycetidae, Agaricomycetes, Agaricomycotina, Basidiomycota, Fungi
Dhingra, G. S., et al. 2012. *Radulomycetopsis* (Agaricomycetes), a new corticioid genus from India. *Mycotaxon 119*:133-136.
From genus *Radulomyces* M. P. Christ.+ Gr. suf. *-ópsis*, for resemblance with this genus. Basidiocarps resupinate, adnate, effused, membranous-ceraceous; hymenial surface smooth to slightly tuberculate, orange to brownish-orange to reddish-brown; margins fibrillose, concolorous but paler; hyphal system monomitic; generative hyphae branched at wide angles, without clamps, basal hyphae covered with a thick sheath of brownish red matter, which dissolves in 3% KOH; cystidia projecting, thin- to slightly thick-walled; simple to somewhat branched hyphoid structures present in hymenium; basidia clavate to subclavate, 4-sterigmata, without a basal clamp; basidiospores broadly ellipsoid to subglobose, smooth, thin- to slightly thick-walled, inamyloid, acyanophilous; both basidia and basidiospores rich in oil drops. On decaying angiospermous branches in Bombila, West Kameng, Arunachal Pradesh, India.

Ramalina Ach.
Ramalinaceae, Lecanorales, Lecanoromycetidae, Lecanoromycetes, Pezizomycotina, Ascomycota, Fungi
Acharius, E. 1809. In Luyken, *Tent. Hist. Lich.*:95.
From L. *ramus*, branch > *ramalis*, pertaining to a branch, with branches, or from *ramale* (plural *ramalia*), small branch, shoot + suf. *-ina*, which denotes similarity. Thallus fruticose, highly branched, branches flattened, arranged more or less dichotomously. Apothecia small or large, lateral or terminal, sessile or pedicellate, with a concave or convex disc, excipulum of same color as thallus, or rarely black. On trees, shrubs, old wood, and maritime rocks.

Ramaria Fr. ex Bonord.
Clavariaceae, Agaricales, Agaricomycetidae, Agaricomycetes, Agaricomycotina, Basidiomycota, Fungi
Bonorden, H. F. 1851. *Handb. Allgem. mykol.* (Stuttgart):166.
From L. *ramus*, branch + L. suf. *-aria*, which indicates connection or possession, for having a branched basidioma. Genus with more than 100 species; terricolous, lignicolous or humicolous; several species edible and/or ectomycorrhizogenous, with only a few species reported as toxic. *Ramaria flava* (Fr.) Quél. and *R. botrytis* (Fr.)

Rick are edible of good quality, while *R. formosa* (Fr.) Quél. has a bitter taste, is toxic, although not fatal.

Ramalina menziesii Taylor: branched, fruticose, green thallus, with flattened branches, on tree bark, x 0.3.

Ramaria botrytis (Pers.) Bourdot: terricolous, yellowish, branched basidioma, x 0.5.

Ramularia Unger (syns. **Acrotheca** Fuckel, **Mycosphaerella** Johanson)
Mycosphaerellaceae, Capnodiales, Dothideomycetidae, Dothideomycetes, Pezizomycotina, Ascomycota, Fungi
Unger, F. 1833. *Die Exantheme der Pflanzen und einige mit diesen verwandte Krankheiten der Gewächse, pathogenetisch und nosographisch dargestellt.* 119. Wien.
From L. *ramus*, a branch + suf. *-aria*, which indicates connection or possession. Conidiophores growing out through stomata of host leaves, clustered, short, hyaline or subhyaline; simple, frequently curved or bent, with prominent conidial scars; conidia (sympodulospores) hyaline, cylindrical, typically 2-celled, but many 1-celled and a few 3-celled, frequently in short chains; parasitic on plants, causing leaf spots, responsible for yield losses to many important crops, including barley, sugar beet and strawberry.

Rattania Prabhug. & Bhat
Inc. sed., Inc. sed., Inc. sed., Inc. sed., Pezizomycotina, Ascomycota, Fungi
Prabhugaonkar, A., D. J. Bhat. 2009. *Rattania setulifera*, an undescribed endophytic hyphomycete on rattans from Western Ghats, India. *Mycotaxon* 108:217-222.
From Malay *rotan*, a type of palm, cane > E. *rattan*, rattan + L. des. *-ia*. Colonies in culture forming superficial, gregarious, dark brown, setose sporodochia with a small stromal base. Setae erect, straight to flexuous, unbranched, tapering to a pointed apex, septate, smooth, thick-walled, brown. Conidiophores branched, hyaline, smooth, arising in a layer from a pseudoparenchymatous stroma. Conidiogenous cells terminal, integrated or discrete, usually monoblastic, sometimes extending sympodially to form successive holoblastic conidia. Conidia slimy, solitary, fusiform, curved, hyaline, smooth, 0-multiseptate, thin-walled, truncate at base and acuminate at tip, setulate at both ends. Isolated as an endophyte from leaves of *Calamus thwaiesii* (rattan) collected in the Western Ghats, India.

Rattania setulifera Prabhug. & Bhat: setulose, dark brown sporodochium, with a small stromal base, x 175; conidiogenous cells with slimy, fusiform, 0-multiseptate, hyaline conidia, with setulae at both ends, x 460.

Redeckera C. Walker & A. Schüßler
Diversisporaceae, Diversisporales, Inc. sed., Glomeromycetes, Glomeromycotina, Glomeromycota, Fungi
Schüßler, A., C. Walker. 2010. *The Glomeromycota,* A species list with new families and new genera (Gloucester):44.
Named in honor of the German mycologist *Dirk Redecker* + L. des. *-a-*, for his pioneering work in the molecular phylogeny of the Glomeromycota. Spore formation disorganized in large, compact sporocarps, containing hundreds to thousands of spores per sporocarp; spores with 2- to rarely 3-wall layers; subtending hyphae generally broad at spore base, with a conspicuous, thick, broad

septum that arises from inner lamina of bi-laminated, structural wall layer; structural wall layer 2 generally continues over short distances into subtending hypha; structural wall 1 fragile, usually inflating in a short distance to spore base where structural wall layer 2 becomes invisible in subtending hypha.

Remispora Linder
Halosphaeriaceae, Microascales, Hypocreomycetidae, Sordariomycetes, Pezizomycotina, Ascomycota, Fungi
Linder, D. L. 1944. I. Classification of the marine fungi, pp. 401-433, *In*: Barghoorn, E. S., D. L. Linder, Marine fungi: their taxonomy and biology. *Farlowia* 1:395-467.
From L. *remus*, oar + *spora* < Gr. *sporá*, spore, seed, for the shape of the ascospores. Perithecia solitary or somewhat gregarious, imbedded in substratum, light colored, isabellinous or cream-colored, subglobose, membranous, with a pronounced truncate-conoid, eccentric ostiole. Paraphyses absent. Asci broadly fusoid with a slight apiculus, tardily deliquescent, 3-8-spored, usually with some spores aborted. Ascospores ovoid, ellipsoid or elongate-ellipsoid, hyaline, 1-septate, with two hyaline, broad, tapering subgelatinous appendages at each end, appendages diverging, nearly at right-angles to long axis of spore, later deliquescing or dropping off. On immersed test blocks.

Repetoblastiella R.F. Castañeda, et al.
Inc. sed., Inc. sed., Inc. sed., Inc. sed., Pezizomycotina, Ascomycota, Fungi
Castañeda Ruiz, R. F., et al. 2010. Two new anamorphic fungi from Cuba: *Endophragmiella profusa* sp. nov. and *Repetoblastiella olivacea* gen. & sp. nov. *Mycotaxon* 113:415-422.
From L. *repetere*, to repeat > *repeto*, repeat + Gr. *blastós*, bud, sprout + L. dim. suf. *-ella*, referring to the blastic mode of conidium ontogeny. Colonies on natural substratum hairy, caespitose, funiculose to arachnoid, effuse, dark green, olivaceous or brown. Mycelium superficial, immersed. Conidiophores micronematous, mononematous, simple or branched, septate, brown or olivaceous, smooth or verrucose. Conidiogenous cells monoblastic, terminal, determinate; sometimes polyblastic with sympodial proliferations. Conidial secession schizolytic. Conidia cylindrical, oblong to bacilliform, multi-septate, olivaceous or brown, blastocatenate, forming irregular chains from indeterminate cells across length of conidial body. Sexual morph unknown. On bark of decaying nuts of *Couroupita guianensis* in Santiago de Las Vegas, Havana, Cuba.

Retiarius D. L. Olivier
Inc. sed., Inc. sed., Inc. sed., Inc. sed., Pezizomycotina, Ascomycota, Fungi
Olivier, D. L. 1978. *Retiarius* gen. nov.: phyllosphere fungi which capture wind-borne pollen grains. *Trans. Br. mycol. Soc.* 71(2):193-201.
From L. *retiarius*, one who fights with a net, in reference to the manner of capture of pollen grains by the fungus. Fungus with septate, branching, anastomosing hyphae, hyaline, on adaxial surface of leaves, producing short erect hyphae by which it traps and parasitizes anemophilous pollen grains. Conidiogenous cells integrated, intercalary in both prostrate hyphae and erect short branches, conidia developing sympodially on short, lateral, indeterminate branches. Isolated from the phyllosphere of evergreen trees and shrubs in the Western Cape region of South Africa.

Reticulosphaeria Sivan. & Bahekar
Phlogicylindriaceae, Amphisphaeriales, Xylariomycetidae, Sordariomycetes, Pezizomycotina, Ascomycota, Fungi
Sivanesan, A., V. Bahekar. 1982. *Reticulosphaeria indica* gen. et sp. nov. from India. *Trans. Br. mycol. Soc.* 78 (3):547-551.
From L. *reticulum*, a small net + genus *Sphaeria* Haller, referring to the reticulate ascospores. Perithecia solitary, immersed, black, flask-shaped. Asci cylindrical, short-stalked, 6-8 spored, unitunicate, with a distinct, large, complex, amyloid, apical apparatus. Ascospores globose, brown, thick-walled, reticulate, one-celled. Paraphyses filiform, hyaline, septate, simple to sparsely branched. On stems of *Achyranthes aspera* in Aurangabad, India.

Reticulosphaeria indica Sivan. & Bahekar: vertical section of an immersed perithecium in the stem of *Achyrantes aspera*, showing the peridium, unitunicate asci and filiform paraphyses, x 60; one ascus with its apical amyloid apparatus, x 400, and one brown, globose ascospore, with reticulate wall, x 640. Collected in Maharashtra, Aurangabad, India.

Rhabdocline Syd. (syn. **Meria** Vuill.)
Hemiphacidiaceae, Helotiales, Leotiomycetidae, Leotiomycetes, Pezizomycotina, Ascomycota, Fungi
Sydow, H. *In*: H. Sydow, F. Petrak, 1922. Ein Beitrag zur Kenntniss der Pilzflora Nordamerikas, insbesondere der nordwestlichen Staaten. *Annls mycol.* 20(3/4):194.
From Gr. *rhábdos*, rod, stick, walking stick, septre +

klíne, bed, inclined, from *klíno*, incline, referring to the disposition of the apothecia on the plant leaves. Stromata subepidermal bearing apothecia that erupt and undergo dehiscence. Asci clavate. Ascospores cylindrical or elliptical, sometimes with one end widened. Conidia obclavate, unicellular, hyaline, catenulate, with basipetal maturation, accumulating in mucilage. Traps and destroys nematodes in soil, also inhabiting living leaves. Causing needle fall on fir trees.

Rhabdocline pseudotsugae Syd.: apothecia in subepidermal stromata, of this parasitic discomycete causing needle fall on fir trees (*Pseudotsuga menziesii*), x 0.5.

Rhexoampullifera P. M. Kirk
Inc. sed., Inc. sed., Inc. sed., Inc. sed., Pezizomycotina, Ascomycota, Fungi
Kirk, P. M. 1982. New or interesting microfungi V. Microfungi colonizing *Laurus nobilis* leaf litter. *Trans. Br. mycol. Soc.* 78(2):293-303.
From Gr. *rhéxis*, rupture + genus *Ampullifera* Deighton, i.e., an *Ampullifera* with rhexolytic separation of the conidia. Mycelium partly superficial, partly aerial, superficial mycelium of smooth, branched, septate, pale brown to brown hyphae bearing hyphopodia, aerial mycelium of smooth, branched, septate, brown to dark brown conidiogenous hyphae. Hyphopodia stipitate, cylindrical, globose or hemispherical to clavate, sometimes curved, pale brown to brown, smooth. Conidiophores micronematous. Conidiogenous cells determinate, holoblastic, monoblastic or polyblastic, conidiogenous loci acropleurogenous. Conidia formed in simple or branched, rhexolytically fragmenting chains, delimited acropetally but maturing basipetally, dry, smooth, septate, dark brown, cylindrical to doliiform. On leaf litter of *Laurus nobilis* near Brodick Castle on the Isle of Arran, Scotland.

Rhexodenticula W. A. Baker & Morgan-Jones
Inc. sed., Inc. sed., Inc. sed., Inc. sed., Pezizomycotina, Ascomycota, Fungi

Baker, W. A., et al. 2001. Notes on hyphomycetes. LXXXIV. *Pseudotrichoconis* and *Rhexodenticula*, two new monotypic genera with rhexolytically disarticulating conidial separating cells. *Mycotaxon* 79:361-373.
From Gr. *rhéxis*, break, rupture + L. *denticulatus*, with small teeth, for the presence of denticulate conidia that separate rhexolytically from the conidigenous cell. Colonies effuse, hairy, brown on substrate. Mycelium partly immersed, composed of branched, septate, pale brown to brown, smooth hyphae. Conidiophores solitary or caespitose, arising from small clusters of dark-brown, globose cells, simple, erect, cylindrical, straight or subflexuous, sympodial, septate, smooth, brown, with a paler apex, macronematous and mononematous. Conidiogenous cells integrated, terminal, polyblastic, denticulate. Conidia attached by a separating cell, solitary, dry, acropleurogenous, cylindrical, septate, verruculose, pale brown, with darker middle cells. On decaying leaves in Cuba and North America.

Rhinocladiella Nannf.
Herpotrichiellaceae, Chaetothyriales, Chaetothyriomycetidae, Eurotiomycetes, Pezizomycotina, Ascomycota, Fungi
Nannfeldt, J. A. 1934. *In*: E. Melin and J. A. Nannfeldt. Researches into the blueing of ground wood pulp. *Svensk Skogsvårdsförening Tidskr.* 3-4:461-462.
From Gr. *rhíne*, file, sandpaper + *kládos*, branch + L. dim. suf. *-ella*, referring to the branches and the principal axis of the conidiophores. Conidiophores terminating in a sympodial rachis with conidiogenous cell. Conidiogenous cells cylindrical or somewhat inflated, with prominent, dark scars giving them a rough appearance. Conidia globose or ovoid, uniseptate, truncate at base. Saprobic in soil, on straw, seeds, and other vegetable substrates.

Rhizocarpon Ramond ex DC.
Rhizocarpaceae, Rhizocarpales, Lecanoromycetidae, Lecanoromycetes, Pezizomycotina, Ascomycota, Fungi
De Candolle, A. P. *In*: Lamarck, J. de., A. P. De Candolle. 1815. *Fl. franç.*, Edn. 3 (París) 2:365.
From Gr. *rhíza*, root > pref. *rhizo-* + *karpós*, fruit. Thallus crustose or subsquamulose, commonly areolate or verrucose, attached to substrate by rhizoids. Apothecia circular or flexuous, generally black, with hymenium hyaline or brown on upper part. Photobiont *Protococcus*.

Rhizoctonia DC. (syn. **Thanatephorus** Donk)
Ceratobasidiaceae, Cantharellales, Inc. sed., Agaricomycetes, Agaricomycotina, Basidiomycota, Fungi
Lamarck, J. de, A. P. De Candolle. 1805. Mémorie sur les rhizoctones nouveau genre de champignons qui attaque les racines des plantes et en particulier celle de la luzerne cultiveé. *Fl. franç.*, Edn 3 (Paris) 5/6:110-111.

Rhizomucor

From Gr. *rhíza*, root + *któnos*, mortality, butchery < *ktonéo*, to assassinate, kill, i.e., producing the death of the roots of plants. Forms a loose basidial hymenium on which develop basidia that produce basidiospores on apices of long sterigmata. Attacks various plant parts near surface of soil. Pathogenic on plants of economic importance throughout the world, also in soil.

Rhizocarpon geographicum (L.) DC.: crustose or subsquamulose thallus, with black, circular or flexuous apothecia, bearing the superior dark hymenium, x 5. On rock, high rock cliffs, Europe.

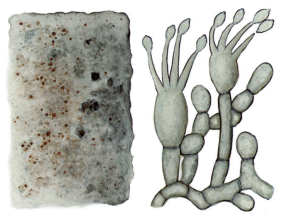

Rhizoctonia fusispora (J. Schröt.) Oberw., et al. (syn. **Thanatephorus fusisporus** (J. Schröt.) Hauerslev & P. Roberts): loose basidial hymenium in chestnut wood, x 0.5; basidia with basidiospores on the apices of long sterigmata, x 350.

Rhizoctonia solani J. G. Kühn.: pine root plantules parasitized by this fungus, x 0.3.

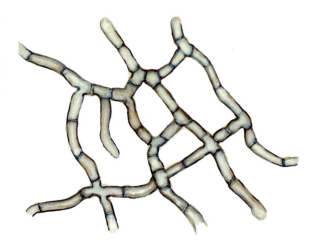

Rhizoctonia solani J. G. Kühn.: mycelium with right angle branches; note the constriction on the base of the hyphal branches, x 525.

Rhizomucor Lucet & Costantin

Lichtheimiaceae, Mucorales, Inc. sed., Inc. sed., Mucoromycotina, Zygomycota, Fungi

Lucet, A., J. Costantin. 1900. *Rhizomucor parasiticus*. Espice pathogene de l'homme. *Rev. gén. Bot.* 12:81-98.

From Gr. *rhíza*, root + L. *mucor*, mold, referring to the presence of rhizoids on the base of the sporangiophores, by which it is differentiated from the genus *Mucor* Fresen. Thermophilic, principally in manure and similar fermenting substrates, widely distributed.

Rhizomucor pusillus (Lindt) Schipper: sporangiophores with sporangia, rhizoids at the base of the sporangiophores, and a stolon connecting two groups of rhizoids, on agar culture, x 100.

Rhizophagus P. A. Dang.

Glomeraceae, Glomerales, Inc. sed., Glomeromycetes, Glomeromycotina, Glomeromycota, Fungi

Dangeard, P. A. 1896. Une maladie du peuplier dans le ouest de la France. *Botaniste* 5:38-43.

From Gr. *rhíza*, root + *phágos*, that which eats < *phágomai*, to eat. Endotrophic mycorrhizal fungus associated with the roots of plants.

Rhizopogon

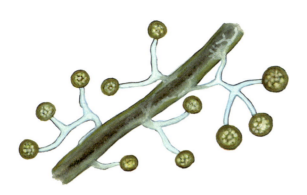

Rhizophagus clarus (T. H. Nicolson & N. C. Schenck) C. Walker & A. Schüssler: hyaline spores of this mycorrhizal fungus, containing globules of variable size, growing from a host root, x 30. Collected in Florida, U.S.A.

Rhizophlyctis A. Fisch.

Rhizophlyctidaceae, Rhizophlyctidales, Chytridiomycetidae, Chytridiomycetes, Inc. sed., Chytridiomycota, Fungi

Fischer, A. 1892. Phycomycetes. Die Pilze Deutschlands, Oesterreichs und der Schweiz, p. 114, *In: Rabenh. Krypt.-Fl.*, Edn 2 (Leipzig) *1*(4):1-490.

From Gr. *rhíza*, root + *phlyctis* < *phlýctaina*, tumor, scab, blister, pimple, i.e., a tumor or blister with roots. Thallus endobiotic with rhizoids. On algae, insect integuments and other substrates.

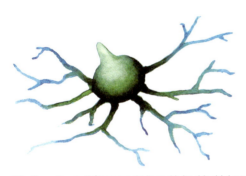

Rhizophlyctis petersenii Sparrow: thallus with its rhizoidal system and papillate sporangium, x 220.

Rhizophydium Schenk (syn. **Phlyctidium** (A. Braun) Rabenh.)

Rhizophydiaceae, Rhizophydiales, Chytridiomycetidae, Chytridiomycetes, Inc. sed., Chytridiomycota, Fungi

Schenk, A. 1858. Algologische Mitteilungen. *Verh. Phys.-Med. Ges. Würzburg* 8:235-259.

From Gr. *rhizophýeo* or *rhizophyés*, to engender or put out roots + L. dim. suf. *-idium*. Thallus endobiotic with rhizoids attached to substrate. Parasitic or saprobic on plants including freshwater algae and animal substrates.

Rhizoplaca Zopf

Lecanoraceae, Lecanorales, Lecanoromycetidae, Lecanoromycetes, Pezizomycotina, Ascomycota, Fungi

Zopf, W. 1905. Zur Kenntnis der Flechtenstoffe. 14. Mitteilung. *Liebigs Annalen der Chemie*. 340:276-309

From Gr. *rhíza*, root > pref. *rhiz-*, *rhizo-* + *pláx*, plaque, sheet, plate, plain > *plakódes*, tabular, foliate. Thallus foliose, closely attached to substrate, umbilicate. Apothecia conspicuously greenish, yellowish, grayish or pink. Spores simple, hyaline. Distinguished from *Umbilicaria* Hoffm. by yellow-green thallus. On exposed rocks in alpine and subalpine forests.

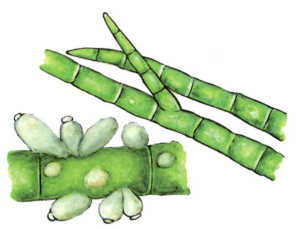

Rhizophydium anatropum (A. Brawn) Karling (syn. **Phlyctidium anatropum** (A. Braun) Sparrow): operculate zoosporangia, on thallus of the freshwater alga (*Stigeoclonium*), x 1,300.

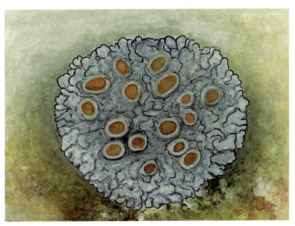

Rhizoplaca chrysoleuca (Sm.) Zopf: foliose thallus with grayish apothecia, which have a reddish hymenium, on exposed rock, x 2.

Rhizopogon Fr.

Rhizopogonaceae, Boletales, Agaricomycetidae, Agaricomycetes, Agaricomycotina, Basidiomycota, Fungi

Fries, E. M., J. Nordholm. 1817. *Symbolae gasteromycetum ad illustrandum floram Suecicam* (Lund) 1:5.

From Gr. *rhíza*, root + *pógon*, beard, lock. Rhizomorphs filiform, adherent, dark-colored, at times numerous on surface of fructification. Fructifications globose or tuberose; peridium of one or two layers; gleba of persistent, anastomosed sheets, with labyrinthiform cavities and spores; spores hyaline or slightly colored, smooth. Hypogeous in pine forests with sandy soils.

Rhizopus

Rhizopogon luteolus Fr.: globose, brownish fructifications, covered with filiform, adherent, dark-colored rhizomorphs, x 0.7. Collected in gravelly ground, Sweden.

Rhizopus Ehrenb. (syn. **Amylomyces** Calmette)
Rhizopodaceae, Mucorales, Inc. sed., Inc. sed., Mucoromycotina, Zygomycota, Fungi
Ehrenberg, C.G. 1820. De mycetogenesi as Acad. C. L.C. N. C. praesidem Epistola. *Nova Acta Phys.-Med. Acad. Caes. Leop.-Carol. Nat. Cur.* 10:159-222.
From Gr. *rhíza*, root + *poús*, foot, support. Sporangiophores with rhizoids at base of them. Cosmopolitan, saprobic, pathogenic, and of industrial importance.
Microbial component of *ragi* and Chinese yeast, used in Asia as inoculum for fermentation of amylaceous substrates such as rice and cassava to obtain traditional fermented foods. Another example is *tempeh*, obtained with the fermatation of soya beans.

Rhodocybe Maire
Entolomataceae, Agaricales, Agaricomycetidae, Agaricomycetes, Agaricomycotina, Basidiomycota, Fungi
Maire, R. 1924. Études mycologiques (fascicule 2). *Bull. trimest. Soc. mycol. Fr.* 40(4):298.
From Gr. *rhódon*, rose-colored + *kýbe* > L. *cybe*, head, in reference to the color of the pileus. Pileus generally pigmented, lamellae rounded to sinuate, adnate, adnexed or decurrent; stipe generally central; veil none; context inamyloid, hyphae not clamped, rarely fibulate. Spore print pink or gray; spores hyaline to pale yellow, angular to rounded, almost smooth to warty spinulose. On the ground or dead woody debris in temperate, subtropical, and tropical regions. Some species are edible.

Rhodophyllus Quél.—see **Entoloma** Fr. ex P. Kumm.
Entolomataceae, Agaricales, Agaricomycetidae, Agaricomycetes, Agaricomycotina, Basidiomycota, Fungi
Quélet, L. 1886. *Enchir. fung.* (Paris):57.
From Gr. *rhódon*, pink + *phýllon*, lamina, leaf, with reference to the pink color of the gills.

Rhodosporidiobolus Q. M. Wang, et al.
Sporidiobolaceae, Sporidiobolales, Inc. sed., Microbotryomycetes, Pucciniomycotina, Basidiomycota, Fungi
Wang, Q. M., et al. 2015. Phylogenetic classification of yeasts and related taxa within Pucciniomycotina. *Stud. Mycol.* 81:149-189.
From Gr. *rhódon*, rose, red + *sporá*, spore + *bállo* > L. *bolus*, a throw, stroke. The genus name refers to the fact that the species were hitherto classified in the genera *Rhodosporidium* I. Banno or *Sporidiobolus* Nyland. Colonies pink to red and butyrous. Budding cells present; pseudohyphae or true hyphae present or not; ballistoconidia formed or not, ellipsoidal, allantoid or amygdaliform. Sexual reproduction observed in some species; clamp connections present; teliospores may be formed and produce transversely septate basidia.

Rhizopus arrhizus A. Fisch. (syn. **Amylomyces rouxii** Calmette): apophysate sporangiophores, with aborted sporangia, x 175.

Rhizopus stolonifer (Ehrenb.) Vuill.: rhizoids on the base of the sporangiophores connecting the stolons, x 100. Isolated from fruit and calyx of *Cornus mas* and *Morus alba*, Germany.

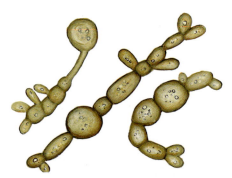

Rhodosporidiobolus lusitaniae (Ä Fonseca & J. P. Samp.) Q. M. Wang, et al. (syn. **Rhodosporidium lusitaniae** Á. Fonseca & J. P. Samp.): mycelial phase with teliospore and basidiospores, and yeast phase with budding pseudomycelium; colonies in solid media produce carotenoid pigments, x 1,100. Isolated from soil, Portugal.

Rhynchosporium

Rhodosporidium I. Banno—see **Rhodotorula** F. C. Harrison

Inc. sed., Sporidiobolales, Inc. sed., Microbotryomycetes, Pucciniomycotina, Basidiomycota, Fungi

Banno, I. 1967. Studies on the sexuality of *Rhodotorula*. *J. gen. appl. Microbiol.* (Tokio) 13:192.

From Gr. *rhódon*, pink + L. *sporidium*, from Gr. *sporá*, spore + L. dim. suf. *-ium*, in reference to the small spores and because the colonies on solid media have yellow, orange, red or pink tones due to containing carotenoid pigments.

Rhodotorula F. C. Harrison (syn. **Rhodosporidium** I. Banno)

Inc. sed., Sporidiobolales, Inc. sed., Microbotryomycetes, Pucciniomycotina, Basidiomycota, Fungi

Harrison, F. C. 1927. A systematic study of some *Torulas*. *Proc. & Trans. Roy. Soc. Canada*, ser. 3, 21 (5):349.

From Gr. *rhódon*, rose-colored + *torula* < L. *torulus*, dim. of *torus*, slender rope or cord, but with constrictions. Colony rose-colored. Yeast cells oval or elongate, reproduce by budding; also forming pseudomycelium. Mycelial phase resulting from sexual reproduction; mycelium with or without clamp connections, with terminal or intercalary teliospores. Basidiospores formed terminally or laterally on holometabasidium or phragmometabasidium. Homothallic or heterothallic. Saprobic on a wide variety of substrates.

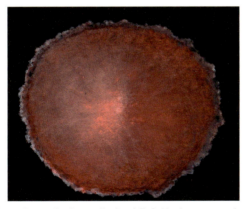

Rhodotorula mucilaginosa (A. Jörg.) F. C. Harrison: rose-colored, mucilaginous giant colony on culture in solid medium, x 1.

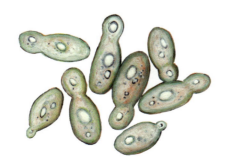

Rhodotorula mucilaginosa (A. Jörg.) F. C. Harrison: vegetative yeast cells with blastospores, x 4,000.

Rhodotus Maire

Physalacriaceae, Agaricales, Agaricomycetidae, Agaricomycetes, Agaricomycotina, Basidiomycota, Fungi

Maire, R. 1924. Études mycologiques (fascicule 2). *Bull. trimest. Soc. mycol. Fr.* 40(3):293-317.

From Gr. *rhódon*, rose, red + *otus*, an ear, for the appearance of the basidiome. Basidiomata small to medium sized. Pileus convex with an incurved margin when young, becoming broadly convex or flat; surface orange, reddish to peach-colored, gelatinous or viscid when wet, decorated with obvious network of raised, round-edged, reticulate ridges; margin incurved, with apricot yellow to cream wrinkle. Lamellae sinuate to adnexed, crowded to subdistant, pale meat-colored, edge even; lamellulae 2-3 tiers, not forked. Stipe eccentric to nearly central, subcylindric or slightly attenuate upwards; nearly white to pale pink; base slightly enlarged. Context whitish to pink, unchanging. Basidiospores globose to subglobose, nearly hyaline, thin-walled, covered with obtuse warts, without germ pore, non-amyloid, non-dextrinoid. Cheilocystidia scarce to abundant, lageniform to ventricose to fusiform, nearly hyaline, thin-walled. On dead trunks of deciduous trees (*Acer*, *Populus* or *Ulmus*), clustered or occasionally solitary; fruiting period extends from summer to autumn in the northern temperate region.

Rhodotus palmatus (Bull.) Maire: basidioma on soil, x 0.5.

Rhynchosporium Heinsen ex A. B. Frank

Inc. sed., Helotiales, Leotiomycetidae, Leotiomycetes, Pezizomycotina, Ascomycota, Fungi

Heinsen, E. 1897. *Wochenschrift für Brauerei.* 14:518.

From Gr. *rhýnchos*, beak, nose + *spóros*, spore + L. dim. suf. *-ium*. Conidia bicellular, with beak on apex, produced on phialides. Causes a necrosis known as scald on Poaceae (barley, oats, wheat, etc.); also saprobic on plant remains.

Rhytisma

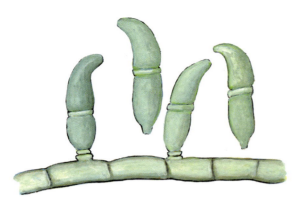

Rhynchosporium secalis (Oudem.) Davis: bicellular conidia (rhynchospores), with a beak on the apex, produced by phialides, x 1,500.

Rhytisma Fr.
Rhytismataceae, Rhytismatales, Leotiomycetidae, Leotiomycetes, Pezizomycotina, Ascomycota, Fungi
Fries, E. M. 1818. Uppställning af de I Sverige funne Värtsvampar (Scleromyci). *K. svenska Vetensk-Akad. Handl.*, ser. 3. 39:104.
From Gr. *rhytís*, genit. *rhytídos*, wrinkle, or from *rhytídóo*, to wrinkle, contract in wrinkles + suf. *-ma*, which indicates the result of an action. Stroma with wrinkled surface in radial furrows subtended by apothecia. Causes tar spot on leaves of maple and willow.

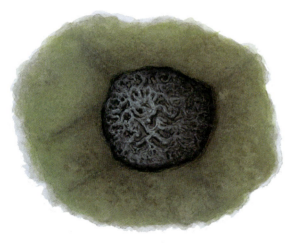

Rhytisma acerinum (Pers.) Fr.: wrinkled stroma, with radial furrows, that represent apothecia of this parasitic fungus on the leaf of maple (*Acer pseudoplatanus*), causing the disease called tar spot, x 3. Collected in Germany.

Rigidoporopsis I. Johans. & Ryvarden—see Amylosporus Ryvarden
Bondarzewiaceae, Russulales, Inc. sed., Agaricomycetes, Agaricomycotina, Basidiomycota, Fungi
Johansen, I., L. Ryvarden. 1979. Studies in the Aphyllophorales of Africa VII. Some new genera and species in the Polyporaceae. *Trans. Br. mycol. Soc.* 77(2):189-199.
From genus *Rigidoporus* + Gr. suf. *-ópsis*, aspect, appearance, for the similarity to this genus.

Roccella DC.
Roccellaceae, Arthoniales, Arthoniomycetidae, Arthoniomycetes, Pezizomycotina, Ascomycota, Fungi
De Candolle, A. P. 1805. *In*: Lamarck, J. de and A. P. De Candolle, 1805. *Fl. franç.*, Edn. 3 (Paris) 2:334.
Probably from It. *roccella*, common name for a lichen; from L. *rocca*, rock + L. dim. suf. *-ella*. Thallus fruticose, erect, with flat branches, smooth or with soralia, more or less twisted. Apothecia lateral, sessile, with a flat to convex disc, excipulum of same color as disc, surrounded by a thalloid cover that is sometimes fugaceous. On rocks and shrubs.

Roccella phycopsis Ach.: fruticose thallus, with flat branches provided with soralia and sessile apothecia, growing on rock, x 0.5.

Rosellinia de Not.
Xylariaceae, Xylariales, Xylariomycetidae, Sordariomycetes, Pezizomycotina, Ascomycota, Fungi
De Notaris, G. 1844. Cenni sulla tribu dei Pirenomiceti sferiacei e descrizione di alcuni generi spettani alla medesima. *G. bot. ital.* 1:322-255.
Named in honor of the Italian botanist F. P. Rosellini + L. des. *-a*. Ascomata ostiolate perithecia, borne on a felt-like subiculum of dark hyphae, single or gregarious, often covering surface of subiculum; perithecia globose or broadly obpyriform, large, black, glabrous, often seated on a definite hypostroma of angular, thick-walled cells. Ostiolar neck papillate or conical, ostiole lined with periphyses. Perithecial wall of several layers of flattened cells. Asci unitunicate, cylindrical, stalked, with an amyloid apical apparatus, lining inside of perithecial wall, 8-spored. Ascospores 1-celled, dark brown, ellipsoidal, often laterally compressed, with a longitudinal germ slit. On wood, roots or similar cellulosic substrates.

Royoungia Castellano, et al. (syn. Australopilus Halling & N. A. Fechner)
Boletaceae, Boletales, Agaricomycetidae, Agaricomycetes, Agaricomycotina, Basidiomycota, Fungi

Ruhlandiella

Castellano, M., et al. 1992. Australasian truffle-like Fungi. III. *Royoungia* gen. nov. and *Mycoamaranthus* gen. nov. (Basidiomycotina). *Australian Syst. Bot.* 5:614.
Named in honor of Mr. *Roy Young*, accomplished collector of hypogeous fungi + L. *-ia*, ending of generic names. Basidiomata flattened to subglobose, bright golden yellow when fresh and as dried, finally tomentose to nearly glabrous. Gleba brown when young, dark chocolate brown when mature; locules sublamellate, irregularly shaped, empty. Rhizomorphs two or three, small, attached at sporocarp base, concolorous with peridium. Columella cartilaginous, abruptly truncate, stout, concolorous with peridium, when cut slowly staining bright red. Peridium with yellow pigment immediately leached from peridial tissue when mounted in KOH. Basidiospores smooth, fusiform to suballantoid, apex obtuse, base mostly symmetrical. Hypogeous, in a community of some species of *Eucalyptus*, *Melaleuca*, *Allocasuarina* and *Leptospermum*, in Australia.

Rubikia splendida Dulym., et al.: vertical section of a unilocular, immersed pycnidium on dead leaf of *Pandanus rigidifolius* in Mauritius, opening by a circular, papillate ostiole, x 85, frontal and lateral views of mature, vertically flattened conidia, with four pigmented cells in the center surrounded by 36 cells devoid of contents, x 670.

elliptical in end view, vertically flattened, consisting of about 12 strongly cyanophilous, hyaline, thin-walled central cells and about 36 brown-walled exterior cells devoid of contents. On dead needles of *Pinus caribaea* in Atlántida, Honduras.

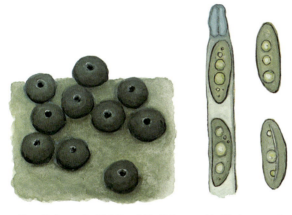

Rosellinia aquila (Fr.) Ces & De Not.: saprobic, black ascomata (ostiolate perithecia), on dead wood, with papillate ostiolar neck, x 2; unitunicate, 8-spored ascus, with an amyloid apical apparatus, x 750; liberated, ellipsoidal 1-celled ascospores, guttulate, with longitudinal germ slit, x 1,000.

Rubikia H. C. Evans & Minter
Inc. sed., Inc. sed., Inc. sed., Inc. sed., Pezizomycotina, Ascomycota, Fungi
Evans, H. C., D. W. Minter. 1985. Two remarkable new fungi on pine from Central America. *Trans. Br. mycol. Soc.* 84(1):57-78.
From E. *Rubik's Cube™* + L. des. *-ia*, for the resemblance of the spores to this puzzle. Mycelium of hyaline, thin-walled, smooth, septate, branched hyphae in a thick mucous matrix. Fruitbodies inconspicuous, only ostioles observable, pycnidial, separate, hyaline, globose, unilocular, deeply immersed like mycelium, opening by a circular, ostiolate papilla. Papilla of a mucilaginous ball of hyphae. Pycnidial wall multilayered. Sporophores hyaline, thin-walled, smooth, branching, constricted at base, each branch producing only one spore. Spores when mature, circular to squarish in face view, elongated

Rugosospora Heinem.
Agaricaceae, Agaricales, Agaricomycetidae, Agaricomycetes, Agaricomycotina, Basidiomycota, Fungi
Heinemann, P. 1973. Leucocoprinées nouvelles d'Afrique centrale. *Bull. Jard. Bot. natn. Belg.* 43 (1-2):12.
From L. *rugosus*, rugose + *spora*, spore, in reference to the rugose or wrinkled spores. Pileus submembranaceous. Spores rugose or verrucose. Possibly a synonym of *Leucocoprinus* Pat.

Ruhlandiella Henn. (syn. **Muciturbo** P. H. B. Talbot)
Pezizaceae, Pezizales, Pezizomycetidae, Pezizomycetes, Pezizomycotina, Ascomycota, Fungi
Hennings, P. 1903. *Ruhlandiella berolinensis* P. Henn. n. gen. et n. sp., eine neue deutsche Rhizinacee. *Hedwigia* (Beibl.) 42:22-24.
Dedicated to the German botanical and mycologist W. O. E. *Ruhland* + L. *-ella*, dim. suf. added to noun stems. Exothecial hypogeus fungi (truffles that lack the outer layer or peridium); the color of ascocarp varies ranging from white to brownish lilac, but typically becomes black with age or when exposed. Asci do not contain opercula, and paraphyses covered with gelatinous sheaths that greatly exceed asci in length. The ascospores are hyaline, globose, with ornamentation (reticulate or truncate). The species of this genus have been found on several continents across the globe; are widely distributed in Nothofagaceae forests in South America and near *Eucalyptus* or *Melaleuca* plants in Australia, North America, and Europe, where they can probably form ectomycorrhizas.

Russula

Ruhlandiella reticulata (P. H. B. Talbot) E. Rubio, et al. (syn. **Muciturbo reticulatus** P. H. B. Talbot): turbinate apothecia, on soil, forming ectomycorrhizae with *Eucalyptus maculata*, South Australia, x 7; vertical section of an apothecium, showing paraphyses, asci with eight ornamented ascospores, x 100.

Russula Pers.

Russulaceae, Russulales, Inc. sed., Agaricomycetes, Agaricomycotina, Basidiomycota, Fungi

Persoon, C. H. 1796. *Observ. mycol.* (Lipsiae) *1*:100. From L. *russulus*, reddish, for the presence of this color in the pileus of species such as *R. emetica*, the type species. Similar to *Lactarius* Pers., but without exudation of latex. Cosmopolitan, although absent in unforested regions, since it is obligately ectomycorrhizogenous, principally on conifers. Several of the species are edible and a few are considered somewhat toxic.

Russula xerampelina (Schaeff.) Fr.: reddish basidioma of this obligate ectomycorrhizal species, principally on conifers, x 0.5.

S

Schizophyllum commune

S

Saccharomyces Meyen

Saccharomycetaceae, Saccharomycetales, Saccharomycetidae, Saccharomycetes, Saccharomycotina, Ascomycota, Fungi

Meyen, F. J. F. 1883. *Arch. Naturgesch.* 4(2):100.

From Gr. *sákcharon*, sugar + *mýkes* > L. *myces*, fungus, in reference to its common occurrence in sugary substrates where it undergoes vigorous fermentation. Asexual reproduction by multilateral budding, also forming pseudomycelium but rarely true mycelium. Pellicle rarely formed in liquid media or, if so, not pulverulent or ascendent. Asci more or less persistent when mature. Asci each with 1-4 ascospores. Ascospores spheroidal or prolate-ellipsoid, sometimes originating by conjugation between mother cell and budding or between individual cells. Able to tolerate high osmotic pressures, also high concentrations of ethanol. On substrates with high sugar, such as nectar of flowers, sap of plants, syrups, jellies, condensed and sweetened milk, and oriental condiments with elevated salt content, as well as products of the fermentation industries, principally wineries, breweries and bakeries.

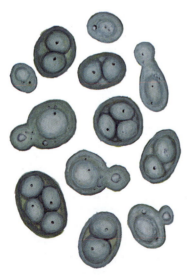

Saccharomyces cerevisiae (Desm.) Meyen: multilateral budding of yeast cells, and asci with 1-4 ellipsoidal ascospores, x 2,500.

Saccharomycodes E. C. Hansen

Saccharomycodaceae, Saccharomycetales, Saccharomycetidae, Saccharomycetes, Saccharomycotina, Ascomycota, Fungi

Hansen, E. C. 1904. Grundlinien zur Systematik der Saccharomyceten. *Centbl. Bakt. ParasitKde,* Abt. I 12 (19-21):537.

From genus *Saccharomyces* < Gr. *sákcharon*, sugar + *mýkes* > L. *myces*, fungus + Gr. suf. *-odes*, which indicates similarity, i.e., similar to the genus *Saccharomyces*. Hyphal cells originating by means of bipolar bud-fission. Asci with four ascospores fusing in pairs inside ascus before germinating. Ascospores spherical, smooth, thin-walled. Isolated from viscous exudates of oaks; capable of fermenting sugars.

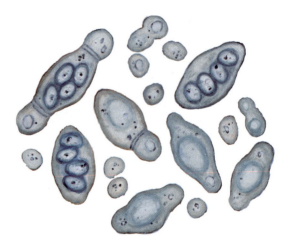

Saccharomycodes ludwigii (E. C. Hansen) E. C. Hansen: bipolar bud-fission yeast cells, and asci with four spheroidal ascospores, x 1,700. Isolated from exudate on trunk of *Quercus*, Europe.

Saccharomycopsis Schiönning (syn. **Endomycopsis** Stell.-Dekk.)

Saccharomycopsidaceae, Saccharomycetales, Saccharomycetidae, Saccharomycetes, Saccharomycotina, Ascomycota, Fungi

Samarosporella

Schiönning, H. 1903. Nouveau genre de la famille des saccharomycètes. *Comptes rendu Trav. Laboratoire d. Carlsberg* 6:101-125.

From genus *Saccharomyces* Meyen < Gr. *sákcharon*, sugar + *mýkes* > L. *myces*, fungus + suf. *-opsis* < Gr. *-ópsis*, appearance, aspect, i.e., similar to *Saccharomyces*. Produces true mycelium and arthrospores. True mycelium fragments into arthrospores, pseudomycelium and yeast-like cells. Blastospores generally formed by plasmodesmata in hyphal septa and intercalary asci. Ascospores spherical, hat-shaped or saturnoid, smooth or verrucose with a border. Ferments glucose and other sugars slowly and weakly. Isolated from fruits of *Crataegus*, bread, decomposed margarine, olives, insect tunnels in trees, and other substrates.

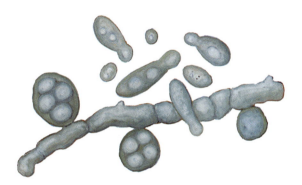

Saccharomycopsis vini (Kreger-van Rij) Van der Walt & D. B. Scott: true mycelium with arthrospores and blastospores, and intercalary asci with four spherical ascospores, x 2,000. Isolated in Chile.

Saccobolus Boud.

Ascobolaceae, Pezizales, Pezizomycetidae, Pezizomycetes, Pezizomycotina, Ascomycota, Fungi

Boudier, J. L. E., 1869. Memoire sur les Ascobolés. *Annls Sci. Nat., Bot.*, sér. 5, 10:191-268.

From L. *saccus*, sac, bag + *bolus* < Gr. *bállo*, to emit, throw, hurl, discharge. Apothecia small, globose, subglobose or discoid. Eight ascospores united in each ascus and actively discharged as a group. Ascospores purple, brown or almost black. Coprophilous, also on vegetable debris or in soil.

Saccobolus beckii Heimerl: sac-shaped apothecia of this coprophilous fungus, from which are actively discharged balls of eight, brown ascospores, x 45.

Sacculospora Oehl, et al.—see **Entrophospora** R. N. Ames & R. W. Schneid.

Acaulosporaceae, Diversisporales, Inc. sed., Glomeromycetes, Glomeromycotina, Glomeromycota, Fungi

Oehl, F. et al. 2011. Revision of Glomeromycetes with entrophosporoid and glomoid spore formation with three new gen. *Mycotaxon* 117:297-316.

From Gr. *sákkos*, sac > L. *saccus*, dim. *sacculus*, saccule + Gr. *sporá*, spore, seed > L. *spora*, spore, referring to the spore formation within the neck of sporiferous saccules.

Saksenaea S. B. Saksena

Saksenaeaceae, Mucorales, Inc. sed., Inc. sed., Mucoromycotina, Zygomycota, Fungi

Saksena, S. B. 1953. A new genus of the Mucorales. *Mycologia* 45(3):426-436.

Named in honor of another mycologist from India, R. K. Saksena, by S. B. Saksena, who isolated this mucoraceous fungus from soil. Sporangium lecythiform. Saprobic in soil, rarely causal agent of zygomycosis in humans.

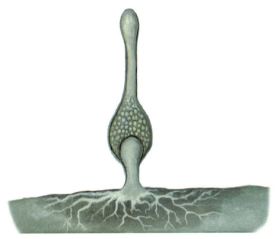

Saksenaea vasiformis S. B. Saksena: lecythiform sporangium with columella and spores, on soil, x 400. Isolated from soil, Madhya Pradesh, India.

Samarosporella Linder

Inc. sed., Inc. sed., Inc. sed., Inc. sed., Inc. sed., Inc. sed., Fungi

Linder, D. L. 1944. I. Classification of the marine fungi, pp. 401-433, *In*: Barghoorn, E. S. and D. L. Linder, Marine fungi: their taxonomy and biology. *Farlowia* 1 (3):395-467.

From genus *Samarospora* Rostr. + L. dim. suf. *-ella*, for the similarity to that genus. Perithecia immersed, black, carbonaceous, globose, with papilliform, slightly exerted ostiole. Paraphyses absent. Asci fusoid with rounded apices, deliquescing, 4-8-spored, usually with some spores aborted. Ascospores hyaline, elongate-ellipsoid or somewhat fusoid, with irregular hyaline wings; aborted ascospores smaller, ovoid or ellipsoid, devoid of hyaline wings or with rudimentary ones. On driftwood.

Samuelsia

Samuelsia P. Chaverri & K. T. Hodge
Clavicipitaceae, Hypocreales, Hypocreomycetidae, Sordariomycetes, Pezizomycotina, Ascomycota, Fungi
Chaverri, P., et al. 2008. A monograph of the entomopathogenic gen. *Hypocrella*, *Moelleriella*, and *Samuelsia*, gen. nov. (Ascomycota, Hypocreales, Clavicipitaceae), and their aschersonia-like anamorphs in Neotropics. *Stud. Mycol.* 60:1-66.

Named in honor of American mycologist *Gary J. Samuels* + L. des. -ia, for his contributions to the biology of the hypocreaceous fungi. Stromata pulvinate, generally hard, pale yellow, orange to brownish-yellow or brown, hypothallus lacking, stromatal surface smooth. Perithecia immersed in stroma, subglobose, scattered. Asci cylindrical to clavate, with a cap. Ascospores hyaline, multiseptate, smooth, long fusiform to filiform. Conidiomata as shallow or deep depressions, U-shaped. Conidiophores in a compact palisade, short, monoverticillate, with 2-5 phialides per verticil. Phialides cylindrical, long, slender, tapering at apex. Conidia small, hyaline, unicellular, smooth, allantoid. On scale insects and whiteflies on leaves.

Samuelsia rufobrunnea P. Chaverri & K. T. Hodge: pulvinate, yellow stroma in scale insect (Coccidae and Lacaniidae, Homoptera) living on leaves of *Geonoma* sp., in Bolivia and Peru, x 7; vertical section of a stroma containing perithecia and conidiomata, x 15; ascus with young ascospores, and mature, multiseptate, fusiform ascospores, x 550; phialides producing allantoid, 1-celled conidia, x 1,000.

Sarcodon Quél. ex P. Karst.
Bankeraceae, Thelephorales, Inc. sed., Agaricomycetes, Agaricomycotina, Basidiomycota, Fungi
Karsten, P. A. 1881. Enumeratio Thelephorearum Fr. et Clavariarum Fr. Fennicarum, systemate novo dispositarum. *Rev. Mycol.* (Toulouse) 3(9):20.

From Gr. *sárx*, genit. *sarkós*, flesh + *odoús*, genit. *odóntos*, tooth. Basidiomata pileate-stipitate, fleshy, generally dark, hymenophore with fleshy teeth. Terricolous or lignicolous on fallen wood. Some species edible, mycorrhizal. One of the most common species in pine forests is *S. imbricatum* (L.) Karst., whose fructifications have a wide, dark brown, scaly pileus, at times with pink to purple tints; stipe brownish-gray, lighter or whitish on the base; hymenophore of brown teeth that hang from the lower part of the pileus and also cover the upper part of the stipe; the flesh is bitter.

Sarcodon squamosus (Schaeff.) Quél.: pileate-stipitate, dark-colored basidioma, with the hymenophore on fleshy teeth, on soil, among mosses, x 0.5.

Sarcoscypha (Fr.) Boud.
Sarcoscyphaceae, Pezizales, Pezizomycetidae, Pezizomycetes, Pezizomycotina, Ascomycota, Fungi
Boudier, J. L. E. 1885. Nouvelle classification des discomycètes charnus. *Bull. Soc. mycol. Fr.* 1:103.

From Gr. *sárx*, genit. *sarkós*, meat + *skýphos*, cup, for the fleshy consistency and shape of the apothecium. Fructifications sessile or pedicellate, generally brilliant red. Gregarious in large or small groups in coniferous and oak forests, on branches partially covered by humus and fallen leaves.

Sarcoscypha coccinea (Gray) Boud.: pedicellate apothecia, brilliant red in color, on soil of coniferous forest, x 1.

Sarcosoma Rehm
Sarcosomataceae, Pezizales, Pezizomycetidae, Pezizomycetes, Pezizomycotina, Ascomycota, Fungi

Sartorya

Rehm, H. 1891. Rabenhorst's Kryptogamen-Flora, Pilze - Ascomyceten 1(3):401-608.

From Gr. *sárx*, genit. *sarkós*, flesh + *sōma*, body, for the fleshy-cartilaginous consistency of the apothecium. Apothecia globose or subglobose, sessile, with small pedicel, shape of deep wine glass, brown or blackish, up to 6 cm diam, 8 cm high. Hymenium with gelatinous material at base, fructification swells, increases considerably in weight. On humid soil in forests, among mosses, or on branches and pieces of wood.

Sarcosoma globosum (Schmidel) Casp.: fleshy-cartilaginous, globose, blackish-colored apothecium, with violaceous hymenium, on humid forest soil, x 0.5.

Sarcosphaera Auersw.

Pezizaceae, Pezizales, Pezizomycetidae, Pezizomycetes, Pezizomycotina, Ascomycota, Fungi

Auerswald, B. 1869. Synopsis pyrenomycetum europaeorum. *Hedwigia* 8:82.

From Gr. *sárx*, genit. *sarkós*, flesh + *sphaíra* > L. *sphaera*, sphere, for the globose, fleshy fructifications. Apothecia subterranean during its young stage, hollow, cartilaginous or fleshy, fragile. At maturity fructifications break through soil surface and open, forming triangular radial lobes like a crown; at base white, fleshy foot remains buried; interior surface smooth, violaceous, lined with hymenium. In coniferous forests.

Sarcosphaera coronaria (Jacq.) J. Schröt.: globose, mature, fleshy, pinkish-colored apothecia, with triangular radial lobes, on soil in coniferous forest, x 0.3.

Sarea Fr.

Trapeliaceae, Baeomycetales, Ostropomycetidae, Lecanoromycetes, Pezizomycotina, Ascomycota, Fungi

Fries, E. M. 1825. *Systema orbis vegetailis* (Lundae) 1:86. Etymology not determined. Apothecia sessile to substipitate, discoid, with a raised margin when young, becoming convex at maturity, up to 1.5 cm diam, fleshy, gelatinous, coriaceous when dry, pale orange to black. Excipulum of widely spaced, short-celled hyphae immersed in a hyaline gel. Subhymenium thin, of interwoven hyphae. Hymenium of asci interspersed with paraphyses, turning blue in iodine. Paraphyses filiform, septate, longer than asci, with a single row of prominent oil globules. Tips of paraphyses narrowly clavate, encrusted with pigmented granules and united in a gel to form an epithecial layer. Asci broadly clavate to obovoid, thick-walled, appearing bitunicate, but inner wall layer not extensible, with a thick apical cap traversed by a narrow canal, staining dark blue in iodine, polysporous. Ascospores globose, 1-celled, hyaline, smooth. Asexual morphs pycnidiella-like. On the resinous exudates of conifers.

Sarocladium W. Gams & D. Hawksw. (syn. Cephalosporium Corda)

Inc. sed., Hypocreales, Hypocreomycetidae, Sordariomycetes, Pezizomycotina, Ascomycota, Fungi

Gams, W., D. L. Hawksworth. 1975. The identity of *Acrocylindrium oryzae* Sawada and a similar fungus causing sheath-rot of rice. *Kavaka* 3:57-61.

From Gr. *sáron*, a broom + *kládos*, branch, sprout + -*ídion* > E. -*idium*, dim. suf., in reference to the broom-like shape of the conidiophores. Elongated phialide rising solitary on vegetative hyphae or on conidiophores that are sparsely or repeatedly branched, the production of abundant phialides and conidia hyaline, cylindrical, arranged in slimy heads. These species are frequently reported as plant pathogens or as saprobes associated with grasses, and mutualistic endophytes; also involved in human infections.

Sarocladium kiliense (Grütz) Summerb.: hyphal strings with phialides and unicellular, fusiform conidia, that agglutinate within a mucous sheath, x 600. Saprobic fungus that can attack plants, and humans with immunological disorders, causing endocarditis, peritonitis and queratities.

Sartorya Vuillemin—see Aspergillus P. Micheli

Inc. sed., Inc. sed., Inc. sed., Sordariomycetes, Pezizomycotina, Ascomycota, Fungi

Vuillemin, P. 1927. *Sartorya*, nouveau genre de plectascinées angiocarpes. *C. r. hebd. Séanc. Acad. Sci., Paris* 184:136-137.

Savoryella

Named in honor of *A. Sartory* + the des. *-a*. It is applied to the cleistothecial state of some species of the *Aspergillus fumigatus* group, such as *A. fischeri* Wehmer.

Savoryella E. B. G. Jones & R. A. Eaton
Savoryellaceae, Savoryellales, Sordariomycetidae, Sordariomycetes, Pezizomycotina, Ascomycota, Fungi
Jones, E. B. G., R. A. Eaton. 1969. *Savoryella lignicola* gen. et sp. nov. from water-cooling towers. *Trans. Br. mycol. Soc.* 52(1):161-165.
Named for the English microbiologist *J. G. Savory* + L. dim. suf. *-ella*. Perithecia solitary to aggregated, immersed or partly immersed in substratum, pale to dark brown, with a well-defined wall, and a short neck lined with periphyses. Asci long-cylindrical with a short foot, 8-spored, unitunicate, lacking paraphyses. Ascospores triseptate with brown central cells and hyaline end cells. On Scots pine test-blocks in a water-cooling tower at Connah's Quay, Flintshire, North Wales, UK.

Savoryella melanospora Abdel-Wahab & E. B. G.Jones: vertical section of a dark-brown perithecium in submerged driftwood, partially buried in a sandy dune, in Mornington Peninsula, Rye, Victoria, Australia, x 110; young and mature asci with eight ascospores, x 240; liberated ascospores, triseptate, with brown central cells and hyaline end cells, x 370.

Scedosporium Sacc. ex Castell. & Chalm.
Microascaceae, Microascales, Hypocreomycetidae, Sordariomycetes, Pezizomycotina, Ascomycota, Fungi
Castellani, A., A. J. Chalmers. 1919. *Manual of tropical medicine*, 3th. ed. (London), pp. 967-1125.
From Gr. *skedáo, skedázo*, to dissipate, disperse + *spóros*, spore + L. dim. suf. *-ium*, referring to the rapid growth of the colony. Hyphae floccose, smoky grey. Conidiophores simple or branched. Conidia aleuriospores, oval or clavate conidia with base conspicuously truncate. Sexual morph petriellidium-like. Causes eumycotic mycetoma in humans; saprobic phase in soil.

Schizophyllum Fr.
Schizophyllaceae, Agaricales, Agaricomycetidae, Agaricomycetes, Agaricomycotina, Basidiomycota, Fungi
Fries, E. M. 1815. *Observationes mycologicae praecipue ad illustrandam floram Suecicam*. (Havniae) 1:103.
From Gr. *schízo*, to split, divide + *phýllon*, leaf, gill, for the presence of "gills" deeply split or divided. Fructifications cupuliform at first, attached to substrate by narrow base, later becoming applanate, discoidal or flabelliform. Gills fused in pairs thus appearing split. Spores hyaline, smooth; spore print white. Lignicolous in temperate, subtropical and tropical regions. *Schizophyllum commune* Fr. has been used in biological investigations, especially genetics.

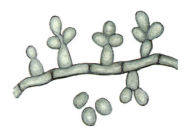

Scedosporium prolificans (Hennebert & B. G. Desai) E. Guého & de Hoog) (syn. **Lomentospora prolificans** Hennebert & B. G. Desai): conidiophore, conidiogenous cells with one-celled conidia, x 1,000.

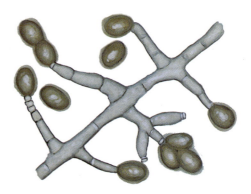

Scedosporium prolificans (Hennebert & B. G. Desai) E. Guého & de Hoog: mycelium with simple or branched conidiophores, with phialides that produce oval or clavate conidia (aleuriospores), x 1,500. Isolated in Belgium.

Schizophyllum commune Fr.: lignicolous, applanate fructifications, with the border of "gills" deeply split or divided, but not considered homologous with those in the Agaricales, x 1.On trunks of *Alnus, Betula, Fagus, Populus* and *Tilia,* Sweden.

Schizosaccharomyces Lindner
Schizosaccharomycetaceae, Schizosaccharomycetales, Schizosaccharomycetidae, Schizosaccharomycetes, Taphrinomycotina, Ascomycota, Fungi
Lindner, P. 1893. *Schizosaccharomyces pombe* n. sp. neuer Gärungsrreger. *Wochenschr. Brau.* 10:1298-1300.
From Gr. *schízo*, to split, divide + genus *Saccharomyces* Meyen < Gr. *sákcharon*, sugar + *mýkes* > L. *myces*, fungus, in reference to the schizogenesis or fission that the vegetative cells and mycelium undergo which results in the formation of arthrospores. Budding absent. Asci formed by somatogamic conjugation. Ascospores round, oval or reniform, with distinctive warts. Ferments sugars in grapes, wine, sugar, and molasses.

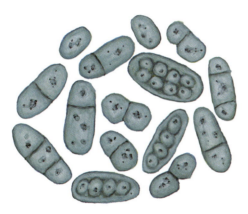

Schizosaccharomyces octosporus Beij.: schizogenesis or fission of vegetative yeast cells which results in the formation of arthrospores, and asci by means of somatogamy conjugation, with four or eight round or oval ascospores, x 2,000.

Schizostoma Ehrenb. ex Lév.
Agaricaceae, Agaricales, Agaricomycetidae, Agaricomycetes, Agaricomycotina, Basidiomycota, Fungi
Léveillé, J.H. 1846. Descriptions des champignons de l'herbier du Muséum de Paris. *Annls Sci. Nat.*, sér. 3, 5:165.
From Gr. *schízo*, to crack, to split, to cleave + *stóma*, mouth, pore, in reference to the dehiscence by cracking from the apex of the pileus. Peridium depressed-globose, borne at apex of a long, slender stem. Exoperidium fragile, fugacious, of hyphae and sand. Endoperidium tough, smooth, dehiscing by irregular rupture from apex, stem inserted into a socket at base of peridium. Gleba of non-septate capillitium and pulverulent spores. Spores globose to subglobose, verrucose or smooth. In sandy soils of Africa, Australia, India and North America.

Schizothyrioma Höhn.
Dermateaceae, Helotiales, Leotiomycetidae, Leotiomycetes, Pezizomycotina, Ascomycota, Fungi
Höhnel, F. von. 1917. Mykologische Fragmente. Nrn. 120-190. *Annls mycol.* 15(5):293-383.
From Gr. *schizo*, to split, divide + *thyrís*, window + suf. *-oma*, which implies the idea of entirety. Lessions on leaves randomly distributed, black, shining, oblong, slightly raised. Apothecia sessile, present on the upper and lower surfaces of leaves. Lateral and bottom excipulum composed of polyhedral cells. Asci clavate, mainly with four didymospores, occasionally with 2-6. These fungi are hosted only by *Achillea ptarmica* and are superficially indistinguishable. Occurs in Europe, Asia and North America.

Schizothyrioma ptarmicae (Desm.) Höhn
(syn. *Schizothyrium ptarmicae* (Desm.) Desm.): dark, mature pseudothecia (apothecioid), which collapse or split in a disorderly manner to expose the asci, on host plant leaf, x 30.

Schizothyrium Desm.
Schizothyriaceae, Inc. sed., Inc. sed., Dothideomycetes, Pezizomycotina, Ascomycota, Fungi
Desmazières, J. B. H. J. 1849. *Annls Sci. Nat.*, Bot., sér.3, 11:360.
From Gr. *schízo*, cutter, to divide, to split, to separate, cleave + *thyreós*, shield + dim. suf. *-ium*, in reference to the manner in which the shield or scutellum of the pseudothecia opens. Pseudothecia apothecioid, dark, at maturity collapsing or splitting in a disorderly manner to expose asci. Asci ovoid or broadly clavate. Ascospores uniseptate, hyaline. On leaves and branches of various plants.

Sclerocystis Berk. & Broome—see **Glomus** Tul. & C. Tul.
Glomeraceae, Glomerales, Inc. sed., Glomeromycetes, Glomeromycotina, Glomeromycota, Fungi
Berkeley, M. J., C. E. Broome. 1873. Fungi of Ceylon. *J. Linn. Soc., Bot.* 14:137.
From Gr. *sklerós*, hard + *kýstis*, bladder, vesicle, cell, in reference to the sporocarps, which are small, hard and rounded.

Scleroderma Pers.
Sclerodermataceae, Boletales, Agaricomycetidae, Agaricomycetes, Agaricomycotina, Basidiomycota, Fungi
Persoon, C. H. 1801. *Synopsis methodica fungorum* (Göttingen) 1:150.

Sclerographiopsis

From Gr. *sklerós*, hard + *dérma*, skin, for the thick, hard peridium of the fructification. Fructifications globose, subturbinate or pyriform, sessile or pseudostipitate. Gleba powdery at maturity, devoid of a capillitium. Spores globose, ornamented, frequently reticulate or echuinulate. In conifer and oak forests, meadows and semiarid scrub land, and clay soils rich in humus, as well as in sandy soils.

Scleroderma citrinum Pers.: globose, sessile fructifications, with hard peridium, on soil of conifer forest, x 0.5.

Sclerographiopsis Deighton
Inc. sed., Inc. sed., Inc. sed., Inc. sed., Pezizomycotina, Ascomycota, Fungi
Deighton, F. C. 1973. *Sclerographiopsis and Spinulospora, two new monotypic hyphomycetous genera from Sierra Leone. Trans. Br. mycol. Soc.* 61(1):193-196.
From genus *Sclerographium* Berk. + Gr. suf. *-ópsis*, aspect, appearance, for the similarity to this genus. Leaf spot none. Synnemata hypophyllous, solitary, sparsely distributed over wide areas. Mycelium superficial, evanescent; hyphae at base of synnemata olivaceous, sparingly septate, branched, smooth. Synnemata erect, substraight, dark brown, diminishing slightly towards long club-shaped head. Conidiogenous cells integrated, terminal or intercalary. Conidiophores olivaceous, on branches of synnema, numerous, cylindric, mostly simple, continuous, longer ones sometimes 1-2 septate with a short lateral branch, smooth, covered with short, strongly cicatricized, denticles. Conidial scars conspicuously thickened. Conidia pale olivaceous, slightly obclavate, often catenulate, terminal ones obtuse at apex, usually slightly curved, smooth, 1-3 septate, not constricted. On living leaves of *Dalbergia heudelotii* in Sierra Leone.

Scleromitrula S. Imai
Rutstroemiaceae, Helotiales, Leotiomycetidae, Leotiomycetes, Pezizomycotina, Ascomycota, Fungi
Imai, S. 1941. Geoglossaceae Japoniae. *J. Fac. agric., Hokkaido Imp. Univ., Sapporo* 45:155-264.
From Gr. *sklerós*, hard + *mítra*, headdress, miter + L. dim. suf. *-ula*, i.e., a small headdress or miter. Stromata an elongated, black sclerotium, foliicolous, formed beneath cuticle of host. Apothecia arising from sclerotia, single or in pairs, stipitate, campanulate to cylindric or subturbinate. Hymenial surface pitted or often longitudinally furrowed. Asci 8-spored. Ascospores ellipsoid to fusiform. Asexual morph lacking. Pathogenic on tree leaves and other plant parts.

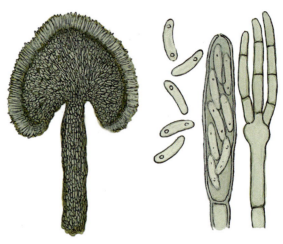

Scleromitrula rubicola T. Schumach. & Holst-Jensen: vertical section of a fertile head of the apothecium, on leaves of *Rubus chamaemorus* in Norway, showing the hymenium, subhymenium and excipulum, x 35; one ascus with eight ascospores, and free, 1-2 guttulated ascopores, x 500.

Sclerotinia Fuckel
Sclerotiniaceae, Helotiales, Leotiomycetidae, Leotiomycetes, Pezizomycotina, Ascomycota, Fungi
Fuckel, K. W. G. L. 1870. *Symbolae mycologicae. Jb. nassau. Ver. Naturk.* 23-24:330.
From Gr. *sklerótes*, hardness + L. des. *-inia*, derived from the Gr. suf. *-ínos*, material, nature or source of something. Sclerotia tuberous in tissues of host. Apothecial arising from sclerotia, pedicellate. Causing rot of rye seeds and other vegetables, in particular lettuce.

Sclerotinia sclerotiorum (Lib.) de Bary: tuberous sclerotium, from which arise pedicellate, yellowish apothecia, x 1.5.

Sclerotium Tode—see **Athelia** Pers.
Inc. sed., Inc. sed., Inc. sed., Inc. sed., Inc. sed., Inc. sed., Fungi
Tode, H. J. 1790. *Fung. mecklenb. sel.* (Lüneburg) 1:2.
From Gr. *sklerós*, hard, dry > *sklerótes*, hardness + L. dim. suf. *-ium*, in reference to the usually dark sclerotia that develop on a white mycelium.

Scoleconectria Seaver—see **Nectria** (Fr.) Fr.
Nectriaceae, Hypocreales, Hypocreomycetidae, Sordariomycetes, Pezizomycotina, Ascomycota, Fungi
Seaver, F. J. 1909. The Hypocreales of North America-II. *Mycologia* 1(5):177-207.
From Gr. *skólekos*, worm > L. *scolecos* + genus *Nectria* (Fr.) Fr., for the shape of the ascospores.

Scopulariopsis Bainier—see **Microascus** Zukal
Microascaceae, Microascales, Hypocreomycetidae, Sordariomycetes, Pezizomycotina, Ascomycota, Fungi
Bainier, G. 1907. Mycothèque de l'Ecole de Pharmacie. XIV. Scopulariopsis (*Penicillium* pro parte) genre nouveau de mucédinées. *Bull. Soc. mycol. Fr.* 23(2):98-99.
From L. *scopula*, dim. of *scopa*, broom + L. suf. *-aris*, belonging to + suf. *-opsis* < Gr. *-ópsis*, aspect, appearance, i.e., like a small broom, referring to the conidiophores, whose phialides generally are arranged with the appearance of a small paint brush or broom, which have catenulae, basally truncate conidia.

Scorias Fr.
Capnodiaceae, Capnodiales, Dothideomycetidae, Dothideomycetes, Pezizomycotina, Ascomycota, Fungi
Fries, E. M. 1832. *Syst. mycol.* (Lundae) 3(2):269.
From Gr. *scórias*, dross, refuse > L. *scoria*, referring to the irregular, black masses of hyphae the fungus produces on the host. Mycelium forming a black subiculum over surface of host; subiculum in some species large and sponge-like, of branching bundles of united hyphae; pseudothecia and pycnidia borne on surface of subiculum, entire mass covered by a thin gelatinous sheath. Pseudothecia black, borne singly, but often crowded, subglobose to ovate or broadly ellipsoid, with an elongate sterile base, glabrous. Asci bitunicate, oblong to saccate. Ascospores phragmosporous, subclavate to clavate, hyaline, but often becoming olivaceous, smooth or finely roughened at maturity. On branches of trees and shrubs.

Scortechiniellopsis Sivan.—see **Nitschkia** G. H. Otth ex P. Karst.
Nitschkiaceae, Coronophorales, Hypocreomycetidae, Sordariomycetes, Pezizomycotina, Ascomycota, Fungi
Sivanesan, A. 1974. Two new genera of Coronophorales with descriptions and key. *Trans. Br. mycol. Soc.* 62(1):35-43.
From genus *Scortechiniella* Arx & E. Müll. + Gr. suf. *-ópsis*, aspect, appearance, for the similarity to this genus.

Scotiosphaeria Sivan.
Inc. sed., Inc. sed., Inc. sed., Sordariomycetes, Pezizomycotina, Ascomycota, Fungi
Sivanesan, A. 1977. British Ascomycetes: *Endoxylina pini* sp. nov., *Scotiosphaeria endoxylinae* gen. et sp. nov. and *Didymosphaeria superapplanata* sp. nov. *Trans. Br. mycol. Soc.* 69:117-123.
From L. *scoticus*, Scottish + genus *Sphaeria* Haller, for the country of origin. Perithecia densely aggregated, superficial or with their bases slightly immersed in a stromatic complex around and in between adjacent ostiolar necks of *Endoxylina pini* Sivan., black, globose, with a short papillate ostiole lined on inside by hyaline, filiform periphyses. Asci cylindrical, pedicellate, unitunicate, 8-spored. Ascospores ellipsoid, brown, 1-septate. Paraphyses numerous, filiform, hyaline. On *Endoxylina pini* on decorticated wood of pine in Perth, Scotland.

Scotiosphaeria endoxylinae Sivan.: black, globose perithecium, on wood, with its base slightly immersed in a stromatic complex adjacent to ostiolar necks of **Endoxylina pini** Sivan., x 180; ascus with eight two-celled ascospores, and three free ascospores, x 1,000.

Scutellinia (Cooke) Lambotte (syn. **Patella** F. H. Wigg.)
Pyronemataceae, Pezizales, Pezizomycetidae, Pezizomycetes, Pezizomycotina, Ascomycota, Fungi
Lambotte, J. B. E. 1887. Flore mycologique belge. *Mém. Soc. roy. Sci. Liège*, Série 2 14:299.
From L. *scutella*, little cup, or *scutellum*, small shield + L. suf. *-inia*, which indicates similarity or possession, for the small size and the almost flat shape of the ascomata, resembling a small shield. Apothecia sessile, discoid with bright red or orange rooted hairs, at times with black or brown hairs on border. On soil or dead wood in coniferous forests.

Scutellospora C. Walker & F. E. Sanders (syn. **Orbispora** Oehl, et al.)

Scutiger

Gigasporaceae, Diversisporales, Inc. sed., Glomeromycetes, Glomeromycotina, Glomeromycota, Fungi
Walker C., F. E. Sanders. 1986. Taxonomic concepts in the Endogonaceae: III. The separation of *Scutellospora* gen. nov. from *Gigaspora* Gerd. & Trappe. *Mycotaxon* 27:179.
From L. *scutellum*, small shield + Gr. *sporá*, spore, referring to the production of germination shields in spores of members of the genus. Spores produced singly in soil (or rarely in cortical cells of roots), large, variable in shape, usually globose or subglobose, but often ovoid, obovoid, pyriform, or irregular especially when constrained during formation; borne on a bulbous suspensor-like cell, usually with a narrow hypha extending from one or more peg-like projections towards the spore. Spore wall structure of at least two wall groups, with one or more flexible membranous or coriaceous walls in the inner group or groups. Germination by means of one or more germ tubes produced near the spore base from a germination shield fomed upon or within a flexible inner wall. Thin-walled, knobby or broadly papillate auxiliary cells borne in soil, on straight or coiled hyphae, formed singly or in clusters. Forming endomycorrhizas with arbuscules and hyphal coils, but without vesicles.

Scutiger Paulet
Albatrellaceae, Russulales, Inc. sed., Agaricomycetes, Agaricomycotina, Basidiomycota, Fungi
Paulet. 1812. *Prosp. Traité Champ.* (Paris):49.
From L. *scuta*, a flat dish + *gero*, to bear, carry, for the shape of the basidiocarps. Annual basidiomata, variously colored (yellow, brown), fan-shaped, stem eccentric to lateral, sometimes concrescent, scaly, fleshy; hymenophore with angular and wide pores. Hyphal system monomitic. Basidia claviform. Basidiospores ellipsoidal or lacrimoid, guttulate, thin-walled, hyaline, inamyloid, cyanophilic. Weak pleasant smell and hazelnut flavor. Ectomycorrhizal fungus.

Scutiger pes-caprae (Pers.) Bondartzev & Singer (syn. **Albatrellus pes-caprae** (Pers.) Pouzar): basidiocarp on soil, x 0.3.

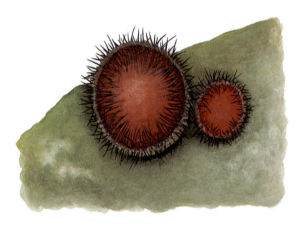

Scutellinia pilatii (Velen.) Svrček: sessile, discoid, bright red apothecia, with black hairs on the border, on dead wood, x 25.

Scutellinia ulloae Izquierdo-San Agustin, et al.: SEM view of a tuberculate-wall ascospore, x 2,300. In pine forest soil in Temascaltepec, Mexico State, near the Nevado de Toluca volcano.

Scytalidium Pesante (syn. **Xylogone** Arx & T. Nilsson)
Inc. sed., Helotiales, Leotiomycetidae, Leotiomycetes, Pezizomycotina, Ascomycota, Fungi
Pesante, A. 1957. Osservazioni su una carie del platana. *Annali Sper. agr., N.S.* 11(suppl.):249-266.
From Gr. *skytalé*, a staff; a serpent of uniform roundness and thickness, a cylinder + *-ídion* > E. *-idium*, dim. suf., in reference to the shape of the spores. Colonies white velvety or floccose, becoming light brown or dark brown. Mycelium partly superficial, partly immersed. Hyphae sub-hyaline to brown, with one or several oil globules, septate, smooth, branched. Conidiogenesis thallic. Conidia cylindrical, rectangular, of variable length, catenate, truncated at both ends. Chlamydospores pale brown to brown, barrel-shaped to oblong, thick walled, catenate. Ascomata nonostiolate, dark brown at maturity; asci dispersed in centrum subglobose or globose, thin-walled, quickly evanescent, 8-spored; ascospores hyaline, smooth, with refractive walls, subglobose to globose. Pathogens of cultivated mushrooms, causing slippery scar on *Auricularia nigricans* (Sw.) Birkenak, et al. (*S. auriculariicola* W. H. Peng, et al.), and causes yellow rot on *Ganoderma* P. Karst. (*S. ganodermophthorum* Kang, et al.).

Scytalidium ganodermophthorum Kang, et al. (syn. **Xylogone ganodermophthora** Kang, et al.): lignicolous fungus, which develops its dark cleistothecia in woody substrates, x 200. Isolated from diseased fruitbody of cultivated **Ganoderma lucidum** (Curtis) P. Karst.,

Scytinopogon Singer

Clavariaceae, Agaricales, Agaricomycetidae, Agaricomycetes, Agaricomycotina, Basidiomycota, Fungi

Singer, R. 1945. New genera of fungi. *Lloydia* 8 (3):139-144.

From Gr. *skýtos*, leather, *skytínos*, leathern + *pógon*, beard, in reference to the leathery, branched fruitbody. Basidiomata clavariaceous, branched, subfleshy-tough, glabrous, pigmented or without color. Hyphae of branch apices thin with rounded tips, hyphae of context beneath subparallel, little interwoven, hyaline. Basidia clavate or attenuated upwards, with usually four apical sterigmata; spores inamyloid, ellipsoid to oblong-subangular, asymmetric. Cystidia absent, clamps frequent. Hymenium not continuous; basidia scattered.

Seaverinia Whetzel

Sclerotiniaceae, Helotiales, Leotiomycetidae, Leotiomycetes, Pezizomycotina, Ascomycota, Fungi

Whetzel, H. H. 1945. A synopsis of the genera and species of the Sclerotiniaceae, a family of stromatic inoperculate Discomycetes. *Mycologia* 37(6):648-714.

Named in honor of the American mycologist *Fred J. Seaver* + L. des. -*inia* < Gr. -*ínos*, nature or source of something, for his extensive work on discomycetes. Stromata substratal, poorly developed, not a definite sclerotium, formed in host rhizomes, visible usually as a narrow, black line. Spermatia not observed. Asexual morph botrytis-like. Conidiophores botryose, pale brown, sparingly septate, formed in tufts on rhizome and roots, under moist conditions, profusely developed, bearing conidia in dense clusters. Conidia pale brown, minutely tuberculate, subglobose, tapering somewhat to attachment. Apothecia arising from partially decayed rhizome, stipitate, length depending on depth of buried rhizome, shallow cup-shaped. Asci cylindric or subcylindric, 8-spored. Ascospores ellipsoid, hyaline. On rootstocks of *Geranium maculatum*.

Sebacina Tul. & Tul.

Sebacinaceae, Sebacinales, Inc. sed., Agaricomycetes, Agaricomycotina, Basidiomycota, Fungi

Tulasne, L.-R., C. Tulasne. 1871. New notes upon the Tremellineous fungi and their analogues. *J. Linn. Soc. Bot.* 13:31-42.

From L. *sebaceous*, like lumps of tallow < *sebum*, grease, tallow, wax + suf. -*ina*, which indicates possession or similarity, for its waxy appearance. Basidiomata resupinate or encrusting, sometimes with free lobes, usually thick; texture tough, coriaceous or gelatinous; hymenium smooth or undulate; basidia tremelloid (ellipsoid and vertically septate), giving rise to long, sinuous sterigmata or epibasidia on which the spores are produced; basidiospores typically ellipsoid to oblong, white in mass, germinating by repetition or by the production of conidia; gloeocystidia and clamp-connections lacking. On soil and litter, sometimes partly encrusting stems of living plants; its species are mycorrhizal, forming a range of associations with trees, orchids, and other plants.

Sebacina pululahuana (Pat.) D. P. Rogers (syn. **Ductifera pululahuana** (Pat.) Donk): cartilaginous-gelatinous basidioma, on dead wood, x 1.5.

Secotium Kunze

Agaricaceae, Agaricales, Agaricomycetidae, Agaricomycetes, Agaricomycotina, Basidiomycota, Fungi

Kunze, J. 1840. *Secotium*, eine neue Gattung der Gasteromycetes Trichogastres. *Flora*, Regensburg 23:321-327.

From L. *sectus*, cut, from *seco*, to cut, to split, to divide + suf. -*ium*, characteristic of or indicates connection or similarity to a cracked or cut object, in reference to the gleba which is divided into superimposed or anastomosed gills in the interior of the fructification. Fructifications sessile or with short stipe; sporiferous sac subglobose, conic, globose, depressed or agaricoid, with a central columella that is a continuation of stipe.

Seimatosporiopsis

Spores colored, smooth, with an apical pore. In forests and pastures.

Seimatosporiopsis B. Sutton, et al.
Inc. sed., Inc. sed., Inc. sed., Inc. sed., Pezizomycotina, Ascomycota, Fungi
Sutton, B. C., et al. 1972. *Seimatosporiopsis* gen. nov. (Sphaeropsidales). *Trans. Br. mycol. Soc.* 59 (2):295-300.
From genus *Seimatosporium* Corda + Gr. suf. *-ópsis*, aspect, appearance, for the similarity to this genus. Pycnidia immersed, dark brown, solitary, globose, glabrous, cavity unilocular, with a single distinct central ostiole. Pycnidial wall several cells thick, of outer sclerotioid cells, inner thin-walled pseudoparenchyma. Conidiogenous cells holoblastic, annellidic, lining inside of pycnidial wall, hyaline, bearing single conidia from apex and from successive proliferations. Conidia blastospores, dark brown, smooth-walled, apex obtuse, base truncate, cylindrical to slightly curved, transversely distoseptate, with one or more apical or subapical, and one or more exogenous basal, occasionally branched, filiform appendages. On stems of *Salvadora oleioidis* in Karachi, Pakistan.

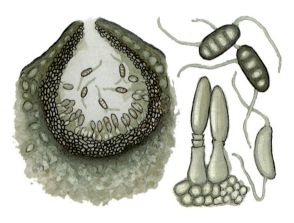

Seimatosporiopsis salvadorae B. Sutton, et al.: vertical section of an immersed pycnidium in host tissues, with conidiogenous cells and conidia, x 160; a close-up of annellidic conidiogenous cells and transversely distoseptate conidia (blastospores), in various development phases, with filiform appendages, x 720.

Seimatosporium Corda (syns. **Discostromopsis** H. J. Swart, **Vermisporium** H. J. Swart & M. A. Will.)
Discosiaceae, Amphisphaeriales, Xylariomycetidae, Sordariomycetes, Pezizomycotina, Ascomycota, Fungi
Corda A. C. J. 1833 *In*: J. Sturm, *Deutschl. Fl.*, 3 Abt. (Pilze Deutschl.) 3(13):79.
From Gr. *seismos*, an earthquake, a shaking + *sporá*, spore. Perithecia single, or in small groups, with sparse stroma, embedded in host tissue, covered by a small, dark clypeus, inverted pear-shaped to irregular. Perithecial wall of several layers of flattened, thin-walled cells. Asci arising from perithecial base, surrounded by paraphyses that degenerate at maturity, long elliptical, unitunicate with thickened apex, with indistinct apical structures. Ascospores partly biseriate, hyaline, 3-septate, elliptical, straight or inaequilateral. Conidiomata closed when young, at maturity wide open due to disintegration of upper cell layer, subepidermal or intra-epidermal, of pale brown pseudoparenchymatous cells. Conidiophores short or absent. Conidiogenous cells holoblastic, annellidic, discrete, cylindrical or flask-shaped. Conidia hyaline, with pigmented median cells, white or pale pink in mass, long, narrow, 3- or usually 4-celled, with thin smooth walls, not or slightly constricted at septa, acerose to falcate, slightly curved or flexuous, with an exogenous basal appendage. Causing leaf spots on dead and living leaves.

Semifissispora H. J. Swart
Inc. sed., Inc. sed., Inc. sed., Dothideomycetes, Pezizomycotina, Ascomycota, Fungi
Swart, H. J. 1982. Australian leaf-inhabiting fungi XII. *Semifissispora* gen. nov. on dead *Eucalyptus* leaves. *Trans. Br. mycol. Soc.* 78(2):259-264.
From L. pref. *semi*, half + *fissus*, split + Gr. *sporá*, spore, seed, referring to the 2-celled ascospores which split at the septum. Ascocarps immersed, scattered in dead host leaf, stromatic, globose or slightly flattened at apex, wall of several layers of dark, thick-walled, slightly flattened cells; ostiole under a stoma. Asci arising from base of the ascocarp surrounded by thread-like paraphysoids, clavate, without distinct stalks, bitunicate. Ascospores partly biseriate, hyaline, 2-celled, at maturity partly splitting at median septum, bending. On dead leaves of *Eucalyptus behriana* in Central and South Australia.

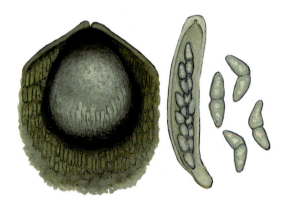

Semifissispora fusiformis H. J. Swart: vertical section of an immersed ascoma in host dead leaves, x 80; a bitunicate ascus with eight 2-celled ascospores, x 450; four liberated, mature ascospores splitting at the median septum and bending, x 600.

Semisphaeria K. Holm & L. Holm
Inc. sed., Inc. sed., Inc. sed., Dothideomycetes, Pezizomycotina, Ascomycota, Fungi
Holm, K., L. Holm. 1991. Ascomycetes on *Myrica gale* in Sweden. *Nordic Jl Bot.* 11(6):675-687.
From L. *semi*, half + genus *Sphaeria* Haller, i.e., like

Sphaeria but with a shield. Ascomata densely to thinly scattered in outermost periderm, scutate. Scutellum of a few layers of strongly incrusted cells; at center small, isodiametric, more elongated, irregularly lobed, meandering at margin, with many cells extending into brownish hyphae. Hamathecium of branched, indistinctly septate filaments. Asci ventricose, almost sessile, 8-spored. Ascospores fusiform, with rather pointed ends, often curved, with one median septum, sometimes with 2 additional septa, hyaline. On bark of *Myrica gale*.

Sepedonium Link—see **Hypomyces** (Fr.) Tul. & C. Tul.
Hypocreaceae, Hypocreales, Hypocreomycetidae, Sordariomycetes, Pezizomycotina, Ascomycota, Fungi
Link, J. H. F. 1809. Observationes in ordines plantarum naturales. *Mag. Gesell. naturf. Freunde, Berlin* 3 (1-2):18.
From Gr. *sepedón*, genit. *sepedónos*, rot, decay + L. dim. suf. *-ium*, because this fungus parasitizes and rots the fruiting bodies of agarics and boletes, on which it develops its conidiophores with aleuriospores in clusters.

Septobasidium Pat.
Septobasidiaceae, Septobasidiales, Inc. sed., Pucciniomycetes, Pucciniomycotina, Basidiomycota, Fungi
Patouillard, N. T. 1892. *Septobasidium*, nouveau genre d'hyménomycètes hétérobasidiés. *J. Bot.*, Paris 6 (4):63.
From L. *septum*, septum, partition + *basidium* < Gr. *basídion*, dim. of *básis*, base, basidium, for the presence of transverse septa in the basidia. Mycelial mats covering insects, mycelium of fungus penetrates insect body from which it absorbs nutritive material. Parasitic on homopterous insects called scales or woodlice in the families Diaspididae and Coccidae, which are parasites that attack wild and cultivated plants.

Septobasidium burtii Lloyd: basidia with basidiospores, emerging from the mycelial mat parasitizing the scale insect host (Diaspididae and Coccidae). This insect is a parasite that feeds on the living tissues of the host plant, an example of hyperparasitism, x 1,000.

Septoglomus Sieverd., et al. (syn. **Viscospora** Sieverd., et al.)
Glomeraceae, Glomerales, Inc. sed., Glomeromycetes, Glomeromycotina, Glomeromycota, Fungi
Oehl, F., et al. 2011. Glomeromycota: three new genera and glomoid species reorganized. *Mycotaxon* 116:75-120.
From L. *septum*, partition, septum + genus *Glomus* Tul. & C. Tul., cluster; referring to the relation with the genus *Glomus*. Spores generally formed in loose clusters; subtending hypha hyaline to white, rarely subhyaline, often thick-walled. Spores with 1-4 wall layers; outer wall layer exuding a mucigale-like substance. Pore closure at spore base often open, or semi-closed by wall thickening.

Septoglomus altomontanum Palenz., et al.: spores, dark-brown, supported by hyphae, x 160. This mycorrhizogenous fungus was isolated from rhizosphere of *Ophioglossum vulgatum* and other plants in grassland in alpine areas of Andalucia, Spain, and propagated on *Sorghum vulgare* and *Trifolium pratense*.

Septoria Sacc.
Mycosphaerellaceae, Capnodiales, Dothideomycetidae, Dothideomycetes, Pezizomycotina, Ascomycota, Fungi
Saccardo, P. A. 1884. *Syll. fung.* (Abellini) 3:474.
From Gr. *septós*, rotten, subject to rotting < *sepo*, to rot + L. suf. *-oria*, which indicates capacity, action or function. Pycnidia small, black, similar to those of *Phoma* Sacc. Pathogenic causing rots, leaf spots and stains of plants.

Septoria lycopersici Speg.: leaf spots with pycnidia, on the surface of tomato plant leaf, x 0.5, which produce narrowly elongate, hyaline, septate conidia, x 75.

Septotinia

Septotinia Whetzel ex J. W. Groves & M. E. Elliott (syn. **Septotis** N.F. Buchw. ex Arx)
Sclerotiniaceae, Helotiales, Leotiomycetidae, Leotiomycetes, Pezizomycotina, Ascomycota, Fungi
Groves, J. W., M. E. Elliott. 1961. Self-fertility in the Sclerotiniaceae. *Can. J. Bot.* 39(1):215-231.
From L. *septa*, partition + L. des. *-inia* < Gr. *-inos*, nature or source, or nature of something, for the septate conidia. Conidiophores branched. Conidia hyaline, elongate, septate, apex and base truncate, except for apical conidia which are tapered towards the acute apices. Sclerotia angular, elongate or circular, thin, black, formed in tissues of host usually on ground. Spermatia ovoid, minute, produced on short, club-shaped spermatiophores, clustered to form minute spermadochia on decaying tissues, accompanying formation of sclerotia. Apothecia contorted, spirally twisted stipes; asymmetrical ascospores. Lesions circular, margin scalloped, brown, zonate, with radiating subcuticular hypomycelium. On leaves of *Podophyllum peltatum* and *Populus* spp.

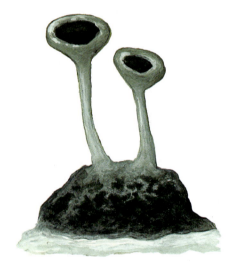

Septotinia populiperda (Moesz & Smarods) Waterman & E. K. Cash (syn. **Septotis populiperda** (Moesz & Smarods) B. Sutton): circular, black apothecia emerging from a sclerotium on agar, x 2. The sclerotia were producing leaf blotch on poplar leaves, Eastern

Septotis N.F. Buchw. ex Arx—see **Septotinia** Whetzel ex J. W. Groves & M. E. Elliott)
Sclerotiniaceae, Helotiales, Leotiomycetidae, Leotiomycetes, Pezizomycotina, Ascomycota, Fungi
Arx, J. A. von 1970 A revision of the fungi classified in *Gloeosporium. Bibliotheca Mycol.* 24:158.
From L. *septa*, partition, for the septate conidia + suf. *-is*, which indicates narrow connection.

Sepultariella van Vooren, et al.
Pyronemataceae, Pezizales, Pezizomycetidae, Pezizomycetes, Pezizomycotina, Ascomycota, Fungi
Van Vooren, N. et al. 2017. Emendation of the genus *Tricharina* (Pezizales) based on phylogenetic, morphological and ecological data. *Ascomycete.org.*9(4):101-123.
From L. *sepultus*, buried, sleeping < *sepelio*, to bury, bury in sleep + dim. suf. *-ella*. Apothecium 2-5 mm across, sessile, cup-shaped when young, later almost flat. Hymenium orange, outer surface concolorous with the hymenium and tomentose, edges crenulate with small whitish teeth. Presents a fine subiculum hypha with which it adheres to the substrate. Asci cylindrical, 8-spored; paraphyses cylindrical, forked, septate, and slightly thickened at the apex; ascospores ellipsoidal to slightly fusiform, smooth, biguttulate, sometimes with a large guttule accompanied by other smaller ones. Several to gregarious on soil, under hardwoods and conifers.

Sepultariella patavina (Cooke & Sacc.) Van Vooren, et al. (syn. **Pustularia patavina** (Cooke & Sacc.) Boud.): gregarious, cup-shaped ascomata, whose external surface has vesicles that resemble pustules, on sandy soil, x 4.

Serpula (Pers.) Gray
Serpulaceae, Boletales, Agaricomycetidae, Agaricomycetes, Agaricomycotina, Basidiomycota, Fungi
Gray, S. F. 1821. *Nat. Arr. Brit. Pl.* (London) 1:637.
From L. *serpula*, small snake, from *serpens*, serpent, crawling, winding, that which wiggles or slithers + dim. suf. *-ula*, for the hymenophore with undulations similar to shape of a moving snake. Fructifications resupinate, effuse, forming undulations with a merulioid to hydnoid hymenial surface, with irregular reticulate folds. *Serpula lacrymans* (Wulf. ex fr.) J. Schröt. (=*Merulius lacrymans* Fr.) is the dry rot fungus that frequently destroys houses and other wood structures.

Setocampanula Sivan. & W. H. Hsieh
Trichosphaeriaceae, Trichosphaeriales, Inc. sed., Sordariomycetes, Pezizomycotina, Ascomycota, Fungi
Sivanesan, A., W. H. Hsieh. 1989. *Kentingia* and *Setocampanula*, two new ascomycete genera. *Mycol. Res.* 93 (1):83-90.
From L. *seta*, bristle + *campanula*, dim. of *campana*, a bell, for the bell-shaped ascomata covered with bristles. Ascomata more or less campanulate, becoming depressed at apex, superficial on a thin subiculum, setose, hairy. Hamathecium of paraphyses and periphyses. Asci cylin-

drical, unitunicate, short-stalked, 8-spored. Ascospores hyaline, oblong, 0-1-septate, not constricted, uniseriate in ascus, smooth. On stems of *Yushania niitakayamensis* in Kaohsiung, Taiwan.

Serpula lacrymans (Wulfen) J. Schröt.: effuse, reticulate, folded fructification on fallen wood, principally of conifers, causing dry rot decay, x 0.2.

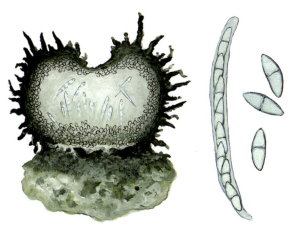

Setocampanula taiwanensis Sivan. & W. H. Hsieh: vertical section of a bell-shaped, superficial ascoma covered with bristles, on host bamboo stems, x 140; unitunicate ascus, with 8 ascospores, x 430; liberated, mature ascospores, hyaline, 1-septate, not constricted, x 520.

Setosphaeria K. J. Leonard & Suggs—see **Exserohilum** K. J. Leonard & Suggs

Pleosporaceae, Pleosporales, Pleosporomycetidae, Dothideomycetes, Pezizomycotina, Ascomycota, Fungi
Leonard, K. J., E. G. Suggs. 1974. *Setosphaeria prolata*, the ascigerous state of *Exserohilum prolatum*. *Mycologia* 66(2):281-297.
From L. *seto*, bristle, hair + genus *Sphaeria* Haller, for the setose hairs on the ascomata.

Setosporella Moustafa & Abdul-Wahid

Inc. sed., Inc. sed., Inc. sed., Inc. sed., Pezizomycotina, Ascomycota, Fungi
Moustafa, A. F., O. A. Abdul-Wahid. 1989. *Setosporella*, a new genus of Hyphomycetes from Egyptian soils. *Mycol. Res.* 93(2):227-229.
From L. *seta*, bristle + Gr. *sporá*, spore, seed + L. dim. suf. *-ella*, for the spores with bristle-like protuberances. Mycelium superficial, immersed, subhyaline, smooth, producing micronematous conidiophores, discrete or penicillate, ampulliform to lageniform, branched. Conidia holoblastic, subspherical, pale brown, covered with long, sometimes branched, columnar protuberances with rounded ends. Isolated from cultivated soil in Ismalia Governorate, Egypt.

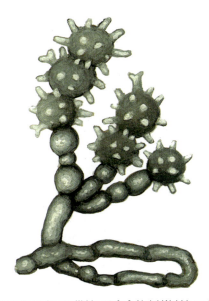

Setosporella mahmoudii Moustafa & Abdul-Wahid: conidiophores with penicillate conidiogenous cells, which retrogressively produce subspherical, brown, catenulate conidia covered with bristle-like protuberances, x 1,100.

Shanoria Subram. & K. Ramakr.—see **Strigula** Fr.

Strigulaceae, Strigulales, Dothideomycetidae, Dothideomycetes, Pezizomycotina, Ascomycota, Fungi
Subramanian, C. V., K. Ramakrishnan. 1956. *Ciliochorella* Sydow, *Plagionema* Subram. & Ramakr., and *Shanoria* gen. nov. *Trans. Br. mycol. Soc.* 39(3):314-318.
Named after the American mycologist *Leland Shanor* + L. suf. *-ia*.

Siamia V. Robert, et al.

Inc. sed., Inc. sed., Inc. sed., Inc. sed., Pezizomycotina, Ascomycota, Fungi
Robert, V., et al. 2000. *Siamia luxuriosa* gen. et sp. nov., a new synnematous hyphomycete from Thailand. *Mycol. Res.* 104(7):893-895.
From *Siam*, the original name for Thailand + L. des. *-ia*, for the type locality. Colonies on natural substrata effuse, brown-olivaceous to black, hairy. Mycelium partly superficial, partly immersed. Stromata absent. Conidiomata macronematous, synnematous, indeterminate, erect. Stipe tightly compacted hyphae,

Siemaszkoa

parallel, septate, brown to dark-brown, becoming progressively paler towards divergent fertile apex. Conidiogenous cells integrated, hyaline or subhyaline, terminal, intercalary, sympodial, monoblastic or polyblastic. Conidial secession schizolytic. Conidia holoblastic, solitary, obclavate to cylindrical, straight to slightly curved, truncate at base, rostrate at apex, multiseptate, hyaline to sub-hyaline, smooth. On dead leaves in Khao Yai National Park, Thailand.

Siamia luxuriosa V. Robert, et al.: saprobic fungus, with black mycelium, on dead leaves of an unidentified angiosperm, producing synnematous, erect conidiomata, x 275, which give rise to cylindrical, hyaline, multiseptate conidia, x 50.

Siemaszkoa I. I. Tav. & T. Majewski
Laboulbeniaceae, Laboulbeniales, Laboulbeniomycetidae, Laboulbeniomycetes, Pezizomycotina, Ascomycota, Fungi
Tavares, I. I., T. Majewski. 1976. *Siemaszkoa* and *Botryandromyces*, two segregates of *Misgomyces* (Laboulbeniales). *Mycotaxon* 3(2):193-208.
Named in honor of the Polish mycologists *Janina* and *Wincenty Siemaszko* + L. des. -a, for their work on laboulbeniaceous fungi. Perithecium with three vertical rows of outer wall cells of three cells of unequal length and one row of four cells of unequal length, fourth cell of latter row protruding apically. Receptacle uniseriate, unbranched, with three or more cells subtending stalk cell of perithecium. Secondary stalk cell and basal cells of perithecium not evident at maturity because their walls do not thicken. Antheridia short, sessile, with a broad base, arising in a series from successive cells of proximal part of unbranched primary appendage. On insects.

Simiglomus Sieverd., et al.—see **Glomus** Tul. & C. Tul.
Glomeraceae, Glomerales, Inc. sed., Glomeromycetes, Glomeromycotina, Glomeromycota, Fungi
Oehl, F., et al. 2011. Glomeromycota: three new genera and glomoid species reorganized. *Mycotaxon* 116:75-120.
From L. *similis*, like, similar + *glomus*, cluster; referring to the relation with the genus *Glomus*.

Simuliomyces Lichtw.
Legeriomycetaceae, Harpellales, Inc. sed., Inc. sed., Kickxellomycotina, Zygomycota, Fungi
Lichtwardt, R. W. 1972. Undescribed genera and species of Harpellales (Trichomycetes) from the guts of aquatic insects. *Mycologia* 64(1):167-197.
Any member of the family Simuliidae (dipteran insects) + Gr. *mýkes* > L. *myces*, fungus. The name means fungus on simulids, that serve as hosts. *Simuliomyces* lives attached to the cuticle that lines the proctodeum of the larvae of these insects.

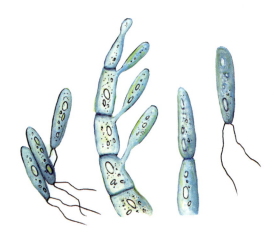

Simuliomyces microsporus Lichtw.: thallus with biflagellate spores, x 750. It lives attached to the proctodeum cuticle of dipteran larvae (*Simulium tuberosum*), Wyoming, U.S.A.

Singerocybe Harmaja
Tricholomataceae, Agaricales, Agaricomycetidae, Agaricomycetes, Agaricomycotina, Basidiomycota, Fungi
Harmaja, H. 1988. Studies on the agaric genera *Singerocybe* n. gen. and *Squamanita*. *Karstenia* 27(2):71-75.
Named in honor of well-known mycologist/agaricologist *Rolf Singer* + the suf. -*cybe* < Gr. *kýbe*, head, in reference to the head-shaped pileus of the agaric genus. Pileus shallowly infundibuliform, hygrophanous, non-pruinose, weakly translucent-striate near margin when moist, surface slightly viscid or dry, pale gray-brown when moist. Stipe equal, curved, white with scanty white mycelial tomentum at base. Lamellae fairly long, decurrent, white, moderately close, elastic; odor slightly acidulose when fresh; taste mild. Spores ellipsoid, lacrymoid, obovoid, subfusiform, smooth, inamyloid. On decaying angiosperm leaves, and spruce needles or mesotrophic *Picea abies* woods in the boreal or hemiboreal regions of Europe in calcareous soils.

Singeromyces M. M. Moser

Boletaceae, Boletales, Agaricomycetidae, Agaricomycetes, Agaricomycotina, Basidiomycota, Fungi

Horak, E., M. Moser. 1965 (1966). Fungi Austroamericani VIII. Über neue Gastrokoletaceae aus Patagonien: *Singeromyces* Moser, *Paxillogaster* Horak und *Gymnopaxillus* Horak. *Nova Hedwigia* 10:329-338.

Dedicated to *Rolf Singer*, prominent mycologist wellknown for his important contributions to science and particularly to Agaricology. The word is formed from *Singer* + L. suf. *-myces* < Gr. *mýkes*, *mýketos*, fungus. Pileus with rudimentary or absent peridium. Gleba ferruginous, irregularly lacunose, stipe cylindric, brown. Spores yellow, ellipsoid or subcylindric, symmetric or subsymmetric, minutely punctate-perforate. Hyphae without clamp connections. The type species *Singeromyces ferrugineus* M. M. Moser was collected under *Nothofagus pumilio* in soil, with hypogeous habit in Prov. Neuquen, Argentina.

Sistotrema Fr.

Hydnaceae, Cantharellales, Inc. sed., Agaricomycetes, Agaricomycotina, Basidiomycota, Fungi

Fries, E. M. 1821. *Syst. mycol.* (Lundae) 1:425.

From Gr. *systáo*, to join, to stick, to sew, to unite closely + *tréma*, hole, opening, because the pores of the hymenophore are closely united to the context of the basidiocarp. Basidiocarps resupinate, exceptionally stipitate, arachnoid, pelliculose or waxy; hyphal system monomitic; generative hyphae nodose-septate, often ampullate. Hymenium smooth to hydnaceous or poroid. Basidia urniform, 6-8 sterigmata. Basidiospores small, smooth, hyaline, ellipsoid, thin-walled, usually with one oil drop. Causes white rot; widely distributed in North America, cosmopolitan.

Skyttea Sherwood, et al.

Inc. sed., Helotiales, Leotiomycetidae, Leotiomycetes, Pezizomycotina, Ascomycota, Fungi

Sherwood, M. A., et al. 1980. *Skyttea*, a new genus of odontotremoid lichenicolous fungi. *Trans. Br. mycol. Soc.* 75(3):479-490.

Named in honor of the Danish lichenologist *M. Skytte Christiansen* + L. des. *-a*, for his contributions to our knowledge of lichenicolous fungi. Apothecia ascohymenial, lichenicolous, immersed, becoming erumpent, small, initially closed, opening by a pore, remaining deeply urceolate, perithecioid even when mature; excipulum dark brown or olive green, noncarbonized, of small-celled pseudoparenchyma, lined near summit with hyaline to brown, thin-walled, smooth hairs. Paraphyses filiform, simple to sparingly branched, septate, sometimes enlarged above. Asci thick-walled, functionally unitunicate, not reacting with iodine, apex thickened but lacking any obvious apical apparatus or pore, 8-spored. Ascospores hyaline, smooth, oval to narrowly ellipsoid, 0-3-septate. Asexual morph unknown. On living thalli of lichens.

Smithiomyces Singer

Agaricaceae, Agaricales, Agaricomycetidae, Agaricomycetes, Agaricomycotina, Basidiomycota, Fungi

Singer, R. 1944. New genera of fungi. I. *Mycologia* 36 (4):366-367.

Genus dedicated "to Dr. *Alexander H. Smith*, whose contributions to American Agaricology during the last decade are among the most outstanding advances in this particular field", according to Singer + L. suf. *-myces* < Gr. *mýkes*, fungus or mushroom. Distinguished from Amanitaceae in having regular trama. Spore print white; clamp connections present; cystidia absent; with free lamellae. On humus and rotten wood in tropical and subtropical forests in Mexico and Florida, U.S.A.

Smittium R. A. Poiss.

Legeriomycetaceae, Harpellales, Inc. sed., Inc. sed., Kickxellomycotina, Zygomycota, Fungi

Poisson, R. 1936. Sur un endomycète nouveau: *Smittium arvernense* n. g., n. s., parasite intestinal des larves de *Smittia* sp. (Diptères Chironomides) et description d'une nouvelle espèce *Stachylina* Lég. et Gauth., 1932. *Bull. Soc. sci. Bretagne* 14:30.

The name derives from *Smittia*, which is the genus of dipteran insects that serve as host + L. dim. suf. *-ium*. Thallus produces trichospores. Endocommensal, attached to the proctodeum of the larvae of these insects.

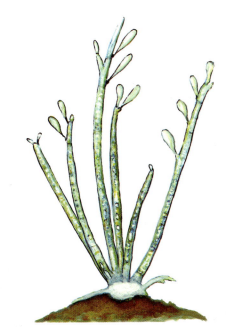

Smittium bulbosporophorum L. G. Valle & Santam.: thallus with trichospores, x 1,160. An endocommensal adhered to the proctodeum of dipteran insect larvae.

Soleella

Soleella Darker
Rhytismataceae, Rhytismatales, Leotiomycetidae, Leotiomycetes, Pezizomycotina, Ascomycota, Fungi
Darker, G. D. 1967. A revision of the genera of the Hypodermataceae. *Can. J. Bot. 45*:1399-1444.
From L. *solea*, sandal + dim. suf. *-ella*, in apparent reference to the flat, elongated shape of the ascomata. Apothecia black, elliptical to elongated, linear, along stomatal lines, subepidermal, opening by a longitudinal fissure. Hymenium flat; subhymenium thin. Paraphyses simple, filiform. Asci subclavate, unitunicate. Ascospores bifusiform. Causing lesions on conifer needles.

Soleella cunninghamiae Saho & Zinno: vertical section of a subepidermal hysterothecium, on conifer needles of *Cunninghamia lanceolata* in Japan, x 80; filiform paraphyses and unitunicate asci with bifusiform ascospores, x 850.

Solorina Ach.
Peltigeraceae, Peltigerales, Lecanoromycetidae, Lecanoromycetes, Pezizomycotina, Ascomycota, Fungi
Acharius, E. 1808. Förteckning på de i Sverige våxande arter af Lafvarnas Familj. *K. Vetensk-Acad. Nya Handl.* 29:228-237.
From Gr. *sólos*, iron disk + *orinós*, mountain, inhabitant of the mountains, reference to the presence of discoidal apothecia on the surface of the thallus, and the frequency of the species in mountain places. Thallus foliose, more or less lobulate, loosely attached to substrate by rhizoids. Apothecia medium sized, 2-6 mm, with a flat, convex or somewhat concave disc, generally brown, blackish or reddish. Spores brown, oblong, ellipsoidal or fusiform, uniseptate. On soil or mosses.

Sordaria Ces. & de Not.
Sordariaceae, Sordariales, Sordariomycetidae, Sordariomycetes, Pezizomycotina, Ascomycota, Fungi
de Notaris, G. 1863. Proposte di alcune rettificazioni al profilo dei discomiceti. *Comm. Soc. crittog. Ital. 1* (fasc. 4):225.
From L. *sordes*, *sordidus*, with a dirty, muddy, or stained appearance + suf. *-aria*, which denotes possession, referring to the dark colored perithecia, giving the substrate a dark appearance. Perithecia with asci swelling at maturity to fill upper part of perithecium. Ascospores expelled explosively through ostiole. Cosmopolitan saprobe on dung of various animals. Some species used in physiological and genetic studies due to the ease with which they can be manipulated experimentally.

Solorina saccata (L.) Ach.: foliose, more or less lobulate thallus with brown apothecia, on mountain soil, x 0.5; vertical section of an apothecium showing the hymenium, the spreading green algal layer, spherical blue-green alga, and rhizoids, x 100. On a green alga, *Cladophora*, Bulgaria.

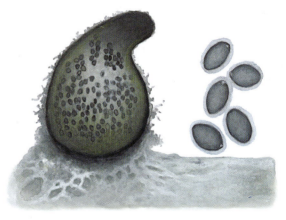

Sordaria fimicola (Roberge ex Desm.) Ces. & De Not.: dark-colored perithecium, with the ostiolar neck bent towards the incident light; dark ascospores, with the sheath and a germ pore on one pole, are expelled explosively through the ostiole of the perithecium, x 150; free ascospores, x 500. Isolated from horse dung, France.

Sparassis Fr.
Sparassidaceae, Polyporales, Inc. sed., Agaricomycetes, Agaricomycotina, Basidiomycota, Fungi
Fries, E. M., 1819. *Novitiae florae Sueciae 5* (cont.):80.
From Gr. *spáraxis*, laceration, tear, from *sparásso*, lacerate, to tear, to chop, for the shape of the basidioma. Basidiomata cerebriform or similar to cauliflower, fleshy, erect, much branched; branches applanate, narrow or wide, petaloid or curly, more or less confluent, giving appearance of a lacerated object. On soil in oak woods, also in subtropical forests. Includes edible species, such as *S. crispa* Wulf & Fr. and *S. radicata* Weir.

Sphacelotheca

Sparassis crispa (Wulfen) Fr.: cerebriform, lacerated basidioma, with petaloid or curly branches, on soil in oak woods, x 0.2.

Spathularia Pers.
Cudoniaceae, Rhytismatales, Leotiomycetidae, Leotiomycetes, Pezizomycotina, Ascomycota, Fungi
Persoon, C. H. 1797. *Tentamen dispositionis methodicae fungorum.* (Lipsiae):36.
From L. *spatha* < Gr. *spathé*, spatula, sheet, a flat, wide piece of wood, a long, wide sword > dim. *spathula*, spatula + L. suf. *-aria*, which denotes similarity or connection to something. Fructifications erect, fleshy, with slender cylindrical stipes; fertile portion bearing hymenium, shaped like a spatula or spoon, yellow or brownish-yellow. Asci clavate, with eight hyaline, multiseptate ascospores. Saprobic, especially in pine forests, in humus or on decaying trunks and branches.

Spathularia flavida Pers.: saprobic, fleshy, yellowish ascocarps, with spatula fertile portion, on humus of pine forest, x 1.

Spathulospora A. R. Caval. & T. W. Johnson
Spathulosporaceae, Lulworthiales, Inc. sed., Sordariomycetes, Pezizomycotina, Ascomycota, Fungi
Cavaliere, A. R., T. W. Johnson. 1965. A new marine ascomycete from Australia. *Mycologia* 57(6):927.
From L. *spathula*, dim. of *spatha*, spatula + Gr. *sporá*, spore. Perithecia with thick wall, lanose, ostiolate, periphysate. Ascospores bicellular. Parasitic on marine Rhodophyceae.

Spathulospora lanata Kohlm.: vertical section of a perithecium, with the bristly peridium, parasitizing the thallus of red alga (*Camontagnea hirsuta*, New Zealand), containing fusoid ascospores,

Spermophthora S. F. Ashby & W. Nowell—see Eremothecium Borzi
Eremotheciaceae, Saccharomycetales, Saccharomycetidae, Saccharomycetes, Saccharomycotina, Ascomycota, Fungi
Ashby, S. F., W. Nowell. 1926. The fungi of stigmatomycosis. *Ann. Bot.*, Lond. *40*:72.
From Gr. *spérma*, seed + *phthorá*, corruption, spot, death, destruction, because it is a parasite of higher plants, especially causing damage on cotton bolls and hazelnut seeds.

Sphacelia Lév.—see Claviceps (Fr.) Tul.
Clavicipitaceae, Hypocreales, Hypocreomycetidae, Sordariomycetes, Pezizomycotina, Ascomycota, Fungi
Léveillé, J. H. 1827. Mémoire sur l'ergot, ou nouvelles recherches sur la cause et les effets de l'ergot. *Mém. Soc. Linn. Paris* 5:578-579.
From L. *sphacelus* < Gr. *sphákelos*, gangrene, inflammation, because of the effect it causes in parasitzed heads of cereals.

Sphaceloma de Bary—see Elsinoë Racib.
Elsinoaceae, Myriangiales, Dothideomycetidae, Dothideomycetes, Pezizomycotina, Ascomycota, Fungi
de Bary, A. 1874. Über den sogenannten Brenner der Reben. *Ann. Oenol. 4*:165-167.
From Gr. *sphákelos*, gangrene, inflamation + suf. *-oma*, tumor, pathologic state.

Sphacelotheca de Bary
Microbotryaceae, Microbotryales, Inc. sed., Microbotryomycetes, Pucciniomycotina, Basidiomycota, Fungi

Sphaeria

de Bary, H. A. 1884. *Vergleichende Morphologie und Biologie der Pilze, Mycetozoen und Bakterien* (Leipzig):187.
From Gr. *sphákelos*, gangrene + *théke*, sheath, covering, for having the mass of spores protected by a membrane. Teliospores or ustilospores originating from complete division of mycelium to form a powdery mass in which spores remain isolated, i.e., they do not unite in pairs or in spheroidal groups. Spore mass covered by a delicate membrane of sterile mycelium forming a sorus with a central columella formed partly of host plant tissue, generally a grain, frequently sorghum or maize; e.g., *S. sorghi* (Link) Clinton, produces covered smut of sorghum.

Sphaerobolus stellatus Tode: mature basidomata on rotted wood, which forcefully hurls the globose peridiole that contains the viscous, dark-colored gleba, x 10.

Sphacelotheca polygoni-serrulati Maire: habit of the powdery masses (sori) in the head of the host (*Persicaria*), x 1.5; magnification of the teliospores of this smut fungus, x 500.

Sphaeria Haller—see **Hypoxylon** Bull.

Hypoxylaceae, Xylariales, Xylariomycetidae, Sordariomycetes, Pezizomycotina, Ascomycota, Fungi
Haller, V. B. von. 1768. *Hist. stirp. Helv.* 3:120.
From Gr. *sphaîra* > L. *sphaera*, sphere, globe, referring to the shape of the fruiting body. This name was originally applied to most fungi with a sphaerical fruiting body, whether perithecia or pycnidia. The name is no longer considered valid. However, it has been widely used to form part of the generic names of numerous ascomycetes with sphaerical ascomata.

Sphaerobolus Tode

Geastraceae, Geastrales, Phallomycetidae, Agaricomycetes, Agaricomycotina, Basidiomycota, Fungi
Tode, H. J. 1790. *Fung. mecklenb. sel.* (Lüneburg) 1:43.
From Gr. *sphaîra* > L. *sphaera*, sphere + *bállo* > L. *bolus*, to hurl, because the mature fructification forcefully hurls the peridiole. Fructification with peridium that fragments at apex into several recurved segments, so that at maturity it becomes star shaped. Gleba viscous, dark in color. Peridioles solitary, globose body with spores covered in mucilage. Spores hyaline, smooth. Gregarious on rotted wood and horse dung.

Sphaerodermatella Seaver

Coniochaetaceae, Coniochaetales, Sordariomycetidae, Sordariomycetes, Pezizomycotina, Ascomycota, Fungi
Seaver, F. J. 1909. The Hypocreales of North America-II. *Mycologia* 1(5):177-207.
From genus *Sphaeroderma* Fuckel + connective -*t*- + L. dim. suf. -*ella*. Stromata erumpent, fleshy. Perithecia in dense cespitose clusters seated on stroma, entirely obscured at maturity, more or less rough, furfuraceous. Asci broad-clavate to ovoid, 4-8-spored. Ascospores simple, becoming dark colored, opaque. On tree bark.

Sphaerodes Clem.

Ceratostomataceae, Melanosporales, Hypocreomycetidae, Sordariomycetes, Pezizomycotina, Ascomycota, Fungi
Clements, F. E. 1909. The Genera of Fungi. H. W. Wilson Co., Minneapolis:44.
From Gr. *sphaîra* > L. *sphaera*, globe, sphere + Gr. suf. -*odes*, like, like a sphere. Ascomata perithecia, ostiole present or lacking, globose to subglobose, scattered to gregarious, superficial or immersed, glabrous or with sparse hairs, yellowish to orange, golden-brown, ochraceous, or brown, but appearing black when filled with mature spores. If present, ostiolar neck short, with coronal setae sparse or absent. Perithecial wall of somewhat flattened, angular pseudoparenchymatous cells, outermost cell walls pigmented, inner cells hyaline, membraneous, translucent. Centrum of thin-walled pseudoparenchymatous cells. Asci arranged in a basal fascicle in perithecium. Asci broadly clavate to clavate or ellipsoid, pyriform, or oblong, thin-walled, undifferentiated at apex, evanescent, 4-8-spored. Ascospores 1-celled, citriform, sometimes laterally compressed, olivaceous-brown to dark brown or black, with coarsely reticulate walls with spiculate to umbonate or tuberculate ends, each with a germ pore.

Sphaerodes mycoparasitica Vujan.: globose, brown perithecium, with sparse hairs and a short ostiolar neck, x 120; reticulate, 1-celled, citriform, dark brown ascospores, x 850, with bipolar germination (one triangular ascospore), and curly branches of mycelium parasitizing the living hyphae of **Fusarium oxysporum** Schltdl. or **F. avenaceum** (Fr.) Sacc.

Sphaerophorus Pers.
Sphaerophoraceae, Lecanorales, Lecanoromycetidae, Lecanoromycetes, Pezizomycotina, Ascomycota, Fungi
Persoon, C. H. 1794. Einige Bemerkungen über die Flechten. *Ann. Bot.* (Usteri) 7:23.
From L. *sphaera* < Gr. *sphaíra*, sphere + Gr. suf. *-phóros* < *phéro*, to bear, produce. Thallus fruticose, produces globose, sessile apothecia on tips of branches. Apothecia open irregularly near apex, at times formed in coralloid branchlets. Photobiont *Protococcus*. On soil, rocks and trees.

Sphaerophorus globosus (Huds.) Vain: fruticose thallus on mountain soil, with globose, sessile, dark-colored apothecia on the tips of coralloid branchlets, x 1.

Sphaerostilbe Tul. & C. Tul.—see **Nectria** (Fr.) Fr.
Nectriaceae, Hypocreales, Hypocreomycetidae, Sordariomycetes, Pezizomycotina, Ascomycota, Fungi
Tulasne, L. R., C. Tulasne. 1865. *Select. fung. carpol.* (Paris) 3:24.
From Gr. *sphaíra* > L. *sphaera*, sphere + Gr. *stílbe*, a lamp, shining, perhaps in reference to the brightly colored ascomata.

Sphaerostilbella (Henn.) Sacc. & D. Sacc. (syn. **Gliocladium** Corda)
Hypocreaceae, Hypocreales, Hypocreomycetidae, Sordariomycetes, Pezizomycotina, Ascomycota, Fungi
Saccardo P.A., D. Saccardo. 1905. *Syll. fung.* (Abellini) 17:778.
From Gr. *sphaíra* > L. *sphaera*, ball, sphere + genus *Stilbella* Lindau. Species of this genus are characterised by a tomentose layer of hairy perithecia formed in a scanty subiculum, and spinulose, naviculate and nonapiculate ascospores. Form gliocladium-type anamorphs, characterised by usually aseptate conidia held in liquid heads at the top penicillately branched conidiophores; sometimes form synnemata. Occur on basidiocarps of various wood-decaying fungi (*Ganoderma* P. Karst., *Gloeophyllum* P. Karst., *Heterobasidion* Bref., and *Stereum* Hill ex Pers.).

Sphaerotheca Lév.—see **Podosphaera** Kunze
Erysiphaceae, Erysiphales, Leotiomycetidae, Leotiomycetes, Pezizomycotina, Ascomycota, Fungi
Léveillé, J. H. 1851. Organisation et disposition méthodique des espèces qui composent le genre *Erysiphe*. *Annls Sci. Nat., Bot.*, sér. 3, 15:138.
From Gr. *sphaíra* > L. *sphaera*, sphere + *théke* > L. *theca*, box, for the cleistothecioid perithecia.

Sphaerulina Sacc.
Mycosphaerellaceae, Capnodiales, Dothideomycetidae, Dothideomycetes, Pezizomycotina, Ascomycota, Fungi
Saccardo, P. A. 1878. Fungi nonnulli extra-Italici novi. *Michelia* 1:399.
From Gr. *sphaîra*, globe, sphere > L. *sphaera*, dim. *sphaerula* + des. *-ina*, likeness, for the similarity of the shape of the ascocarp to a sphere. Ascomata uniloculate, perithecioid pseudothecia, scattered or grouped, sometimes confluent, appearing multiloculate, immersed or erumpent; pseudothecia brown, globose or conical, ostiolate, with a papilla. Ascomatal wall of brown, polygonal cells, thick- or thin-walled. Asci bitunicate, oblong, clavate or cylindric-clavate, fasciculate, stalked or sessile, 8-spored. Ascospores phragmosporous, cylindric to filiform, straight or curved, guttulate, smooth, hyaline, sometimes becoming pale brown upon discharge. Asexual morph cercospora-like and septoria-like. Causing leaf spots on leaves of dicotyledonous plants.

Sphinctrina Fr.
Sphinctrinaceae, Mycocaliciales, Mycocaliciomycetidae, Eurotiomycetes, Pezizomycotina, Ascomycota, Fungi
Fries, E. M. 1825. *Syst. orb. veg.* (Lundae) 1:120.
From L. *sphincter, sphincteris* < Gr. *sphinctér*, sphincter, circular muscle of the anus and of the bladder + suf. *-ina* < *-inus*, which indicates similarity or possession.

Spilocaea

Apothecia spheroid, closed or open, a small black, disk. Photobiont *Protococcus*. Parasitic on other lichens, such as *Cyphelium* Ach. and *Pertusaria* Lam. & D.C.

Sphinctrina turbinata (Pers.) De Not.: thallus of this lichen parasitic on other lichen thalli, such as **Pertusaria** (Germany), with spheroid apothecia, x 30.

Spilocaea Fr.—see Venturia Sacc.
Venturiaceae, Venturiales, Pleosporomycetidae, Dothideomycetes, Pezizomycotina, Ascomycota, Fungi
Fries, E. M. 1819. *Novit. Flo. Svec.* 5:79.
From Gr. *spiló*, contaminated, ugly < *spilás*, *spílos*, spot, contagion, leprosy + suf. *-cea* < *keázo*, to split, break, because this fungus causes a disease of apple and related species, known as scab, characterized by the formation of necrotic, corky and chlorotic spots on leaves and fruits.

Spinalia Vuill.
Inc. sed., Dimargaritales, Inc. sed., Inc. sed., Kickxellomycotina, Zygomycota, Fungi
Vuillemin, P. 1904. Le *Spinalia radians* g. et sp. nov. et la série des Dispirées. *Bull. Soc. mycol. Fr.* 20:26-33.
From L. *spina*, spine + suf. *-alis*, belonging to or referring to. Sporangiola spiny, bispored, borne on a terminal vesicle. On dung, mycoparasitic on mucoraceous fungi and *Chaetomium* Kunze.

Spinalia radians Vuill.: this mucoraceous fungus forms groups of bispored sporangiola that are borne on a terminal vesicle, on herbivorous dung, x 400.

Spinellus Tiegh.
Phycomycetaceae, Mucorales, Inc. sed., Inc. sed., Mucoromycotina, Zygomycota, Fungi
Tieghem, P. van. 1875. Nouvelles recherches sur les mucorinées. *Annls Sci. Nat., Bot.*, sér. 6, 1:66.
From L. *spina*, spine + dim. suf. *-ellus*, i.e., with little spines. Aerial mycelium spiny. Parasitic on agaric basidiocarps, especially *Collybia* Fr. and *Mycena* Fr. in cold regions.

Spinellus fusiger (Link) Tiegh.: aerial mycelium, with little spines, x 600 (shown below), that develops sporangia as a parasite of agaric basidiocarps, especially **Mycena**, x 2.

Spiniger Stalpers—see Heterobasidion Bref.
Bondarzewiaceae, Russulales, Inc. sed., Agaricomycetes, Agaricomycotina, Basidiomycota, Fungi
Stalpers, J. A. 1974. *Spiniger*, a new genus for imperfect states of Basidiomycetes. *Proc. K. Ned. Akad. Wet., Ser. C, Biol. Med. Sci.* 77(4):402-407.
From L. *spiniger*, spiny < *spina*, spine + *gero*, to carry on or with, referring to the conical spines or denticles on the surface of the terminal ampullae of the conidiophores, and on which the dacryoid conidia arise synchronously.

Spinulosphaeria Sivan.
Inc. sed., Inc. sed., Sordariomycetidae, Sordariomycetes, Pezizomycotina, Ascomycota, Fungi
Sivanesan, A. 1974. Two new genera of Coronophorales with descriptions and key. *Trans. Br. mycol. Soc.* 62(1):35-43.
From L. *spinula*, small thorn, spine + genus *Sphaeria* Haller, for the sharp, tooth-like spines on the ascomatal wall. Ascomata superficial, scattered to densely crowded,

on a dense subiculum of brown, septate, branched hyphae, black with a metallic iridescence, turbinate to clavate, ornamented with short, tooth-like often furcate spines, lacking a definite ostiole but have "Quellkorper" of mucilaginous cells at the region below apex. Asci clavate, long stalked, unitunicate, 8-spored, evanescent. Ascospores brown, one septate. Paraphyses filiform and hyaline. On dead wood and bark.

Spinulospora Deighton

Inc. sed., Inc. sed., Inc. sed., Inc. sed., Pezizomycotina, Ascomycota, Fungi

Deighton, F. C. 1973. *Sclerographiopsis* and *Spinulospora*, two new monotypic hyphomycetous genera from Sierra Leone. *Trans. Br. mycol. Soc.* 61(1):193-196.

From L. *spinula*, small thorn, spine + Gr. *sporá* > L. *spora*, spore, for the densely spiny conidia. Mycelium evanescent. Sporodochia hypophyllous, developing over rust sori and leaf surface, discrete, widely scattered or more or less gregarious, pale yellowish to pale greenish, thin-walled, of densely crowded, much branched hyphae divergent from base, constricted at septa, terminal cells each bearing a solitary terminal conidium. Conidia not caducous, mostly clavate-cylindric, sometimes ellipsoid, broadly rounded at apex, continuous, coarsely and densely spinulose except near extreme base where spines become fewer. Conidia germinate *in situ*, each apparently producing a single germ tube, smooth, septate, almost hyaline. On leaves of *Smilax kraussiana* in conjunction with *Puccinia kraussiana* Cooke and *Scolecobasidium acanthacearum* (Cooke) M. B. Ellis in Sierra Leone.

Spinulospora pucciniiphila Deighton: pale yellowish sporodochium, developed over the leaf surface, covered by conidiogenous cells and spiny, not caducous conidia, x 400; two liberated clavate-cylindric, spinulose conidia, that germinate *in situ*, x 600. On leaves of *Smilax kraussiana* (Sierra Leona) infested with the rust **Puccinia kraussiana** Cooke.

Spiralum J. L. Mulder

Inc. sed., Inc. sed., Inc. sed., Inc. sed., Pezizomycotina, Ascomycota, Fungi

Mulder, J. L. 1975. *Spiralum*, a new genus from Asia. *Trans. Br. mycol. Soc.* 65(3):517-521.

From Gr. *speíra* > L. *spira*, coil, for the spirally twisted conidiophores. Primary mycelium immersed in leaf tissue. Hyphae emerging through stomata, giving rise to secondary superficial mycelium, which forms a network or reticulum on which conidiophores are borne. Conidiophores dematiaceous, mostly in groups or rarely single, not in fascicles, unbranched, septate, thin to slightly thickened wall, spirally twisted, apical region swollen, verrucose, producing conidia singly, leaving a terminal scar pushed to one side by continued growth of conidiogenous cell, scars often conspicuous, slightly raised. Conidia hyaline, occasionally olivaceous, smooth, thin-walled, pluriseptate, cylindrical, straight, curved or helicoid. On leaves of *Ficus* in the Solomon Islands.

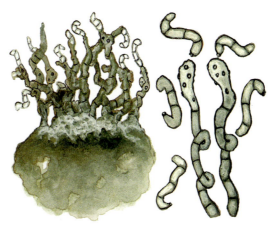

Spiralum helicosporum J. L. Mulder: primary mycelium, immersed in the leaf tissue, giving rise to the network of dark-colored secondary mycelium, x 200, that forms spirally twisted conidiophores, and multiseptate conidia, with a large hilum, x 600. On leaves of *Ficus copiosa*.

Spiroplana Voglmayr, et al.

Inc. sed., Inc. sed., Inc. sed., Inc. sed., Pezizomycotina, Ascomycota, Fungi

Voglmayr, H., et al. 2011. *Spiroplana centripeta* gen. & sp. nov., a leaf parasite of *Philadelphus* and *Deutzia* with a remarkable aeroaquatic conidium morphology. *Mycotaxon* 116(1):203-216.

From Gr. *speíra*, something wound or wrapped around > L. *spira*, coil + *planus*, flat, referring to the laterally flattened conidia made up of conidial filaments branching and coiling in one plane. Mycelium partly superficial, partly endoparasitic. Hyphae septate, branching. Conidiophores erect, mononematous, septate. Conidiogenous cells terminally integrated, holoblastic. Conidia irregularly globose to elongated, laterally flattened, formed by branched, tightly spirally interwoven, septate conidial filaments; primary conidial filament giving rise to centripetally growing, coiled filaments at inner side of coil. On *Philadelphus schrenkii* in Bongmyeong-ri, Chuncheon, Korea.

Sporidesmiella

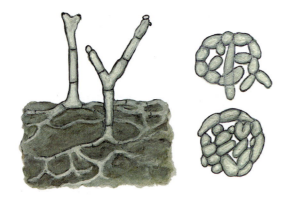

Spiroplana centripeta Voglmayr, et al.: parasitic mycelium on the host leaf, giving rise to simple or branched conidiophores, x 50; conidiogenous cells form globose to elongate, flattened spirally interwoven conidia, x 600. Parasitic on leaves of *Philadelphus schrenkii* and *Deutzia*, South Korea.

Sporidesmiella P. M. Kirk

Melanommataceae, Pleosporales, Pleosporomycetidae, Dothideomycetes, Pezizomycotina, Ascomycota, Fungi

Kirk, P. M. 1982. New or interesting microfungi VI. *Sporidesmiella* gen. nov. (Hyphomycetes). *Trans. Br. mycol. Soc.* 79(3):479-489.

From genus *Sporidesmium* Link + L. dim. suf. -*ella*. Mycelium partly superficial, partly immersed, of branched, septate, pale brown to brown hyphae. Conidiophores macronematous, mononematous, solitary, simple, straight or slightly flexuous, septate, brown to dark brown. Conidiogenous cells holoblastic, monoblastic, integrated, terminal, proliferating percurrently, rarely sympodially. Conidia acrogenous, solitary, dry, seceding schizolytically, cylindrical, narrowly clavate, obovoid to broadly obovoid or cuneiform, truncate at base, rounded or rarely coronate at apex, 1-5-distoseptate, cell lumina reduced, pale olivaceous to olivaceous-brown or brown. On dead twigs of *Ulex europaeus* in Surrey, U.K.

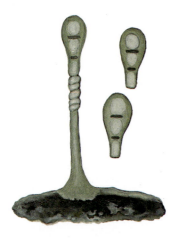

Sporidesmiella claviformis P. M. Kirk: dark brown conidiophore with proliferating percurrently conidiogenous cell, which gives rise to acrogenous, cuneiform, 1-5-distoseptate, cell lumina reduced, olivaceous conidia, x 650. On dead stem and twigs of *Rubus fruticosus* and dead branches of *Ulex europaeus*, Surrey, U.K.

Sporobolomyces Kluyver & C. B. Niel (syn. **Aessosporon** Van der Walt)

Inc. sed., Sporidiobolales, Inc. sed., Microbotryomycetes, Pucciniomycotina, Basidiomycota, Fungi

Kluyver, A. J., C. B. van Niel. 1924. Über Spiegelbilder erzeugende Hefenarten und die neue Hefengattung *Sporobolomyces*. *Centbl. Bakt. ParasitKde*, Abt. II, 63:1-20. From Gr. *spóros*, spore + *bállo* > L. *bolus*, to throw, hurl + Gr. *mýkes* > L. *myces*, fungus, because it actively liberates its spores (ballistospores). Vegetative reproduction by budding, pseudomycelium and true mycelium, or by ballistospores. Ballistospores forcefully ejected by isolated or mycelial cells. Sexual reproduction by production of teliospores that germinate to form metabasidium or promycelium that gives rise to basidiospores. Produces carotenoid pigments. In phyllosphere of plants, especially on leaves parasitized by rusts, in air and seawater.

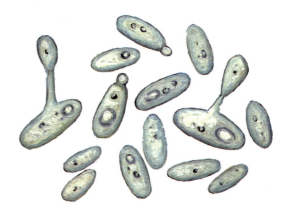

Sporobolomyces roseus Kluyver & C. B. Niel: vegetative yeast cells with ballistospores, x 1,900.

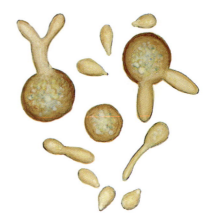

Sporobolomyces salmonicolor Van der Walt: teliospores originating basidiospores, in agar culture, x 1,500. Isolated in Netherlands.

Sporophormis Malloch & Cain—see **Aspergillus** P. Micheli ex Link

Trichocomaceae, Eurotiales, Eurotiomycetidae, Eurotiomycetes, Pezizomycotina, Ascomycota, Fungi

Sporothrix

Malloch, D., R. F. Cain. 1972. The Trichocomataceae: Ascomycetes with *Aspergillus*, *Paecilomyces* and *Penicillium* imperfect states. *Can. J. Bot.* 50:2613-2628.
From Gr. *sporá*, seed > L. *spora*, spore + Gr. *phórmos*, a woven basket, mat, referring to the loosely woven ascocarps bearing asci.

Sporormia De Not.
Sporormiaceae, Pleosporales, Pleosporomycetidae, Dothideomycetes, Pezizomycotina, Ascomycota, Fungi
De Notaris, G. 845. Micromycetes italici novi vel minus cogniti. *Mém. R. Accad. Sci. Torino*, Ser. 2, 10:342.
From Gr. *sporá*, *spóros*, spore + *órmos*, string, chain or necklace. Perithecia ostiolate. Asci bitunicate, cylindrical, eight ascospores in each ascus surrounded by a common mucilaginous covering. Ascospores multiseptate, 4-18 cells arranged in a row. Saprobic on dung.

Sporormia fimetaria (Rabenh.) De Not.: bitunicate ascus of this coprophilous fungus, containing eight multiseptate ascospores, which have 4-18 cells arranged in a row, like a necklace, x 1,200.

Sporormiella Ellis & Everh.
Sporormiaceae, Pleosporales, Pleosporomycetidae, Dothideomycetes, Pezizomycotina, Ascomycota, Fungi
Ellis, J. B., B. M. Everhart. 1892. *N. Amer. Pyren.* (Newfield):136.
From genus *Sporormia* de Not. + L. dim. suf. *-ella*, the first from Gr. *sporá*, *spóros*, spore + *órmos*, string, chain or necklace. Perithecia ostiolate. Asci cylindrical, bitunicate. Ascospores triseptate, dark brown, each surrounded by a mucilaginous covering, forcefully discharged through ascal apex. Coprophilous.

Spororminula Arx & Aa
Sporormiaceae, Pleosporales, Pleosporomycetidae, Dothideomycetes, Pezizomycotina, Ascomycota, Fungi
Arx, J. A. von, H. A. Van der Aa. 1987. *Spororminula tenerifae* gen. et sp. nov. *Trans. Br. mycol. Soc.* 89 (1):117-120.
From genus *Sporormia* De Not. + L. dim. suf. *-ula*, for a sporormia-like fungus. Ascomata uniloculate pseudothecia, immersed in substrate when young, becoming erumpent, subglobose to ovate-ampulliform, papillate, ostiolate, black. Pseudoparaphyses numerous, filamentous, hyaline, evanescent. Asci cylindrical, bitunicate, 8-spored. Ascospores cylindrical, long, phragmosporous, with a gelatinous sheath, triangular in cross-section, lacking a germ slit, brown, readily separating into part-spores. Isolated from rabbit dung, in Tenerife, Canary Islands.

Sporormiella capybarae (Speg.) S. I. Ahmed & Cain: cylindrical, bitunicate asci of this coprophilous fungus, with dark brown triseptate ascospores, each one with a germ slit and surrounded by a mucilaginous covering, x 400.

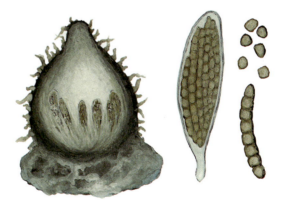

Spororminula tenerifae Arx & Aa: vertical section of a black, erumpent, papillate, uniloculate pseudothecium x 100, with bitunicate, 8-spored asci, with phragmosporus, lacking a germ slit, ascospores, x 270; one ascospore triangular in cross-section, which separates into part-spores, x 350. Isolated on wet paper in Petri dishes, from rabbit dung, Tenerife, Canary Islands.

Sporothrix Hektoen & C. F. Perkins
Ophiostomataceae, Ophiostomatales, Sordariomycetidae, Sordariomycetes, Pezizomycotina, Ascomycota, Fungi

Stachybotrys

Hektoen, L., C. F. Perkins. 1901. Refractory subcutaneous abscesses caused by *Sporothrix schenkii*, a new pathogenic fungus. *J. Exp. Med.* 5:80.

From Gr. *spóros*, spore + *thríx*, hair, in reference to the hyphae that produce the conidia, which arise more or less directly from the vegetative hyphae. Sporogenous cells denticulate. Conidia non-septate, borne on denticles in acropetal succession at apex of these cells. Saprobic in soil, on vegetable remains, also causes sporotrichosis in humans and higher animals; this is a localized subcutaneous mycosis acquired by traumatic implantation of the conidia.

Stachybotrys chartarum (Ehrenb.) S. Hughes: this cellulolytic saprobic fungus forms mycelial aggregations of hyphae which produce conidiophores that terminate in an apical cluster of 3-7 inflated phialides, which give rise to the dark-colored, oval, unicellular conidia, at times in chains, surrounded by a mucilaginous covering, x 500.

Sporothrix schenckii Hektoen & C. F. Perkins: this fungus produces denticulate sporogenous cells (saprobic state in soil plant remains), giving rise to the conidia (amerospores) which are borne on the denticles in acropetal succession on the apex of these cells, x 1,000. At right, the budding yeast cells (parasitic state), x 1,800. Isolated from a human patient.

Stachybotrys Corda (syn. **Ornatispora** K. D. Hyde, et al.)

Stachybotryaceae, Hypocreales, Hypocreomycetidae, Sordariomycetes, Pezizomycotina, Ascomycota, Fungi
Corda, A. C. J. 1837. *Icon. fung.* (Prague) 1:21.
From Gr. *stáchys*, ear (of grain) + *bótrys*, cluster. Ascomata superficial, globose, collabent when dry, black, coriaceous, lacking or covered in numerous setae, papillate. Papilla short, beak-like, black, shiny, periphysate. Sterile tissue filiform, aseptate, flexuose, deliquescing in dried material. Asci 8-spored, clavate, pedicellate, thin-walled, unitunicate, lacking an apical apparatus, deliquescent at maturity. Ascospores 2-3-seriate, ellipsoidal, 1-septate, hyaline, verrucose and surrounded by a mucilaginous sheath. Conidiophores in an apical cluster of 3-7 inflated phialides that produce conidia, at times in chains, or more commonly, in gloeoid heads. Conidia globose oval or cylindrical, unicellular, pigmented. Common in soil as a cellulolytic saprobe; also known to produce satratoxins, mycotoxins that affect humans and cattle, causing stachybotryotoxicosis. Cause of indoor mold or black mold that develops on gypsum board and other areas that have become wet repeatedly. Inhaling the spores can result in serious illness.

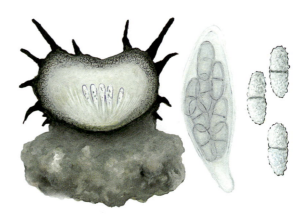

Stachybotrys punctatus (Dulym., et al.) Yong Wang bis, et al. (syn. **Ornatispora punctata** Dulym., et al.): longitudinal section of an ascoma, on dead leaves of *Pandanus rigidifolius* and *P. barkleyi*, x 220; ascus with eight, one-septate ascospores, x 750; free, verrucose ascospores, x 780. Collected in Mauritius.

Staninwardia B. Sutton

Inc. sed., Chaetothyriales, Chaetothyriomycetidae, Eurotiomycetes, Pezizomycotina, Ascomycota, Fungi
Sutton, B. C. 1971. *Staninwardia* gen. nov. (Melanconiales) on *Eucalyptus*. *Trans. Br. mycol. Soc.* 57(3):539-542.
Named in honor of the English botanist *Stanley F. Inward* + L. suf. *-ia*. Immersed mycelium sparse, of pale brown septate hyphae. Fructifications acervular, formed in epidermis and hypodermis, erumpent, with basal wall formed of pale brown, smooth-walled pseudoparenchymatic cells. Conidia holothallospores, catenate, basipetal, 1-septate, pale brown, verruculose, both ends truncate except for terminal conidium which is obtuse at apex, formed in a mucilaginous sheath. On living leaves of *Eucalyptus* in Nouvelle Decouverte, Mauritius.

Starkeyomyces Agnihothr.—see **Myrothecium** Tode
Inc. sed., Hypocreales, Hypocreomycetidae, Sordariomycetes, Pezizomycotina, Ascomycota, Fungi

Stephanotheca

Agnihothrudu, V. 1956. Fungi isolated from the rhizosphere II. *Starkeyomyces*, a new genus of the Tuberculariaceae. *J. Indian bot. Soc.* 35:40.

Dedicated to the American microbiologist *R. L. Starkey*, for his outstanding studies on fungi of the rhizosphere + L. *-myces* < Gr. *mýkes*, fungus. Sporodochia brightly colored, covered by a fertile layer of irregularly branched conidiophores. Conidia hyaline with a membranous appendage at apex. In rhizosphere of plants.

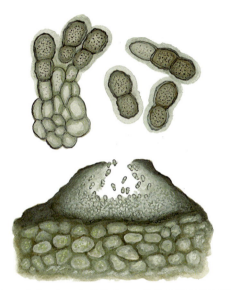

Staninwardia breviuscula B. Sutton: vertical section of an erumpent acervulus, in living leaves of *Eucalyptus* in Nouvelle Decourvete, Mauritius, x 350; conidiogenous cells with catenate, 1-2-septate, pale brown, verruculose conidia (holothallospores), formed in a mucilaginous sheath, x 1,000.

Starkeyomyces koorchalomoides Agnihothr.: black, shiny sporodochium, covered by a fertile layer of conidiophores, x 20; hyaline, unicellular conidia with characteristic membranous appendage on the apical part, x 1,000. Isolated in Tamil Nadu, India.

Starmerella C. A. Rosa & Lachance
Inc. sed., Saccharomycetales, Saccharomycetidae, Saccharomycetes, Saccharomycotina, Ascomycota, Fungi
Rosa, C. A., M. A. Lachance. 1998. The yeast genus *Starmerella* gen. nov. and *Starmerella bombicola* sp. nov., the teleomorph of *Candida bombicola* (Spencer, Gorin & Tullock) Meyer & Yarrow. *Int. J. Syst. Bacteriol.* 48 (4):1413-1417.

Named in honor of the American microbiologist *William T. Starmer* + dim. suf. *-ella*, in recognition of his major contributions to the ecology and evolution of yeasts associated with plants and insects. Vegetative reproduction by multilateral budding. Vegetative cells ovoid to ellipsoidal. Asci evanescent, conjugated, forming one spheroidal ascospore with a convoluted surface and a membranous basal ledge, released terminally and tend to agglutinate. Pseudomycelium or true mycelium not formed. Fermentative. Isolated from hedge bindweed, *Calystegia sepium*.

Stegophora Syd. & P. Syd.
Sydowiellaceae, Diaporthales, Sordariomycetidae, Sordariomycetes, Pezizomycotina, Ascomycota, Fungi
Sydow, H., P. Sydow. 1916. Weitere Diagnosen neuer philippinischer Pilze. *Annls mycol.* 14(5):353-375.

From Gr. *stégos*, a covering, roof + *phóra*, bearing, referring to the clypeus covering the perithecia. Ascomata ostiolate perithecia formed in host tissues beneath a black, crustose clypeus in upper epidermis; perithecia upright, immersed, often several clustered beneath a common clypeus, ostioles short, erumpent through lower epidermis. Ascomatal wall brown, of flattened cells. Asci unitunicate, ellipsoid, with an apical ring. Ascospores hyaline, 2-celled, with septum near base, ellipsoid to ovoid or straight, with rounded ends. On leaves of deciduous trees.

Stemphylium Wallr. (syn. **Pleospora** Rabenh. ex Ces. & De Not.)
Pleosporaceae, Pleosporales, Pleosporomycetidae, Dothideomycetes, Pezizomycotina, Ascomycota, Fungi
Wallroth, K. F. W., 1833. *Fl. crypt. Germ.* (Nüremberg) 2:300.

From Gr. *stémphyila*, genit. *stémphylon*, sediment of wine or olive oil, or < *stemphylídes*, black olives + L. dim. suf. *-ium*, for the dark, almost black color of the conidiophores and conidia. Pseudothecia with bitunicate, cylindrical asci and pseudoparaphyses. Ascospores muriform, pigmented, generally dark brown. Conidiophores proliferating through conidial scars giving rise to new conidia through pores in wall. Conidia dark, muriform, of variable shape, with a constriction at median septum. Causing leaf spots on herbaceous stems of plants, also a saprobe frequently isolated from soil.

Stephanotheca Syd. & P. Syd.
Elsinoaceae, Myriangiales, Dothideomycetidae, Dothideomycetes, Pezizomycotina, Ascomycota, Fungi
Sydow, H., P. Sydow. 1914. *Philipp. J. Sci.*, C, Bot. 9 (2):178.

From Gr. *stéphanos*, or *stéphane*, crown, circumference + *théke*, case, box. Pseudothecia apothecioid, superficial, with a thin, flat, radial shield. Asci ovoid, embedded in parenchymatous tissue. Ascospores dictyosporous, hyaline.

Stereocaulon

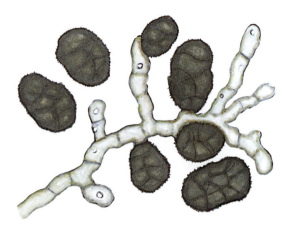

Stemphylium botryosum Wallr. (syn. **Pleospora tarda** E. G. Simmons): conidiophore and polytretic conidiogenous cells that give rise to muriform, verruculose, dark conidia (porospores), x 600.

Stemphylium vesicarium (Wallr.) E. G. Simmons (syn. **Pleospora herbarum** (Pers.) Rabenh.): vertical section of a pseudothecium, showing the cylindrical asci and muriform, dark brown ascospores, in tissues of an herbaceous stem, x 100; free ascospore, x 1,000, Collected in Germany.

Stereocaulon Hoffm.

Stereocaulaceae, Lecanorales, Lecanoromycetidae, Lecanoromycetes, Pezizomycotina, Ascomycota, Fungi
Hoffman, G. F. 1796. *Deutschl. Fl.*, Zweiter Theil (Erlangen):128.
From Gr. *stereós*, solid, hard, firm + *kaulós*, a plant stem. Thallus inconspicuous with secondary thallus of pseudopodetia, branches or solid stems, much branched, twisted and irregular. Apothecia lateral or terminal, small or medium, with a flat or convex disc, reddish-brown or black. On soil or rocks.

Stereum Hill ex Pers.

Stereaceae, Russulales, Inc. sed., Agaricomycetes, Agaricomycotina, Basidiomycota, Fungi
Persoon, C. H. 1794. Neuer Versuch einer systematischen Einteilung der Schwämme. *Neues Mag. Bot.* 1:110.
From Gr. *stereón*, firm, hard, solid, for the consistency of the fructification. Fructifications coriaceous, resupinate, pileate, bracket-shaped or infundibuliform, sessile or with a short central or lateral stipe with smooth hymenium. Spore print white. On wood or soil. *Stereum purpureum* Fr. has purple-brown resupinate or imbricate basidiocarps. It causes wood decay in living and fallen trees, e.g., willow, poplar, beech, birch, elm and occasionally conifers, as well as the disease called silver leaf of fruit trees, such as plum, apple and peach.

Stereocaulon ramulosum Raeusch.: solid branches of pseudopodetia (secondary thallus), developed from the inconspicuous primary thallus, with black, lateral or terminal apothecia, on soil, x 15.

Stereum ostrea (Blume & T. Nees) Fr.: coriaceous, resupinate, zonate, bracket-shaped fructifications, on host wood, x 2.

Sticta (Schreb.) Ach.

Lobariaceae, Peltigerales, Lecanoromycetidae, Lecanoromycetes, Pezizomycotina, Ascomycota, Fungi
Acharius, E. 1803. *Methodus qua Omnes Detectos Lichenes Secundum Organa Carpomorpha ad Genera, Species et Varietates Redigere atque Observationibus Illustrare Tentavit Erik Acharius*:275.
From L. *stictus*, punctate < Gr. *stiktós*, marked with dots, to brand. Thallus foliose, brown, broadly lobulate with

large, white, depressed pores on lower surface. Photobiont cyanophyceae. On bases of trees or on mosses that grow on rocks and trees.

Sticta latifrons A. Rich.: foliose, brown, broadly lobulate thallus, with large, white, depressed pores dispersed on the lower surface of the thallus, x 1.

Stictis Pers.

Stictidaceae, Ostropales, Ostropomycetidae, Lecanoromycetes, Pezizomycotina, Ascomycota, Fungi
Persoon, C. H. 1800. *Observ. mycol.* (Lipsiae) 2:73.
From Gr. *stiktós*, marked, distinguished with dots, mottled, from *stízo*, sting, mark, stigmatize, in reference to the ascocarps which this fungus forms partially or completely immersed in the substrate. Ascocarps apothecioid, at maturity opening widely exposing a wide, white margin that frequently splits into lobes. Asci cylindrical with a thickened apex and a small pore through which it violently shoots filiform, multiseptate ascospores. On leaves of plants.

Stictis stellata Wallr.: mature, lignicolous ascocarps on a plant twig, which open widely and expose the white margin that splits into lobes, x 10.

Stigmatomyces H. Karst.

Laboulbeniaceae, Laboulbeniales, Laboulbeniomycetidae, Laboulbeniomycetes, Pezizomycotina, Ascomycota, Fungi
Karsten, H. 1869. *Chemismus Pfl.-Zelle* (Viena):1-90.
From Gr. *stigma*, genit. *stígmatos*, mark, sign, spot + *mýkes* > L. *myces*, fungus, due to the granulose marks, nodulose borders or transverse lines that the perithecium has on its external wall. Perithecia with basal cells with well developed walls and an unbranched appendage with a long unilateral series of phialides. Black cell basal reduced in size. Parasitizes free-living flies as well as associated acarids and coleopterans.

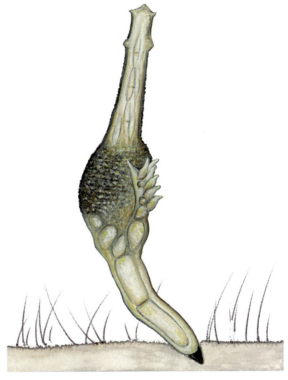

Stigmatomyces psilopae Thaxt.: receptacle of the thallus of this fungus, which is a parasite of diverse species of insects, showing the granulose marks of the perithecium, and an unbranched appendage provided with a long unilateral series of phialides, x 250.

Stilbella Lindau

Inc. sed., Inc. sed., Hypocreomycetidae, Sordariomycetes, Pezizomycotina, Ascomycota, Fungi
Lindau, G. 1900. Fungi imperfecti. *In*: A. Engler & K. Prantl, *Nat. Pflanzenfam.*, Teil. 1 (Leipzig) 1:489-491.
From genus *Stilbum* Tode < L. *stilbus*, sharp < Gr. *stilbo*, to shine + L. dim. suf. *-ella*, for the hyaline or brightly colored synnemata with mucoid heads that are also brightly colored. In soil and on insects.

Stilbocrea Pat.

Bionectriaceae, Hypocreales, Hypocreomycetidae, Sordariomycetes, Pezizomycotina, Ascomycota, Fungi
Patouillard, N. T. 1900. Champignons de la Guadaloupe. *Bull. Soc. mycol. Fr.* 16(4):186.
From Gr. *stilbo*, shining + *kreas* > L. *creas*, flesh, an apparent reference to the brightly colored stroma. Stromata of a hypocreoid base with several erect stilbum-like outgrowths, fleshy, bright colored. Perithecia globose or ovate, immersed or with necks slightly protruding. Asci 8-spored. Ascospores hyaline or subhyaline, 1-septate, smooth or rough.

Stipella

Stilbocrea macrostoma (Berk. & M. A. Curtis) Höhn. (syn. **Hypocreopsis macrostoma** (Berk. & M. A. Curtis) E. Müll.): macroscopic details of the bionectriaceous ascomata in a natural substrate, on bark of leaf plants in Taiwan, x 12; ascus with eight, unicellular, hyaline ascospores, x 820.

Stirtonia latispora Seavey & J. Seavey: foliicolous, crustous, greenish thallus, with carbonaceous pseudothecia, ascostromata called hysterothecia, x 20.

Stipella L. Léger & Gauthier—see **Stypomyces** Doweld
Legeriomycetaceae, Harpellales, Inc. sed., Inc. sed., Kickxellomycotina, Zygomycota, Fungi
Léger, L., M. Gauthier. 1932. Endomycetes nouveaux des larves aquatiques d'insectes. *C. r. hebd. Séanc. Acad. Sci., Paris* 194(26):2262-2265.
From L. *stipes*, stake, trunk, foot, peduncle + dim. suf. *-ella*, i.e., little foot, referring to the basal portion of the axial hypha of the thallus, which secretes an adhesive mucilage to attach it to the proctodeum of the dipteran insect larvae that serve as hosts.

Stipitochaete Ryvarden—see **Hymenochaete** Lév.
Hymenochaetaceae, Hymenochaetales, Inc. sed., Agaricomycetes, Agaricomycotina, Basidiomycota, Fungi
Ryvarden, L. 1985. *Stipitochaete* gen. nov. (Hymenochaetaceae, Basidiomycotina). *Trans. Br. mycol. Soc.* 85 (3):535-539.
From L. *stipes*, genit. *stipitis*, stipe + NL. *chaeta*, bristle, referring to a stipitate fungus with bristles (setae).

Stirtonia A. L. Sm.
Arthoniaceae, Arthoniales, Arthoniomycetidae, Arthoniomycetes, Pezizomycotina, Ascomycota, Fungi
Smith, A. L. 1926. Cryptotheciaceae, a family of primitive Lichens. *Trans. Br. mycol. Soc.* 11(3-4):195.
Dedicated to the Scottish physician and lichenologist *James Stirton* + L. suf. *-ia*, which denotes pertaining to. Thallus crustous. Fructifications pseudothecia lacking an excipulum, carbonaceous, hysterothecia. Asci aggregated in fertile areas differentiated from thallus. On living leaves.

Stomiopeltis Theiss.
Micropeltidaceae, Microthyriales, Inc. sed., Dothideomycetes, Pezizomycotina, Ascomycota, Fungi
Theissen, F. 1914. De Hemisphaerialibus Notae Supplende. *Brotéria, sér. bot.* 12(1):73-96.
From Gr. *stóma*, mouth, opening + *pélte*, a small shield, a shield-shaped ascomata with a pore. Ascomata dimidiate-scutate, ostiolate thyriothecia, uniloculate or multiloculate, formed on surface of mycelium. Mycelium brown, septate, reticulately branched, superficial, appressed to cuticle of host, not penetrating host cells. Shield of thyriothecium brown, shallow-convex, rounded or somewhat irregular in top view, with a central ostiole, covering a thin layer of hyaline hyphae giving rise to asci. Wall of shield of a thin layer of laterally united, pigmented hyphae that radiate outward; hyphae of wall sinuous, lobed and branched, lighter near margin of shield. Centrum of a single layer of asci among short, filiform pseudoparaphyses. Asci clavate to ovoid or subglobose, bitunicate, arranged around periphery of locule, with apices inclined toward ostiole, 8-spored. Ascospores hyaline, 2-celled, septum median to supramedian, narrowly obovoid, smooth. On living leaves and stems of herbaceous plants.

Streptotinia Whetzel
Sclerotiniaceae, Helotiales, Leotiomycetidae, Leotiomycetes, Pezizomycotina, Ascomycota, Fungi
Whetzel, H. H. 1945. A synopsis of the genera and species of the Sclerotiniaceae, a family of stromatic inoperculate Discomycetes. *Mycologia* 37(6):648-714.
From Gr. *streptós*, twisted + L. des. *-inia*, nature of something, for the twisted conidiophore branches. Stromata small, black sclerotia, flattened or hemispherical, firmly attached to substratum, flat or flattish on attached surface. Conidiophores with branches strikingly and characteristically twisted tightly. Conidia smooth, hyaline or nearly so.

Stropharia

Stomiopeltis betulae J. P. Ellis: shield-shaped, brown thyriothecia, with clear ostioles, formed on the surface of the mycelium, on dead stems of *Betula* and *Sorbus*, in England, x 220; bitunicate ascus with bicellular ascospores; x 900, free hyaline, 2-celled ascospores, x 1,200.

Strigula Fr. (syn. **Shanoria** Subram. & K. Ramakr.)
Strigulaceae, Strigulales, Dothideomycetidae, Dothideomycetes, Pezizomycotina, Ascomycota, Fungi
Fries, E. M. 1823. *Syst. mycol.* (Lundae) 2(2):535.
From Gr. *strix*, genit. *strigós*, groove, canal > L. *striga*, furrow, groove, crease, ray, or *stría*, stripe, at times formed from hard, closely united hairs, from bristles or scales, often of unequal length, + L. dim. suf. *-ula*. Thallus finely striate-crustous. Perithecia small or minute, more or less immersed in thallus, with a dimidiate wall and inconspicuous ostiole. Stromata black, carbonaceous, with one or more loculi, lined with conidiophores, at maturity dehiscing by irregular, longitudinal rupture. Conidiophores cylindrical or clavate. Conidia hyaline, continuous, with one subapical appendage at each end. On leaves of tropical and subtropical plants in association with algae of the genera *Cephaleuros*, *Heterothallus*, and *Phyllactidium*.

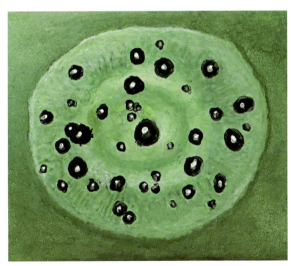

Strigula smaragdula Fr.: striate, crustose, greenish thallus, with minute, ostiolate perithecia, x 10, on leaves of *Loranthus ligustroides*, India and Nepal.

Strobilomyces Berk.
Boletaceae, Boletales, Agaricomycetidae, Agaricomycetes, Agaricomycotina, Basidiomycota, Fungi
Berkeley, M. J. 1851. Decades of fungi. *Hooker's J. Bot. Kew Gard. Misc.* 3:78.
From L. *strobilus*, a pine cone + suf. *-myces* < Gr. *mýkes*, fungus, referring to the scales on the pileus, similar to those of a pine cone. Basidiomata usually dark, context light, stains pink when exposed. Hymenophore tubular. Ectomycorrhizal in temperate forests, edible.

Strobilomyces strobilaceus (Scop.) Berk.: mature, dark basidiome of this ectomycorrhizal fungus, growing on soil in temperate forest, with the scaly pileus and tubular hymenophore, x 0.5.

Stromatinia (Boud.) Boud.
Sclerotiniaceae, Helotiales, Leotiomycetidae, Leotiomycetes, Pezizomycotina, Ascomycota, Fungi
Boudier, J. L. E. B. 1907. *Hist. Class. Discom. Eur.* (Paris):108.
From Gr. *strōma*, genit. *strōmatos*, bed, rug, mattress, cushion + L. des. *-inia* < Gr. suf. *-ínos*, material, nature or source of something, due to the formation of the apothecia on stromata that develop, in the manner of layers or cloaks, in the tissues of the host. Parasitic on cultivated plants, causes dry rot of bulbs, rhizomes and corms, i.e., gladiolus and narcissus.

Stropharia (Fr.) Quél.
Hymenogastraceae, Agaricales, Agaricomycetidae, Agaricomycetes, Agaricomycotina, Basidiomycota, Fungi
Quélet, L. 1872. Les champignons du Jura et les Voges. *Mém. Soc. Émul. Montbéliard*, Sér. 2, 5:141.
From L. *strophium*, girdle or chest sash, from Gr. *stróphos*, belt, sword, from *strépho*, to twist + L. suf. *-aria*, which

Strumella

indicates possession, for the presence of a subapical ring on the stipe. Fructifications solitary or cespitose, small to medium. Pileus fleshy, regular, continuous with stipe, ring persistent or fugaceous; gills adnate; spores purple to brownish-gray, rarely blackish. Common in forests and grasslands. Includes both edible and toxic species.

From L. *strumella*, dim. of *struma*, small tumor or protuberance, referring to the canker that it causes in trees. Sporodochia verruciform. Bulbils dark, solitary, and give rise to helicospores. Saprobic on wood.

Strumella griseola Höhn.: greenish, verruciform sporodochia in the bark of a tree host, x 90; it produces hyaline helicospores, x 1,000.

Stylopage Drechsler
Zoopagaceae, Zoopagales, Inc. sed., Inc. sed., Zoopagomycotina, Zygomycota, Fungi
Drechsler, C. 1935. Some non-catenulate conidial phycomycetes preying on terricolous amoebae. *Mycologia* 27(2):197-198.
From Gr. *stýlos*, column, punch + *páge*, loop, snare, because the hyphae of this predatory fungus of terricolous amoebae develop internal haustoria and distinctive external sporangiophores. Sporangiophores long, produce a single unispored merosporangium on apex or a succession of merosporangia on a sympodial sporangiophore.

Stromatinia cryptomeriae Kubono & Hosoya: brown, stipitate apothecia, formed on stromata that develop in the branches of *Cryptomeria japonica*, the Japanese cedar, x 1.5.

Stropharia aeruginosa (Curtis) Quél.: bluish basidiome, with subapical ring on the stipe, common in forest grass and in woodland, Great Britain, x 0.5.

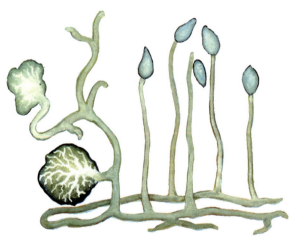

Stylopage rhynchospora Drechsler: hyphae, with haustoria inside the cells of terricolous amoebae, from which develop external sporangiophores with unispored merosporangium, x 200.

Strumella Fr.
Sarcosomataceae, Pezizales, Pezizomycetidae, Pezizomycetes, Pezizomycotina, Ascomycota, Fungi
Fries, E. M. 1849. *Summa veg. Scand.* (Stockholm):482.

Stypomyces Doweld (syn. **Stipella** L. Léger & Gauthier)
Legeriomycetaceae, Harpellales, Inc. sed., Inc. sed., Kickxellomycotina, Zygomycota, Fungi

Doweld A. B. 2014. *Index Fungorum* 112:1.

From Gr. *stypo*, a stump, stem + *mýkes* > L. *myces*, fungus, referring to the mode of living of this fungus. Endocommensal within the hindguts of *Simulium equinum* larvae. Thallus consists of a long, unbranched non-septate, multinucleate axial hyphae, with shorter, unbranched non-septate lateral hyphae. The axial hypha is terminated basally by a dichotomously branched, cellular holdfast, closely applied to, though not penetrating, the cuticle. Every holdfast was surrounded by a mucilaginous sheath. Spore production occurs basipetally and in an unilateral series along the hyphae; each generative cell producing a conidium at its distal end. Zygospore production appears to be limited to those thalli inhabiting the penultimate and ultimate larval stages.

Suillus luteus (L.) Roussel: terricolous, reddish-brown basidioma, with the viscous or glutinous character of the pileus, on soil of conifer forest, Europe, x 0.5.

Stypomyces vigilans (L. Léger & M. Gauthier) Doweld (syn. **Stipella vigilans** L. Léger & M. Gauthier): axial hypha at the base of the branching thallus, which produce trichospores, x 150. This thallus secretes an adhesive mucilage to attach to the proctodeum of *Simulium*, diptera insects.

Suillus Gray
Suillaceae, Boletales, Agaricomycetidae, Agaricomycetes, Agaricomycotina, Basidiomycota, Fungi
Gray, S. F. 1821. *Nat. Arr. Brit. Pl.* (London) 1:646.

From L. *sus*, *suis*, pig, from this *suillus*, pertaining to the pig, in reference to its flavor, or perhaps to the viscous or glutinous character of the pileus. They were called "porcien fungi" by the medieval ancients, such as Vidus (1540). Fructifications gregarious, medium-sized, with viscous pileus, generally smooth, reddish-brown or yellowish, with stipe short or medium length, whitish or yellow-grayish, more or less granulose-verrucose. Common on soil among fallen leaves in conifer forests and oak woods in temperate forests. Mycorrhizal; none toxic.

Sutorius Halling, et al.
Boletaceae, Boletales, Agaricomycetidae, Agaricomycetes, Agaricomycotina, Basidiomycota, Fungi
Halling, R.E., et al. 2012. *Sutorius*: a new genus for *Boletus eximius*. *Mycologia* 104(4):951-961.

From L. *sutorius*, shoemaker, the profession of Ch. C. Frost, a Vermont shoemaker, who described *Boletus robustus* from Eastern North America; later changed by C.H. Peck to *Boletus eximius*, which is the basis of the new genus *Sutorius*. Pileus dry, viscous or humid, fuscolilaceo-brunneous or violaceo-brunneous. Hymenophore adnexed, vinaceo-brunneous. Stipe dry, squamulose. Spore print ruber to brunneous, spores smooth, inamyloid, oblong. Grows among litter on soil in forests associated with *Fagus*, *Quercus* and *Tsuga*, in North America, Eastern Canada to Georgia, Wisconsin, (U.S.A.), Costa Rica and Indonesia. Possibly Japan, China and Guyana.

Syncephalastrum J. Schröt.
Syncephalastraceae, Mucorales, Inc. sed., Inc. sed., Mucoromycotina, Zygomycota, Fungi
Schröter, J. 1886. Mucorineae. *In*: J. F. Cohn, *Krypt.-Fl. Schlesien* (Breslau) 3.1(9-16):217.

From Gr. *sýn*, with, together + *kephalé*, head + L. *astrum*, astro, star, i.e., with stellate heads, in reference to the globose or ovoid vesicles on which are borne radially cylindrical sporangiola or merosporangia. Saprobic in soil, decomposing vegetation and dung.

Syncephalis Tiegh. & Le Monn.
Piptocephalidaceae, Zoopagales, Inc. sed., Inc. sed., Zoopagomycotina, Zygomycota, Fungi
Tieghem, P. van, G. Le Monnier. 1873. Recherches sur les mucorinées. *Annls Sci. Nat., Bot.*, sér. 5, 17:372.

Synchytrium

From Gr. *sýn*, with, together + *kephalé*, head + L. suf. *-alis*, belonging to or relative to, referring to the vesicles or heads that bear the sporangiola, which are formed on the apices of clavate, simple or sparsely branched sporangiophores. Mycoparasite of Mucorales.

Synchytrium endobioticum (Schilb.) Percival: group of sporangia or sori, endoparasitic on cells of various higher plants, x 450.

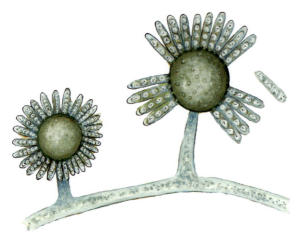

Synchytrium longispinosum (Couch) Karling (syn. **Micromyces longispinosus** Couch): endoparasitic thalli inside a cell of the alga *Spirogyra*, x 250.

Syncephalastrum racemosum Cohn ex J. Schröt.: sporangiophores of this saprobic fungus on rice and bread, Poland, with stellate heads of globose or ovoid vesicles, on which are borne radially cylindrical sporangiola or merosporangia, over all the surface, x 400.

Syzygites Ehrenb.

Inc. sed., Mucorales, Inc. sed., Inc. sed., Mucoromycotina, Zygomycota, Fungi

Ehrenberg, C. G. 1818. *Sylvae mycologicae berolinenses.* (Berlin):25.

From Gr. *sy-*, together, with, united + *zygíe*, coupling, nuptial union + suf. *-ítes*, which indicates a close connection. Zygosporangia borne between opposed suspensors almost equal in shape and size. Common, saprobic on decomposing fleshy fruiting bodies of basidiomycetes.

Syncephalis nodosa Tiegh.: clavate sporangiophore of this mycoparasite of Mucorales, which forms on the apex a vesicle or head that bear the sporangiola, x 600.

Syzygites megalocarpus Ehrenb.: zygosporangia between suspensors that are opposed and almost equal in shape and size, x 60.

Synchytrium de Bary & Woronin

Synchytriaceae, Chytridiales, Chytridiomycetidae, Chytridiomycetes, Inc. sed., Chytridiomycota, Fungi
de Bary, A., M. Woronin. 1863. Beitrag zur Kenntniss der Chytridieen. *Verh. Naturf. Ges. Freiburg* 3(2):46.
From Gr. *sýn*, together, a prep. used to give the idea of solidarity or concrescence + *chytrís*, kettle, receptacle, which in this case corresponds to the sporangium + L. dim. suf. *-ium*. Sporangia or sori in groups. Endoparasitic on higher plants.

Syzygospora G.W. Martin

Carcinomycetaceae, Tremellales, Inc. sed., Tremellomycetes, Agaricomycotina, Basidiomycota, Fungi
Martin, G.W. 1937. A new type of heterobasidiomycete. *J. Wash. Acad. Sci.* 27:112-114.
From Gr. *sýzygos*, yoked together, joined in pairs + *sporá*, spore, for the fused basidiospores. Basidiocarps sessile, gelatinous; hymenium covering exposed surface; basidia bluntly clavate, transversely septate into two cells, each cell producing a single basidiospore on a short sterigma; basidiospores fusing in pairs before detachment.

T

Tricholomopsis humboldtii

T

Taiwanascus Sivan. & H. S. Chang
Niessliaceae, Hypocreales, Hypocreomycetidae, Sordariomycetes, Pezizomycotina, Ascomycota, Fungi
Sivanesan, A., H. S. Chang. 1997. A lignicolous ascomycete, *Taiwanascus tetrasporus* gen. et sp. nov., and a new family, Taiwanascaceae. *Mycol. Res.* 101(2):176-178.
From *Taiwan*, the type country + L. *ascus*, ascus < Gr. *askós*, wine skin. Ascomata cleistothecia, appendaged, superficial, globose. Appendages aseptate, brown, thick-walled, straight, more or less dichotomously branched at apex. Peridium thin-walled, of a single layer of pseudoparenchymatous cells. Hamathecium absent. Asci broadly cylindrical, long stalked, unitunicate, 4-spored, deliquescent. Ascospores hyaline, aseptate to 1-septate, filiform, parallel in ascus. On angiospermous wood in Taipei, Taiwan.

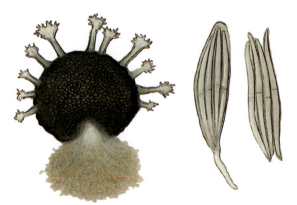

Taiwanascus tetrasporus Sivan. & H. S. Chang: globose cleistothecium, x 134, with dichotomously branched appendages at the apex, and, at the base, open peridium to liberate the asci; pedicellate ascus with four filiform ascospores, x 1,400; 1-septate ascospores, x 1,500. On dead wood from an unidentified angiosperm in Taipei, Taiwan.

Taiwanoporia T. T. Chang & W. N. Chou
Inc. sed., Inc. sed., Agaricomycetidae, Agaricomycetes, Agaricomycotina, Basidiomycota, Fungi
Chang, T. T., W. N. Chou. 2003. *Taiwanoporia*, a new aphyllophorean gen. *Mycologia* 95(6):1215-1217.
From the country *Taiwan* + genus *Poria* Pers., referring to the type locality. Basidiomata annual, resupinate, effused-reflexed to pileate, soft when fresh, whitish to creamy. Pores angular to rounded. Hyphal system monomitic; generative hyphae with simple septa (clampless). Basidiospores tear-shaped to subglobose, smooth, hyaline, amyloid. On decaying wood of *Trochodendron aralioides*.

Talaromyces Benjamin
Trichocomaceae, Eurotiales, Eurotiomycetidae, Eurotiomycetes, Pezizomycotina, Ascomycota, Fungi
Benjamin, C. R. 1955. Ascocarps of *Aspergillus* and *Penicillium*. *Mycologia* 47(5):681-685.
From Gr. *tálaros*, *tálaron*, basket, small basket + *mýkes* > L. *myces*, fungus. Gymnothecia with peripheral hyphae loosely interwoven, forming a small basket-like structure around asci. Conidiophores with conidia, penicillium-like. Common fungi of soil and other diverse habitats.

Talaromyces flavus (Klöcker) Stolk & Samson: basket-like gymnothecium, covered by loose interwoven hyphae, x 300; tuberculate, ellipsoidal ascospores, x 3,600.

Tandonea M. D. Mehotra
Inc. sed., Inc. sed., Inc. sed., Inc. sed., Pezizomycotina, Ascomycota, Fungi
Mehotra, M. D. 1991. *Tandonea*, a new Coelomycete gen. causing leaf spotting and blight in *Eubucklandia populnea* from India. *Mycol. Res.* 95(9):1074-1076.
Named in honor of the Indian plant pathologist *R. N. Tandon* + L. des. -ia. Mycelium immersed, branched, septate, hyaline to pale brown. Conidiomata pycnidial, immersed,

subglobose to globose, dark brown, thin-walled. Ostiole single, central, circular. Paraphyses straight to flexuous, 1-2-septate. Conidiogenous cells phialidic, integrated or discrete, determinate, hyaline, straight to slightly curved, cylindrical smooth, aperture apical. Conidia holoblastic, hyaline, guttulate, aseptate, ellipsoidal to acerose. On living leaves in Darjeeling, West Bengal, India.

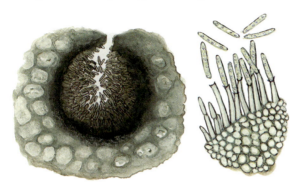

Tandonea dargentiana M. D. Mehrotra: vertical section of a pycnidial conidioma, with a central ostiole and conidia, in living host leaf of *Eubucklandia populnea*, in Darjeeling, West Bengal, India, x 136; conidiogenous cells with cylindrical, unicellular conidia, x 500.

Tapesia (Pers.) Fuckel

Inc. sed., Helotiales, Leotiomycetidae, Leotiomycetes, Pezizomycotina, Ascomycota, Fungi

Fuckel, L. 1870. Symbolae Mycologicae. Beiträge zur Kenntniss der Reinischen Pilze. *Jb. nassau. Ver. Naturk.* 23-24:1-456.

From L. *tapete* > NL. *tapesium*, carpet, referring to the thick, tomentose subiculum. Apothecia superficial, seated on a conspicuous web of white or dark-colored mycelium, minute, soft, waxy, scutellate or discoid, excipulum dark. Hymenium often at first light-colored, becoming dark. Asci cylindric or clavate, 4-8-spored. Ascospores ellipsoid, fusoid or elongated, never or rarely septate, hyaline. Paraphyses filiform. On plant materials.

Taphridium Lagerh. & Juel

Taphrinaceae, Taphrinales, Taphrinomycetidae, Taphrinomycetes, Taphrinomycotina, Ascomycota, Fungi

Lagerheim, G., H. O. Juel. 1902. *Bih. K. svenska VetenskAkad. Handl.*, Afd. 3 27 (no. 16):7.

From Gr. *taphría*, act of opening a pit or trench < *táphre*, genit. *táphros*, grave + L. dim. suf. *-idium*. Mycelium with chlamydospores and synasci or compound asci. Parasitic on umbelliferous plants.

Taphrina Fr.

Taphrinaceae, Taphrinales, Taphrinomycetidae, Taphrinomycetes, Taphrinomycotina, Ascomycota, Fungi

Fries, E. M. 1815. *Observ. mycol.* (Havniae) 1:217.

From Gr. *taphría*, act of opening a grave or trench, from *táphre*, genit. *táphros*, sepulchre, with the suf. *-ina*, which indicates possession by or belonging to. Galls or deformations with blisters that open to expose palisade of naked asci appearing in depressions or pits in deformed and hypertrophied tissues of leaves. On living leaves and other organs of plants, such as peach and plum.

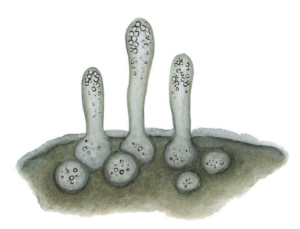

Taphridium umbelliferarum (Rostr.) Lagerh. & Juel: intracellular mycelium, chlamydospores and synasci (or compound asci) of this fungus that parasitizes umbelliferous plants, x 700.

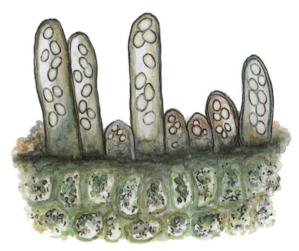

Taphrina pruni (Fuckel) Tul.: infected leaf of plum, which has blisters that open to expose the palisade of naked asci with ascospores, x 560.

Tarzetta (Cooke) Lambotte (syn. Pustularia Fuckel)

Pyronemataceae, Pezizales, Pezizomycetidae, Pezizomycetes, Pezizomycotina, Ascomycota, Fungi

Lambotte, E. 1887 (1888). La Flore mycologique de la Beguique. I. Supplément comprenant les Hyménomycètes, Pyrénomycètes, Discomycètes. *Mém. Soc. roy. Sci. Liège,* Série 2, *14*:325.

The etymology could not be determined. Fruitbody deeply cup-shaped to urn-shaped, sometimes splitting into lobes or becoming expanded when old, with a distinct stalk; margin dentate to slightly notched; apothecia, without or with carotenoids, apothecia not clothed with hairs or setae. Asci inamyloid, 8-spored; paraphyses slender, septate, forked toward the base, tips not clavate but often lobed; ascospores elliptic,

Tectimyces

smooth, biguttulate, hyaline, with carminophilic nuclei. Several to gregarious on the ground, under hardwoods and conifers.

Tarzetta cupularis (L.) Svrcek: sessile apothecium in dead wood, slightly flattened, with the margin crenulated, smooth hymenium, yellowish ochre, x 1. This fungus lives in the north of Europe, with occasional records in south Spain and Morocco.

Tectimyces L. G. Valle & Santam.
Legeriomycetaceae, Harpellales, Inc. sed., Inc. sed., Kickxellomycotina, Zygomycota, Fungi
Valle, L. G., S. Santamaria. 2002. *Tectimyces*, a new genus of Harpellales on mayfly nymphs (Leptophlebiidae) in Spain. *Mycol. Res. 106*(7):841-847.
From L. *tectus*, hidden + *myces* < Gr. *mýkes*, fungus, due to the fungus being hidden among gut debris. Trichospores without appendages bearing an inconspicuous collar upon release. Generative cells elongate, broadened distally. Biconical zygospores of type II, submedially and obliquely attached to zygosporophore. Upon release zygospores bear a lateral collar and a single mucilaginous appendage-like structure. Thalli irregularly pinnate or umbellate. Attached to the hindgut cuticle of Leptophlebiidae nymphs.

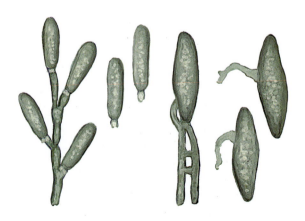

Tectimyces leptophlebiidarum L. G. Valle & Santam.: fertile thallic branch, giving rise to four trichospores, x 290, with an inconspicuous basal collar, a zygospore obliquely attached to the zygosporophore, and upon release the zygospores bear a lateral, single mucilaginous appendage-like structure, x 510. On hindgut cuticle of nymph of *Habroleptoides confusa* (Leptophlebiidae), Barcelona, Spain.

Tectonidula Réblová—see **Barbatosphaeria** Réblová
Inc. sed., Inc. sed., Inc. sed., Sordariomycetes, Pezizomycotina, Ascomycota, Fungi
Réblová, M., V. Štĭpánek. 2009. New fungal genera, *Tectonidula* gen. nov. for *Calosphaeria*-like fungi with holoblastic-denticulate conidiogenesis and *Natantiella* gen. nov. for three species segregated from *Ceratostomella*. *Mycol. Res. 113*(9):991-1002.
From L. *tectus*, covered, hidden + *nidus*, nest + dim. suf. *-ula*, referring to groups of perithecia hidden under the bark.

Telimenochora Sivan.
Phyllachoraceae, Phyllachorales, Inc. sed., Sordariomycetes, Pezizomycotina, Ascomycota, Fungi
Sivanesan, A. 1987. *Telimena*, *Telimenopsis*, and a new ascomycete genus *Telimenochora* of the Phyllachorales. *Trans. Br. mycol. Soc. 88*(4):473-477.
From genus *Telimena* Racib. + Gr. *chóros*, place, region. Ascomata solitary to aggregated, immersed in a stroma, amphigenous, clypeate, scattered, black, globose, unilocular, ostiolate. Hamathecium with periphyses in ostiole, persistent periphysoids and evanescent paraphyses. Asci unitunicate, deliquescent, cylindrical to broadly ellipsoidal, 8-spored. Ascospores hyaline to pale brown when mature, fusiform, straight to curved, smooth, 2-septate in middle, with a dark, thick band at each end when mature. On *Cordia spinescens* in Veracruz, Mexico.

Telimenochora abortiva (F. Stevens) Sivan.: vertical section of an immersed ascostroma, showing the clypeus and the hamathecium, x 45; unitunicate ascus with eight young ascospores, x 650, and four, mature ascospores with two or more septa, some with a dark, thick band at each end, x 750. On living leaves of *Cordia spinescens* in Veracruz, Mexico.

Teloschistes Norman
Teloschistaceae, Teloschistales, Lecanoromycetidae, Lecanoromycetes, Pezizomycotina, Ascomycota, Fungi
Norman, J. M. 1853. Connatus praemissus redactionis novae generum nonnullorum lichenum. *Nytt Mag. Natur. 7*(3):228.
From Gr. *télos*, end, final, term + *schistós*, divided, split, separated. Thallus fruticose, divided into slender,

round or flattened branches, often spinulose, with terminal apothecia when fertile. Occasionally thallus foliose, of more or less imbricate lobules; apothecia lateral or dispersed. Thallus fruticose, generally bright orange or yellow. Photobiont *Protococcus*. On wood, trees and rocks.

Teloschistes chrysophthalmus (L.) Norman ex Tuck.: bright orange or yellow, fruticose thallus on wood, with round or flattened branches, spinulose, with terminal apothecia, x 1.

Teratosphaeria Syd. & P. Syd. (syn. Kirramyces J. Walker, et al.)

Mycosphaerellaceae, Mycosphaerellales, Dothideomycetidae, Dothideomycetes, Pezizomycotina, Ascomycota, Fungi

Sydow, H., P. Sydow. 1912. Beschreibungen neuer südafrikanisher Pilze. *Annales Mycologici* 10:39.

From Gr. *téras*, genit, *tératos*, a monster, wonder + *spháira*, ball, sphere, referring to the anomalous aspect of some ascomata, which are inconsistent with or deviating from normal or expected. Important foliicolous pathogens of plants, causes leaf spots, leaf blotch, and shoot blight and stem cankers, causes reduction of wood volume and in severe cases tree death. Some distinctive characters of this genus are the presence of superficial stromatic tissue, ascospores that darken in their asci, remnants of the hamathecial tissue, ascospores that are frequently covered by a mucoid sheath, asci with a multi-layered endotunica, and the presence of ostiolar periphyses.

Terfezia Tul. & C. Tul.

Pezizaceae, Pezizales, Pezizomycetidae, Pezizomycetes, Pezizomycotina, Ascomycota, Fungi

Tulasne, L. R., C. Tulasne. 1851. *Fungi hypog.* (Paris):172.

From Arabic *terfez* or *terfás*, a name applied in North Africa to the fructifications of the edible fungus *Terfezia leonis* (Tul. & C. Tul.) Tul. & C. Tul.[=*T. arenaria* (Moris) Trappe], also called *kamés* or poor man's truffle. Ascocarps hypogeous, similar to truffles of genus *Tuber* P. Micheli ex F. H. Wigg., globose, fleshy, compact, characterized by absence of veins that traverse flesh from surface. Asci with five to eight ascospores that do not constitute a typical hymenium, arranged in spherical alveoli. Associated with plants of the families Cistaceae and Plantaginaceae, which they colonize.

Teratosphaeria angophorae (Andjic, et al.) Andjic, et al. (syn. **Kirramyces angophorae** Andjic, et al.): pycnidial conidiomata, with a central ostiole, x 3; conidiogenous cells and euseptate, 3- or 4-septate conidia, x 640. On leaves of *Angophora floribunda* in New South Wales.

Terfezia arenaria (Moris) Trappe: globose, fleshy, and brown ascocarps, x 0.5.

Termitomyces R. Heim

Lyophyllaceae, Agaricales, Agaricomycetidae, Agaricomycetes, Agaricomycotina, Basidiomycota, Fungi

Heim, R. 1942. Nouvelles études descriptives sur les agarics termitophiles d'Afrique tropicale. *Arch. Mus. Hist. Nat. Paris*, ser. 6, 18:147.

From L. *termitarium*, mounds made by termites + suf. *-myces* < Gr. *mýkes*, fungus, mushroom, for the habitat of the fungus. Basidiocarps usually rather fleshy and large with a prominent umbo; spore print pink; lamellae free or adnexed; stipe central, with a simple or double veil, sometimes without veil; spores hyaline, inamyloid, ellipsoid, smooth; hyphae afibulate. Primordia developing in holes of termite nests in Africa and Asia, also in the South Pacific. Most species are edible and considered of good quality in Africa and Asia.

Thamnidium

Termitomyces titanicus Pegler & Piearce: brownish basidiome growing symbiotically on termite nest (genus *Macrotermes*); the cap may reach 1 m in diameter on the stipe with more than 57 cm in length, with the creamy densely crowded gills, x 0.1.
It is the largest edible mushroom in the world, and this can be found throughout Zambia, Africa. **T. titanicus** is packed with a wide array of antioxidants including phenols, flavonoids, and betacarotene, which enhance immunity of humans, improving the body's ability to fight hypertension, arthritis, diabetes and other diseases.

Thamnidium Link

Mucoraceae, Mucorales, Inc. sed., Inc. sed., Mucoromycotina, Zygomycota, Fungi
Link, H. F. 1809. Observationes in ordine plantarum naturales. *Mag. Gesell. naturf. Freunde, Berlin* 3:31.
From Gr. *thámnos*, shrub + L. dim. suf. *-idium*. Sporangiophores with shrub-like branching bearing sporangia and sporangiola. Saprobic in soil.

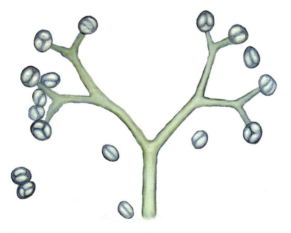

Thamnidium elegans Link: shrub-like sporangiophore of this saprobic fungus in soil, bearing sporangia and sporangiola, x 450.

Thamnostylum Arx & H.P. Upadhyay (syn. **Helicostylum** Corda)

Syncephalastraceae, Mucorales, Inc. sed., Mucoromycetes, Mucoromycotina, Zygomycota, Fungi
Arx, J. A. von. 1970. The genera of fungi sporulating in pure culture:247.
From Gr. *thámnos*, shrub + *stýlos* > L. *stylum*, style, pillar, for the shrub shape of the sporangiophores. Mycelium branched, forming stolons and rhizoids; sporangiophores erect, often nodose, terminating in a sporangium or a blunt spine or a node; branches arising in whorls, simple or forked and ending in recurved stalks bearing sporangiola; sporangia spherical or pyriform, apophysate; sporangiola pyriform, with or without a columella; sporangiospores ellipsoidal, smooth, hyaline; zygospores borne aerially between opposed suspensors, dark, rough-walled.

Thamnostylum piriforme (Bainier) Arx & H. P. Upadhyay (syn. **Helicostylum piriforme** Bainier): sporangiophore of this saprobic fungus, in soil and in dung, with helicoid branches, curved like a helicoid, that bear sporangiola, x 1,000.

Thanatephorus Donk—see **Rhizoctonia** DC.
Ceratobasidiaceae, Cantharellales, Inc. sed., Agaricomycetes, Agaricomycotina, Basidiomycota, Fungi
Donk, M. A. 1956. The generic names proposed for Hymenomycetes. *Reinwardtia* 3:376.
From Gr. *thánatos*, death, from *thanatóo*, to kill + suf. *-phóros*, from *phéro*, to carry, sustain, produce, bring.

Thaptospora B. Sutton & Pascoe
Inc. sed., Inc. sed., Inc. sed., Inc. sed., Pezizomycotina, Ascomycota, Fungi
Sutton, B. C., I. G. Pascoe. 1987. Some cupulate Coelomycetes from native Australian plants. *Trans. Br. mycol. Soc.* 88(2):169-180.
From Gr. *thaptó*, burial + *sporá*, spore, seed > L. *spora*, spore, referring to the conidiomata forming on the leaves at the base of the leaf hairs. Mycelium branched, septate, pale brown, anastomosing, superficial. Conidiomata superficial, eustromatic, cylindrical, lageniform to campanulate, dark brown above, paler below; wall of brown textura angularis to intricata, 1-2 cells thick; marginal

cells dark brown, projecting, irregularly. Conidiophores absent. Conidiogenous cells formed from inner cells of lower conidiomatal wall, discrete, hyaline, ampulliform to lageniform or cylindrical, producing a succession of conidia at same level by percurrent enteroblastic proliferation; collarette absent, periclinal thickening and cytoplasmic channel present. Conidia holoblastic, hyaline, smooth, guttulate, aseptate, fusiform. On leaves of *Olearia argophylla*, Victoria, Australia.

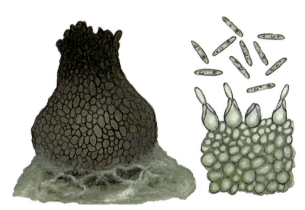

Thaptospora atrobrunnea B. Sutton & Pascoe: dark brown conidioma on living leaf of *Olearia agophylla*, Victoria, Australia, x 530; ampulliform conidiogenous cells with hyaline, fusiform, guttulate conidia, x 1,000.

Thaxterina Sivan., et al.
Tubeufiaceae, Tubeufiales, Pleosporomycetidae, Dothideomycetes, Pezizomycotina, Ascomycota, Fungi
Sivanesan, A., et al. 1988. *Thaxterina*, a new tubeufiaceous genus with multispored asci from India. *Trans. Br. mycol. Soc.* 90(4):662-665.
Named in honor of the American mycologist *Roland Thaxter* + L. suf. -ina, likeness, for the similarity to the genus *Thaxteriella* Petr. Ascomata superficial on a hyphal subiculum, scattered or aggregated, black, setose, at first globose later collapsing inwards at apex. Hamathecium of numerous filamentous, septate, hyaline, branched, distinct pseudoparaphyses. Asci multispored, bitunicate, thick-walled, short-stalked, obovoid, clavate or broadly cylindrical. Ascospores cylindric to oblong, transversely 1-3 septate, hyaline, straight to curved, with obtuse ends. Asexual morph moristroma-like. On dead wood of *Terminalia arjuna* in Jabalpur, Madhya Pradesh, India.

Thermoascus Miehe (syn. Dactylomyces Sopp.)
Trichocomaceae, Eurotiales, Eurotiomycetidae, Eurotiomycetes, Pezizomycotina, Ascomycota, Fungi
Miehe, H. 1907. *Selbsterhitz. Heus.* G. Fischer, Jena:70-73.
From Gr. *thermós*, warm, hot, or *thérme*, heat + *askós* > L. *ascus*, wine-skin, sac, ascus. Ascomata cleistothecial. Conidiophores finger-like branches producing chains of conidia. Asexual morph dactylomyces-like. Thermophilc, developing on fermenting vegetable substrates such as straw, cacao and guayule, during which the temperature rises up to 60°C.

Thaxterina multispora Sivan., et al.: vertical section of a black, setose ascoma in dead wood of *Terminalia arjuna*, in Madhya Pradesh, India, x 100; bitunicate ascus with numerous, young ascospores, x 260, and cylindric, hyaline ascospores, with 1-3 septa, x 830.

Thermoascus aurantiacus Miehe: agar culture of this thermophilic fungus, with developing cleistothecia, x 0.3; cleistothecia, with irregular shape and size, x 8; asci with eight, dark ascospores, x 1,600.

Thielavia Zopf
Chaetomiaceae, Sordariales, Sordariomycetidae, Sordariomycetes, Pezizomycotina, Ascomycota, Fungi
Zopf, W. 1876. Genus novum Perisporiacearum.*Verh. bot. Ver. Prov. Brandenb.* 18:101.
Dedicated to F. V. Thielau, with the L. suf. -ia, which denotes pertaining to. Perithecia lacking an ostiole or monostiolate, pale brown, semitransparent or dark brown, with or without hairs. Saprobic in soil.

Thielaviella Arx & T. Mahmood—see Boothiella Lodhi & Mirza
Chaetomiaceae, Sordariales, Sordariomycetidae, Sordariomycetes, Pezizomycotina, Ascomycota, Fungi
Arx, J. A. von, T. Mahmood. 1968. *Thielaviella humicola* gen. et sp. nov. from Pakistan. *Trans. Br. mycol. Soc.* 51:611-613.
From genus *Thielavia* Zopf + L. dim. suf. -ella.

Thielaviopsis

Thielavia aurantiaca Tad. Ito, et al.: nonostiolate, pale brown perithecium of this saprobic fungus; vertical section of the perithecium showing the asci and ascospores, x 500. Isolated from field soil, Japan.

Thielaviopsis Went— see **Ceratocystis** Ellis & Halst. Ceratocystidaceae, Microascales, Hypocreomycetidae, Sordariomycetes, Pezizomycotina, Ascomycota, Fungi
Went, F. A. F. C. 1893. De ananasziekte van het suikerriet (*Thielaviopsis*). *Meded. Proefstat. Suikerriert W. Java* 5:4.
From genus *Thielavia* Zopf, which was named in honor of *F. V. Thielau*, + L. suf. *-ia*, which denotes belonging to + Gr. suf. *-ópsis*, appearance, aspect, i.e., it has characters similar to those of *Thielavia*. Conidiophores producing phialides and phialospores; also forming thick-walled aleuriospores which eventually break apart; parasitic or saprophytic.

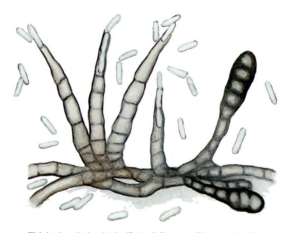

Thielaviopsis basicola (Berk. & Broome) Ferraris: phialides with endoconidia, and aleuriospores, x 350.

Thuemenella Penz. & Sacc. (syn. **Chromocreopsis** Seaver)
Hypoxylaceae, Xylariales, Xylariomycetidae, Sordariomycetes, Pezizomycotina, Ascomycota, Fungi
Penzig, O., P. A. Saccardo. 1897. Diagnoses fungorum novorum in insula Java collectorum. *Malpighia* 11(11-12): 518.
Named in honor of the German mycologist *Felix von Thümen* + L. dim. suf. *-ella*. Stromata erumpent through the bark, lobed or tuberculate, solitary, scattered and somewhat hemispheric or crowded and tending to coalesce into larger, elongate to irregularly shaped compound stromata. Stroma surface bright lemon-yellow, later darkening and turning brown when dry or with age, becoming uniformly dotted with brown punctations at maturity due to the presence of the dark ascospores extruded through the ostioles. Endostroma yellowish, prosenchymatous. Perithecia immersed, in a single layer at the periphery of the stroma, obpyriform to subglobose. Asci unitunicate, cylindrical, with ascospores ellipsoidal, short-cylindrical, cuboid to angular, initially yellowish to gray brown, becoming reddish brown at maturity, smooth or finely wrinkled (longitudinally striate). On dead wood and bark.

Thuemenella cubispora (Ellis & Holw.) Boedijn: yellowish stromata with the immersed perithecia, growing on dead wood in Iowa, U.S.A., and Java, Indonesia, x 1. Ascus with 8-ascospores, cuboid, x 1,000.

Thyronectria Sacc.
Inc. sed., Inc. sed., Inc. sed., Sordariomycetes, Pezizomycotina, Ascomycota, Fungi
Saccardo, P. A. 1875. Nova Ascomycetum Genera. *Grevillea* 4(no. 29):21-22.
From Gr. *thýra*, door + genus *Nectria* (Fr.) Fr. Stromata erumpent-superficial or sub-immersed, with perithecia in dense cespitose clusters. Perithecia subglobose, smooth, roughened, often clothed with a yellowish-green, furfuraceous coat that may disappear with age, perithecia dark colored, red to brown, collapsing or entire. Asci 8-spored, cylindrical to clavate. Ascospores hyaline, becoming dark brown, elliptical, becoming muriform, often with minute spore-like bodies. On tree bark.

Thyronectroidea Seaver
Thyridiaceae, Inc. sed., Inc. sed., Sordariomycetes, Pezizomycotina, Ascomycota, Fungi
Seaver, F. J. 1909. The Hypocreales of North America-II. *Mycologia* 1(5):177-207.
From genus *Thyronectria* Sacc. + NL. suf. *-oides*, like-

ness. Perithecia cespitose in erumpent clusters. Asci clavate-cylindrical, 8-spored. Ascospores elliptical, many-septate, becoming muriform, at first hyaline, becoming dark brown. On tree bark.

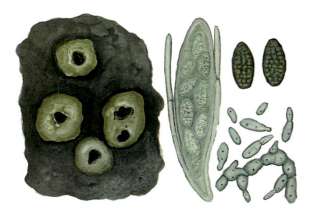

Thyronectroidea chrysogramma (Ellis & Everh.) Seaver (syn. **Thyronectria chrysogramma** Ellis & Everh.): superficial stroma in the bark of *Ulmus americana* (seen in specimens of Ontario and Ottawa, Canada), with yellowish-green perithecia, x 22; ascus with eight young ascospores, x 270, and mature, muriform ascospores, x 300. Lower right, the asexual morph showing phialides with conidia, some of them budding, x 400.

Ticosynnema R. F. Castañeda, et al.

Inc. sed., Inc. sed., Inc. sed., Inc. sed., Pezizomycotina, Ascomycota, Fungi

Castañeda, R. F., et al. 2012. A microfungus from Costa Rica: *Ticosynnema* gen. nov. *Mycotaxon* 122:255-259.

From *Tico*, a local name for Costa Rica + L. *synnema* < Gr. *sýn*, together + *néma*, filament, for the type of conidiophore. Mycelium superficial, immersed. Conidiomata on natural substratum synnematous, scattered, determinate, dark brown to black. Conidiophores macronematous, mononematous, erect, septate, loosely packed or compact, brown to dark brown. Conidiogenous cells monoblastic, integrated, determinate, terminal. Conidial secession rhexolytic. Conidia solitary, acrogenous, cylindrical, vermiform to oblong, with a conspicuous basal frill produced by rhexolytic fracture of wall of conidiogenous cells, septate, foveate, smooth or verruculose, pale brown to brown. On twigs in Guanacaste, Costa Rica.

Tieghemiomyces R. K. Benj.

Dimargaritaceae, Dimargaritales, Inc. sed., Inc. sed., Kickxellomycotina, Zygomycota, Fungi

Benjamin, R. K. 1959. The merosporangiferous Mucorales. *Aliso* 4(2):321-433.

Named in honor of the French mycologist *P. van Tieghem* + Gr. *mýkes* > L. *myces*, fungus. It is a dimargaritaceous parasite of Mucorales, which is characterized by producing bispored sporangiola (merosporangia) on numerous fertile branches of the sporangiophore.

Ticosynnema carranzae R. F. Castañeda, et al.: synnematous conidiomata, on decomposing twig in Costa Rica, showing conidiogenous cells and conidia, x 300; three mature, multiseptate, cylindrical conidia, x 750.

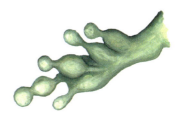

Tieghemiomyces californicus R. K. Benj.: fertile branches of the sporangiophores of this dimargaritaceous parasite of Mucorales, that produces bispored sporangiola (merosporangia), x 100; a close-up of bispored sporangiola, x 1,200. Collected in California, U.S.A.

Tilletia Tul. & C. Tul.

Tilletiaceae, Tilletiales, Exobasidiomycetidae, Exobasidiomycetes, Ustilaginomycotina, Basidiomycota, Fungi

Tulasne, L. R. 1847. Second mémoire sur les Uredinées

Tilletiopsis

et les Ustilaginées. *Annls Sci. Nat., Bot.,* sér. 3, 7:77-196.
Dedicated to the French botanist *Mathieu Tillet* (1730-1791), who conducted important field experiments on wheat smut, now called *Tilletia tritici* G. Winter. Sori grayish, more or less powdery due to large quantity of spores. Spores small, smooth or reticulate, upon germination forming an aseptate germ tube that produces primary sporidia (basidiospores) arranged in an apical crown, which fuse in pairs, forming secondary spores (also called secondary sporidia or conidia). Pathogenic of cereals.

Tilletia caries (DC.) Tul. & C. Tul.: grayish masses of teliospores of this pathogenic species causing stinking bunt of wheat, x 4; reticulate teliospores upon germinating form an aseptate germ tube that produces basidiospores arranged in an apical crown, which then fuse in pairs (secondary sporidia), x 900.

Tilletiopsis Derx

Inc. sed., Inc. sed., Exobasidiomycetidae, Exobasidiomycetes, Ustilaginomycotina, Basidiomycota, Fungi
Derx, H. G. 1948. *Itersonilia*, a nouveau genre de sporobolomycetes a mycélium bouclé. *Bull. bot. Gdns Buitenz.* 17(4):465-472.
From genus *Tilletia* Tul. & C. Tul. + L. suf. *-opsis* < Gr. *-ópsis*, appearance, aspect, due to the similarity to this genus. In early stages yeast phase formed and ballistospores, later giving rise to an apparently monokaryotic mycelium lacking clamps.

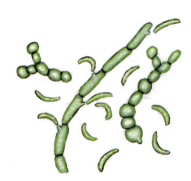

Tilletiopsis washingtonensis Nyland: budding yeast phase, which later gives rise to a monokaryotic mycelium producing ballistospores, x 2,200. Isolated in Washington, U.S.A.

Tirmania Chatin

Pezizaceae, Pezizales, Pezizomycetidae, Pezizomycetes, Pezizomycotina, Ascomycota, Fungi
Chatin, A. 1892. *Truffe,* Edn 2, Paris:80.
Dedicated to *M. Tirman,* Governor General of Algeria, who sent the type collection to Chatin + L. suf. *-ia,* which denotes pertaining to. Fructifications hypogeous, ascocarps solid, globose, slender toward lower part with basal mycelium; gleba venose with eight-spored asci and a wide pedicel. Mycorrhizal with *Helianthmum* spp. (Cistaceae), principally in North Africa.

Tolypocladium W. Gams

Ophiocordycipitaceae, Hypocreales, Hypocreomycetidae, Sordariomycetes, Pezizomycotina, Ascomycota, Fungi
Gams, W. 1971. *Tolypocladium*, eine Hyphomycetengattung mit geschwollenen Phialiden. *Persoonia* 6(2):185-191.
From Gr. *tolyp,* wind up, something wound up, a ball + *kládos,* a branch, sprout, dim. *kladion* > L. *cladium,* a club baton, referring to the wound up aspect of the conidiophores. Is a genus of soil-borne moniliales, characterized by slow-growing, upholstery-shaped, whitish cultures, terminal and lateral, partly on short lateral branches, whorled arranged phialides consisting of a swollen base and a thread-shaped, and small unicellular conidia. The sexual reproductive states include robust stipitate stroma with clavate to capitate clava to highly reduced stroma comprising rhizomorphs and aggregated perithecia; perithecia may be immersed and ordinal to the long axis of the stroma, or superficial and produced on a highly reduced stromatic pad; asci are single-walled long and cylindrical with a pronounced apical cap; ascospores are filiform, approximately as long as asci, septate and typically disarticulate into part-spores. Includes species that are parasites of other fungi, insect pathogens, rotifer pathogens and soil inhabiting species with uncertain ecological roles.

Torrentispora K. D. Hyde, et al.

Annulatascaceae, Inc. sed., Sordariomycetidae, Sordariomycetes, Pezizomycotina, Ascomycota, Fungi
Hyde, K. D., et al. 2000. *Torrentispora fibrosa* gen. sp. nov. (Annulatascaceae) from freshwater habitats. *Mycol. Res.* 104(11):1399-1403.
From L. *torrens,* torrent + *spora,* spore < Gr. *sporá,* seed, spore, referring to the running water where the fungus was found. Ascomata globose or subglobose, immersed to superficial, black, coriaceous, papillate, ostiolate, paraphysate, solitary. Neck long, black. Peridium of an outer layer of dark-brown, angular cells; an inner layer of hyaline, compressed cells. Paraphyses wide, septate, tapering distally. Asci 8-spored, long-cylindrical, uni-

Trametes

tunicate, pedicellate, with a large refractive apical ring. Ascospores uniseriate, oval to fusiform, occasionally flattened on one side, unicellular, hyaline, surrounded by a narrow fibrillar sheath. On submersed, decayed wood in Hong Kong.

Tolypocladium capitatum (Holmsk.) Quandt, Kepler & Spatafora: perithecial stroma parasitizing the semifleshy ascocarp of **Elaphomyces** T. Nees, x 1.

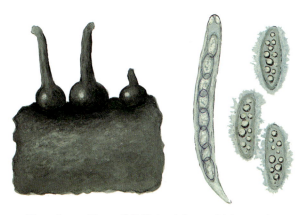

Torrentispora fibrosa K. D. Hyde, et al.: superficial ascomata, growing on submersed, decayed wood in a stream in Hong Kong, x 30; unitunicate ascus, with massive refractive apical ring, with eight young, unicellular ascospores, x 200; and three mature, fusiform, guttulate ascospores, covered by a fibrillar sheath, x 1,000.

Torula Pers.

Torulaceae, Pleosporales, Pleosporomycetidae, Dothideomycetes, Pezizomycotina, Ascomycota, Fungi
Persoon, C. H. 1794. Observationes mycologicae. *Ann. Bot. (Usteri)* 15:25.
From L. *torulus*, small string, dim. of *torus*, rope, cord. Porospores dark, in simple or branched chains, which break up easily giving rise to phragmospores or amerospores, formed more or less directly from vegetative hyphae. Saprobic in straw and other vegetable substrates.

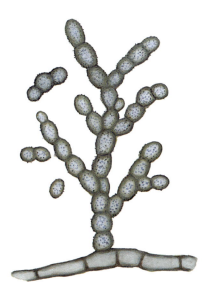

Torula herbarum (Pers.) Link: this saprobic fungus grows in straw and other vegetable substrates, and forms simple and branched chains of dark porospores, x 870.

Torulopsis Berlese

Inc. sed., Saccharomycetales, Saccharomycetidae, Saccharomycetes, Saccharomycotina, Ascomycota, Fungi
Berlese, A. N. 1895. I funghi diversi dai saccaromiceti e capaci di determinare la fermentazione alcoolica. *Giorn. Vitic. Enol.* 3:52-55.
From genus *Torula* Pers. < L. *torulus*, dim. of *torus*, slender rope or cord, but with constrictions + L. suf. *-opsis* < Gr. *-ópsis*, appearance, aspect, due to the similarity of the pseudomycelium of this yeast to the conidia of the dematiaceous genus *Torula* Pers.

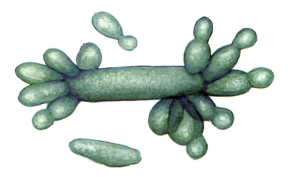

Torulopsis taboadae Ulloa & T. Herrera: monopolar and multipolar budding of vegetative yeast cells, x 2,300. Isolated from a sample of *colonche*, a fermented drink made from tuna fruits, in Zacatecas, Mexico.

Trametes Fr. (syn. **Coriolus** Quél.)

Polyporaceae, Polyporales, Inc. sed., Agaricomycetes, Agaricomycotina, Basidiomycota, Fungi
Fries, E. M. 1835. Corpus Florarum provincialium suecicae I. *Fl. Scan.* 1-339.
From L. *trama*, trama, referring to the trama, as a well-developed tissue of the fructification under the subhyme-

Trechispora

nium + suf. *-etes*, which means presence or possession. Fruitbody annual, pileate, broadly attached, solitary or imbricate, coriaceous, tough when fresh and dry. Pileus smooth to hirsute, white, cream, greyish or brownish; pore surface white cream or grey, with age becoming pale brown to ochraceous; pores round to angular; tubes non-stratified, regular; context tough to woody hard, in some species with a thin black zone between context and upper tomentum. Hyphal system trimitic; generative hyphae thin-walled and with clamps at septa; binding and skeletal hyphae thick-walled. Spores ellipsoid, cylindrical to allantoid, smooth. Mostly in deciduous woods, also in coniferous forests. Cosmopolitan.

Trametes versicolor (L.) Lloyd (syn. **Coriolus versicolor** (L.) Quél.): zonate basidiomata, on wood, x 1.

Trechispora P. Karst. (syn. **Echinotrema** Park.-Rhodes)

Hydnodontaceae, Trechisporales, Inc. sed., Agaricomycetes, Agaricomycotina, Basidiomycota, Fungi

Karsten, P. 1890. Fragmenta mycologica. *Hedwigia* 29:147.

From Gr. *trechýs*, *trecheía*, rough, rugged + *sporá*, spore, for the texture of the surface of the spores. Fructifications resupinate, irregularly effused, corticioid, granulose, spinulose or porose, of a pruinose, arachnoid or membranous-fragile texture, dingy whitish, of much-branched hyphae, with often ampulose clamp connections on borders of septa. Hymenophore sistotremoid, composed of parallel sinuous plates. Sterile margin of receptacle lacking. Trama composed of uniform hyphae without typical clamp connections, distinct, hyaline. Basidia with 3-6 sterigmata diverging from apex. Spores globose, hyaline, strongly echinulate. Cystidia or other specialized endings lacking. On decaying wood.

Tremella Pers.

Tremellaceae, Tremellales, Inc. sed., Tremellomycetes, Agaricomycotina, Basidiomycota, Fungi

Persoon, C. H. 1794. Dispositio methodica fungorum. *Neues Mag. Bot.* 1:111.

From L. *tremulus*, trembling, from *tremo*, to tremble + dim. suf. *-ella*, for the small movements that are produced when the basidiomata are touched or shaken, due to their gelatinous consistency. Fructifications cerebriform, foliose or lobulate, with variable colorations (dark, light or bright). They grow as parasites on other fungi, especially on wood rotting fungi on branches.

Trechispora clancularis (Park.-Rhodes) K. H. Larss. (syn. **Echinotrema clanculare** Park.-Rhodes): basidia and echinulate basidiospores, in hollows under *Armeria*, Great Britain, x 1,700.

Trechispora mollusca (Pers.) Liberta: ellipsoidal basidium on whose apex are formed four, peripheral sterigmata with roughened basidiospores, x 3,000.

Tremelloscypha D. A. Reid

Sebacinaceae, Sebacinales, Inc. sed., Agaricomycetes, Agaricomycotina, Basidiomycota, Fungi

Trichoderma

Reid, D. A. 1979. *Tremelloscypha* and *Papyrodiscus* - two new genera of Basidiomycetes from Australasia. *Beih. Sydowia* 8:332-334.

From genus *Tremella* Pers. + Gr. *skýphus*, *skýpha*, cup, referring to the shape of the fructification. Sporophores small, varying from monopodal and either infundibuliform or umbilicate to pleuropodal and then either pseudo-infundibuliform or with a lateral stipe gradually expanding into the ascending flabellate pileus. Pileus coarsely strigose and somewhat spongy. Hymenial surface inferior, decurrent, smooth, covered with a dark slate grey pruina. Stipe small, buff, with a conspicuous basal ball of earth. Hyphal structure monomitic, hyphae loosely interweaved with a firm distinct wall, but lacking clamp connections. Hymenium with hyphal branching just below the basidia. Spores subcylindric, narrowly elliptic to allantoid. On sandy soil in Australia.

Tretovularia Deighton

Inc. sed., Inc. sed., Inc. sed., Inc. sed., Pezizomycotina, Ascomycota, Fungi

Deighton, F. C. 1984. *Tretovularia*, a new hyphomycetous genus. *Trans. Br. mycol. Soc.* 82 (4):743-745.

From Gr. *tretós*, pierced, perforated + genus *Ovularia* Sacc., for the tretic conidiogenous cells. Foliicolous. Caespituli white, effuse. Mycelium immersed. Stroma small. Conidiophores fasciculate, emerging through the stomata, macronematous. Conidiogenous cells integrated, polytretic, sympodial, colorless. Conidia subglobose, colorless, non-septate, with unthickened hilum. Collected on leaves of *Vicia cassubica* in northern Europe.

Trichaptum Murrill

Inc. sed., Hymenochaetales, Inc. sed., Agaricomycetes, Agaricomycotina, Basidiomycota, Fungi

Murrill, W. A. 1904. The Polyporaceae of North America: IX. *Inonotus*, *Sesia* and monotypic genera. *Bull. Torrey bot. Club* 31(11):608.

From Gr. *trichápton*, with a texture of hairs < Gr. *thríx*, *thrichós*, hair, referring to the hairs covering the basidiocarps. Fruitbodies annual, resupinate, effused, reflexed to pileate; upper surface hispid to adpressed tomentose, gray to dirty white; hymenophore irpicoid, lamellate to poroid; pore surface purplish to violet, pale brown in age and on drying; hyphal system dimitic to trimitic, generative hyphae with clamps, skeletal hyphae dominant in the basidiocarp, binding hyphae rarely present; spores cylindrical, smooth hyaline. On both conifers and hard woods, causing a white rot. Cosmopolitan.

Trichobelonium (Sacc.) Rehm. See Belonopsis (Sacc.) Rehm.

Dermateaceae, Helotiales, Leotiomycetidae, Leotiomycetes, Pezizomycotina, Ascomycota, Fungi

Rehm, H. F. 1891. Ascomyceten: Hysteriaceen und Discomyceten. On L. Rabenhorst, Kryptogamen-Flora von Deutschland, *Krypt.-Fl.,* Edn 2 (Leipzig) 1.3(lief. 36):401-608.

From Gr. *trichós*, hair + genus *Belonium* Sacc.

Trichoceridium R. A. Poiss.

Legeriomycetaceae, Harpellales, Inc. sed., Inc. sed., Kickxellomycotina, Zygomycota, Fungi

Poisson, R. A. 1932. Sur deux entophytes parasites intestinaux de larves de diptères. *Annls Parasit. hum. comp.* 10:435-443.

The name derives from *Trichocera*, the genus of dipterids to which the larvae of the host belong + L. dim. suf. *-idium*. This organism lives attached to the chitinous lining of the intestine of the host by means of an adhesive basal organ, from which grows a thallus that produces arthrospores.

Trichoceridium ramosum R. A. Poiss.: thallus with arthrospores, attached by means of an adhesive basal organ to the chitinous lining of the intestine of dipterid insect larva, x 30.

Trichoderma Pers. (syns. Chromocrea Seaver, Hypocrea Fr.)

Hypocreaceae, Hypocreales, Hypocreomycetidae, Sordariomycetes, Pezizomycotina, Ascomycota, Fungi

Persoon, C. H. 1794. Neuer Versuch einer systematischen Eintheilung der Schwämme. *Neues Mag. Bot.* 1:92.

From Gr. *thríx*, genit. *trichós*, hair + *dérma*, skin, bark. Stromata patellate or subpatellate, whitish, yellowish or reddish to greenish-black, more or less variable in a given species, fleshy. Perithecia entirely immersed with necks only slightly prominent. Asci cylindrical, 8-spored, becoming 16-spored by the separation of each original spore into 2 subglobose spores. Ascospores colored, greenish or brownish. Conidiophores erect, solitary or aggregated in loose clusters bearing phialides. Conidia globose, green or hyaline, accumulating in mucose balls.

Trichoglossum

Common in soil and decomposing detritus, also on cellulosic substrates, antagonistic to other fungi including some plant pathogens.

Trichoderma flavipes (Peck) Seifert, et al. (syn. **Dendrostilbella hanlinii** Hammill & Shipman): synnemata with small heads on which are produced hyaline conidia, on soil, x 16. Collected in forest soil, Georgia, U.S.A.

Trichoderma gelatinosum P. Chaverri & Samuels (syns. **Hypocrea gelatinosa** (Tode) Fr., **Chromocrea gelatinosa** (Tode) Seaver): sagittal section of a stroma, with immersed perithecia, and ascospores, x 20.

Trichoderma strictipile Bissett (syn. **Hypocrea strictipilosa** P. Chaverri & Samuels): erumpent perithecial stromata, on rotting log, Québec, Canada, x 5.

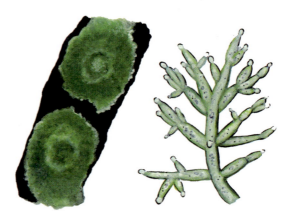

Trichoderma viride Pers.: velvety, green colonies growing on cellulosic substrate, x 0.5; conidiophore and phialides with globose, green, mucose conidia, x 1,000.

Trichoglossum Boudier
Geoglossaceae, Geoglossales, Inc. sed., Geoglossomycetes, Pezizomycotina, Ascomycota, Fungi
Boudier, J. L. E. 1885. Nouvelle classification des discomycètes charnus. *Bull. Soc. mycol. Fr.* 1:110.
From Gr. *thríx*, genit. *trichós*, hair + *glõssa*, tongue, for the shape of the setose hymenial, pedicellate ascophore or ascoma, which is setose. Ascomata similar to *Geoglossum* Pers., from which it differs mainly by having abundant dark brown, acuminate setae in the hymenium and on the stipe, as indicated in the etymology.

Trichoglossum hirsutum (Pers.) Boud.: pedicellate ascomata, with dark brown, acuminate setae in the hymenium and on the stipe, on soil, x 1.

Tricholoma (Fr.) Staude (syn. **Cortinellus** Roze)
Tricholomataceae, Agaricales, Agaricomycetidae, Agaricomycetes, Agaricomycotina, Basidiomycota, Fungi
Staude, F. 1857. *Schwämme Mitteldeutschl.*:xxviii, 125.
From Gr. *thríx*, genit. *trichós*, hair, tomentum, down, filament + *lōma*, fringe, margin, border, referring to the fimbriate ornamentation of the margin of the pileus of several of the species. Pileus fleshy and regular, with

Trichosporon

an incurved margin. Stipe central and also fleshy. Gills sinuate, adnate or somewhat decurrent, and spores generally white. Terricolous in habitat; mostly in temperate forests, and in open places in tropical regions. Several of the species are edible and/or ectomycorrhizogenous with some species of conifers or broad-leaved trees, and in certain cases they form fairy rings. Some of the edible species could be propagated or cultivated commercially.

Tricholoma vaccinum (Schaeff.) P. Kumm.: terricolous basidioma, with fimbriate ornamentation of the margin of the pileus, x 0.5.

Tricholomopsis Singer
Tricholomataceae, Agaricales, Agaricomycetidae, Agaricomycetes, Agaricomycotina, Basidiomycota, Fungi
Singer, R. 1939. Phylogenie und Taxonomie der Agaricales. *Schweiz. Z. Pilzk.*17:52-57.
From genus *Tricholoma* (Fr.) Staude < Gr. *thríx*, genit. *trichós* hair, filament, tomentum + *lōma*, fringe, margin, border, + *-ópsis*, appearance, aspect. Is a genus of saprophytic agaricoid fungi with yellow lamellae, a fibrillose or squamulose dry pileus with red or yellow tones, and it is usually associated with decaying wood. The spores are smooth and inamyloid, the hyphae are clamped, and the lamellae have a sterile edge, and cheilocystidia.

Tricholomopsis humboldtii Singer, et al.: basidiomes, x 0.3. It grows solitary to clustered on conifer stumps and logs (*Quercus humboldtii*).

Trichopeltum Bat., et al.
Microthyriaceae, Microthyriales, Inc. sed., Dothideomycetes, Pezizomycotina, Ascomycota, Fungi
Batista, A. C., et al. 1957. Organogênese e sistemática dos fungos Trichopeltinaceae (Theiss.) emend nobis. *Publicações Inst. Micol. Recife* 90:1-24.
From Gr. *thríx*, genit. *trichós*, hair + *pélte*, shield, referring to the presence of hairs or bristles on the shield or scutellum of the hemiperithecioid pseudothecium. Pseudothecia develop as circular enlargements beneath the thallus, which is formed by a superficial, branched and striped mycelium. It lives on leaves and stems, principally in tropical and subtropical places. The asci contain hyaline ascospores with several septa.

Trichophaea Boud.
Pyronemataceae, Pezizales, Pezizomycetidae, Pezizomycetes, Pezizomycotina, Ascomycota, Fungi
Boudier, J. L. L. É. 1885. Nouvelle classification naturelle des Discomycètes charnus. *Bull. Soc. mycol. Fr.* 1:105.
From Gr. *thríx*, genit. *trichós*, hair + *pháios* > L. *phaeo*, dark-colored, i.e., with dark-colored hairs. Ascoma an apothecium, sessile, cupulate to slightly concave or flat, outer surface dark gray to light or dark brown, clothed with prominent hairs of various types. Hairs stiff, thick-walled, septate, arising from a bulbous base formed in the excipular tissues, and tapering to a point, subhyaline to brown. Hymenium white, gray, cream, to blackish-brown, composed of asci interspersed with paraphyses. Paraphyses filiform, with clavate apices, septate. Asci unitunicate, operculate, cylindrical, tapered toward the base, apex not blueing in iodine, 8-spored. Ascospores 1-celled, hyaline, elliptical, with one or two oil droplets, smooth. Asexual state in *Dichobotrys* Hennebert. On bare or burned soil and wood, often among mosses.

Trichophyton Malmsten
Arthrodermataceae, Onygenales, Eurotiomycetidae, Eurotiomycetes, Pezizomycotina, Ascomycota, Fungi
Malmsten, P. H. 1848. *Arch. Anat. Physiol. Wiss. Med.*:14.
From Gr. *thríx*, genit. *trichós*, hair + *phytón*, plant, i.e., a vegetable with hair, since initially the fungi were considered plants. Dermatophyte or fungi causing tineas in humans and other mammals. Due to its capacity to utilize keratin, the species of this genus attack hair, skin and nails. Macroconidia cylindrical or clavate, septate, and microconidia unicellular, ovoid, and truncate. Common in soil, on keratinous substrates.

Trichosporon Behrend
Trichosporonaceae, Trichosporonales, Inc. sed., Tremellomycetes, Agaricomycotina, Basidiomycota, Fungi
Behrend, G. 1890. Über *Trichomycosis nodosa* (Juhel-Rénoy); piedra (Osorio). *Berliner Klin. Wochenschr.* 21:464-467.

Trichothecium

From Gr. *thríx*, genit. *trichós*, hair, filament + *spóron*, seed, spore, due to the fact that the hyphae formed by this yeast. Blastospores, septate, disarticulate and form arthrospores. Saprobic and pathogenic species, the latter causing cutaneous lesions, "white piedra" in humans and vertebrate animals.

From Gr. *thríx*, genit. *trichós*, hair + *thyréos*, oblong shield + *mykés* > L. *myces*, fungus. Mycelium composed of brown hyphae, anastomosing, becoming membranaceous-reticulate, ribbon-like, adhered to the host fungus, lacking hyphopodia and setae. Thyriothecia scutate, radial, prosenchymatous, brownish-black, round, with a central pore, without setae. Basal wall distinct, composed of radiating hyphae. Paraphyses septate, thick, simple or branched, hyaline. Asci clavate, obclavate or ellipsoidal, sessile or short-pedicellate, 8-spored, in fascicles. Ascospores 1-septate, brown. Conidiophores simple, subhyaline to brown, cylindrical; conidia isthmoid, oblong, multicellular, dark brown, echinulate. Collected in Recife, Brazil, parasitic on *Asteridiella melastomatacearum* (Speg.) Hansf.

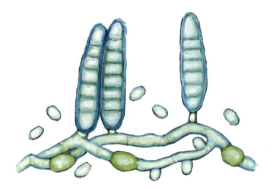

Trichophyton ajelloi (Vanbreus.) Ajello: cylindrical, septate macroconidia, and unicellular, ovoid and truncate microconidia developed in agar culture, x 1,100. This keratinophilous fungus causes tineas in humans and other mammals.

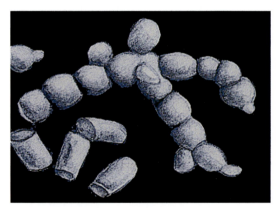

Trichosporon beigelii (Küchenm. & Rabenh.) Vuill.: hyphae with arthrospores and blastospores, x 3,000.

Trichothecium Link
Inc. sed., Hypocreales, Hypocreomycetidae, Sordariomycetes, Pezizomycotina, Ascomycota, Fungi
Link, J. H. F. 1809. Observationes in ordines plantarum naturales. *Mag. Gesell. naturf. Freunde, Berlin* 3 (1-2):18.
From Gr. *thríx*, genit. *trichós*, hair + NL. *thecium* < Gr. *théke*, box, case. Cellular conidia retrogressively, which are arranged in basipetal chains, supported by mucus. Worldwide, common on vegetable remains and fruiting bodies and sclerotia of higher fungi, which it covers with a pink conidial powder.

Trichothyriomyces Bat. & H. Maia
Microthyriaceae, Microthyriales, Inc. sed., Dothideomycetes, Pezizomycotina, Ascomycota, Fungi
Batista, A. C., et al. 1955. Alguns fungos hiperparasitas. *Anais Soc. Biol. Pernambuco* 13(2):94-107.

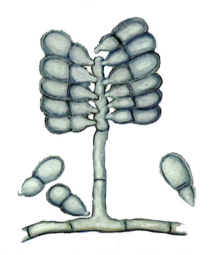

Trichothecium roseum (Pers.) Link: conidiophore that produces basipetal, retrogressive chains of bicellular conidia, supported by mucus, x 420.

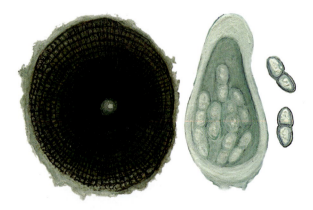

Trichothyriomyces notatus Bat. & H. Maia: thyriothecium scutate, radial, brownish-black, with a central pore, of this mycoparasite developed in the leaf underside of **Asteridiella melastomatacearum** (Speg.) Hansf., in Recife, Brazil, x 400; bitunicate ascus with eight, bicellular ascospores, and two free, mature, brownish ascospores, x 1,500.

Tricispora Oehl, et al.
Acaulosporaceae, Diversisporales, Inc. sed., Glomeromycetes, Glomeromycotina, Glomeromycota, Fungi

Oehl, F., et al. 2011. Revision of Glomeromycetes with entrophosporoid and glomoid spore formation with three new genera. *Mycotaxon* 117:297-316.

Formed as an anagram, *trici*, from L. *cicatrix*, cicatrix, scar + Gr. *sporá*, spore, seed > L. *spora*, spore, referring to the two conspicuous cicatrices left on the structural wall layer of the spores, even when the sporiferous saccules and the hyphal neck distal to the saccule have detached completely from the spores. Sporocarps unknown. Spores formed within the hyphal neck of closely adherent terminal or intercalary sporiferous saccules. The globose saccule is substantially smaller than the attached mature spore. Spores have an outer and an inner wall. At least two layers (including the outer wall structural layer) are continuous with the sporiferous saccule wall. The outer layer of the outer wall is evanescent; the inner layers are permanent. After the hyphal neck connections break off, spores show two, often opposite, cicatrices that are closed by the permanent sublayers of the outer wall structural layer. The inner wall forms *de novo*, consists of several layers without granular, beaded, appearance and does not stain with Melzer's reagent. The fungal structures in the roots stain blue to dark blue with trypan blue; forming vesicular-arbuscular mycorrhiza.

Tricladiospora Nawawi & Kuthub.
Inc. sed., Inc. sed., Inc. sed., Inc. sed., Pezizomycotina, Ascomycota, Fungi
Nawawi, A., A. J. Kuthubutheen. 1988. *Tricladiospora*, a new genus of dematiaceous Hyphomycetes with staurosporous conidia from submerged decaying leaves. *Trans. Br. mycol. Soc.* 90(3):482-487.

From genus *Tricladium* Ingold + Gr. *sporá*, spore, seed > L. *spora*, spore, for the staurosporus conidia. Conidiophores macronematous, mononematous, single to caespitous, indeterminate, simple, smooth, brown, erect to flexuous. Conidiogenous cells integrated, terminal, monoblastic, proliferations percurrent. Conidia holoblastic, solitary, acrogenous, colored, multiseptate, staurosporous; main axis elongated, truncate at the base, tapered towards the apex, with 1-3 lateral branches, single, in succession, with basal constrictions, liberated by dissolution of basal septum. Collected on decaying immersed angiospermous leaves in Pasoh Forest Reserve, Negeri Sembilan, Malaysia.

Tripospora Sacc.
Coryneliaceae, Coryneliales, Coryneliomycetidae, Eurotiomycetes, Pezizomycotina, Ascomycota, Fungi
Saccardo, P. A., 1886. *Syll. fung., Addit.* I-IV (Abellini):194.

From Gr. *trípos*, with three feet, tripodal + *sporá*, spore, for the markedly lobed, or more or less star-shaped ascospores, which are produced in pedicellate asci within ascostromata that are transitional between the Euascomycetes and the Loculoascomycetes. Parasite on the bark of tropical conifers.

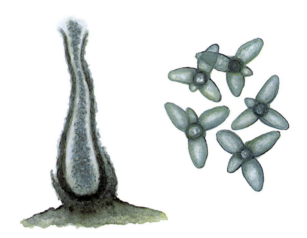

Tripospora macrospora Fitzp.: vertical section of a superficial perithecium, on leaves of *Podocarpus lambertii*, Brazil, showing the accumulation of ascospores, x 50; free, 3-celled (or 5-celled) ascospores, x 600.

Triscelophorus Ingold
Inc. sed., Inc. sed., Inc. sed., Inc. sed., Pezizomycotina, Ascomycota, Fungi
Ingold, C. T. 1943. *Triscelophorus monosporus*, n. gen., n. sp., an aquatic hyphomycete. *Trans. Br. mycol. Soc.* 26(3-4):148-152.

From Gr. *trís*, thrice + L. *cella*, cell + Gr. *phórus*, to bear, for the spores with three cells. Submerged aquatic fungi with branched septate mycelium. Conidia (aleuriospores) terminal, branched, consisting of an elongated main axis continuous with the conidiophore, and elongated secondary ramuli forming a whorl of three branches arising from the lower part of the main axis.

Tryssglobulus B. Sutton & Pascoe
Inc. sed., Inc. sed., Inc. sed., Inc. sed., Pezizomycotina, Ascomycota, Fungi
Sutton B. C., I. G. Pascoe. 1987. *Argopericonia* and *Tryssglobulus*, new Hyphomycete genera from *Banksia* leaves. *Trans. Br. mycol. Soc.* 88(1):41-46.

From Gr. *tryssós*, dainty + L. *globus*, dim. *globulus*, a little ball, referring to the small, cellular conidial head. Colonies saprotrophic. Mycelium superficial, brown, branched, euseptate. Stromata absent. Conidiophores arising from the vegetative mycelium, mononematous, determinate or indeterminate, erect, dark brown, smooth, euseptate, producing an apical globose, multicellular conidiogenous region, sometimes proliferating through the conidiogenous head to form additional shorter conidiophores each with an apical conidiogenous region. Conidiogenous cells integrated, restricted to and comprising the peripheral cells of the conidiogenous

Tuber

head, non protuberant, brown, smooth, each with protuberant, unthickened denticulate conidiogenous loci; denticles cylindrical. Conidia holoblastic, dry, brown, smooth, subglobose to lenticular, upper wall thicker and flatter than the lower wall; denticles sometimes persisting. On living leaves of *Banksia marginata* Cav., in Brisbane Ranges, Victoria, Australia.

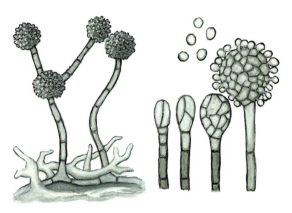

Tryssglobulus aspergilloides B. Sutton & Pascoe: saprotrophic conidiophores with small, cellular conidial heads, growing among the foliar hairs in the leaf of *Banksia marginata*, Yaugher, Victoria, Australia, x 75; conidiophores showing the development of conidiogenous cells, and brown, subglobose conidia, x 1,300.

Tuber P. Micheli ex F. H. Wigg.
Tuberaceae, Pezizales, Pezizomycetidae, Pezizomycetes, Pezizomycotina, Ascomycota, Fungi
Wiggers, F. H. 1780. *Prim. fl. holsat.* (Kiliae):1-112.
From L. *tuber*, tumor, tubercle, protuberance, for the globose shape of the ascocarps, resembling protuberances, tumors or tubercules. Hypogeous, commonly known as truffles, valuable in gastronomy and silviculture, since besides being edible they are mycorrhizogenous. They grow in coniferous and broad-leaved forests, especially of oak. Globose or subglobose ascocarps, dark brown, from 2-8 cm in diam, with the appearance of potato tubers, whose surface is smooth or warted. Interior of the fructification with numerous canals and chambers lined by hymenia with asci and paraphyses. Asci with usually from two to four oval ascospores, with prolongations that resemble spines.

Tuber melanosporum Vittad.: globose, warted, dark brown ascoma, on soil, x 1.

Tubercularia Tode—see **Nectria** (Fr.) Fr.
Nectriaceae, Hypocreales, Hypocreomycetidae, Sordariomycetes, Pezizomycotina, Ascomycota, Fungi
Tode, H. J. 1790. *Fung. mecklenb. sel.* (Lüneburg) 1:18. From L. *tuberculum*, dim. of *tuber*, tumor, lump, gall + suf. *-aria*, connection, possession, recipient, receptacle, in reference to the sessile cushion-shaped or tuberculate sporodochia.

Tulasnella J. Schröt.
Tulasnellaceae, Cantharellales, Inc. sed., Agaricomycetes, Agaricomycotina, Basidiomycota, Fungi
Schröter, J. 1888. Pilze Schlesiens. On J. F. Cohn, *Krypt.-Fl. Schlesien* (Breslau) 3.1(25-32):397.
Dedicated to the French mycologists *R. L. Tulasne* (1815-1885) and *C. Tulasne* (1816-1884) + L. dim. suf. *-ella*. The fructification is fleshy-membranaceous or gelatinous-ceraceous at first, then cartilaginous, resupinate, i.e., with the hymenium on the upper part, since the lower part remains against the substrate. The hymenium is smooth or folded; it has globose basidia with two to four sterigmata, apical or lateral, thick, at first obtuse, then long and filiform, on which are formed white, globose, ovoid, ellipsoid or pyriform, smooth spores, with the capacity to produce conidia or a mycelium at the time of germination. On humus or on wood.

Tulasnella violea (Quél.) Bourdot & Galzin: cartilaginous, resupinate, violaceous fructification, on dead wood, x 2.

Tulostoma Pers.
Agaricaceae, Agaricales, Agaricomycetidae, Agaricomycetes, Agaricomycotina, Basidiomycota, Fungi
Persoon, C. H., 1794. *Neues Mag. Bot.* 1:86.
From Gr. *týlos*, hump, callus, hardness, peg + *stóma*, mouth, pore, for the manner of dehiscence, by means of a frequently umbonate apical pore. Fructification a globose pileus supported by a slender, fibrous, sub-woody or scaly stipe. Gleba of abundant true capillitium and globose or irregular, smooth or ornamented spores. On sandy soils in arid and semiarid zones with scrub vegetation.

Tumularia Descals & Marvanová
Inc. sed., Inc. sed., Inc. sed., Inc. sed., Pezizomycotina, Ascomycota, Fungi

Marvanová, L., E. Descals. 1987. New taxa and new combinations of 'aquatic Hyphomycetes'. *Trans. Br. mycol. Soc. 89*:499-507.

From L. *tumulus*, hill, mound, protuberance + suf. *-ia*, state of being. Colony dark, hyphae hyaline to dark. Conidiophores single, apical or lateral, semimacronematous, simple or sparsely branched, hyaline. Conidiogenous cells single, apical, integrated, percurrent or sympodial. Conidia single, apical, fusoid to clavate, hyaline to fuscous, 1-few septate; one cell larger, central or apical, often protuberate; scar truncate to convex. Secession schizolytic.

Tulostoma brumale Pers.: mature basidiomata, with stipe and one-ostiolate pileus, on soil, x 1.

Tunicopsora Suj. Singh & P. C. Pandey—see **Kweilingia** Teng
Phakopsoraceae, Pucciniales, Inc. sed., Pucciniomycetes, Pucciniomycotina, Basidiomycota, Fungi
Singh, S., P. C. Pandey. 1971. *Tunicopsora*, a new rust genus on bamboo. *Trans. Br. mycol. Soc. 56*:301-303.
From L. *tunica*, covering + Gr. *psóra*, itch, scab.

Turbinellus Earle
Gomphaceae, Gomphales, Phallomycetidae, Agaricomycetes, Agaricomycotina, Basidiomycota, Fungi
Earle, F. S. 1909. The genera of North American gill fungi. *Bull. New York Bot. Gard. 5*:373-451.
From L. *turbo, turbinis*, anything that whirls around, such as a top; *turbinatus*, cone or top shaped + dim. suf. *-ellus*, i.e., little top. Pileus turbinate, fleshy-tuberous, thick, rugose infundibuliform, putrescent. Hymenium covering irregular, forking and reticulating folds; spores white or hyaline. Stipe central, short, thick, similar to club shaped species of *Craterellus* Pers. Common in North America.

Turturconchata J. L. Chen, et al.
Inc. sed., Inc. sed., Inc. sed., Pezizomycotina, Ascomycota, Fungi
Chen, J. L., et al. 1999. *Turturconchata*, a new genus of hyphomycetes from Taiwan. *Mycol. Res. 103*(7):830-832.
From L. *turtur*, turtle-dove + *concha*, shell + des. *-ata*, for the resemblance of the spore to a turtle shell. Colonies growing slowly, plane, dark grey to brownish-grey. Conidiophores macronematous, mononematous, single to clustered, simple or branched, flexuous, curved to spiral, septate, smooth, pale brown. Conidiogenous cells integrated, monoblastic or polyblastic, terminal to intercalary. Conidia solitary, dry, smooth, flattened dorsiventrally, borne horizontally, surface view globose, subglobose, broadly ellipsoidal to rectangular, margin irregular, somewhat crenate, lateral view fusiform, muriform, thick-walled, pale brown to olive brown, with a pale brown, eccentric protuberate pedicel. Isolated from decaying herbaceous stem from Shansia, Taiwan.

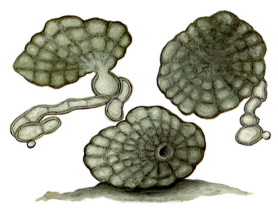

Turturconchata reticulata J. L. Chen, et al.: flexuous conidiophores of this fungus, isolated from decaying stem of herbaceous dicotyledon plant, in Taiwan, showing conidiogenous cells and flattened dorsiventrally ellipsoidal, muriform, conidia with 19-58 cells, x 1,500.

Tylopilus P. Karst.
Boletaceae, Boletales, Agaricomycetidae, Agaricomycetes, Agaricomycotina, Basidiomycota, Fungi
Karsten, P. A. 1881. Enumeratio boletinearum et polyporearum. Fennicum systemate novo dispositarum. *Rev. Mycol.* (Toulouse) *3*(no. 9):16.
From Gr. *týlos*, wooden nail, callus, pillow + *pílos*, hat, pileus, for the generally convex shape of the pileus. Fructification nail-shaped with pileus and stipe together; pores white to pink at openings of tubular hymenophore. Terricolous, solitary or in small groups in oak forests; important in forestry as ectomycorrhizal fungus. Some species are edible while others are somewhat toxic.

Tyrannosorus Unter. & Malloch
Inc. sed., Inc. sed., Inc. sed., Dothideomycetes, Pezizomycotina, Ascomycota, Fungi
Untereiner, W. A., et al. 1995. A molecular-morphotaxonomic approach to the systematics of the Herpotrichiellaceae and allied black yeasts. *Mycol. Res. 99*(8):897-913.
From Gr. *týrannos*, tyrant, terrible + *sorós*, sorus. Ascoma

Tyrannosorus

a uniloculate pseudothecium, solitary, scattered on surface of substrate, dark brown to black. Pseudothecia ostiolate, ovate to obpyriform, covered with stiff, pointed, dark brown setae. Wall of pseudothecium composed of several layers of pseudoparenchymatous cells, outermost cells thick-walled and pigmented, inner cells thinner-walled and subhyaline to hyaline. Centrum of young pseudothecium containing anastomosed pseudoparaphyses. Asci bitunicate, saccate to narrowly clavate, short-stipitate, 8-spored. Ascospores 2-celled, brown, fusoid to ellipsoid, with longitudinal striations, and each cell with 3-5 germ slits, smooth. Asexual morph helicodendron-like. Growing on wood of conifers (*Pinus* spp.).

Tylopilus plumbeoviolaceus (Snell & E. A. Dick) Snell & E. A. Dick: pink-colored basidiome, with the convex-shaped pileus, on soil in oak forest, x 0.5.

U

Usnea rubicunda

U

Uleomyces Henn. (syn. **Kusanoopsis** F. Stevens & Weedon)

Cookellaceae, Inc. sed., Inc. sed., Dothideomycetes, Pezizomycotina, Ascomycota, Fungi

Hennings, P. 1895. Fungi goyazenses. *Hedwigia* 34:107. Named in honor of the German botanist *E. H. G. Ule* + L. *myces*, fungus. Ascomata superficial, dull black, arising from an intramatricular mycelium, erumpent, epiphyllous, cushion-shaped, attached by a central foot, outer layer not differentiated. Asci formed at different levels, ovoid, thick-walled, hyaline, 4-spored. Ascospores hyaline, muriform, with usually 3 transverse septa and 1, 2 or 3 longitudinal septa. On unknown dicotyledonous host. British Guiana.

Uleomyces tapirirae (F. Stevens & Weedon) Petr. (syn. **Kusanoopsis guianensis** F. Stevens & Weedon): sagittal section of an erumpent, epiphyllous ascoma, x 120; hyaline, muriform ascospore, x 850. Collected in Guyana.

Umbelopsis Amos & H. L. Barnet (syn. **Mortierella** W. Gams)

Umbelopsidaceae, Umbelopsidales, Inc. sed., Inc. sed., Mucoromycotina, Zygomycota, Fungi

Amos, R. E., H. L. Barnett. 1966. *Umbelopsis versiformis*, a new genus and species of the Imperfects. *Mycologia* 58:805-808.

L. *umbella*, parasol, umbel + . suf. *-opsis*, from Gr. *-ópsis*, appearance, aspect, referring to the typical manner in which sporogenous branches all arise from the swollen head. Sporangiophores hyaline, often septate, bearing swollen head with sporogenous branches at the apex; sporangia globose or elongate, often reddish or ochraceous, multispored or single-spored, often with more or less conspicuous columella; spores 1-celled, hyaline, globose or ellipsoidal, rounded or angular. Isolated from the roots of red oak (*Quercus borealis*), and subsequently of soil near roots of oak trees, in Eastern West Virginia, U.S.A.

Umbelopsis vinacea (Dixon-Stew.) Arx (syn. **Mortierella ramanniana** var. **angulispora** (Naumov) Linnem.): sporangiophore with terminal sporangium, showing columella and polygonal spores, x 3,800.

Umbilicaria Hoffm. (syn. **Gyrophora** Ach.)

Umbilicariaceae, Umbilicariales, Umbilicariomycetidae, Lecanoromycetes, Pezizomycotina, Ascomycota, Fungi

Hoffmann, G. F. 1789. *Descr. Adumb. Plant. Lich.* 1(1):8. From L. *umbilicus*, navel + suf. *-aria*, which indicates possession or connection, for the umbilicate thallus. Thallus foliose, more or less concave, united to substrate by a central umbilicate depression, at times with outline circular, entire or regularly or irregularly lobed. Apothecia small or medium, immersed or short pedunculate, black or blackish, with a flat or convex disc and a proper excipulum. Photobiont *Pleurococcus*. On rocks, principally in high mountain areas.

Uncispora

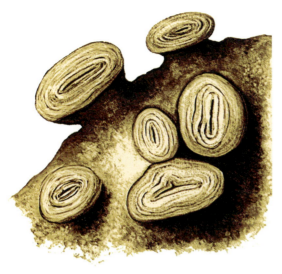

Umbilicaria cylindrica (L.) Delise: brownish, foliose thallus, on rock, united to the substrate by the funiculus (umbilical cord), x 15.

Umbilicaria vellea (L.) Ach.: thallus foliose-umbilicate, rigid, 4-5 (-20) cm diameter with entire to incised margins, attached by a thick, dark, central holdfast (umbilicus), pale brown, x 1. The photobiont is a green alga, the reproductive strategy is by soredia or soredia-like structures (blastidia). Found on artic alpine on exposed surfaces of siliceous rocks with some water seepage.

Uncinocarpus Sigler & J. W. Carmich.
Onygenaceae, Onygenales, Eurotiomycetidae, Eurotiomycetes, Pezizomycotina, Ascomycota, Fungi
Sigler, L., J. W. Carmichael. 1976. Taxonomy of *Malbranchea* and some other hyphomycetes with arthroconidia. *Mycotaxon* 4(2):349-488.

From L. *uncinus*, hook + *carpos*, fruit < Gr. *karpós*, for the hooked tips of the peridial hyphae. Gymnothecia more or less spherical, reddish-brown. Ascomatal initials on short stalks, bulbous. Peridial hyphae smooth, aseptate, of elongate appendages loosely intertwined. Free ends uncinate, sometimes spiral. Asci subglobose, evanescent, 8-spored. Ascospores oblate, smooth, yellow to reddish-brown. Asexual morph malbranchea-like. On keratinous substrates.

Uncinocarpus reesii Sigler & G. F. Orr: spherical, reddish-brown gymnothecium of this keratinophilic fungus, with elongated, spiral appendages of the peridial hyphae, loosely intertwined, covering the asci and ascospores, x 1.

Uncinula Lév.—see **Erysiphe** R. Hedw. ex DC.
Erysiphaceae, Erysiphales, Leotiomycetidae, Leotiomycetes, Pezizomycotina, Ascomycota, Fungi
Léveillé, J. H. 1851. Organisation et disposition méthodique des espèces qui composent le genre *Erysiphe*. *Annls Sci. Nat., Bot.*, sér. 3, 15:151.

From L. *uncinulus*, dim. of *uncinus*, fish-hook, hook, referring to the uncinate peridial appendages of the chasmothecia (cleistothecioid perithecia), which are simple or branched, but with the apex uncinate. Obligate parasite of grapevine and other plants, on which it causes a powdery mildew. The asexual state or conidial state is oidium-like.

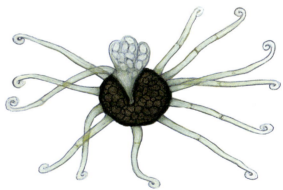

Uncinula longispora R. Y. Zheng & G. Q. Chen: cleistothecioid perithecium, with uncinate peridial appendages, x 100. An obligate parasite of grapevine and other plants causing powdery mildew, collected in Hebei, China.

Uncispora R. C. Sinclair & Morgan-Jones
Inc. sed., Inc. sed., Inc. sed., Inc. sed., Pezizomycotina, Ascomycota, Fungi
Sinclair, R. C., G. Morgan-Jones. 1979. Notes on Hyphomycetes. XXVI. *Uncispora harroldii* gen. et sp. nov. *Mycotaxon* 8(1):140-143.

Underwoodia

From L. *uncus*, hook + Gr. *sporá*, spore, for the hooked apex of the spore. Mycelium mostly immersed in substrate, of branched, septate, hyaline to pale brown hyphae. Conidiophores macronematous, synnematous or arranged in fascicles, erect, straight or flexuous, simple or branched, brown, smooth, septate. Conidiogenous cells integrated, monoblastic, terminal, determinate, cylindrical. Conidia solitary, simple, septate, subhyaline to pale brown, smooth or verruculose at maturity, obclavate, hooked towards apex, truncate at base. On dead branches of *Betula nigra* in Alabama, U.S.A.

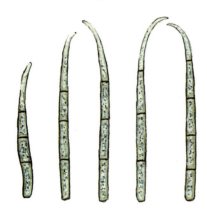

Uncispora hainanensis Jian Y. Li & Z. F. Yu: septate, subhyaline conidia, hooked towards the apex, truncate at the base, x 1,170. Isolated from decomposing leaves in a river of Hainan Island, China.

Underwoodia Peck

Inc. sed., Pezizales, Pezizomycetidae, Pezizomycetes, Pezizomycotina, Ascomycota, Fungi
Peck, C. H. 1890. *Ann. Rep. Reg. N.Y. St. Mus.* 43:32.
Dedicated to the American mycologist L. M. Underwood + L. suf. *-ia*, which denotes belonging to. Fructifications columnar, differentiated into a pileus and stipe; pileus saddle-shaped with two thick, equal branches, separated by a deep central fissure. Uncommon in humid soils of coniferous and mixed woods.

Underwoodia columnaris Peck.: columnar fructifications, with the saddle-shaped pileus, on humid soil of coniferous forest, x 0.7.

Unguiculariopsis Rehm (syn. **Mollisiella** (W. Phillips) Massee)

Dermateaceae, Helotiales, Leotiomycetidae, Leotiomycetes, Pezizomycotina, Ascomycota, Fungi
Rehm, H. 1909. Ascomycetes exs. Fasc. 44. *Annls mycol.* 7(5):400.
From L. *unguis*, dim. *ungiculus*, nail, claw + suf. *-ópsis* < Gr. *-ópsis*, aspect, for the appearance of the apothecia. Apothecia small cupulate, becoming expanded, externally dark-colored, brownish, tomentose, or clothed with poorly developed hairs. Asci clavate or cylindric, usually 8-spored. Ascospores at maturity 1-seriate, simple, globose. Paraphyses filiform, slightly enlarged above. On other fungi.

Unguiculariopsis lettaui (Grummann) Coppins: thallus of **Evernia prunastri** (L.) Ach. being parasitized by this helotiaceous fungus, showing the brown to blackish apothecia, urceolate, with hairy exciples, x 3. Mature biguttulate ascospores, x 220.

Ungulina Pat.—see **Fomes** (Fr.) Fr.

Polyporaceae, Polyporales, Inc. sed., Agaricomycetes, Agaricomycotina, Basidiomycota, Fungi
Patouillard, N. T. 1900. *Essai Tax. Hyménomyc.* (Lons-le-Saunier):102.
From L. *ungula*, hoof + dim. suf. *-ina*, which also indicates similarity, for the shape of the fructifications of this genus, which resemble horse hooves.

Urnula Fr.

Sarcosomataceae, Pezizales, Pezizomycetidae, Pezizomycetes, Pezizomycotina, Ascomycota, Fungi
Fries, E. M. 1849. *Summa veg. Scand.*, Sectio Post. (Stockholm):364.
From *urnula*, L. dim. of *urna*, jug for water, urn, for the shape of the apothecium, similar to a small water pitcher. Apothecia opening by small fissures from apex, margin becoming crenate; exterior surface tomentose, brownish-black, interior surface lighter, whitish or yellowish; consistency coriaceous, cartilaginous or gelatinous. Frequent on humid soil with decomposing wood in temperate and cold forests.

Urosporellopsis

Urnula craterium (Schwein.) Fr.: cartilaginous, brownish-black apothecia, with the crenate margin, on decomposing wood in temperate forest, x 0.5.

Uromyces appendiculatus (Pers.) Link: unicellular, orange-reddish colored teliospores, causing a rust in the tissues of *Phaseolus vulgaris*, x 1,000.

Urocystis Rabenh. ex Fuckel
Urocystidaceae, Urocystidales, Inc. sed., Ustilaginomycetes, Ustilaginomycotina, Basidiomycota, Fungi
Rabenhorst, G. L. 1870. *Jb. nassau. Ver. Naturk.* 23-24:41. From Gr. *ourá*, tail + *kýstis*, covering, bladder, for the manner in which the teliospores germinate. Ustilospores dark, surrounded by lighter sterile cells, grouped into balls of two or more spores; germinate in a manner similar to the teliospores of *Tilletia* Tul. & C. Tul.

Urophlyctis J. Schröt.—see **Physoderma** Wallr. Physodermataceae, Blastocladiales, Inc. sed., Blastocladiomycetes, Inc. sed., Chytridiomycota, Fungi
Schröter, J. 1886. Die Pilze Schliesiens, p. 196. *In*: F. Cohn, *Krypt.-Fl. Schlesien* (Breslau) *3.1*(9-16):1-814. From Gr. *ourá*, tail + *phlyctís* < *phlýctaina*, tumor, scab, blister, pimple, due to the resting spore having a swollen, turbinate cell in the basal or posterior part.

Urosporellopsis W. H. Hsieh, et al.
Phlogicylindriaceae, Amphisphaeriales, Xylariomycetidae, Sordariomycetes, Pezizomycotina, Ascomycota, Fungi
Hsieh, W. H., et al. 1994. *Urosporellopsis taiwanensis* gen. et sp. nov., a new amphisphaeriaceous ascomycete on *Sassafras* from Taiwan. *Mycol. Res.* 98(1):101-104. From genus *Urosporella* G. F. Atk. + Gr. suf. *-ópsis*, aspect, appearance, for the similarity to this genus. Stromata thin, pseudoparenchymatous. Ascomata solitary or in pairs, brown to dark brown, dispersed, immersed to erumpent. Ostiole absent. Asci clavate, 8-spored, short-stalked, thin-walled, evanescent. Paraphyses deliquescent. Ascospores fasciculate, hyaline, fusiform, 1-septate, upper cell tapering toward apex and lower cell extending to a slender, long, filiform, simple appendage. On dead stems of *Sassafras randaiense* in Taichung Hsien, Taiwan.

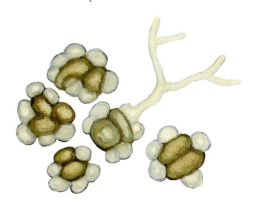

Urocystis ranunculi (Lib.) Moesz: dark teliospores, surrounded by sterile cells, lighter in color, of this pathogen (smut) of *Ranunculus*, x 1,050.

Uromyces (Link) Unger
Pucciniaceae, Pucciniales, Inc. sed., Pucciniomycetes, Pucciniomycotina, Basidiomycota, Fungi
Unger, F. J. A. N. 1833. *Exanth. Pflanzen* (Wien):277. From Gr. *ourá*, tail + *mýkes* > L. *myces*, fungus, for the pedicellate urediniospores and teliospores, or from L. *uro*, to burn, to light, consume, inflame + *mýkes* > L. *myces*, fungus, for the appearance of the lesions that this rust causes on the plants that it parasitizes. It is a genus similar to *Puccinia* Pers., from which it differs principally in having unicellular teliospores. Aecia subepidermal, eruptive. Urediniospores borne individually on pedicels. Telia exposed or covered by epidermis of host plant; teliospores pedicellate with a generally pigmented wall. Causing rust diseases.

Urosporellopsis taiwanensis W. H. Hsieh, et al.: vertical section of an immersed, brown ascostroma, isolated from dead stems of *Sassafras randaiense*, Taiwan, showing the asci inside, x 212; 8-spored ascus, x 440, and one fusiform ascospore, with a septum and, at the base, a filiform, long appendage, x 620.

Urupe

Urupe Viégas

Meliolaceae, Meliolales, Inc. sed., Sordariomycetes, Pezizomycotina, Ascomycota, Fungi

Viégas, A. P. 1944. Alguns fungos de Brasil II. Ascomicetos. *Bragantia* 4(1-6):1-392.

From Tupi *Urupê*, fungus. Mycelium intercellular, subtorulose, septate, with irregular haustoria, forming a subepidermal stroma that erupts through epidermis to form ascomata. Ascomata globose, formed in clusters on stroma, black, with an apical pore. Asci clavate, short-pedicellate, 8-spored. Ascospores fusiform, hyaline, 3-septate, with end cells curved at tips. On leaves of *Guadua* sp. in Santa Catarina State, Brazil.

Urupe guaduae Viégas: epiphyte ascomata on the surface of a host plant (*Guadua* sp.), x 5; vertical section of an ascoma showing the peridium, x 260; ascus with 8 developing ascospores, x 1,400; one released ascospore, fusiform, 2-septate, x 1,500. Described in the Morro de Aipium, Brazil.

Usnea Dill. ex Adans.

Parmeliaceae, Lecanorales, Lecanoromycetidae, Lecanoromycetes, Pezizomycotina, Ascomycota, Fungi

Adanson, M. 1763. *Fam. Pl.* (Paris) 2:7.

From Arabian *oshnab*, moss. Thallus fruticose, erect or pendulous, commonly branched, similar to certain mosses. Apothecia small or large, lateral or terminal, sessile, with disc concave or cyathiform, at times flat, greenish or yellowish, with excipulum the same color as thallus, more or less irregular. On trees, shrubs, and sometimes on old wood and rocks.

Usnea rubicunda Stirt.: fruticose, branched, yellowish-green to reddish thallus, on a tree branch, x 1.

Ustilago (Pers.) Roussel

Ustilaginaceae, Ustilaginales, Ustilaginomycetidae, Ustilaginomycetes, Ustilaginomycotina, Basidiomycota, Fungi

Roussel, H. F. A. 1806. *Fl. Calvados, Edn* 2:47.

From L. *ustus, ustulatus*, burned, scorched, cauterized, from *ustulo*, to burn, for the black color of the mass of spores in the sorus. Fructifications fragment completely into teliospores. Sori formed principally in inflorescence, lacking columella, naked or covered by a peridium. Spores echinulate, not arranged in chains or balls; a large quantity produced within each sorus, constituting black, powdery masses at maturity; germinating by forming a short, triseptate basidium that produces basidiospores (sporidia), one per cell. Cause smut diseases, principally on grasses. *Ustilago maydis* (D.C.) Corda causes the smut of maize. In Mexico the immature smut galls are called *cuitlacoche* or *huitlacoche* and are eaten by indigenous and mestizo peoples.

Ustilago maydis (DC.) Corda: black colored, powdery sori of teliospores, on a maize cob, x 0.4.

Utharomyces Boedijn

Pilobolaceae, Mucorales, Inc. sed., Inc. sed., Mucoromycotina, Zygomycota, Fungi

Boedijn, K. B. 1959. Notes on Mucorales of Indonesia. *Sydowia* 12(1-6):321-362.

From Gr. *oúthar*, genit. *outháratos*, udder, breast + *mýkes*

> L. *myces*, fungus, for the shape of the sporangium. Sporangia globose, columellate, with a subsporangial vesicle on apophysate or nonapophysate sporangiophores supported by trophocysts. Sporangial wall with stellate dehiscence wall. Saprobic, isolated from mouse dung.

Utharomyces epallocaulus Boedijn ex P. M. Kirk & Benny: breast-shaped, columellate sporangia, with stellate dehiscence of the sporangial wall; note the subsporangial vesicle, x 250. Isolated from dung of *Rattus losea*, Taiwan.

Utriascus Réblová

Inc. sed., Sordariales, Sordariomycetidae, Sordariomycetes, Pezizomycotina, Ascomycota, Fungi

Réblová, M. 2003. *Utriascus*, a new ascomycetous genus in the Sordariales. *Mycologia* 95(1):128-133.

From L. *uter*, bag + ML. *ascus* < Gr. *askós*, wine skin, for the saccate shape of the ascus. Ascomata superficial, solitary or in small groups, subglobose to conical, papillate, setose when old, with a pore at top. Ascomata attached to substratum with long, brown, flexuous, unbranched hyphae. Setae short, apically pointed to blunt, 0-1-septate, dark brown, covering whole perithecium. Perithecial wall hard, fragile, two-layered. Interthecial filaments septate, filiform, hyaline fragments among asci, deliquescing. Periphyses not observed. Asci unitunicate, thin-walled, saccate, formed in a basal fascicle, long-stipitate, apically narrowly to broadly rounded depending on arrangement of ascospores within ascus, 8-spored, persistent at maturity. Ascal apex non-amyloid, lacking apical ring. Ascospores bilaterally flattened with an elliptical outline, circular in front view, 1-celled, olivaceous to pale brown when young, brown center at maturity, enclosed in a firm, hyaline, finely verruculose sheath. Sheath present, ornamented in earliest stages of ascospore development. Ascospores filled with many droplets. Germ pores and germ slit lacking. On soft decayed wood of *Ulmus glabra* in the Šumava Mts. National Park, Czech Republic.

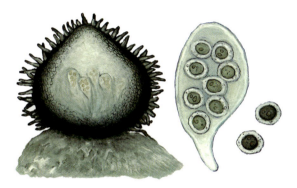

Utriascus gabretae Réblová: vertical section of a superficial, subglobose-conical, papillate, setose perithecium, on soft decayed wood of trunk of *Ulmus glabra*, x 90; unitunicate, saccate, 8-spored ascus, x 500, and free ascospores bilaterally flattened, 1-celled, brown center, enclosed in a hyaline, verruculose sheath, x 600.

Uyucamyces H. C. Evans & Minter—see Ocotomyces H. C. Evans & Minter

Inc. sed., Rhytismatales, Leotiomycetidae, Leotiomycetes, Pezizomycotina, Ascomycota, Fungi

Evans, H. C, D. W. Minter 1985. Two remarkable new fungi on pine from Central America. *Trans. Br. mycol. Soc.* 84(1):57-78.

Named after Cerro *Uyuca* + L. *-myces*, fungus, for the type locality of the fungus in Honduras.

V

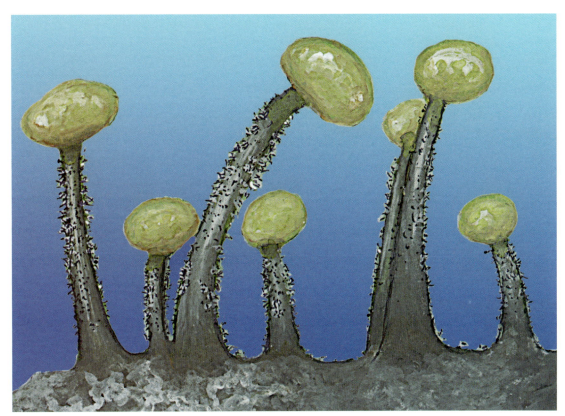
Vibrissea truncorum

V

Valsa Fr.—see **Cytospora** Ehrenb.
Valsaceae, Diaporthales, Sordariomycetidae, Sordariomycetes, Pezizomycotina, Ascomycota, Fungi
Fries, E. M. 1849. *Summa veg. Scand.,* Sectio Post. (Stockholm):410.
The etymology could not be determined.

Vascellum F. Šmarda
Agaricaceae, Agaricales, Agaricomycetidae, Agaricomycetes, Agaricomycotina, Basidiomycota, Fungi
Šmarda, F. 1958. Lycoperdaceae. *In:* A. Pilat, *Fl. CSR,* B-1, Gasteromycetes:760-761.
From L. *vas,* urn, vessel + suf. *-cellum,* which indicates dim., due to the shape of the fructification. Gleba and sterile base separated by a conspicuous membrane, a diaphragm. In grasslands in tropical regions, conifer forests, arid and subarid regions.

Vascellum lloydianum A. H. Sm.: vessel-shaped, white fructifications growing in wood in conifer forest, x 1. Described in Washington, U.S.A.

Velutarina Korf ex Korf
Hyaloscyphaceae, Helotiales, Leotiomycetidae, Leotiomycetes, Pezizomycotina, Ascomycota, Fungi
Korf, R. P. 1971. Some new discomycete names. *Phytologia* 21(4):201.
From L. *vellus,* wool + suf. *-aria,* denotes likeness, for the woolly ascomata. Apothecia erumpent, becoming superficial, single or in dense cespitose clusters, at first closed, opening but remaining cupulate, externally clothed with a dense woolly growth, tan or pale brown. Hymenium concave or nearly plane. Asci subcylindric, inoperculate, 8-spored. Ascospores ellipsoid, simple, hyaline or slightly colored. Paraphyses filiform to clavate. On dead branches.

Velutarina rufo-olivacea (Alb. & Schwein.) Korf (syn. **Velutaria rufo-olivacea** (Alb. & Schwein.) Fuckel): two rehydrated apothecia growing in the bark of the natural host, branches of *Rubus fruticosus,* Brandenburg, Germany; the apothecial hymenium is green-olivaceous to brownish, externally clothed with a dense woolly growth, brick-brownish, x 15; inoperculate ascus with ellipsoid, hyaline ascospores, x 400.

Venturia Sacc. (syn. **Spilocaea** Fr.)
Venturiaceae, Venturiales, Pleosporomycetidae, Dothideomycetes, Pezizomycotina, Ascomycota, Fungi
Saccardo, P. A. 1882. *Syll. fung.* (Abellini) 1:586.
Dedicated to the Italian mycologist A. Venturi + *-a,* L. fem. ending. Ascocarps sunken in tissues of dead leaves. Ascospores bicellular, yellowish, with upper cell shorter than lower one. Forcibly ejected through opening of ascocarps. Conidiophores with annellidic phialides. Parasitizes apple, pear, loquat and other plants of economic importance, on which it produces scab or canker. Forms necrotic, corky and chlorotic spots on leaves and fruits.

Verpa

Venturia inaequalis (Cooke) G. Winter (syn. **Spilocaea pomi** Fr.): vertical section of a pseudothecium, sunken in the tissues of dead leaf of apple parasitized by the fungus, showing the asci and bicellular, yellowish ascospores, x 1,200.

Venturia inaequalis (Cooke) G. Winter (syn. **Spilocaea pomi** Fr.): erumpent conidiophores, which cause apple scab, with annellidic phialides that produce flame-shaped, greenish-brown conidia, x 690.

Veramycella G. Delgado

Inc. sed., Inc. sed., Inc. sed., Inc. sed., Pezizomycotina, Ascomycota, Fungi

Delgado, G. 2009. South Florida microfungi: *Veramycella bispora*, a new palmicolous anamorphic genus and species, with some new records for the continental U.S.A. *Mycotaxon* 107:358.

From genus *Veramyces* Matsush. + L. dim. suf. *-ella*, i.e., like a small *Veramyces*. Colonies effuse, hairy, brown with mycelium mostly immersed, composed of septate, branched, smooth-walled, brown hyphae. Dark stromata usually present. Condiophores macronematous, mononematous, single or grouped, erect, unbranched, straight or subflexuous, cylindrical, smooth-walled, brown, paler toward apex, regenerating percurrently. Conidiogenous cells polyblastic, integrated, terminal, intercalary, sympodial, with loci flattened, non-protuberant, apical and lateral. Conidial secession schizolytic. Conidia holoblastic, acropleurogenous, catenate in simple, acropetal chains, polymorphic, mitrate, fusiform, clavate, pyriform, distoseptate, subhyaline to pale olivaceous, smooth-walled. On dead leaves of *Sabal palmetto* in Miami-Dade Co., Florida, U.S.A.

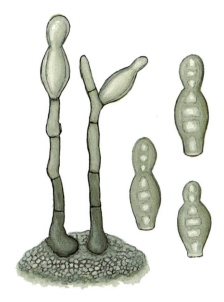

Veramycella bispora G. Delgado: saprobic fungus, which has the stroma at the base of the conidiophores; the conidiophores have enteroblastic proliferation, with straight or slightly sympodial conidiogenous cells that give rise to acropetal, distoseptate, 2-celled conidia, x 900. On rachides of dead leaves of *Sabal palmetto*, Florida, U.S.A.

Vermisporium H. J. Swart & M. A. Will.—see Seimatosporium Corda

Inc. sed., Inc. sed., Inc. sed., Inc. sed., Pezizomycotina, Ascomycota, Fungi

Swart, H. J., M. A. Williamson. 1983. Australian leaf-inhabiting fungi XVI. *Vermisporium*, a new genus of Coelomycetes on *Eucalyptus* leaves. *Trans. Br. mycol. Soc.* 81(3):491-502.

From L. *vermis*, worm + *spora*, spore + NL. dim. suf. *-ium*, for the shape of the conidia. Conidiomata in necrotic leaf spots, apparently closed when young, at maturity wide open (by disintegration of most or all of the upper cell layer), subepidermal or intra-epidermal, composed of a few layers of light brown pseudoparenchymatous cells. Conidiophores short or absent. Conidiogenous cells holoblastic, annellidic, discrete, cylindrical or flask-shaped. Conidia hyaline, white or pale pink in mass, long and narrow, 3- or usually 4-celled, with thin smooth walls, not or slightly constricted at the septa, acerose to falcate, slightly curved or flexuous, with an exogenous basal appendage. Collected on leaves of *Eucalyptus melliodora*, Box Hill, Australia.

Verpa Swartz (syn. Ptychoverpa Boud.)

Morchellaceae, Pezizales, Pezizomycetidae, Pezizomycetes, Pezizomycotina, Ascomycota, Fungi

Swartz, O. 1815. Svampar saknade i Fl. Sv. L., funne i Sverige, och anteknade. *K. svenska Vetensk-Akad. Handl.*, ser. 3, 36:129.

From L. *verpa*, penis, phallus, for the shape of the fructification. Ascocarps with pileus campanulate or

Verrucaria

conical, smooth or slightly pleated, pendulous on apex of a cylindrical stipe, so that it resembles a thimble on stipe, which is long and hollow. Hymenium lines outer surface of smooth or somewhat pleated longitudinally pileus. In grasslands and roadsides or forests.

Verpa bohemica (Krombh.) J. Schröt.: apothecium with the pileus conical, lightly pleated, on soil, x 0.7.

Verrucaria Schrad.

Verrucariaceae, Verrucariales, Inc. sed., Eurotiomycetes, Pezizomycotina, Ascomycota, Fungi
Schrader, H. A. 1794. *Spicil. fl. germ.* 1:108.
From L. *verruca*, wart + suf. *-aria*, which indicates belonging to or possession. Thallus crustose with a warted or areolated surface. Perithecia small or minute, more or less immersed one to several in each wart or areola of thallus; wall dimidiate or complete; ostiole inconspicuous. On wet or submerged rocks along the marine coasts and in fresh water or on rocks in dry places.

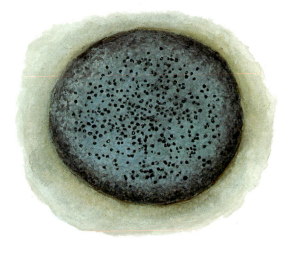

Verrucaria ditmarsica Erichsen: bluish, crustous thallus, with a warted or areolated surface, which possess immersed minute perithecia in each wart or areola, on rock in dry places, x 1.5.

Verticicladiella S. Hughes—see Leptographium Lagerb. & Melin

Ophiostomataceae, Ophiostomatales, Sordariomycetidae, Sordariomycetes, Pezizomycotina, Ascomycota, Fungi
Hughes, S. J. 1953. Conidiophores, conidia and classification. *Can. J. Bot.* 31:577-659.
From L. *verticillus*, verticil, disk + Gr. *kládos*, branch + L. dim. suf. *-ella*, i.e., having verticillate branches, referring to the conidiophores.

Verticillium Nees

Plectosphaerellaceae, Inc. sed., Hypocreomycetidae, Sordariomycetes, Pezizomycotina, Ascomycota, Fungi
Nees von Esenbeck, C. G. 1816. *Syst. Pilze* (Würzburg):57.
From L. *verticillus*, verticil, slice + dim. suf. *-ium*, referring to the verticillate arrangement of the phialides. Conidiophores arranged in primary, secondary or a higher order of verticils along principal axis or branches. Conidia phialospores, unicellular, hyaline, embedded in mucilage, forming gloeoid balls on apices of phialides. Common in soil as a saprobe, although there are plant pathogenic species that attack the vascular system of plants.

Vibrissea Fr. (syn. Apostemidium (P. Karst.) P. Karst.)

Vibrisseaceae, Helotiales, Leotiomycetidae, Leotiomycetes, Pezizomycotina, Ascomycota, Fungi
Fries, E. M. 1822. *Syst. mycol.* (Lund) 2:4, 31.
From L. *vibrissae*, small hairs in the nose, from *vibro*, *vibras*, to vibrate, wiggle, to shake with a tremulous movement. Apothecia sessile, turbinate or convex, soft-wavy or subgelatinous with hairs or filaments on pedicel. Hairs moved by water, appear to have tremulous movement. Pedicel widens toward apex and terminates in a small expanded, compressed head or cap, where asci formed. Hymenium spread on upper, convex or surface of substrate, sterile below. Asci narrowly cylindric, opening by a pore, 8-spored. Ascospores in a parallel fascicle in ascus, hyaline, filiform, many-septate, nearly as long as the ascus. Paraphyses present. On sawn wood soaked in water and other moist plant materials.

Viegasella Inácio & P. F. Cannon

Vibrisseaceae, Helotiales, Leotiomycetidae, Leotiomycetes, Pezizomycotina, Ascomycota, Fungi
Inácio, C. A., P. F. Cannon. 2003. *Viegasella* and *Mintera*, two new genera of Parmulariaceae (Ascomycota), with notes on the species referred to *Schneepia*. *Mycol. Res.* 107(1):82-92.
Named in honor of the Brazilian mycologist *Ahmés Pinto Viégas* + L. dim. suf. *-ella*, for his work on Brazilian fungi. Colonies superficial on host leaves. Ascostromata black, circular or stellate, multilocular, opening by

longitudinal slits. External mycelium and conidiomata absent. Ascomatal locules round, ellipsoidal or elongate. Asci cylindric-clavate, thick-walled, not blueing in iodine, 6-8-spored, with rostrate dehiscence. Ascospores cylindric-ellipsoidal, pale brown, verrucose, 1-septate. On leaves of Sapotaceae in Paraguay and Brazil.

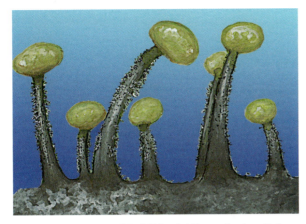

Vibrissea truncorum (Alb. & Schwein.) Fr.: apothecia, with the black pedicel, covered by hairs or filaments that vibrate with the movements of the water; the pedicel expands and terminates in a yellowish compressed cap, x 10. This fungus develops on sawn submerged wood.

Viegasella pulchella (Speg.) Inácio & P. F. Cannon: black ascomata, sometimes confluent, circular to star-shaped, irregular, light brown to reddish with a diffuse edge, growing on the surface of plant host (Sapotaceae), x 50; vertical section of an ascoma with bitunicate asci, x 225; one ascus with 8 ascospores, which are 1-septate, guttulate, x 300.

Vinculum R. Y. Roy, et al.—see **Barnettella** D. Rao & P. Rag. Rao
Inc. sed., Inc. sed., Inc. sed., Inc. sed., Pezizomycotina, Ascomycota, Fungi
Roy, R. Y., et al. 1965. *Vinculum*, a new genus of Deuteromycetes. *Trans. Br. mycol. Soc.* 48(1):113-115.
From L. *vinculum*, cord, chain, for the chains of conidia and connecting cells.

Virgella Darker
Rhytismataceae, Rhytismatales, Leotiomycetidae, Leotiomycetes, Pezizomycotina, Ascomycota, Fungi

Darker, G. D. 1967. A revision of the genera of the Hypodermataceae. *Can. J. Bot.* 45:1399-1444.
From L. *virga*, streak, stripe of color on cloth + dim. suf. *-ella*, because the ascomata appear as dark streaks on the host. Ascoma a linear apothecium, hypophyllous, dark brown, immersed in host tissues, opening by a longitudinal fissure. Covering layer of dark pseudoparenchyma tissue; basal layer plectenchymatous and hyaline. Asci saccate-clavate, unitunicate. Ascospores rod-shaped, 1-celled, with a gelatinous sheath. Pycnidia epiphyllous, spermatia minute, bacillar. On needles of conifers, causing lesions.

Viscospora Sieverd., et al.—see **Septoglomus** Sieverd., et al.
Glomeraceae, Glomerales, Inc. sed., Glomeromycetes, Glomeromycotina, Glomeromycota, Fungi
Oehl, F., et al. 2011. Glomeromycota: three new genera and glomoid species reorganized. *Mycotaxon* 116:75-120.
From L. *viscosus*, sticky + *spora*, spore; referring to the adhesive nature of the spore surface of the type species of the genus.

Vivantia J. D. Rogers, et al.
Xylariaceae, Xylariales, Xylariomycetidae, Sordariomycetes, Pezizomycotina, Ascomycota, Fungi
Rogers, J. D., et al. 1996. *Biscogniauxia anceps* comb. nov. and *Vivantia guadalupensis* gen. et sp. nov. *Mycol. Res.* 100(6):669-674.
Named for the French mycologist *Jean Vivant*, who collected the fungus + L. des. *-ia*. Stromata applano-pulvinate, probably bipartite when immature, wide spreading; externally and internally black, fairly carbonaceous. Perithecia embedded in stroma. Ostioles umbilicate. Asci cylindrical, shortstipitate, with apical ring blueing in Melzer's reagent. Ascospores hyaline to subhyaline, smooth, two-celled, with septum median, submedian, or strongly eccentric, without discernible germination site. Paraphyses broad, longer than asci, numerous. On wood in Guadeloupe.

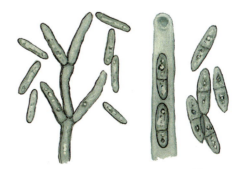

Vivantia guadalupensis J. D. Rogers, et al.: conidiogenous cells of the nodulisporium-like asexual morph with unicellular conidia, x 750; ascus with an apical ring blueing in Melzer's reagent, x 1,000, and free, hyaline, two-celled, with the septum median, ascospores, x 950. Isolated from wood, Guadeloupe.

Vizella

Vizella Sacc.
Vizellaceae, Inc. sed., Dothideomycetidae, Dothideomycetes, Pezizomycotina, Ascomycota, Fungi
Saccardo, P. H. 1883. *Syll. fung.* (Abellini) 2:662.
Named in honor of the English mycologist Reverend *J. Vize* + L. dim. suf. *-ella*. Pseudothecia black, carbonaceous, subcuticular, hemiperithecioid with a small ostiole in the center of the shield and obclavate or cylindrical asci with muriform, hyaline, oblong ascospores. On living branches and stems of plants in temperate regions.

From genus *Volvaria* Fr. (=*Volvariella* Speg.), a name that is composed of L. *volva*, wrapping, covering, cover, matrix + suf. *-aria*, which indicates connection or possession + L. dim. suf. *-ella*, due to the presence of a universal veil. Basidioma fleshy, medium-sized, with covering, which, on breaking, remains at base of stipe as a membranous cup. Free gills, pink-colored due to color of spores. Lignicolous, humicolous or coprophilous. The species are edible and exploited commercially, principally in Asiatic countries where they are cultivated on an industrial scale.

Vizella oleariae H. J. Swart: vertical section of a black, carbonaceous, subcuticular pseudothecium, with a small ostiole, on leaf of *Olearia argophylla*, in Victoria, Australia, showing the cylindrical asci with muriform ascospores, x 380.

Volvariella Speg.
Pluteaceae, Agaricales, Agaricomycetidae, Agaricomycetes, Agaricomycotina, Basidiomycota, Fungi
Spegazzini, C. 1898. Fungi Argentini novi V. critici. *Anal. Mus. nac. Hist. nat. B. Aires* 6:118.

Volvariella bombycina (Schaeff.) Singer: pinkish basidioma of this lignicolous fungus, with a membranous universal veil at the base of the stipe, x 0.5.

W

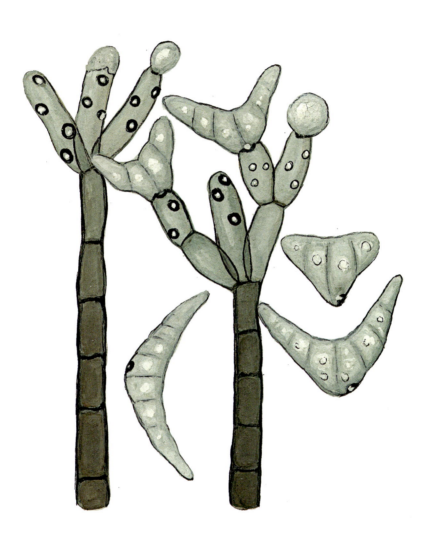

Weufia tewoldei

W

Wardinella Bat. & Peres

Inc. sed., Inc. sed., Inc. sed., Inc. sed., Pezizomycotina, Ascomycota, Fungi

Batista, A. C., et al. 1960. Novos fungos Asterinothyriaceae e Plenotrichaceae. *Publicações Inst. Micol. Recife* 221:1-22.

From genus *Wardina* G. Arnaud + L. dim. suf. *-ella*. Mycelium superficial, reticulate, smooth, with intercalary hyphopodia. Pycnostromata superficial, dimidiate-scutate, smooth, brown, membraneous, unilocular, with a central circular ostiole. Conidiophores short, formed on underside of upper wall. Conidia cylindrical, phragmosporous, subhyaline at first, then olivaceous. On leaves of an unknown plant in Jabatão, Pernambuco, Brazil.

Wardomyces F. T. Brooks & Hansf.

Microascaceae, Microascales, Hypocreomycetidae, Sordariomycetes, Pezizomycotina, Ascomycota, Fungi

Brooks, F. T., C. G. Hansford. 1922. Mould growth upon cold-store meat. *Trans. Br. mycol. Soc.* 8(3):135-137.

In memory of the British mycologist *Marshall Ward* + Gr. *mýkes* > L. *myces*. Conidiophores with terminal, inflated conidiogenous cells. Conidia unicellular, ovoid, dark with a germ slit. Isolated from chicken eggs, rabbit meat, mangrove mud, and soils cultivated with cereals.

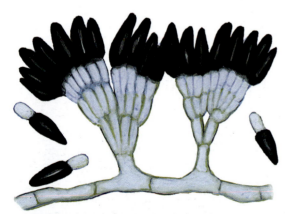

Wardomyces humicola Hennebert & G. L. Barron: conidiophores, with terminal conidiogenous cells that form unicellular, ovoid, dark conidia with a germ slit, x 1,000. Described in Ontario, Canada.

Wenyingia Zheng Wang & Pfister

Pyronemataceae, Pezizales, Pezizomycetidae, Pezizomycetes, Pezizomycotina, Ascomycota, Fungi

Wang, Z., D. H. Pfister. 2001. *Wenyingia*, a new genus in Pezizales (Otideaceae). *Mycotaxon* 79:397-399.

Named in honor of the Chinese mycologist *Wen-ying Zhuang* + L. des. *-ia*. Ascomata cupulate, stipitate apothecia, pale brown to buff, strongly concave when dry. Hymenium dark brown when dry, covered by a thin, white membrane originating from parallel hyphae at margin. Asci operculate, thick-walled, cylindrical, tapering toward base, J-, 8-spored. Ascospores ellipsoid to ovoid, smooth, with two oil droplets. On soil in Sichuan Province, China.

Westea H. J. Swart

Inc. sed., Inc. sed., Inc. sed., Dothideomycetes, Pezizomycotina, Ascomycota, Fungi

Swart, H. J. 1988. Australian leaf-inhabiting fungi. XXIX. Some Ascomycetes on *Banksia*. *Trans. Br. mycol. Soc.* 91(3):453-465.

Named in honor of the Australian mycologist *Gretna M. Weste* + euphonic des. *-a*, who pioneered study of *Phytophthora cinnamomi* in Victoria. Mycelium both in and on host leaf; hyphae thin, hyaline to straw-colored. Ascomata hypophyllous, seated among epidermal hairs, small, globose or broadly ellipsoidal. Ascomatal wall of an outer zone of 2-3 layers of firm-walled, brown pseudoparenchyma and an inner zone of flattened hyaline cells. Asci bitunicate, obovoid with a short stalk, 8-spored. Ascospores hyaline, phragmosporous, with a pronounced near-median constriction at which they easily break. On living leaves of *Banksia intregrifolia*, Mallacoota Inlet, Victoria, Australia.

Westerdykella Stolk

Sporormiaceae, Pleosporales, Pleosporomycetidae, Dothideomycetes, Pezizomycotina, Ascomycota, Fungi

Stolk, A. C. 1955. *Emericellopsis minima* sp. nov. and *Westerdykella ornata* gen. nov., sp. nov. *Trans. Br. mycol. Soc.* 38(4):419-424.

Named in honor of the Dutch mycologist *Johanna West-*

Wickerhamia

erdijk, director of the Centraal Bureau voor Schimmelcultures + L. dim. suf. *-ella*. Perithecia globose, astomous, black. Perithecial wall membranaceous, of one layer of brown to black thick-walled cells. Asci numerous, subglobose to elliptical, stalked, many-spored, evanescent. Ascospores globose to subglobose, continuous, brown, with spiral bands. No germ pores. Isolated from mangrove mud on the island of Inhaca, Mozambique, East Africa.

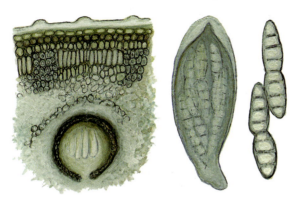

Westea banksiae H. J. Swart: vertical section of a leaf of *Banksia integrifolia* from Victoria, Australia; note the hypophyllous ascostroma, among the epidermic hairs, with four asci, x 100; bitunicate ascus with eight young ascospores, x 450, and free mature hyaline ascospores (phragmospores), with a near-median constriction at which they easily break, x 500.

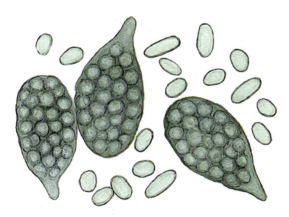

Westerdykella cylindrica (Malloch & Cain) Arx: coprophilous pyrenomycete, with subglobose, stalked asci containing 32 brown, oblong-ellipsoidal ascospores, x 1,100. Described in horse dung in the Tuscan Archipelago, Italy.

Weufia Bhat & B. Sutton

Inc. sed., Inc. sed., Inc. sed., Inc. sed., Pezizomycotina, Ascomycota, Fungi

Bhat, D. J., B. C. Sutton. 1985. New and interesting Hyphomycetes from Ethiopia. *Trans. Br. mycol. Soc.* 85 (1):107-122.

From Amharic *weuf*, bird-like appearance + L. des. *-ia*, for the shape of the conidia. Colonies effuse, black, velvety. Conidiophores mononematous, septate, brown, branched only at apex. Conidiogenous cells integrated, terminal, with several thickened conidiogenous loci; proliferation at loci enteroblastic and simultaneous with conidial ontogeny. Conidia holoblastic, produced through predetermined pores in conidiogenous cell wall, solitary, dry, V-shaped, with 2 divergent arms, brown, smooth, distoseptate, bilaterally symmetrical. On dead branches, Kaffa, Ethiopia.

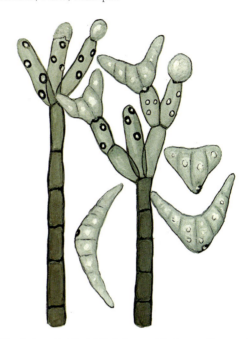

Weufia tewoldei Bhat & B. Sutton: conidiophores with several thickened conidiogenous loci; proliferation at the loci is enteroblastic and simultaneous with conidial ontogeny, to give rise to "V"-shaped conidia, with 2-divergent arms, brown, distoseptate (6-14 distosepta), bilaterally symmetrical, x 480. On dead twigs, Ethiopia.

Whalleya J. D. Rogers, et al.

Xylariaceae, Xylariales, Xylariomycetidae, Sordariomycetes, Pezizomycotina, Ascomycota, Fungi

Rogers, J. D., et al. 1997. *Jumillera* and *Whalleya*, new genera segregated from *Biscogniauxia*. *Mycotaxon* 64:39-50.

Named in honor of the British mycologist *Anthony J. S. Whalley* + euphonic des. *-a-*, for his studies of the xylariaceous fungi. Stromata applanate, solitary or confluent, lacking KOH-extractable pigments. Outer dehising layer dark brown, thin, exposing mature black layer. Perithecia globose, monostichous. Ostioles depressed, appearing punctate. Asci 8-spored, cylindrical, short-stipitate, persistent, with a discoid apical ring, amyloid. Ascospores pale brown, unicellular, ellipsoid-inaequilateral, with narrowly rounded ends and a straight germ slit, smooth. Asexual morph geniculosporiun-like, but with denticulate secession scars.

Wickerhamia Soneda

Saccharomycodaceae, Saccharomycetales, Saccharomycetidae, Saccharomycetes, Saccharomycotina, Ascomycota, Fungi

Wickerhamomyces

Soneda, M. 1960. On a new yeast genus *Wickerhamia*. *Nagaoa* 7:9.

Named in honor of the American zymologist *Lynferd J. Wickerham* + L. des. *-ia*, for his contributions to yeast taxonomy. Ascomata and hyphae lacking. Somatic cells unicellular, biapiculate, sometimes forming short chains, ovoid, elongate, slightly curved, often swollen in middle. Sugar fermentative; nitrate not assimilated. Asci form directly from somatic cells, 1-2-spored, occasionally with up to 16 spores. Ascospores 1-celled, hyaline, cap-shaped, with a lateral brim, smooth. Asexual morph kloeckera-like. On dung.

Wickerhamomyces Kurtzman, et al.
Wickerhamomycetaceae, Saccharomycetales, Saccharomycetidae, Saccharomycetes, Saccharomycotina, Ascomycota, Fungi

Kurtzman, C. P., et al. 2008. Phylogenetic relationships among species of *Pichia*, *Issatchenkia* and *Williopsis* determined from multigene sequence analysis, and the proposal of *Barnettozyma* gen. nov., *Lindnera* gen. nov. and *Wickerhamomyces* gen. nov. *FEMS Yeast Res.* 8 (6):939-954.

Named in honor of the American zymologist *Lynferd J.Wickerham* + connective *-o-* + L. *myces*, fungus, for his extensive work on yeast taxonomy and ecology. Asci globose to ellipsoid, unconjugated or arise from conjugation between a cell and its bud or between independent cells. Some species heterothallic. Asci deliquescent or persistent, forming one to four ascospores. Ascospores hat-shaped or spherical with an equatorial ledge. Cell division by multilateral budding on a narrow base and budded cells spherical, ovoid or elongate. Pseudohyphae and true hyphae are formed by some species. Glucose fermented by most species; some species ferment other sugars. A variety of sugars, polyols and other carbon sources are assimilated, but not methanol and hexadecane. Nitrate is utilized by some species. The predominant ubiquinone is CoQ-7. The diazonium blue B reaction is negative.

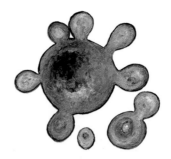

Wickerhamomyces anomalus (E. C. Hansen) Kurtzman, et al.: multilateral budding of yeast cells in agar culture, x 4,500.

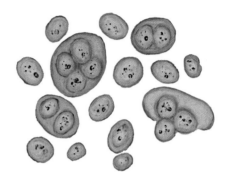

Wickerhamomyces anomalus (E. C. Hansen) Kurtzman, et al.: yeast cells with active budding in the male gonoducts of mosquito *Anopheles stephensi* in Italy, the malary vector, not shown, and asci which produce one to four spherical ascospores, x 1,500.

Wolfiporia Ryvarden & Gilb.
Polyporaceae, Polyporales, Inc. sed., Agaricomycetes, Agaricomycotina, Basidiomycota, Fungi

Ryvarden, L., R. L. Gilbertson. 1984. Type studies in the Polyporaceae, 15 species described by L. O. Overholts, either alone or with J. L. Lowe. *Mycotaxon* 19:141.

Named in honor or the American mycologist *Frederick A. Wolf* + *-i-* connective + genus *Poria* Pers., who first described the fungus. Fructifications resupinate, effuse; pores cream, ochraceous or brownish. Hyphal system monomitic or subdimitic. Generative hyphae thin-walled, lacking clamps. Skeletal hyphae branched. Basidiospores thin-walled, hyaline, cylindric to ellipsoidal, nonamyloid.

X

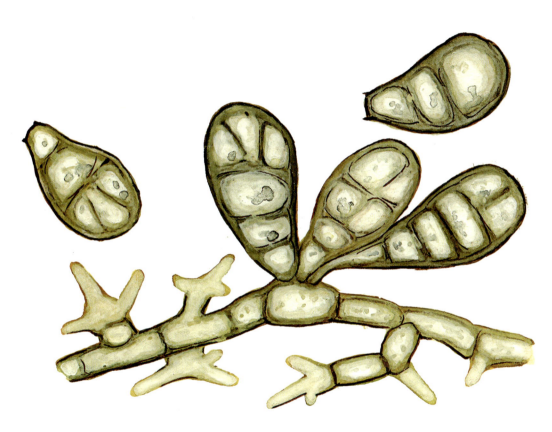

Xylochia stegonsporioides

X

Xanthophyllomyces Golubev (syn. **Phaffia** M. W. Mill., et al.)

Mrakiaceae, Cystofilobasidiales, Inc. sed., Tremellomycetes, Agaricomycotina, Basidiomycota, Fungi
Golubev, V. I. 1995. Perfect state of *Rhodomyces dendrorhous* (*Phaffia rhodozyma*). *Yeast* 11(2):105.

From Gr. *xánthos*, yellow + *phýllon*, leaf + L. *myces*, fungus. Anascosporogenic yeast, carotenogenic, fermentative. Isolated from viscus exudates of plants. Because of its astaxanthin content, it is used as a source pigment incorporated into the diet of fish and crustaceans in hatcheries.

Xanthoria (Fr.) Th. Fr.

Teloschistaceae, Teloschistales, Lecanoromycetidae, Lecanoromycetes, Pezizomycotina, Ascomycota, Fungi
Fries, T. M. 1860. Lichenes arctoi Europae Groenlandiae que hactenus cogniti. *Nova Acta R. Soc. Scient. Upsal.*, Ser. 3:166.

From Gr. *xánthos*, yellow + suf. -*oria*, which indicates capacity, action, function or characteristic, for the yellow color of the thallus in the majority of the species in the genus. Thallus foliose, yellow, red or orange, with small, narrow lobules. Some species with soredia, others with only apothecia. On tree trunks, branches or rocks.

Xanthoria parietina var. **parietina** (L.) Th. Fr.: foliose, yellow thallus, with small narrow lobules and reddish apothecia, on tree trunk, x 1.

Xanthoriicola D. Hawksw.

Inc. sed., Inc. sed., Inc. sed., Inc. sed., Pezizomycotina, Ascomycota, Fungi
Hawksworth, D. L., E. Punithalingam. 1973. New and interesting microfungi from Slapton, South Devonshire: Deuteromycotina. *Trans. Br. mycol. Soc.* 61:57-69.

From genus *Xanthoria* (Fr.) Th. Fr. + L. suf. -*icola*, inhabitant, i.e., growing on *Xanthoria*. Fungus lichenicolous. Stromata, setae, and hyphopodia absent. Conidiophores semi-macronematous, densely aggregated, mononematous, erect, branched, irregularly penicillate, short, smooth, brown. Conidiogenous cells monophialidic, terminal, determinate, oblong to ampulliform, with a short apical collarette, brown. Conidia produced singly, dry, globose, simple, with minutely echinate walls, olivaceous-brown to brown. Growing on the ascocarps of *Xanthoria*.

Xenocylindrocladium Decock, et al.

Nectriaceae, Hypocreales, Hypocreomycetidae, Sordariomycetes, Pezizomycotina, Ascomycota, Fungi
Decock, C., et al. 1997. *Nectria serpens* sp. nov. and its hyphomycetous anamorph *Xenocylindrocladium* gen. nov. *Mycol. Res.* 101(7):786-790.

From Gr. *xénos*, foreign + genus *Cylindrocladium* Morgan, referring to the distinctive characters that separate it from this genus. Macroconidiophores penicillate. Stipe septate, hyaline, straight, becoming sinuous, coiled in upper part; avesiculate, tapering towards a terminal cell; stipe long. Conidiophore branches: primary branches non-septate or rarely 1-septate, secondary branches non-septate. Phialides doliiform to reniform, hyaline, non-septate. Conidia cylindrical, hyaline, straight with rounded ends, 1-septate. Colony reverse amber brown. Chlamydospores in extensive numbers, with medium to extensive sporulation on aerial mycelium. On bark of fallen tree trunk, Cuyabeno, Ecuador.

Xerocomellus Šutara

Boletaceae, Boletales, Agaricomycetidae, Agaricomycetes, Agaricomycotina, Basidiomycota, Fungi

Xylobotryon

Šutara, J. 2008. *Xerocomus* s.l. in the light of the present state of knowledge. *Czech. Mycol. 60*:44.

From Gr. *xerós*, dry + *kóme*, hair L. dim. suf. *-ellus*, referring to the dry, tomentose pileus of the majority of the species. Basidiocarps small to medium-size; pileus velvety, at times cracked, slightly viscous; hymenophore tubular with yellow pores, flesh turning bluish-green when exposed to air or damaged. Ectomycorrhizal with trees in particular pines and fir in temperate forests. Non-toxic but seldom eaten.

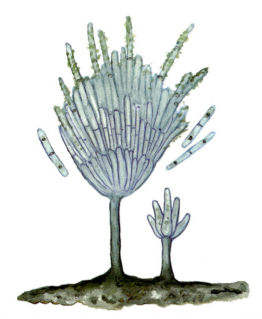

Xenocylindrocladium guianense Crous & Decock: penicillate conidiophores, with greenish exudate in the wall, terminating in a group of hyaline phialides bearing cylindrical, 1-septate conidia, x 450. On plant debris in rain-forest litter, French Guiana.

Xerocomellus pruinatus (Fr. & Hök) Šutara: dry, tomentose basidiocarps, with velvety pileus, on soil of temperate forest, x 0.5.

Xeromphalina Kühner & Maire

Mycenaceae, Agaricales, Agaricomycetidae, Agaricomycetes, Agaricomycotina, Basidiomycota, Fungi
Konrad, P., A. Maublanc. 1934. *Icon. Select. Fung.* 6:236.
From Gr. *xerós*, dry + genus *Omphalina* Quel.; a small and dry *Omphalina*. Pileus with initially incurved margin, yellow, vinaceous, or fulvous; spores hyaline, smooth, amyloid, ellipsoid, oblong or cylindric; spore print white; trama of pileus with clamp connections, inamyloid; stipe generally central, but often eccentric; veil none. In temperate zones of both hemispheres and tropical montane regions. Used by foresters and ecologists as a reliable indicator of coniferous wood.

Xerula Maire

Physalacriaceae, Agaricales, Agaricomycetidae, Agaricomycetes, Agaricomycotina, Basidiomycota, Fungi
Maire, R. 1933. Fungi Catalaunici: Contributions à l'étude de la Flore Mycologique de la Catalogne. *Treb. Mus. Ciènc. nat. Barcelona*, sér. bot., 15(2):66.
From Gr. *xéros*, dry, in reference to the absence of viscosity characteristic of the related genus *Mucidula* Pat. + dim. suf. *-ula*, i.e., a small and dry mushroom. Pileus convex, conic or bell-shaped, frequently becoming flat, generally pilose; gills broadly adnate or uncinate, not decurrent; stipe pilous, often hairy or furfuraceous. Spores smooth, veil poorly developed. Singly or in groups in mixed forests. Some species are edible.

Xylaria Hill ex Schrank

Xylariaceae, Xylariales, Xylariomycetidae, Sordariomycetes, Pezizomycotina, Ascomycota, Fungi
Hill, J. 1773. *Baier. Fl.* (München) 1:200.
From Gr. *xýlon*, wood + L. suf. *-aria*, similarity, resembling, due to its hard consistency. Perithecia embedded in a stroma, generally dark, globose, erect, clavate or irregular. Saprobe or weak parasite on wood in temperate forests and tropical zones.

Xylaria hypoxylon (L.) Grev.: hard, dark colored, woody ascostromata, with perithecia embedded, in this lignicolous fungus on dead wood, x 1.5.

Xylobotryon Pat.

Xylobotriaceae, Xylobotrales, Xylobotryomycetes, Pezizomycotina, Ascomycota, Fungi
Patouillard, N. T., G. De Lagerheim. 1895. Champignons

Xylochia

de l'Equateur (Pugillus IV). *Bulletin de l'Herbier Boissier* 3(1):53-74.

From Gr. *xýlon*, log, wood + *bótrys*, cluster of, grapes Stromata large, dark brown to black, upright, stipitate, branched or unbranched, bearing numerous superficial ascomata. Ascomata perithecioid, subglobose to ellipsoid or oval, sessile or short-stipitate; ostioles inconspicuous to papillate; ostiolar canal densely lined with hyaline periphyses with bluntly rounded ends. Paraphyses abundant, filiform, with free ends, hyaline. Asci bitunicate, narrowly clavate to slightly fusiform, 8-spored. Ascospores ellipsoid to slightly fusiform, equilateral, with broadly rounded ends, 2-celled, pale brown, with several longitudinal germ slits in each cell. Saprobic or possibly parasitic on wood or bark.

Xylobotryum andinum Pat.: a corymbose stroma with black perithecia, growing as saprobe on decorticated wood in Ecuador, x 15; vertical section of a perithecium showing its peridium of three layers, and the developing asci with 8 ascospores in the interior, x 175; two released ascospores, ellipsoidal to fusiform, bicellular, brown olivous, x 500.

Xylochia B. Sutton

Inc. sed., Inc. sed., Inc. sed., Inc. sed., Pezizomycotina, Ascomycota, Fungi

Sutton, B. C., N. D. Sharma. 1983. *Xylochia* gen. nov. and *Murogenella*, two Deuteromycete genera with distoseptate conidia. *Trans. Br. mycol. Soc.* 80(2):255-262.
From Gr. *xylochos*, a thicket + L. des. *-ia*, in reference to the conidiomatal structure. Mycelium branched, septate, brown, smooth. Conidiomata globose, pale brown, indehiscent, of a loose network of irregularly branched, septate, smooth hyphae with numerous small projections along length and on short lateral branches. Conidiophores micronematous, irregularly branched, pale brown, septate, smooth, individual cells function conidiogenously. Conidia holoblastic, solitary or in groups of 2-7, dry, smooth, pale golden-brown, obovoid to pyriform, with transverse, oblique or longitudinal distosepta, sometimes curved or inclined to one side, basal septum often thickened, lumina reduced; secession rhexolytic. On *Saccharum officinarum* in India.

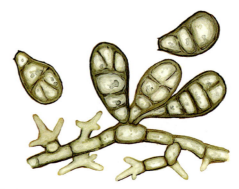

Xylochia stegonsporioides B. Sutton: holoblastic, pale golden, obovoid to pyriform, distoseptate, rhexolytic conidia, x 1,200. Isolated from *Saccharum officinarum*, India.

Xylocladium P. Syd. ex Lindau

Xylariaceae, Xylariales, Xylariomycetidae, Sordariomycetes, Pezizomycotina, Ascomycota, Fungi

Engler, A., K. Prantl. 1900. *Die natürlichen Pflanzenfamilien nebst ihren Gattungen und wichtigeren Arten insbesondere den Nutzpflanzen*: I. Tl., 1. Abt.: Fungi (Eumycetes):1-570.
From Gr. *xýlon*, wood + *kládos*, branch + L. dim. suf. *-ium*. Conidiophores with ampullae or radulae that give rise to unicellular, elliptical, hyaline conidia. On wood, branches and fallen leaves.

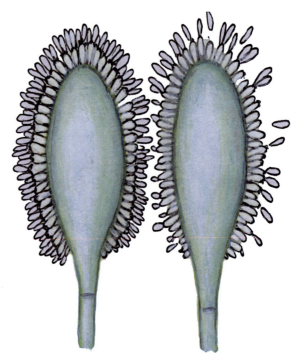

Xylocladium claviforme (J. L. Crane & Dumont) Arx: lignicolous fungus that forms conidiophores with radulae that produce unicellular, elliptical, hyaline conidia, x 1,800.

Xylogone Arx & T. Nilsson—see **Scytalidium** Pesante

Inc. sed., Helotiales, Leotiomycetidae, Leotiomycetes, Pezizomycotina, Ascomycota, Fungi

Arx, J. A. von, T. Nilsson. 1969. *Xylogone sphaerospora*, a new ascomycete from stored pulpwood chips. *Svensk bot. Tidskr.* 63:345-348.

From Gr. *xýlon*, log, wood + *gónos*, the engendered, offspring, in reference to the lignicolous habitat of this fungus.

Xylosphaera Dumort.—see **Xylaria** Hill ex Schrank
Xylariaceae, Xylariales, Xylariomycetidae, Sordariomycetes, Pezizomycotina, Ascomycota, Fungi
Dumortier, B. C. J. 1822. *Comment. bot.* (Tournay):91.
From Gr. *xýlos*, wood + *sphaíra* > L. *sphaera*, sphere, for the consistency and shape of the perithecial stroma of this fungus.

Xylostroma Tode
Fomitopsidaceae, Polyporales, Inc. sed., Agaricomycetes, Agaricomycotina, Basidiomycota, Fungi
Tode, H. J. 1790. *Fung. mecklenb. sel.* (Lüneburg) 1:36.
From Gr. *xýlon*, wood + *strōma*, bed, mattress. Mycelium a dense, interwoven, membranous, suberose or coriaceous, flexible. In fissures of bark of trees such as oaks, eucalyptus and others.

Y

Ypsilomyces elegans

Y

Yinmingella Goh, et al.
Inc. sed., Inc. sed., Inc. sed., Inc. sed., Pezizomycotina, Ascomycota, Fungi
Goh, T. K., et al. 1998. *Yinmingella mitriformis* gen. et sp. nov., a new sporodochial hyphomycete from submerged wood in Hong Kong. *Can. J. Bot.* 76(10):1693-1697.
Derived from the first name of the senior author's wife, *Yin-Ming Leong* + L. dim. suf. *-ella*, for her appreciation and support of his work in mycology. Colonies on natural substratum punctiform, black, glistening. Mycelium superficial, immersed in substratum. Setae and hyphopodia absent. Stromata well developed, hemispherical, dark brown to black. Conidiophores absent. Conidiogenous cells discrete, determinate, monoblastic, medium brown to black, thick-walled, smooth, lageniform. Conidia holoblastic, acrogenous, catenulate, dematiaceous, unicellular, mitriform to limoniform, asetulate, seceding by schizolysis.

From L. *ypsilon*, letter Y + *myces*, fungus, referring to the Y-shaped conidia. Stromata pseudoparenchymatous, brown to dark brown. Conidiophores macronematous, mononematous, unbranched or branched, cylindrical, erect, straight or flexuous, septate, smooth, pale brown to brown. Conidiogenous cells terminal, cylindrical, smooth, pale brown, with enteroblastic percurrent extensions producing fertile branches. Conidia thallic-arthric after disarticulation, single, dry, smooth, septate, usually more or less dichotomously branched, Y-shaped, furcated or cylindrical, subhyaline.

Ypsilomyces elegans D. A. C. Almeida & Gusmão: group of conidiophores with Y-shaped, multiseptate conidia, on decomposing plant leaves, Brazil, x 250.

Yinmingella mitriformis Goh, et al.: punctiform, black, glistening conidioma on wood submerged in Tung Chung River, Lantau Island, Hong Kong, x 300; conidiogenous cells with dematiaceous, unicellular, mitriform conidia, x 450.

Ypsilomyces D. A. C. Almeida & Gusmão
Inc. sed., Inc. sed., Inc. sed., Inc. sed., Pezizomycotina, Ascomycota, Fungi
Almeida D. A. C., L. F. P. Gusmão. 2014. *Ypsilomyces*, a new thallic genus of conidial fungi from the semi-arid Caatinga biome of Brazil. *Mycotaxon* 129(1):182.

Yuccamyces Gour, et al.
Inc. sed., Inc. sed., Inc. sed., Inc. sed., Pezizomycotina, Ascomycota, Fungi
Dyko, B. J., B. C. Sutton. 1979. Two new and unusual Deuteromycetes. *Trans. Br. mycol. Soc.* 72(3):411-417.
From plant genus *Yucca* + L. *myces*, fungus, for the resemblance of the conidiomata to small purple *Yucca* plants. Conidiomata sporodochial, of textura intricata and textura oblita. Ostiole absent. Conidiophores absent.

Conidiogenous cells integrated, cylindrical, terminal, arising from basal hyphae of conidiomata. Conidia holoblastic, hyaline, cylindrical, euseptate, formed acropetally in long, branched chains. Isolated from *Flacourtia indica*, University of Udaipur, India.

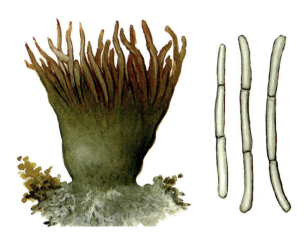

Yuccamyces purpureus Gour, et al.: sporodochial conidioma, with the appearance of purple *Yucca* plants, isolated from *Flacourtia indica*, University of Udaipur, Rajasthan, India, x 200, with pedicellate conidiogenous cells which give rise to cirrhi of cylindrical, euseptate, 2-septate conidia arranged in long and branched chains, x 750.

Yuea O. E. Erikss.
Xylariaceae, Xylariales, Xylariomycetidae, Sordariomycetes, Pezizomycotina, Ascomycota, Fungi
Eriksson, O. E. 2003. *Yuea*, a new genus in Xylariales. *Mycotaxon* 85:313-317.
Named in honor of the American mycologist Dr. *Yue Jing-zhu*. Perithecia scattered, each seated in a separate cavity in the wood, flattened subglobose. Asci cylindrical, 8-spored. Ascospores ellipsoid-subcylindrical, brown, smooth, with a germ slit spiralling one full circumference. On culms of *Chusquea cummingii*.

Yunnania H. Z. Kong
Inc. sed., Inc. sed., Inc. sed., Inc. sed., Pezizomycotina, Ascomycota, Fungi
Kong, H. Z. 1998. *Yunnania* gen. nov. of Hyphomycetes. *Mycotaxon* 69:319-326.
From *Yunnan*, a Chinese province name. Colony fuscous-grey or fuscous; conidiophores borne from substratum, funicular or aerial hyphae, with short stipes, smooth-walled, lightly grey or lightly brown; penicilli biverticillate or monoverticillate, occasionally terverticillate; phialides ampulliform, with short collula (neck), no annellation, conidia ellipsoidal, truncate at the base.

Z

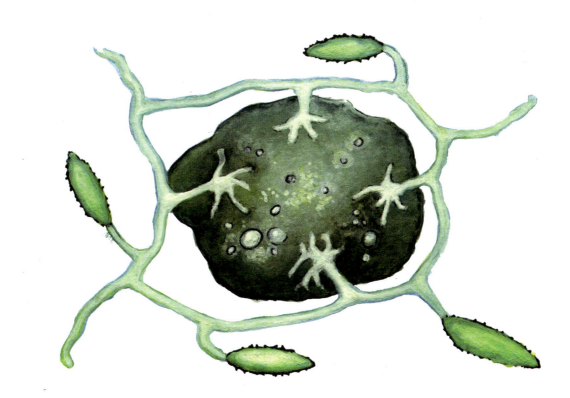

Zoopage phanera

Z

Zasmidium Fr.
Mycosphaerellaceae, Mycosphaerellales, Dothideomycetidae, Dothideomycetes, Pezizomycotina, Ascomycota, Fungi
Fries, E. 1849. *Summa veg. Scand., Sectio Post.* (Stockholm):407.
The etymology was not determined. Colonies villose, dark green, without powdery conidial areas. Conidial structures sparse. Conidiophores subhyaline to brown, a rachis or sympodial conidiogenous axis on apex of each branch, which produces ellipsoid conidia with a truncate base. Saprobic in dark, humid sites, such as basements, where it grows on wine or vinegar casks, and in caves.

From Abyssinian *zibra* > Fr. *zebre* and NL. *zebra*, zebra + L. *spora*, spore, for the light and dark bands on the spores. Hyphae immersed, pale yellowish. Conidiophores macronematous, mononematous, single or caespitose, unbranched, erect, straight or flexuous, smooth, septate, dark reddish-brown, swollen at apex. Conidiogenous cells polyblastic, integrated, terminal, sympodial, cylindrical or clavate, conidial scars reddish-brown, prominent. Conidia solitary, acropleurogenous, straight, cylindrical with rounded ends, 3-septate, dark reddish-brown, frequently with a dark band at each septum and a hyaline band on outer edge of each septum. On dead leaves of *Freycinetia* spp. in the South Pacific area.

Zasmidium cellare (Pers.) Fr.: dark green, villose colony, without powdery conidial areas; the conidial areas are sparse on the mycelium, x 1.

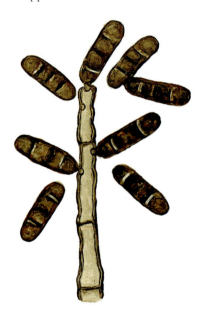

Zebrospora bicolor McKenzie: dark reddish-brown conidiophore, producing cylindrical, 3-septate conidia, with a dark band at each septum, x 720. Collected on dead leaves of *Freycinetia baueriana* subsp. *banksii*, Cook Islands, New Caledonia, New Zealand and Western Samoa.

Zebrospora McKenzie
Inc. sed., Inc. sed., Inc. sed., Inc. sed., Pezizomycotina, Ascomycota, Fungi
McKenzie, E. H. C. 1991. Dematiaceous hyphomycetes on *Freycinetia* (Pandanaceae). 2. *Zebrospora* gen. nov. *Mycotaxon* 41(1):189-193.

Zelodactylaria A. C. Cruz, et al.
Inc. sed., Inc. sed., Inc. sed., Inc. sed., Pezizomycotina, Ascomycota, Fungi

Cruz, A. C., et al. 2012. *Zelodactylaria*, an interesting new genus from semi-arid northeast Brazil. *Mycotaxon* 119:241-248.

From Gr. *zélo*, emulation + genus *Dactylaria* Sacc., for the similarity to this genus. Colonies on natural substratum effuse, hairy, brown to olivaceous or black. Mycelium superficial, immersed. Conidiophores macronematous, mononematous, erect, branched, septate, brown or black, smooth or verruculose. Conidiogenous cells polyblastic, discrete, sympodially proliferating, denticulate. Conidial secession schizolytic. Conidia solitary, clavate, obovoid to globose, septate, smooth or verruculose, hyaline. On decaying bark of unidentified plant in "Serra da Maravilha", Senhor do Bonfim, Bahia, Brazil.

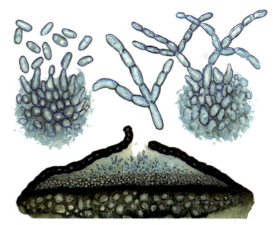

Zelopelta thrinacospora B. Sutton & R. D. Gaur: vertical section of unilocular, pellicular pycnothyrial conidioma on dead leaves of *Hedera nepalensis*, Garhwal, Uttar Pradesh, Himalaya, India, x 350; conidiogenous cells with microconidia or macroconidia, x 800.

Zelodactylaria verticillata A. C. Cruz, et al.: conidiophore with polyblastic, denticulate, conidiogenous cells, and obovoid, verruculose, hyaline conidia, x 1,200. Collected from decaying bark of unidentified plant in the semiarid Caatinga biome, Bahia, Brazil.

Zelopelta B. Sutton & R. D. Gaur

Inc. sed., Inc. sed., Inc. sed., Inc. sed., Pezizomycotina, Ascomycota, Fungi

Sutton, B. C., R. D. Gaur. 1984. *Zelopelta thrinacospora* gen. et sp. nov. (Pycnothyriales). *Trans. Br. mycol. Soc.* 82(3):556-559.

From Gr. *zélos*, rival + *pélte* > L. *pelta*, shield. Conidiomata pycnothyrial, superficial, septate, scutate, circular, pale brown, reticulate, unilocular; margin entire, pellicular; upper wall one cell thick, brown, of textura epidermoidea; lower wall up to four cells thick, hyaline, of textura globulosa; ostiole absent, dehiscence irregular. Conidiogenous cells determinate, discrete, hyaline, restricted to lower wall. Conidia holoblastic, acrogenous, hyaline, consisting of a basal cell and three divergent, septate arms. Microconidia formed from minute phialides, hyaline, aseptate, cylindrical to fusiform. On dead leaves of *Hedera nepalensis* in Garhwal, Himalaya, India.

Zeus Minter & Diam.

Rhytismataceae, Rhytismatales, Leotiomycetidae, Leotiomycetes, Pezizomycotina, Ascomycota, Fungi

Minter, D. W., et al. 1987. *Zeus olympius* gen. et sp. nov. and *Nectria ganymede* sp. nov. from Mount Olympus, Greece. *Trans. Br. mycol. Soc.* 88(1):55-61.

Named after *Zeus*, the king of gods in ancient Greek mythology, who was believed to inhabit Mount Olympus. Ascomata apothecial, scattered or in small groups, not associated with scars of short-shoot needle bundles; when immature immersed below bark and a black fungal covering layer; when mature circular or rather angular, becoming erumpent, throwing back bark breaking black covering layer by irregular radial splits to reveal dark fawn-colored hymenium (fragments of bark often remaining attached to black covering layer hiding its dark appearance), raising substratal surface, but lacking a stalk, not protruding beyond level of dislodged bark. On dead twigs and small branches of *Pinus leucodermis*.

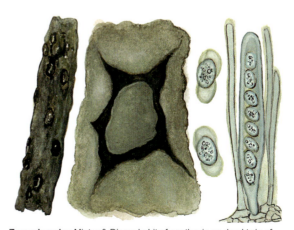

Zeus olympius Minter & Diam.: habit of apothecia on dead twig of pinus (*Pinus leucodermis*), in the northeast of Greece, x 1; detail of an erumpent apothecium in pinus bark, x 30; ascus with young, unicellular ascospores, x 450, and two mature ascospores surrounded by a mucous sheath, x 470.

Zoopage

Zoopage Drechsler
Zoopagaceae, Zoopagales, Inc. sed., Inc. sed., Zoopagomycotina, Zygomycota, Fungi
Drechsler, C. 1935. Some conidial phycomycetes destructive to terricolous amoebae. *Mycologia* 27(1):30.
From Gr. *zōon*, animal + *páge*, loop, snare, i.e., a trap for animals, which in the case of this fungus are terricolous amoebae, which they trap by means of adhesive hyphae and destroy with intracellular haustoria. Hyphae forming external to host, producing sporangiophores and multispored merosporangia.

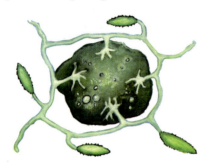

Zoopage phanera Drechsler: external hyphae with sporangiospores, and adhesive branch to capture a cell of *Amoeba terricola*, with intracellular, destroying haustoria, x 400. Described in Washington, U.S.A.

Zoophagus Sommerst.
Zoopagaceae, Zoopagales, Inc. sed., Inc. sed., Zoopagomycotina, Zygomycota, Fungi
Sommerstorff, H. 1911. Ein Tiere fangender Pilz (*Zoophagus insidians*, nov. gen., nov. sp.). *Öst. bot. Z.* 61:361-373.
From Gr. *zōon*, animal + *phágos*, that which eats < *phágomai*, to eat, i.e., that which eats animals, which in this case are freshwater rotifers, which they trap by means of specialized organs.

Zoophagus insidians Sommerst.: rotifer captured by one of the adhesives, trapping hyphal branches of this fungus, x 180. Described in Austria.

Zygorhynchus Vuill.—see Mucor Fresen.
Mucoraceae, Mucorales, Inc. sed., Inc. sed., Mucoromycotina, Zygomycota, Fungi
Vuillemin, P. 1903. Importance taxonomique de l'appareil zygospore des mucorinées. *Bull. Soc. mycol. Fr.* 19:116.
From Gr. *zygón*, yoke, pair, sexual union + *rhýnchos*, pick, referring to the suspensors of the zygospore.

Zygosaccharomyces B. T. P. Barker
Saccharomycetaceae, Saccharomycetales, Saccharomycetidae, Saccharomycetes, Saccharomycotina, Ascomycota, Fungi
Barker, B. T. P. 1901. A conjugating yeast. *Phil. Trans. Roy. Soc. London, Ser. B, Biolog. Sci.* 194:467-485.
From Gr. *zygón*, yoke, pair + genus *Saccharomyces* Meyen ex E. C. Hansen < Gr. *sákcharon*, sugar + *mýkes* > L. *myces*, fungus. Conjugation immediately precedes ascus formation between mother cell and bud or between two individual cells. Ascospores 1-4 per ascus, globose or ellipsoidal. Asexual reproduction by multilateral budding. Pseudomycelium, no true mycelium. Vigorous fermentation of sugars. Isolated from wine, vinegar, fruit juices, pickled gherkins, feces, tea fungus, bark, soil, gaseous drinks and other substrates.

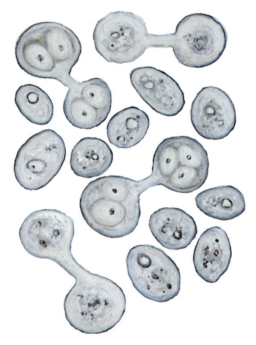

Zygosaccharomyces bailii (Lindner) Guilllierm.: conjugation of asci between mother cell and the bud, or between two different individual cells, to produce 1-4 spherical ascospores per ascus, x 2,000.

Kingdom Chromista

Hyphochytrium, Aqualinderella, Pontisma, Peronospora, Plasmopara, Ectrogella

Myzocytium, Sphaerita, Monorhizochytrium, Bremia, Dictyuchus, Pythium

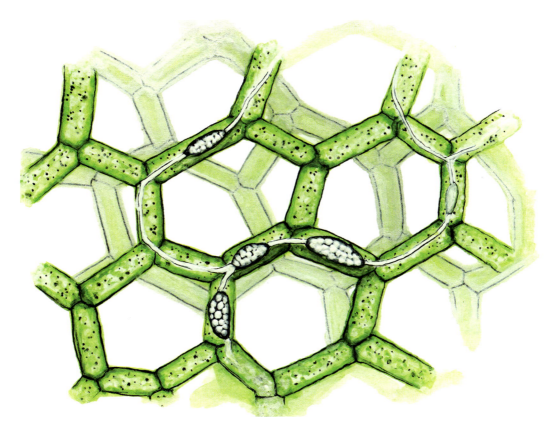

Hyphochytrium hydrodictii

Chromista

Achlya Nees

Saprolegniaceae, Saprolegniales, Saprolegniidae, Peronosporea, Inc. sed., Oomycota, Chromista

Nees von Esenbeck, C. G. 1823. *In*: C. G. Carus, Beitrag zur Geschichte der unter Wasser an verwesenden Thierkörpern sich erzeugenden Schimmel-oder Algen-Gattungen. *Nova Acta Phys.-Med. Acad. Caes. Leop.-Carol. Nat. Cur.* 11:514.

From Gr. *achlýo*, to darken, or < *achlýs*, darkness, haze, + L. suf. *-ia*, which denotes quality or state of something. Sporangia and oogonia dark grey. Saprobic on the remains of plants and animals in water or soil.

Achlya americana Humphrey: oogonium with oospheres, and antheridia, x 750.

Albugo (Pers.) Roussel

Albuginaceae, Albuginales, Albuginidae, Peronosporea, Inc. sed., Oomycota, Chromista

Gray, S. F. 1806. *Nat. Arr. Brit. Pl.* (London) 1:47.

From L. *albus*, white, particularly opaque white rather than brilliant or lustrous white + L. suf. *-ugo*, which indicates the possession of a substance or property. Pustules white, resulting from formation of conidiosporangia beneath epiderm. Causes a disease known as white rust on plants (mainly Amaranthaceae, Brassicaeae, and Solanaceae).

Albugo candida (Pers.) Roussel: conidiosporangia within a pustule, on the living host, x 200.

Anisolpidium Karling

Anisolpidiaceae, Anisolpidiales, Olpidiopsididae, Peronosporea, Inc. sed., Oomycota, Chromista

Karling, J. S. 1943. The life history of *Anisolpidium ectocarpi* gen. nov. et sp. nov., and a synopsis and classification of other fungi with anteriorly uniflagellate zoospores. *Am. J. Bot.* 30:637-648.

From Gr. *ánisos*, unequal + genus *Olpidium* (Braun) Rabenh. < Gr. *olpís*, oil bottle, small glass + dim. *-ídion* > L. *-idium*. Thallus similar to that of *Olpidium* (olpidioid), but with zoospores having one mastigonemate flagellum on the anterior pole, instead of one whiplash type on the posterior pole. A parasite of marine algae.

Aphanomyces de Bary

Leptolegniaceae, Saprolegniales, Saprolegniidae, Peronosporea, Inc. sed., Oomycota, Chromista

Apodachlya

de Bary, A. 1860. Einige neue Saprolegnieen. *Jb. Wiss. Bot.* 2:178.

From Gr. *aphanés*, invisible, secret, unknown + *mýkes* > L. *myces*, fungus. Hyphae delicate, long, scantily branched, difficult to detect. Saprobic on plant and animal remains or as a parasite of *Daphnia* and other small animals.

From genus *Aphanomyces* de Bary (< *aphanés*, invisible + *mykes* > L. *myces*, fungus) + L. suf *-opsis* < Gr. *-ópsis*, appearance, aspect, i.e., similar to the fungus *Aphanomyces* de Bary. An endobiotic parasite of diatoms and desmids.

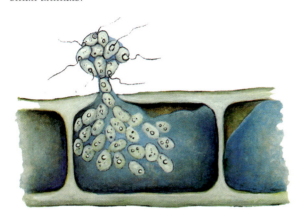

Anisolpidium ectocarpi Karling: zoosporangium with uniflagellate zoospores, in the cells of *Ectocarpus mitchellae* and *E. siliculosus*, in North Carolina, U.S.A., x 800.

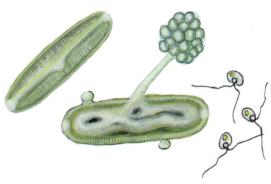

Aphanomycopsis bacillariacearum Scherff.: zoosporangium in a cell of *Pinnularia*, and biflagellate zoospores, x 475.

Apodachlya Pringsh.

Leptomitaceae, Leptomitales, Saprolegniidae, Peronosporea, Inc. sed., Oomycota, Chromista

Pringsheim, N. 1883. Über Cellulinkörner, eine Modification der Cellulose in Körnerform. *Ber. dt. bot. Ges.* 1:289.

From genus *Achlya* Nees (< Gr. *achlýo*, darkened) + prep. *apó*, which is employed to give the idea of something that is lacking, separation, transformation, i.e., that differs, in this case with respect to the hyphae, which in *Apodachlya* are more slender, slightly refractile or of a brilliant gray color, whereas in *Achlya*, with which it often coexists on submerged branches and fruits, the hyphae are coarser.

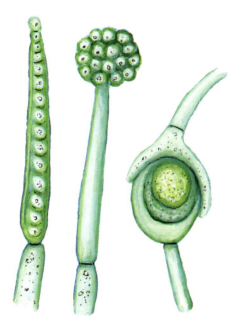

Aphanomyces laevis de Bary: two zoosporangia, one with internal zoospores, and another releasing zoospores. At right, oogonium and antheridium, x 500.

Aphanomycopsis Scherff.

Leptolegniellaceae, Leptomitales, Saprolegniidae, Peronosporea, Inc. sed., Oomycota, Chromista

Scherffel, A. 1925. Endophytische Phycomyceten Parasiten der Bacillariaceen und einige neue Monadinen. Ein Beitrag zur Phylogenie der Oomyceten (Schröter). *Arch. Protistenk.* 52:11.

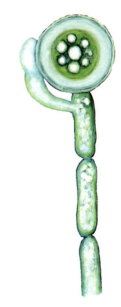

Apodachlya brachynema (Hildebr.) Pringsh.: a hypha with oogonium and antheridium, x 600.

Apodachlyella

Apodachlyella Indoh
Apodachlyellaceae, Leptomitales, Saprolegniidae, Peronosporea, Inc. sed., Oomycota, Chromista

Indoh, H. 1939. Studies on the Japanese aquatic fungi. I. On *Apodachlyella completa*, sp. nov., with revision of the Leptomitaceae. *Sci. Rep. Tokyo Bunrika Daig., Sect. B*, 4:43-50.

From genus *Apodachlya* Pringsh. < *apó*, separation + *Achlya* Ness + des. *-ia*, which denotes quality or state of something + L. dim. suf. *-ella*. Similar to *Apodachlya*, but differs in having multispored oogonia and lacking zoosporangia. Saprobic in soil and water.

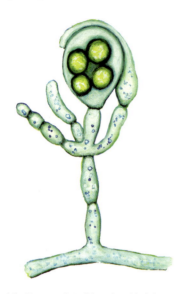

Apodachlyella completa (Humphrey) Indoh: oogonium with four oospheres, and external antheridium, x 420.

Aqualinderella R. Emers. & W. Weston
Rhipidiaceae, Rhipidiales, Peronosporidae, Peronosporea, Inc. sed., Oomycota, Chromista

Emerson, R., W. H. Weston. 1967. *Aqualinderella fermentans* gen. et sp. nov., a phycomycete adapted to stagnant waters. I. Morphology and occurrence in nature. *Am. J. Bot.* 54:702-719.

From L. *aqua*, water + *linder*, in honor of the American mycologist *David H. Linder*, who was the first to see and collect this organism + L. dim. suf. *-ella*. Living in stagnant water on submerged fruits, free or almost free of oxygen, with high concentrations of CO_2 and fermentable organic material.

Araiospora Thaxt.
Rhipidiaceae, Rhipidiales, Peronosporidae, Peronosporea, Inc. sed., Oomycota, Chromista

Thaxter, R. 1896. New or peculiar aquatic fungi. 4. *Rhipidium, Sapromyces*, and *Araiospora*, nov. gen. *Bot. Gaz.* 21:326.

From Gr. *araíos*, rare, tenous, or < *araiós*, delicate, slender + *sporá*, spore. Oospore thick-walled, with a periplasmic covering that form attractive reticulations, frequently of a golden color. Saprobic on plant remains, particularly branches in water.

Aqualinderella fermentans R. Emers. & W. Weston: thallus with zoosporangia, x 50.

Araiospora pulchra Thaxt.: thallus with golden oospores, growing on aquatic plant debris, x 180.

Basidiophora Roze & Cornu
Peronosporaceae, Peronosporales, Peronosporidae, Peronosporea, Inc. sed., Oomycota, Chromista

Roze, E., M. Cornu. 1869. Sur deux nouveaux tyres génériques pour les fammes des Saprolegniées et Péronosporées. *Annls. Sci. Nat., Bot. Sér.* 5, 11:84.

From Gr. *basídion*, dim. of *básis*, base + suf. *-phóros* < *phéro*, to carry, bear. Sporangiophore cylindrical to clavate; apex with small sterigmata producing sporangia. An obligate parasite of vascular plants.

Basidiophora entospora Roze & Cornu: sporangiophores and sporangia, coming through a stoma of a leaf of *Aster*, x 120.

Bremia Regel

Peronosporaceae, Peronosporales, Peronosporidae, Peronosporea, Inc. sed., Oomycota, Chromista

Regel, E. 1843. Beiträge zur Kenntnis einiger Blattpilze. *Botan. Ztg.* (Berlin) 1:665-667.

A genus dedicated in honor of the cleric *Bremi* + L. suf. *-ia*, which denotes belonging to. Sporangiophores projecting from stomata of leaves, branching dichotomously, each branch terminating in an flattened enlargement from whose edges arise sterigmata bearing spheroid sporangia (conidiosporangia), with apex depressed, papillate. Oospores small, globose with a subrugose, yellow-brown episporium. Parasitizes lettuce.

Bremia lactucae Regel: sporangiophore with sporangia, x 120.

Brevilegnia Coker & Couch

Saprolegniaceae, Saprolegniales, Saprolegniidae, Peronosporea, Inc. sed., Oomycota, Chromista

Coker, W. C., 1927. Other water molds from the soil. *J. Elisha Mitchell Sci. Soc.* 42:212.

Combresomyces

From L. *brevis*, short, of short duration + Gr. *légnon*, *légne*, border, slender filament, fringe, edge + L. suf. *-ia*, quality or state of something. Mycelium sparse. Sporangial wall disintegrates to liberate zoospores. Saprobic in detritus in soil.

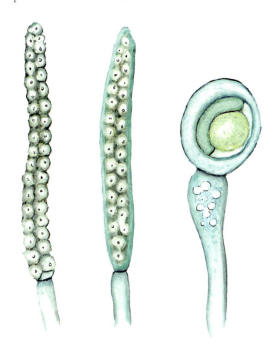

Brevilegnia subclavata Couch: sporangia with zoospores, and a hypha with masculine and feminine gametangia, x 300.

Brevilegniella M. W. Dick

Leptolegniellaceae, Leptomitales, Saprolegniidae, Peronosporea, Inc. sed., Oomycota, Chromista

Dick, M. W. 1961. *Brevilegniella keratinophila* gen. nov. sp. nov. *Pap. Mich. Acad. Sci.* 46:195-204.

From genus *Brevilegnia* Coker & Couch + L. dim. suf. *-ella*. Saprobic on keratinous subtrates in water and soil.

Combresomyces Dotzler, et al.

Fossil Oomycota

Dotzler, N., et al. 2008. *Combresomyces cornifer* gen. sp. nov., an endophytic peronosporomycete in *Lepidodendron* from the Carboniferous of central France. *Mycol. Res.* 112:1107-1114.

Named for *Combres* (or *Combre*) + L. *myces*, fungus, a village in the Massif Central, central France, where the fossil was found. Pyriform to subglobose terminal oogonium, thin-walled, subtended by wide hypha (oogonial stalk); simple septum present between oogonium and oogonial stalk; oogonium with prominent antler-like surface extensions positioned on hollow, column-like or broadly triangular papillations of oogonial wall; extensions once or twice dichotomized, densely spaced, regularly distributed on entire surface except in neck region; oogonium empty, sometimes

Cornumyces

containing several (usually 3-7) small spherules or a single aplerotic or nearly plerotic oospore; antheridium clavate, paragynous. Endophytic in *Lepidodendron*, intracellular in periderm.

Cornumyces M.W. Dick
Inc. sed., Leptomitales, Saprolegniidae, Peronosporea, Inc. sed., Oomycota, Chromista

Dick, M. W. 2001. Straminipilous Fungi: Systematics of the Peronosporomycetes including accounts of the marine straminipilous protists, the plasmodiophorids and similar organisms (Dordrecht):327.

Named in honor of *Marie Maxime Cornu*, a French botanist, algologist and mycologist + L. *myces*, fungus. Zoosporangium irregularly tubular with short contorted branches or lobes of varying length, occasionally ellipsoidal or narrowly reniform, causing pronounced swelling of the infected apex of the hypha, with a single short sessile or slightly protruding discharge tube; reniform and biflagellate zoospores. Endobiotic on other zoosporic fugi such as *Saprolegnia* Nees and *Achlya* Nees.

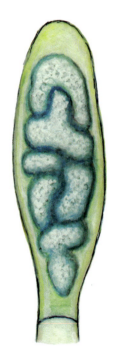

Cornumyces irregularis (Const.) M. W. Dick (syn. **Petersenia irregularis** (Const.) Sparrow: endobiotic zoosporangia inside the hyphae of **Achlya** Nees, x 100.

Dictyuchus Leitg.
Saprolegniaceae, Saprolegniales, Saprolegniidae, Peronosporea, Inc. sed., Oomycota, Chromista

Leitgeb, H. 1868. Zwei neue Saprolegnieen. *Bot. Ztg.* 26:502-503.

From Gr. *díktyon*, net, reticulum + *ychóo*, to weave, or *ycheos*, fabric. Sporangia with encysted spores remaining attached, thus retaining sporangial shape. When encysted spores germinate and liberate zoospores, walls of empty cysts give appearance of a net. Saprobic on plant detritus in water and soil.

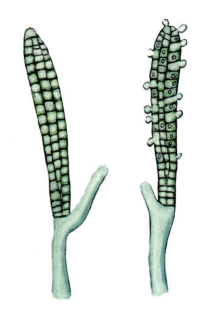

Dictyuchus monosporus Leitg.: sporangium with the encysted spores, and another releasing zoospores, x 30.

Ectrogella Zopf
Ectrogellaceae, Peronosporales, Peronosporidae, Peronosporea, Inc. sed., Oomycota, Chromista

Zopf, W. 1884. Zur Kenntniss der Phycomyceten. I. Zur Morphologie und Biologie der Ancylisteen und Chytridiaceen. *Nova Acta Acad. Caes. Leop.-Carol. German Nat. Cur.* 47:143-236.

From Gr. *ektro*, to make abort + L. *gelatinum*, gel, gelatin, or < *ektrétho*, inflame, swell + L. dim. suf. *-ella*. Thallus in host expands and causes separation of valves of host cell. Exit tubes of zoospores remain partially extramatrical, not perforating siliceous wall of diatom. Endoparasitic on freshwater and marine diatoms.

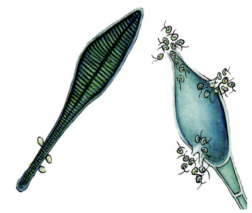

Ectrogella perforans H. E. Petersen) (syn. **Ectrogella licmophorae** Scherff.): zoosporangia emerging from the diatomaceous host (*Licmophora*), x 50; sporangium releasing biflagellate zoospores, x 320.

Labyrinthula

Eurychasma Magnus
Eurychasmataceae, Inc. sed., Inc. sed., Peronosporea, Inc. sed., Oomycota, Chromista
Magnus, P. 1905. Über die Gattung, zu der *Rhizophydium dicksonii* Wright gehört. *Hedwigia* 44:347-349.
From Gr. *eurýs*, wide + *chásma*, opening. Zoospores with wide, short discharge tubes that project from host cells. Endobiotic parasite of marine algae.

Eurychasma dicksonii (E. P. Wright) Magnus: zoosporangium discharging tubes of zoospores, on thallus of a marine alga, *Ectocarpus granulosus*, Great Britain, x 500.

Eurychasmidium Sparrow
Eurychasmataceae, Inc. sed., Inc. sed., Peronosporea, Inc. sed., Oomycota, Chromista
Sparrow, F. K., Jr. 1936. Biological observations on the marine fungi of Woods Hole waters. *Biol. Bull. Mar. biol. Lab. Woods Hole* 70:236-263.
From genus *Eurychasma* Magnus + L. dim. *-idium*. Differs from *Eurychasma* principally in thallus remaining endobiotic with only tips of zoospore discharge tubes protruding. Endoparasite of marine Rhodophyceae.

Eurychasmidium tumefaciens (Magnus) Sparrow: endobiotic zoosporangium with numerous tubes for discharging zoospores, on thallus of marine red alga, *Ceramium*, x 250.

Geolegnia Coker
Saprolegniaceae, Saprolegniales, Saprolegniidae, Peronosporea, Inc. sed., Oomycota, Chromista
Harvey, J. V. 1925. A study of the water molds and pythiums occurring in the soil of Chapel Hill. *J. Elisha Mitchell scient. Soc.* 41:153.
From Gr. pref. *geo-* < *ge*, the earth, as a planet, and soil or land + *légnon*, *légne*, edge, fringe, slender filament + L. suf. *-ia*, which denotes quality or state of something. Hyphae slender, lacking flagellate cells; spores encysting inside sporangium, liberated by decomposition of sporangial wall, then germinating by a germ tube. Saprobic in soil.

Geolegnia intermedia Höhnk: encysted spores inside the sporangium, of this saprolegniaceous saprobic fungus, x 1,700. Described in Germany.

Hyphochytrium Zopf
Hyphochytriaceae, Hyphochytriales, Inc. sed., Hyphochytrea, Inc. sed., Oomycota, Chromista
Zopf, W. 1884. Zur Kenntniss der Phycomyceten. I. Zur Morphologie und Biologie der Ancylisteen und Chytridiaceen. *Nova Acta Acad. Caes. Leop.-Carol. German Nat. Cur.* 47:143-236.
From Gr. *hyphé*, hypha, tissue, spider web + *chytrís*, pot, receptacle + L. dim. suf. *-ium*. Sporangia originate from terminal or intercalary swellings of mycelium. Parasitic in freshwater algae, ascocarps of discomycetes, and dead maize stems.

Labyrinthula Cienk.
Labyrinthulaceae, Labyrinthulida, Inc. sed., Labyrinthulea, Inc. sed., Bigyra, Chromista
Cienkowski, L. 1867. Über den Bau und die Entwicklung der Labyrinthuleen. *Archs. Anat. microsc.* 3:274-310.
From Gr. *labýrinthos*, labyrinth, tortuous path + dim. suf.

Lagena

-ula. Forming networks of viscous, reticulate tubes, up to several centimeters long, individual cells, generally fusiform. Cells move, grow, and reproduce within the transparent mucilage that cells elaborate. Pathogenic in marine plants, in particular *Zostera marina* and the green alga, *Ulva*.

Lagena radicicola Vanterp. & Ledingham: endobiotic zoosporangia, isolated from living roots of *Triticum*, in U.S.A., x 650.

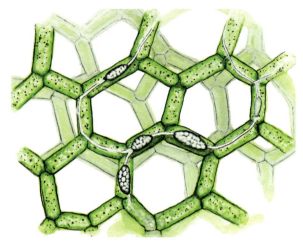

Hyphochytrium hydrodictyi Valkanov: sporangia inside cells of freshwater alga, *Hydrodictyon*, x 700.

Lagenidium Schenk

Pythiaceae, Peronosporales, Peronosporidae, Peronosporea, Inc. sed., Oomycota, Chromista
Schenk, A. 1859. Algologische Mitteilungen. *Verh. Phys.- Med. Ges. Würzburg* 9:27.
From L. *lagena* < Gr. *lágenos*, bottle, flask, vial, jar + L. dim. suf. *-idium*, i.e., a little bottle, because of the shape of the zoosporangia which develop parasitically. Endobiotic in the cells of diverse hosts (freshwater algae, other zoosporic fungi, rotifer eggs, microscopic animals and others) or as saprobes in substrates of animal origin.

Lagenocystis H. F. Copel.—see Lagena Vanterp. & Ledingham

Lagenaceae, Inc. sed., Inc. sed., Inc. sed., Inc. sed., Oomycota, Chromista
Copeland, H. F. 1956. *Classific. of Lower Organisms*:82.
From Gr. *lágenos*, bottle, flask, jar + *kýstis*, bladder, vesicle, for the shape of the zoosporangia.

Leptomitus C. Agardh

Leptomitaceae, Leptomitales, Saprolegniidae, Peronosporea, Inc. sed., Oomycota, Chromista
Agardh, C. A. 1824. *Syst. alg.* Lund, i-xxxviii + 1-312.
From Gr. *leptós*, small, fine, delicate, graceful, slender + *mítos*, filament in reference to the hyphae of the thallus. Saprobic in contaminated fresh water and on submerged substrates, principally fruits and branches.

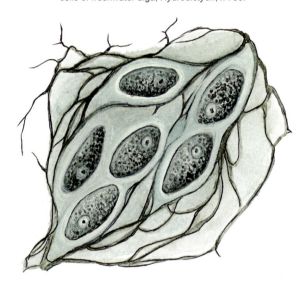

Labyrinthula terrestris D. M. Bigelow, M. W. Olsen & Gilb.: reticulate plasmodium, formed by individual and fusiform cells, x 600. Isolated from *Lolium perenne* on golf courses in Arizona, U.S.A..

Lagena Vanterp. & Ledingham (syn. Lagenocystis H. F. Copel.)

Lagenaceae, Inc. sed., Inc. sed., Inc. sed., Inc. sed., Oomycota, Chromista
Vanterpool, T. C., G. A. Ledingham. 1930. Studies on "browning" root rot of cereals. I. The association of *Lagena radicicola* n. gen. n. sp. with root injury of wheat. *Canadian Journal of Research, Section C* 2:171-194.
From L. *lagena* < Gr. *lágenos*, bottle, flask, jar. Zoosporangium bottle-shaped. Endobiotic in the root cells of cereals.

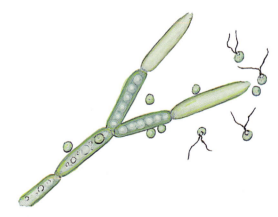

Leptomitus lacteus C. Agardh: slender hyphae of the thallus, with empty zoosporangia after discharging biflagellate zoospores, x 700.

Mindeniella Kanouse
Rhipidiaceae, Rhipidiales, Peronosporidae, Peronosporea, Inc. sed., Oomycota, Chromista
Kanouse, B. B. 1927. A monographic study of special groups of the water molds. I. Blastocladiaceae. II. Leptomitaceae and Pythiomorphaceae. *Am. J. Bot. 14*:301.
Named in honor of the mycologist M. von Minden + L. dim. *-ella*. Saprobic on decomposing rosaceous fruits, in freshwater habitats.

Mindeniella spinospora Kanouse: thallus with spinulose zoosporangia, on rotting fruit of *Crataegus* and *Pyrus malus*, Michigan, U.S.A., x 400.

Monorhizochytrium K. Doi & D. Honda
Thraustochytriidae, Thraustochytrida, Inc. sed., Labyrinthulea, Sagenista, Bigyra, Chromista
Doi, K., D. Honda. 2017. Proposal of *Monorhizochytrium globosum* gen. nov., comb. nov. (Stramenopiles, Labyrinthulomycetes) for former T*hraustochytrium globosum* based on morphological features and phylogenetic relationships. *Phycol. Res. 65(3)*:198.
From Gr. *mónos*, one, only + *rhíza*, root + *chytrís*, pot, earthern jar, flower vase + L. dim. suf. *-ium*. Vegetative cells globose or subglobose, monocentric, eucarpic. Mature zoosporangium divides into 8-32 daughter cells through cleavage of the multinucleate protoplasm. Reniform zoospores, containing two to three granules at the ventral side of the flagellar apparatus. Colony cells exhibit angular shapes such as tetragons or hexagons. Isolated from Futomi Beach, Kamogawa, Chiba, Japan.

Myzocytium Schenk
Pythiaceae, Peronosporales, Peronosporidae, Peronosporea, Inc. sed., Oomycota, Chromista
Schenk, A. 1858. *Über das Vorkommen kontraktiler Zellen im Pflanzenreiche*. Würzburg:70.
From Gr. *mýzo*, suck, sip + *kýtos*, cavity, cell + L. dim. suf. *-ium*, i.e., a small cell that is nourished by absorbing its food. Perhaps the name also derives from the Gr. *myxós*, cave, grotto, an interior or intimate place, in reference to its endobiotic habitat. Endoparasitic in freshwater algae and microscopic animals.

Monorhizochytrium globosum Kobayasi & M. Ôkubo: fragile sporangium, epibiotic in marine algae, which undergoes an explosion and dissolution of the distal part of the wall, x 800, that liberates the aflagellate spores, which later they become biflagellate. Collected in Japan. Vegetative cell with branched ectoplasmic nets, x 150. Zoosporangium with spores, x 150. Releasing biflagellate zoospores, x 180.

Myzocytium rabenhorstii (Zopf) M. W. Dick (syn. **Lagenidium rabenhorstii** Zopf): zoosporangia in a cell of *Spirogyra*, with the exit tubes releasing biflagellate zoospores, x 375.

Myzocytiopsis

Myzocytiopsis M. W. Dick
Myzocytiopsidaceae, Peronosporales, Peronosporidae, Peronosporea, Inc. sed., Oomycota, Chromista
Dick, M. W. 1997. The Myzocytiopsidaceae. *Mycol. Res.* 101(7):878.
From genus *Myzocytium* Schenk + L. suf. *-opsis* < Gr. *ópsis*, appearance, aspect. Resembling *Myzocytium*. Thallus holocarpic, unbranched, sometimes septate. Zoosporangia with zoopores hemispherical with flagella inserted subapically or laterally. Parasite or rarely saprotrophic, terrestrial, freshwater or marine.

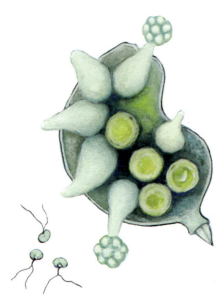

Myzocytiopsis zoophthora (Sparrow) M. W. Dick (syn. **Myzocytium zoophthorum** Sparrow): endoparasitic zoosporangia, releasing biflagellate zoospores, and gametangia, inside a rotifer body, x 360.

Nematophthora Kerry & D. H. Crump
Leptolegniellaceae, Leptomitales, Saprolegniidae, Peronosporea, Inc. sed., Oomycota, Chromista
Kerry, B. R., D. H. Crump. 1980. Two fungi parasitic on females of cyst nematodes (*Heterodera* spp.). *Trans. Br. mycol. Soc.* 74:119-125.
From Gr. *nématos*, thread + *phthorá*, destruction, referring to the fungus killing nematodes. Mycelium filamentous, holocarpic, mainly intramatrical, hyphae irregular, much branched. Extramatrical hyphae narrow, rarely branched, at maturity functioning as sporangia. Zoospores laterally biflagellate, completing development within sporangia, encysting inside or outside zoosporangia. Oospores produced laterally on undifferentiated hyphae, septate at sporogenesis, one spore produced on each hyphal segment. Morphologically distinct gametangia absent. Oospores containing reserve globules in granular cytoplasm. Oospore wall very thick, two-layered with a separable endospore membrane. Parasitic on females of cyst nematodes (*Heterodera* spp.), widely distributed in England and Wales.

Olpidiopsis Cornu (syn. **Pseudolpidium** A. Fisch.)
Olpidiopsidaceae, Olpidiopsidales, Olpidiopsididae, Peronosporea, Inc. sed., Oomycota, Chromista
Cornu, M. 1872. Monographie des Saprolégniées; étude physiologique et systématique. *Ann. Sci. Nat. Bot.* 15:127.
From genus *Olpidium* (A. Braun) J. Schrot. + L. suf. *-opsis* < Gr. *-ópsis*, appearance, aspect. Similar to *Olpidium* but thallus endobiotic, holocarpic, and zoospores biflagellate. Parasitic on other zoosporic fungi and freshwater or marine algae.

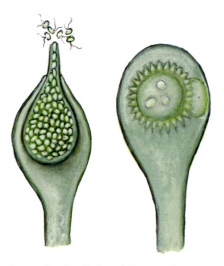

Olpidiopsis saprolegniae (A. Braun) Cornu: endobiotic, holocarpic thallus, releasing biflagellate zoospores, x 100; at right, spiny resting spore, with companion cell, x 100. This fungus is a parasite of **Saprolegnia** Nees.

Peronophythora C. C. Chen—see **Phytopthora** de Bary
Pythiaceae, Peronosporales, Peronosporidae, Peronosporea, Inc. sed., Oomycota, Chromista
Chen, C. C. 1961. A species of *Peronophythora*, gen. nov. parasitic on lichi fruit in Taiwan. *Special Publications of the National Museum of Natural Sciences* 10:32.
The name from the combination of *Peronospora* Corda and *Phytophthora* de Bary, whose etymologies are given elsewhere.

Peronoplasmopara (Berl.) G. P. Clinton—see **Pseudoperonospora** Rostovzev
Peronosporaceae, Peronosporales, Peronosporidae, Peronosporea, Inc. sed., Oomycota, Chromista
Clinton, G. P. 1905. Report of the Botanist. *Rept. Conn. Exp. Sta.*:329.
The name derives from the combination of the generic names *Peronospora* Corda and *Plasmopara* Schröt.

Peronospora Corda
Peronosporaceae, Peronosporales, Peronosporidae, Peronosporea, Inc. sed., Oomycota, Chromista
Corda, A. C. J. 1837. *Icon. fung.* (Prague), Vol. 1:20.

Plasmopara

From Gr. *péra*, *peré*, purse, knapsack, sack, or < *peróne*, bubble, ampoule, tassel + *sporá*, spore. Conidiosporangia spheroid or obovoid, germinate, liberating zoospores. Obligate parasites of higher plants.

Peronospora hyoscyami de Bary: sporangiophore with sporangia (conidiosporangia), x 180.

Petersenia Sparrow

Pontismataceae, Peronosporales, Peronosporidae, Peronosporea, Inc. sed., Oomycota, Chromista

Sparrow, F. K., Jr. 1934. Observations on marine Phycomycetes collected in Denmark. *Dansk bot. Ark.* 8(6):13.

Named in honor of the mycologist H. E. Petersen + L. suf. -ia, belonging to. Endobiotic on other zoosporic fungi or growing as saprobes or parasites of marine algae.

Phytophthora de Bary

Peronosporaceae, Peronosporales, Peronosporidae, Peronosporea, Inc. sed., Oomycota, Chromista

de Bary, A. 1876. Researches into the nature of the potato-fungus-*Phytophthora infestans*. *J. Roy. Agric. Soc. England, ser.* 2 12:239-269.

From Gr. *phytón*, plant + *phthorá*, destruction. Sporangiophores branched, with determinate growth, limoniform; periplasm in oogonium inconspicuous. Necrotic spots in tissues. Attacks and destroys solanaceous, citrus, cacao, lychee, and many other plants.

Plasmopara J. Schröt.

Peronosporaceae, Peronosporales, Peronosporidae, Peronosporea, Inc. sed., Oomycota, Chromista

Schröter, J. 1886. Die Pilze Schlesiens. *In: Cohn, Krypt.-Fl. Schlesien (Breslau)* 3.1:236.

From L. *plasma* < Gr. *plásma*, formation, bland material with which an organ or living being is formed + L. *parere*, to engender. Sporangial wall breaks and protoplasmic contents escape as a nonmotile protoplast (perhaps enveloped by the endospore), that later germinates by a tube and forms mycelium. Downy mildews of higher plants.

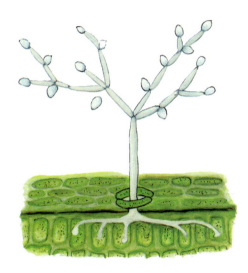

Phytophthora infestans (Mont.) de Bary: sporangiophore with limoniform sporangia, emerging from a stoma of leaf of *Solanum*, France, x 150.

Phytophthora litchii (C. C. Chen ex W. H. Ko, et al. & L. S. Leu) Voglmayr, et al. & Oberw. (syn. Peronophythora litchii C. C. Chen ex W. H. Ko, et al. & L. S. Leu): sporangiophore, whose branches give rise to limoniform sporangia, x 200. Collected in lychees, Taiwan.

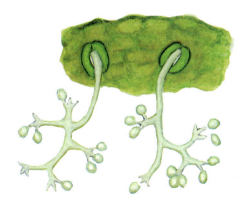

Plasmopara viticola (Berk. & M. A. Curtis) Berl. & De Toni: sporangiophores with ovoid conidiosporangia, emerging from two stomata in the leaf of vine (*Vitis vinifera*), Missouri, U.S.A., x 150.

Plasmoverna

Plasmoverna Constant., et al.
Peronosporaceae, Peronosporales, Peronosporidae, Peronosporea, Inc. sed., Oomycota, Chromista
Constantinescu, O., et al. 2005. *Plasmoverna* gen. nov., and the taxonomy and nomenclature of *Plasmopara* (Chromista, Peronosporales). *Taxon* 54:813-1121.
From L. *plasma* < Gr. *plásma*, formation, bland material with which an organ or living being is formed + L. *verno*, pertaining to spring, when these organisms are most abundant. Hyphae intercellular; haustoria intracellular, as obpyriform, globose, or slightly elongated vesicles, often surrounded by a callose sheath. Sporangiophores hyaline, sparsely branched in upper part, branching monopodially, branches more or less divergent, ending in a number of elongated, often cylindrical, ultimate branchlets, tips remain open after sporangium discharge. Sporangiogenesis holoblastic. Sporangia produced synchronously. Sporangia obovoidal, subglobose to ellipsoidal; wall hyaline, surface appearing smooth in light microscopy. Oogonia irregularly shaped, wall folded, thickened, colored; oospores globose, almost aplerotic, wall yellowish, of uniform thickness. Parasitic on plants.

Plectospira Drechsler
Leptolegniaceae, Saprolegniales, Saprolegniidae, Peronosporea, Inc. sed., Oomycota, Chromista
Drechsler, C. 1927. Two water molds causing tomato rootlet injury. *J. Agric. Res.* 34:294.
From Gr. *plektós*, braided, interwoven + *speíra* > L. *spira*, spirally, coil, each of the spiral turns of a helix. Zoosporangia arranged spiral, of a complex of inflated, interwoven elements giving rise to a slender, elongated zoosporangium. Zoospores differentiated into two or more rows inside zoosporangium remaining at tip of zoosporangium when evacuated. Parasitic on roots of vascular plants and crustaceans.

Plectospira myriandra Drechsler: slender, elongated zoosporangium; a group of zoospores are differentiated in various rows inside the zoosporangium, and they remain at the tip of the zoosporangium when it is evacuated, x 250.

Pontisma H. E. Petersen
Pontismataceae, Peronosporales, Peronosporidae, Peronosporea, Inc. sed., Oomycota, Chromista
Petersen, H. E. 1905. Contributions à la connaissance des Phycomycètes marins (Chytridineae Fischer). *Overs. K. danske Vidensk. Selsk. Forh.* 5:482.
From Gr. *póntisma*, that which has been put in the sea < *póntios, pontía*, marine. Principally or exclusively an endobiotic parasite of *Ceramium*, a marine alga.

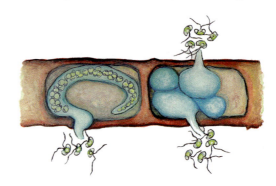

Pontisma lagenidioides H. E. Petersen: endobiotic zoosporangia inside cells of *Ceramium rubrum*, a red marine alga, some zoosporangia discharging biflagellate zoospores, x 1,300. Collected in Denmark.

Pseudolpidium A. Fisch.—see **Olpidiopsis** Cornu
Olpidiopsidaceae, Olpidiopsidales, Inc. sed., Peronosporea, Inc. sed., Oomycota, Chromista
Fischer, A. 1891. Phycomycetes. *Rabenh. Krypt.-Fl., Pilze* 1(4):33.
From Gr. *pseúdos*, false + genus *Olpidium* (Braun) Rabenh. < Gr. *olpís*, oil-jar + L. dim. suf. *-ium*. Endoparasite that differs from *Olpidium* because its zoospores are biflagellate.

Pseudoperonospora Rostovzev (syn. **Peronoplasmopara** (Berl.) G. P. Clinton)
Peronosporaceae, Peronosporales, Peronosporidae, Peronosporea, Inc. sed., Oomycota, Chromista
Rostovsev, S. J. 1903. Beiträge zur Kenntnis der Peronosporeen. *Flora, Regensburg* 92:424.
From Gr. *pseúdos*, false + genus *Peronospora* Corda < Gr. *péra, peré*, purse, knapsack, sack, or < *peróne*, bubble, ampoule, tassel + *sporá*, spore. Characteristics intermediate between *Peronospora* and *Plasmopara* J. Schröt. Sporangiophores subdichotomously branched with branches at a sharp angle, branchlets with subacute apices. Sporangia large with a prominent papilla through which zoospores liberated as in *Plasmopara*.

Pseudosphaerita P. A. Dang.
Pseudosphaeritaceae, Rozellopsidales, Olpidiopsididae, Peronosporea, Inc. sed., Oomycota, Chromista
Dangeard, P. A. 1895. Mémoire sur les parasites du noyau et du protoplasma. *Botaniste* 4:242.

Pythium

From Gr. *pseúdos*, false + genus *Sphaerita* Dang. < L. *sphaera*, sphere + adverb *ita*, thus, in this manner. Lagenidial fungus endoparasitic on *Euglena* and *Cryptomonas*, which superficially resemble the thallus of *Sphaerita*. The two genera differ principally in the method of zoospore formation and in their flagellation.

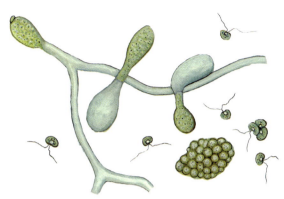

Pythiogeton transversum Minden: zoosporangia with biflagellate zoospores of this saprobic species, x 600.

Pythiopsis de Bary

Saprolegniaceae, Saprolegniales, Saprolegniidae, Peronosporea, Inc. sed., Oomycota, Chromista

de Bary, A. 1888. Species der Saprolegnieen. *Bot. Ztg.* 46:597-610.

From genus *Pythium* Pringsh. (< Gr. *pýtho*, decay, to make rot) + L. suf. *-opsis* < Gr. *-ópsis*, appearance, aspect, i.e., similar, in the shape of the sporangia, to the fungus of the genus *Pythium*. Differs in zoospore shape, in number of oospores in each oogonium, in having wider hyphae, and in the manner in which the gemmae germinate. Saprobic in plant and animal detritus and in water and soil.

Pseudoperonospora cannabina (G. H. Otth) Curzi (syn. **Peronoplasmopara cannabina** (G. H. Otth) Peglion): subdichotomously branched sporangiophore, with large sporangia, x 150.

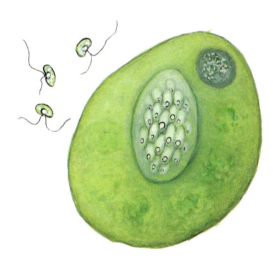

Pseudosphaerita euglenae P. A. Dang.: lagenidial fungus, endoparasitic on *Euglena viridis*, France, forming biflagellate zoospores, x 800.

Pythiogeton Minden

Pythiogetonaceae, Peronosporales, Peronosporidae, Peronosporea, Inc. sed., Oomycota, Chromista

von Minden, M. D. 1916. Beitrage zur Biologie und Systematik einheimischcr submerser Phycomyceten. *In*: R. Falck, *Mykol. Untersuch. Ber.* 2:241.

From Gr. *pýtho*, rot, to make rot + *géion* < *géiton*, inhabitant, neighbor. Saprobic on plant remains decomposing in water.

Pythiopsis cymosa de Bary: papillate sporangia, supported by wider hyphae, when they are compared to the hyphae of **Pythium** Pringsh, x 550.

Pythium Pringsh.

Pythiaceae, Peronosporales, Peronosporidae, Peronosporea, Inc. sed., Oomycota, Chromista

Rhipidium

Pringsheim, N. 1858. Beiträge zur Morphologie und Systematik der Algen. II. Die Saprolegnieen. *Jb. wiss. Bot.* 1:304.

From Gr. *pýtho*, rot, to make rot + L. dim. suf. *-ium*. Saprobic or parasitic on plant and animal parts in water and in soil. Causing necrosis and rotting of host tissues.

Pythium oligandrum Drechsler: zoosporangia with spores, and spiny oogonia with oospores, x 400.

Rhipidium Cornu

Rhipidiaceae, Rhipidiales, Peronosporidae, Peronosporea, Inc. sed., Oomycota, Chromista

Cornu, M. 1871. Note sur deux genres nouveaux de la famille des saprolégniées. *Bull. Soc. bot. Fr.* 18:58-59. From L. *rhipidium* < Gr. *rhipídion*, dim. of *rhipís*, *rhipídos*, fan + L. dim. suf. *-ium*. Thallus branches from basal cell; branches bear zoosporangia; gametangia originate individually or in umbellate racemes thus appearing fan-like. Develops as white, gummy pustules on branches and fruits of rosaceous plants in water.

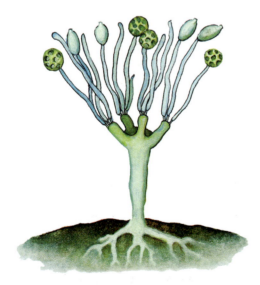

Rhipidium americanum Thaxt.: basal cell of a thallus, with umbellate racemes of branches bearing zoosporangia or gametangia, that give the appearance of a fan, on rosaceous plant in water, x 160.

Rhizidiomyces Zopf

Rhizidiomycetaceae, Hyphochytriales, Inc. sed., Hyphochytrea, Inc. sed., Oomycota, Chromista

Zopf, W. 1884. Zur Kenntniss der Phycomyceten. I. Zur Morphologie und Biology der Ancylisteen und Chytridiaceen. *Nova Acta Acad. Caes. Leop.-Carol. German Nat. Cur.* 47:143-236.

From Gr. *rhíza*, root + dim. *-ídion* + *mýkes* > L. *myces*, fungus. Thallus produces intramatrical rhizoidal system. Parasitic on oospheres and oospores of other fungi and aquatic algae.

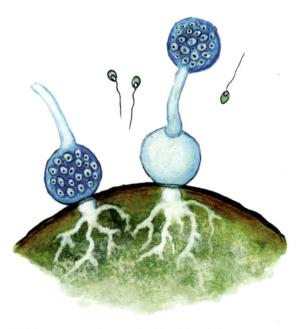

Rhizidiomyces apophysatus Zopf: interbiotic thallus, with the intramatrical rhizoidal system, and epibiotic zoosporangia with uniflagellate zoospores, on oospore of aquatic fungus, x 650. Described in Germany.

Saprolegnia Nees

Saprolegniaceae, Saprolegniales, Saprolegniidae, Peronosporea, Inc. sed., Oomycota, Chromista

Nees von Esenbeck, C. G. 1823. Beitrag zur Geschichte der unter Wasser an verwesenden Thierkörpen sich erzeugenden Schimmel-oder Algen-Gattungen. *Nova Acta Acad. Caes. Leop.-Carol. German Nat. Cur.* 11:513. From Gr. *saprós*, rotten + *légnon*, *légne*, colored border or edge, edge, fringe, slender filament + L. suf. *-ia*, which denotes quality or state of something. Hyphae develop as saprobes in plant and animal remains. Some species parasitic on aquatic animals, including fish, frog eggs, and others.

Sapromyces K. Fritsch

Rhipidiaceae, Rhipidiales, Peronosporidae, Peronosporea, Inc. sed., Oomycota, Chromista

Fritsch, K. 1893. Nomenclatorische Bemerkungen. VI. *Naegeliella* Schröt. *Öest. bot. Z.* 43:420-421.

From Gr. *saprós*, rotten + *mýkes* > L. *myces*, fungus. Saprobic on plant remains, especially branches in fresh, cold water.

Saprolegnia ferax (Gruith.) Kütz.: filamentous sporangia with biflagellate zoospores; at left, one sporangium with intrasporangial proliferation; at right, oogonium with oospheres, and three antheridia performing gametangial contact with the oospheres, x 400.

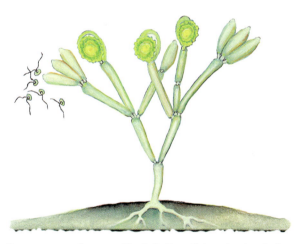

Sapromyces androgynus Thaxt.: thallus with branches terminating in zoosporangia, with biflagellate zoospores, and gametangia (oogonia and antheridia), on branches in fresh, cold water, x 100.

Sclerospora J. Schröt.
Peronosporaceae, Peronosporales, Peronosporidae, Peronosporea, Inc. sed., Oomycota, Chromista
Schröter, J. 1879. *Protomyces graminicola* Saccardo. *Hedwigia* 18:83-87.
From Gr. *sklerós*, hard + *sporá*, spore. Oospore with exospore confluent with wall of oogonium, forming a hard resistant or latent spore. Downy mildews, obligate parasites of grasses (Poaceae).

Sclerospora graminicola (Sacc.) Schroet.: sporangiophore; each branch terminates in a very hard or latent spore, x 200.

Sirolpidium H. E. Petersen
Sirolpidiaceae, Peronosporales, Peronosporidae, Peronosporea, Inc. sed., Oomycota, Chromista
Petersen, H. E. 1905. Contributions à la connaissance des Phycomycètes marins (Chytridineae Fischer). *Overs. K. danske Vidensk. Selsk. Forh.* 5:478.
From Gr. *seirá*, filament, chain + genus *Olpidium* (Braun) Rabenh. < Gr. *olpís*, oil-jar + L. dim. suf. *-ium*. Thallus initially tubular but later becomes septate, separates to form a chain or linear series of olpidioid sporangia. Endobiotic, parasitic on marine organisms.

Sommerstorffia Arnaudov
Saprolegniaceae, Saprolegniales, Saprolegniidae, Peronosporea, Inc. sed., Oomycota, Chromista
Arnaudov, N. 1923. Ein neuer Rädertiere (Rotatoria)-fangender Pilz (*Sommerstorffia spinosa*, nov. gen., nov. sp.). *Flora, Regensburg* 116:109-113.
Genus dedicated to *H. Sommerstorff* + L. suf. *-ia*, belonging to. Parasitic on rotifers, which they capture by means of adhesive hyphal branches.

Sommerstorffia spinosa Arnaudov: saprolegniaceous fungus, with a zoosporangium, parasitic on rotifers which they capture by means of adhesive hyphal branches, x 180.

Sphaerita

Sphaerita P. A. Dang.
Pseudosphaeritaceae, Rozellopsidales, Olpidiopsididae, Peronosporea, Inc. sed., Oomycota, Chromista
Dangeard, P. A. 1886. Recherches sur les organismes inferieurs. *Annls. Sci. Nat. Bot.*, sér. 7,4:240-341.
From L. *sphaera* < Gr. *sphaíra*, sphere + adverb *ita*. Sporangium globose. Endobiotic in rhizopods and euglenas.

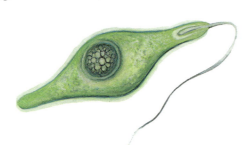

Sphaerita dangeardii Chatton & Brodskii: endobiotic sporangium inside the cell of *Euglena*, x 800.

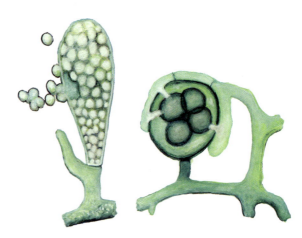

Thraustotheca clavata (de Bary) Humphrey: at left, sporangium liberating the primary spores, but later these spores germinate by forming zoospores, x 350; at right, oogonium and antheridia performing gametangial contact of this saprobic species in the soil, x 380.

Thraustochytrium Sparrow
Thraustochytriidae, Thraustochytrida, Inc. sed., Labyrinthulea, Sagenista, Bigyra, Chromista
Sparrow, F. K., Jr. 1936. Biological observations on the marine fungi of Woods Hole waters. *Biol. Bull. Mar. biol. Lab. Woods Hole* 70:236-263.
From Gr. *thraustós*, fragile, brittle, or < *threustés*, fragmented + *chytrís*, pot, receptacle + L. dim. suf. *-ium*. Sporangium liberating spores by explosion and dissolution of distal part of wall; liberated spores initially aflagellate but later become biflagellate. Epibiotic in marine algae.

Thraustotheca Humphrey
Saprolegniaceae, Saprolegniales, Saprolegniidae, Peronosporea, Inc. sed., Oomycota, Chromista
Humphrey, J. E. 1893. The Saprolegniaceae of the United States, with notes on other species. *Trans. Am. Phil. Soc.* 17:131.
From Gr. *thraustós*, fragil, brittle or < *thraustés*, fragmented + *théke*, case. Sporangium breaks irregularly or disintegrates to liberate spores, which germinate by forming zoospores. Saprobic in soil.

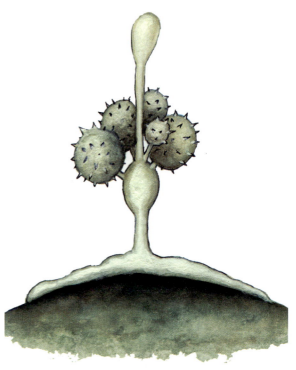

Trachysphaera fructigena Tabor & Bunting: prominent wart-like protuberances in the oogonium, x 500, a parasite of cacao and coffee fruits.

Trachysphaera Tabor & Bunting
Pythiaceae, Peronosporales, Peronosporidae, Peronosporea, Inc. sed., Oomycota, Chromista
Tabor, R. J., R. H. Bunting. 1923. On a disease of cocoa and coffee fruits caused by a fungus hitherto undescribed. *Ann. Bot. Lond.* 37:153-157.
From Gr. *trachýs*, rough, scabrous + *sphaíra* > L. *sphaera*, sphere. Oogonial wall ornamented with prominent wart-like or ampulliform protuberances; sporangia spherical, echinulate. Parasite on cacao and coffee fruits.

Woronina Cornu
Plasmodiophoridae, Plasmodiophorida, Inc. sed., Phytomyxea, Endomyxa, Cercozoa, Chromista
Cornu, M. 1872. Monographie des saprolégniées; étude physiologique et systématique. *Annls Sci. Nat., Bot.* 15:176.
Dedicated to the Russian microbiologist M. Woronin (1838-1903) + L. fem. termination *-a*. Parasitic on zoosporic fungi (*Achlya* Nees, *Saprolegnia* Nees, *Pythium* Pringsh.) and freshwater algae (*Oedogonium*, *Vaucheria*).

Kingdom Protozoa

Acrasis, Physarum, Leocarpus, Cribraria, Lamproderma, Arcyria

Fuligo, Didymium, Stemonitopsis, Hemitrichia, Diachea, Physarella

Leocarpus fragilis

Protozoa

Acrasis Tiegh.
Acrasiaceae, Acrasida, Inc. sed., Heterolobosea, Tetramitia, Percolozoa, Protozoa
van Tieghem, M. P. 1880. Sur quelques myxomycètes à plasmode agrégé. *Bull. Soc. bot. Fr.* 27:317-322.
From Gr. *akrásia*, bad mixture or < *akrátos*, immiscible. Myxamoebae aggregate to form a pseudoplasmodium, but retain their individuality, no fusing. On pods, capsules, and old or moribund fruits still attached to the plant.

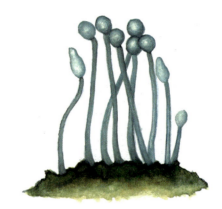

Acytostelium leptosomum Raper: sorocarps on soil, x 75. Isolated from leaf mould and surface soil from deciduous forest, in Illinois, Michigan, and Wisconsin, U.S.A.

Aethalium Link—see **Fuligo** Haller
Physaraceae, Stemonitiida, Myxogastria, Myxogastrea, Mycetozoa, Amoebozoa, Protozoa
Link, J. H. F. 1809. Observationes in ordines plantarum naturales. *Mag. Gesell. naturf. Freunde, Berlin* 3(1-2):24.
From Gr. *aíthalos*, soot, for the black color of the spores en masse + L. dim. suf. -*ium*.

Amaurochaete Rostaf.
Stemonitidaceae, Stemonitiida, Myxogastria, Myxogastrea, Mycetozoa, Amoebozoa, Protozoa
Rostafinsky, J. T. 1873. *Vers. Syst. Mycetozoen* (Strassburg):1-21.
From Gr. *amaurós*, dark + *chaíte*, crest, head of hair. Fructification with dark brown or black capillitium. On trunks and branches of living conifers, fallen leaves, and mosses.

Acrasis rosea L. S. Olive & Stoian.: sorocarp composed of pedicellar and prespore cells, x 160. Described in New Jersey and New York, U.S.A.

Acytostelium Raper
Acytosteliaceae, Acytostelida, Dictyostelia, Stelamoeba, Mycetozoa, Amoebozoa, Protozoa
Raper, K. B. 1956. Factors affecting growth and differentiation in simple slime molds. *Mycologia* 48(2):169-205.
From Gr. *a-*, without + *kýtos*, cavity, cell + *stéle*, column, stalk, + L. dim. suf. -*ium*, i.e., an acellular column. Sporocarp with column or pedicel of hollow tube lacking cells. In soil, humus, insect remains and decomposing agaric basidiocarps.

Amoebidium Cienk.
Amoebidiidae, Eccrinida, Inc. sed., Ichthyosporea, Choanofila, Choanozoa, Protozoa
Cienkowski, L. 1861. Über parasitische Schläuche auf Krustaceen und einigen Insektenlarven. *Bot. Ztg* 19:169.
From Gr. *amoibé*, change, transformation + L. dim. suf. -*idium*. Sporangiospores amoeboid. Ectocommensals on diverse families of crustaceans and insects.

Astreptonema

Amaurochaete atra (Alb. & Schwein.) Rostaf.: aethalia growing on the conifer twig, x 15.

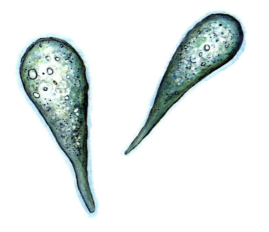

Amoebidium parasiticum Cienk.: two parasitized amoebae, with sporangiospores inside, x 1,300.

Arcyria Hill & F. H. Wigg.

Arcyriaceae, Trichiida, Myxogastria, Myxogastrea, Mycetozoa, Amoebozoa, Protozoa

Wiggers, F. H. 1780. *Prim. fl. holsat.* (Kiliae):109.

From Gr. *árkys, árkyos*, net + L. suf. *-ia*, which denotes quality of or state of a being. Sporangia with exposed capillitium when peridium disintegrates. On dead wood, fallen leaves, plant remains, and dung of herbivorous animals.

Arcyria cinerea (Bull.) Pers.: sporangia on dead wood, x 5.

Arundinula L. Léger & Duboscq

Eccriniidae, Eccrinida, Inc. sed., Ichthyosporea, Choanofila, Choanozoa, Protozoa

Léger, L., O. Duboscq. 1906. L' evolution des *Eccrina* des *Glomeris*. *Compt. Rend. Hebd. Séances Mém. Acad Sci., Paris* 42:590-592.

From L. *arundina*, cane, reed grass + dim. suf. *-ula*. Thallus unbranched. Endocommensal attached to the stomodeum and proctodeum of marine and freshwater decapod crustaceans.

Arundinula galatheae Manier & Ormières ex Manier: thalli, two with internal spores, adhered to the guts of aquatic crustaceans, x 170.

Astreptonema Hauptfl.

Eccriniidae, Eccrinida, Inc. sed., Ichthyosporea, Choanofila, Choanozoa, Protozoa

Hauptfleisch, P. 1895. *Astreptonema longispora* n. g., n. sp., eine neue Saprolegniaceae. *Ber. Deutsch. Bol. Ges.* 13:83-88.

From Gr. *a*, without + *streptós*, necklace, rosary, coiled, curved + *néma*, filament, i.e., a filament without spores arranged in chains. Dimorphic with microthalli of distal uninucleate cells beneath cells of apical generative spores; macrothalli larger, with curved base and generative spores formed at apex. Spores with filamentous prolongations at maturity. Endocommensal attached to the proctodeum of marine and freshwater amphipods.

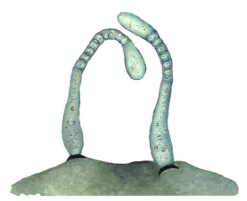

Astreptonema gammari (L. Léger & Duboscq) Manier ex Manier) (syn. **Eccrinella gammari** L. Léger & Duboscq): sporulating microthalli, adhered to the proctodeum of *Gammarus pulex* (amphipod crustacean), France, x 310.

Badhamia

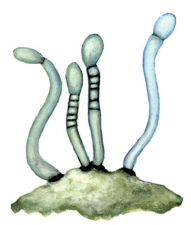

Astreptonema longisporum Hauptfl.: endocomensal macrothalli adhered to the host (aquatic amphipods), x 310.

Calomyxa metallica (Berk.) Nieuwl.: sporangia on rotten wood, x 35.

Badhamia Berk.

Physaraceae, Stemonitiida, Myxogastria, Myxogastrea, Mycetozoa, Amoebozoa, Protozoa

Berkeley, M. J. 1853. On two new genera of fungi. *Trans. Linn. Soc. London* 21:149-154.

Dedicated to the English mycologist *Ch. D. Badham* (1806-1857) + L. suf. *-ia*, which denotes belonging to. Sporangia sessile or pedunculate with a thin peridium, at times with a thick incrustation of calcium salts. Capillitium of a network of calcareous tubes, on occasion also with hyaline tubes lacking calcareous salts. Columella present or absent. Spores black en masse, frequently disposed in groups. On bark, generally moss-covered, of living or dead trees, leaves, and vegetable detritus.

Cavostelium L. S. Olive

Cavosteliida, Protostelida, Inc. sed., Protostelea, Mycetozoa, Amoebozoa, Protozoa

Olive, L. S. 1965. A new member of the Mycetozoa. *Mycologia* 56(6):885-896.

From L. *cavus*, hole, swollen + Gr. *stéle*, column, stalk + L. dim. suf. *-ium*. Sporocarp with stalk or pedicel that swells in water before liberating spores. On dead plant structures attached to the plant; worldwide.

Cavostelium apophysatum L. S. Olive: sporocarps, on dead plant, x 1,050.

Ceratiomyxa J. Schröt.

Ceratiomyxaceae, Protostelida, Inc. sed., Protostelea, Mycetozoa, Amoebozoa, Protozoa.

Schröter, J. 1889. Myxogasteres (eigentliche Myxomyceten). *In*: A. Engler and K. Prantl, *Die natürlichen Pflanzenfamilien*. Engelmann, Leipzig, Vol. 1(1):16.

From Gr. *kerátion* < *kéras*, horn + *mýxa*, slime. Sporocarps horn-shaped originating from plasmodial columns. On decomposing wood.

Badhamia gracilis (T. Macbr.) T. Macbr.: sporangia on tree bark, x 15.

Calomyxa Nieuwl.

Dianemataceae, Trichiida, Myxogastria, Myxogastrea, Mycetozoa, Amoebozoa, Protozoa.

Nieuwland, J. A. 1916. Critical notes of new and old genera of plants. *Am. Midl. Nat.* 4:333-335.

From Gr. *kalós*, beautiful + *mýxa*, slime. Sporangial peridium copper-colored or of iridescent shades. On rotten wood and the bark of living trees.

Ceratiomyxella L. S. Olive & Stoian.

Cavosteliida, Protostelida, Inc. sed., Protostelea, Mycetozoa, Amoebozoa, Protozoa

Olive, L. S., C. Stoianovitch. 1971. A new genus of protostelids showing affinities with *Ceratiomyxa*. *Amer. J. Bot.* 58:32-40.

Comatricha

From Gr. *kerátion* < *kéras*, leather + *mýxa*, slime + L. dim. suf. *-ella*; the generic name indicates its affinity with *Ceratiomyxa* Schroet. On pedicels of citrus fruits, legume pods, and dead grass.

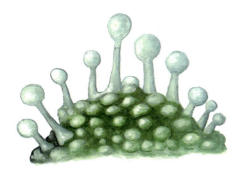

Ceratiomyxa fruticulosa (O. F. Müll.) T. Macbr.: sporocarps, growing from decomposing wood, x 4.

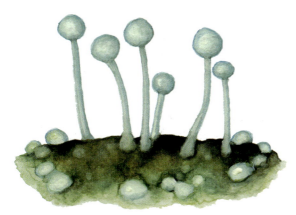

Ceratiomyxella tahitiensis L. S. Olive & Stoian.: sporocarps, described in Tahiti, x 210.

Clastoderma A. Blytt

Clastodermataceae, Echinostelida, Myxogastria, Myxogastrea, Mycetozoa, Amoebozoa, Protozoa

Blytt, A. G. 1880. *Clastoderma* A. Blytt, novem Myxomycetum genus. *Bot. Ztg. (Berlin)* 38:343.

From Gr. *klastós*, broken into various pieces + *dérma*, genit. *dérmatos*, skin, rind. Sporangium with peridium, which fragments to liberate spores at maturity. On dead wood, bark of living and dead trees, and other organic substrates.

Clastostelium L. S. Olive & Stoian.

Protosteliaceae, Protostelida, Inc. sed., Protostelea, Mycetozoa, Amoebozoa, Protozoa

Olive, L. S., C. Stoianovitch. 1977. *Clastostelium*, a new ballistosporous protostelid (Mycetozoa) with flagellate cells. *Trans. Br. mycol. Soc.* 69(1):83-88.

From Gr. *klastós*, broken in pieces > L. *clastos* + Gr. *stele*, support + NL. suf. *-ium*, ending of generic names. Sporocarp with two stalk segments, upper inflated segment rupturing and forcibly discharging spores; spores producing flagellate, or sometimes amoeboid, cells during germination; trophic stage amoeboid, uninucleate to plurinucleate, producing filose pseudopodia. On dead grass inflorescences and leguminous pods on the Micronesian islands of Guam and Truk.

Clastoderma debaryanum A. Blytt.: sporangia, the first on the left has the peridium broken to liberate the spores, on dead wood, in pine forest on lower surface of dead polypore, in Norway, x 90.

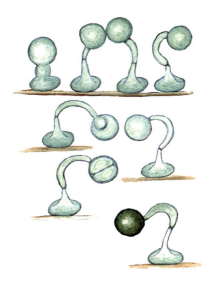

Clastostelium recurvatum L. S. Olive & Stoian.: development of a sporocarp, which has two stalk segments, the upper inflated segment rupturing and forcibly discharging the spore, x 1,000. Described in Guam.

Comatricha Preuss

Stemonitidaceae, Stemonitiida, Myxogastria, Myxogastrea, Mycetozoa, Amoebozoa, Protozoa

Preuss, C. G. T. 1851. Übersicht untersuchter Pilze, besonders aus der Umgegend von Hoyerswerda. *Linnaea* 24:99-153.

From Gr. *komáo*, *kóme*, head of hair, crest + *thríx*, genit. *trichós*, hair + L. fem. ending *-a*. Capillitium ending in small, free branches without anastomosing to form a network. On dead wood, mosses, lichens, and leaves.

Craterium

Craterium Trentep.
Physaraceae, Stemonitiida, Myxogastria, Myxogastrea, Mycetozoa, Amoebozoa, Protozoa
Roth, A. W. 1797. *Catalecta botanica*. Leipzig, 1:224.
From L. *craterium* < Gr. *kraterídion*, goblet or small cup. Sporangium cup-shaped. On leaves and dead branches, occasionally on wood or bark.

Craterium minutum (Leers) Fr.: sporangia, growing on wood, x 12.

Cribraria Pers.
Cribrariaceae, Liceida, Myxogastria, Myxogastrea, Mycetozoa, Amoebozoa, Protozoa
Persoon, C. H. 1794. Neuer Versuch einer systematischen Einteilung der Schwämme. *Neues Mag. Bot.* 1:63-128.
From L. *cribrum*, sieve + suf. *-aria*, which denotes similarity to or connection with something. Sporangium with peridium fugacious at maturity leaving a visible net-like mesh. On dead wood.

Cribraria cancellata var. **cancellata** (Batsch) Nann.-Bremek.: cancellate sporangia, on dead wood, x 20.

Dermocystidium Pérez
Inc. sed., Dermocystida, Inc. sed., Ichthyosporea, Choanofila, Choanozoa, Protozoa
Pérez, C. 1908. *Crypt. Fr. Exs.* 64:738.
From Gr. *dérma*, genit. *dérmatos*, skin, hide + *kýstis*, bladder, vesicle, sac + dim. suf. *-ídion* > L. *-idium*. Small vesicles or cysts forming in tissues of host. Parasitizes oysters causing serious problems in the oyster industry.

Dermocystidium marinum Mackin, et al.: cyst with spores, in tissues of the fish host, x 1,000. Described in Louisiana, U.S.A.

Diachea Fr.
Stemonitidaceae, Stemonitiida, Myxogastria, Myxogastrea, Mycetozoa, Amoebozoa, Protozoa
Fries, E. M. 1825. *Syst. orb. veg. (Lundae)* 1:143.
From Gr. *diachéo*, to diffuse, or < *diá*, with + *chéo*, heaped. Sporangia gregarious, with iridescent peridium. On leaves, branches, and bark of living and dead trees with mosses.

Diachea leucopoda (Bull.) Rostaf.: sporangia on plant remains; one sporangium has the broken peridium to show the columella and capillitium, x 30.

Dictyostelium Bref.
Dictyosteliaceae, Dictyostelida, Dictyostelia, Stelamoeba, Mycetozoa, Amoebozoa, Protozoa
Brefeld, O. 1870. *Dictyostelium mucoroides*. Ein neuer Organismus aus der Verwandtschaft der Myxomyceten. *Abh. senckenb. naturforsch. Ges.* 7:85-107.
From Gr. *díktyon*, net + *stéle*, column, stalk + L. dim. suf. *-ium*. Sporocarps with stalk or pedicel of a framework of empty cells appearing net-like. Widely distributed in soils and dung of animals.

Dictyostelium discoideum Raper: sorogenesis to form sorocarps primordia, on soil, x 150.

Didymium crustaceum Fr.: pedicellate sporangia with calcareous peridium, on dead wood, x 90.

Dictyostelium mucoroides Bref.: cellular net of the sorocarp pedicel, x 200. Collected on dung of rabbit and horse, France.

Diderma Pers.

Didymiaceae, Stemonitiida, Myxogastria, Myxogastrea, Mycetozoa, Amoebozoa, Protozoa

Persoon, C. H. 1794. Neuer Versuch einer systematischen Einteilung der Schwämme. *Neues Mag. Bot. 1*:89.

From Gr. *dís*, two times + *dérma*, skin. Sporangial peridium typically double, with an external calcareous or cartilaginous layer and an internal membranaceous layer. On dead wood, bark, leaves, lichens, and mosses.

Diderma radiatum (Rostaf.) Morgan: sporangia, with the cracked peridium, on dead wood, x 15.

Didymium Schrad.

Didymiaceae, Stemonitiida, Myxogastria, Myxogastrea, Mycetozoa, Amoebozoa, Protozoa

Schrader, H. A. 1797. *Nov. gen. pl.* (Lipsiae):20.

From Gr. *dídymos*, double + L. dim. suf. *-ium*. Sporangial peridium double. Differs from *Diderma* in having calcium carbonate of peridium of natural crystals. On wood and dead leaves, rarely old dung.

Eccrinidus

Eccrinella Léger & Duboscq—see Astreptonema Hauptfl.

Eccriniidae, Eccrinida, Inc. sed., Ichthyosporea, Choanofila, Choanozoa, Protozoa

Léger, L., O. Duboscq. 1933. *Eccrinella* (*Astreptonema*?) Gammari Leg. et Dub. Eccrinidae des Gammares d'eau Douce. *Archs. Zool. Exp. Gén.* 75:283-292.

From Gr. *ekkríno*, to separate, expel + L. dim. suf. *-ella*. Filamentous thallus expels sporangiospores.

Eccrinidus Léger & Duboscq

Eccriniidae, Eccrinida, Inc. sed., Ichthyosporea, Choanofila, Choanozoa, Protozoa

Léger, L., O. Duboscq. 1933. *Eccrinella* (*Astreptonema*?) Gammari Leg. et Dub. Eccrinidae des Gammares d'eu Douce. *Archs. Zool. Exp. Gén.* 75:283-292.

From Gr. *ekkríno*, to separate, expell + L. suf. *-idus*, which indicates a state or an action in progress. Filamentous thallus expels sporangiospores. Endocommensal in the proctodeum of terrestrial myriapods and diplopods.

Eccrinidus flexilis (L. Léger & Duboscq) Manier: endocommensal thalli with sporangiospores, adhered to the proctodeum of myriapods and diplopods, x 30. Described from intestine of *Glomeris marginata* (pill millipede), France.

Eccrinoides

Eccrinoides Léger & Duboscq
Eccriniidae, Eccrinida, Inc. sed., Ichthyosporea, Choanofila, Choanozoa, Protozoa

Léger, L., O. Duboscq. 1929. *Eccrinoides henneguyi* n. g., n. sp. et la systématique des Eccrinides. *Arch. Anat. Microscop.* 25:309-324.

From genus *Eccrina* Leidy < Gr. *ekkríno*, to separate, expel + L. suf. *-oides* < Gr. *-oeídes*, similar to, i.e., similar to *Eccrina* Leidy, now a synonym of *Enterobryus* Liedy.

Eccrinoides henneguyi L. Léger & Duboscq: thalli with uninucleate spores, endocommensals of terrestrial isopods gut, x 500.

Echinostelium de Bary
Echinosteliida, Echinostelida, Myxogastria, Myxogastrea, Mycetozoa, Amoebozoa, Protozoa

Rostafinsky, J. 1873. *Versuch eines Systems der Mycetozoen* (Strassburg):7.

From L. *echinus*, bristle + Gr. *stéle*, column, stalk + L. dim. suf. *-ium*. Sporangium with stalk or pedicel; spores spiny. On tree bark, fallen leaves, and dung.

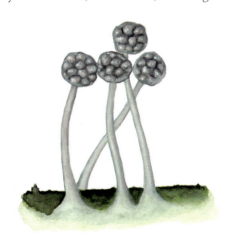

Echinostelium minutum de Bary: multisporate sporangia, on tree bark, x 185.

Enterobryus Leidy
Eccriniidae, Eccrinida, Inc. sed., Ichthyosporea, Choanofila, Choanozoa, Protozoa

Leidy, J. 1850. *Enterobryus*, a new genus of Confervaceae. *Proc. Acad. Nat. Sci. Philadelphia* 4:225-227.

From Gr. *énteron*, intestine + *brýon*, moss < *brýo*, to spring from, germinate. Sporangiospores multinucleate, attached to intestinal cuticle of host, later uninucleate sporangiospores exit digestive tract of host. Endocommensal attached to the proctodeum of terrestrial and freshwater insects, terrestrial myriapods, and marine crustaceans.

Enterobryus borariae Lichtw.: multinucleate sporangiospores attached to the intestinal cuticle of the host (endocommensal of terrestrial myriapods and marine crustaceans), x 380. Described in North Carolina, U.S.A.

Fuligo Haller
Physaraceae, Stemonitiida, Myxogastria, Myxogastrea, Mycetozoa, Amoebozoa, Protozoa

Haller, A. von. 1768. *Historia stirpium indigenarum Helvetiae inchoata*:110.

From L. *fuligo*, soot. Aethalia with black spores en masse. On decomposing dead wood, straw, dung, fallen leaves, and other moist organic substrates.

Fuligo septica (L.) F. H. Wigg.: aethalium with black spores en masse, on bark, x 1.5.

Guttulina Cienkow.
Guttulinaceae, Acrasida, Inc. sed., Heterolobosea, Tetramitia, Percolozoa, Protozoa

Cienkowsky, L. 1873. *Guttulina rosea. Trans. Bot. Sect. 4th Meeting Russian Naturalists* at Kazan. Also described

by A. Batalin in German in *Botanischer Jahresbericht*, pp. 61-62, 1874.
From L. *guttula*, dim. of *gutta*, drop + *-ina*, which denotes similarity. Sorus viscid, refringent, on sorocarp. On bark, lichenized or not, of trees.

Hemitrichia Rostaf.
Trichiaceae, Trichiida, Myxogastria, Myxogastrea, Mycetozoa, Amoebozoa, Protozoa
Rostafinski, J. 1873. *Versuch eines Systems der Mycetozoen. Dissertation inaugural.* IV. Strasburg:14.
From Gr. *hémi*, half, intermediate + genus *Trichia* Haller < Gr. *thríx*, genit. *trichós*, hair + L. suf. *-ia*, which denotes quality of, or state of a being, referring to capillitial filaments, which show a certain variation, because of which there are intermediate species that can be assigned equally to the genera *Trichia* Haller and *Arcyria* Wigg. On dead wood and fallen leaves.

Hemitrichia calyculata (Speg.) M. L. Farr (syn. **Hemitrichia stipitata** (Massee) T. Macbr.): sporangia showing the capillitial filaments, on dead wood, x 17.

Kelleromyxa Eliasson
Inc. sed., Inc. sed.., Inc. sed., Myxogastrea, Mycetozoa, Amoebozoa, Protozoa
Eliasson, U. H., et al. 1991. *Kelleromyxa*, a new generic name for *Licea fimicola* (Myxomycetes). *Mycol. Res.* 95(10):1201-1207.
Named in honor of the American mycologist *Harold W. Keller* + L. *myces* < Gr. *mýkes*, fungus, for his extensive work on myxomyxetes. Sporangia scattered, gregarious or tightly clustered in groups, fusiform, black, erect on a constricted base. Peridium closely adhering as two layers, cartilaginous, outer layer shining, smooth, thickened, inner layer membranous, slightly roughened, papillose; dehiscence irregular, lacking preformed lines of dehiscence or sutures. Capillitium present, arising from inner peridium as a sparse system of branching, anastomosing threads or reduced to simple, unbranched threads with dark nodules sometimes present. Spores black en masse, dark brown by transmitted light, lacking smoky colors, globose, free, surface markings of long, thin, truncate processes uniformly covering over two thirds of spore surface, remainder smooth with two or three cup-like depressions, wall evenly thickened.

Kelleromyxa fimicola (Dearn. & Bisby) Eliasson: phaneroplasmodium with mature sporangia of this obligate coprophile, on horse dung, Canada, x 15.

Lajassiella Tuzet & Manier ex Manier.
Parataeniellaceae, Eccrinida, Inc. sed., Ichthyosporea, Choanofila, Choanozoa, Protozoa
Tuzet, O., J. F. Manier. 1968. *Lajasiella aphodii*, n. g., n. sp., Palavascide parasite d'une larve d'*Aphodius* (coleoptère Scarabaeidae). *Annls. Sci. Nat., Bot. Biol. Vég.*, sér. 12, 9:96.
The generic name derives from the name of the locality "La Jasse-du-Maurin", in the south of France, where the host larvae of this Trichomycete were collected + L. dim. suf. *-ella*. Sporangia producing round spores in distal chains, oval spores arranged in two series in proximal part of thallus. Thallus attached to intestine of beetle larvae of the genus *Aphodius*.

Lajassiella aphodii Tuzet & Manier ex Manier: thallus with resistant spores, adhered to the intestine of the larvae of beetles (*Aphodium*), in France, x 175.

Lamproderma Rostaf.
Stemonitidaceae, Stemonitiida, Myxogastria, Myxogastrea, Mycetozoa, Amoebozoa, Protozoa
Rostafinski, J. 1873. *Versuch eines Systems der Mycetozoen. Dissertation inaugural,* IV. Strassburg, 21 pp.

Leocarpus

From Gr. *lamprós*, brilliant, beautiful + *dérma*, skin. Fructifications with brilliant peridium, generally iridescent, with blue, purple, silver or bronze tones. On dead wood and fallen leaves with mosses.

Lamproderma pulchellum Meyl.: sporangia on dead wood, with a brilliant and iridescent peridium, x 25.

Leocarpus Link

Physaraceae, Stemonitiida, Myxogastria, Myxogastrea, Mycetozoa, Amoebozoa, Protozoa

Link, J. H. F. 1809. Observationes in ordines plantarum naturales. *Magazin Ges. Naturf. Freunde* 3:25.

From L. *leo*, lion + Gr. *karpós*, fruit. Fructification yellow-ochraceous. On remains of plants and at times on living herbaceous plants.

Leocarpus fragilis (Dicks.) Rostaf.: yellow-ochraceous sporangia, growing on plant remains, x 35.

Licea Schrad.

Liceaceae, Liceida, Myxogastria, Myxogastrea, Mycetozoa, Amoebozoa, Protozoa

Schrader, H. A. 1797. *Nov. gen. pl.* (Lipsiae):16.

From L. *lix, licis*, ashes. Aethalium with peridium which breaks to expose dark, ash-colored spores en masse. On tree bark or dead wood.

Liceopsis Torrend—see Reticularia Bull.

Tubiferaceae, Liceida, Myxogastria, Myxogastrea, Mycetozoa, Amoebozoa, Protozoa

Torrend, C. 1908. Catalogue raissonné des Myxomycetes du PortugaL *Bull. Soc. Pon Sci. Nat.* 2:55-73.

From genus *Licea* Schrad. + Gr. suf. *-ópsis*, appearance, aspect. Spores ash-colored en masse similar to those of *Licea*.

Licea pusilla Schrad.: mature aethalia, on rotten wood in pine forest, Germany, x 35.

Lycogala Pers.

Tubiferaceae, Liceida, Myxogastria, Myxogastrea, Mycetozoa, Amoebozoa, Protozoa

Persoon, C. H. 1794. Neuer Versuch einer systematischen Einteilung der Schwämme. *Neues Magazin für die Botanik* 1:63-128.

From Gr. *lýkos*, wolf + *gála*, milk; or < Gr. *lýkos*, hood, cover. Aethalium with peridium covered by a creamy mass of spores. On dead wood or on the bark of living trees.

Lycogala epidendrum (J. C. Buxb. ex L.) Fr.: mature aethalia, on dead wood, x 4.5.

Metatrichia Ing

Trichiaceae, Trichiida, Myxogastria, Myxogastrea, Mycetozoa, Amoebozoa, Protozoa

Ing, B. 1964. Myxomycetes from Nigeria. *Trans. Br. mycol. Soc.* 47:49-55.

From Gr. *metá*, later, in the sense of advanced evolutionary condition + genus *Trichia* Haller < Gr. *thríx*, genit. *trichós*, hair + L. suf. *-ia*, which denotes quality of, or state of being. Sporangia with a double-layered peridium that dehisces by a preformed operculum. On rotted wood and fallen leaves.

Metatrichia vesparium (Batsch) Nann.-Bremek. ex G. W. Martin & Alexop.: sporangia that dehisce by a preformed operculum, on rotted wood, x 30.

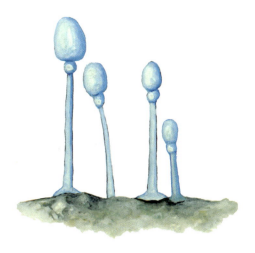

Nematostelium ovatum (L. S. Olive & Stoian.) L. S. Olive & Stoian.: sporocaps with slender stalk, on soil, x 210. Isolated in Florida, U.S.A.

Microglomus L. S. Olive & Stoian.
Protosteliida, Protostelida, Inc. sed., Protostelea, Mycetozoa, Amoebozoa, Protozoa
Olive, L. S., C. Stoianovitch. 1983. Redescription of the protostelid genus *Microglomus*, its type species and a new variety. *Trans. Br. mycol. Soc.* 81:449-454.
From Gr. *mikros* > L. *micro*, small + L. *glomus*, ball, round body, referring to the globose spore. Sporocarp with a single globose spore with spinulose wall containing 2 or 4 amoeboid cells; synaptonemal complexes present in nuclei of developing spores prior to nuclear and cell divisions; trophic stage consisting of holozoic, mostly uninucleate amoeboid cells typically with filose, rarely lobose, pseudopodia, lacking flagella; microcysts uninucleate. Isolated from bark of *Casuarina equisetifolia* in Kahana, Oahu, Hawaii, U.S.A.

Nematostelium L. S. Olive & Stoian.
Protosteliida, Protostelida, Inc. sed., Protostelea, Mycetozoa, Amoebozoa, Protozoa
Olive, L. S. 1970. The Mycetozoa: A revised classification. *Bot. Rev.* 36:59-87.
From Gr. *néma, némmatos*, thread + *stéle*, column, stalk + L. dim. suf. *-ium*. Sporocarp with a long, slender stalk or pedicel. In soil and humus as well as on still attached dead plant parts.

Octomyxa Couch, et al.
Plasmodiophoridae, Plasmodiophorida, Inc. sed., Phytomyxea, Endomyxa, Cercozoa, Protozoa
Couch, J. N., et al. 1939. A new genus of the Plasmodiophoraceae. *J. Elisha Mitchell Sci. Soc.* 55:399-408.
From Gr. *októ*, eight + *mýxa*, slime. Resting spores aggregate in groups of eight inside thalli of *Achlya* Nees and other genera of Saprolegniaceae.

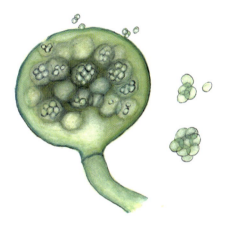

Octomyxa achlyae Couch, et al.: resting spores in groups of eight, inside the thallus of **Achlya glomerata**, x 450. Collected in North Carolina, U.S.A.

Orthotricha Wingate
Clastodermataceae, Echinostelida, Inc. sed., Myxogastrea, Mycetozoa, Amoebozoa, Protozoa
Wingate, H. 1886. A new genus of Myxomycetes. *J. Mycol.* 2:125-126.
From Gr. *orthós*, straight, upright, in the sense of correct, normal or regular + genus. *Trichia* Haller < Gr. *thrix, trichós*, hair + L. suf. *-ia*, which denotes quality of or state of being. Capillitium filamentous.

Palavascia Tuzet & Manier ex Lichtw.
Palavasciaceae, Eccrinida, Inc. sed., Ichthyosporea, Choanofila, Choanozoa, Protozoa
Lichtwardt, R. W. 1964. Validation of the genus *Palavascia* (Trichomycetes). *Mycologia* 56:318-319.
Named for *Palavas*, Hérault, France, the type locality for the genus + L. suf. *-ia*, which denotes belonging to. Sporangiferous thalli with terminal cells giving rise to one or several microthalli upon germination. Endocommensal in the intestine of isopod crustaceans.

Paramoebidium

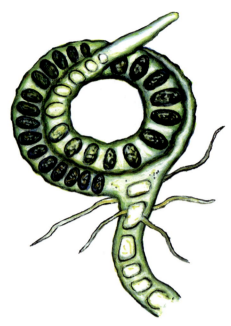

Palavascia sphaeromatis Tuzet & Manier ex Manier: sporangiferous thallus with terminal cells giving rise to microthalli, x 850. Collected in France.

Paramoebidium Léger & Duboscq
Amoebidiidae, Eccrinida, Inc. sed., Ichthyosporea, Choanofila, Choanozoa, Protozoa
Léger, L., O. Duboscq. 1929. L'evolution des *Paramoebidium*, nouveau genre d' eccrinides, parasite des larves aquatiques d' insectes. *Compt. Rend. Hebd. Séances Mém. Acad Sci., Paris* 189:75-77.
From Gr. *pará*, at the side, almost + genus *Amoebidium* Cienk. < Gr. *amoibé*, change, transformation + L. dim. suf. *-idium*. Sporangiospores amoeboid, differs from *Amoebidium* in having thalli that adhere to cuticle of posterior digestive tube of larvae and nymphs of ephemeropteran and plecopteran insects.

Parataeniella R. A. Poiss.
Parataeniellaceae, Eccrinida, Inc. sed., Ichthyosporea, Choanofila, Choanozoa, Protozoa
Poisson, R. 1929. Recherches sur quelques eccrlnides parasites de crustacés amphipodes et isopodes. *Archs Zool. exp. gén.* 69:179-216.
From Gr. *pará*, at the side, almost + genus *Taeniella* Léger & Duboscq, from L. *taenia*, sash, band, belt, tapeworm + dim. suf. *-ella*. Shares many of the characteristics of the genus *Taeniella* L. Léger & Duboscq, another eccrinal endocommensal of the proctodeum of marine crustaceans, which produces unispored sporangia.

Physarella Peck
Physaraceae, Stemonitiida, Myxogastria, Myxogastrea, Mycetozoa, Amoebozoa, Protozoa
Peck, C. H. 1982. New species of fungi. *Bull. Torrey Bot. Club* 9:61-62.

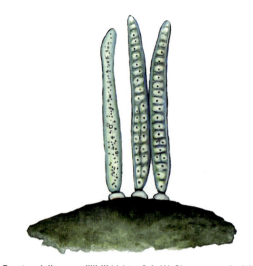

Parataeniella armadillidii Lichtw. & A. W. Chen: an eccrinal thallus that produces unispored sporangia, adhered to the proctodeum of marine crustaceans, Kansas, U.S.A., x 185.

From genus *Physarum* Pers. + L. dim. suf. *-ella*, which indicates a similarity. On wood and dead leaves.

Physarella oblonga (Berk. & M. A. Curtis) Morgan: ampulliform, tuberculate sporangia, with gyrose stipe, on dead wood, x 9.

Physarum Pers.
Physaraceae, Physarida, Inc. sed., Myxogastrea, Mycetozoa, Amoebozoa, Protozoa
Persoon, C. H. 1794. Neuer Versuch einer systematischen Einteilung der Schwämme. *Neues Magazin für die Bot.* 1:88.
From Gr. *phýsa*, blister, bubble, bladder + L. suf. *-arum*, similar to. Sporangia globose. On woody plant remains, leaves, fallen leaves, and mosses.

Plasmodiophora Woronin
Plasmodiophoridae, Plasmodiophorida, Inc. sed., Phytomyxea, Endomyxa, Cercozoa, Protozoa
Woronin, M. 1877. *Plasmodiophora brassicae*, der die unter dem Namen Hernie bekannte Krankheit der Kohlpflanzen verursacht. *Arbeiten der Sankt Petersburguer Naturforschender Gesellschaft* 8:169.

From L. *plasmodium*, plasmodio < Gr. *plásma*, bland material with which a living being is formed + *-oeídes*, similar to > L. *-odium*, + Gr. *phóros*, bearer. Produces plasmodium. Parasitic on terrestrial vascular plants in the Brassicaceae and marine aquatics.

Physarum polycephalum Schwein.: globose, mature sporangia, arising from the phaneroplasmodium, with the peridium black and gyrose, on woody plant remains, x 10.

Physarum pusillum (Berk. & M. A. Curtis) G. Lister: globose, pedicellate sporangia, with whitish peridium, growing on a twig, x 23.

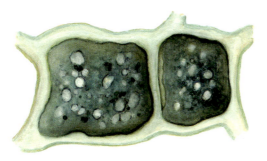

Plasmodiophora brassicae Woronin: plasmodia inside root cells of cabbage (*Brassica oleracea*), x 820. This pathogen grows on roots of Brassicaceae, mainly *Brassica*, rarely *Iberida umbellata*, Germany.

Pocheina A. R. Loeblich & Tappan

Guttulinaceae, Acrasida, Inc. sed., Heterolobosea, Tetramitia, Percolozoa, Protozoa

Loeblich, A. R., H. Tappan. 1961. Suprageneric classification of the Rhizopoda. *J. Paleontol.* 35:245-330.
Dedicated to the protozoologist F. Poche + L. suf. *-ina*, which denotes similarity. Sporocarps pedicellate, each with a single terminal spherical or subspherical sorus. On lichenized bark of trees.

Pocheina rosea (Cienk.) A. R. Loebl. & Tappan: pedicellate sporocarp, with a single, subspherical sorus, on rotten wood, Germany, x 325.

Polymyxa Ledingham

Plasmodiophoridae, Plasmodiophorida, Inc. sed., Phytomyxea, Endomyxa, Cercozoa, Protozoa

Ledingham, G. A. 1939. Studies on *Polymyxa graminis*, n. gen. n. sp., a plasmodiophoraceous root parasite of wheat. *Can. J. Res.* 17:38-51.

From Gr. *polýs*, many + *mýxa*, slime. Resting spores forming in interior of host cell from protoplasts or myxamoebae lacking a cell wall. Parasitic on grasses, such as wheat.

Polymyxa graminis Ledingham: zoosporangial thalli and zoosporangia inside root cells of wheat (*Triticum sativum*), x 550.

Polysphondylium Bref.

Dictyosteliaceae, Dictyostelida, Inc. sed., Dictyostelea, Mycetozoa, Amoebozoa, Protozoa

Brefeld, O. 1884. *Polysphondylium violaceum* und *Dictyostelium mucoroides*. *Untersuchungen aus dem Gesammtgebiete de Mykologie* 6:1-34.

From Gr. *polýs*, many + *sphóndylos*, vertebra, dorsal spine, verticil + L. dim. suf. *-ium*. Sorocarp with intercalary verticils along its axis. In soil and dung of rodents and other animals.

Protosporangium

Polysphondylium violaceum Bref.: sorocarp with intercalary verticils of sori along the axis, on agar culture, x 15.

Protosporangium L. S. Olive & Stoian.
Cavosteliaceae, Protostelida, Inc. sed., Protostelea, Mycetozoa, Amoebozoa, Protozoa
Olive, L. S., C. Stoianovitch. 1972. *Protosporangium*: A new genus of protostelids. *J. Protozool.* 19:563-571.
From Gr. *prōtos*, first, ahead + *sporá*, spore + *angeíon*, receptacle. Microsporangia diminutive. On tree bark.

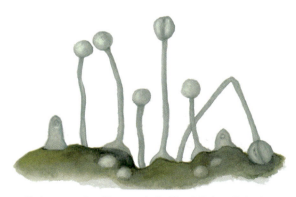

Protosporangium bisporum L. S. Olive & Stoian.: diminutive microsporangia, double spored in the terminal part of two sporangia, on tree bark, x 350. Isolated in North Carolina, U.S.A.

Protosteliopsis L. S Olive & Stoian.
Protosteliaceae, Protostelida, Inc. sed., Protostelea, Mycetozoa, Amoebozoa, Protozoa
Olive, L. S., C. Stoianovitch. 1966. *Protosteliopsis*, a new genus of the Protostelida. *Mycologia* 58:452-455.
From Gr. *prōtos*, first, forward + *stéle*, column, stalk + L. suf. *-opsis* < Gr. *-ópsis*, appearance, aspect, due to the similarity of the sporocarp to that of the genus *Protostelium* Olive & Stoian. Common, cosmopolitan organism in dung, soil, and dead, still-attached plant structures.

Protosteliopsis fimicola (L. S. Olive) L. S. Olive & Stoian.: sporocarps of a cosmopolitan organism, on soil, x 1,000. Isolated in North Carolina, U.S.A.

Protostelium L. S. Olive & Stoian.
Protosteliaceae, Protostelida, Inc. sed., Protostelea, Mycetozoa, Amoebozoa, Protozoa
Olive, L. S., C. Stoianovitch. 1960. Two new members of the Acrasiales. *Bull. Torrey Bot. Club* 87:1-20.
From Gr. *prōtos*, first, primitive, forward + *stéle*, column, pillar, stallk + L. dim. suf. *-ium*. Sporocarp with spores typically borne at apex of a filiform, acellular stalk. On dead, and still attached plant parts.

Protostelium mycophagum L. S. Olive & Stoian.: filiform sporocaps, with acellular stalk, x 350. Isolated from cultures of **Rhodotorula mucilaginosa** (A. Jörg.) F. C. Harrison and **Phoma** Sacc., New Jersey, U.S.A.

Ramacrinella Manier & Ormières
Eccriniidae, Eccrinida, Inc. sed., Ichthyosporea, Choanofila, Choanozoa, Protozoa
Manier, J. F., R. Ormières. 1962. *Ramacrinella raibauti* n. g., n. sp., Eccrinide ramifié commensal de l'intestin postérieur de *Microdeutopus gryllotalpa* A. Costa (Amphipodes-Aoridae). *Annls. Sci. Nat., Bot. Biol. Vég., sér.* 12, 2:625-634.

From L. *rama*, branch + Gr. *ekkríno*, to separate, expel + L. dim. suf. *-ella*. Thallus branched from which sporangiospores separate. Endocommensal in the proctodeum of marine amphipod crustaceans.

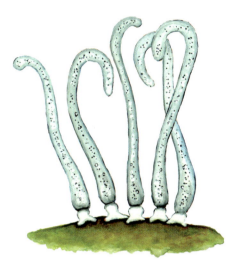

Ramacrinella raibautii Manier & Ormières ex Manier: endocommensal thalli with sporangiospores, adhered to the proctodeum of marine amphipod crustaceans, x 220. Described in France.

Reticularia Bull. (syn. **Liceopsis** Torrend)

Tubiferaceae, Liceida, Myxogastrea, Mycetozoa, Amoebozoa, Protozoa

Bulliard, J. B. F. 1791. *Histoire des champignons de la France*. Paris, Vol. 1, 368 pp.

From L. *reticulatus*, reticulate, made up of a net + suf. *-aria*, which denotes similarity or connection with something. Pseudocapillitium with filaments, which unite to form a net. Spores with reticulate wall. On dead wood.

Reticularia lobata Lister: above, the pseudocapillitium of a sporangium, formed by reticulate filaments, x 250; below, sporangia on dead wood, x 50.

Rhinosporidium Minchin & Fantham

Inc. sed., Inc. sed., Inc. sed., Inc. sed., Inc. sed., Choanozoa, Protozoa

Minchin, E. A., H. B. Fantham. 1905. *Rhinosporidium kinealyi* n. g. n. sp. A new sporozoon from the mucous membrane of the septum nasi of man. *Quart. J. Microbiol. Sci.* 49:521-532.

From Gr. *rhín*, genit. *rhinós*, nose + *sporá*, spore, seed + dim. suf. *-ídion* > L. dim. suf. *-idium*. Sporangia large, with small (7-9 μm diam) spores, spherical or ovoid. In the nasal polyps of patients with the mycosis known as rhinosporidiosis.

Rhinosporidium seeberi (Wernicke) Seeber: sporangium with endospores, in a nasal polyp of a human patient with rhinosporidiosis, x 160.

Rozella Cornu

Rozelliidae, Rozellida, Inc. sed., Rozellidea, Paramycia, Choanozoa, Protozoa

Cornu, M. 1872. Monographie des saprolégniées; étude physiologique et systématique. *Annls. Sci. Nat. Bot.*, sér. 5, 15:1-89.

From genus *Rozia* Cornu, dedicated to M. *Roze* + L. dim. suf. *-ella*. Endoparasitic on hyphae and reproductive organs of aquatic zoosporic fungi such as *Allomyces* Butler, *Blastocladia* Reinsch and *Chytridium* A. Braun.

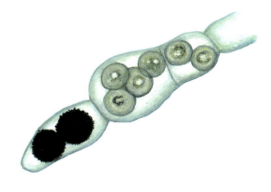

Rozella allomycis Foust: resting spores and young thalli of this chytridiaceous fungus inside the hyphae of **Allomyces javanicus**, x 700.

Schizoplasmodium

Schizoplasmodium L. S. Olive & Stoian.
Protosteliaceae, Protostelida, Inc. sed., Protostelea, Mycetozoa, Amoebozoa, Protozoa
Olive, L. S., C. Stoianovitch. 1966. A simple new mycetozoan with ballistospores. *Amer. J. Bot.* 53:344-349.
From Gr. *schýzo*, to split, divide + L. *plasmodium*, plasmodium < Gr. *plásma*, bland material with which a living being is formed + *-oeídes*, similar to > L. *-odium*. Plasmodium divides into prespoaral cells. Uncommon on dead flowers and still attached fruits of plants.

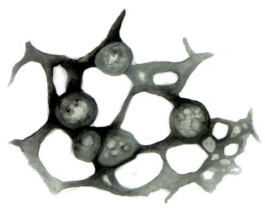

Schizoplasmodium cavostelioides L. S. Olive & Stoian.: reticulate plasmodium that segments just before sporulation into plurinucleate prespore cells, on agar surface, x 1,050.

Schizoplasmodium cavostelioides L. S. Olive & Stoian.: sporocarps on moribund parts attached to flowers, with the short stalk subtended by a distinct apophysis, and the sporocarp in the center with a ballistospore liberated, x 1,100. Isolated from pods of *Cytisus scoparius*, New Zealand.

Sorodiscus Lagerh. & Winge
Plasmodiophoridae, Plasmodiophorida, Inc. sed., Phytomyxea, Endomyxa, Cercozoa, Protozoa
Winge, O. 1912. Cytological studies in the Plasmodiophoraceae. *Ark. Bot.* 12:23.
From Gr. *sorós*, pile, heap + *dískos*, disc. Cystosori or resting spores flat oval disks, of one or two layers of spores. Parasitic on aquatic vascular plants, algae (*Chara*) and filamentous Oomycetes (*Pythium* Pringsh.).

Sorosphaerula Neuh. & Kirchm.
Plasmodiophoridae, Plasmodiophorida, Inc. sed., Phytomyxea, Endomyxa, Cercozoa, Protozoa
Neuhauser, S., M. Kirchmair. 2011. *Sorosphaerula* nom. n. for the Plasmodiophorid genus *Sorosphaera* J. Schröter 1886 (Rhizaria: Endomyxa: Phytomyxea: Plasmodiophorida). *J. Eukaryot. Microbiol.* 58(5):469-470.
From Gr. *sorós*, cask, reservoir + *sphairula*, little ball < *sphaíra* > L. *sphaera*, sphere, referred to the spherical sporosori. Cystosori or resting spores intracellular, predominantly hollow, globose or ellipsoid. Parasitic on aquatic vascular plants, such as *Veronica*.

Sorodiscus radicicola Ivimey Cook: flat, oval disk of a cystosorus, or resting spores, which forms inside the host cell (*Chara*, an alga, or **Pythium** Pringsh, a member of kingdom Chromista), x 1,500.

Sorosphaerula veronicae (J. Schröt.) Neuh. & Kirchm. (syn. **Sorosphaera veronicae** (J. Schröt.) J. Schröt.): intracellular, hollow spheres of the cystosorus or resting spores, parasitizing aquatic vascular plants, such as *Veronica*, x 100.

Spongospora Brunch.
Plasmodiophoridae, Plasmodiophorida, Inc. sed., Phytomyxea, Endomyxa, Cercozoa, Protozoa
Brunchorst, J. 1887. Ueber eine sehr verbreite Krankheit der Kartoffelknollen. *Bergens Museums Årsberetning* 1886:219.
From L. *spongia* < Gr. *spóngos*, sponge + *sporá*, spore. Resting spores more or less solid, spongy masses or aggregations, developing inside host cells. Parasitic on potato tubers.

Stemonitis Roth
Stemonitidaceae, Stemonitiida, Inc. sed., Myxogastrea, Mycetozoa, Amoebozoa, Protozoa

Roth, A. W. 1787. Verschiedene Abhandlungen: *Stemonitis*. *Bot. Mag. (Römer & Usteri)* 1(2):25-26.
From Gr. *stémon*, genit. *stémonos*, thread, yarn + suf. *-itis*, like. Sporangia fibrous. On dead wood and vegetable detritus.

Spongospora subterranea (Wallr.) Lagerh.: resting spores, which adopt the shape of spongy masses or aggregations, parasitizing potato tubers, x 800.

Stemonitis axifera (Bull.) T. Macbr.: fibrous shape of the sporangia, which develop on dead wood and vegetable remains, x 40.

Stemonitopsis (Nann.-Bremek.) Nann.-Bremek.
Stemonitidaceae, Stemonitiida, Inc. sed., Myxogastrea, Mycetozoa, Amoebozoa, Protozoa
Nannenga-Bremekamp, N.E. 1974. De Nederlandse Myxomyceten:203.
From Gr. *stémon*, genit. *stémonos*, thread, yarn + suf. *-ópsis*, appearance, aspect. Sporangia scattered to gregarious, stipitate, cylindric to narrowly ovoid, obtuse, erect, lilac-gray with a silvery shine, after its loss then brown. Hypothallus membranous, red-brown. Capillitium densely reticulate arising from the entire columella. On dead wood and vegetable detritus.

Stemonitopsis typhina (F. H. Wigg.) Nann.-Bremek. (syn. **Comatricha typhoides** (Bull.) Rostaf.): sporangia, some with broken peridium to see the inside capillitium, on dead wood, x 40.

Taeniella L. Léger & Duboscq
Eccriniidae, Eccrinida, Inc. sed., Ichthyosporea, Choanofila, Choanozoa, Protozoa
Léger, L., O. Duboscq. 1911. Sur les eccrinides des crustacées décapodes. *Annls. Univ. Grenoble* 23:139-141.
From L. *taenia*, sash, band, belt, tapeworm + dim. suf. *-ella*. Thallus elongated. Endocommensal, attached to the proctodeum of marine decapod crustaceans.

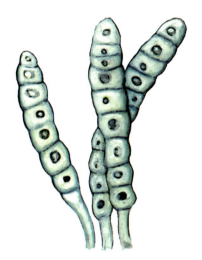

Taeniella carcini L. Léger & Duboscq: belt-shaped thallus with spores, of this endocommensal adhered in the rectum of marine decapod crustaceans, *Carcinus moenas*, in France, x 360.

Tetramyxa K. I. Göbel
Plasmodiophoridae, Plasmodiophorida, Inc. sed., Phytomyxea, Endomyxa, Cercozoa, Protozoa
Göbel, K. 1884. *Tetramyxa parasitica*. *Flora* 67:517-521.
From Gr. *tétra*, four + *mýxa*, slime. Myxamoebae divide twice forming four resting spores, which generally remain united in tetrads. Parasitic on marine vascular plants (*Ruppia*, *Zannichellia* and others).

Trichia

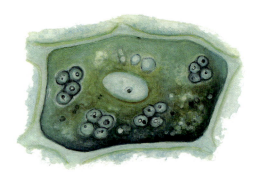

Tetramyxa parasitica K. I. Goebel: tetrads of resting spores inside the cell of a marine vascular plant, x 800.

Trichia favoginea (Batsch) Pers.: mature sporangia on dead wood, two of them showing the hairs or filaments of the capillitium, x 15.

Trichia Haller

Trichiaceae, Trichiida, Inc. sed., Myxogastrea, Mycetozoa, Amoebozoa, Protozoa

Haller, A. von. 1768. *Historia stirpium indigenarum helvetiae*. Bern 3:114.

From Gr. *thríx*, genit. *trichós*, hair + L. suf. *-ia*, which denotes quality of, or state of a being. Sporangia with hair-like or filamentous capillitium. On dead wood, tree bark, and fallen leaves.

Tubifera J. F. Gmel.

Tubiferaceae, Liceida, Inc. sed., Myxogastrea, Mycetozoa, Amoebozoa, Protozoa

Gmelin, J. F. 1792. *Syst. Nat.* 2(2):1472.

From L. *tubifer* < *tuba*, trompet, or < *tubus*, tube, pipe + *-fer* < *fero*, to bear, carry. Sporangia tubular shaped, of the sporangia that aggregate to form pseudaethalia. On dead wood and fallen leaves.

Abbreviations and symbols

Am. Sp.	-	American Spanish	OG. -	Old German
cm	-	centimeter	pl. -	plural
des.	-	desinence (ending)	Por. -	Portuguese
E.	-	English	pref. -	prefix
et al.	-	and others (<L. *et alii, aliorum*)	prep. -	preposition
fem.	-	feminine	Skt. -	Sanskrit
Fr.	-	French	sp. -	species
G.	-	German	spp. -	species (plural)
genit.	-	genitive	suf. -	suffix
Gr.	-	Greek	syn., syns.	synonym, synonyms
i.e.	-	that is (<L. *id est*)	var. -	variation, variety
Inc. sed.	-	uncertain position (<L. *Incertae sedis*)	< -	derived from, less than
			> -	giving rise to, greater than
It.	-	Italian	& -	and
L.	-	Latin		
LL.	-	Late Latin		
m	-	meter (<L. *metrum*, measure)		
ML.	-	Middle Latin		
mm	-	millimeter		
μm	-	micrometer		
NL.	-	Neo-Latin		

Bibliography

Borror, D. J. 1960. *Dictionary of Word Roots and Combining Forms*. Mayfield, Palo Alto, CA, U.S.A., 134 pg.

Cannon, P. F., and P. M. Kirk. 2007. *Fungal Families of the World*. CAB International, Wallingford, U.K., 456 pg.

Clements, F. E., and C. L. Shear. 1957. *The Genera of Fungi*. Hafner, New York, U.S.A., 496 pg.

Cuyás, A. 1939. *Appleton's New English-Spanish and Spanish-English Dictionary*. D. Appleton-Century, New York, U.S.A., 554 pg. (part I), 521 pg. (part II).

Donk, M. A. 1960. The generic names proposed for Polyporaceae. *Persoonia 1(2)*:173-302.

Donk, M. A. 1962. The generic names proposed for Agaricaceae. *Nova Hedwigia, Beihefte 5*:1-320.

Font Quer, P. 1963. *Diccionario de Botánica*. Labor, Barcelona, Spain, 1,244 pg.

Index Fungorum. 2019. Index Fungorum Partnership. www.indexfungorum.org

Jaeger, E. 1950. *A Source-book of Biological Names and Terms*, 2nd ed. Charles C Thomas, Springfield, IL, U.S.A., 287 pg.

Kirk, P. M., P. F. Cannon, D. W. Minter, and J. A. Stalpers, eds. 2008. *Ainsworth & Bisby's Dictionary of the Fungi*, 10th ed. CAB International, Wallingford, U.K., 784 pg.

MycoBank. 2019. International Mycological Association. www.mycobank.org

Oltra, M. 1991. *Origen etimológico de los nombres científicos de los hongos*. Monografía 1 de la Sociedad Micológica de Madrid, Real Jardín Botánico, Madrid, Spain, 136 pg.

Ortega Pedraza, E. 1980. *Etimologías. Lenguaje culto y científico*. Diana, México, D.F., 286 pg.

Pabón S. de Urbina, J. M., and E. Echauri Martínez. 1959. *Diccionario griego-español*. Publicaciones y Ediciones Spes, Barcelona, Spain, 633 pg.

Padres Escolapios. 1943. *Diccionario manual griego-latino-español*, 2nd ed. Albatros, Buenos Aires, Argentina, 966 pg.

Quintana-Cabanas, J. M. 1989. *Introducción etimológica al léxico de la biología*. Dykinson, Madrid, Spain, 153 pg.

Rodríguez Castro, S. 2004. *Diccionario etimológico griego-latín del español,* 21st ed. Esfinge, México, D.F., 248 pg.

Snell, W. H., and E. A. Dick. 1971. *A Glossary of Mycology*, rev. ed. Harvard University Press, Cambridge, MA, U.S.A., 181 pg.

Stearn, W. T. 1983. *Botanical Latin*. David & Charles, London, U.K., 566 pg.

Ulloa, M. 1997. Imágenes y palabras, una dualidad dinámica de la comunicación científica. *Revista Mexicana de Micología 13*:12-27.

Ulloa, M., and R. T. Hanlin. 2012. *Illustrated Dictionary of Mycology,* 2nd ed. American Phytopathological Society, St. Paul, MN, U.S.A., 762 pg.

Ulloa, M., and T. Herrera. 1994. *Etimología e iconografía de géneros de hongos*. Cuadernos del Instituto de Biología Núm. 21, UNAM, México, D.F., 304 pg.

Webster's Ninth New Collegiate Dictionary. 1989. Merriam-Webster, Springfield, MA, U.S.A., 1,562 pg.